EVOLUTION

EVOLUTION
The Origins and Mechanisms of Diversity

JONATHAN BARD

CRC Press
Taylor & Francis Group
Boca Raton London New York

CRC Press is an imprint of the
Taylor & Francis Group, an **informa** business

First edition published 2022
by CRC Press
6000 Broken Sound Parkway NW, Suite 300, Boca Raton, FL 33487-2742

and by CRC Press
2 Park Square, Milton Park, Abingdon, Oxon, OX14 4RN

© 2022 Taylor & Francis Group, LLC

CRC Press is an imprint of Taylor & Francis Group, LLC

ISBN: 978-1-032-13848-0 (hbk)
ISBN: 978-0-367-35701-6 (pbk)
ISBN: 978-0-429-34621-7 (ebk)

DOI: 10.1201/9780429346217

Typeset in Utopia
by KnowledgeWorks Global Ltd.

Access the Support Material: <Link to be confirmed with Editorial in a few months, please>

For Gillian

To explore evolution is to celebrate life

Contents

Preface

Biology, unlike any other science, has a single unifying principle at its core: this is that the evolutionary path of every living organism can be traced back almost four billion years to a very early and simple bacterium known as the Last Universal Common Ancestor (or LUCA). The basis for this principle was seeded by Lamarck in 1809: he was the first person to prove on the basis of data that evolution meant branching descent rather than climbing up the ladder of the complexity of life. Lamarck's insight did, however, mean little until 1859 when Darwin, on the basis of 20 years of research, proposed a mechanism of how new species would evolve from existing ones as a result of natural variation and selection. In 1953, almost a century later, Theodosius Dobzhansky was able to write 'Nothing in biology makes sense except in the light of evolution.' That statement was right then, and it is right now.

The purpose of this book is to explore that principle of evolutionary decent, and it is based on earlier one (*Principles of Evolution*). This book has, however, been completely reorganised, rewritten and brought up to date with ~700 new references; it also includes two new sections. One is on the *History of Life* and describes the major events that led from the LUCA to the glorious diversity of life today. The other is on *Human Evolution* and details how a spur off the great ape line in East Africa some seven million years ago led to the clade of hominins, now represented across Earth by *Homo sapiens* alone. I hope that these sections together with those on the evidence for evolution and the mechanisms of evolutionary change will enable the reader to feel that the text, as a whole, provides an integrated view of evolutionary biology.

What makes this book different from others on evolution is its perspective. Most focus on genetics, palaeontology, or taxonomy (now known as systematics). This book, while, of course, covering that material, is written from the perspectives of molecular genetics and developmental anatomy. This enables the reader to take, in particular, a close look at the mechanistic origins of anatomical diversity, which is, of course, how we recognise evolutionary change.

As the study of evolution involves every aspect of biology from genomics to ecology, basic material that senior biology students should know has been excluded so as to keep the size of the book manageable. The aim is to focus on the core principles of evolution rather than to cover the vast amount of supporting evidence (PubMed includes more than 600,000 papers that touch on evolution). Two further aims are to point the reader to a selection of particularly interesting primary research and review papers and to provide the background language and knowledge needed to read them. These papers serve several purposes:

- They pay tribute to important early work.
- They provide background and supporting information.
- They expand areas that, for reasons of space, are only lightly discussed here.
- They highlight the many aspects of evolution that still need investigation.
- They are a starting point for essays and discussions—suggestions for these are included on the Book Website.

Another, immediately accessible resource is Wikipedia with its interlinked and highly referenced entries that cover all aspects of biology. Its anonymous reviews on, for example, evolutionary clades, show details of taxonomic relationships that are too extensive to include on a printed page. I would like to thank all who have helped write and maintain those pages—they are a generous gift to students and academics alike.

The achievements of evolutionary biologists since Darwin have been remarkable. By the 1980s, generations of palaeontologists, evolutionary zoologists, and classical geneticists had provided us with a broad picture of the evolution of post-Cambrian animal and plant life based on the fossil record, shown how changes in the gene flow through a population would lead to phenotypic change and explained novel speciation in terms of the genetic properties of small groups that broke away from parent populations.

Since then, our understanding of evolution has been greatly expanded by scientists from two other disciplines. Evolutionary informaticians have revealed the taxonomic history of early clades for which there is no fossil record and shown how the eukaryotic cell derived from several bacteria through the process of endosymbiosis, as well as providing computational tools for others to analyse DNA sequence data. Molecular developmental biologists have explored the effects of mutation on phenotypes, shown how ancient proteins have maintained common functions across very different clades, particularly during embryogenesis (the area known as evo-devo) and revealed the likely anatomy of the earliest bilaterian ancestor. In addition, palaeontologists, archaeologists, and informaticians have greatly expanded our knowledge of human evolution and the timings and sizes of the small groups of early humans who left Africa and spread across world. All of these topics are discussed in this book.

My hope is that the reader will want still more. Lamarck and Darwin opened doors that revealed the complex landscape of evolution, and generations of biologists have since passed through them to investigate this terrain. I can assure future evolutionary biologists that there is still plenty of uncharted territory awaiting exploration.

Jonathan Bard
Oxford

Timing convention. There are several abbreviations for historic time in the literature. Those used here are By and Bya, My and Mya, and Ky and Kya for billions, millions, and thousand of years, with the additional 'a' meaning 'ago.'

Website Support Materials

Students and instructors have access to several Support Materials which should be seen as an integral part of the book. These Support Materials can be downloaded by visiting the following link, under Support Materials: https://www.routledge.com/9780367357016

The book's Support Materials include:

- Further Reading and Website Resources, with links
- References, with links
- The full version of Appendix 3 from the book
- Copy of the Glossary

Instructors can register to gain access to two additional resources. To register, they must request access at the following location: https://routledgetextbooks.com/textbooks/instructor_downloads/.

The two additional resources available to registered instructors are:

- Discussion points for each chapter
- Figure slides — all the figures of the book, downloadable in either PDF or PowerPoint format

Acknowledgements

No one can write a book as wide-ranging as this without a considerable amount of help. I would like to thank my editor, Chuck Crumly, for his unceasing encouragement and feedback, while he, Jordan Wearing, and the CRC Press/Taylor & Francis production team have made the production stages far less painful than I had feared. I am also grateful to the host of authors and publishers who gave me permission to use their images. Their names are included in the picture acknowledgements.

Many experts have read chapters and made helpful suggestions as well as pointing out errors in early drafts. Here, I would like to express my gratitude to Julia Clarke, Nick Colgrave, Darren Curnoe, Robert Desalle, Vernon French, Martin Goldway, Brian Hall, Sariel Hübner, Tom Kemp, Kevin Padian, Idan Perelman, Kathy Pitirri, Susannah Porter, Aston Saino, Sebastian Shimeld, Alastair Simpson, Bernard Wood, and several anonymous reviewers. I would particularly like to thank Per Ahlberg: a request to read one chapter on palaeontology turned into a willingness to read six; this was accompanied by a long, interesting and educational email correspondence on modern systematics. I, of course, remain responsible for the remaining errors and would appreciate being informed of them.

Finally, I owe an immeasurable debt to my wife, Gillian Morriss-Kay. Few authors on evolution can have the good fortune to be married to someone who is not only an expert mammalian developmental anatomist, but who also edited a major anatomical journal for a decade. Not only did we have many discussions of one or another aspect of evolution over breakfast, but she read and criticised drafts of every chapter. I do not even dare to think what the book would have been like without her help.

About the Author

Professor Jonathan Bard is a British biologist whose main experimental work has explored vertebrate developmental anatomy and whose theoretical research has been in the areas of developmental modelling, computational anatomy, and evolution. He is the author of more than 100 research papers and four books

After a degree in physics at Cambridge and a Ph.D. in biophysics at Manchester, he became a developmental biologist, with postdocs at Edinburgh and Harvard. He then worked for the MRC Hunan Genetics Unit in Edinburgh and at the University of Edinburgh where he was also Director of the Graduate School of Biosciences. He is currently a graduate advisor at Balliol College, Oxford.

JB is particularly known for his work on:

- Morphogenesis in vertebrate embryos
- Models of the development of zebra striping and other animal patterns
- Anatomical and other ontologies (making biological knowledge computer accessible)
- Systems biology approaches to development and evolution

SECTION ONE

AN INTRODUCTION TO EVOLUTION

This first section aims to provide sufficient background for the reader who has not previously studied evolution in any detail to understand the evidence for evolutionary change (Section 2) and the work that has unraveled the mechanisms by which that change occurs (Section 4). The section also aims to provide the timelines within which the history of life developed (Section 3). Readers should note that although some secondary reading is given at the end of each chapter in this section, there are few references. This is because these chapters are meant to be browsed through rather than to be studied in detail. The associated literature is supplied in the subsequent chapters, where the various topics discussed in this introductory section are treated in far more detail.

Chapter 1 is an introduction that sets the context for approaching the study of evolution and, in particular, focuses on the perspective from which this book is written. This includes the use of our knowledge of how protein networks operate during development to help explain how mutation leads to changes in the anatomical phenotype (more information on this area, known as systems biology, is given in Appendix 1).

Chapter 2 gives a very brief history of evolutionary thought. The aim is to show how our current knowledge of evolution has been achieved, why different problems were worth working on at different times, and how each area of the subject built on others. Interested readers can find more detail on this fascinating subject in Appendix 2.

Chapter 3 gives a brief history of the prehistoric world, mainly on the basis of fossil and geological records. More detail on each of the geological periods is given in Appendix 3 and on the book's website, while explanations of how fossils were produced and how they are dated are given in Appendix 4.

Chapter 4 describes today's biosphere with its remarkable variety. In a sense, this chapter asks the questions about the origins of organism diversity that the rest of the book aims to answer.

DOI: 10.1201/9780429346217-1

Approaching Evolution

1

Our planet teems with life. In the sky, on the land, and in the sea, there are some millions of species of animals, plants, algae, fungi, bacteria, and viruses and *almost* all thrive as they develop and reproduce using energy directly or indirectly derived from the sun. This is possible because each species is appropriately adapted to the environment in which it lives: its ecosystem provides it with food, and it, in turn, provides nourishment and a context for other local organisms, sometimes while it lives and always when it dies. Unicellular organisms, be they prokaryotic or eukaryotic, can usually reproduce by simple division, whereas most multicellular organisms have the ability to reproduce by mating – no multicellular individual survives or reproduces in the long term unless it is part of a population.

The word *almost* is italicised in the first sentence above because, under all circumstances, some organisms are declining, with their niche in the ecosphere being occupied by a variant of that species or perhaps another species that does a little better in that environment. This competition is generally known as natural selection, and it is fueled by mutations that start off by affecting the phenotype of an individual that then slowly spread across a population through reproduction. So it is in every niche and for every species. The result is a continuing but very slow turnover of species, as the fossil record clearly shows. The bioworld is always in a state of flux and the wonderful diversity that we see today represents just a single frame in the movie of Earth's life.

The key purpose of the science of evolution is to make sense of this richness, to understand how it arose, to track the diversity of life backwards in time to its origins and to study the mechanisms by which that diversity evolved and continues to evolve; these mechanisms ensure that all species adapt to their environment and continue to do so as it changes. The data required for this enterprise come from across the whole of biology and include information from contemporary and ancient organisms, molecular biology and cytogenetics, comparative and molecular embryology, *evolutionary population genetics*, comparative genomics, and even from the comparison of the two gene sequences in a particular pair of chromosomes in a diploid individual. It is worth noting that other than this last example, evidence from an individual organism or even the most wonderful of fossils, rarely represents more than just a single, interesting fact. The science of evolution mainly depends on interpreting data from across populations and species.

Such comparisons are, however, just the start: understanding evolutionary change requires a much broader approach because all organisms live and evolve in a rich and complex biological and physical environment and have always done so. Whether a species thrives or dies out depends on its interactions with that environment. What we know about an individual species in particular and evolutionary biology, in general, depends as much on ecology and even the climate as much as it does on genomics.

We would, of course, love to be able to study evolutionary change directly, but we cannot: it can take a million or more generations for enough novel mutations

DOI: 10.1201/9780429346217-2

to be generated and to spread across a population for the environment to select a new subpopulation that is recognised as a novel and distinct species (Chapter 22). Heritable, selectable change to an organism's phenotype, other than through selective breeding of a rare phenotype, is usually far too slow to be appreciated in the short term; research work on speciation has, therefore, to be indirect. In general, the most that we can usually do in experimental evolutionary research is to investigate how the balance of traits and gene polymorphisms change as a result of mutation and selection pressures.

It is sometimes claimed that, unlike physics, biology doesn't have any profound theories. Physics, however, requires four forces and many, many laws about how they operate to explain the world of particles and inanimate objects. What is unusual about biology, as compared to physics and indeed all other areas of science, is that a single underlying principle integrates so much of its subject matter. This principle is that novel species evolve from existing ones through selection acting on heritable variants, with the result being that new species form through *descent with modification* – this is the heart of evolution. Its corollary is that the evolutionary path of every living organism can be traced back to a very early and simple bacterium known as the Last Universal Common Ancestor.

Darwin had this deep insight around 1840 and acquired evidence to support the idea over successive years. Nevertheless. he only published his work in 1858, after Alfred Russell Wallace had written to ask him to pass on his similar idea for publication. Darwin then rapidly wrote *The Origin of Species*, an 1859 classic that is still worth reading. Darwin's insight provides the framework within which it is possible to discuss the evolutionary change from its simplest beginning to now. Work since then has built on Darwin's view of evolution and little of this, other than his approach to how variation happens, has been shown to be wrong. It does, however, need to be emphasised that since the mid-19th century, a great deal of work has been needed to flesh out the rather bare skeleton that Darwin bequeathed us.

It should also be noted that it was not Darwin who invented the idea of evolution in the 1850s. The idea had been loosely discussed for several hundred years, but only as an idea and not as a scientific inquiry built on data and analysis. The credit for the first successful scientific work on evolution goes to Jean-Baptiste Lamarck (Chapter 2): in 1809, he realised on the basis of comparative anatomy that life did not, as was then suggested, climb a ladder of complexity with humans on the top step but evolved by descent and branching. It was Lamarck who, working almost 50 years before Darwin, provided the context within which we still discuss the long history of life.

PERSPECTIVES

A helpful starting point for making sense of the complexity of evolutionary change is through considering its time-scale (Metz, 2011). Microevolution focuses on changes in DNA sequences, proteins, and gene frequencies—this is essentially the domain of molecular mutation and population genetics and mainly reflects the short term (perhaps up to hundreds of generations). Mesoevolution looks at changes in traits within a population under the richness of selection, which covers environmental effects in the very widest of senses—this is the domain of the mathematical theory of evolutionary genetics (up to perhaps thousands of generations). Macroevolution considers changes in major anatomical features, whole-genome effects, and speciation—this is the domain of gross anatomy, embryology, coalescence theory, and palaeontology, and here tens of thousands of generations can be a short time. Novel speciation in which breeding between new and parent populations irreversibly break down, mainly because of chromosomal rearrangements, can take a million or more generations and is reflected in the fossil record. What makes the study of evolution difficult is that it covers almost all of biology over periods of time from the immediate to the essentially infinite.

Over the years since Darwin published *The Origins of Species*, successive generations have focused on particular aspects of the biology of evolution. In the decades immediately after Darwin, palaeontologists investigated the fossil record while naturalists explored how selection worked. Once Mendelian genetics had been rediscovered early in the 20th century, mathematically inclined biologists set out to produce a formal theory of evolutionary change that incorporated ideas of selection into population genetics.

Once the structure of DNA and the molecular basis of mutation had begun to be understood in the 1950s, proper studies of molecular change could be investigated. However, this area of research only really took off in the 1970s, once techniques for sequencing proteins and DNA had been established and started to come into general use. This new sequence data allowed computational biologists to work out molecular phylogenies of groups of organisms for which the fossil record was inadequate or even absent and trace them back to the origins of life. It was this ability, for example, that allowed us for the first time to understand how the first eukaryotic cell evolved through endosymbiosis, a rare event in which one cell engulfs another without then breaking it down; this enables the two cells to maintain a stable symbiotic relationship (Chapter 10).

Since the 1980s, the new emphases in evolutionary biology have particularly been on coalescence theory and on evo-devo. The former is a mathematical approach that integrates population genetics with DNA-sequence variation and is based on the idea that homologous sequences evolved through mutation from a common ancestor. This sophisticated computational technique allows us to track populations back in time and has, for example, allowed us to study how founder populations of early *Homo sapiens* spread across the world.

Evo-devo is the study of the roles of homologous proteins (i.e., those whose sequences show them to derive from a common ancestor sequence) in the development of very different organisms; it was a surprising discovery in the 1980s that they often have very similar roles during embryogenesis. A classic example is the role of the transcription factor Pax6 in the development of the eye, first identified in the mouse (Hill et al., 1991). Over the following few years, it became clear that a Pax6 homologue initiated the development of every known eye, be it the camera eye of vertebrates and molluscs such as the squid or the compound eye of invertebrates such as the fruit fly, *Drosophila melanogaster*. Proof that all these Pax6 variants really play such a role comes from the experimental observation that a mouse Pax6 protein can substitute for its homologue in *Drosophila* and initiate a compound eye, irrespective of where in the fly's ectoderm it is expressed (Gehring, 1996; Chapter 8).

EVOLUTIONARY RESEARCH TODAY

Now that molecular biology has been incorporated into our understanding of evolution, we can wonder if there are further perspectives left for this generation of evolutionary biologists to investigate that go beyond filling in the gaps of those approaches just described. One is certainly the molecular mechanisms that facilitate the evolution of symbionts and holobionts (Chapter 19). The former occurs when two organisms evolve to become mutually dependent (e.g. lichens are a symbiont of cyanobacteria and fungi) and so become able to colonise ecological niches in which neither could survive alone. Holobionts are assemblages of species that are normally observed as an individual: humans have almost as many bacteria in their microbiome as they have eukaryotic cells, with many helping in normal physiology; humans also carry many small parasitic invertebrates. These minor organisms are not inherited via the zygote but are acquired after birth and are present until the human dies. It is now clear that, although the many organisms in a holobiont are capable of leading independent lives, the assemblage has cooperative properties that allow the major organism to develop traits that it would not have on its own.

A classic example here is herbivory, the ability to survive by eating plants, a property that vertebrate herbivores still acquire through hosting cellulose-degrading bacteria (Chapter 19; Gilbert, 2019). Once a carnivorous vertebrate species could start to survive on plants, there were many, many new niches that its descendants could colonise. The fossil record shows how carnivorous anatomy slowly evolved into a vegetarian anatomy, with the most obvious changes being to the teeth, jaws, and gut.

A further perspective comes from the recent realisation that few proteins work alone; the great majority cooperate within complex networks to produce a single, functional output. In biochemistry, one thinks of the Krebs and fatty acid cycles together with other such pathways; in physiology, life depends on the outputs of a myriad of protein pathways that include those that drive muscle contraction, neuronal signaling, and the cardiac cycle. They are also particularly important in development where a host of protein networks drives cell proliferation and apoptosis, the various

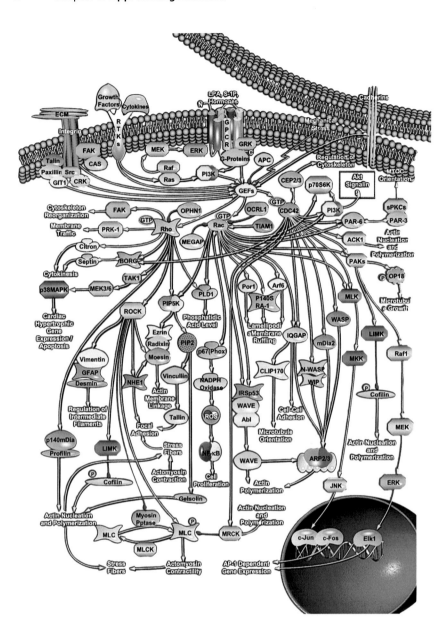

Figure 1.1 An illustration of complexity.
The rho-GTPase protein network controls cytoskeletal behaviour and hence much of morphogenesis. It contains almost a hundred proteins and includes several alternative pathways that produce different outputs. The details of the activity of this network of proteins are still not understood.

(From https://proteinlounge.com/pathway. php. Courtesy of ProteinLounge.)

morphogenetic mechanisms, and the many modes of cell differentiation (e.g. Fig. 1.1). The functional outputs of all of these networks depend on the cooperative properties of their constituent proteins, each of whose outputs can only rarely be predicted on the basis of what we know about the network structure and its component proteins. When we think about the sorts of traits that are particularly associated with selection, we are almost always considering the outputs of complex networks.

Our understanding of how these networks operate is still very thin, and we know very little about how mutation might modulate their outputs, be it for neuronal activity, cell proliferation rates, degree of skin pigmentation or for biochemical network activity. Minor changes to these outputs do however provide the molecular basis of the minor trait variants that together define the spread of individuals within a population. Over the longer term, successful variants flourish under natural selection and, for example, lead to the anatomical changes that can be followed in fossil sequences.

A major focus of this book is the molecular basis of phenotypic variation and particularly of anatomical change. This focus is not to underplay the importance of other aspects of evolutionary biology but reflects the great advances that have been made in the last few years in our understanding of the molecular genetics of embryology and of developmental mechanisms. The evolutionary link here is that almost all the anatomical changes that are seen as organisms evolve reflect changes that took place during development. If we are to understand these changes, we need to understand developmental mechanisms, the networks that underpin them and how mutation changes their outputs.

SYSTEMS BIOLOGY

The study of protein networks is a key aspect of systems biology, a topic that has attracted considerable attention over the last decade or so (Noble, 2008; Appendix 1). The realisation that proteins cooperate within networks to generate an output capability has changed how we think about the relationship between the genotype and the phenotype, particularly with respect to the effects of mutation (Capra & Luisi, 2014). This focus on networks is thus important, but does, however, reflect a narrow area of systems biology

There is the second aspect of systems biology, and this is the analysis of multilevel complexity, a topic of particular importance in the context of evolutionary change. The effect of a heritable mutation in an organism has to work its way up from a change in a DNA sequence, through to a change in protein function, to a change in the output of a network in a particular tissue system, and upwards to the change of phenotype that is affected by selection (Table 1.1). It is only rarely, however, that this progression of change is understood in any detail. As if this multilevel complexity was not hard enough to understand, the situation is made further complicated by there usually being feedback controls both up and down this hierarchy (Appendix 1).

An important example of evolutionary complexity comes in the study of selection: while the fitness of a variant in a particular environment can sometimes be measured as a numerical constant, rarely if ever can that constant be calculated or predicted or even expected to stay constant over long periods of time. This is because the extent to which one variant organism leaves more or fewer offspring than its peers in a particular environment depends on many factors, some of which may interact. Worse, we cannot assume that selection is the result of pressures on a single trait. This is because every trait in an organism can show variation and is subject to selection; although it is possible to focus on a single trait, it is usually unrealistic to do so except under conditions of selective breeding. The study of such complexity, underlain by multilevel causality, can be viewed as systems biology in the broader sense.

Although much of this book discusses where we are today in the field of evolution, its viewpoint is as much from a molecular and systems biology perspective as from

TABLE 1.1 BIOLOGICAL LEVELS AND EVOLUTIONARY CHANGE

Level	Some evolutionary roles
Environment	This provides both ecological niches and selection pressures, both of which are subject to change.
Population	This is the context within which individual variation and selection take place. They include variants capable of adapting to new environments.
Individuals	Mutations in the networks that affect development (e.g. patterning) will change adult anatomy. Adults are, of course, the unit of breeding and success here will affect the genotype of future populations.
Tissues	Tissue properties reflect the totality of cell properties. If mutations affect the growth networks, for example, tissue size will be affected.
Cells	The properties of a cell reflect the sum of its network outputs, and these are affected by mutation. Interactions between networks make their exact properties hard to predict.
Protein networks	The basic generators of biochemical, physiological and developmental properties, the outputs of which, because of their intrinsic complexity, are affected by mutation in unpredictable ways.
Proteins	The effect of mutation on a protein is to affect its binding specificities and rate constants.
Chromosomes	Chromosomal rearrangements underpin irreversible speciation but take a long time to form and spread through a population.
DNA sequences	Many types of information are stored in DNA sequences, each of which can vary through mutation, in turn affecting protein and cell properties.

more traditional ones. This can particularly be seen in its analysis of variation, in its approach to evolutionary genetics and in its analysis of evolutionary change – here, mathematical graphs (the natural language of systems biology, see Appendix 1) help, for example, to articulate the logic that underpins cladistics, the formal taxonomic system that integrates inheritance with anatomical modification.

Although it cannot be claimed that a systems biology approach can explain all the unsolved problems of evolutionary biology, it can certainly help push the subject forward. A mediaeval rabbi wrote that "it is not your duty to complete the work, but neither are you free to avoid it." So it is with evolution: every generation adds their perspective to a great tradition, but there will always be unsolved evolutionary problems awaiting future generations of biologists who will bring new techniques and approaches to help solve them.

FURTHER READINGS

Bard J (2013) Systems biology – The broader perspective. *Cells*; 19:413–431.

Barresi MJF, Gilbert SF (2019) *Developmental Biology* (12th ed). Sinauer Ass.

Darwin C (1859) *On the Origin of Species by Means of Natural Selection, or the Preservation of Favoured Races in the Struggle for Life*. John Murray. http://darwin-online.org.uk/converted/pdf/1859_Origin_F373.pdf

Noble D (2008) *The Music of Life: Biology Beyond Genes*. Oxford University Press.

WEBSITES

- The Wikipedia entries on the *History of Evolutionary Thought* and *Systems Biology*.

- This article on evolutionary biology: https://evolution.berkeley.edu/evolibrary/article/0_0_0/evo_toc_01

2

A Potted History of Evolutionary Science

This chapter briefly summarises the key steps and the major figures that, over the last two hundred or so years, were responsible for the modern understanding of evolution. It provides the context for appreciating the many strands that make up our contemporary understanding of evolution and that are discussed further in the subsequent chapters. A much fuller history of this fascinating story, together with references, is given in Appendix 2. Readers interested in some of the key texts should consult Ayala (Ayala & Avise, 2014).

THE PRE-DARWINIAN ERA

Towards the end of the 18th century, the general belief across Europe and the rest of the western world, insofar as people cared, was that the bible provided a true account of the origins of life. Over the centuries, however, a few people had suggested that humans had not been divinely created, but all their arguments were based on thought rather than evidence. (Appendix 2). The first serious step that led the way to modern views of evolution came from Charles Bonnet, a Swiss biologist, who originally trained and practiced as a lawyer. Around 1770, he suggested that there had been step-by-step progress up the long-established ladder of life (or great chain of being) that graded organisms from basal moulds through plants, insects, worms, shellfish, and on to fish, birds, and quadrupeds before finally achieving the ultimate perfection of humans.

There were actually quite good reasons then for believing the bible: the biological world seemed constant and static, animals were perfectly suited to their environments, one species could not breed with another, and there was little or no evidence in favour of any other account. The bible story was credible because there was no alternative rational explanation (albeit that the bible gives different creation stories in Genesis 1 and 2). The one oddity was the presence of fossils, but these were mainly viewed as organisms either lost in or displaced by Noah's flood.

The first modern evolutionary scientist in the sense that his work was based on the analysis of hard evidence was Jean-Baptiste Lamarck (1744–1829). Lamarck realised that because the annelid and parasitic worms were so anatomically different, they couldn't be treated as being on the same step of the ladder, as Bonnet had suggested. The key point here was that the internal anatomy of the former was far more complicated than that of the latter and, in particular, had a coelom or body cavity. In 1809, Lamarck suggested that evolution occurred not by ascending a ladder but through successive descending branches from an original organism. Lamarck was thus the first person to put forward the idea of descent based on change and did so on the basis of scientific evidence, although it is generally agreed that his writing was less than clear. He did not, of course, know how change happened or was inherited but suggested that organisms had two intrinsic abilities which were, in essence, to become more complicated and to adapt, with the resultant changes being inherited. These general ideas were accepted for the best part of the next century by most biologists, including Darwin.

DOI: 10.1201/9780429346217-3

Lamarck's dreadful reputation today derives from the obituary written by Jean Cuvier, his fellow professor at the Muséum National d'Histoire Naturelle in Paris and the major comparative vertebrate anatomist and palaeontologist of his day. Cuvier did not believe in evolution (or transmutation, the term then used) because, as he interpreted the fossil evidence seen in the different rock layers, the data pointed to successive extinctions followed by new bouts of creation. He used his grossly unfair and misleading obituary in 1829 to say that Lamarck was no more than a foolish philosopher, so destroying his work and his reputation.

In spite of Cuvier's attempts, the idea that life had evolved rather than had been created was becoming widely discussed in scientific circles in the 1830s. One obvious reason for this was the increasing number of vertebrate fossils being discovered that obviously bore little relationship to modern animals. While it was not possible to date them, it was clear as early as the 1820s that the deeper and hence older were the rock strata in which they were found, the more primitive were the fossils that they contained. The biblical stories were thus becoming rather less credible than they had been. Palaeontology and geology soon became sister subjects, and the insights that derived from them were a driving force for establishing the understanding that evolution could be responsible for the diversity of modern life. What was, of course, missing was any idea of how change could happen.

THE DARWINIAN ERA

Around 1837, Charles Darwin (1809–1882) came to realise that Malthus' idea that exponential growth of the human population would be held in check by food availability also applied to the rest of the living world. Competition among variant individuals for food would lead to the success of the fittest, in the sense that they would leave offspring that were best adapted for future breeding, and this was the key to understanding how evolution occurred. Darwin then spent more than 20 years assembling data on variation and selection but did not publish his work, mainly because he didn't want the inevitable public argument as his health was poor.

His hand was, however, forced by Alfred Russell Wallace (1858), who was working in Indonesia on what is now called biogeography, an area that he invented. Wallace, a professional collector of exotic animals and plants, had developed similar ideas on how species form as a result of variation and selection, also after reading Malthus. In 1858, he set these down in an essay that he sent to Darwin to pass on for publication. After much soul-searching and discussion with friends, Darwin then rapidly abstracted some of his own unpublished writing, and both papers were then submitted to the Linnaean Society and soon published. Darwin then rapidly wrote *The Origin of Species* (1859), a book that looks at how new species form, a shrewd way of discussing evolution without having to be too explicit. The book itself focuses on a single question: what are the evolutionary implications of the variation that any observer can see within a species?

Darwin's answer, buttressed by his own experiments and a wide range of examples from across the then known biological spectrum, starts with the fact that, left to themselves, populations grow exponentially, but such growth is constrained by pressures from the environment, such as food availability and predator activity. Darwin then saw new species as arising from the variants that naturally formed within the populations and suggested that these would disappear or flourish, even to the extent of replacing their parent population as selection pressures from their environment dictated (*the struggle for existence*). Such Malthusian competition, now interpreted as natural selection leading to the *survival of the fittest* (i.e., those leaving the most successful offspring), would, if the new variant flourished, result in descent with modification, which is of course evolution. The balance of variation and selection, two of the many ideas that underpin the book, is thus the key to understanding the origin of new species. In 1871, Darwin produced his second book on evolution, *The Descent of Man, and Selection in Relation to Sex*, in which he discussed sexual selection and human evolution. A year later, he published *The Expression of the Emotions in Man and Animals*, in which he sought to show that faculties that might seem special to humans had, in fact, a basis in animal behaviour.

It is worth noting that Darwin never explicitly mentions evolution, and this was for three reasons. First, his wife was a devout Christian, and he didn't want to upset her. Second, evolution had a second meaning then, which was embryonic development, and its use could have been confusing. Third, as mentioned, Darwin was acutely aware of the public arguments that had been caused by an earlier book that talked about evolution (see Appendix 2), and he wanted to avoid such problems. This was the reason why he used the phrase *descent with modification*. This key perception was, however, a far deeper explanation of evolution than Lamarck's view of descent as it showed how change could be recognised: derived organisms are much like their ancestors but with small modifications. Today, such changes in anatomy are key to making sense of the fossil record (Chapters 5 and 6), while the small mutational changes in genomes (i.e., mutations) allow us to reconstruct evolutionary trees in the absence of fossil records (Chapter 7). Darwin opened the doors to understanding both the history of life and the mechanisms of evolutionary change.

This wonderful work left several questions unanswered, the most important being about the mechanisms that created variants and their inheritance. There were also questions about the relative importance of hard inheritance, the effect of intrinsic variation, and of soft inheritance (often called Lamarckian inheritance) due to environmentally determined change that became heritable, and whether variation proceeded by slow continuous change or substantial discontinuous jumps (saltations). The basic question of inheritance was solved in the 1880s, mainly by August Weismann (1889), who proposed the theory of the continuity of the germplasm: this stated that inheritance is carried by sperm and eggs in their nuclei and that the key factor in generating variation is the crossing-over of chromosomes during meiosis. His view was that soft inheritance was impossible because there was no mechanism by which it could work, and even today, there is little evidence so far to support this possibility over anything except the short term (Chapter 17). Darwinian evolution combined with Weismannian cell biology formed the basis of *Neo-Darwinism*.

THE ERA OF EVOLUTIONARY GENETICS

A deeper understanding of variation became possible once Mendel's work on genetic inheritance, which had been published in 1865 and then forgotten, had been rediscovered around 1900. Soon afterwards, de Vries proposed his mutation theory of evolution. This was soon followed by the work of Hardy and Weinberg, who independently showed in 1908 that in the absence of selection and other factors, gene populations would be stable over time. A decade later, Ronald Fisher produced the basic theory of *evolutionary population genetics* that included ideas of selection whose effects would be to break down Hardy-Weinberg equilibrium. This remarkable and original work was extended and deepened over the next few decades by a series of major figures that included Chetverikov, Dobzhansky, Haldane, Huxley, Mayr, and Wright. The resulting *Modern Evolutionary Synthesis* combined Neo-Darwinist ideas on evolution with models of population genetics to produce a mathematical theory of how gene and trait frequencies could change over time and so lead to evolutionary change.

The strength of the synthesis was that it integrated evolutionary change and population genetics in a formal mathematical framework that was based on four key points.

1. Within a population, there is a slow accumulation of mutations that have small effects, and these lead to genetic and phenotypic (trait) variation.
2. Natural selection (Malthusian competition) acts on these variants, and some will do well, while others will be lost.
3. Evolutionary change can be modelled by adding selection and other factors to the mathematical theory of population genetics.
4. The mechanism by which natural selection works is that the subgroup within any population that produces the largest number of best-adapted offspring in the sense that they too will leave disproportionately large numbers of offspring, the technical meaning of survival of the fittest, will eventually come to dominate the population.

It was not, however, until the 1940s that Mayr and others realised that the formation of new species always started with small groups of organisms becoming reproductively isolated from their parent populations. If this happened in an environment with different selection pressures from those of its parents, variants within the small group that had alleles and traits that enabled them to thrive in the new environment would eventually come to dominate. Further mutations in both parent and daughter groups would drive them further apart genetically and eventually lead to reproductive isolation. In due course, the descendant of the original small group would have become a new species.

In the late 1960s, Motoo Kimura pointed out that neutral genes (those whose phenotypes were not subject to selection) would be spread through a population through genetic drift, hence playing a part in the evolutionary genetics of speciation. (Drift reflects the way in which gene frequencies can alter in a population due to random breeding.) In a population where a new allele of a gene arises through mutation, genetic drift plays an important role in its distribution across the group. In a large population, distribution is slow, but drift in a small population can produce quite large changes over a relatively small number of generations (Chapter 21). This realisation had two important implications. First, genetic drift can be as important as the mutation in creating phenotypic variants and often more important in the short term as mutation rates are low. Second, in small, segregated groups of organisms, the asymmetric gene distributions resulting from genetic drift are key to producing new variants rapidly. It is from these variants that new species eventually arise (Chapter 22).

A further point implicit in this synthesis is that organisms are engaged in a battle for survival. This idea, summarised in Tennyson's phrase "Nature, red in tooth and claw" from his 1850 poem *In memoriam,* had been argued against as early as 1902 by the Russian biologist (and anarchist) Pyotr Kropotkin in his book *Mutual Aid,* which focused on how organisms depend on each other to survive and flourish. It nevertheless took until the 1960s before kin selection, which incorporated altruism (organisms acting in a way that was beneficial to others, particularly relatives, at the expense of their own interests), was accepted and explained within the modern evolutionary synthesis. This synthesis has since been extremely useful in analysing population changes, in working through how selection can operate and in seeing how speciation can occur. It currently underpins thinking about how symbionts and holobionts evolve and flourish (Chapter 19).

These ideas were, however, based on the idea that speciation takes place at a slow and uniform rate that depends on the gradual incorporation into a population of trait-changing mutations that survived selection (this is phyletic gradualism or anagenesis). In 1972, however, Eldredge and Gould suggested that novel speciation did not simply occur through the slow accumulation of small changes in a species but could proceed through periods of stasis alternating with short periods of rapid change, a process that they called punctuated equilibrium. This was not to say that change occurred through a single mutation that had a major effect (saltation), but that the weight of accumulated mutations eventually led to a relatively rapid change.

As ever, the answer to whether this view was correct or not lay in analysing data and, by this time, the fossil record was very rich. Detailed examination of long-lived families of fossilised organisms extending from small bryozoans (coral-like sea organisms) to large dinosaurs revealed examples of punctuated equilibrium that reflected changes in the environment and hence changes in selection pressures. Other organism sequences, however, showed phyletic gradualism or slow, roughly continuous change. In other words, there turned out to be a spectrum of change dynamics.

Meanwhile, a major change was taking place in ideas on how best to group species for discussions of evolution. Traditionally, this had been on the basis of embedded classes, each with their own common anatomical properties, with higher classes having fewer than lower ones (Linnaean taxonomy, see Chapter 4 and Appendix 2) so that a species was a member of a family, families were grouped in orders that were grouped in classes, and so on, irrespective of when that species lived. There had been some dissatisfaction with such a hierarchy because it failed to capture evolutionary relationships in any coherent way. In the 1950s, Willi Hennig produced such a system that he called phylogenetic systematics, an approach now called *cladistics*. In this, species are linked by the relationship *A descends with modification from B* on the basis of shared and derived characteristics (Chapter 5), and relationships are traced back to a succession of common ancestors. So successful has this tree-of-life approach been in integrating living and extinct organisms that Linnaean taxonomy has had to be modified to include its results (Chapter 4).

THE MOLECULAR ERA

These classical ideas were all based on observable traits and behaviours (the *phenotype*) which, of course, could say nothing tangible about their then unknown heritable underpinnings (collectively grouped in the *genotype*). The discovery of the DNA structure in the 1950s and of the molecular apparatus for turning

DNA sequences into proteins in the 1960s were landmarks for our understanding of evolutionary mechanisms: they not only demonstrated that the genotype was encoded in specific DNA sequences but showed that variation was achieved through mutation of those sequences. Work over the last thirty years, in particular, has explained a great deal about the nature of the genotype, the sorts of mutation that have occurred and their evolutionary implications for generating anatomical and physiological change (Chapters 7 and 8).

One particularly fruitful area of research that has resulted from this knowledge has been the use of computational approaches to analyse DNA sequences for proteins with similar functions in very different organisms (e.g. cytochrome C). This work was based on analysing the DNA-nucleotide or amino-acid sequences on the basis of the mutational differences between them to see if it was possible to generate a hierarchy of the most likely paths of descent from an original candidate sequence. The computations showed that the information would indeed form a tree, and the differences were thus shown to have formed by mutation from an ancestor sequence; this result had been expected but not hitherto demonstrated. This early success has been widely extended; we now know that there is a very wide range of homologous proteins (i.e., based on common descent) with similar functions across a wide range of organisms. Each set can be integrated into an evolutionary hierarchy on the basis of descent with mutational modification and traced pack to an original sequence possessed by a last common ancestor.

Such phylogenetic trees or molecular phylograms not only show how contemporary organisms should be grouped but, because they use much more precise data, do so with greater accuracy than is possible with cladistics analysis based on anatomical differences (Chapter 7). Computational sequence analysis has also allowed us to reconstruct the likely genetic composition of early prokaryotic organisms from the Archaea and Eubacteria, and to explore the ways in which eukaryotic life evolved from them through endosymbiosis (Chapters 9 and 10), even though there is only minimal fossil record.

Phylogenetic analysis can now be extended using coalescent theory to model the mutational changes of current DNA variants backwards in time to a last common ancestral sequence within a population. The technique for doing this takes, for example, the two sequences of the many genes on an individual organism's pair of chromosomes to conduct a phylogenetic analysis within the context of a model of population genetics and a likely mutation rate. The analysis produces the estimates of the original sequence, how many generations back it was present and the size of the population over time since that last common ancestor started to vary (Chapter 7). Such approaches have, in particular, highlighted how and where successive founder groups of *Homo sapiens* colonised the world (Chapter 23).

Further understanding of evolutionary change has come from the application of molecular genetics to investigating an old insight, that the generation of novel or changed anatomical features in adults reflect the effect of changes taking place during embryogenesis (it seems that this obvious point was first pointed out to Darwin by Thomas Huxley). There had always been some interest in similarities in the development of different embryos, but it was not until the 1980s that it became possible to compare the activity of homologous proteins in very different organisms. This work opened up the field now known as *evo-devo* (Chapter 8 and Appendices 6 and 7), and it soon became clear that very similar genes which clearly had an early common ancestor and so were homologues had very similar roles across a wide range of organisms. One well-known example here, in addition to the Pax6 homologues mentioned in the last chapter, is the set of Hox genes that organise body patterns in animals as different as mouse and *Drosophila*. Evo-devo has provided some of the most powerful evidence on how evolutionary change occurs.

The most recent insights in this area have come from the realisation that much of embryonic development results from the outputs of the complex protein networks that drive growth, differentiation, and cell movement. It has now been shown that there is a great deal of protein homology in the networks that drive similar events in very different organisms, and this observation is clarifying further evolutionary relationships among very different organisms (Appendix 1).

Such work has also highlighted a longstanding distinction between the use of the word *gene* by classical and by molecular geneticists. The former, following Mendel, defines the term by its effects on the phenotype (e.g. Mendel's gene alleles for wrinkled and smooth peas and the genes underlying complex and quantitative traits such as altruism and height), whereas the latter defines a gene as a DNA sequence with some function. The former has the advantage that mutations in what can be called a *trait gene* are immediately obvious, whereas many mutations in *sequence genes* have no noticeable effect on the phenotype. The difficulty with the trait gene today is that its molecular basis is often hard to work out. It is only in relatively few cases that individual mutations in such a gene can be directly linked to changes in the phenotype (these are now called *Mendelian genes*). Although the relationship between the two definitions for a gene has not yet been clarified, it is becoming clear that the classical gene is often far closer to representing the effect of a network of proteins than that of a single protein (Chapters 17 and 18).

Finally, it is worth noting that the molecular revolution has not so much changed our basic understanding of evolution as it has opened up great areas of research into the previously hidden genotype and, in particular, explained much about the mechanisms of variation. The net result is that some 200 years of anatomical, embryological, genetic, and molecular research has established beyond doubt that the richness of today's biosphere evolved from a single and simple bacterium (Chapter 9).

A very great deal is now understood about the evolutionary history of life on Earth and the nature of the mechanisms that have driven the associated explosion in species. There are, however, still gaps in our understanding: we do not know how that first bacterium formed, and we know almost nothing about the evolution of neuronal activity and cognition. We also need to identify the details of the evolutionary changes that, over a period of ~7 My, led a minor spur off the great-ape line to become modern humans with unique brains. These became so neuronally powerful that they could produce speech and then capture it in written language, the major factor responsible for cultural inheritance and our current dominance.

FURTHER READINGS

Ayala FJ, Avis JC (eds) (2014) *Essential Readings in Evolutionary Biology.* Johns Hopkins University Press.

Darwin CR (1859) *On the Origin of Species by Means of Natural Selection or the Preservation of Favoured Races in the Struggle for Life.* John Murray. http://darwin-onlineorguk/converted/pdf/1859_Origin_F373pdf.

Gould SJ (2000) A Tree Grows in Paris: Lamarck's Division of Worms and Revision of Nature. In *The Lying Stones of Marrakech: Penultimate Reflections in Natural History*, pp. 115–114. Harmony Books. http://wwwfacstaffbucknelledu/sdjordan/PDFs/Gould_Lamarck'sTree.pdf.

Mayr E (1982) *The Growth of Biological Thought.* Belknapp Press.

Wallace AR (1858) On the tendency of varieties to depart indefinitely from the original type. *Proc. Linn Soc*; 3:53–62. people.wku.edu/charles.smith/wallace/S043.htm

WEBSITE

- The Wikipedia entry on Evolution and the people mentioned in this chapter.

The Ancient World

The purpose of this chapter is to give a basic, introductory picture of the history of life, particularly as told by the fossil record and the associated rock datings. Supplementary information on the geological and evolutionary events during each of the geological periods is given in Appendix 3 and online in the book's website, while core information on the formation of rocks and fossils, together with some of the techniques for dating them, are in Appendix 4. Note that the fossil data alone are not enough to provide the full history of life that is detailed in Section 3a as there are few fossils representing pre-Ediacaran life or of small organism. Building a full picture of life's history also requires insights from phylogenetic analysis based on anatomy and genomics (Chapters 5–7) and from evolutionary developmental biology (eve-devo – Chapter 8). Together, this information not only supplements the fossil record of the last ~600 My but provides a window into the distant past before that record was laid down.

Our direct knowledge of the history of the Earth and the animals and plants that have lived on it comes from analysing fossilised organisms and the sedimentary rocks in which they were found. Here, the rocks provide information about when and where organisms lived (Appendix 4), while fossils give us anatomical information, albeit mainly limited to hard tissues. This is because fossilisation depends on organisms dying under conditions in which their tissues degrade slowly enough for mineralisation to take place before they have been eaten by predators or broken down by the environment. As soft tissues are usually lost fairly rapidly, this means that organisms lacking a hard internal or external skeleton will rarely have had the chance to be preserved. Even for large organisms, fossilisation is a random and rare process. The result is that the fossil record says little about small, soft-bodied organisms, be they prokaryotic or eukaryotic, although they and other organisms may have left traces of their behaviour.

Occasionally, however, the fossil record is remarkably good and includes organisms of astounding quality because their soft tissue structure has been preserved. The extent of the tissue integrity and anatomical detail seen, for example, in the fossils of early marine animals from the Cambrian (541 Mya, Chapter 13), is almost unbelievable. The rare sites in which fossils showing fine anatomical detail have been found are known as *Lagerstätte* (see Websites at the end of chapter). As to prokaryotic fossils, even they have occasionally been preserved, albeit as assemblages rather than individuals (e.g. stromatolites; Chapter 9).

The low chances of small soft-bodied organisms being fossilised explains why little direct evidence of any life has been found before the Ediacaran, when macroscopic organisms first evolved, and particularly before the Cambrian, when animals with exoskeletons first appeared. Similarly, there is little evidence of plant life before the Mid-Ordovician period (~470 Mya), when bryophytes (e.g. mosses) that evolved from green algae became able to live on land because a waxy cuticle had evolved to protect these primitive organisms from desiccation. The early evidence for algae and fungi, the other two major groups of multicellular phyla, is even weaker as few, if any, have

DOI: 10.1201/9780429346217-4

robust tissues. The fossil record, other than for animals with shells, thus mainly represents a series of almost random snapshots of the hard tissues of relatively large and physically robust organisms.

The situation today is that many tens of thousands of fossilised organisms from the Ediacaran (~635 Mya) onwards have been catalogued and dated. This record of extinct organisms is impressive and still improving as more fossils, and new dating techniques are discovered. Such information as we now have is sufficient to describe much of the detail of the history of multicellular life. In some cases, it is even possible to follow the process of transition as one structure evolved into another (Chapter 6 and websites). Cumulatively, the fossil record shows how life evolved from very primitive organisms to the rich diversity of the biosphere today, although its primary focus is on vertebrate and invertebrate animals with only a secondary focus on plants.

It should be noted that the information that can be obtained from a specific fossil is more limited than might be apparent when one first looks at a beautifully preserved and mounted skeleton. Without many more examples of the species, it can be hard to know whether it was male or female, whether it is typical, and what were the start and end dates of the species, unless, of course, it died out during a major extinction. Its evolutionary timing can also be opaque because the fossil evidence often confirms that similar organisms survived in parallel for long periods. This is particularly clear for human evolution: we know that several human subspecies lived at the same time in different parts of the world until relatively recently (Chapter 23).

It is therefore important to emphasise that a single or even a few similar fossils, no matter how spectacular, give limited insight into the origins and descendants of that organism unless the details of its anatomy can be analysed in the context of the anatomy of other similar organisms (i.e., through cladistic analysis, Chapter 5). Although this can generally be done, it is worth noting the anomalous status of the "Tully monster" (see Box 14.1).

THE EVOLVING EARTH

Earth formed ~4.5 Bya, with the crust forming soon after (the Hadean Eon). The earliest rocks date to ~4 Bya (the arbitrary beginning of the Archaean Eon – 4.0–2.5 Bya), and there is evidence of liquid water being present soon after. Although the first tectonic plates probably date to ~3.9 Bya (Næraa et al., 2012), there are very few examples of early rocks. One reason may be that in the Archaean Eon, when the Earth was hot, the movement of tectonic plates was faster than it is now, and this was accompanied by substantial subduction as plates were lost as others moved over them. During the Proterozoic Eon (2500–541 Mya), which started with the appearance of atmospheric oxygen, things seemed to have become more settled, and the geologic record was much better. From then onwards, we have an ever-improving picture of the movements of the early continental masses and the intervening seas in which early life was very slowly evolving.

The simplest way to view the Earth's surface over geological time is through the movement of a succession of supercontinents formed from tectonic plates (Table 3.1), each of which eventually broke up. Although the earliest, Vaalbara, dates back to more than 3 Bya, those relevant to evolutionary history start with Rodinia, which lasted from ~1.1–0.75 Bya and covered a period when all life was probably marine.

TABLE 3.1 SUPERCONTINENTS OVER THE PAST TWO BILLION YEARS

Name	Approx. age (Mya)	Period/Era Range
Columbia (Nuna)	1,820–1,350	Paleo- + Meso-Proterozoic
Rodinia	1,130–750	Mesoproterozoic-Tonian
Pannotia	633–573	Ediacaran
Gondwana	596–180	Ediacaran-Jurassic
Laurasia	472–200	Ordovician-Jurassic
Pangaea	335–173	Carboniferous-Jurassic

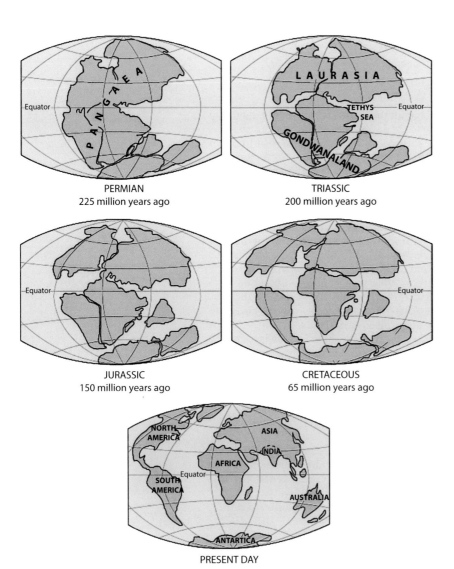

Figure 3.1 World maps from the
beginnings of the breakup of Pangea into
Laurasia and Gondwana, from ~250 Mya to
modern times.

In due course, Pannotia formed (~633 Mya), and its breakup (~573 Mya) came towards the end of the Ediacaran, an event that was accompanied by major climatic change. Part of Pannotia became Gondwana, a major southern landmass, while the remaining fragments eventually reassembled to form the more northerly Laurasia (~425 Mya). ~336 Mya, Gondwana and Laurasia coalesced to form Pangaea, the last of the supercontinents. Pangaea lasted for ~160 Mya before breaking up into slightly different versions of Gondwana and Laurasia ~173 Mya during the Jurassic.

Following this breakup, Gondwana and Laurasia started to fragment, and the separate drifting of their constituent plates eventually led to the continents of the modern world (Fig. 3.1). It is worth noting that the early tetrapods began to live on land a little before the formation of Pangaea and that the subsequent history of terrestrial life is thus linked to plate tectonics and continent movements. Thus, for example, Gondwana included the future Australian, Antarctic, and South American plates, and it was across this region that the Metatheria (marsupial mammals) migrated during the early Cenozoic Era.

In general, we have some understanding of the relationship between time, rocks, and fossils over the last two billion years or so; an understanding that improves as time approaches the present. There are, however, a few exceptions where erosion and complex plate activity have led to layers of rock whose structure is chaotic; these are called *unconformities*. The best known of these is the *Great Unconformity* (525–515 Mya) and was probably due to substantial tectonic activity crushing already formed rocks. This unconformity took place near the beginning of the Cambrian (Keller et al., 2019) and is particularly noticeable in the layers of the Grand Canyon rock face. This unconformity means that finding early Cambrian animal fossils is particularly hard, although those sites that remain have yielded spectacular fossils.

Extinctions

In addition to the effects of tectonic plate movements, the second set of events helped shape the evolution of life. This was the sequence of dramatic climate changes that led to substantial and virtually immediate extinctions of great swathes of life, particularly over the last ~500 Mya (Table 3.2). Each of these was followed by further radiations of the surviving organisms, with different groups becoming predominant in different periods.

The first of the major population-devastating extinctions seems to have occurred ~2.4 Bya and was probably due to the build-up of oxygen by cyanobacteria which led to the poisoning of anaerobic bacteria. Some, such as the great Permian extinction, appear to have been due to the build-up of volcanic gases or have origins that are still not understood. Many, however, seem to have been the result of large asteroids colliding with the Earth. Such collisions on land would have caused impact craters and the release of vaporised rock and sufficient atmospheric soot and ash to block out the Sun. This would have caused both a temperature drop for substantial periods, and perhaps for the atmosphere, to become poisonous. Asteroid collisions at sea would have led to megatsunamis, whose waves would have devastated nearby lands.

The net effect of major asteroid collisions would thus have been to devastate living organisms of all types, particularly land-based animal fauna. The resulting extinctions can now be seen as abrupt changes in the fossil record in the rock layers, and they mark the transition from one geological period to the next (Table 3.1). Many of these extinctions have been analysed on the basis of geological data, and we now understand the nature of some of the events that caused such dramatic changes to life at the time. The origins of others still remain obscure and are only recognised through dramatic changes in fossil populations.

Extinctions are not, however, limited to those caused by global catastrophes: the fossil record shows that almost every species becomes extinct in due course and is often replaced by others that, on morphological grounds, seem to have derived from it or its near ancestors. In contrast, a few, such as the horseshoe crab and velvet worms, have changed relatively little over very long periods; these species are referred to as living fossils (Fortey, 2011).

The net result of two centuries of palaeogeographical research is that we can now identify rock layers around the world on the basis of the age of the layer and the fossils that they contain (Appendices 3 and 4). Together, they give us a clear picture of how life in all of its variety evolved. Table 3.3 gives core information detail about these periods, while Hoyal et al. (2020) provide a modern, quantitative perspective on their effects.

TABLE 3.2 SOME MAJOR EXTINCTIONS

Name	Mya	Likely cause	% Genera lost	Main results
Ordovician-Silurian	~445	Gondwana moved south, global cooling.	60%	Loss of most brachiopods, bivalves, corals, echinoderms.
Late Devonian	374	Unclear.	~25% marine	Marine losses: most trilobites, brachiopods and corals. Land insects & plants seem to have been relatively unaffected.
Permian-Triassic	252.2	Volcanic activity and complex environmental effects, perhaps amplified by an asteroid collision.	~83%	Every group was affected. On land, most mammal-like reptiles, labyrinthodont amphibian and large insects were lost. At sea, all spiny sharks, trilobites, sea scorpions and most snails, echinoderms, brachiopods, bivalves, ammonites and crinoids were lost.
Triassic-Jurassic	201.3	Volcanic activity + global warming.	~20% marine	Loss of conodonts in sea. On land, many large non-dinosaur archosaurs, some therapsids and large amphibians were lost.
Cretaceous-Paleogene (K-T)	66	Asteroid collision and perhaps other factors.	~75%	All dinosaurs (except birds) and ~75% of other animal and plant species were lost on land together with the giant marine lizards.

TABLE 3.3 GEOLOGICAL TIME AND EVENTS *(FROM THE WIKI ON GEOLOGICAL TIME)*

Eon	Era	Period	Epoch	Starts at	Major biological events
	Cenozoic	Quaternary	Anthopocene	1945	Starts with the first atom bomb; it marks the effects of humans on Earth's rocks
			Holocene	11.7 Kya	Starts after last ice age—modern times
			Pleistocene	2.59 Mya	Modern *Homo* species
		Neogene	Pliocene	5.33 Mya	*Homo habilis* (at end of epoch)
			Miocene	23.03 Mya	Modern mammals, first apes, modern birds
		Paleogene	Oligocene	33.9 Mya	Further mammals, anthropoid primates, plant diversification
			Eocene	56 Mya	More mammal diversification; grasses appear
			Paleocene	66 Mya	Placental-mammal diversification; modern flowering plants
	Mesozoic	Cretaceous		145 Mya	Further dinosaur radiation, marsupial and placental mammals split
		Jurassic		201.3 Mya	Dinosaur radiation, first birds and mammals
		Triassic		252.2 Mya	Archosaur radiation, earliest small mammals, teleost fishes, flowering plants
	Paleozoic	Permian		298.9 Mya	Mammal-like-reptile radiation, therapsids, cycads, conifers
		Carboniferous		358.9 Mya	First land reptiles, amphibian radiation, winged insects, forests, gymnosperms
		Devonian		419.2 Mya	Jawed fish radiation, amphibia, first trees, club-mosses, horsetails, land insects
		Silurian		443.4 Mya	Bony fish, first vascular plants on land, first land invertebrates
		Ordovician		485.4 Mya	Invertebrate diversification, jawless then jawed fishes, first land plants and fungi
Phanerozoic		Cambrian		541 Mya	Explosion of phyla in the fossil record
	Neoproterozoic	Ediacaran		635 Mya	Ediacaran fauna and sponges, trace fossils of worms
		Cryogenian		850 Mya	This seems to have been a period of extreme cold - fossils are rare
		Tonian		1000 Mya	Trace fossils of multicellular organisms such as plankton (e.g. dinoflagelates)
	Mesoprerozoic			1600 Mya	Green algae colonies
Proterozoic	Paleoproterozoic			2500 Mya	First eukaryotic cells & algae; oxygen from cyanobacteria change atmosphere
Archean				4000 Mya	Archaea, bacteria, stromatolites, oxygen-producing, blue-green cyanobacteria
Hadean				4567 Mya	No good evidence of life

FOUR BILLION YEARS OF LIFE

Prokaryotes

Initially, the world's atmosphere seems to have been composed mainly of nitrogen and carbon dioxide together with the gases that resulted from volcanic activity. The effects of these gases on the ocean would have been to make it slightly acid and reducing with a pH of about 6.6. It was in this environment that a primitive bacterium known as the First Universal Common Ancestor (FUCA) evolved almost four Bya (Chapter 9). By ~2.5 Bya, some of these eubacteria evolved into cyanobacteria capable of photosynthesis, a process that harnessed carbon dioxide to synthesise glucose and released oxygen, so raising the ocean pH to about 7.0 (Krissansen-Totton et al., 2018) and slowly changing the atmosphere. Today, the ocean pH is about 8.1 but is slowly declining as a result of today's carbon dioxide build-up.

Both geological and fossil evidence (e.g. from stromatolites) suggest that there was a major change in the Earth's climate ~2.4 Bya, when carbon dioxide levels dropped and oxygen levels rose; this was the *Great Oxygen Event*. It followed ~100 Mya of cyanobacterial activity and marks the time when sufficient amounts of oxygen were being produced in the oceans to change them from being weakly reductive to mildly oxidising and defines the Archean–Proterozoic boundary. These events seem to have resulted in the poisoning of anaerobic bacteria so lowering the amount of carbon dioxide in the atmosphere. This drop in greenhouse gases in the atmosphere led to a lowering of temperature and eventually to the Huronian glaciation that ended ~2.1 Bya.

It seems that, a little later, archaebacteria evolved from earlier bacteria in ways that are still not understood. These prokaryotic cells are very different from the original eubacteria: they have a membrane based on glycerol-ether lipids rather than the glycerol-ester lipids of contemporary eubacteria, while their proteins for cell division and protein synthesis are closer to those of eukaryotic cells than are those of eubacteria (Chapter 9). They are also far more robust than eubacteria in that they can live under far more extreme conditions (which is where they were first identified and so were originally known as extremophiles).

Unicellular eukaryotes

It is only recently that the origins of eukaryotic cells have become understood through the use of phylogenetic analysis on DNA sequences: there is now an overwhelming amount of evidence that eukaryotic cells formed as a result of several endosymbiotic events over the Proterozoic era (2.5–1 Bya). These very rare processes occur when one cell engulfs another but, instead of breaking it down, lets it maintain a semiautonomous existence (Chapter 10). The exact details of what actually happened are still argued about (Chapters 9 and 10) but, by ~1.7 Bya, the Last Eukaryotic Common Ancestor or LECA had evolved. This unicellular eukaryote had a nucleus that was predominantly archaebacterial, membranes that reflected a eubacterial origin, and a mitochondrion that probably derived from a *Rickettsia* eubacterium. This cell was the ancestor of animals, fungi, and amoebae. There was then a further endosymbiotic event ~1.1 Bya: in this, a LECA descendent endosymbiosed a cyanobacterium that became its chloroplast; this was the last common ancestor of green algae and, in due course, plants. Further algae formed later when unicellular protists from other phyla endosymbiosed a unicellular alga, so secondarily acquiring a chloroplast.

Over the ~1.7 Bya since the LECA first appeared, its descendants evolved in many directions, so leading to the many phyla that now inhabit the world. In parallel with these events, early eukaryotic cells within many phyla seem to have become social, sometimes forming stable colonies of cells that later evolved into small organisms; the development of gamete-forming cells within them was probably an early step in cell differentiation. Sadly, the fossil record of these small, soft organisms is very limited indeed (Chapter 10).

Early large eukaryotes

Large-scale organisms up to perhaps a metre across are first seen in the Ediacaran period (635–541 Mya). These had frond- and disc-like morphologies and bear little resemblance to today's organisms. Simple taxa that were clearly metazoan also evolved during the Ediacaran or perhaps even earlier (Turner, 2021): sponges were probably first on the basis that they have the most simple organization, but soon, the earliest organisms with an ectoderm and an endoderm germ layer appeared. These are now the diploblast coelenterates that have radial symmetry and include the Ctenophera (e.g. comb jellies) and Cnidaria (e.g. jellyfish). Later in the Ediacaran, the first primitive triploblasts evolved: these were characterised by a mesoderm layer and mirror symmetry along their anterior–posterior axis. The resulting worm-like organisms were the root organisms of the Bilateria, the category that today includes almost all animals. It now seems that the stem species of many of today's major groups of phyla had evolved by the end of the Ediacaran.

A second dramatic increase in atmospheric oxygen levels occurred for unknown reasons in the Neoproterozoic Era, probably leading to modern levels by the end of the Ediacaran (541 Mya), the period before the start of the Cambrian (Liu et al., 2019b). This, it is thought, might have been the event that triggered the line of evolution that resulted in the massive increase in metazoan diversity, the only kingdom with an absolute requirement for oxygen. The period of cold that ended the Ediacaran was followed by the beginning of the Paleozoic Era, which started with the Cambrian (541–485 Mya), during which there was a spectacular burst of mutational activity known as the Cambrian explosion. This resulted in the evolution of a host of new body patterns that are seen in the fossils of these mainly small marine organisms; some were lost, but most survived and further evolved to give the animal phyla that we see today (Chapter 13). The dominant phylum was that of arthropods, but there were other invertebrate clades. The Cambrian fossil record also includes primitive marine chordates (e.g. *pikaia* and *haikouella*) that eventually evolved into fish (Chapter 14; see Fig. 3.2 for a diagram of post-Cambrian evolution).

There is some fossil evidence to suggest that limited numbers of primitive plants and fungi were established on land by the Ordovician (485–443 Mya), while a few invertebrates, such as scorpions, were terrestrial by the early Silurian (443–419 Mya). The earliest land plants that evolved from algae were probably bryophytes such as mosses. These have minimal roots and lack vascular tissues and proper leaves but are covered with a water-impermeable, waxy cuticle that stops them from drying out (Chapter 11). By the Devonian (419–359 Mya), ferns with wood and roots had evolved, and they were soon followed by gymnosperms. Early species produced seeds from leaves but, in due course, these seed-producing leaves were modified to become cones. Angiosperms (flowering plants) did not, however, evolve until the Jurassic

Figure 3.2 A diagram showing the progression of the richness of life on Earth, focusing on events since the Cambrian Period.

(201–145 Mya); they then diversified rapidly to the extent that flowering plants of all sizes now dominate the plant world (Chapter 11).

All vertebrate life was marine until ~365 Mya, late in the Devonian, when one at least of the inshore species with fins that had evolved into limbs and whose ancestors had separated off the sarcopterygian line (today's lungfishes and coelocanths), clambered up onto the shore. It was able to do this because its fins had extended to become limbs, and these had been strengthened by a pelvic and pectoral girdle. These innovations had evolved to allow it to walk on the sea bottom in shallow waters and to wade through dense inshore seaweed in search of food (Chapter 14). Such an organism was a stem tetrapod, and its amphibian descendants dominated land for some 50 Mya, by which time another line of descendants had evolved to become stem amniotes whose eggs could survive away from water (late Carboniferous, 359–299 Mya).

These stem amniotes rapidly diversified during the Permian (299–252 Mya), forming groups of tetrapods that could be differentiated by skull morphology. The synapsids, amniotes characterised by a single temporal fenestra (a space at the side of the cranium behind the eye that was occupied by a large muscle used for biting) on either side of the head then became dominant. Perhaps the most famous synapsid was *Dimetrodon*, a large crawling animal that had a substantial bony sail on its back. Meanwhile, a further group of early amniotes that had two temporal fenestrae (the diapsids) diversified to become the reptiles. In due course, some of these diapsids lost their fenestrae and eventually became anapsids (today's turtles and tortoises). A particularly important evolutionary change occurred towards the end of the Permian when a group of diapsid reptiles evolved that were characterised by an antorbital fenestra (a space in the bones between the eye and the nostrils). This group gave rise to dinosaurs, crocodiles, and eventually birds (Chapter 15).

Animal life on land was almost destroyed in the Permian extinction (Table 3.1) and took several million years to recover. When, as it was, the dust had cleared, and the Mesozoic Era had started, the previously dominant synapsids had been reduced to a relatively minor group that slowly evolved into stem mammals and then mammaliaforms, many of which were nocturnal and lived in burrows. At the same time, the synapsid archosaurs became dominant and diversified widely, eventually evolving into a clade that included crocodilians, flying pterosaurs, and a broad range of carnivorous and herbivorous dinosaurs, with the largest weighing more than 50 tons (Chapter 15).

Of particular interest are the small carnivorous dinosaurs (theropods) that were less than a metre long - it was within this clade that birds evolved. Many earlier dinosaurs had feathers, presumably for insulation and perhaps display, but these small theropods seem to have been the only taxon that used them for gliding and later for flight. *Archaeopteryx* was the first such gliding dinosaur whose fossils were discovered, but many more have since been found so that the pathway to flight is now well documented in the fossil record. Meanwhile, mammaliaforms had, over the Mesozoic, evolved into the clade of monotreme, marsupial, and placental mammals, most of which were relatively small.

It was, however, the dinosaurs that were the dominant taxon over the whole of the Mesozoic Era (252–66 Mya) until an asteroid, which crashed into what is now the Yucatán Peninsula in Mexico, led to the K-T extinction and ended the Cretaceous. This catastrophe destroyed almost all animal life, particularly affecting the synapsid species. When the world recovered, all the large sea lizards and most of the archosaurs had been lost, leaving only the Crocodilia and non-flying birds, the smaller of which evolved to become capable of flight. The non-archosaur synapsids were reduced to a few taxa, the most important of which now are snakes and lizards. It was in this environment that early mammals flourished.

Soon after the beginning of the Cenozoic era, the mammalian taxa had become predominant (Chapter 16): they diversified widely, occupying most of the land ecological niches and eventually taking to the air (bats) and the sea (whales, seals, and manatees). The primate line originated quite early in the Paleocene epoch (66–56 Mya), but it took a further ~50 Mya until a spur that branched off from the primate branch in Africa ~7 Mya led to humans. The earliest group anatomically recognisable as a *Homo* taxon was *Homo erectus*, whose earliest fossil remains date to ~1.9 Mya (Chapter 23), and this gave rise to many subsequent subspecies, some of which left Africa. The best known of these are *Homo sapiens* and *Homo Neanderthalensis* that diverged some 350 Kya (they could still interbreed when they met up again in the Near East and Europe some 40 Kya), but there were many more (Table 23.2).

Meanwhile, a line that remained in Africa slowly gave rise *H. sapiens* who, by ~200 Kya, could be considered as an anatomically modern human. All have been lost apart from *H. sapiens*, a single species that now dominates the world to an extent that is unique, although its DNA still carries genetic contributions from its earlier relatives.

FURTHER READINGS

Benton MJ (ed) (2019) *Cowen's History of Life* (6th ed). Wiley Blackwell.

Cocks LRM, Torsvik TH (2016) *Earth History and Palaeogeography*. Cambridge University Press.

Hoyal Cuthill JF, Guttenberg N, Budd GE (2020) Impacts of speciation and extinction measured by an evolutionary decay clock. *Nature*; 588:636–641. doi: 10.1038/s41586-020-3003-4.

WEBSITES

- The Wikipedia entries on *Extinction Events and History of Earth*.
- *Lagerstätte* (sites of particularly fine fossils): en.wikipedia.org/wiki/Lagerst%C3%A4tte www.fieldmuseum.org/science/blog/mazon-creek-fossil-invertebrates
- Burgess Shale fossils: https://burgess-shale.rom.on.ca/en/science/burgess-shale/03-fossils.php
- The movement of plates over the last 3.3 By: https://www.youtube.com/watch?v=UwWWuttntio

Life Today: Species, Diversity, and Classification

4

Studying the origins of the diversity of organisms across the biosphere starts with examining the complexity of life today. The first step here is to organize current knowledge of living organisms in a way that lends itself to evolutionary investigation. The first part of this chapter discusses many of the elements of this rich field that illuminate the diversity of life; these include the basics of speciation, some of the difficulties in defining and recognising species and the extent of individual variation within a population.

The latter part of the chapter discusses how species can be grouped, and this can be done in two very different ways. The classical approach is on the basis of their anatomical similarities and differences. This produces the traditional Linnaean taxonomy, which includes a hierarchy of classes. Higher-level classes have fewer defining features so that a phylum, for example, is defined solely by a body pattern: arthropods are invertebrates with appendages that are paired and segmented. Insects are a class within the arthropods that have three pairs of limbs and a pair of antennae. The key grouping for recognizing individuals is the species, and these are marked by unique features, such as the details of skin patterning. In the Equidae, for example, *Equus zebra* has about 40 stripes, and *Equus grevyi burchelli* has about seventy stripes. Linnaean taxonomies say nothing about evolution as inheritance is not a property that it uses in classifying organisms.

The second way of classifying organisms explicitly incorporates evolutionary relationships and is based on the idea that organisms can be grouped because the evidence shows that they share a common ancestor. Such classifications generate historical trees, known as phylogenies, and are constructed on the basis of inherited similarities and novel differences. If these reflect anatomy, the phylogenies are known as cladograms: a clade is a branch of the tree that starts with an ancestor species with some new feature and includes all the descendants with this feature. Trees can also be constructed on the basis of a set of homologous gene sequences associated with a group of living organisms that share a common ancestor (i.e. the sequences are homologous): computational techniques that analyse mutational differences allow these sequences to be traced back to a common ancestral sequence. Such gene trees are known as molecular phylograms. In practice, cladograms only help integrate the fossil record, while molecular phylograms can be used across the biosphere as every contemporary organism has sequenceable genes. It is worth noting that, even though they are based on very different properties, Linnaean taxonomies, cladograms, and molecular phylograms group contemporary species in the same way.

The key purpose of this chapter is to introduce many of the themes that underpin the analysis of evolutionary change and that are discussed in subsequent chapters. It also points to some of the specific terminologies that is used for this, although full discussion of cladistic definitions is postponed until Chapter 5.

DOI: 10.1201/9780429346217-5

Over the last two hundred or so years, biologists have identified vast numbers of species. Some are prokaryotes, lacking a nucleus, but most are eukaryotes, having one or more cells with a nucleus. Two hundred years ago, this diversity was seen as reflecting divine creation and the wish of God to place organisms in environments in which they would thrive. Mainly because of Darwin's writings, however, this view was generally seen to be wrong by the end of the 19th century, and it is now clear that this diversity arose through evolution.

It is, however, one thing to acknowledge evolutionary change and another to work out how today's organisms evolved from ancestral ones. This chapter serves two purposes in this context. The first part introduces the reader to the complexity of life today and how sense can be made of it by grouping species on the basis of shared features. The latter part of the chapter discusses how species are integrated within taxonomies that have a formal structure. These have an importance beyond providing a means of organising species. Linnaean taxonomies group anatomically similar classes of organisms and these classifications provide the raw material for constructing phylogenies based on common evolutionary descent. These, in turn, pose the questions about *when* and *how* diversity was generated. Section 3 of this book is concerned with *when* and so considers the history of life (the study of this is now known as *Biological systematics*). Section 4 focuses on *how* by discussing the mechanisms of change and the ways in which new species evolve from existing ones.

THE DIVERSITY OF LIFE TODAY

Grouping species begins with identifying those that share common properties, and then organising these groups within a formal structure. Some indication of the difficulty of making sense of the diversity of life today is clear when the remarkable range of sizes of multicellular organisms now living is considered (Fig. 4.1). Vertebrates, for example, extend from *Paedophryne amauensis*, a small frog less than 8 mm in length and ~0.5 g in weight, to the blue whale, *Balaenoptera musculus*, up to 30 m long and weighing up to ~170 tonnes (the volume ratio is ~$1:3 \times 10^8$). Flowering plants extend from around 1–2 mm (*Wolffia globosa*, a duckweed that weighs a few milligrams) to almost 100 m in height (e.g. the giant sequoia, *Sequoiadendron giganteum*, and the fig, *Ficus benghalensis*, both of which can weigh well over 1000 tons). The length ratio of these two trees to *Wolffia* is ~$1:10^7$ and the weight ratio about ~$1:10^{20}$. Single-celled organisms are, of course, very much smaller: *E. coli* is a cylinder ~2 μm in length and ~1 μm in diameter and weighs ~2×10^{-12} gm, while a small unicellular eukaryotic yeast can be as little as ~4 μm in size and weighs ~6×10^{-11} gm. These examples show that the scale of life is almost inconceivably large.

When the smallest vertebrate is compared with the largest, the differences are more than obvious, but the parallels shouldn't be ignored. The organ systems in the frog and the whale are remarkably similar, and the range of cell types and basic morphologies are much the same. Their key tissues are epithelia, essentially 2-D sheets of cells that cover external surfaces and line internal tubes, and mesenchyme, the embryonic precursor of most 3-D tissues (e.g. dermis, muscles, bones, tendons, and cartilage), together with the neuronal system and the many subtypes of blood cells. This is, of course, because their last common ancestor was a Permian stem tetrapod with the same set of organ systems.

As to the plants, the similarities between duckweed and the fig tree are particularly apparent in their respective reproductive systems: both have the flowers, pistils, stamens, and seeds characteristic of flowering plants. The differences are in the supporting tissues. Comparison of the fig or sequoia with a small woody shrub called *Salix herbacea*, which is only a few cm high, shows that it is hard to discover any differences in the anatomical entities or the cell types that make up the plant—the differences are merely of scale and minor morphological detail.

Our eye naturally looks for differences and inevitably focuses on the exterior rather than on the organism as a whole. Most of an organism's tissues are, of course, internal and, if scale is set aside, can usually be seen to be similar to most other organisms in the same phylum, at least developmentally. It is these common features that allow us to group organisms into species and then into families of species.

Figure 4.1 (a): The frog *paedophryne amauensis* is the smallest vertebrate known (~8 mm). (b): The skull of the blue whale, *balaenoptera musculus*, the largest and heaviest contemporary vertebrate (29 m, 180 tons). (c): Hundreds of plants of the duckweed, *Wolffia globosa*, the smallest known flowering plant (2 mm). (d): An example of *salix herbacea*, the smallest woody plant (~4 cm). (e): The redwood, *Sequoiadendron giganteum*, the tallest tree known (90 m, ~1000 tons).

(a Courtesy of Dr. Chris Austin.

b Published under a Creative Commons CCO 1.0 license.

c Courtesy of Wayne Armstrong.

d Published under a GNU free documentation license.

e Courtesy of Mike Murphy. Published under a Creative Commons Attribution-Share alike 2.0 generic license.)

DEFINING A SPECIES

In distinguishing one population of organisms from other similar groups, it is universally agreed that the obvious unit for classifying populations of eukaryotic, sexually reproducing organisms is the *species*. This is ideally defined by the interbreeding criterion: if males and females from two different groups cannot interbreed to produce fertile offspring, they belong to different species; if they can, they are viewed as members of the same species. Such a functional test works quite well in the context of contemporary speciation and is particularly important in theoretical discussions on the evolution of novel species. This is because the reproductive isolation that derives from a new species' inability to breed with related species drives future diversity.

The interbreeding criterion is not, however, usually appropriate. First, the test can only be applied to living organisms; second, cross-group breeding experiments are generally hard to do with animals but are easier in plants, although the results can be difficult to interpret as plants form fertile hybrids far more readily than animals—plants seem to be particularly tolerant of polyploidy and genetic variation for reasons that are still not fully clear. The boundaries between plant species are thus often more blurred than those between animal ones.

Second, the criterion can lead to apparently counterintuitive results: all canines, for example, can interbreed to produce fertile offspring. On this definition, therefore, each group within the genus Canis, which includes domestic dogs, jackals, dingoes, coyotes, and wolves, would have to be viewed as members of the same species. In fact, of course, an ability between two distantly related species to crossbreed and produce fertile offspring normally only reflects that fact that the two have the same number of chromosomes. The interbreeding criterion is inappropriate for canine taxa because, for all their anatomical and other differences, they each have 78 chromosomes.

There are other criteria for demarcating species boundaries: these include morphology, habitat requirements, and niche adaptation. The last of these is particularly important for exploring the early stages of separation (Coyne & Orr, 2004; Chapter 22), but morphological criteria are the most commonly used. Even these are not simple for two good reasons. First, one has to be sure that the differences are substantial and do not simply reflect natural variation. There is a lesson here from domestic dogs. It is not obvious that, if the only data for analysis were their fossilised skeletons, an adult pug that has a very short snout and is ~30 cm high would be considered as being in the same species as a great dane with a very long snout and a height of more than 100 cm, or even a terrier that is also ~30 cm high but with a very different skull (Fig. 4.2).

Second, one has to be certain that any similar features are homologous. This term means that they reflect descent from a common ancestor that had those or similar features. It is important to exclude the possibility that the features arose independently and do not just reflect independent adaptive responses to similar habitats. Thus, for example, cacti and desert euphorbias can appear to be of the same family because both are succulents with spikes and thick layers of cuticle; other features, together with DNA phylogenetic analysis, however, show that they are only distantly related. Similarly, both wings and sabre teeth have arisen independently many times (Chapter 5), and there are many other examples of similar anatomical features in different organisms that have evolved independently as adaptations to a similar environmental niche rather than through common descent. These are known as *homoplasies* and reflect what is known as convergent evolution (McGhee, 2011); their existence emphasizes the importance of using a broad range of phenotypic characteristics in making taxon assignments.

Figure 4.2 The skull of a terrier with an elongated snout and a pug with a much shorter one.

(From Almén MS et al. (2016). *Bioessays*; 38: 14–20. With permission from John Wiley & Sons.)

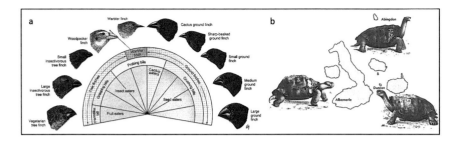

Fig. 4.3 Variation in the Galapogos Islands. (a): The various finch species showing how beak morphology adapted to a particular food. (b): Three species of tortoises. The Abingdon and Duncan forms feed on cactus on dry islands, whereas the Albemarle form feeds on vegetation closer to the ground on a wetter island. The former have upturned 'saddles', longer necks, and straighter backs. The latter has a domed shell and shorter neck.

(a From Patel N (2006) How to build a longer beak. *Nature*; 442:515–516. With permission of Springer Nature.
b Text by Steven M. Carr, © 2019. With thanks to Steven Carr for supplying the image.)

When assigning living organisms to distinct species, the normal anatomical criteria are the external characteristics (e.g. external anatomical features, size, and the details of skin morphology, such as patterns, appendages, and colour), which can be interpreted on the basis of habitat requirements. An example of this is Darwin's classic work on the different finches that flourished on each of the Galapogos islands: it is now clear that a key distinguishing feature here is beak morphology and its adaptation to island-specific food resources (Fig. 4.3). As the finches are still on their respective islands, this example of speciation is still accessible for study (Abzhanov et al., 2006; Abzhanov, 2010; Almén et al., 2016). It is worth noting that Darwin's original interest in variation was in the Galapogos tortoises as there were different species on the different islands, but he seems not to have analysed them carefully (Fig. 4.3b; Browne, 1995).

Extinct taxa can, however, only be assigned to specific species on the basis of fossilised tissues: these mainly reflect the tough exoskeleton of invertebrates, the hard tissues of plants and the bones and teeth of vertebrates. Palaeontologists have to know anatomy in very great detail if they are to recognise the differences between sets of apparently similar fossilised remains that enable them to be assigned to different species.

It used to be difficult to decide whether organisms that looked very similar should be assigned to the same taxon as it was not always clear whether the similarity derived from convergence or close inheritance. Today, DNA analysis usually resolves such problems (Chapter 6) and is widely used to confirm the validity of Linnaean classifications (see below). Comparative DNA sequencing of a particular gene across a group of organisms shows which of its members are closely related and which only distantly.

Such molecular phylogenetic analysis can be used to clarify detail at taxonomic levels much higher than that of species. Thus, for example, comparative sequence analysis of the various 18s ribosome RNA genes was important in showing that there was an early and high-level subdivision of the chordate phylum into protostomes and deuterostomes (Fig. 13.7; Peterson & Eernisse, 2001; Giribet, 2008). It was also used to confirm that the traditional subdivision of protostome animals (those whose first entry into the interior of the early embryo becomes the mouth) into moulting *Ecdysoa* and nonmoulting *Lopotrochozoa* was correct and reflected a high-level, and hence a very old evolutionary separation. (The origins of the phyla are discussed in Chapters 10 and 11.)

VARIATION WITHIN A SPECIES

No two organisms within a species are identical; this is a result of mutation, and its extent in successive generations is generally increased as a result of sexual reproduction. Even identical twins, who start off with identical genomes, soon develop small genetic differences and eventually come to look less similar than they once did. This variation is due to the fact that, for all the DNA-error-correcting molecular machinery in cells, a few such somatic mutations are incorporated into the genome of the daughters every time a cell divides. Such mutations are, of course, only inherited if they occur in the germline.

Sexual reproduction involves mating, and this implies the local presence of members of the opposite sex. Hence, when one speaks of a species, one is implicitly talking about a population of individuals that are slightly different and such variation is the key to the formation of new species. It is worth noting that the extent of variation is much less in organisms that reproduce asexually by *parthenogenesis* (e.g. water fleas, such as *Daphnia* and many single-cell eukaryotes) or are *hermaphroditic* and are able to self-fertilise (this large group of organisms carries both male and female reproductive cells and includes plants, slugs, and other molluscs, together with worms, such as nematodes and platyhelminths).

The major, immediate sources of variation within a sexually reproducing population come from the almost random combinations of genes from two individuals that occur during reproduction. This randomness has two sources: first, the homologous recombination between pairs of chromosomes that occurs during meiosis randomly assorts genes, after allowing for linkage effects, so that each gamete chromosome is different from those of both parents. Second, mating brings together two haploid genomes, each of which have own set of minor mutations. For the next generation, further mutations will accrue during gamete development and meiosis, and these will add a further degree of randomness to be inherited by further generations. Breeding thus increases variation.

Such resulting phenotypic variation within a population mainly reflects the effects of the mutations that lead to minor changes in protein function and most tend to have minimal effects on individuals and their reproductive abilities—they just increase the amount of genetic variation within a population. Cases where a heritable or germline mutation have a more dramatic effect fall into two classes. Some have effects that are deleterious to the individual and are likely to be lost, particularly if they generate congenital abnormalities. Those, on the other hand, that enhance the ability of the organism to thrive and reproduce in a particular environment will be kept and may slowly spread through the population. They may even be candidate genes for facilitating novel speciation. It should, however, be said that the effects of such mutations are just about impossible to identify when they first form, unless reproduction takes place under extreme selective pressures designed to identify them. These tests can usually only be applied to bacteria but have been used occasionally for *Drosophila* (Chapters 20 and 22).

Within any population, there is thus a broad spectrum of unique genotypes with the phenotype of each individual, other than of very young identical twins, being slightly different. If we just look at humans, the major features that mark an individual as being distinct include overall size, pigmentation patterns (skin, hair, and eyes) and minor differences in the relative size of particular features, notably in the face; in terms of distinct anatomical tissues, there are few if any. The molecular basis for the origins of the minor differences in phenotype among individuals is discussed in Chapters 17 and 18.

THE NUMBERS OF SPECIES TODAY

Investigations by naturalists from across the world over the last few centuries have led to the identification of the very great majority of the larger species (Table 4.1) and many of the smaller ones. However, as one goes down in scale to the smallest invertebrates, tiny plants, and fungi that are barely visible to the naked eye, there will inevitably be many yet to be discovered as each habitat can be a source of speciation. A recent estimate of eukaryotic biodiversity, or the number of such

TABLE 4.1 APPROXIMATE NUMBERS OF CATALOGUED SPECIES IN SELECTED GROUPS OF EUKARYOTES

Monocotyledonous plants	~5K
Flowering plants	~300K
Nematode worms	25-1,000K
Beetles	~450K
Fishes	~28K
Birds	~10K
Mammals	~5K

species is 8.7 ± 1.3 million; if this number is even roughly correct, less than 20% of the distinct organisms with which we share the planet have so far been identified (Mora et al., 2011). This is particularly so in the deepest parts of the sea: these have barely been assessed for animal and other lifeforms and are thought to be the habitat for many unknown species.

The numbers of living species that have so far been found is very large: the catalogue of life database (see Websites) includes information on about 1.1 million living species, of which the very great majority (~920,000) are arthropods. Each species is a member of a particular phylum, and the numbers in some phyla are very large (Table 3.1). The key reason for this is that natural variation allows organisms to explore new habitats. Every small region on the planet affords an opportunity for organisms to invade it and, once there, not only change themselves and form a new, local species but to change that habitat (e.g. it may deplete one source of food while offering itself as another). This may allow further organisms to invade the territory, to adapt and survive there—and so further speciation occurs.

One reason why the possible number of species is so large is that we have little idea of the number of parasites that are specific to each of the larger ones. There is a standard joke that goes: "Which eukaryote phylum has the most species?" and the answer is the nematode worms because every sizeable animal has a nematode parasite, although most have yet to be catalogued—hence the wide range in Table 4.1. There are, however, many, many other species-specific parasitic organisms that remain unknown, and their identification will increase the total. Another reason for the uncertainty is that we are unlikely to have catalogued all the single-celled and other very small eukaryotic organisms, and indeed it is unlikely that we ever will: there will always be odd niches that we will have neither the resources nor the imagination to investigate. Furthermore, evolution has not ceased, and groups of existing variants are always making the slow transition to becoming new species.

Such reasons also help explain the unexpectedly low number of prokaryotic species so far identified: the PubMed taxonomy database includes only ~18,000 eubacterial and ~800 archaeabacterial species. Since the prokaryotic populations of most habitats and organisms, other than humans, have not been systematically investigated, there are bound to be many more prokaryotic species to be discovered. It is however, harder to define a prokaryotic than a eukaryotic species: because bacteria divide asexually, a morphological rather than a breeding criterion is primarily used. There is, however, only a limited number of possible shapes for a single-celled organism; individual species in each morphological group have to be distinguished by their DNA differences, which are hard to quantify, and by the hosts to which they are restricted, and of course, there can be overlaps here (Wayne, 1988; Oren & Garrity, 2014).

We may, therefore, have to accept that the differences within groups of bacteria reflect a continuum rather than a set of distinct species (note that a similar problem occurs in eukaryotes in what are known as ring species, Chapter 22). As to the viruses, their effects are usually organism-specific and can only be identified if they lead to disease. As we care little about the diseases that affect the great majority of organisms unless they become able to infect us (e.g. COVID-19 that seems to have made the transition from a bat host to a human host in 2019) or our food organisms, the ~3400 virus species catalogued in 2014 are bound to be a gross underestimate of the full number.

TAXONOMIES: ORGANISING DIVERSITY

A classification of species based on anatomical or other similarities is known as a *taxonomy*. Although such classifications go back to Aristotle, the credit for modern approaches to the subject is given to Linnaeus through his book *Systema Naturae*, published in 1735. His taxonomy is a graded hierarchy of classes: similar taxa or groups of organisms can be put into broader taxa on the basis of common similarities and so on upwards (Box 4.2). In systems terminology, such a taxonomy is a series of embedded classes linked by the relationship <A> *is a subclass of* <B>, normally shortened to <A> *is a* <B> and can be visualized as a hierarchical tree or as sets of embedded circles (Chapter 5, Appendix 1).

Membership of a particular taxon at a particular level is defined by the possession of a set of properties, with higher ranks having a smaller number of defining properties than lower ones and so including more organisms (Box 4.2). Species are, following Linnaean practice, given a double name written in italics: the former is the genus or family name, the latter the species name; thus, for example, *Pan paniscus* and *Homo sapiens* are the species names for bonobos and humans. Subspecies have a third name so that, for example, the Sumatran tiger is *Panthera tigris sondaica*.

The highest level in Linnaean taxonomies is the domain. These are the Eubacteria (anuclear prokaryotes with a membrane containing glycerol-ester lipids), the Archaebacteria (anuclear prokaryotes with a membrane containing glycerol-ether lipids), and the Eukaryota (cells with nuclei). Below these domains is the taxon of kingdoms which particularly subdivides the eukaryotes. The two major ones are the Archaeplastida (subkingdoms: red algae, green algae, plants, and the unicellular glaucophytes, all of which have chloroplasts with double membranes and whose motile cells have two flagella) and the Amorphea (subkingdoms: animals, fungi, and amoebae, all of whose motile cells, such as spermatazoa have a single flagellum); there are also several very minor kingdoms, most of which are single-celled or very simple social eukaryotes (see Chapter 10).

The next important level down is the *phylum*. This term has a significant biological meaning: each phylum includes all species possessing the same body plan; the taxon currently includes 35 animal, 6 fungal, 11 plant, one moss, 52 bacterial, and 5 archaeal members. Table 4.2 shows twenty-two of the main animal phyla (the omitted 15 or so minor phyla are protostomes). The primitive Metazoa (the animal clade) include the Poriphera (sponges) and Placozoa (the simplest multicellular animal) as well as the Cnidaria and Ctenophera that have a gut, radial symmetry and are diploblastic (this means that their early embryos have two layers, an epithelial ectoderm and an epithelial ectoderm, with any muscles or neurons forming from ectoderm later in development).

The rest of the taxa in Table 4.2 are members of the Bilateria, which have basic mirror-image symmetry and are all triploblasts: they have a layer of mesoderm between the embryonic ectoderm and endoderm that makes muscles and other cell types not seen in diploblasts (for details, see Chapter 10). The Bilateria are divided into two major groups on the basis of their embryonic development: most are protostomes (the mouth originally derived from the first cavity in the early blastula with the second becoming the anus), and only three are deuterostomes (the order of formation of these cavities was, originally at least, reversed; note that the later evolution of the amniotic egg altered the way in which the amniote gut forms). The *chordates* are the major deuterostome taxon, and their most important subgroup is the vertebrates.

Inclusion within a phylum was originally defined solely on the basis of adult morphology, but, as the example of the echinoderms shows, this can be ambiguous. Adult echinoderms such as starfish have essentially radial symmetry, normally with five arms, but their larvae, which are triploblastic, initially show bilateral symmetry with their early development being characteristic of deuterostomes. Later in their development, these larvae undergo a major and very rapid reorganisation generating radial morphology and rebuilding the gut (Wray, 1997; Cary & Hinman, 2017). For the purposes of classification, the larval form with its three layers and bilateral morphology is a better indication of the origins of the organism than the adult morphology with its radial symmetry. Echinoderms are, therefore, included in the Bilateria within the deuterostome clade.

Today, the most successful phylum on the basis of catalogued numbers is that of the arthropods (Table 3.1). These protostome invertebrates, which possess a chitin exoskeleton and jointed limbs, have flourished from the Cambrian onwards. The famous evolutionary biologist J.B.S Haldane is said to have replied to a question from a theologian about what the living world tells us about the nature of God by saying, "If one could conclude as to the nature of the Creator from a study of creation, it would appear that God has an inordinate fondness for stars and beetles."

As to the prokaryotes, they are currently subdivided into two domains, Eubacteria and Archaebacteria, and subdivided into about 50 phyla on the basis of their morphology and composition. The more common Eubacteria form the majority: they mainly have outer membranes that contain glycerol-ester lipids and peptidoglycan (the cyanobacteria, an important bacterial phylum, contain cellulose rather than peptidoglycan). The rarer Archaebacteria have cell membranes that contain glycerol-ether lipids and are subdivided into four phyla. Archaebacteria were originally known as extremophiles because they were found in what were considered inhospitable environments, such those with high salt concentrations (the haplophiles), high temperatures (up to 105°C—the thermophiles), high acidity or alkalinity, and those lacking oxygen (the methanogens). More recently, however, as the search for Archaebacteria has widened, these prokaryotes have been found in almost every

TABLE 4.2 THE MAIN ANIMAL PHYLA. MOST OF THOSE EXCLUDED ARE PROTOSTOMIC WORM AND PROTIST PHYLA, ALL OF WHICH HAVE RELATIVELY FEW SPECIES. (ADAPTED FROM EN.WIKIPEDIA.ORG/WIKI/PHYLUM).

	Superphylum	Phylum	Common name	Distinguishing characteristic	Species number
Coelenterata		Placozoa		Epithelioid disc-like structures ~1mm in diameter	3 genera
		Poriphera	Sponges	No tissues, water channels	~10,000
		Cnidaria	Jellyfish, Corals	Nematocysts (stinging cells)	~11,000
		Ctenophera	Comb jellies	8 "comb rows" of fused cilia	~ 100
	Deuterostomes	Vertebrata	Vertebrates	Chordates with vertebrae, neural-crest-derived sense tissues and a tail	~45,000
		Cephalochordata	Amhioxus	Simple chordates with a notochord and gill slits but no neural-crest-derived tissues	23
		Tunicata	Ascidians	Larva with notochord; adults lose most tissue and have a sac-like body with syphons	~3,000
		Echinodermata	Echinoderms	Fivefold radial symmetrys, mesodermal calcified spines	~7,000
		Hemichordata	Acorn worms	Stomochord in collar, pharyngeal slits	~ 100
		Arthropoda	Arthropods	Chitin exoskeleton, jointed limbs	>1,130,000
	Ecdysozoa (Animals that shed their exoskeleton)	Kinorhyncha	Mud dragons	Eleven segments, each with a dorsal plate	~ 150
		Loricifera	Brush heads	Umbrella-like scales at each end	~122
Bilateria		Nematoda	Round worms	Round cross section, keratin cuticle	25-1,000K
		Nematomorpha	Horsehair worms	Non-functional gut	~ 320
		Onychophora	Velvet worms	Legs with chitinous claws	~ 200
		Priapulida	Penis worms	Eversible proboscis	16
	Protostomes	Tardigrada	Water bears	Head + body with 4 segments, insemented limbs	~1,000
		Annelida	Segmented worms	Multiple circular segments	~17,000
		Brachiopoda	Lamp shells	Lophophore and pedicle, shells	300-500
	Lophotrochozoa	Bryozoa	Moss animals, Sea mats	Lophophore, no pedicle, ciliated tentacles; some have shells	5,000
	Ciliated larvae	Entoprocta	Goblet worm	Anus inside ring of cilia	~150
		Mollusca	Molluscs	Muscular foot and mantle	60-100,000
		Nemertea	Ribbon worms	Eversible proboscis	~1,200
		Phoronida	Horseshoe worms	U-shaped gut	11

In Linnaean taxonomies, substantial anatomical differences are reflected in the criteria used for high-level taxa, such as phyla and classes, while lesser differences distinguish low-level ones, such as species and subspecies. These differences actually reflect the different ways that embryos and larvae develop; relatively few anatomical features form during the growth phase that follows morphological development in the embryo and larva.

The basis for this insight came from von Baer almost 200 years ago (see Appendix 2), although it seems obvious today. He realised that basic features, such as the body plan, emerge early in embryonic development while more minor interspecies differences appear much later, building on what is already there. Although embryological stages are not precisely linked to taxonomic classifications, they provide the basis for them: early events reflect higher taxonomic levels and later events, lower ones (Davidson & Erwin, 2009).

In the case of the vertebrates, the main subphylum of the chordates: all go through a series of early stages that lead to an elongated bilateral embryo with a notochord (a cartilaginous rod just ventral to the neural tube) and mesodermal segments (somites) that, in particular, become vertebrae and muscles. An additional developmental structure, the neural crest (Chapter 14), then allows animals to develop a proper head with eyes. The cephalochordates are a sister group to vertebrates that lack vertebrae and

neural crest; as a result, they have few head tissues other than a mouth and gills. The very few members of this group are all branchiostomes such as *Amphioxus*, also called the lancelet.

The embryos of vertebrates soon develop vertebrae, their defining feature, and most go on to produce paired lateral outgrowths. If these become fins, the animals are fish but if these outgrowths develop into limbs, they are tetrapods, with subsequent features of development distinguishing amphibians, reptiles, birds and mammals and their subgroups and families. It is the later, minor development differences that distinguish the various species within a family. Typical areas where such terminal differences occur includes size, tooth and bone morphology, and the details of skin patterns.

The class of mammals illustrates this. They are divided into the subclasses of Prototheria (monotremes that lay eggs) and Theria whose young are born alive. The latter are subdivided into two infraclasses: the Metatheria (marsupials) and the Eutheria (placental mammals). The essential differences among them are not particularly impressive in the greater scheme of things: they all have, to a close approximation, the same body form and the same cell types. Each group is then subdivided on the basis of later-forming characteristics. The discussion in Chapter 16 shows the importance of geographical isolation in the evolution of the different mammalian families.

environment that has been studied—we can eat them in fermented seafood (Roh et al., 2010). The relationship between Archaea and Eubacteria is discussed in more detail in Chapter 9.

The taxonomic levels between the phylum, defined by body plan, and the species, defined by interbreeding or morphological criteria, are essentially constructs often based on artifical criteria criteria that indirectly reflect the development of the organisms within a phylum (Box 4.1). The Integrated Taxonomic Information System (see Websites) shows that the domestic cat hierarchy has 27, the honeybee has 18, and the garden daffodil has 12 such levels between kingdom and species. There is little that is contentious in these levels between the phylum and the species, albeit that names occasionally change (e.g. the family of legumes used to be called the Leguminosae but are now called the Fabaceae).

Today's Linnaean taxonomies, which are based on common anatomical properties, are now so large and complicated (see Box 4.2) that they can only be held in an online database and searched computationally.

According to http://www.ncbi.nlm.nih.gov/Taxonomy, the full taxonomic hierarchy (with *taxonomic* levels where known) for the domestic cat is:

Cellular organisms; Eukaryota (*domain*); Opisthokonta; Metazoa (*kingdom*); Eumetazoa; Bilateria (*subkingdom*); Deuterostomia (*infrakingdom*); Chordata (*phylum*); Craniata; Vertebrata (*subphylum*); Gnathostomata (*infraphylum*); Dipnotetrapodomorpha; Tetrapoda (*superclass*); Amniota; Mammalia (*class*); Theria (*subclass*);

Eutheria (*infraclass*); Laurasiatheria; Carnivora (*order*); Feliformia (*suborder*); Felidae (*family*); Felinae (*subfamily*); Felis (*genus*); Felis catus (*species*)

In more formal, systems terminology, this 24-level hierarchy is a series of embedded classes or sets linked by the relationship <*is a subclass of*> (abbreviation: *is a*). Thus, <my cat, Coco> <*is a*> <*Felis catus*> <*is a*> <Felis> <*is a*> <Felinae> <*is a*> <Felidae> <*is a*> <Feliformia> etc.

PHYLOGENIES AND EVOLUTIONARY HISTORY

Linnaean taxonomies can, of course, include extinct organisms on the basis of their anatomical features, but these placings say nothing about evolutionary descent. This is because the age and ancestry of a species (even when they are known) are not anatomical properties. Actually, age is a very difficult property to handle taxonomically. One reason for this is that the dates associated with a particular fossilised species are rarely precise as to when that species evolved, what were its descendants and when it became extinct, unless the latter date coincided with a major extinction. The more important reason, however, is that taxon age is not a relationship that describes evolutionary descent (parent and daughter species can easily coexist). What marks species as distinct in the standard taxonomies are only the details of the anatomical differences.

Modern Linnaean taxonomies often include species with common ancestors within the same high-level groupings, but this can lead to some unexpected results. The superorder of ungulates originally included mammals that walked on their toes and whose nails evolved into hooves. It has now become clear, mainly on the basis of DNA analysis, that not only have hooves evolved more than once but that the Cetacea (whales, dolphins, and other sea mammals) have the same common ancestor as modern even-toed ungulates, such as pigs, sheep, and camels (Chapter 16). Although they were originally considered as unrelated groups, the Linnaean vertebrate taxonomy has had to be changed so as to include both within a single class. Today, the cetaceans are now grouped with the hippos in the Whippomorpha, a suborder within the Artiodactyla (even-toed ungulates) superorder.

Far more useful in an evolutionary context are taxonomies explicitly based on inheritance (these are phylogenies), and there are two ways of making these. The first uses cladistics analysis to construct trees on the basis of shared anatomical properties that derive from a common ancestor; these trees are known as cladograms and are discussed in detail in Chapter 5. Cladistics provides a means of structuring the fossil record on the basis of comparative anatomy and is particularly important in tracing the history of vertebrates because this clade has the strongest fossil record.

The second type of phylogeny is usually based on homologous DNA and amino acid sequences from groups of contemporary organisms (e.g. protein-coding sequences with the same function in a set of related organisms) and is known as molecular phylogeny. Phylogram construction uses computational methods to produce the most likely tree of descent on the basis of tracing mutational differences back to a common ancestor (Chapter 7). This approach is the most widely used way of grouping organisms because it can be used for every organism based on its genome. If enough gene sequences are used in this process, the cumulative and averaged phylogenetic tree approaches a species tree that goes back to a common-ancestor root.

One major achievement of such sequence analysis is the proof that eukaryotic cells derive from endosymbiosis whereby one bacterium engulfed another, allowing it to maintain a semiautonomous presence in the composite. Such is the strength of the approach that it has been able to pinpoint the likely bacterial ancestors that together led to the LECA (Last Eukaryotic Common Ancestor) from which all subsequent eukaryotic cells descended (Chapter 9).

Cladogram trees are more intuitive than molecular phylogenies as they have the advantage that the anatomical criteria for determining intermediate nodes in the tree are explicit as they come from analysing fossilised material (many examples will be found in subsequent chapters). Phylograms are, however, important for tracing evolutionary relationships in the absence of a fossil record and allow us to probe the early stages of life on the basis of common proteins sequences that can be identified computationally (Chapters 9 and 10). Short-term phylograms are easy to visualise, but those integrating the inheritance of different groups can be so large that they can only be constructed computationally and visualised online.

Cladograms and phylograms differ from Linnaean hierarchies in two important ways. First, the link that connects terms is *descended with modification from* rather than *is a* (Chapter 5). Second, in evolutionary hierarchies, every link connects, in principle at least, two species of organisms, the parent and the child, although it is usually hard to identify precisely the exact parent species. In Linnaean classifications, all higher levels reflect groups of species, alive or extinct, that share an ever-restricted group of anatomical features. These three very different taxonomies share one feature that reflects the fact that they all describe the same set of organisms: they should each group contemporary species in the same way because any such group reflects both genomic and anatomical descent from a common ancestor together with shared anatomical features. This prediction is found to be true.

KEY POINTS

- There are more than a million known animal species distributed across about 50 phyla (basic body forms), with vertebrates varying in size from less than a cm (a frog) to more than 30m long (the blue whale). Plants likewise show a similar size range.

- There is no definition of a species that is suitable in all contexts, but, where appropriate, the best is that a species is a group of organisms that do not breed with any other group to produce fertile offspring.

- Where the definition of species based on breeding cannot be used (e.g. where the organisms will not mate or in cases where an organism is extinct), species are usually distinguished on the basis of morphology, habitat or DNA profiles.

- Within any population, variation is normal and mainly derives from random assortment of gene variants during meiosis and mating, with new mutations providing a slow background increase in diversity.

- Novel speciation builds on this variation.

- In Linnaean taxonomies, organisms can be grouped within an anatomy-based hierarchy in which widely shared features are placed in higher classes (e.g. phyla and orders) and more restricted ones in lower classes (e.g. families and species). This hierarchy roughly reflects the order in which developmental events takes place.

- In an evolutionary context, phylogenies based on descent with modification (homology) are far more useful than Linnaean ones. Cladograms for a group are constructed on the basis of shared and derived anatomical properties while phylograms for a group are constructed on the basis of the particular mutations in homologous DNA sequences.

FURTHER READINGS

Gaston KJ, Spicer JI (2004) *Biodiversity: An Introduction.* Blackwell Science Ltd.

Jeffrey C (1986) *An Introduction to Plant Taxonomy.* Cambridge University Press.

Polaszek A (ed) (2010) *Systema Naturae 250 - The Linnaean Ark.* CRC Press.

Ruse M, Travis J (eds) (2009) *Evolution: The First Four Billion Years.* Harvard University Press.

Verma A (2015) *Principles of Animal Taxonomy.* Alpha Science International.

WEBSITES

- The Wikipedia entry on the Phylocode: www.ncbi.nlm.nih.gov/taxonomy/
- The Integrated Taxonomic Information System: www.itis.gov/

- The Catalogue of Life: www.catalogueoflife.org

SECTION TWO
THE EVIDENCE FOR EVOLUTION

The evidence for evolution set out in this section comes from comparative anatomy of both living and extinct species, together with genetic-sequence analysis and comparative embryology of extant species. The fossil record, briefly discussed in Chapter 4, is the most obvious source as it demonstrates the major lines of species radiations and extinctions, particularly over the past ~600 My. Although this record provides a historical context for evolution, individual fossil descriptions are only a start: direct evidence for evolutionary descent comes only from a close comparison of the anatomy of similar fossils to identify those features that are common and those features that might indicate descent with modification. The formal approach for analysing such differences is known as cladistics and will be discussed in Chapter 5.

In terms of the direct evidence for evolution on the basis of the fossil record, the most important data comes from fossil sequences in which the details of transitions from one anatomical feature to another can be closely followed, revealing the ways in which descent with modification happened. Three examples are discussed in Chapter 6: the formation of pentadactyl limbs from the pectoral and pelvic fins of fishes, the changes that led from the skull of stem mammals to one with the basic features of modern mammals, and the evolution of the modern horse (*Equus*) taxon, whose feet have a single hoofed toe, from its multitoed ancestors.

A second means of elucidating evolutionary relationships among living species comes from computational analysis of homologous DNA and protein sequences on the basis of their mutational differences. Once homology (common ancestry) has been established on the basis of the similarity of sequence, computational and statistical methods can be used to generate a hierarchy known as a phylogram (Chapter 7). This is done by identifying the ancestral sequence that would lead to all the contemporary sequences with the minimum number of mutations.

A further use of such computational methods is in exploring the early stages of life, before the fossil record became substantial. Such are the richness of the sequence database and the sophistication of the computational tools that it is straightforward to identify related sequences in, for example, prokaryotes and eukaryotes and across very different phyla that lack a fossil record. The analysis of such related sequences has uncovered the origin of eukaryotes and the roots of the major groups of organisms. This hidden history of the first three billion years of life will be discussed in Chapters 9-11.

The third line of evidence comes from work showing that the development of very different organisms involves the activities of homologous proteins and protein networks that have very similar functions. This is the area known as evo-devo

DOI: 10.1201/9780429346217-6

(Chapter 8), but its appreciation requires some background knowledge about embryonic development; this is provided in Appendices 6 and 7. One of the successes of evo-devo methodology has been to identify a set of proteins, protein networks and likely phenotypic properties that would have been possessed by the last common ancestor of all animal species with basic bilateral (mirror) symmetry. This small worm-like organism is known as *Urbilateria* and probably lived towards the end of the Ediacaran Period. The fossil data and likely anatomy of this organism will be discussed in Chapter 12.

It is worth noting that these three lines of evidence, from anatomical comparison of fossils, from DNA-sequence analysis and from comparative molecular embryology (evo-devo), are completely independent. Each, of course, probes its own area of data with different degrees of temporal and species resolution, but where their insights overlap, it is a prediction that they should turn out to support rather than contradict one another. That this prediction is borne out is not, of course, surprising: each approach examines a specific but very different facet of a single history.

Analysing Evolutionary Change

5

This chapter provides the framework for analysing evolutionary descent on the basis of anatomical change. It shows that the resulting taxonomic formalism known as cladistics, which is based on this insight, provides a natural language for discussing evolutionary history and phylogenies (descent trees). The first part of the chapter provides the formal derivation of cladistics and explains its terminology. Application of the approach is not always straightforward as various anatomical properties may have changed dramatically over time and so do not at first sight fit within the framework. The latter part of the chapter considers three types of such phenomena. The first are homoplasies in which very similar adaptations have arisen independently in very different species (e.g. wings and sabre teeth). The second are lost plesiomorphies in which characteristic traits that would be expected to be present in a taxon have been modified dramatically or even been lost (e.g. the legs of snakes). The third, and perhaps most intriguing, are those tissues that apparently appear in the fossil record in a fully formed and functioning state with no apparent antecedents. Such events are known as exaptations and examples are feathers, which are considered here, and the camera eye, which is discussed in detail in Chapter 20.

When Darwin wrote about "descent with modification" rather than evolution in *The Origin of Species*, he articulated the precise way in which evolutionary changes in anatomical features can first be recognised and then maintained or further modified over time. This phrase is also the key to understanding how to track features back in time: the more widely spread is a particular feature, the earlier it must have evolved. It is thus, as this chapter discusses, the theme that underpins the procedures that enable us to produce a species taxonomy based on evolutionary change (this is known as a phylogeny) rather than one based on common anatomical features (Linnaean taxonomy).

Such an approach can be viewed as explaining the broad brush of evolution, tracing lines of descent and highlighting the relationships that enable today's organisms to be grouped on the basis of evolutionary descent. This approach says nothing, however, about the details of the processes and events that determine when, where and how change happens, whether a new species will form or how one possible trait change might be maintained and another lost. Discussion on these aspects of the mechanisms of novel speciation is the subject matter of Section 4.

DESCENT WITH MODIFICATION IS THE IDENTIFIER FOR EVOLUTION

New species start to evolve as the result of novel, heritable changes arising in a distinct group of organisms belonging to an existing species that have become reproductively isolated from its parent group. A new species become established from the members of this group if, in due course, those changes are sufficient to ensure that they can no

DOI: 10.1201/9780429346217-7

longer interbreed with the descendants of their parent species to produce fertile offspring (Chapter 22). This may be through choice, through inaccessibility or through genetic incompatibility (e.g. chromosomal reorganisations have taken place). These changes, of course, imply that any new species carries most of the properties of its parent species, but has novel, heritable features, often recognised visually, that facilitate the adaptation of the descendants of the separated group to its new niche.

This expanded version of Darwin's *descent with modification* can be expressed more formally: if species **B** shares many features with another species **C**, but has some features that are different, then species **B** and species **C** are both likely to have descended from a common ancestor, species **A**. This chapter shows how such an idea can be formalized in cladograms (phylogenies based on anatomical features): these are hierarchies that link species through the relationship *<derives with modification from>*. The formalism of cladistics, together with its unfortunately complex but logical terminology (Box 5.1; Figs. 5.2 and 5.3), provides a strong framework within which to articulate anatomical change and evolutionary relationships.

Descent with modification in a succession of species can be expressed more precisely. Suppose that an ancestral species **A** has a set of anatomical properties or traits (**i j k ... p q r...**), the expectation is that any descendant, but different species **B** has most of these ancestral properties (which are known as *plesiomorphies*), but also has at least one novel feature. This will have resulted from the loss or change of an existing feature (e.g. feature **p** is modified to become feature **x**) so that **B** now has the features (**i j k... q r...x**). In due course, descendants of **B** will form a new species **C** characterized by a further novelty **y** which derives from **q** and so has features (**i j k ...r...x y**). Later, descendants of **C** will evolve to give species **D** with novel feature **z** from **r** and hence with features (**i j k...q...x z**).

If there is a contemporary descendent, **A′**, of the original ancestor species **A**, it would be expected to lack the **x y z** innovations, although other changes in **A′** might have evolved since the time when **A** lived. The key point is that evolutionary change can only be recognized through examining the similarities and differences among a group of species; a single species on its own says nothing in this context.

All this phenotypic information can be expressed as a diagram known as a *cladogram* (from the Greek: κλαδος (klados) = branch). This is a hierarchical model of the evolutionary history of a set of species (Fig. 5.1) constructed on the basis of simple triplet relationships of the form:

< species **G** >< *derives with modification from* >< species **F** >

Together, these triplet relationships can be assembled into a hierarchy giving the lines of descent. It is worth noting that any tree composed of such triplets is an example of a *mathematical graph*, and its more general use is discussed in Appendix 1. It should also be emphasised that species hierarchies are normally constructed on the basis of many anatomical features rather than the very few discussed here.

Such a triplet relationship immediately raises a problem because, for a fossil species, one can never be sure about its precise ancestor, only its approximate location on the diagram. In the case of species **B** in the example above, all one can say is that its parent species had to be a new species whose phenotype included

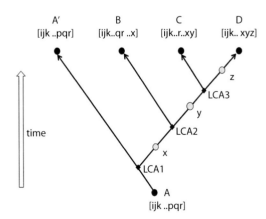

Figure 5.1 A cladogram illustrating Darwinian descent with modification. This has been drawn to illustrate how an ancestral organism **A** has given rise to four different organisms **A′,B,C,D**, any of which could be alive or extinct today. If this cladogram contains **A** and all of its descendants, it is the crown group of **A** (see Box 5.1 and Fig. 5.3). Note also that lines of descent are not shown as stemming directly from known species, but from last common ancestors (LCA); this is because the exact species from which a descendant evolved is not generally known.

property **x** that separated it from the inheritance line of species **A**. This parent species is considered to be the *last common ancestor* (LCA) of all organisms with property **x,** and the historical facts are rewritten in the form.

$$< species\,\mathbf{G}> < derives\ with\ modification\ from > < LCA\ with\ property\ \mathbf{x} >$$

The cladogram for the example above is shown in Fig. 5.1, and the terms are of the form:

$$< species\,\mathbf{C}> < derives\ with\ modification\ from > < \mathbf{LCA3} >$$
$$< species\,\mathbf{D}> < derives\ with\ modification\ from > < \mathbf{LCA3} >$$
$$< species\,\mathbf{LCA3}> < derives\ with\ modification\ from > < \mathbf{LCA2} >$$

Such a cladogram can, of course, be extended to include further species (or indeed any taxonomic class or *taxon*) as the data demand. The key point, however, is that every taxon included in a branch of the hierarchy should share with the others on that branch a single last common ancestor at the base of the branch (species **A** in this case). A branch that includes all and only the descendants of a single root species is known as a *monophyletic taxon*.

The cladogram in Fig. 5.1 includes an important convention: the grey circle labelled **x** signifies that all and only species on its far side possess property **x**. Occasionally, there may be a further downstream species in which **x** has been lost. In this case, all descendant species will also lack **x**, but these species should maintain a sufficient number of other ancestral and clade-defining features (plesiomorphies and synapomorphies) to demonstrate that their classification is not ambiguous (see the example of snakes later in the chapter). They will thus still share the same last common ancestor with descendant species that have retained **x**.

An example here is the ancient clade of flying insects that had two pairs of wings and evolved from unwinged ancestors. In the Diptera (gnats, mosquitoes, and flies, such as *Drosophila melanogaster)*, the posterior wings later evolved to become halteres (small and vibrating balancing organs). In contrast, the Strepsiptera (small "twisted wing" parasitic insects) retain their posterior flying wings, and it was the original anterior wings that evolved to become halteres. Because both recent groups share many other properties, such as limb number, they are all insects with the same common ancestor and part of the same clade.

It is important to realise that the formalism represented in Fig. 5.1 says nothing explicit about time, other than its direction. One might immediately think that the framework applies to fossilised organisms where timings are known, and it of course does. Such timings are, however, irrelevant to a cladogram, even where there is apparently good fossil evidence: this is because the dates when such organisms appeared and became extinct are rarely precise enough to confirm that one derived from the other. The only properties considered in making a cladogram based on anatomy are the structural features inherent within similar organisms, independent of whether they are living or extinct. In fact, timings can be misleading as many organisms carried on with little change long after other species had evolved from their line. Perhaps the best-known exceptions here are evolutionary changes that follow global extinctions that have been accurately dated.

Fig. 5.1 also clarifies the two sorts of prediction that derive from descent with modification. First, whenever a novel property arises as the result of change to an ancestral property (e.g. **x**, which derives from **p**), it should normally be present in all subsequent downstream taxa; an example of this is the amniote egg, a clade-defining synapomorphy (this is a shared and derived property for a clade; a summary of cladistics terminology is given in Box 5.1) and maintained in all descendant vertebrates. Since synapomorphies are shared by all downstream taxa in a clade, they represent strong and clear evidence for common evolutionary descent of that clade from a last common ancestor. In cases where a property is lost (e.g. humans, unlike other primates, have minimal amounts of thick body hair), the cladogram can be viewed as accurate provided that the apparently anomalous species possesses sufficient number of other synapomorphies and plesiomorphies to sustain its position in the tree. Decisions here require detailed knowledge of comparative anatomy.

BOX 5.1 THE MAIN CLADISTICS TERMS

The use of the terms below is illustrated in Figs. 5.2 *and* 5.3

Taxon: any named group of related organisms, usually of the same Linnaean rank (e.g. the bryophyte taxon includes mosses, hornworts, and liverworts)

Cladistics deals with the inheritance of properties (**...morphies**): as evolution proceeds, one of more *pre-existing or ancestral* properties (**plesiomorphies**) in a species changes in some heritable way to produce innovations or *novel, derived* properties (**apomorphies**). If the species with a particular apomorphy then gives rise to a group of further organisms, each should inherit that property, so forming a new **clade**. The original apomorphy thus becomes a *shared, derived* property for the clade (a **synapomorphy**—syn means together) and an example is milk, which is produced by all mammalian clades. A novel feature that is specific to the terminal leaf of a branch, whether living or extinct, is known as an **autapomorphy** (e.g. the ~40 black and white stripes on the mountain zebra).

Clade: this is the complete group of living and extinct species that share a last common ancestor (its visualization is a **cladogram**); it is also known as a **monophyletic clade** to distinguish it from a **paraphyletic clade** or **grade** in which some of the taxa are missing (e.g. a dinosaur clade that excludes birds). A cladogram has at its base **a last common root ancestor species** that gives rise to a spread of intermediate species and terminates in a group of **leaf species**.

Crown group: this is a clade that includes all the living members of the group and its ancestors back to a last common ancestor in which the synapomorphy that characterizes the clade originated.

Stem group: this group includes all the early organisms between the last common ancestor of the crown group back to the branch that led to the sister group (below). The stem group provides a location for early organisms closer to the crown group than the sister group (e.g. the line of stem mammals that diverged from the reptile clade and eventually gave rise to the mammalian clade).

Total (or **Pan**) **group**: this includes the crown group and the stem group.

Grade: this is a subgroup within a particular clade that excludes some branches (e.g. fishes comprise a grade within the vertebrate clade).

Sister group: if a total group reflects a subclade within a large clade, the sister group is the immediately basal spur (i.e., it separated immediately before the last common ancestor of the total group). An **outgroup** is a lineage less closely related to the clade than the sister group.

If the synapomorphies in different organisms later evolve to give different structures, these structures are, because they derive from a common ancestor, known as **homologues**: for example, the forewings of early flying insects evolved to give halteres in *Strepsiptera* and elytra or wing covers in beetles. Properties that appear similar but arise independently and not from plesiomorphies are known as **homoplasies**; examples are the wings of bats, pterosaurs, birds, and insects. This is sometimes called homoplastic or convergent evolution.

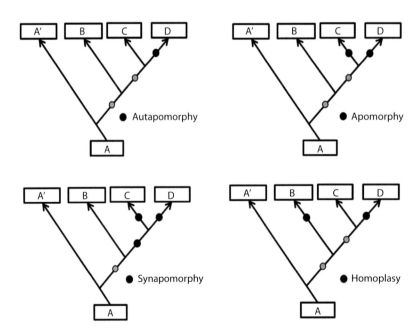

Figure 5.2 Diagram illustrating the various types of novel property that can be identified in cladograms (see Box 5.1). (Adapted from Page & Holmes, 1998.) Note that the cladistic term assigned to some property is defined by its location in a particular cladogram. If that location changes, the term may also change (see text).

BOX 5.1 (*Continued*) THE MAIN CLADISTICS TERMS

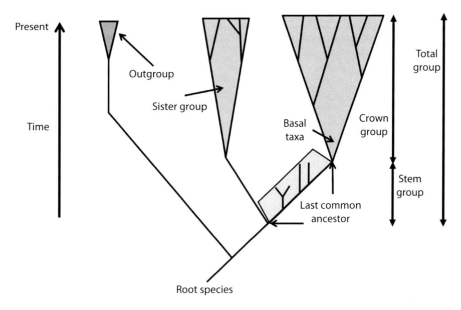

Figure 5.3 Diagram illustrating the relationships between the various types of groups used in cladistic analysis.

The term **last common ancestor** is used in the context of groups of species, while the term **most recent common ancestor** is used for discussing individuals within a species.

Some examples help clarify the use of cladistic terminology. Consider the tetrapod clade that is defined by the possession of jointed digits and whose origins are discussed in Chapter 6.

- The *crown group* includes all tetrapods, alive or dead, back to a node that represents the last common ancestor that had jointed digits and no fin rays, together with other tetrapod-like features (Chapter 16). The two branches from this node lead to the amphibians and the amniotes.
- The lower part of the *stem group* of tetrapods includes all extinct fish with tetrapod-like fins that lacked an autopod with jointed bones (e.g. the elpistostegids) but are clearly more tetrapod-like than sarcopterygian fish. The upper part of the stem group includes early tetrapods with an incomplete autopod and lacking

other features that one would expect of a member of the tetrapod clade.
- The modern sarcopterygian fishes, which lack limb-like fins, are thus within the *sister group* to the tetrapods.
- The echinoderms form an *outgroup* to the tetrapods.
- The reptiles, a group that excludes their avian descendants, form a *grade* within the tetrapod clade.
- Jointed digits in autopods are a *synapomorphy* for the tetrapods, although they can occasionally be lost secondarily (e.g. whales have no hindlimbs).
- Bony vertebrae in tetrapods are a *plesiomorphy* that they inherited from bony fish.
- Dorso-ventral stripes are an *apomorphy* for zebras.
- The pattern of ~70 dorso-ventral stripes is an *autapomorphy* for Grevyi's zebra.
- Sabre teeth, which arose independently in a few members of the tetrapod clade, are a *homoplasy*. The group of animals with sabre teeth is thus *polyphyletic* because the trait cannot be followed back to a common ancestor.

Second, very different and widely separated species that share a common ancestor and so are clearly related (such as the different vertebrate families) would still be expected to maintain many common, ancestral properties or plesiomorphies (these are the **ijk** tissues in Fig. 5.1). To repeat the example from Chapter 3, the tiny frog, *Paedophryne amauensis*, and the enormous blue whale, *Balaenoptera musculus*, both have standard vertebrate eyes. Similarly the fact that nerve cells in squid and humans are essentially the same in both structure and function indicates that this cell type is a plesiomorphy and hence that vertebrates and molluscs share a very ancient common ancestor (a stem bilaterian) that had such cells. Such ancient, shared properties are particularly studied in evo-devo work (Chapter 8).

CLADISTIC LINKS ORGANISMS BY ANATOMICAL INHERITANCE

The idea of grouping organisms on the basis of shared, derived characteristics had long been thought about, but it was not until the 1950s that it was formalised by Willi Hennig who called his approach "phylogenetic systematics" (Hennig, 1966). It is now known as cladistics (Kitching et al., 1998; Lipscombe, 1998) and has become the standard way of analysing evolutionary change and interspecies relationships. It should, however, be said that its use is not always straightforward and can occasionally give ambiguous results in cases where there are many characteristics to consider, some of which may be homoplasies (i.e., similar features that evolved independently). In addition, as in all attempts to formalize differences, it has difficulties with quantitative parameters, such as size, while its terminology, summarized in Box 5.1, is not intuitively obvious. Nevertheless, because it captures the nature of evolutionary relationships so well, cladistics is now universally used as a primary descriptor for evolutionary descent based on anatomical features, and it will provide the language for doing so here and in all subsequent chapters.

Constructing a cladogram for a set of related species is generally straightforward, in principle at least. The process starts by listing a set of characters or traits for each species: some will be species-specific, and others will be shared by two or more of the species, with those that are shared by a wider set of species being viewed as plesiomorphies or ancestral properties. These characters are then integrated into a taxon matrix of species and their traits; this is then used to construct a tree that describes the inheritance relationships among all the species in the most parsimonious way (it was originally known as a Wagner Tree). In essence, the first step is to identify a set of last common ancestors in which innovations (apomorphies) first appear and to identify the species that carry them together with other traits, some of which will be novel; if there are several such species, they are all in the same clade and share what is now a synapomorphy. The next step is to group species on the basis of such shared properties. Eventually, the analysis produces a set of statements of the general form:

$<$ Species **A** $><$ *carries trait x and derives from* $><$ Last Common Ancestor **LCA$_A$** $>$

$<$ Last Common Ancestor **LCA$_A$** $><$ *derives from* $><$ species **B** $>$

where all modifications are explicit. This information is sufficient to construct a hierarchical cladogram.

For simple cases, producing this tree can be done by hand, but, for more complicated examples where species have many traits, the tree needs to be constructed computationally and there are readily available algorithms and programs for this. One such program is PHYLIP (see Websites at the end of chapter); this tries out all combinations of possible lines of evolution to work out the tree that requires the smallest number of assumptions and hence is the most likely to represent the evolutionary history of the clade. Cladistics cannot, of course, say what actually happened, only suggest the most probable way that it did happen. In many cases, however, the most likely version of a cladogram has a very much greater likelihood of being correct than any other version, and in simple cases, is the only plausible version (see below).

It is worth repeating that the one aspect of evolution that is not directly included in cladograms is time. This means that the length of the branch in a cladogram carries no implication either about the degree of differences or the time between species. The most that can be done was assign a date when the earliest fossil showing an apomorphy first lived, but this date is, of course, provisional, and may need to be revised in the light of any future data. The corollary of this is that the sequence in which organisms evolved emerges from the analysis as a testable prediction for the fossil record and this, in turn, may point to likely gaps in the fossil record.

There is a further important aspect of cladistics that needs to be mentioned and that is scale. Cladograms are useful both for broad-brush information and for detailed phylogenies. An example of the former is given by the general lines of the evolution of stem tetrapods (Fig. 15.1) which gives little detail about species, while an example of the latter is the evolution of modern horses. Here, a broad group of organisms, many of which are known, narrowed to a single species that diverged to give the seven species of Equidae that alive today (Figs. 6.10 and 6.11). Complete cladograms bear some resemblance to Mandelbrot patterns in that, as one focuses in, more and more detail appears. In those few cases where it is possible to explore the detail of a node in a cladogram, it is usually seen to be a bush of variant species and subspecies, only one of which survived selection. The route by which this happens cannot be predicted and may even include hybrids which, on the graph, are seen as links between clades. An example of this is anatomically modern humans who mated with Neanderthals and Denisovans (Fig. 24.1).

Fortunately, there are a few taxa where sufficient numbers of species are represented in the fossil record for us to be able to get some insight into the details of this process of transition. Examples that are discussed in Chapter 6 include the evolution of the pentadactyl limb, the mammalian skull and the modern horse. A further important example is the evolution of *Homo sapiens*: our species is the sole survivor of a host of hominin species that lived over the last few million years, many of which were contemporaneous and with several surviving for up to a million or more years. The net result, however, was that all but our species were lost (Chapter 23). Such evolutionary bushes of species (technically, networks if they include hybrids) may be complicated, but they represent the reality of the process of evolutionary change.

LINNAEAN AND CLADISTIC TAXONOMIES ARE DIFFERENT

Cladistic taxonomies organise species on the basis of *descent with modification* and so include a sense of time. They are thus fundamentally different from the classical Linnaean taxonomies which structure species on the basis of the relationship <*is a member of the class of*> (Chapter 3, Box 3.1). The formal structure of Linnaean taxonomy is, like that of cladograms, a mathematical graph but one based on shared morphological features. Low-level Linnaean categories, such as families can be defined using details, such as tooth number (member species of the Canidae, for instance, have 42 teeth, while member species of the Felidae have 30 teeth), while higher level categories have fewer and broader properties: phyla, for example, are defined by body plans.

What Linnaean and cladistic taxonomies should have in common is that the groupings of terminal (or leaf) species in both hierarchies will be the same because they both cluster such living species on the basis of common anatomical properties. It is the higher-level groupings that are very different, and this is because the former represents a set of groups with common anatomical properties while the latter is based on ancestral history.

It is worth noting that cladistic hierarchies implicitly contain much of the anatomical information included in Linnaean classifications. This is because ancestral taxa all have the more general features of the clade (plesiomorphies) but lack the derived properties seen in later groups. The dinosaurs provide an obvious example; the last common ancestor of all the ornithiscian dinosaurs was herbivorous and can be identified by the tooth-bearing dentary bone of the lower jaw terminating in a predentary bone, a feature seen in all of its descendants. However, the taxon lacked the later anatomical features that would characterize descendant clades, such as the armoured plates (osteoderms) seen in the ankylosaurs, the hollow head crests of hadrosaurs, the heavily ossified skulls of the pachycephalosaurs, and the facial horns and frills of the ceratopsian dinosaurs. It is the sharing of such anatomical features that provide the grouping framework for Linnaean taxonomies.

Finally, it needs to be emphasised that the various cladistic -*morphies* are not fixed properties of an individual feature but reflect the position of the organisms within a cladogram that possess the feature. Thus, for example, the amniote egg was an autapomorphy for the first amniote; it is now a synapomorphy for the amniote clade and a plesiomorphy for mammals.

APPARENT ANOMALIES

In constructing taxonomies based on anatomy, there can appear to be contradictory or difficult features. Three well-known examples are (i) similar tissues seen in very different organisms that do not share a recent common ancestor, (ii) tissues, even plesiomorphies, that should be present throughout a clade but are missing in some taxa, and (iii) structures that appear in the record apparently fully formed and functional with no obvious ancestral species, and that could not have worked had they been incomplete. These tissues are respectively known as *homoplasies*, *lost plesiomorphies* (which are, in effect, new apomorphies) and *exaptations*.

Homoplasies

Just because two groups of organisms share a novel feature, it does not necessarily mean that they share a recent evolutionary relationship: that particular feature could have evolved independently. Consider the vertebrate wing seen in pterosaurs and birds that have diapsid skulls, and bats that have synapsid skulls: the fossil record shows that each wing has its own characteristic skeletal details, and each evolved separately from vertebrate forelimbs originally used for walking (bats and pterodactyls were originally quadrupedal) or grasping (birds evolved from bipedal dinosaurs). Each, of course, also reflects an adaptation that originally enabled a species to fly (for more detailed discussion, see Chapter 13). It is thus clear that the vertebrate wing is a structure defined by function rather than by common descent and so is not a helpful property for analyzing evolutionary relationships.

In traditional evolutionary analysis, a similar character that has evolved separately in different species (i.e., it is shared, but *not* derived from the same common ancestor) is described as resulting from convergent evolution. In cladistics, such a character is called a *homoplasy* (Box 5.1). Another well-known example of a homoplasy is the sabre tooth: although this no longer present in any extant animal, the fossil record shows that this dental feature occurred in many extinct species, only some of which were closely related to the cat family (Van Valkenburgh, 2007; Goswami et al., 2011). The earliest known example of a species with sabre teeth is *Tiarajudens*, an anomodontid therapsid from the mid-Permian (~260 Mya). It was probably a herbivore that used its sabre teeth for defense (Cisneros et al., 2011; this paper also includes what is probably the first recorded example of tooth decay). A more recent animal with sabre teeth was *Thylacosmilus*, a marsupial mammal that lived in S America and died out ~3 Mya. The other examples were members of Felidae-related taxa, but they were not closely related as they do not share a last common ancestor possessing a sabre tooth.

The camera-type eye is another classic example of a homoplasy: this evolved completely separately in the vertebrates and the cephalopods, taxa whose last common ancestor lived in the latter part of the Ediacaran Period. These eyes are identical in function and similar in morphology but differ in structural detail. The two types of eye have very different crystallin proteins in their lenses and different retinal-nerve organization. In vertebrates, the early nerve fibres from the retinal nerve cells migrate *in front of* the early retina and coalesce centrally before moving through the pigmented epithelium (thus forming a blind spot) and then extending to the brain. In contrast, the cephalopod retinal nerve fibres migrate *behind* the retina on their way to the brain and there is no blind spot (Koenig et al., 2016). The question of how such a complex organ could evolve when it required such a detailed structure to function properly is mentioned below and discussed more fully in Chapter 20.

It is intriguing to note that these two very different eye types both use rhodopsin homologues as light-sensitive proteins. Other homologues of these proteins are also responsible for light sensitivity in the compound eyes of arthropods, and even for light sensitivity in eubacteria and archaebacteria (Ernst et al., 2014). All this suggests that both camera and compound eyes originally evolved from light-sensitive patches on very early organisms whose original proteins have mutated but retained their function (Yoshida & Ogura, 2011; see also Chapter 7).

Lost plesiomorphies

A member of a species carries its deep, historical features (plesiomorphies) as well as its clade-specific (synapomorphies) and species-specific novelties (autapomorphies). The interesting question arises as to what is happening when the chain is broken because a plesiomorphy is lost, as in the case of the substantial diminution in the amounts of human body hair mentioned above. A classic example of this is the taxon of snakes: all the fossil, morphological and molecular evidence points to snakes being reptilian, but a defining plesiomorphy of the reptilian clade is tetrapod morphology. The most obvious apomorphy of the snakes is their lack of limbs, a loss that, on the basis of the fossil record, started about 100 Mya and occurred progressively (Martill et al., 2015).

This fossil data strongly suggests that the ancestors of today's snakes originally had normal limbs, a view strengthened by the fact that boas and pythons have tiny

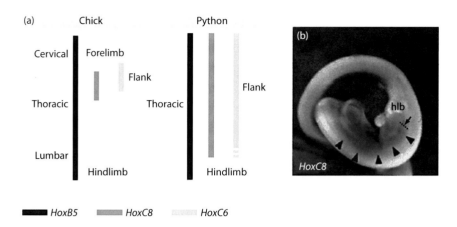

Figure 5.4 Hindlimb development in pythons. (a): Diagram comparing expression domains of *HOXB5*, *HOXC8*, and *HOXC6* genes (the broken end indicates lack of certainty about precise expression limits) in chick and python embryos. In the chick, forelimbs develop in the posterior part of the cervical region that is marked by HoxB5 expression alone. This domain is anterior to the expression of the other hox genes that pattern a thoracic identity (marked by rib production). In the snake, the thoracic region (and ribcage) extends to the head because all three genes are expressed in what would have been the cervical region. (b): HOXC8 expression (arrowheads) extends posteriorly to level of hindlimb bud (hlb). The dashed line and arrow mark the posterior expression boundary of Gene expression).

(From Cohn MJ & Tickle C (1999). *Nature* 399:474–479. With permission from Springer Nature.)

hindlimbs (pelvic spurs), but not forelimbs. The explanation for the fact that snakes once had limbs and have now lost them comes from comparative evo-devo work on their molecular development by Cohn and Tickle (1999). For this, they examined the patterning of Hox genes in the somites of early chick and python embryos that lie on either side of the neural tube (Fig. 5.4a; Appendix 6). This was because ribs and limb bones derive from the sclerotome components of these somites, with the details of the various Hox genes expressed in each being responsible for its future development (Chapter 8; Appendix 7).

The pattern of Hox gene expression in the chick shows how the arrangement of three genes, in particular, specify the cervical and thoracic regions. The cervical region is specified by HOXB5 protein alone. The immediately posterior thoracic (rib-forming) region is essentially defined by the added expression of the *HOXC6* and *HOXC8* genes (note: Hox homeobox genes in the chick are capitalized). The forelimb bud forms in the posterior region of the cervical region just anteriorly to the thoracic domain. In the python, the whole of the anterior region of the embryo carries the expression of the three Hox genes associated with thoracic tissue. As a result, there are ribs all the way up to the head region and, because there is no cervical region, there is consequently no forelimb.

The situation in the hindlimb is different. The Hox patterning is such that early hindlimb buds form in the python as they do in the chick (Figure 5.4b). Cohn and Tickle showed that the reason why the python limb buds do not develop further is that the signaling systems in the overlying ectoderm that activate cell proliferation and tissue patterning in the early limb mesenchyme of the chick are not present in the snake; their buds, therefore, remain as rudiments. Python mesenchymal cells in the limb bud do, however, retain their ancestral abilities: if their limb bud mesenchyme is grown in the presence of chick limb-bud ectoderm, the mesenchyme can be activated to participate in normal proliferation and patterning.

The implication is clear: modern snakes carry their evolutionary history in their genes, still maintaining those needed for limb formation; their expression is, however, repressed by an evolutionary modification that blocks expression of the signaling system in their limb bud ectoderm. The historical line of synapomorphies has been tinkered with, but not fully lost. It is also intriguing that a similar, but reciprocal effect has been found in cetaceans (whales and dolphins): here, forelimbs form while hindlimb buds cease development shortly after they have formed (Chapters 7 and 16). The implications of such patterning changes in the context of variation are discussed in Chapter 13.

Exaptations

It sometimes happens that a structural feature with a novel and specific function just seems to appear in the evolutionary record apparently fully functional. In practice, however, such exaptations are always found to derive from a similar but usually simpler pre-existing tissue fulfilling a different function in an earlier organism within the clade. One classic example of a tissue that appeared fully formed is, as mentioned, the camera eye but it now seems to have evolved from a flat photosensitive plaque over a relatively short time, by evolutionary standards (Chapter 20). Another is the feathers that early birds needed for flight. These were first identified in *Archaeopteryx*, a fossil of an organism with both dinosaur and birdlike features that was discovered in the late 19th century at the Solnhofen Lagerstätte in Bavaria (Southern Germany). Given that such feathers would have had no use in flight unless they were fully functional, it is at first sight hard to understand how they evolved.

Feathers, it has now turned out, did not evolve for the purpose of flight but were a common feature seen in dinosaurs that that lived long before the earliest birds evolved, and that evolved from skin scales through a complex series of changes (Wu et al., 2018). Their initial role seems to have been to help maintain body temperature, much like hair in mammals (Chen et al., 2015; Fig. 15.10). Persons & Currie (2019) have suggested that the next step was that some of these simple feathers evolved to become patterned and complex so allowing them to be used for display and hence for sexual selection. It was only later that these more sophisticated feathers further evolved to become first gliding and then flight feathers.

This shift in role is now called an *exaptation*, a term that replaces the older one of *preadaptation*, a word which carries overtones of teleology. Another example of an exaptation is the walking limb of land vertebrates. This evolved from the boned fins of late sarcopterygian fishes whose role was to facilitate swimming maneuverability. These fins eventually became robust enough to support the fish: fossilised trackways show that they walked on the bottom of shallow seas near the shore (Ahlberg, 2019). Over time, these swimming appendages strengthened to produce primitive autopods with radial bones, as opposed to fin rays (as in the elpistostegids). Eventually, these early limbs became strong enough drag the "fishapods" (e.g. Tiktaalik; Shubin, 2007; the more formal name is epistosteglids) up beaches, presumably in search of food (Chapter 6; Clack, 2012; Ahlberg, 2019). In due course, a branch in this clade evolved to produce stem tetrapods whose limbs had jointed digits and girdles strong enough for them to walk properly without dragging their body. An appendage whose initial role was to facilitate swimming had become one whose role was in walking on the sea bottom and then on land.

THE BROADER IMPORTANCE OF DARWIN'S IDEAS

The focus of this chapter has been on how Darwin's original idea, that novel speciation derived from heritable descent with modification, can be used to construct anatomical phylogenies. Darwin thus integrated two important ideas: first, only heritable changes are important in the context of evolution thus implying that new species derive from earlier ones and, second, there need to be mechanisms that generate heritable change. Together, they explain why we are able to group organisms in a hierarchy based on ancestry using the formalism of cladistics. What this chapter has also shown is that implementing what seems a simple thought in a way that is useful turns out to be rather more complicated than one might have hoped.

One apparent problem here is that cladistic analysis tends to assume that, because one structure could have evolved from another, it actually did so. In many cases, the fossil evidence to support the details of such a contention is missing—the fossil record is mainly a very low-resolution movie of past life whose detail decreases as one goes further back in time. Hence the importance of transitional fossils that allow us to track some of the details of the changes by which new tissues evolved from old. The next chapter focuses on three of these transitions, which provide good examples of how evolutionary change takes place.

At a broader level, one way of recognising the importance of a scientific idea is that, although it might have been produced to help explain a specific area of ignorance, it turns out to have a much wider significance. In Darwin's time, the key assay for discussing evolutionary change through descent with modification was anatomical detail; the formalism of cladistics with its various "-morphies" sets this idea in a contemporary anatomical context.

It is, however, important to note that the application of the term <*descends with modification*> is not in principle restricted to discussions of anatomy alone but can be applied to any heritable property of an organism. Another obvious example here is the inheritance of genomic mutations, an area in which we now have vast amounts of data from a host of species across the phyla. Comparative sequencing of DNA across taxa (Chapter 7) has the great advantage that investigating evolutionary closeness in a way that does not depend on the fossil record: it uses sequence data from living organisms (and occasionally ancient DNA). A third area is the cross-species analysis of proteins with specific functions, particularly during development. Experimental evidence has now shown that that very different organisms control their very different development using very similar molecular mechanisms (Chapter 8).

These three approaches construct evolutionary history on the basis of completely independent types of data: anatomical tissues, DNA sequences, and the function of homologous proteins and networks, often during embryogenesis. Any scientific theory should make predictions that can be tested; hence, each of these approaches, which represent different windows on the same evolutionary events, should thus group living species in the same way. These hierarchies, in turn, should mesh with what is known about the evolutionary time-scale as inferred from the fossil record. In cases where the amounts of data are not the same, lower-resolution groupings should be compatible with higher-resolution ones. A further prediction is that very different organisms with homologous proteins should use them in very similar ways, allowing for their different structural characteristics and functional abilities. It is hard to identify even a single case where, if there is data, such predictions are not met.

It might be thought unnecessary to make this point about testing Darwinian ideas against predictions. It is, however, a commonly held view, even amongst some scientists (although not biologists) that the theory of evolution is not a "proper" theory because it does not make predictions. This view is completely wrong, and readers might like to think about other predictions that the theory makes and how they might be tested.

KEY POINTS

- The evidence on evolutionary relationships can be organized within hierarchical trees, known as cladograms, that reflect descent with modification from last common ancestors.

- Cladistics provides the formalism for constructing and studying such trees and is normally based on anatomical homologies and innovations.

- Cladograms are mathematical graphs that organise groups of species into hierarchies on the basis of shared and derived features. Typical relationships are:

< species **B** >< *descends with modification from* >< species **A** >
< species **C** >< *descends with modification from* >< species **B** >

- *Synapomorphies* are characteristics that are common to a clade as they are shared and derived from those seen in a common ancestor species.

- *Plesiomorphies* represent ancient features: i.e. those present in species **A, B,** and **C** together with their ancestral taxa.

- If an expected plesiomorphy is missing in some taxon within a clade (e.g. snakes should have limbs, a plesiomorphy for the reptile clade), there should be a molecular explanation for this that is accessible to investigation. The correct location of the taxon within a clade can be stablished through a wide range of other predicted synapomorphies. Snake synapomorphies should include other reptilian plesiomorphies, such as scales, amniote eggs, and skull morphology.

- The presence of similar properties in two species is not of itself unequivocal evidence of the species being related as the property may have arisen independently in each. Such properties are known as *homoplasies;* examples are wings and sabre teeth.

- Whenever a new anatomical feature appears fully formed in the fossil record, it will be found to have derived from a simpler structure with another function. These are known as *exaptations* and examples include flight feathers and vertebrate limbs.

FURTHER READING

Kitching IJ, Forey PL, Humphries CJ, Williams DM (1997). *Cladistics: Theory and Practice of Parsimony Analysis* (2nd ed). Oxford: University Press.

WEBSITES

- The Wikipedia entries on *Cladistics, Homoplasy* and *Synapomorphy* and *Apomorphy*
- PHYLIP (the PHYLogeny Inference Package) is a set of programs for inferring phylogenies (evolutionary trees). http://evolution.gs.washington.edu/phylip/general.html

- More historical and other details on cladistics are given in: https://www.newworldencyclopedia.org/entry/Cladistics: https://www2.gwu.edu/~clade/faculty/lipscomb/Cladistics.pdf

The Anatomical Evidence for Evolutionary Change

6

The key to understanding evolution is not just a matter of placing fossils in temporal order—this provides little direct insight into the process of changes that led one taxon to evolve slowly into another. As both taxa may then have continued to survive, although probably in slightly different habitats, fossil dates are of themselves little help in determining the timings of relationships. Evolutionary descent can only be established by careful analysis of anatomical differences between closely related species, identifying the differences and using cladistic analysis to establish a phylogeny. The fossil record is generally good enough to enable the major lines of post-Cambrian animal and plant evolution to be worked out—these are detailed in Section 3. Harder to discover, because the evidence is much thinner, are the details of how evolutionary change occurred as one taxon transitioned into another.

This chapter examines the evidence on three important anatomical transitions where the fossil record is rich if not complete. First: the evolution of the pentadactyl walking limbs from the fins of sarcopterygian fishes, a change that took place in the Devonian seas over a period of ~30 My. Second: the evolution of the mammaliaform skull from that of a stem mammal. This required a series of changes that took place over ~55 My between the mid-Permian and the early Jurassic. Third: the evolution of the Equidae, with a particular focus on how the multi-toed foot of Eocene horses was reduced a single hoofed toe over a period of ~50 My.

Together, these examples show the types of anatomical change that characterise evolutionary descent in vertebrates and the complexity of the events that occur at evolutionary nodes. There is one other message that emerges from these examples: it is that evolutionary change is slow, at the anatomical level at least, with the amount of change happening over a million years sometimes being surprisingly small.

Darwin's proposal that evolutionary change arises through *descent with modification* makes two types of prediction that can be examined in the fossil record. The first is that any organism that ever lived can, where the fossil record is adequate, be traced back to earlier ancestors with more primitive features. The second is that change should arise from selection acting on variants and hence that, for major changes in anatomical features, there will be a set of transitional organisms that show how that change happened. The formalism of cladistics given in the last chapter provides a means of articulating the first prediction, and its use will be demonstrated in this and future chapters. Given the vagaries and limitations of the fossil record, however, the second is harder to test, but there are a few examples where the fossil record is rich enough to include sufficient transitional species for us to be able to follow the process of change. Three of these provide the main focus of this chapter (see Websites for other examples).

The key requirement for the first prediction is being able to establish that similar contemporary species share properties that can be traced back in time to a node at which such properties evolved, and then identifying all other species that share

DOI: 10.1201/9780429346217-8

some at least of these derived properties (synapomorphies), albeit that it may have become modified in some way over time. One obvious example is the evolution of all bony fishes that followed the ossification modification to the cartilaginous skeleton of earlier fishes. Another is that the evolution of stem tetrapods to produce amniote eggs with shells and internal membranes that could develop in nonaqueous conditions. The early amniotes, in turn, gave rise to diapsids and synapsids: the former were the reptiles and the latter were stem mammals that eventually evolved to give mammals (Chapter 15).

This first prediction has been fairly straightforward to satisfy wherever the fossil record is good enough. Those relatively few ambiguities in the history of animal and plant evolution where the fossil detail is inadequate to explain the origins of a particular contemporary taxon are usually resolved by using phylogenetic approaches (Chapter 7). These were, for example, used to untangle the phylogenetic history of whales at a time when the fossil record was inadequate (Chapter 16). The anatomical history of almost every animal and plant taxon can now be included in a cladogram that details anatomical change. The Tully monster (Chapter 12) is one of those few ancient organisms for which the fossil record so far provides no obvious antecedents or descendants.

The second prediction is harder to meet because such anatomy-based cladograms are mainly broad-brush and say little about the details of what is going on at the nodes that mark the emergence of a new feature. Obtaining detail here requires close analysis of as many fossils as possible that lived over what was, by evolutionary standards, a relatively short time. Even in the few cases for which this criterion is met, any explanation can only be at a macroscopic level. Probing more deeply into how the changes in phenotype were achieved almost always requires molecular-genetic comparisons of morphogenetic events in a range of contemporary embryos—this is evo-devo (Chapters 8 and 18). Further analysis of the roles of variation and selection in speciation can only be explored experimentally, and these approaches are considered in Section 4.

Understanding evolutionary transitions at a fine degree of temporal and anatomical resolution requires many fossils from a range of similar organisms. The difficulty here is that the fossil record, as discussed in Chapter 3, is stochastic by nature and, of the many millions of transitions that have happened over time, only a few are adequately represented in the rocks. Even when it possible to find some fossilised organisms that contributed to a transition, the evidence generally turns out to be inadequate to reconstruct the details of what happened at the nodes within the formalism of a cladogram.

Fortunately, there are enough examples of fossil sequences to illustrate the details of a few of these transitions and so meet, in part at least, the expectations of the second prediction. The interested reader is particularly referred to the wiki on *Transitional fossils* that discusses more than 20 such cases (see Websites). Here, we will consider three of the better-known examples in vertebrate evolution, the clade that, because of the robustness of its bony skeleton, is particularly well represented in the fossil record. These are (i) the transition from the rayed pectoral and pelvic fins of bony fish to the pentadactyl limbs of early tetrapods in the Devonian (419–359 Mya), (ii) the evolution of the changes that led to an early mammalian skull from that of a stem mammal, events that started in the late Permian (~245 Mya), and (iii) the changes in the multi-toed feet of early equids that led to an early horse in the lower Eocene whose feet had a single hoof, and the subsequent diversification of the clade during the Miocene. In Section 5, we will also consider the skeletal transitions that led to the evolution of humans from an early great ape ancestor.

THE EVOLUTION OF THE PENTADACTYL LIMB

In Chapter 14, we will look in more detail at the many anatomical modifications that had to evolve before a member of the line that evolved from bony fishes acquired limbs that allowed it to clamber onto land, and its descendants to become stem tetrapods, the earliest vertebrates with jointed toes. This section will just focus on the evolution of the pentadactyl limb in Devonian fish. The bony fishes then and now are grouped in two taxa, the Actinopterygii and the Sarcopterygii; members of the former have nonfleshy pectoral and pelvic fins that include some proximal bones and terminal bony rays (lepidotrichia) that support the web, while the latter have fleshy fins whose

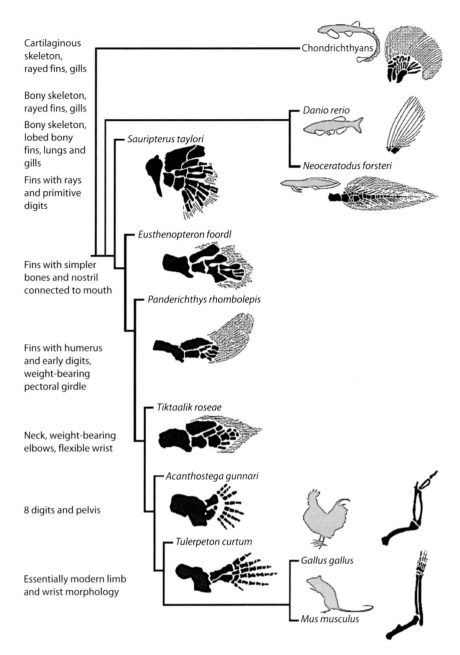

Figure 6.1 An evogram based on well-known fossils showing details of the evolution of the tetrapod limb from the fin of an early fish.

(From Schneider et al. (2011) *Proc Natl Acad Sci U S A*; 108: 12782–12786. With permission.)

proximal region included bones homologous to those in tetrapods and whose distal region again has bony rays within the fin web (Fig. 6.1). Today, almost all bony fish (e.g. trout) are members of the Actinopterygians, whereas the only surviving taxa of the Sarcopterygii are lungfish (Dipnoi) and the very rare coelacanths (Latimeriae).

The anatomical evidence shows unequivocally that tetrapods with limbs evolved from sarcopterygian fish with bony fins. Apart from these fins, sarcopterygian fish, such as *Eusthenopteron* (~350 Mya) also had the labyrinthodont teeth, nostrils that connected with the oral cavity and a flexible neck joint that allowed the head to turn, all features that are seen in the fossils of stem tetrapods. Although only hard tissues have been preserved, *Eusthenopteron* would, on the basis of these and other features, probably also have had primitive paired lungs and some breathing capacity, as seen in today's lungfish.

Over the last decade or so, the analysis of the anatomy of a wonderful series of late-Devonian fossils, some recently discovered and others reexamined, has illuminated the details of the transition from water to land and the evolution of the tetrapod limbs that took place during the Devonian. This work has been supplemented by the analysis of trace-fossil trackways showing evidence of tetrapod activity that date

TABLE 6.1 THE TRANSITION FROM SEA TO LAND—THE FOSSIL EVIDENCE

Name	Mya	Class	Apomorphies
Kenichthys	~395	Sarcopterygian fish	Earliest known fish with nostrils linking to the oral cavity (Zhu & Ahlberg, 2004)
Eusthenopteron	~385	Sarcopterygian fish	Aquatic, internal nasal adaptations, labyrinthodont teeth; the bones of pectoral and pelvic fins had growth plates for lengthening (Meunier & Laurin, 2012).
Sauripterus	416-360	Sarcopterygian fish	Fins with digits and rays, scapulocoracoid bones, and pelvic girdle (Davis et al., 2004).
Pentlandia macroptera	~385	Sarcopterygian fish	Jointed radial bones, cleithrum, clavicle, and ossification centres (Jude et al., 2014)
Panderychthys	~380	Elpistostegid	Four unjointed digit-like bones, tetrapod cranium (Boisvert et al., 2008)
Tiktaalik	~375	Elpistostegid	Basic wrist with rays, flexible neck, lungs, basic pectoral and pelvic girdles (Shubin et al., 2014).
Elpistostege	~380	Elpistostegid	Pectoral girdle, limbs with clear homologues of modern long bones. fins with unbranched radials and rays; basic girdles (Cloutier, 2020)
Acanthostega	~365	Stem tetrapod	8 forelimb & 7 hindlimb jointed digits, non-weight-bearing forelimbs, complete pelvic girdle (Coates, 1996).
Tulerpeton	~365	Stem tetrapod	6 jointed digits, powerful "wading" limbs, pectoral girdle, lungs, no gills (Lebedev & Coates, 1995)
Ichthyostega	~365	Stem tetrapod	Amphibian-type lungs, strong ribs and fore and hind limbs (7 jointed toes) + fish gills and tail. *This animal was able to clamber on land* (Pierce et al., 2012)
Perdepes	~348	Early land stem tetrapod	5 (+1?) digits. Land-adapted feet (Clack, 2002)

Note: The dates reflect only the age of the fossil and the organisms are ranked by their closeness to tetrapod morphology (Cloutier et al., 2020).

back to ~390 Mya (Ahlberg, 2019) and demonstrate that tetrapod organisms were able to walk on the sea bottom in shallow waters. Slightly later trackways are found, for example, near the beach of Valentia Island off the coast of Ireland that date to ~385 Mya; these show that it was not long before stem tetrapods had acquired sufficient limb strength and girdle support to enable them to walk on land without dragging their body (Table 6.1; Fig. 6.1; see Websites). These trackway dates precede all the well-known fossils of stem tetrapods making it clear that the line to the tetrapods was initiated rather earlier than is indicated by the fossil record.

Limbs differ from fins in two important ways: fins are shorter and include a proximal set of bony elements and a distal rayed fin. Limbs are longer and are patterned into three domains rather than two: the stylopod with a humerus or femur; the zeugopod with a radius and ulna or a tibia and fibula, and the autopod, an apomorphy, with wrist or ankle bones and jointed digits. Fig. 6.1 shows that it is possible to identify homologies between early sarcopterygian fin bones and modern limb bones.

Sarcopterygian fishes have proximal fin bones, some of which are homologous to but much shorter than modern long bones; the key to their elongation was the presence of growth plates in these bones, a feature seen in sarcopterygian fish, such as *Eusthenopteron*. The transition to modern limbs is shown in the elpistostegids (e.g. *Tiktaalik*), a branch off the sarcopterygian line that has now been lost. Their long bones had started to lengthen and their autopods included both bony rays and unjointed bones (radials) in the autopod region. Ancient fossilised trackways suggest that the elpistostegids were capable of clambering onto land (Ahlberg, 2019). It also seems likely that such an ability in these stem tetrapods may, through developmental plasticity, have strengthened the bones of the ambulatory system (Chapter 19; Molnar et al., 2017; Standen et al., 2014).

Figure 6.2 (a): Fossilised fin of *Sauripterus* showing perhaps seven unjointed digits together with a dozen or more rays (arrow). (b): Reconstruction of a skeleton of *Ichthyostega* showing its relatively massive head. Note that the model forelimb phalange bones have been given five digits, although the exact number is still not known.

(a Courtesy of Ghedo. Published under a GNU Free Documentation License 1.2.
b Courtesy of Oleg Taranov. Published under a CC Attribution-Share Alike 3.0 License.)

The loss of lepidotrichia in fins and their eventual replacement with jointed digits in autopods was a key feature of the stem tetrapods that were the first predominantly land vertebrates and that in time gave rise to both amphibians and amniotes (Stewart et al., 2020). The early stages of change are seen in Devonian sarcopterygian fish, such as *Sauripterus*, which had well-developed girdles and fins containing bony radials as well as rays (Fig. 6.2a; Davis et al., 2004). This addition reflects an important developmental event: bony rays are made of dermal bone that forms from mesenchyme that lies subjacent to the external epithelium of the fin web and is derived from embryonic dermatome (Appendix 7). Digits, in contrast, are made of endochondral bone, and this is produced by mesoderm from embryonic lateral plate. The formation of radials and eventually jointed digits required the migration of cells from the lateral plate mesoderm into the distal part of the limb and the evolution of a novel patterning system for its development (Appendix 7).

The transition from a fin to an autopod can be followed in a series of fishes and elpistostegids, such as *Panderichthys* and *Tiktaalik*, which had both rays and unjointed radials. Jointed digits were, however, an apomorphy of the later stem tetrapods such as *Ichthyostega* (Table 6.1; Fig. 6.2b). It is interesting to note that early stem tetrapods had anything up to eight digits and, on the basis of the fossil record to date, this number didn't stabilise to five until ~20 My had passed.

Ideally, there would be sufficient fossil material to produce a detailed picture of tetrapod evolution, but our current sample of organisms is probably only adequate for an analysis of limb evolution (Fig. 6.1). The comparison of the elpistostegids, *Panderychthys,* and *Tiktaalik*, for example, shows that the former had bony radials (unjointed digit bones) but a rigid neck, whereas the latter had a mobile neck but lacked radials. Similarly, a comparison of the stem tetrapods *Ichthyostega* (374 Mya) and *Acanthostega* (365 Mya) shows that the former had stronger forelimbs than the latter. As there are no unique lines of descent, these various fossils have to be seen as a sample of the rich fauna of increasingly tetrapod-like fishes that existed in shallow water during Devonian times. The question of whether the early tetrapods developed in fresh or salt water is discussed in Chapter 15.

THE EVOLUTION OF THE MAMMALIAN SKULL

The skulls of early amniotes were anapsid as they lacked temporal fenestrae: these are the openings in the side of the skull positioned posterior to the orbit (the fenestra within which the eye is located) and used for muscle attachments. By the early Permian, two taxa had emerged that could be distinguished by temporal fenestrae: these were the synapsids with a single lower opening (the stem mammals) and the diapsids with an upper and a lower fenestra (the reptiles). The diapsids gave rise to the Euryapsida with a single upper opening (the future ichthyosaur sea lizards), the Testudines (turtles) that lost both fenestrae and, in due course, the diapsid Archosauria, which would include the crocodiles, pterosaurs, dinosaurs, and eventually the birds (see Figure 15.1 for detailed cladogram).

During the Permian, the synapsids were the major tetrapod clade; early taxa were known as pelycosaurs and later ones as therapsids (note: these animals used to be known as mammal-like reptiles or protomammals but are now called stem mammals). The Permian extinction (252 Mya) led to a major loss of stem mammals, which were eventually reduced to just the cynodont and a very few other taxa that died out (Chapter 16). As the world slowly recovered from this trauma, the other taxon of land animals, the diapsid reptiles, rapidly radiated and evolved over the Triassic to give the wide range of archosaur taxa (Chapter 15). During the Triassic, the stem mammals were reduced to small, shrew-like animals known as mammaliaforms that were mainly nocturnal and probably lived in burrows. By the early Jurassic (~195 Mya), the mammaliaforms had evolved to become mammals (see Chapter 16 for more details).

It thus seems to have taken until the early Jurassic (~195 Mya) for the various skull synapomorphies that characterise modern mammals to have evolved from the more basal properties shown in the skulls of late-Permian stem mammals (Fig. 6.3). The fossil record of these changes is surprisingly good considering that the skulls of small animals are not particularly robust. There are now sufficient crania, jaws, and particularly teeth to follow some of the key changes and nine examples of skulls that cover the transition from a mammaliaform to an early mammal are given in Table 6.2 and Fig. 6.5. Some of these changes reflect alterations or repatterning of the cranial bones and are rather technical in nature. The following four are, however, more anatomically straightforward and can be readily followed in the fossil record.

1. The formation of two of the middle-ear ossicles (malleus and incus) from the quadrate and articular bones that formed the jaw joint.
2. The formation of the secondary palate, an event that facilitated later endothermy.

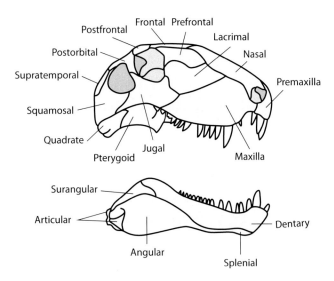

Figure 6.3 The skull of dimetrodon. This Permian pelycosaur (stem mammal) lacks a zygomatic arch and has a simple jaw hinge (the temporomandibular joint). All of its teeth are of protoconid morphology (note the absence of cusped molars) and differ mainly in size. The black region posterior to the orbit (grey) is the temporal fenestra.

(Courtesy of "Smokeybjb." Published under a CC Attribution-Share Alike 3.0 Unported License.)

TABLE 6.2 THE EVOLUTION OF THE MAMMALIAN JAW DURING THE EARLY MESOZOIC ERA

Species	Period span	Details
Dimetrodon	Early-Mid Permian	A pelycosaur with basal amniote jaws, teeth, and skull.
Dinogorgon	Upper Permian	A gorgonopsid stem mammal whose jaws show some mammaliaform features. It had incisors and sabre teeth but no secondary palate.
Moschorhinus	Upper Permian Lower Triassic	A large therocephalian stem mammal with reptilian tooth morphology (including sabre teeth) and a partial palate.
Thrinaxodon	Lower-Mid Triassic	A cynodont (stem mammal) with a single jaw joint, an early zygomatic arch, and a range of tooth forms.
Cynognathus	Lower Triassic	A large cynodont with a primitive zygomatic arch, a secondary palate, and early molars.
Probainognathus	Late Triassic	A typical cynodont with a well-formed zygomatic arch showing muscle attachments. It had a range of tooth forms and a secondary palate.
Diarthrognathus	Late Triassic—Early Jurassic	This mammaliaform had a double jaw hinge: a synapsid jaw joint between the quadrate and articular bones and a future mammalian jaw joint.
Morganucodon	Late-Triassic—Mid-Jurassic	A shrew-sized mammaliaform with a mammalian jaw hinge but a multibone lower jaw.
Hadrocodium	Early Jurassic	A mouse-sized mammaliaform with a jaw hinge formed by the dentary and squamosal alone, with the malleus and incus now detached from both the jaw and Meckel's cartilage. As it had a modern jaw too, this was probably the first known well-defined mammal.

3. Changes in the morphology of the lower bones forming the temporal fenestra, particularly the jugal, to give the zygomatic arch (the modern cheek bone).
4. The evolution of complex teeth from simple, reptilian ones.

The evolution of middle-ear bones and the reorganization of the jaw

The basis of the evolutionary changes that led from the jaw and ear of a stem mammal to that of a modern mammal was first identified in the middle of the 19th century and clarified over the following years (e.g. Hopson, 1950; Allin, 1975; Lautenschlager et al., 2017). Stem mammals had a single auditory ossicle, the columella auris (homologous with the stapes), and complex jaws. The lower jaw was composed of the toothed dentary, articular, angular, and several other bones; the upper jaw contained the premaxilla, maxilla, and quadrate (Fig. 6.3). The joint between the lower jaw and the cranium was formed by an articulation between the quadrate and articular bones.

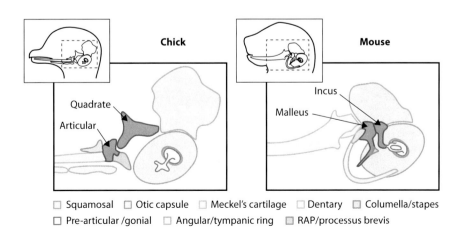

Figure 6.4 Diagram showing the bones of the jaw and middle ear in a chick (this is representative of the anatomy seen in reptiles and stem mammals) and middle ear of a mouse (this is typical of the anatomy seen contemporary mammals). The anatomy of the chick illustrates the origins of the bones adapted to contribute to the mammalian middle ear.

(From Anthwal et al. (2012). *J Anat* 222: 147–160. With permission)

☐ Squamosal ☐ Otic capsule ☐ Meckel's cartilage ☐ Dentary ☐ Columella/stapes
☐ Pre-articular /gonial ☐ Angular/tympanic ring ☐ RAP/processus brevis

The changes that led to mammalian jaw and ear anatomy included (i) the generation of the middle-ear bones (the malleus, incus, and stapes) together with the tympanic ring (these are all endochondral bones that form from cartilage) and (ii) the reduction in complexity of the jawbones to give two upper tooth-carrying jaw bones, the maxilla and premaxilla, and a single lower jawbone, the mandible (these are membrane bones that form directly from mesenchyme). The proximal ends of the half mandibles are located in a fossa (or hollow) within the temporal bone of the skull, so forming the temporomandibular joint; the distal ends meet and fuse at the point of the jaw.

These modifications arose as changes to the development of the upper and lower jaws. These form during embryogenesis from the first pair of pharyngeal (or branchial) arches that appear soon after neurulation, each of which is an epithelial pocket containing mesenchyme that derives from neural crest (Figure 14.6; Appendix 7). The left and right first arches each split into an upper and lower component and each extends, meets and merges with its opposite component. The mesenchyme within the upper pair will form the central premaxilla and small primary palate and the lateral maxilla. In due course, these dermal bones will usually fuse to make the upper jaw bone.

The lower arches fuse in the chin region and the arch mesenchyme then lays down the long thin cylinder of Meckel's cartilage that extends across the lower jaw into the upper jaw region on either side. In early amniotes, the posterior ends of the cartilage then ossified to become the articular and the quadrate (they are thus made of the endochondral bone) which formed the jaw joint. The central and major parts of the cartilage acted as a template for the tooth-bearing dentary. This was made of dermal bone, which was laid down by the surrounding mesenchyme. Meckel's cartilage is still present in adult reptiles and birds but usually lost in mammals.

In later mammaliaforms, the articular and the quadrate became detached, reduced in size, and incorporated into the middle ear as the malleus and incus bones, while the angular, a small bone proximal to the dentary, gave rise to the ectotympanic ring. These events necessitated a change to the jaw joint: as the quadrate and articular bones became part of the inner ear, the jaw articulation became the temporomandibular joint that forms between the dentary and the squamosal, which later became the squamous part of the temporal bone (for further details, see Lautenschlager et al., 2017, 2018).

The evolution of the mammalian jaw took place over a period if ~55 My and is illuminated by a series of fossilised synapsid skulls (Table 6.2; Fig. 6.5). The following

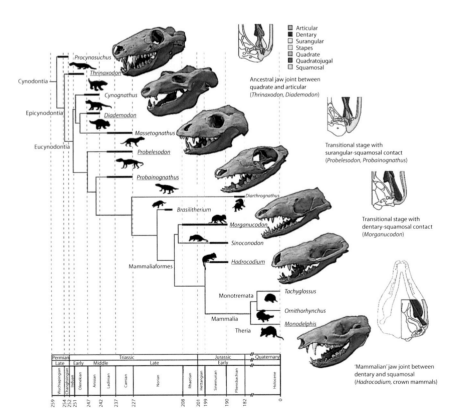

Figure 6.5 Evogram that includes a cladogram of mammaliaform evolution together with datings, skulls, and diagrams of the changes in the jaw that gave rise to the middle ear bones and reshaped the jaw joint during the transition from cynodonts to mammals.

(From Lautenschlager et al. (2017) *Biol Rev Camb Philos Soc*; 92:1910-1940. doi: 10.1111/brv.12314. Published under a Attribution 4.0 International (CC BY 4.0).)

four, in particular, detail aspects of the evolutionary steps that mark the changes to the middle-ear and jaw anatomy.

- *Thrinaxodon* (Early Triassic), a cynodont, had a single jaw joint where the articular and quadrate bones met.
- *Probelesodon* (Middle Triassic), a later cynodont, had a jaw joint in which the position of the squamosal had changed slightly so that it made contact with the surangular.
- *Morganucodon* (Late Triassic), an early mammaliaform, had an intermediate jaw joint in which the dentary makes contact with the squamosal, adjacent to which is the separated stapes.
- *Hadrocodium* (Early Jurassic), a late mammaliaform, had a jaw hinge formed by the dentary and squamosal bones, with the malleus and incus now being detached from both the jaw and Meckel's cartilage.

On the basis of its middle-ear structure and also its brain shape, *Hadrocodium* is probably the earliest mammal so far identified (Rowe et al., 2011). These synapomorphies are seen in all extant mammals, suggesting that their last common ancestor had this form of jaw articulation and three middle ear bones, a feature that enhanced aural sensitivity.

It is worth noting that until at least the Early Cretaceous, an ossified Meckel's cartilage was maintained as part of the mandible in many species, such as *Liaoconodon*, the gobiconodontids, and *Yanoconodon* (Anthwal et al., 2012). In modern mammals, Meckel's cartilage mostly disappears apart from its most posterior region that ossifies and becomes the malleus and the immediately adjacent region that becomes the sphenomandibular ligament (a cable between the mandible and the sphenoid process of the cranium). Curiously, an ossified Meckel's cartilage has very occasionally been seen in modern humans (Shattock, 1880; Keith, 1910).

The evolution of the zygomatic arch

~275 Mya, the therapsid stem mammals, like earlier synapsids, had a single temporal fenestra on each side of the cranium just behind the orbit. Its role was to provide a space through which the external mandibular adductor muscle linked the cranium to the lower jaw, so facilitating biting (Fig. 6.6). The lower bar of this space on each side of the skull was formed by the posterior part of the jugal bone that joined with the anterior part of the squamosal bone, which mainly formed part of the back of the skull. This was the original zygomatic bone.

The zygomatic arch became apparent when the skull enlarged and the squamosal extended outward to form an arch that provided a conduit for the external mandibular adductor muscle (today's temporalis muscle) while its surface provides the superior attachment of the masseter muscle (for mastication), both of which extend to the lower jaw. This change in geometry can be seen in the skull of *Thrixanodon* (~245 Mya; Fig. 6.6). The chewing ability was later strengthened by the evolution of the superficial

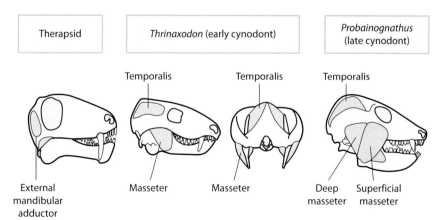

Figure 6.6 The evolution of the zygomatic arch. Early therapsids (~270 Mya) had a single temporal fenestra which was probably an adhesion region for the external mandibular adductor muscle. The skull of *Thrinaxodon*, an early Triassic cynodont (~250 Mya), shows (left view) that the jugal, the bone that forms the base of the temporal fenestra, extends outwards to form the early zygomatic arch so increasing the space available for the temporalis muscle. The right view shows that the temporalis muscle attaches to the medial surface of the mandible, while the masseter muscle links the lateral surface of the mandible to the jugal bone. The skull of *Probainognathus*, a late Triassic cynodont (~220 Mya), shows that the arch has extended further outwards, providing more space and additional adhesion regions for the muscles controlling the jaw.

(From Carroll RL (1988) *Vertebrate Paleontology and Evolution*. WH Freeman & Co.)

masseter muscle, which connects the upper and lower jaws, a feature seen in *Probainognathus* (235-221.5 My; Fig. 6.6).

The evolution of the secondary palate

The early mammaliaforms had a primary palate that formed from the premaxilla, the bone that carries the incisor teeth. This small bony region just behind the upper teeth strengthens the upper jaw but does not divide the oronasal cavity. A secondary palate, which completely separates the oral and nasal cavities, appeared at an early stage of mammalian evolution (~245 Mya; Groenewald et al., 2001). It allowed animals to breathe and chew at the same time, rather than to have to swallow unmasticated food. This had two effects: first, it allowed digestion to begin in the mouth through the production of enzymes within saliva and, second, it allowed the animal to increase the efficiency of oxygen uptake; this turned out to be an exaptation necessary for later endothermy.

The developmental modifications involved in making the secondary palate are now well understood at both the tissue and molecular level. During mammalian embryogenesis, two shelves of tissue, the palatine processes of the maxilla, extend from either side of the embryonic oronasal cavity and are stiffened by extracellular matrix. They meet at the midline, where they seal with each other and to the small primary palate. In due course, all but the posterior part of the initially soft secondary palate becomes ossified (see Gritli-Linde, 2007, and Goodwin et al., 2020, for further details on the molecular mechanisms of secondary palate formation).

The evolutionary steps that led to the hard secondary palate are seen in the fossilised jaws of three synapsids that bridge the Permian extinction (Fig. 6.7; Van Valkenburgh & Jenkins, 2002). In *Dinogorgon*, a mammaliaform that lived during the late Permian (~260 Mya), there are openings in the roof of the mouth (arrows) that show that the palatine processes have yet to evolve. In *Moschorhinus,* another synapsid that lived from the late Permian to the early Triassic, the processes are present but do not yet meet in the midline. In *Cynognathus,* an early Triassic synapsid (~247-37 Mya), the processes have fully formed and fused.

Incidentally, the fossil record shows that *Moschorhinus* fossils are found in rocks both above and below the line marking the Permian extinction. It is noteworthy that skeletons below the line are markedly larger than the Triassic ones above it. This change in size is an example of the Lilliput effect, whereby

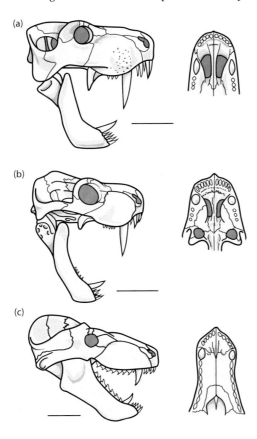

Figure 6.7 Diagram illustrating the geometric differences in skull, jaw, tooth, and palate morphology as the secondary palate evolved. (a): *Dinogorgon* (late Permian); (b): *Moschorhinus* (early Triassic); (c): *Cynognathus* (mid-Triassic). Ventral views of the anterior palates (right) show how the secondary palate evolved, as marked by the progressive closing of the open part (black): *Dinogogon* had no secondary palate; *Moschorhinus* has had a partial secondary palate, while a complete secondary palate has formed in *Cynognathus*. Scale bar = 100 mm.

(From Van Valkenburgh B, Jenkins I (2002) *Paleont Soc Papers*; 8:267–288. With permission from Blaire van Valkenburgh.)

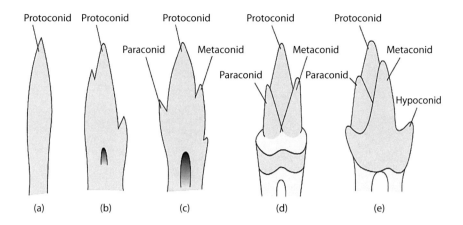

Protoconid   Protoconid   Protoconid   Protoconid   Protoconid

Paraconid   Metaconid   Metaconid   Metaconid

Paraconid   Paraconid

Hypoconid

(a)   (b)   (c)   (d)   (e)

Figure 6.8 Drawings illustrating the evolution of molar (cusped) from conid teeth in a series of stem mammals, mammaliaforms, and early mammals. (a): Permian stem-mammal; (b,c): triassic mammaliaforms; (d): late Triassic mammaliaform; (e): middle Jurassic early mammal.

(From Beddard FE (1902) *The Cambridge Natural History.* Cambridge University Press. (permission assumed on basis of 1st edition). Beddard died in 1925.)

a species that survives an extinction diminishes in size (Twitchett, 2007). The reasons for this diminution are not clear, but the effect may result from limitations on postextinction food availability.

The evolution of mammalian teeth

The teeth of Permian synapsids were all morphologically similar, with a simple, protoconid structure that varied mainly in size (e.g. Fig. 6.3). Teeth with different and more complex morphologies first evolved in the early Triassic cynodonts (the name means dog tooth) and became more sophisticated over time (see Luo, 2007). While these teeth were good for catching and holding prey that would then be swallowed whole, they were not intended for chewing. This requires molars, enameled teeth normally located at the rear of the jaw, that have a flattened and cusped chewing surface; primitive examples of these are first seen in cynodonts. Originally, molars seem to have been used for grinding food, but the organization of the three cusps on matching upper and lower molars across a wide range of early mammaliaforms, such as *Morganucodon* (~205 Mya, early Jurassic) suggests that they were beginning to acquire a shearing function that facilitated more efficient chewing (Schultz & Martin, 2014; Fig. 6.8). The organization of these cusps in fossils, together with the signs of local wear, reveals their function and helps indicate the age of the animal.

Dentition has been important in working out when the early mammals separated into the current three taxa, the key synapomorphies being the structure of the molars and their number in each half jawbone (Williamson et al., 2014; Figure 16.1). It seems that the basic three-cusped or tribosphenic molar was first seen in late mammaliaforms; this feature is maintained in monotremes, although teeth today are only seen in the young of platypuses. Metatherian (marsupial) and eutherian (placental) mammals have more complex molars that differ in detail and number, with the Eutheria having three such molars and Metatheria four. The fossil evidence suggests that tribosphenic teeth evolved independently in the Monotremata and in the Metatheria and Eutheria (Luo et al., 2001).

The molecular basis of tooth development is being actively investigated, and it is now known that cusp patterning is controlled by the downstream effects of complex signaling (Tapaltsyan et al., 2016; Kim et al., 2019), although the details of how the different molar morphologies evolved are not yet clear.

THE EVOLUTION OF THE EQUIDAE

The Equidae clade (horses) evolved ~50 Mya in North America from within Perissodactyla taxon (ungulates with odd numbers of toes), and the fossil evidence includes material from more than 20 extinct species. Species from the Hyracotherium genus (older name: Eohippus) were root members of the clade; these dog-sized animals had four front and three hind toes together with teeth that were adapted for browsing leaves. As the equid line spread widely, it diverged over time; some 20 My later, a successor had only one walking toe (the third digit) and two much smaller lateral toes, a proportionately bigger muzzle and longer teeth adapted for grazing on grass; it could thus be recognised as a member of the Equidae. The oldest known member of the clade with such an autopod was *Merychippus* (~15 Mya), which

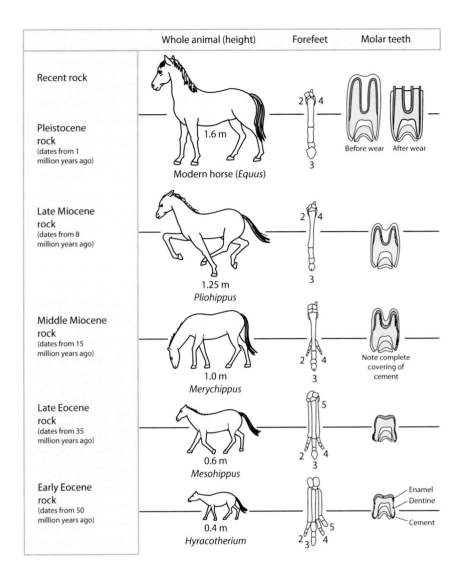

	Whole animal (height)	Forefeet	Molar teeth

Recent rock

Pleistocene rock
(dates from 1 million years ago)

1.6 m
Modern horse (*Equus*)

2 4
3
Before wear   After wear

Late Miocene rock
(dates from 8 million years ago)

1.25 m
Pliohippus

2 4
3

Middle Miocene rock
(dates from 15 million years ago)

1.0 m
Merychippus

2 4
3
Note complete covering of cement

Late Eocene rock
(dates from 35 million years ago)

0.6 m
Mesohippus

5
2 4
3

Early Eocene rock
(dates from 50 million years ago)

0.4 m
Hyracotherium

2 4
3 5

Enamel
Dentine
Cement

Figure 6.9 Key changes in the evolution of the modern horse. AS the Equidae evolved, there were substantial changes in height, foot skeleton, and molars. Note that the foot skeletons showing monodactyly do not include the reduced digits 2 and 4 of *Merychippus*, which became splint bones.

(Courtesy of "Mcy Jerry." Published under a CC Attribution-Share Alike 3.0 Unported License.)

was followed by many subspecies that can be distinguished on the basis of minor differences in the skeleton and teeth.

In the ~10 My that followed the appearance of the taxon, there were changes in height, autopod structure, and molar morphology (Fig. 6.9). These included an increase in the length of legs, the progressive loss of digits, and the strengthening and lengthening of the molars (from < 2 to > 4 cm). Autopod changes started with the loss of digits 1 and 5, a feature of *Merychippus*, and were followed by successive reduction in size of digits 2 and 4 to the extent that these digits today are just splint bones that support the internal structure of the autopods. It is particularly worth noting that the core cladogram of the evolution of the Equidae (Fig. 6.10) includes the many lines of descent that became extinct, particularly those with a triple-digit autopod

The selection pressures responsible for these changes were clearly the new availability of grasslands and the consequent opportunity for grazing on rough but grassy plains, an ability facilitated by the lengthened and strengthened teeth. In such an environment, hooved feet were not only adapted to running but provided a defensive weapon. The developmental mechanism responsible for digit diminution was probably the reduction in width of the limb bud; this was achieved through domains of apoptosis (programmed cell death) being programmed to appear on the medial and lateral sides of the bud (Saxena et al., 2017).

Following examination of some 15 species of Equidae whose fossils were found in North America, Prado & Alberdi (1996) have constructed a cladogram based on detailed analysis of some 20 anatomical features (Fig. 6.10). They mainly reflect digit number and the fine details of tooth, jaw, and cranial morphology. The cladogram

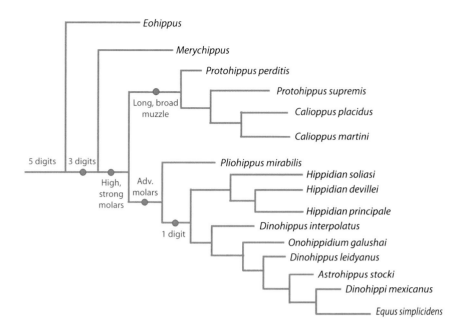

Figure 6.10 A cladogram showing the evolution of the horse based on the data in Prado & Alberti (1996). Some core synapomorphies for the major clades are shown in blue. The synapomorphies for the subclades that are not shown mainly reflect details of molar and skull morphology.

shows that although the early Equidae initially diversified widely to give a broad range of species, all but one eventually became extinct ~3.5 Mya. The species that passed through this bottleneck was *E. simplicidens*, sometimes known as the Hagerman horse, because a skeleton was discovered in Hagerman Valley, Idaho, USA. This animal seems, on the basis of its anatomy, to be close to the last common ancestor of the seven species of the Equidae that are alive today.

Such bottlenecks are frequent in evolutionary descent, as a host of examples demonstrate. Well-known cases include the large, shallows-inhabiting sarcopterygian fish, only one taxon of which seems to have had descendants that clambered onto land; the broad dinosaur clade from which only the ground-living birds survived the Cretaceous extinction, and the mammaliaform clade from which only the cynodonts survived the Permian extinction and its aftermath. The other important example of a bottleneck-surviving species is *Homo sapiens*. Some 7 Mya, a line separated from early great apes that subsequently diverged widely to give a broad range of hominin taxa. All were lost except us, albeit with a little interbreeding on the way with species that are now extinct (Chapter 24).

The closeness of *E. simplicidens* to the seven contemporary horse species is too great for a cladogram to be constructed on the basis of their skeletal differences. Ishida et al. (1995) were, however, able to construct a phylogram using the various mutational differences in their mitochondrial sequences (Fig. 6.11). This shows that modern members of the Equidae fall into three groups: *Equus caballus* (the domestic

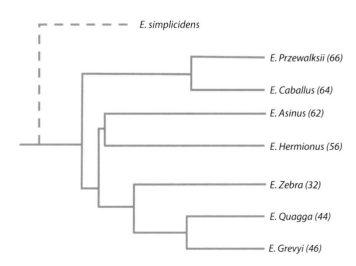

Figure 6.11 A horse phylogeny based on mitochondrial-DNA differences across the modern Equidae (Ishida et al., 1995). The location of *E. simplicidens* (extinct) is shown through a dotted line. The length of a line is an estimate of the amount of mutation that has occurred in mtDNA since a separation node. The chromosome number for each species is given in brackets.

horse) together with Przewalskii's horse, the asses, and the zebras. Particularly striking is the range of chromosome numbers in the different species that has resulted since they diverged some 3.5 Mya (Fig. 6.11; Table 22.3). Nevertheless, and perhaps surprisingly, the different species can interbreed to produce hybrids, albeit that they are almost all sterile (Chapter 22). The range of chromosome numbers also demonstrates the speed with which chromosomal reorganisation can take place. These differences do not, however, yet seem to have been analysed in any detail.

The analysis of the evolution of the Equidae further demonstrates that it is possible to trace apomorphies through time and to see how modern animals evolved from ancient ones through transitional species, often illustrating in some detail how minor changes eventually led to more substantial ones. The fossil evidence also shows how very slow the process of change can be. For the horse, it took >10 My and perhaps half as many generations for three digits to become one, for molar morphology to reach contemporary levels of sophistication and for shoulder height to increase from ~1.0 up to ~1.6 m. The evolution of the stem tetrapod from a sarcopterygian fish seems to have taken 15–20 My, and it required ~55 My for the early mammalian skull to evolve from the skulls of stem mammals. The examples given in this chapter suggest that a million years and almost as many generations of evolutionary change is, in general, unlikely to give rise to anything other than relatively minor anatomical change, although the evolution of the camera eye may be an exception (Chapter 20). The differences in time may represent the strength of the relative selection pressures, but the evidence here is thin.

KEY POINTS

- The strongest anatomical evidence for the process of evolutionary change comes from analysis of the transitions that led to the emergence of new features.

- Although the broad fossil record for vertebrates is strong, there are relatively few cases where it is good enough for detailed insights into how major anatomical changes were achieved.

- The fossil record is particularly good for the evolution of the tetrapod limb with its pentadactyl foot. There is a set of fossils of late Devonian sarcopterygian fish showing how a rayed fin with primitive bones slowly evolved over ~15–20 My to become a walking limb.

- These changes were originally concerned with improving the ability of various taxa to walk on shallow sea floors and to move through dense weed near the shore.

- A second example of a transition where there is good fossil data is the evolution of the mammalian skull from that of

an early mammaliaform, a process that took ~55 My. The fossil record allows us to trace the evolution of the various fenestrae (spaces in the cranium for muscles), the temporo-mandibular joint of the mammalian jaw and ear bones, the secondary palate and the expansion of the range of tooth types.

- The third example is the evolution of the taxon of modern hoofed horses from an early and smaller ancestor with a multi-toed foot. These events started some 35 My in the Oligocene and only reached their current form ~1 Mya.

- The fossil record shows that the progeny of the *Equus* common ancestor initially diversified widely but that the clade was almost lost with only a single successor species surviving (an example of an evolutionary bottleneck). This gave rise to the eight species of Equidae alive today, each of which has a unique chromosome number.

- These examples of transitions emphasise how slow anatomical change tends to be.

FURTHER READINGS

Benton MB (2004) *Vertebrate Palaeontology* (3rd ed). Wiley Blackwell.

Kardong KV (2018) *Vertebrates: Comparative Anatomy, Function, Evolution* (8th ed). McGraw-Hill.

WEBSITE

- The Wikipedia entries on the *Evolution of the Horse, List of Transitional Fossils* and *Early Tetrapod Trackways*.

The Genomic Evidence

7

A set of homologous DNA sequences from a group of related species provides important evolutionary insights because it not only reflects current genetic information about these sequences, but also, through their mutation-derived differences, includes the implicit history of how they diverged over time. Phylogenetic analysis aims to reconstruct that history as a phylogram whose root is the most likely sequence from which that set evolved, and this can be done for any set of homologous sequences, whether or not there is a fossil record. The various ways of constructing such trees differ in the nature of the information that they use, the ways in which they analyse it and the amount of information that the analysis provides. This chapter summarises the various approaches, the language of phylogenetics, the problems that can be encountered in making phylograms and the quality of the resulting evolutionary information. Further computational details are given in Appendix 5.

The theme of *Descent with Modification* applies not only to anatomical properties but to any heritable aspect of life and this is particularly so for DNA sequences. It is an obvious prediction that the closer are two species anatomically, the more similar should be the DNA sequences of any proteins that are homologous, and this can, of course, be checked by comparing those sequences. If they are homologous, then each derives by mutation from a sequence that was present in some undefined common ancestor. Indeed, the fact that all cellular organisms have DNA-based genomes that use essentially the same coding reflects the fact that they all descend from a primitive and very early bacterium known as the Last Universal Common Ancestor (the LUCA—Chapter 9), albeit that there are a very few minor coding variations, particularly in mitochondrial DNA (Bernt et al., 2013; Elzanowski & Ostell, 2013).

Because DNA sequencing is so cheap and fast today, there are now full genome sequences for many hundreds of plants, invertebrates, and vertebrates and for many thousands of prokaryotes, together with protein-coding data for a host of genes in other organisms yet to be fully sequenced. Also beginning to be collected are the details of many sequence variants (polymorphisms). As this information is stored in massive public databases such as ENSEMBL, there is an essentially infinite amount of readily available sequence data from a very wide group of organisms available online, together with a set of sophisticated informatics tools for exploring evolutionary relationships among them (see Websites at the end of the chapter).

The genomic perspective on evolution starts with the hypothesis that contemporary sequences that are similar almost certainly reflect a common-ancestor sequence from which all evolved. The slight doubt here allows for two possibilities: first, that one organism donated a gene-containing DNA sequence to another through horizontal gene transfer (Chapters 6 and 10; Keeling & Palmer, 2008); second, that the similarities in short sequences could have arisen by chance. In practice, a sequence of about 17 bases is more than enough to identify a unique sequence: the odds on two nonrelated sequences of this size being the same are 4^{17}:1 ($>10^{10}$:1). In a human genome of 2×10^9 bases of which ~2% codes for proteins, it is highly unlikely that

DOI: 10.1201/9780429346217-9

there would be even a single repeat by chance in a protein-coding region (*Note*: small, amino-acid coding domains with specific roles, such as transcription-factor motifs, are not present by chance, see below). Nevertheless, and to reduce even further the chances of coincidence and within-species variation, phylogenetic analyses are normally carried out on sequences of at least ~100 nucleotides.

The first role of computational genetics in probing evolution is the identification of sequence homologues. If we have a sequence of any sort, tools exist for probing the databases for similar and potentially homologous sequences that might be present in a very wide range of organisms, be they prokaryotic, eukaryotic, or even viral. In cases of doubt, one important test of homology is that the sequences have the same molecular function, this can often be queried in the gene ontology (GO) database—see Websites. Another test is that a sequence can be placed with high probability within a computed phylogeny (see below). Once such sequences have been identified, they can then be used for phylogenetic analysis and examining evolutionary relationships. This work has had an unexpected bonus: sequence comparisons between prokaryotic and eukaryotic genes have shown that homologues of many in the latter are present in the former. This insight has revolutionized our understanding of the early history of life, the period before there is any meaningful fossil record (Chapters 9 and 10).

Protein-coding genes are long: in humans, they range from 189 to ~2.4M base pairs (Piovesan et al., 2016). In practice, therefore, phylograms can only be constructed computationally but modern programs can do this using as many homologous sequences as are available. In such cases, the result will be a tree with the most likely original-common-ancestor sequence at its root and the current sequences as terminal leaves. It is worth noting that one of the origins of the algorithms used to construct such sequence trees came from work done on reconstructing the scribal history of sets of mediaeval manuscripts on the basis of analysing copying errors (Buneman, 1971). This problem is, of course, formally equivalent to that of constructing phylogenies on the basis of mutation. The use of the algorithms originally designed for linguistic analysis to study gene evolution is an early and even charming example of interdisciplinary computational work!

In any analysis of a group of sequences shown to be homologous, we would expect to see two features in the results. First, the relative closeness of the sequences should mesh with the closeness of the associated contemporary organisms based on cladistic analysis and, second, their relationships back in time should produce a hierarchy analogous to a cladogram in which all the sequences can be tracked back to a common ancestor sequence—this is a phylogram or molecular-phylogenetic tree. Both predictions have been found to be correct and, because so much data underpins the tree, such phylograms can be used to probe areas where relationships among contemporary organisms are unclear because the fossil evidence required to construct a cladogram is incomplete or absent (e.g. Schneider & Sampaio, 2013; see also Chapter 5).

This chapter considers what is involved in phylogenetic analysis, and its general use, Appendix 5 considers the computational methodology in more detail, and examples of its use are given here and in Chapters 16 and 24. The importance of such comparative genomics in probing the early stages in the history of life is particularly discussed in Chapters 9–11.

PHYLOGRAMS AND CLADOGRAMS ARE SUBTLY DIFFERENT

Although phylograms and cladograms are both types of evolutionary tree and look superficially similar, they are not quite the same. Consider two contemporary organisms, **B** and **C** with different anatomical features, that share an unknown last common ancestor **A**. In these, **B** and **C** have homologous sequences **Q** and **R** that derive from sequence **P** possessed by an unknown common ancestor **D**, which may or may not be the same as **A**.

In formal terms, cladograms are composed of triplets of the general type

$<$ species **B** $><$ *descends with modification from* $><$ species **A** $>$

$<$ species **C** $><$ *descends with modification from* $><$ species **A** $>$

while a phylogram is composed of triplets of the general form

< sequence **Q** of species **B** >< *descends through mutation from* >< sequence **P** >

< sequence **R** of species **C** >< *descends through mutation from* >< sequence **P** >

For such phylograms based on sequences from many organisms, the first higher level nodes represent the last-common-ancestor sequences of each pair of current sequences but give no indication of what those ancestor organisms were. Thus, the last common ancestor **D** that carried sequence **P** may have lived very much earlier than species **A** if the sequence that led to **P** underwent more random mutation than expected; alternatively, sequence **P** might have survived for some time on the lines leading to **B** and **C** if it underwent less mutation. Where, however, the sequences are clear homologues and come from species that are in very different phyla, such as mouse and *Drosophila*, we can be sure that this sequence was possessed by the last common ancestor of the two species, an organism that lived in the early Cambrian, at the very latest. Perhaps the most important aspect of a phylogram is that it is independent of the fossil record because, its construction is, with a very few exceptions that use ancient DNA, based only on genes from the contemporary organism.

There are four other major differences between phylograms and cladograms. First, anatomical changes can, in principle, be traced back in time through the fossil record, whereas sequence changes cannot, other than through those very occasional samples from ancient organisms whose DNA has been fortuitously preserved (Der Sarkissian et al., 2015; Orlando et al., 2008). In such cases, it is possible to include their sequences in phylogram analyses. It is worth noting that the study of such ancient DNA is becoming ever more interesting and is starting to give insight into protein stability and change over the Pleistocene period. Thus, for example, recent work on sequencing mitochondrial DNA from a bear dating to ~300 Kya (Dabney et al., 2013) and on horse DNA perhaps twice as old (Orlando et al., 2013) suggests that there may be more to come here from other ancient material containing traces of DNA. It is also possible that robust cross-linked proteins, such as collagen from ancient organisms have maintained their primary sequence well enough for parts at least to be sequenced (Buckley et al., 2008).

Second, DNA sequences are not samples of the phenotype of the organism but of the genotype, and thus of less clear significance than details of anatomical change. One cannot tell whether a mutation changed the phenotype of a past organism or not, although mutations that changed amino acids are more likely to be evolutionarily significant than those that did not (neutral mutations) because their effects may be subject to selection. What this means in practice is that one cannot tell whether the last common ancestor of two contemporary DNA sequences was the same as the last common ancestor of two organisms with slightly different anatomy, be they old or new. That said, phylogenetic trees do establish the relative closeness of a group of current organisms extremely well, and the more genes that are sampled in the study, the better is the estimate. Although ancestor species cannot be identified, if enough sequence trees are generated for a group of species and integrated, one would expect that the resulting tree would be equivalent to the species tree.

Third, the relationships of phylogram triplets are quantitative, whereas those of cladogram triplets are qualitative: the latter are essentially truth statements with the length of a cladogram edge carrying no quantitative meaning—it is hence *unscaled*. In phylograms, on the other hand, the length of an edge or branch has a specific meaning: it is a measure of the minimum number of mutations required to change a parent sequence into daughter ones so that, the fewer the number, the shorter is the branch length. It is difficult to be precise about the exact meaning of this length because one has to decide how to weigh the relative likelihoods of a single-nucleotide change, a back mutation, a deletion or even an insertion. It is, however, sometimes possible to do this and so assign a quantitative mutational meaning to the length of branch in a phylogram (see below and Appendix 5).

Although cladograms explicitly exclude time in their construction, they can be linked to the fossil record through the age of the earliest organisms to include particular synapomorphies; such data gives a minimal estimate of when a feature evolved. Time can also be included in phylograms provided that one knows how long it takes on average for a mutation to become an established feature of the genome. Where this can be done, the length of that branch becomes a measure of time between two contemporary sequences and their last common ancestor. The addition of time to a phylogenetic tree enables it to be used as a history of the evolutionary relationship of the organisms being analysed that is to some extent independent of cladograms based on the fossil record. The hedge here is because estimates of mutation rates often link back to one or another aspect of the fossil record (below). In general, however, it is usually safe to assume that the longer the branch length generated in a phylogenetic tree, the longer the time period that it represents.

Finally, the nodes have different meanings. Those in cladograms represent species that, while they may not be known, do have specific traits. While phylogram nodes between leaves and the root sequence also represent branch points, these cannot be linked to ancient organisms, just to intermediary sequences. In this sense, phylograms of single-gene sequences are less useful than cladograms.

In short, a phylogram indicates the closeness of the relationship amongst homologous DNA sequences, and implicitly among the species possessing these sequences. If the problem of horizontal gene transfer is ignored (next section), a phylogram for a set of homologous sequences for a particular protein will provide an estimate of the closeness of the parent sequences to daughter sequences. The more genes that are included in the analysis, the nearer will be the estimate to the correct relationships and, hence, to the true evolutionary history. Indeed, if the fossil record is weak or absent, the phylogenetic tree may provide the only accurate description of species relationships.

CONSTRUCTING SEQUENCE-BASED PHYLOGENETIC TREES

Phylogenetic trees are always constructed computationally, and there are four different approaches that use increasing amounts of homologous-sequence information for one or more species or even variants within a single species. The result is always a hierarchy, although it can be displayed in various ways that are, of course, topologically equivalent (e.g. below and Fig. 10.4).

- The most basic method is to analyse the presence or absence of sequences (e.g. the analysis of whale evolution discussed in Chapter 16). This approach can also be used for cladistic analysis using, for example, the PAUP (Phylogenetic Analysis Using Parsimony) program, whose most recent versions include bootstrapping and other tests (see below). This approach has been used recently for analysing early amniote evolution (Chapter 15; Ford & Benson, 2020).

- The simplest approach using sequence data is to choose a set of taxa and analyse the differences between homologous sequences. To do this, one first constructs a distance matrix whose terms are numbers that reflect the genetic distance between each pair of sequences. A computer program then uses neighbourhood-joining algorithms to group taxa that are genetically close, constructing the tree whose branch lengths are minimised and so most likely to represent the paths of sequence descent. As this approach is computationally rapid, it was an important method initially when computers were slow (Hall, 2004). Today, they are so fast that techniques that originally took a long time can now be done rapidly; as a result, difference techniques are now rarely used.

- The standard approach today is to construct a phylogram on the basis of nucleotide comparisons across sequences. These methods not only generate trees but can also construct likely ancestral sequences. There are three approaches here.

  - *Maximum parsimony methods:* These set out to find the most likely phylogenetic tree on the basis of identifying those ancestral sequences that generate contemporary ones with the minimum number of mutations.

  - *Maximum likelihood methods:* These are similar to maximum parsimony approaches, but also incorporate a model of evolutionary change based, for example, on the details of the probabilities of different sorts of mutation.

  - *Bayesian methods:* These aim to generate the set of hypothetical phylogenies that is supported by the data, and assign probabilities to each, implicitly recommending that with the highest.

- All these methods use consensus sequences and exclude information on sequence variation or polymorphisms across a population. The final approach is explicitly based on such information and is known as coalescent theory (Chapter 21). This methodology aims to analyse sequence variants within the genome of a species using a model of population genetics and a defined mutation rate. The theory then uses stochastic methods to simulate evolution backwards in time to the genome of the most recent common ancestor (MRCA) that had the original parent sequence. If enough simulations are done (perhaps a million), the most likely ancestor genome can be identified. Coalescent theory has a second and more important use: because it incorporates a model of population genetics, it can generate the past population dynamics of the species carrying the sequences. This approach has been particularly used to study the details of early human evolution and the migrations of *Homo sapiens* across the world (Chapter 23).

In practice, molecular phylogenetics uses a set of homologues of a single gene for analysing sequence evolution and a broad set of such genes for analysing interspecies relationships. In constructing phylograms today, several approaches tend to be used and any slight differences in the trees that they generate are integrated to construct a consensus tree on the basis of their relative likelihoods. It should, however, be emphasised that the different methods each, as expected, give very similar results (e.g. Fig. 7.3). It is also worth mentioning that the field is not staying still and that ever more sophisticated phylogenetic approaches continue to be developed (Szöllősi et al., 2015).

Ancestral sequences

One important strength of tree-searching methods is that they can generate estimates of ancestral sequences, and these can be used as the basis for the synthesis of ancient proteins (Thornton, 2004). If the analysis, for example, uses sequences from both protostome and deuterostome animals and they can be shown to derive from a common root, then the generated last-common-ancestor sequence is an estimate of one that came from a very early bilaterian ancestor, one that probably lived >550 Mya, and gives some insight into the phenotype of that early animal.

Mirabeau and Joly (2013) provide an example of this approach: they showed that some peptide hormones and their receptors co-evolved before the protostome/deuterostome split. At a more physiological level, Shi and Yokoyama (2003) subjected to phylogenetic analysis genetic sequences of short-wavelength-sensitive retina receptors from modern birds, fishes, reptiles, and mammals. This analysis yielded likely ancestral protein sequences that they were able to synthesise and then examined their reactivity to UV and blue light. They were thus able to follow how the sensitivity of the various receptors changed as the various clades evolved. Such work shows how phylogenetic approaches can now be used to study how ancient organisms lived. This intriguing area of research is now casting a light on ancient physiological capabilities.

CHOOSING SEQUENCES FOR PHYLOGENETIC ANALYSIS

For the purpose of making a phylogenetic tree, the essential data is a set of homologous sequences from a group of species (these are orthologous sequences, and an example is the set of genes of the Krebs [citric acid] cycle whose proteins are present in the mitochondria of all eukaryotes) or even within a single species (these are paralogous sequences, and an example is the set of hox genes in the mouse genome). This sequence may code for a protein, an RNA molecule, or may even be untranslated—function is irrelevant here. In practice, the choice of a sequence is less simple than it might seem because there are a series of potential pitfalls.

First, one needs to be clear about what homologous means in this context. The standard meaning is that two sequences are homologues if they derive from a common ancestor sequence. If, however, a sequence was acquired through horizontal gene transfer, where one organism donates to another a piece of DNA (or RNA that can be copied into DNA) that becomes incorporated into the recipient's genome, then the donor and recipient will not share a common ancestor in any immediate sense. Such xenologues certainly occur among bacteria (see Chapter 9) but turn out to be rather less common in multicellular organisms because the incorporation is unlikely to be carried through to the germline. The best-known exception here is *Wolbachia*, a parasitic bacterium that is commonly found in insects; the two organisms lead a symbiotic existence, with the insect genome often including many *Wolbachia* sequences (Le et al., 2014; Chapter 19). There are other examples in holobionts (a host organism together with its associated microbiome—Chapter 19). Any phylogenetic analysis carried out using sequences acquired through horizontal gene transfer will be flawed because the common-descent assumption will have been violated.

Second, there is little point in conducting phylogenetic analyses among a group of closely conserved sequences as the number of mutations required to generate a last common ancestor are so few that it can be hard to decipher relationships; one is merely left with the conclusion that the sequence probably evolved very early. Examples where this sort of problem occurs are in the various DNA-binding motifs, such as homeodomains, zinc-finger domains, and paired-box domains, that characterize the different classes of transcription factors. Such motifs alone show very minor differences across the phyla and are of little use in making phylogenetic trees, although, of course, complete protein sequences that include such motifs are routinely used for phylogenetic analysis.

Third, where the homologous sequences are very different, alignment becomes difficult because of the number of ways that the analysis can be done. Under these circumstances, it is hard for the sorting routines to calculate a unique phylogenetic tree because there are several possible options, with none having a much stronger likelihood than the others. This is the long-branch problem that made analysing the origin of whales so hard (Chapter 16). There is also the interesting question of whether, for protein-coding sequences, one should analyse the genomic or the equivalent amino-acid sequences. DNA-based analyses are more sensitive because of the redundancy in codons and hence more useful for comparisons among closely related species. Amino acid-based analyses are a little coarser and thus more useful in analysing sequences that are from distantly related species.

The fourth complication in making phylogenetic trees derives from the assumption that the sequences that are used computationally represent the consensus sequence of that gene in each organism. The correct choice of sequences is not necessarily obvious, given that even the two alleles on the paired chromosome of a diploid organism would not be expected to be the same, but will include polymorphisms (coalescent analysis can explore these differences). Indeed, one might even differ from the other through deletions or additions and both may differ from the eventual choice of consensus sequence. Identifying the most likely nucleotide for each position within a population may require so much sequencing that compromises may have to be struck on the basis that variation within a species is usually very much less than that between species.

In practice, the types of sequences that particularly lend themselves to phylogenetic analysis are, as has long been realized, those that share considerable lengths that are similar but not excessively so, such as the many homologues of cytochrome C and Pax6 (Fig. 7.1). It thus pays to take some care in the choice of sequences before moving to computational phylogenetic methodologies. In general, the results in which we can

```
93.9% identity in 132 residues overlap; Score 644.0; Gap frequency: 0.0%


Pax6_mouse 5
HSGVNQLGGVFVNGRPLPDSTRQKIVELAHSGARPCDISRILQVSNGCVSKILGRYYETG
Pax6_DrMel 57
HSGVNQLGGVFVGGRPLPDSTRQKIVELAHSGARPCDISRILQVSNGCVSKILGRYYETG
           ************ ********************************************


Pax6_mouse 65
SIRPRAIGGSKPRVATPEVVSKIAQYKRECPSIFAWEIRDRLLSEGVCTNDNIPSVSSIN
Pax6_DrMel 127
SIRPRAIGGSKPRVATAEVVSKISQYKRECPSIFAWEIRDRLLQENVCTNDNIPSVSSIN
           **************** ****** ******************* * **************


Pax6_mouse 125 RVLRNLASEKQQ
Pax6_DrMel 117 RVRNLAAQKEQ
               ******* * *
```

Figure 7.1 A comparison of parts of the Pax6 amino acid sequences from mouse and *Drosophila melanogaster* made using the ExPasy comparison tool. Although the former has 422 and the latter 857 amino acids, there are regions such as that shown where amino-acid matching (*) is almost perfect—there have been virtually no mutational changes here in >500 My of evolutionary divergence.

have the strongest confidence come from analysing rapidly changing sequences for shorter periods of evolution and slowly changing sequences, ideally of amino acids, for longer periods.

THREE EXAMPLES OF MOLECULAR PHYLOGENIES

There are now many thousands of molecular phylogenies in the literature, and it is not practical to consider this collection in any detail. The three examples given here have been chosen to make some explicit points about the uses and limitations of the various molecular phylogenetic approaches.

The evolutionary status of the AmphilLim 1/5 Gene

Langeland et al. (2006) used distance-matrix analysis methods to investigate the evolutionary context of the *AmphilLim 1/5* gene in *Amphioxus*, a nonvertebrate chordate. To do this, they analysed a set of homologous gene sequences that included the Lim1, 3, and 5 genes from a range of vertebrates. To ground the phylogram, they also included the *Amphioxus* islet gene, a distantly related sequence, together with the associated sequences from *Ciona*, an ascidian sea squirt (a deuterostome tunicate), which was chosen as the outgroup. *Amphioxus* is a cephalochordate and hence the sister group to the chordates. It has a dorsal nerve chord and a notochord but lacks a skeleton, eyes, and other sensory tissues. The main reason for these differences is that the *Amphioxus* embryo has no neural crest cells (Appendix 7), and the adult thus lacks all of the neural crest derivatives.

The analysis used the distance-matrix methodology, and the expectation was that the grouping of the genes by species would reflect the same grouping as one would expect to find on the basis of cladistic analysis, with the *Amphioxus* genes branching at a higher level than the vertebrates. Fig. 7.2 shows first, as expected, that the outlier gene lies outside the main phylogeny and, second, that the related genes map to the analogous cladogram, albeit with one exception. The Lim1 and the Lim5 branches directly reflect the standard phylogeny with the AmphilLim1/5 branching at a higher (older) node and the *Ciona* genes branching at an even higher one.

The exception is in the Lim3 hierarchy, which suggests that the zebrafish and *Xenopus* share a common ancestor with the mouse and human rather than the mammalian sequences tracing back to an amphibian ancestor. As this is clearly wrong, one immediately wants to know how much confidence one can have in this analysis, and the authors provide this through bootstrap analysis (given in the numbers adjacent to the nodes). This analysis, which was carried out a thousand times, replaces small amounts of the data randomly. If a node is robust (shown by a

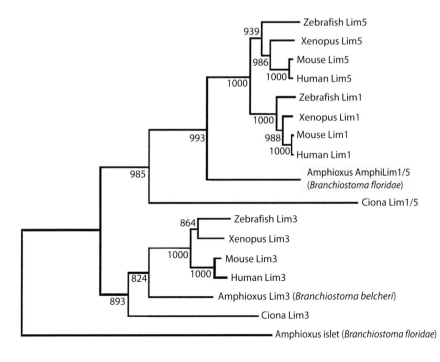

Figure 7.2 A distance-matrix phylogenetic tree designed to show the evolutionary status of the *Amphioxus* Amphilim1/5 Gene which contains a homeodomain. The outlier gene used for rooting the tree was *Amphioxus* islet Gene which also contains a homeodomain. The program used for constructing the tree was clustal x (see websites), and the numbers are bootstrap values based on 1000 trials (From Langeland et al., 2006).

(From Langeland JA et al. (2006) *Int J Biol Sci* 2:110–116. Published under a CC BY-NC License.)

high bootstrap number—see Appendix 5), this should have very little effect on the phylogram; if the node is not robust (shown by a lower bootstrap number), many of the simulations will give a different result.

Most of the bootstrap numbers are very close to 1000, and the associated nodes can thus be viewed as robust. Two of the numbers in the Lim3 branch are considerably lower and we can have less confidence in these nodes which reflect early vertebrate evolution. In practice, the meaning of the low bootstrap values is probably that the sequences didn't have strong enough mutational differences for a slightly crude method like the distance-matrix to provide an accurate analysis. It is for such reasons that the approach is now not generally used, even though it is computationally fast.

Phylogenetic relationships among a group of anemone fish

As the anatomical differences across the various species of anemone fish really only affect skin patterns and there is no relevant fossil record (Fig. 7.3), there is not enough anatomical data to undertake a standard cladistic study. Li et al. (2015) therefore analysed their evolutionary relationships phylogenetically on the basis of some 40 protein-coding mitochondrial sequences. For this analysis, they used both Bayesian and maximum-likelihood tree-searching methods. As measures of their accuracy for each node, the Bayesian analysis automatically assigns a probability of correctness while ML analysis uses the bootstrap methodology (Fig. 7.3). As an outlier to ground the phylogram, they used mitochondrial genes from a butterfly fish (*Chaetodon auripes*).

The resulting phylograms for the two methodologies were identical, with every node having a 1.0 confidence level in the Bayesian analysis. Of the eight nodes in the ML analysis, five have bootstrap number of 100 (every simulation gave the same result) and, of the other three, none were less than 94. We can thus have considerable confidence that the phylogeny is correct. The phylogram includes another feature: the edges have different lengths. These differences reflect the number of mutations between sibling, species, and the programs give a measure of this difference as indicated by the scale bar which is used to calculate them. Here, the scale value of 0.05 represents a 5% difference, and the branch lengths are used to calculate the likelihood of a mutation occurring in any mitochondrial nucleotide downstream of a node.

Phylogenetic relationships across the family of Hox genes

Hox genes are a member of the class of genes that contain a homeobox, a 180-bp sequence that codes for the homeodomain (a helix-turn-helix motif of 60 amino acids) in the associated proteins; this binds to specific DNA sequences, usually in conjunction with other motifs (Bobola & Merabet, 2017). The Hox genes were first recognized through mutants that altered the patterning of the *Drosophila* body plan, and it was then discovered that a homologous set of genes patterned the mouse body. This naturally raised the questions of where else in the Bilateria such patterning could be found, and whether other organisms also possessed homologous genes with a homeodomain. The results presented here reflect phylogenetic analysis by several groups using a range of methods across a wide range of organisms.

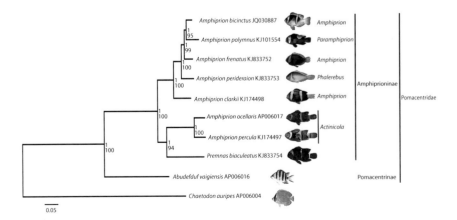

0.05

Figure 7.3 A phylogeny of anemone fish based on Bayesian and maximum likelihood analyses using the butterfly fish *Chaetodon auripes* as an outgroup to root the tree. The analysis used ~40 protein- and non-protein-coding genes, and the Bayesian probabilities for the tree are the upper figures at each node (all equal one), and the bootstrap numbers from the maximum likelihood analysis are the lower figures. The branch lengths are from the Bayesian analysis, and the scale bar represents nucleotide substitutions per site.

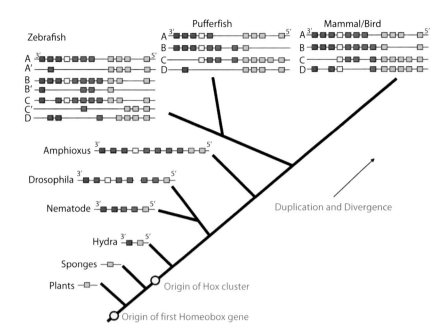

Figure 7.4 Evolution of the hox genes and clusters through duplication and divergence from a primordial homeobox gene estimated to have arisen >1.5 Bya in a common ancestor of plants and animals.

(From Lappin TR et al. (2006) HOX genes: seductive science, mysterious mechanisms. Ulster Med J. 75:23–31. Published under a Creative Commons 4.0 Attribution License (BY/NC/SA).)

Phylogenetic analysis showed that the earliest animal homeodomain-containing protein probably dates back to ancestral Cnidaria (e.g. jellyfish) in the Ediacaran period, perhaps 50 My before the evolution of the Bilateria (Fig. 7.4; Ryan & Baxevanis, 2007). However, the presence of a homeodomain-containing protein in plants (e.g. Costanzo et al., 2014; Meng et al., 2018) implies that the original homeodomain-containing protein was probably present in the last common ancestor of plants and animals, an early unicellular eukaryotic cell that had a nucleus and a mitochondrion but lacked a chloroplast and dates to before ~1.7 Bya. Even today, there are single-cell amoebae, such as *Capsaspora owczarzaki,* that include proteins containing a homeodomain (Holland, 2013).

Hox evolution has, however, mainly been studied in the Bilateria, which fall into two groups: the protostomes and deuterostomes. The former includes almost all of the invertebrates (e.g. *Drosophila*) and the latter includes all the chordates (e.g. the mouse), together with the hemichordates and the echinoderms (e.g. the sea urchins). *Drosophila* has eight homeotic genes that form a group of ordered paralagous sequences (the HOM-C cluster), all on chromosome 3. The mouse has thirty-eight such genes, separated into four clusters on different chromosomes (2, 6, 11, 15). The genes on each cluster are orthologous to those in *Drosophila,* with each group of paralogues being in the same genomic order (Fig. 8.3; Carroll, 1995; Gehring et al., 2009).

Working out the relationships among all these homeotic genes was not simple, and it took a considerable amount of detailed phylogenetic analysis to determine which sequences were closest to one another (Holland & Takahashi, 2005). This analysis shows that the *Drosophila* line maintained a single copy of each of the homeotic genes, but things became far more complicated in the line leading to the mouse. There was first a single duplication of the second gene (the orthologue of Pb in *Drosophila*) and then multiple duplications of the terminal gene (Abdb in *Drosophila*), eventually giving five copies. Later, this group of 13 genes underwent two duplications with three of the groups being translocated to other chromosomes and with some of the possible 52 sequences being lost. (Fig. 7.4; Holland & Takahashi, 2005; Pascual-Anaya et al., 2013). Apart from the technical details of Hox gene evolution, this brief case study makes a further important point: it is possible to use phylogenetic analysis to probe into the deep past, back to the time where the fossil record is too limited to be useful.

GENE TREES, SPECIES TREES, AND PHYLOGENOMICS

It is important to remember that sequence phylogenies are not the same as species phylogenies: the variation that takes place in a single sequence over time is a very limited assay of the extent of the changes in a whole genome, which should, of course, be equivalent to the species phylogenetically. The techniques for constructing species trees from phylogenetic trees are based on finding a tree that represents a minimalization of the differences between the individual gene trees (e.g. Chaudhary et al., 2013) so that as the number of gene trees increases, that tree will converge onto the species tree (Maddison, 1997). The use of large numbers of gene sequences to construct such trees is known as *phylogenomics* (Schraiber & Akey, 2015). For a useful discussion of how this is done in practice, together with some of the difficulties involved, see Choi and Kim (2017).

Such multigene phylogenetic analyses have been useful in, for example, understanding the taxonomy of single-celled eukaryotes (informally known as protists), which are present in most phyla and discussed in more detail in Chapters 10 and 11. There are a large number of these that can only be distinguished by a few morphological characters that include the number of flagella (one, two, or many), the presence of pseudopods, and the presence or absence of chloroplasts. The importance of these organisms is that, in times past, they were the roots of all the major phyla groups. Many of these plesiomorphies have been lost with, for example, the flagella of multicellular organisms only being obvious in sperm cells, with animals having a single flagellum and most plants having two.

It is worth pointing out that all phylogenies, apart from those for the few animal taxa, where there are strong fossil records, are now based on phylogenetic sequencing of several sets of homologous genes. It is these phylogenies that are mainly to be found in the literature and on Wiki sites.

ADDING TIMINGS TO PHYLOGRAMS

In the 1960s, when proteins were first being sequenced, it became clear that the extent of the differences between protein sequences depended strongly on the time that had passed since their host organisms had shared a last common ancestor. The discovery of the genetic code confirmed that some DNA mutations changed the amino acids whereas others did not—these are the neutral mutations. This, in turn, led to the idea of a **molecular clock** (Kumar, 2005) whose regular ticking would represent neutral mutation change becoming established within a sequence, spreading throughout the population as a result of genetic drift (an effect of random mating) at a rate assumed to be constant.

The importance of such a clock was that because it only counted mutations that did not affect protein function, its speed was essentially independent of selection, unlike the fossil record. Such clocks cannot, of course, be linked directly to a cladogram as its branches do not measure time but represent truth values (e.g. a taxon does or does not carry some feature). They can, however, be included within a molecular phylogram as the branch lengths here represent a measure of the quantitative difference between sequences (e.g. Fig. 7.3), albeit that in using the clock for comparisons, distances are not linear: as discussed in Appendix 5, one has to take account of backward and repeated mutations—long-distance branches turn out to represent underestimates of time.

The idea that there was a single, uniform clock rate for all species at all times turned out to be oversimplistic, and for two reasons. First, analysing sequence differences is complicated: not only are there concerns about balancing the relevant weights of the different mutation types (replacements, deletions, and insertions), but there are decisions to be made as to which mutations are neutral and which are active and so subject to selection, particularly when protein-coding sequences are being discussed. Second, the rate at which whole-sequence differences accrue between two taxa depends on the generation times, the degree of selection that may vary over time, and the speed of genetic drift, something that depends on population size (Ayala, 1999).

In short, the speed of the clock cannot be assumed constant even within a clade, at least over long periods. In practice, therefore, the clock always needs to be calibrated for particular applications against the general fossil record or for the species under consideration (van Tuinen & Blair Hedges, 2001; Ho & Duchêne, 2014). An interesting exception here is the work of Shih and Matzke (2013; Chapter 10) who studied the events of early eukaryote evolution. They were able to add evidence from horizontal gene transfer events to calibrate their clocks on the basis that the same events represented in different phylogenetic trees had to have occurred at the same time. They considered that using this in their Bayesian analyses, together with additional information from maintained duplications of ancient genes, improved estimates of the times of evolutionary change by ~20%. This enabled them to estimate that mitochondria and chloroplasts had been endosymbiosed ~1200 and ~900 Mya, respectively. These dates do, however, now seem young as the fossil evidence argues for rather earlier dates (Section 3).

Where the clock idea turns out to be of more general use is over limited periods of time and for groups of related taxa with the same generation time.

Weir and Schluter (2008), for example, analysed the sequences for cytochrome b, a mitochondrial gene, for 12 orders of birds which typically breed after a year. They used detailed fossil data for absolute timings and a set of corrections to allow for known computational difficulties. The clock rates ranged from 1.03 to 3.63 divergences per million years (dpmy), and the combined mean rate was 2.21±0.68% (assuming a generation time of one year). A reasonable working figure for adding divergence timings to vertebrate phylogenies in the absence of any better data is probably about 2 dpmy per generation, but this figure should be taken as no more than indicative.

MOLECULAR PHYLOGENETICS TODAY

This chapter started with the prediction that similar DNA-coding sequences for proteins with similar functions in different organisms are homologous and can be traced back to a common ancestor or shown to have resulted from horizontal gene transfer. The accuracy of this prediction has now been confirmed in many tens of thousands of analyses and tested against cladistic analyses. Its correctness has enabled a generation of evolutionary molecular biologists to group contemporary organisms with an accuracy far beyond what can be achieved using anatomical criteria, the fossil record, and cladistics. The resulting work has clarified the complex evolution of gene sequences, and the species that include them in their genomes. The evidence that has emerged from analysing sequence data computationally has illuminated great swathes of biohistory, unpicking the complex evolution of gene sequences and clarifying ambiguities in the Linnaean taxonomy.

Equally important, as we will see in Section 3 on the History of Life, has been the role that phylogenetic analysis and comparative genomics have played in illuminating the details of early evolution for which there is no fossil record. These areas particularly include the evolution of bacteria, the evolutionary origins of the eukaryotic cell, the origins of the major phyla, and the properties of *Urbilateria*. This organism was the root member of the clade that includes all taxa with, in their early stages at least, bilateral symmetry. This is the subclade of animals that excludes the radial coelenterates and the sponges.

All of this success has been built on substantial mathematics and informatics work, the details of which are normally taken for granted. This is partly because most biologists don't understand the underlying theory, and its implementation and partly because the resulting computer programs are hidden behind computer interfaces that just require a user to input appropriate sequence IDs, quality criteria, and output requirements. It is also worth mentioning that this chapter has not detailed every area of the computational work underlying our understanding of evolution. In Chapter 21, we will consider the topic known as coalescent analysis: this integrates phylogenetics, sequence variation, and population genetics to probe the population dynamics of ancient communities of organisms.

KEY POINTS

- Comparisons across groups of homologous DNA sequences in very different contemporary organisms allow molecular phylogenetic trees (phylograms) to be constructed computationally.

- There are several computational approaches for making such phylograms. The two most widely used are based on Maximum Likelihood methods and on Bayesian statistics. For the former, bootstrapping is used to estimate confidence levels for nodes, while these figures are a natural part of the output of the Bayesian methods.

- These groupings are independent of those from cladograms as they are based on completely different information. Comparisons can, however, be made in the ways that cladistics and molecular phylogenetics group extant species. Where they differ, the latter is more likely to be convincing provided that it uses information from sufficient numbers of gene sequences.

- Tree-searching algorithms can generate likely ancestor sequences. If the range of organisms under analysis spans very different organisms, the resulting sequences represent those of very ancient and basal organisms, whose physiological abilities may therefore be illuminated.

- The branch lengths of phylogenetic trees are a measure of mutation differences. If neutral mutations accumulate at a constant and known rate per generation (this is the molecular clock hypothesis, which often applies over short periods of evolution), then these branch lengths are also a measure of time. Such information is not, however, easy to interpret.

- Phylogenetics and comparative genomics have not only provided information about protein divergence, but have also illuminated early evolution within the prokaryotic kingdoms and the origins of eukaryotic cells.

FURTHER READINGS

For the basics of phylogenetic analysis, see:
Desalle R, Rosenfeld J (2012) *Phylogenomics.* Garland Press.
Felsenstein J (2004) *Inferring Phylogenies.* Sinauer Press.

For more mathematical treatments, see:
Caetano-Anollés G (ed.) (2010) *Evolutionary Genomics and Systems Biology.* Hoboken, NJ: John Wiley.
Koonin EV, Galperin MY (2003) *Sequence—Evolution—Function: Computational Approaches in Comparative Genomics.* Boston: Kluwer Academic. http://www.ncbi.nlm.nih.gov/books/NBK20260/
Lemey P, Salemi M, Vandamme A-M (eds) (2009) *The Phylogenetic Handbook: A Practical Approach to Phylogenetic Analysis and Hypothesis Testing.* Cambridge University Press.
Saitou N (2013) *Introduction to Evolutionary Genetics.* London: Springer.

WEBSITES

- Wikipedia entries for *Computational Phylogenetics, Molecular Clocks,* and *Phylogenomics.*
- The ENSEMBL genome database: http://ensemblgenomes.org/info/genomes
- The Gene Ontology (GO) database of protein function, location and process: http://geneontology.org
- ClustalX: A set of sequence-aligning programs: http://www.clustal.org/clustal2/
- Bioinformatics resources European Bioinformatics Institute: https://www.ebi.ac.uk
- National Centre for Biotechnology Information: https://www.ncbi.nlm.nih.gov
- The ExPASy portal: https://www.expasy.org/resources

The Evo-Devo Evidence

8

This chapter discusses the third main area of evidence about evolution: how proteins that are homologous often have very similar functions in very different organisms, particularly during development. This is the area known as evo-devo, and the chapter focuses on homologies in the mouse (a deuterostome) and *Drosophila* (a protostome). This is partly because they are so different, partly because we know so much about their respective developments and partly because they exemplify much of what evo-devo demonstrates within the context of evolution. Homologies between the mouse and *Drosophila* reflect the behaviour of their last common ancestor, a primitive worm-like organism known as *Urbilateria* that lived >540 Mya in the Ediacaran. The chapter starts with a short section on homologies showing that animals across the deuterostomes and protostomes have inherited a common toolkit of regulatory pathways and networks. It then examines examples of tissue homologies in the mouse and *Drosophila*. These include the development of eyes, body patterning, limbs, and nerve cords. Summaries of mouse and *Drosophila* embryonic development are given in Appendix 6 and some of the principles underlying the mechanisms of embryonic development are given in Appendix 7.

The theme of this chapter is evolutionary developmental biology, or evo-devo, and it is about the ways in which very different organisms use homologous proteins for similar functions, particularly during development, and sometimes go on to re-use them. Where this can be shown to have happened, it implies that such behaviour reflected that in the last common ancestor of the two organisms.

Although molecular work on evo-devo started in the 1980s, the subject has old roots: it seems to have been Thomas Huxley in the 1860s who first made what seems to be the obvious point in discussions with Darwin that the anatomical differences among adult organisms reflect the different ways in which their embryos develop. What no one ever expected was that experimental work in this area of evo-devo would identify molecular homologies in the way that all Bilaterian animals develop and so suggest some of the core anatomical features of *Urbilateria*, their last common ancestor. Those molecular developmental biologists who first started comparing regulatory events in *Drosophila*, and the mouse back in the 1980s opened a door to what may have been the most exhilarating period in the history of evolutionary science since the age of Darwin.

It had been known since the early 1970s that there were sequence and functional homologies in proteins, such as cytochrome C that were involved in biochemical pathways across the phyla. It was then straightforward to identify other structural proteins and enzymes in different organisms that were both sequence and functional homologues. It was not, however, until a decade later, when the molecular regulation of developmental change was first being investigated, that it started to become clear that homologous proteins also had very similar roles in regulating the development of similar tissues in very different organisms.

This had not been an obvious expectation. Indeed, the very idea that organisms that were as different as *Drosophila*, and the mouse had a common anatomical root had barely been considered until then. There were, for example, no reasons to

DOI: 10.1201/9780429346217-10

suppose that there should be any similarity in the means of producing functionally similar but morphologically distinct organs such as the eyes in organisms as different as vertebrates, molluscs, and arthropods. It was for this reason that Salvini-Plawen and Mayr (1977) suggested that eyes had separately evolved some 40 times across the phyla. They would certainly never have predicted that a single transcription factor, Pax6, had a key role in the development of all of them. This was, however, only one example of the extent of protein homology across the phyla that has been revealed by experimental work and computational phylogenetics.

The fact that distantly related organisms express homologous proteins does not of itself imply that they have the same role in each: the mutations that each will have acquired since their host organisms shared a last common ancestor may well have led to changes in function. An enzyme, for example, may have changed its substrate, while the descendants of an original homeodomain-containing transcription factor may have diversified to fulfil different roles in different organisms. The possibility that homologous proteins do share a common function is a hypothesis that has to be tested against the evidence on a case-by-case basis.

Such tests have now been carried out across a wide range of proteins. They have shown that there are many functional homologies in the underlying signals, transcription factors, and protein network systems that regulate the building of tissues in embryos across the phyla (e.g. Barresi & Gilbert, 2019; Held, 2014, 2017). This is so despite the obvious and extensive differences in the external appearances of individual species, their internal tissue anatomy and their modes of development. If one just considers the mouse and *Drosophila*, whose last common ancestor lived before the lower Cambrian, the profound differences between their phenotypes represent >540 My of evolutionary divergence.

It is, of course, obvious that, the more recently the last common ancestor of two groups of species lived, the closer should be both the sequences and the functional homologies. The differences in the molecular underpinnings of the development of mouse and man, whose anatomical tissues are much the same and whose last common ancestor lived ~30 Mya, are relatively minor: they have the same basic anatomy with the most obvious differences being in body proportions, in size and hair coverage and in relative cranial capacity. Experimentation has shown that human proteins with particular developmental functions usually have similar effects when expressed in mouse embryos at the equivalent time and in the appropriate place. It is such observations, together with the ease of breeding, short generation time and ease of genetic engineering, that makes the mouse the primary model system for studying the mutations that underlie human genetic disease (Chapter 17).

Evo-devo insights are important in two contexts. First, the functional evidence is completely independent of, and so complements the evidence for evolution from, anatomical comparisons and genomic sequences. Second, the developmental plesiomorphies shown by embryos as different as those of *Drosophila*, and the mouse illuminate the ways in which tissues must have formed in the early embryos of ancient organisms, for which there is, of course, no fossil data. Such work, in particular, suggests that the forerunners of today's genes and developmental mechanisms were present in *Urbilateria*. The likely anatomy of this early organism is discussed in Chapter 12.

THE MOLECULAR BASIS OF EVO-DEVO HOMOLOGIES

This section discusses some well-known examples of homologous signals and receptors, transcription factors and networks that underly developmental change, particularly in *Drosophila* and the mouse, the key model protostome and deuterostome organisms.

Signal and receptor homologies

Much of *Drosophila* development depends on the activity of a set of intercellular signaling proteins, many of which were identified through analysing mutant flies that were often named on the basis of their odd appearances. Thus, the *hedgehog* gene acquired its name because the mutant embryo looked prickly as it had double the usual number of denticles (small bristles) on its surface and was also dumpy in shape, while the *dishevelled* embryo had denticles that were abnormally oriented. Some of the genes for *Drosophila* signals (their receptors are given in brackets) that are widely used during development include *Decapentaplegic* (*Thickveins* & *Punt*), *Hedgehog* (*Patched*), *Wnt* (*Frizzled*), and *Delta* (*Notch*).

The mouse embryo uses ~30 signaling molecules in driving development and many were identified through searching for homologues of *Drosophila* developmental signals using molecular genetics and bioinformatics approaches. The latter identified mouse homologues of all the early signals seen in *Drosophila*, although there was not a direct one-to-one matching: thus, for example, the mouse was found to have three homologues of the single *Drosophila Hedgehog* gene; these reflect early genetic duplications of the original gene and are named *Sonic*, *Indian*, and *Desert Hedgehog* (Pereira et al., 2014).

The key point, however, is that the mouse homologues of *Drosophila* signals were also signal proteins: sequence homologues turned out to be functional homologues. Closer analysis of the roles of these mouse signal proteins showed that, in many, at least one of their effects was on the growth pathways that are active as the mouse embryo develops but that play only a minor role in *Drosophila* embryogenesis (Appendix 6). Some are, therefore, known as growth factors, with well-known examples being the epidermal, fibroblast, and the vascular endothelial growth factors (EGF, FGF, and VEGF).

It is a similar story for the signal receptors: given that *Drosophila* and mouse signals are homologous in structure and function, it was not surprising to discover that the mouse equivalents of the *Drosophila* membrane receptors for these signals were also homologues. These homologies are also found in many of the internal members of the downstream signal-transduction pathways (below). Note, however, that the homologies in these signaling pathways carry no implications as to the nuclear transcription factors that they eventually activate. In these two very different embryos, their respective signals are used in many different contexts, with their downstream effects on expression depending on the transcription factors expressed in the receptor cells as a result of their lineage.

Transcription factor homologies

This class of proteins is responsible for producing further proteins that change phenotypes, with many *Drosophila* transcription-factor families having homologues in the mouse. Four important examples are considered here and more will be discussed in Chapter 12 in the context of the likely anatomy of *Urbilateria*. These pairs of *Drosophila* and mouse homologues should not be viewed as particularly special since such homologies exist across all organisms; the closer these organisms are in their evolutionary origins, the more are found and the closer, of course, are their genomic sequences. One of the down-stream effects of transcription-factor gene expression is activating the various process networks that drive organogenesis. Homologies in these protein networks are briefly discussed at the end of the chapter and in more detail in Chapters 12 and 18.

Pax6 and eye development

This is the best-known and most striking example of an evo-devo insight into the deep past. The *Pax6* gene, a transcription factor with homeobox and paired-box motif domains, was discovered through the analysis of the mouse *smalleye* mutant: the heterozygote (with only one mutant gene) has small eyes with no iris. This disorder is known as aniridia and a similar phenotype is seen in humans (Hanson et al., 1994). The mouse homozygote phenotype (both gene copies carry mutations) is more profound: the animal has only very rudimentary eyes, together with other facial and brain abnormalities (Grindley et al., 1997). *Drosophila* records were immediately examined to see if there were known mutants that lacked eyes, and it took very little time to identify the *eyeless* mutant, named on the basis of its dramatic phenotype, one that had been recognised many years earlier. Sequencing showed that the underlying mutant fly gene was a homologue of *Pax6* and confirmed that both genes encoded a transcription factor with homeobox and paired box motives, each with regions of similarity to and differences from the other (Fig. 7.1).

This discovery, made in the 1990s, was completely unexpected because the two organisms make their eyes in such different ways. The compound eye of *Drosophila* (Fig. 8.1) forms as a morphogenetic furrow passes over the future eye domain of the

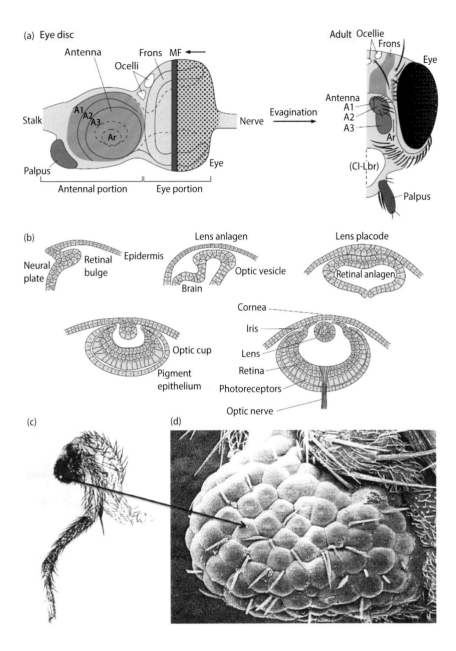

Figure 8.1 **a:** Diagram of the development of the compound eye in *drosophila*. This starts when a wave of activation headed by a morphogenetic furrow (MF) moves across the epithelium of the antenna-eye imaginal disc. The final eye forms during pupation when the disc everts during pupation. **b:** Development of the camera eye of the mouse. **c:** Ectopic eyes resulting from the experimental expression of pax 6 in a *Drosophila* leg. **d:** An enlargement of the ectopic eye shown in **c**.

(a Courtesy of Lewis Held.
b From Harris WA (1997) PNAS 94:2098–2100. With permission from The National Academy of Sciences.
c Halder G, Callaerts P & Gehring WJ (1995) Science 267:1788–1792. With permission from the American Association for the Advancement of Science.)

appropriate head imaginal disc, which derives from ectoderm alone. This is followed by a wave of Hedgehog-dependent activity (below) that leaves in its wake a developing array of ~800 small clusters of cells, with the central one in each regulating the development of its neighbours through signaling. When the eye discs evert during pupation, each cluster becomes an ommatidium with a nerve, a lens, photosensitive cells, pigment, and accessory cells. This compound eye is essentially a hexagonal array of separate, ectoderm-derived optic units, each of which projects back to the optic lobe, the part of fly's brain that undertakes the first steps of integrating the 800 separate images into a single one (Otsuna et al., 2014).

The mouse eye is completely different (Fig. 8.1b): this is partly because it uses a single cornea and lens to focus a single image onto its retina, working much like a camera, and partly because so many different tissues contribute to the organ. The retina, with its pigmented and light-sensitive cell layers, forms as an outgrowth of the future brain region of the neural ectoderm, while the lens and the outer layer of the cornea form from surface ectoderm. The sclera that protects the retina derives from mesoderm, as do some of the muscles while the neural crest contributes further mesenchyme, iris musculature, the posterior layers of the cornea and pericytes (the small muscles in the walls of arteries; Gage et al., 2005). In the mouse eye, the retinal fibres project back to the lateral geniculate nucleus and from there to the visual cortex of the brain which undertakes the interpretation of a single image essentially made of pixels, one from each neuron.

It still seems remarkable that the early development of these two profoundly different eyes is each activated by orthologues of the same transcription factor, Pax6. It is, however, important to emphasise that the mouse and *Drosophila* Pax6 protein sequences are not identical: some regions of the sequences are very similar while others are very different (Fig. 7.1). Moreover, the role of Pax6 in *Drosophila* is much more restricted than it is in vertebrates, where homozygous mutations lead to a wide range of facial and brain abnormalities (Hanson et al., 1994). These differences reflect the effects of >500 years of divergence and mutation that have passed since the two genes inherited a sequence in a last common ancestor.

Sequence homology and protein similarity do not, of course, predict functional homology. Indeed, it was originally thought that, even though the mouse and *Drosophila* Pax6 genes had homologous sequences, each was most unlikely to work in the "wrong" organism. A classic experiment showed that this intuition was wrong. Using some clever genetic tricks, Halder et al. (1995) were able to express the mouse Pax6 protein in *Drosophila* limbs and other external ectodermal tissue in a small region of a toad's head. The results were remarkable: those regions of ectoderm in which the mouse Pax6 protein was expressed developed into compound eyes (Fig. 11.1c,d). Perhaps even more surprising, given that the vertebrate camera eye is so complex, is that ectopic expression of the *Xenopus* Pax6 homologue alone is enough to generate a small but essentially normal eye in the toad (Chow et al., 1999). Given the many cell lineages that contribute to the vertebrate eye, this result clearly means that the descendants of the host cell containing Pax6 can mobilise nearby cells to participate in eye development and differentiate in their own species-specific way.

The production of eyes in *Drosophila* and the mouse, of course, requires the expression and activity of many proteins other than Pax6, which are, of course, under the control of further transcription factors. Analysis has shown that there are both sequence and functional homologies among several other transcription factor genes downstream of Pax6 that help regulate the development of eyes in the two organisms (Kumar & Moses, 2001). These genes include *Sine oculis, Optix, Eyes-absent, Dachsund,* and *Teashirt* in *Drosophila* whose respective mouse homologues include the *Six* family, *Eya1* and *Eya4, Dach1* and *Dach2, Mtsh1* and *Mtsh2*. It is worth noting that homologues of all these genes are also expressed in both the normal and the ectopic toad eye (Chow et al., 1999). This wide use of homologous proteins in these organism makes an important point discussed further below: the regulatory systems that determine a particular tissue are better buffered against the challenges of mutation than are the proteins that contribute to the actual anatomy of that tissue.

There are two further interesting similarities in the very different eyes of *Drosophila* and mouse. First, the wave of ommatidia induction in *Drosophila* is regulated by Hedgehog signaling and the expression effects of Atonal, another transcription factor. Something similar happens with the induction of ganglion cells in the zebrafish retina: a wave of induction passes across the retina propagated by the zebrafish homologue of hedgehog and activated by Ath, the zebrafish homologue of the atonal protein (Jarman, 2000). Patterning here may not, however, be a homology but may be a homoplasy, based on the availability of transcription factors not otherwise being used. Second, and at a more functional level, the key light-sensitive proteins in both types of eyes are members of the opsin family. Such proteins are, in fact, ubiquitously used for light activation: opsin homologues, such as rhodopsin, are present in the visual systems of all diploblast and triploblast organisms (Arendt et al., 2004; Shichida & Matsuyama, 2009).

Although other proteins are important in eye development, the evidence shows that, in every case, Pax6 is the key transcription factor: detailed work across the phyla has demonstrated that, wherever an animal has an eye, upstream of its early development is the expression of a Pax6 homologue (Fig. 8.2). This has been confirmed in mammals, arthropods, molluscs, and even the cnidarian box jellyfish, whose eyes have a cornea, retina and lens—this may be the simplest organism with eyes (Kozmik et al., 2003, Kozmik, 2005). It is worth noting that one animal taxon in Fig. 8.2 that does not express Pax6 is the Anthozoa (corals, anemones, etc.); these animals lacks eyes but do, however, have photoreceptor patches that include opsins (Gornik et al., 2020)

What these results almost certainly mean is that, very early in the evolution of multicellular organisms, a protein with paired-box and homeodomain regions was associated with modulating the organisation of patches of light-sensitive cells. As the

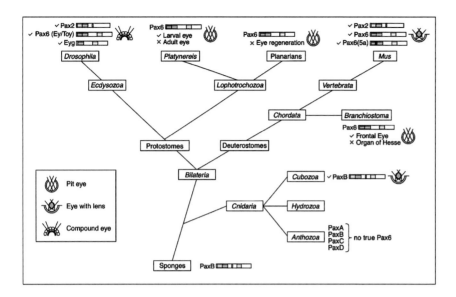

Figure 8.2 Evogram showing the evolution of the eye across the Animal Kingdom, the different eye morphologies and the types of Pax6 expression. It clearly demonstrates that every organism across the phyla that has an eye expresses Pax6.

(From Kozmik Z, Daube M & Frei E (2003). *Dev Cell* 5:773–785. With permission from Elsevier.)

fossil evidence shows that lower Cambrian organisms had eyes (Chapter 13), while worms and jellyfish probably had light-sensitive cells even earlier, the ancestor of Pax6 may first have become involved in light sensitivity almost 600 Mya. That so many organisms use a Pax6 homologue for eye development and an opsin homologue for light detection provides spectacular evidence that animals with light-sensitivity shared a common ancestor which probably lived well before the end of the Ediacaran period.

Patterning the early embryo

Another remarkable homology is the close and completely unexpected similarity in the molecular mechanisms by which invertebrates, such as *Drosophila* and vertebrates such as the mouse pattern their very different antero-posterior tissues during embryogenesis (see Appendix 7 and Barresi & Gilbert, 2019). Consider *Drosophila*: its larvae each have 14 ectodermal segments laid down early in embryogenesis, most of which contain imaginal discs that will produce the external tissues in the adult during pupation. These include location-dependent organs that extend from the antennae at the front to the external reproductive apparatus at the back. The basis of the initial antero-posterior molecular system that will eventually generate this pattern is laid down by nurse cells and is present in the maturing oocyte (Clark et al., 2019). Once the embryo starts to develop, a complex series of molecular interactions takes place which involves many transcription factors together with a great deal of signaling through, among others, Hedgehog and Wnt. The net result of all this molecular activity is that the AP axis is subdivided into segments, each with its own molecular identity and future fate (for details, see Barresi & Gilbert, 2019).

This identity is conferred on the segments by a group of genes that encode transcription factors. The evidence here came from a set of mutations that can change segment identity. These were first recognised more than a century ago and a classic example is the *Antennapedia* mutant, which has a leg on its head instead of an antenna. Such flies are known as homeotic mutants and, in the 1980s, the underlying homeotic patterning genes were shown to encode a set of eight transcription factors. These are known as *Hox* (or homeotic) genes because each includes the 180 bp homeodomain motif within its sequence. In situ hybridization work then showed that, as predicted, each segment expressed its own set of these genes. Much more surprising was the observation that the sequence of *Hox* genes along *Drosophila* chromosome 3 was the same as the sequence of their expression in the embryonic segments along the anterior-posterior axis (a relationship known as collinearity; Fig. 8.3).

The discovery of Hox patterning in *Drosophila* naturally led to an examination of whether anything similar might be happening during the development of the mouse, whose paraxial mesoderm (that adjacent to the neural tube) also undergoes visible segmentation (seen as somites). Genetic analysis soon showed that each somite, together with its adjacent region of neural tube had its pattern specified by a unique set of mouse *Hox* genes, which together comprised the Hox code. The number of *Hox* genes in the mouse was, however, very much greater than that of *Drosophila*

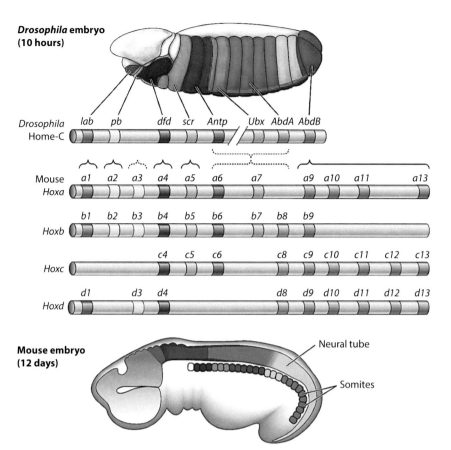

Drosophila embryo (10 hours)

Drosophila Home-C: *lab   pb   dfd   scr   Antp   Ubx   AbdA   AbdB*

Mouse Hoxa: *a1   a2   a3   a4   a5   a6   a7   a9   a10   a11   a13*

Hoxb: *b1   b2   b3   b4   b5   b6   b7   b8   b9*

Hoxc: *c4   c5   c6   c8   c9   c10   c11   c12   c13*

Hoxd: *d1   d3   d4   d8   d9   d10   d11   d12   d13*

Mouse embryo (12 days)

Neural tube

Somites

Figure 8.3 **Hox patterning in *Drosophila* and mouse embryos illustrate homologous colinearity and homology in the two embryos, together with the organisation of paralogues in the mouse.**

(From Carroll SB (1995) *Nature*; 376:479-485. With permission from Springer Nature.)

(see below). Instead of a single set of eight genes, the mouse had four sets of paralagous genes (a,b,c,d), each of which was on a separate chromosome and had up to 13 homologues (Fig. 8.3). There were two further surprises. First, the sequence of genes for each of the four mouse homeotic clusters along their respective chromosomes was the same as the homologous sequence in *Drosophila*. Second was the observation that the expression sequence of the *Hox* gene set along the neural tube and somites of the mouse embryo was the same as in *Drosophila* (they too were colinear). Gene knockout and other mouse work have now shown that the fate of a group of cells depends on the exact subset of *Hox* genes that it expresses.

These similarities were particularly unexpected because the mechanisms for setting up the patterns of homeotic gene expression are very different in the two organisms. In *Drosophila*, Hox patterning is the end result of a complex and multistage sequence of molecular patterning and signaling interactions established soon after fertilisation. In the mouse, Hox patterning is set up rapidly during or soon after gastrulation, a little before segmentation occurs, through signal gradients (FGF8, BMPs, and Wnts) interacting in conjunction with the CDX group of transcription factors. The result is that each segment expresses a unique set of Hox genes that together comprise the Hox code. Each set then, in ways that are still not fully understand, drives the future development of that segment. Hence, although both *Drosophila* and mouse embryos use Hox genes for the antero-posterior patterning of the adult organism, all other aspects of their segmentation and subsequent development are different.

It is worth noting that the process of segmentation in *Drosophila*, where all segments form at about the same time, is different from that in many other arthropods: *Drosophila* is a long germband arthropod whereas most arthropods are short germband and here the segments form sequentially rather than simultaneously (Clark et al., 2019). Such short germband segmentation has some similarities to the mouse in which a newly epithelialised posterior somite forms every two hours from a paraxial mesoderm condensation as the posterior region of the embryo extends, eventually forming the tail (Choe et al., 2006). The mechanism for this in the mouse involves growth, a molecular clock mechanism and a mesenchyme-to-epithelial transition (Hubaud & Pourquié, 2014) but the mechanism by which this happens in short germband arthropods is not yet known.

Since the original discovery of these sets of *Hox* genes, a great deal of work has been done on the wide family of genes that includes a homeobox domain; it now seems that members of this gene family are present across the Bilateria, and phylogenetic analysis shows that they all share a common ancestor. There was clearly an original *protohox* gene that appeared very early in multicellular evolution, going through many rounds of mutation, duplication, and diversification over time and across all the phyla (Fig. 7.4). Even plants have proteins with homeodomains (Costanzo et al., 2014; Meng et al., 2018) which probably evolved from homeodomain-carrying genes first seen in early descendants of the LECA (Derelle et al., 2007) and are still present in single-cell amoebae, such as *Capsaspora owczarzaki* (Holland, 2013).

In invertebrate phyla, *Hox* gene expression seems only to underpin patterning in ectoderm-derived tissues, but Holland (2015) has pointed to two further Hox-related families of genes, the *NK* and the *paraHox* (*Gsx*, *Xlox*, and *Cdx*) clusters, that also have patterning roles in primitive organisms and so may be basal for the clade. The former group has a role in patterning invertebrate mesoderm while the latter can pattern endoderm. Here, *Xlox* is expressed in the anal region, *Cdx* in the midgut and, in molluscs at least, *Xlox* is expressed around the future mouth. The links are far weaker in deuterostomes, but *Pdx1* (a *Cdx* homologue) is needed for the development of the mouse pancreas and the duodenum (Offield et al., 1996).

The key point that Holland (2015) makes is that, associated with the evolution of the early mesoderm layer that resulted in diploblasts becoming triploblasts, was the evolution of three related sets of transcription factors that patterned the ectoderm, mesoderm, and endoderm. Their functional presence was thus a key factor in the evolution of *Urbilateria* (Chapter 12). Today, after >500 My and a host of mutations and duplications, these genes have not necessarily maintained their exact original functions across all the species whose patterning they underpin. It is, however, clear that the various patternings responsible for the anatomy of all living animals have deep and common roots.

Patterning the vertebrate limb and the re-use of proteins

If there is any theme that runs across the whole of evolution, it is that of *opportunism*. This is the idea that, if a genetic change can happen, it will eventually do so through a mix of random mutation, genetic drift and associated random breeding—this is normal variation. Most options will fail, but occasionally a phenotypic variant may thrive in a particular environment—this is normal selection. An interesting aspect of opportunism is seen when a new tissue forms. The development of this tissue needs regulatory systems, and there is little point in, as it were, reinventing the wheel. Once a range of signal transduction and process networks were in place for the development of one set of tissues, they were then available for use in others, provided that their secondary expression could be made tissue specific. One molecular device for this is the enhancer/repressor system: these are small cis-acting regions of the genome that may be upstream or downstream of the gene in question and to which proteins associated with a transcription complex can bind. Activators enhance transcription on a tissue-specific basis while repressors silence it (Nord, 2015). Opportunistic use of such networks required the evolution of new enhancers and repressors that would enable transcription factors to become active in novel environments.

An example of such molecular opportunism is the secondary role adopted by the set of *Hox* patterning genes in vertebrates. Once these genes had been used to establish antero-posterior patterning of the embryo, they were available to be used in new tissues that arose later. This is what happened when the sarcopterygian fishes diverged from the Actinopterygii, with a synapomorphy for the clade being bony pectoral and pelvic fins. It has now been shown that patterning of these bones is regulated by a subset of the *Hox* genes and this regulatory apparatus has been carried over to the limb bones of tetrapods (Amemiya et al., 2013; Fig. 8.4; Chapter 6).

The later evolution of digits in these limb-like fins then required a second round of patterning. With hindsight, it is not surprising that the Hox system, already active in the fin, was re-used to pattern the autopods of amphibians and all its descendant taxa (Fig. 8.4). The details of tetrapod limb organization thus became determined by two rounds of Hox patterning in the embryo that were much less complex than those determining axial organisation. The first is proximal-distal (from the scapula to

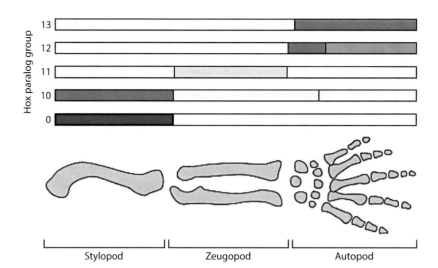

Figure 8.4 **The adult mouse forelimb skeleton showing the domains patterned by the Hoxa and Hoxd paralogues.**

(From Gilbert SF (2014) *Developmental Biology* (10th Ed). Sinauer Press. With permission from Sinauer Press.)

autopod plate in the forelimb) with specific tissue identity being determined by Hox a,d 9-13 (Fig. 8.4); this almost certainly derived from the original patterning of the sarcopterygian fin. The second is the patterning of the early autopod that gives rise to the digits and is the result of antero-posterior patterning of Hoxa13 and Hoxd10-13 (Tickle, 2006); this is established through wnt, FGF and sonic hedgehog signalling (Duboule et al., 2007). This latter system has turned out to be important in the wider context of understanding the molecular basis of variation because the effects of the various mutations that alter digit patterning are easy to identify, particularly in humans. Some of this work will be considered in Chapter 18.

It is worth noting that many of the proteins activated by the Hox system that are involved in patterning vertebrate limbs have homologues that are involved in patterning the very different *Drosophila* limbs, which are predominantly made from ectoderm. Examples include the signals Shh and Hh (anterior-posterior patterning) and Meis/Hth and Dlx/Dll (proximo-distal patterning; Tabin et al., 1999). There are two possible explanations for these coincidences which come down to the question of whether *Urbilateria* had limbs, or at least appendages, such as ectodermal antennae whose development required extant networks. If it did, then the behaviour of these genes in limb formation can be seen as a plesiomorphy and their roles in contemporary vertebrates and invertebrates reflects homology. The alternative view is that *Urbilateria* did not have such appendages and that the similar roles of these genes reflects independent evolution (homoplasy).

The fossil record from the Cambrian is ambiguous: neither *Haikuella* (Fig. 14.4), a very early chordate, nor the pseudocoelomate *Markuelia* (known for its fossilised embryonic forms, see Liu et al., 2014) had limbs, unlike early limbed organisms, such as *Halucigenia* (Fig. 13.1), perhaps an early velvet worm, and *Aysheaia*, which was probably an early arthropod. The Ediacaran record is more helpful here: recent work (Fig. 12.2; Chen et al., 2019) has identified a simple worm-like bilaterian from that period that did not have limbs, although the possibility that it may have had small antennae cannot be excluded. It thus seems likely that limb development is a later event and any apparent homologies probably only represent the likelihood that early organisms had a limited number of potential molecular networks available for limb development.

Such opportunistic reuse of existing proteins and networks has been a parsimonious way of facilitating the evolution of new tissues or of making existing ones more complex; indeed, the gene-expression databases include many examples of the reuse of transcription factors, signals, and process networks, usually under tissue-specific control. This opportunism does, however, carry a medical caveat in the context of building on such information for the treatment and prevention of human disease. When considering possible manipulation of the genes underpinning regulatory systems in the context of drug production, it is important to check work on mouse development to find out whether the proteins have broader roles than expected (Uhl & Warner, 2015). Although rodents are not humans, such work is a first step in avoiding the risk of unpleasant side effects occurring in humans.

Nerve cord location in protostomes and deuterostomes

The final example to be considered here examines a classic problem: why do invertebrates and vertebrates have their dorsal-ventral tissue organisation reversed so that, for example, the nerve cord is ventral (below the gut) in protostomes and dorsal (above the gut) in deuterostomes. This very basic difference had intrigued evolutionary biologists since the 19th century as it seemed to deny the possibility of a common ancestor and so reflected independent origins for the two clades. Comparative molecular genetic analysis of nerve cord determination in mouse and *Drosophila* has greatly clarified the situation: it is now clear that the homologues of the same signals and transcription factors pattern both embryos, which are themselves representative of their wider families of organisms in this context (Fig. 11.3).

The dorso-ventral patterning of the nerve cord has now been shown to be set up in both embryos by a pair of counteracting gradients, one is for a signal and the other is for its antagonist: these are DPP and Sog in *Drosophila* and BMP4 and Chordin in the mouse. The relative concentrations of these gradients at any position in the early embryo activate one of three transcription factors; these are Vnd, Ind, and Msh in *Drosophila* and Nkx2, Gsx, and Msx in the mouse (Fig. 8.4). There are two points of particular interest in these systems. First, as the signals and transcription factors are each homologous pairs, the patterning systems as a whole are also homologous and therefore, reflect common descent (Lichtneckert & Reichert, 2005). Second, the polarity of each system is reversed with respect to the other (Fig. 8.5).

The direct implication of this work is that, very early in the evolution of the Bilateria, there was a polarity reversal in the neuronal patterning system that led to the neural cord becoming relocated with respect to the gut. We have no experimental evidence for why or how this happened or even which was basal. However, the fact that the very great majority of animal phyla had, from the Cambrian, onwards a ventral cord, with only the chordates having a dorsal chord, argues strongly that the protostome organization represented the original morphology (Denes et al., 2007; Lowe et al., 2015b).

Protein network homologies

While signal and transcription-factor activity initiate change in expression during embryogenesis, they do not mediate it; this is mainly the task of protein networks, and these fall into two categories. First, there are the signal transduction pathways triggered by signals binding to their receptors, whose downstream result is to activate transcription-factor-activated gene expression in the nucleus. Second are the process networks that result from this new gene expression and that drive cell differentiation, morphogenesis, proliferation and apoptosis (Appendices 1 and 7). As these networks tend to be very complicated, only the Notch-Delta signal transduction and the key

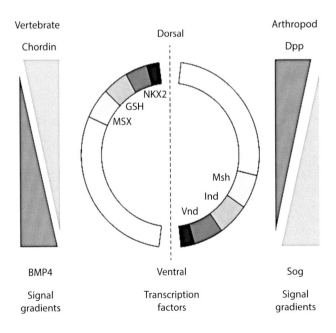

Figure 8.5 Diagram showing patterns of signaling gradients and transcription factor expression in vertebrates and arthropods embryos during neurulation. Chordin/Sog, BMP4/Dpp, Nkx2/vnd, Gsh/ind, and Msx/msh are all homologous pairs of proteins.

(Redrawn from Lichtneckert R & Reichert H (2005) *Heredity* 94:465–477. With permission, from Springer Nature.)

(a) Drosophila Notch signaling pathway

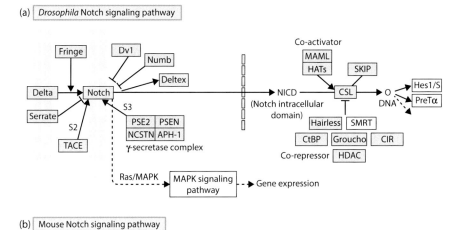

(b) Mouse Notch signaling pathway

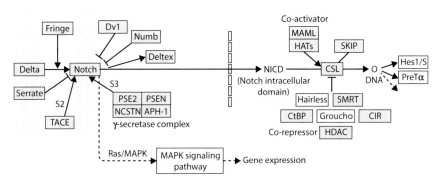

Figure 8.6 The Notch-Delta pathways in *Drosophila* and mouse. Genes in blue boxes are present while those in white boxes are absent. The two networks are surprisingly similar considering that they are separated by >500 My of evolutionary history. Note that proteins in the two diagrams with the same names are not identical, but are organism-specific orthologues, named using the KEGG convention.

(From Kanehisa M, Goto S, Sato Y et al. (2014) *Nucleic Acids Res.* 42:D199–D205. http://www.kegg.jp/pathway/mmu04330. With permission from KEGG.)

features of the apoptosis (caspase) pathways are illustrated here (Figs. 8.6 and 8.7). The effect of mutation on these protein networks will be examined in Chapter 18.

Phylogenetic analysis has shown that most of the signal-transduction pathways seen in mouse and *Drosophila* are closely related, with each pair sharing many homologous proteins that are closely related in sequence and function. Visualising this on the basis of the literature is often quite difficult because the names of homologous proteins tend to be organism-dependent, often because these names were based on the phenotype of the mutant or its embryo. Such interspecies comparisons are usually better done computationally.

Information on signal transduction and other pathways and networks is now readily accessible in the various bioinformatics databases. The key one for making interspecies comparisons is KEGG (Kyoto Encyclopedia of Genes and Genomes, see Websites; Kanehisa et al., 2012); this includes interspecies comparisons among many networks for both embryos and adults, albeit that its focus is more on vertebrates than invertebrates. Fig. 8.6 shows the pathways generated by KEGG for the Notch-Delta signal transduction pathway in mouse and *Drosophila*, and one does not need to understand the details of how these pathways operate to see just how closely related they are.

Similar homologies are seen when comparing the cell proliferation networks in mouse embryos and *Drosophila* larvae: the activation of each can be initiated through several paracrine growth factors that are themselves homologous (e.g. wnt1/Wg, shh/Hh, and BMP4/Dpp) and that excite homologous pathways. *Drosophila* does, however, have two important endocrine factors associated with pupation, juvenile hormone and ecdysone, whose homologues are not present in vertebrates. These, of course, have their own hormonal signals that operate from the growth phase of embryos onwards and are not present in invertebrates. In all pathways associated with proliferation, a key feature is the activation of the output of the cyclin/CDK pathway whose activity results in mitosis, whose molecular basis has roots in ancient unicellular organisms (Morse et al., 2016).

Growth, apart from enlargement due to the swelling of extracellular matrix, reflects the balance between cell proliferation and apoptosis (programmed cell death), with the latter being implemented through signaling that initiates the caspase pathway. This process is not only present across the Bilateria, but even has components in sponges and other very primitive organisms (Sakamaki et al., 2015). The caspase pathway can be activated via *extrinsic* and *intrinsic* signaling: the former

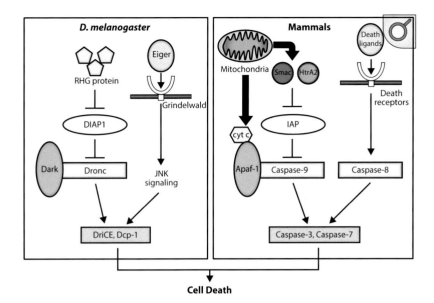

Figure 8.7 **Illustration showing the key similarities between the core features of the apoptosis pathways in *Drosophila* and mammals.** In each diagram, the intrinsic apoptotic pathway is on the left and the extrinsic pathway on the right. Proteins in boxes of the same shape and colour in each diagram represent functional homologies (e.g. DrICE and Dcp-1 are both caspases). In *Drosophila*, Reaper, HiD and Grim (the RHG proteins) are located in the mitochondrion, while their functional equivalents in the mouse are in the cytoplasm (see **Fig. A7.1** for a diagram showing full details of the very complex apoptotic pathway in mice).

(From Kawamoto Y, Nakajima YI, Kuranaga E (2016) Apoptosis in Cellular Society: Communication between Apoptotic Cells and Their Neighbors. *Int J Mol Sci.* 2016; 17:2144. With permission from Erina Kuranaga.)

is used to model tissue details and to remove unneeded cells during development. An example is in the early autopod of animals, such as the mouse, that have separate digits. Here, the embryonic autopod plate initially includes tissue between the future digits. Sometime after digit patterning, TNF and BMP patterning signals activate the extrinsic caspase pathway in the ectoderm and mesenchyme between the digits, so causing their death and separating the digits (Sanz-Ezquerro & Tickle, 2003).

The intrinsic apoptosis pathway is used when a cell suffers internal damage. Here, the caspase pathway is triggered through the release of mitochondrial cytochrome C into the cytoplasm, often through the upstream activity of P53, Bcl-2 and associated proteins. **Fig. 8.7** shows some of the core homologies between the *Drosophila* and mouse caspase pathways (see legend for details; the mouse caspase pathway is shown in more detail in **Fig. A7.2**). It is important to emphasise that, although the pathways are similar, they are not identical (Denton et al., 2013). Indeed, they would not be expected to be so after >500 My of separate evolution. Of particular interest, nevertheless, is the fact that, in both mouse and *Drosophila,* the regulation of both proliferation and apoptosis are interlinked through homologues of the Hippo signalling pathway (Zhao et al., 2011).

The protein networks governing apoptosis and growth (e.g. **Figs. A1.1** and **A7.2**) are not unique in being homologous across many clades; the networks for many other systems also show remarkable degrees of homology across the phyla with respect to both their proteins and their structure. Although the molecular biology of these essentially homologous networks is extremely complicated, many represent contemporary versions of networks with extremely ancient evolutionary provenances. This again demonstrates that all animals inherited the seeds of their organogenesis toolkits from a common ancient ancestor.

If there is so much similarity across the animal phyla in the signals, transcription factors and networks, one is left asking about how the differences across species arise. The core answer seems to be in the details of the mechanisms that pattern the embryo, and in the ways that the cell proliferation and apoptosis networks are used. The most obvious differences between species are in size and the scale differences can be spectacular (Chapters 3, 18), although we do not yet know how the outputs of the proliferation networks are modulated. The patterning mechanisms are even more important, however, as it is these that initially determine first the bauplan (basic body plan for a phylum) and then the map of tissues within an organism. Modulation in these mapping systems determines, for example, the organization of the bones of the vertebrate cranium, whether a fish skeleton is cartilaginous or bony, the number of segments in an arthropod and whether that arthropod has wings.

IMPLICATIONS OF PROTEIN HOMOLOGIES

It is a natural prediction of the *descent with modification* theme that there should be ancient homologies between species in very different animal phyla below the anatomical level. The obvious examples of a common DNA code and basic molecular biology, a similar basic cell structure and various modes of differentiation, such as an epithelial or neuronal cell type had long been recognised. It was, however, completely unexpected in the 1980s to find that the molecular mechanisms that pattern the diversity of early animal embryos and drive their subsequent development would be so similar and would employ so many homologous proteins.

The reasons for this early lack of imagination were partly because so little was known about the molecular basis of development before the 1980s and partly because of the great range of animal morphologies. Even today, it is still surprising that ~500 My of mutation and selection has had so little effect on the signals, the transcription factors, the process networks and the structural proteins that build animal tissues. The conclusion of ~40 years of work are that evolution seems to have tinkered relatively little with the basic regulatory system that underpins embryogenesis. The major differences across the phyla are in how different embryos interpret patterning and timing cues to produce a species-specific adult anatomy.

The question immediately arises as to why these network systems remain so stable, given that both the genomes and phenotypes of organisms can be so very different. If other genes have mutated widely why should their regulatory networks still maintain their integrity? The answer is not known, but there are two possibilities. First, the functional inter-relatedness of the proteins in a network, such as a signal-transduction pathway, conferred a degree of stability against mutation on such systems because it was probably more efficient to achieve change in developing anatomical phenotypes through modulation and the addition of a few new components than through the production of entirely new networks. Second, selection acts on the phenotype and only indirectly on underlying control systems. Provided that these systems maintained an ability (through very minor genetic tinkering) to use a range of old and newly modified genes to produce a modified phenotype that would survive selection, there would have been no pressure for them to be replaced.

The results of almost 40 years of evo-devo work have been as important for developmental biology as they have for evolutionary biology. The latter now has an experimental basis on which to analyse the fossil record and a wide set of functional homologies to strengthen its basis. The implications for developmental biology have been equally substantial: evo-devo approaches have opened up new ways of studying embryo diversity. Perhaps even more important, the work has shown that embryogenesis across the animal phylum is underpinned by a stable regulatory and construction system based on a set of ancestral genes that were present in an organism that lived at the beginning of Bilaterian evolution. It is for this reason that these are known as *deep homologies* (Shubin et al., 2009).

KEY POINTS

- The development of each species reflects its evolutionary history (shown in its plesiomorphies and synapomorphies), and its own novelties (autapomorphies)

- The area of evo-devo is mainly concerned with identifying protein families that have homologues with very similar functions in very different phyla. It is also concerned with the re-use of proteins as organisms evolve.

- There are, in addition, many biochemical and physiological features whose underlying molecular networks show homology (e.g. the Krebs cycle).

- A classic example of such evo-devo relationships is the development of eyes across the animal phyla: in every case, a homologue of the transcription factor Pax6 underpins their early development.

- Across the animal phyla, anterior-posterior patterning of body structures is under the control of homologues of hox transcription factors.

- *Drosophila*, an invertebrate, has a ventral nerve cord while the mouse, a vertebrate, has a dorsal one although the organising proteins are homologues. The reason seems to be that a polarity reversal occurred in the patterning of an invertebrate embryo that lived in the early Cambrian resulting in the evolution of the deuterostome line.

- Many of the protein networks that control gene expression and that drive the development of diverse organisms each include protein homologues to the extent that the networks themselves can be seen as essentially homologous.

- The results as a whole show that the molecular toolkits that drive organogenesis in embryos across the Bilateria are substantially homologous and represent descent from the simpler networks present in *Urbilateria*, the last common ancestor of the Bilateria.

- It is for this reason that the relationships in the constituents of this system are known as deep homologies.

FURTHER READINGS

Bard J (2013) Driving developmental and evolutionary change: A systems biology view. *Prog Biophys Mol Biol*; 111:83–91.

Barresi MF, Gilbert SF (2016) *Developmental Biology* (12th ed). Oxford University Press.

Held LI (2014) *How the Snake Lost Its Legs: Curious Tales from the Frontier of Evo-Devo*. Cambridge University Press.

Shubin N (2007) *Your Inner Fish*. NY: Random House

Todd Streelman J (ed.) (2013) *Advances in Evolutionary Developmental Biology*. Hoboken, NJ: John Wiley.

WEBSITES

- Wikipedia entries for: *Enhancer (genetics), Evolutionary Developmental Biology, Protein-Protein Interactions.*

- Kyoto Encyclopedia of Genes and Genomes: www.genome.jp/kegg

SECTION THREE
THE HISTORY OF LIFE

This section aims to give a summary of our current view of the history of life since its beginning, the area now known as systematics. In the past, this history was based solely on the fossil record and particularly focused on the evolution of animals and later plants as these were the only taxa for which the information was rich enough to provide detail. This record started in the mid-to-late Ediacaran for animals (~570 Mya) and strengthened after the Cambrian explosion; that for plants was later. After the evolution of teeth and particularly bone ~400 Mya, the chordate taxon became particularly well represented in the fossil record, while the record for plants was strengthened by the evolution of lignin >400 Mya. Cladistic analysis of all this data has been used since the 1960s to trace history backwards on the basis of synapomorphies that modern organisms shared with ancient, fossilised ones.

There is, however, no detailed fossil record for the first three billion or so years of life and the early events of evolution could consequently not be explored for many years. The situation changed completely in the 1970s with the introduction of the computational techniques of comparative and phylogenetic analysis that analysed mutational difference in DNA sequences. Use of these techniques enabled ancestral relationships to be reconstructed on the basis of DNA sequences from any group of contemporary organisms. This approach thus provided a quantitative basis for reconstructing the evolutionary history of all contemporary organisms within molecular phylograms (Chapter 7).

Sequence analysis has also proven important in exploring other aspects of the deep past, discussion of which had previously been pure conjecture. These include how eukaryotes first formed and how these early nucleated cells diversified to initiate the

DOI: 10.1201/9780429346217-11

many phyla that now inhabit Earth. Detailed sequence comparisons combined with phylogenetic analysis have pointed towards how eukaryotic cells formed through endosymbiosis of eu- and archaebacterial prokaryotes and how the descendants of those cells evolved in different directions to initiate today's phyla and major clades. Molecular phylogenetics has thus transformed our understanding of the evolution of prokaryotes and the early eukaryotes. Without the details that they provide, very little indeed would be known about the first three billion or years of life.

The early chapters of this section use such informatics approaches to give the current view of the evolution of life forms whose origins predate times before the fossil record becomes useful. It should, however, be mentioned that the details of the relationships and the groupings of very similar tiny organisms are still in a state of flux, but it is to be hoped that the different views of the key research workers in the field are now converging so that the final consensus view will not be very different from that given here.

Our current understanding of the history of life to date thus reflects the cooperative work of several generations of biologists and palaeontologists who have collected and published vast amounts of material. Although every aspect of this history is not yet in place, we now have extensive information on the lines of descent of almost every clade in every phylum. The amount of information on the evolutionary lines of descent for a given species is often so great that it difficult to show it on a printed page. The reader interested in pursuing any aspect of evolutionary history of any taxon in full detail, which is always far lengthier than can be given here, is therefore recommended to start by exploring the appropriate Wikipedia article.

The aim of this section is not, however, to provide a detailed taxonomy of all the major types of organisms but to pick out the key anatomical and molecular changes across the phyla that have enable living organisms to diversify. The question of the mechanisms by which all this change actually happened are discussed in Section 3.

The First Two Billion Years

While the evidence is clear that life evolved in a marine environment almost 4 Bya, we still do not know exactly how this happened, even after more than half a century of experimental work. It is, however, clear that it was based on the availability of relatively small organic molecules. Some came from extra-terrestrial rock and dust while others formed in ways that resulted from lightning discharges in an atmosphere that contained ammonia, methane, hydrogen sulphide, and hydrogen.

The first part of this chapter summarises current thinking on the processes that led to the origin of life, the process known as abiogenesis. The first living organism was a small, primitive prokaryotic cell (the First Universal Common Ancestor or FUCA) that probably had an RNA-based genome. The evidence for this comes mainly from chemical and geological experiments under conditions designed to emulate marine conditions in those early days. This cell slowly evolved and eventually became the Last Universal Common Ancestor or LUCA, the parent cell of all subsequent life with a DNA-based genome. The chapter goes on to consider prokaryote diversification and the evolution of the Eubacteria and the Archaebacteria, the two kingdoms of the prokaryote world.

The last part of the chapter examines the origins of the first eukaryotic cells. There is now incontrovertible evidence on the basis of sequence homologies that this was the result of endosymbiosis. This happened when one bacterium engulfed another, with the internalised one continuing to maintain a degree of independent integrity rather than being degraded. One way in which a eukaryote could have formed was that a eubacterium endosymbiosed an archaebacterium that became its nucleus to give a composite cell known as the First Eukaryotic Common Ancestor (the FECA). This event was then followed by the further endosymbiosis of a eubacterium that included an early citric acid (Krebs) cycle and so was able to produce considerable amounts of ATP; this became its mitochondrion. An alternative possibility is that the eukaryotic cell resulted from a methanogenic archaebacterium endosymbiosing a eubacterium to give a mitochondrion, with the eukaryotic cell eventually forming as a result of major internal reorganisations, particularly of the membrane system. Either way, this cell with its archaebacterium-derived nucleus and eubacterium-derived mitochondrion is known as the Last Eukaryotic Common Ancestor (the LECA) and was the parent cell of all living eukaryotic taxa.

Life started in very early seas almost four billion years ago with the formation of the first cell, known as the First Universal Common Ancestor (FUCA). This very primitive prokaryote slowly evolved, increasing its spectrum of abilities and acquiring a DNA-based genome, so becoming the Last Universal Common Ancestor (LUCA) of all subsequent organisms. Its taxonomic status is still not clear, but evolutionary change led to the existence of two very different types of prokaryotes (cells that lack a defined nucleus), Eubacteria and Archaebacteria, with the latter evolving some time over the period 3.3–2.7 Bya (Table 9.1; Gribaldo & Brochier-Armanet, 2006).

DOI: 10.1201/9780429346217-12

Today, prokaryotes are, in terms of population numbers, by far the most numerous organisms on Earth. Prokaryotes now occupy every niche, even in environments, such as rocks in which it had been thought that life was unsustainable (Wierzchos et al., 2012). Prokaryotes are also present within larger organisms: many animals, for example, are host to almost as many prokaryotic cells as they have eukaryotic ones— these comprise the organisms' microbiome.

Prokaryotes slowly diverged, living essentially independent existences until ~2 Bya when the eukaryotes started to form. This occurred through a series of endosymbiotic events in each of which one prokaryote engulfed another. These events resulted in cells with a nucleus, mitochondrion and in some a chloroplast. The diversification of these cells eventually resulted in today's biosphere.

All of these events, from the start of life to the evolution of cells with chloroplasts, took about three billion years and are the subject of this chapter. Unfortunately, we know nothing about the morphological details of any of these primitive organisms as the fossil record over this period is minimal and completely inadequate for analysis. The only methodology for probing these early events is computational phylogenetic analysis (Chapter 7). This uses homologous gene sequences across contemporary prokaryotes and eukaryotes to identify taxonomic relationships back into the deep past. These techniques illuminate the history of all organisms back to the LUCA, the prokaryote parent of all subsequent species including eukaryotes. They even give some insight into the molecular abilities of the FUCA, although, of course, they say nothing about its origins.

THE ORIGIN OF LIFE

Putting dates on when life originated is not easy, but analysis of the radioactive profile of zircon fragments from Australia that date back to ~4.4 Bya argues that oceans were present around or a little later than 4.2 Bya and so were available to contain a molecular soup of complex organic molecules (see Appendix 3 for details of dating methods). Sleep et al. (2011) have suggested that the Earth would have cooled down enough for life to evolve ~4 Bya and point to evidence that this had happened before ~3.8 Bya, with perhaps even photosynthetic prokaryotes having evolved by then, although this view is contentious (Weiss et al., 2016). Ohtomo et al. (2014) date the origins of life to a little later on the basis that structured carbon deposits laid down in Western Greenland ~3.7 Bya had low levels of the stable ^{13}C isotope, indicative of living organisms.

The first direct evidence of life in the fossil record is the presence of stromatolite fossils in Western Australia that formed ~3.48 Bya and whose morphology is characteristic of contemporary stromatolites that are made from microbial mats (Fig. 9.1). These probably contain a wider range of prokaryotes than the much more primitive prokaryotes that formed the original stromatolites (Noffke et al., 2013). The earliest microfossils (small carbonaceous spherical microstructures <1 mm in diameter), identified on the basis of their likely organic walls, come from rocks in South Africa aged to ~3.2 Bya (Javaux et al., 2010). If these dates are even roughly correct, several hundred million years of chemical jostling in the early oceans were required to produce the FUCA.

Because all the cells of contemporary organisms share a basic structure, genetic code, and molecular machinery, it is a fair deduction that life only evolved once and almost certainly under anaerobic conditions as there seems to have been no free

Figure 9.1. Stromatolites. (a): Stromatolites seen today in Shark Bay, Western Australia. (b): Section through an Australian stromatolite from >3 Bya showing the layered structure.

(a Courtesy of Paul Harrison. Published under a Creative Commons Attribution-Share Alike 3.0 Unported License.

b Courtesy of Didier Descouens. Published under a Creative Commons Attribution-Share Alike 4.0 International License.)

oxygen in the early seas. Although there has been a great deal of discussion about the evolution of the FUCA, its origins and details remain elusive. One fanciful line of speculation has suggested that the FUCA did not evolve on Earth but has an extra-terrestrial origin. It then reached Earth via a meteor from some source outside the solar system (non-scientists have come up with more fanciful explanations), landing in the sea, an environment in which it flourished. This old idea, known as directed panspermia, was resuscitated in the middle of the last century by a group whose most famous member was Francis Crick. The idea has the advantage to its advocates, at least, of not being disprovable, but it is obviously unsatisfactory because it just transfers the problem of how life originated to some other star system rather than trying to solve it.

There was actually no need to invoke the idea of directed panspermia because there was a better, testable theory, which had been put forward in the 1950s. This was that the early seas were a "primordial soup" containing complex chemicals that are known to be the building blocks of life and whose self-assembly over millions of years eventually led to the FUCA. Even then, it was known that part of the organic mix in the early seas came from the many meteors and rocks that bombarded early Earth (Kvenvolden et al., 1970; Bernstein, 2006). As well as inorganic material, these rocks contained organic molecules, which were likely to have included polycyclic aromatic hydrocarbons (Ehrenfreund et al., 2006). There is, however, no evidence that the contribution included viable cells and, indeed, it would be hard to see how any such cell would survive the great rise in temperature such a meteor would have experienced due to the friction in moving through the Earth's atmosphere.

The first real evidence to support this theory came from Stanley Miller and colleagues (1953) who assumed that the Earth's early atmosphere included ammonia, methane, hydrogen, and water vapour, but not oxygen. They ran an electrical discharge through these gases (simulating the effect of lightening) and found that the molecules then synthesized included amino acids, albeit in a racemic mixture of L and D isomers (Parker et al., 2011; for a recent perspective on these experiments, see Bada, 2013). Later, similar experiments with other reducing gases produced nucleic acids, such as adenine, although it is now thought that the early sea was not as acidic as originally considered. The advantage of the "soup" approach is that it provides a good explanation of how large amounts of essential organic chemicals, particularly those requiring nitrogen could have been present in the early sea.

This basic view does not, however, explain either the origins of the energy needed to drive the synthesis of complex molecules or how concentrations of these chemicals would become adequate for the many chemical interactions needed for the formation of the complex macromolecules required for even the simplest of cells There are two obvious sources of energy: electrical energy from lightning bolts near the surface and thermal energy from the sea, particularly in the vicinity of the thermal vents present in deeper water (the current habitat of thermophilic Archaebacteria).

As to the formation of complex macromolecules, Sleep et al. (2011) point to the catalytic role of serpentinite rock, a source of nickel and other minerals, known to be present at least 3.9 Bya, in facilitating the emergence of life. Serpentinite forms from mantle rock in the location of underwater thermal vents where the local pH is slightly alkaline and the water contains sulphur-, iron-, and calcium-containing minerals and sufficient CO_2 to act as a source of carbon. Sleep et al. (2011) argued that submarine hot springs flowing through mineralized chimneys that included serpentinite would provided geochemical free energy, reactive mineral surfaces and protective pore spaces that would serve to concentrate life's "building blocks." Such chimneys would particularly provide a surface that could act as a catalyst for the polymerization and other reactions of and between adherent molecules by lowering the free energy requirement for them to take place (Ferris & Ertem, 1992). It was hence a far more likely environment for life to evolve than the traditional "primaeval soup."

A further problem was ensuring the availability of sufficient nitrate for the production of amino acids. While organic and inorganic molecules containing nitrogen were present in the early seas as a result of lightening, it seems unlikely that there was sufficient of this to sustain and expand life. It thus seems probably that an early property of the FUCA was an ability to fix nitrogen. Today, only a limited range of eubacteria and archaebacteria are able to do this (mainly producing ammonia); all require the presence of molybdenum for their activity. As there was unlikely to have been a great deal of this ion in early seas, it seems likely that nitrogen fixation was first carried out using different enzymes, and the mechanism that used molybdenum evolved later (Mus et al., 2018). Details of how this happened are still not understood.

As to the necessary physical conditions, Mat et al. (2008) have summarized, giving their advantages and disadvantages, five possible temperature zones where the FUCA might have formed. They refer to the most likely model as the *hot cross origin*: in this, the FUCA formed at a tepid temperature and slowly spread, but it only evolved further when it was in a far hotter environment; this additional thermal energy facilitated the formation of more sophisticated chemicals that enabled it to acquire complex functions and eventually to evolve into the LUCA. The details of how these events could have happened are, however, still unclear.

There is a further problem, also still unsolved, and this affects chirality. While the Miller experiments indicated how organic molecules could form, they did not suggest how or why only L-amino acids were incorporated into proteins and nucleotides based on D-ribose into nucleic acids. As to the tertiary structures of such macromolecules, this probably depended on the local roles of hydrophilic and hydrophobic bonds and one can envisage that, as macromolecules started to form, all options were randomly generated and most failed to be useful. Biophysical approaches that may illuminate possible ways in which the larger-scale properties of the FUCA emerged are now being explored, and they may well clarify these early events (Budin & Szostak, 2010; Islam et al., 2017).

FUCA, THE FIRST UNIVERSAL COMMON ANCESTOR

The FUCA clearly included a genome and some associated protein machinery, together with metabolizing and synthesizing proteins within a containing sac. It is now widely accepted that information in the FUCA was stored in RNA rather than DNA chains. This is partly because RNA chains can act as ribozymes with protein-synthesizing and other enzymic properties and partly because of the wide range of functions that RNA molecules undertake today as compared to the very limited, albeit key role taken by DNA chains in all extant organisms except RNA viruses (Moore & Steiz, 2005; Higgs & Lehman, 2015; Kun et al., 2015).

As to the minimum number of proteins required for reproduction, construction and metabolism, it could have been as low as 300 (Mat et al., 2008; Fig. 9.2), although others hold that it was 500–1000 (Weiss et al., 2016). Today, symbiotic bacteria, such as *Nasuia deltocephalinicola* can have as little as 137 protein-coding genes, while the very simple nonsymbiotic eubacterial mycoplasma can have less than 500 protein-coding genes. Insight into the likely protein set of the FUCA comes from a phylogenetic

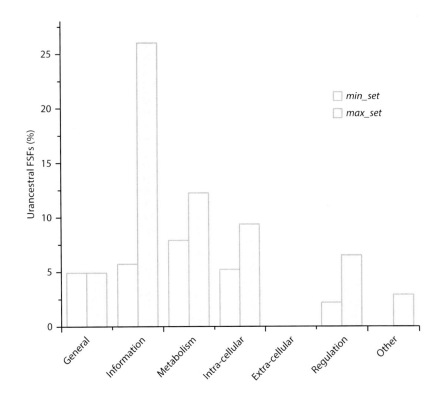

Figure 9.2 The roles of the classes of likely protein possessed by the FUCA. This is based on phylogenetic analysis of fold superfamilies (FSFs) of protein domains. The pairs show upper and lower bounds while the percentages are comparisons with today's organisms, (From Kim & Caetano-Anollés, 2011).

(Courtesy of Gustavo Caetano-Anollés.)

analysis by Kim and Caetano-Anollés (2011) of conserved structural domains of proteins (these are known as fold superfamilies or FSFs) in 420 diverse organisms that included a range of Eubacteria, Archaebacteria and eukaryotes. This analysis was designed to give upper and lower bounds on the relative numbers of these domains in the last common ancestors. The results (Fig. 9.2) indicate that the number of such protein domains in the FUCA was much less than that in contemporary organisms.

A further concern was how a primitive cell obtained energy in a form that could be used to mobilise ATP as there was no free oxygen in the early sea. On the basis of the energy sources used by contemporary prokaryotes operating under anaerobic conditions, Weiss et al. (2016) suggest that this could have been achieved either by using carbon dioxide to produce methane or by using hydrogen to produce acetic acid. Today, the former is employed by methanogens (Archaebacteria) and the latter by homoacetogens (monoderm, gram-positive Eubacteria). Weiss et al. (2016) suggest that such differences marked out the original divergence of the Eubacteria and Archaebacteria, albeit that there was likely to have been a degree of horizontal gene transfer among early representatives of the two clades.

A significant step towards answering the question of how the FUCA's bilayer membrane evolved comes from the work of Mansy et al. (2008) who showed that fatty acid esters can spontaneously self-assemble to form vesicles with many of the expected properties of an early bilayer, particularly in their ability to enlarge and to allow unaided molecular transport across them. Further questions immediately arose about how such vesicles could come to contain the machinery of life and how the various reactions within them became compartmentalized. These are still under investigation (Dziecio & Mann 2012), but there are as yet no generally accepted answers.

It is important to note that the protein systems in the FUCA could not have worked independently: cell division would clearly have failed if, for example, it had been independent of genome reproduction, protein synthesis or membrane expansion. Some degree of molecular communication between the various protein systems within the FUCA would thus have been required for its success. The difficulty of course is that development of such feedback really requires a living, reproducing cell so that its evolution could have been facilitated by the blind hand of selection. It will be interesting to see if a set of steps can be suggested and modeled that start with a membrane containing the sorts of complex organic molecules likely to have been in a primeval soup and ends with a cell with the abilities of the FUCA. This may be the only way to obtain insight into how the FUCA cell evolved. There is, however, little reason to doubt the conclusion of Glansdorff et al. (2008) that life was born complex.

LUCA, THE LAST UNIVERSAL COMMON ANCESTOR

In due course and probably after a fair amount of genetic and molecular enhancement, the LUCA evolved in the early seas. If the FUCA did have an RNA-based genome, the evolution of the LUCA would have required transition to a DNA-based genome. Two nonexclusive drivers for this have been suggested: the enhanced stability against attack from RNA viruses and the improved thermal stability that DNA chains have over RNA ones (Mat et al., 2008). Today, as all cells share a DNA genome with essentially the same genetic code and basic transcriptional machinery and metabolism, it is highly likely that this transition only occurred once. If this was the evolutionary route, the resulting DNA-based cell outcompeted RNA-based cells, all of which were lost. The resulting LUCA probably had a DNA genome, basic protein toolsets for replication, transcription and protein synthesis, and some core biochemical pathways, all enclosed in a within a lipid bilayer membrane.

There has been considerable discussion as to whether the LUCA was (a) a simple monoderm eubacterium with a single bilayer membrane, (b) a diderm eubacterium with a double bilayer (and hence with a morphology similar to that of *Escherichia coli*, below), (c) an archaebacterium, (d) a very primitive eukaryotic forerunner, or (e) something with a mix of such properties, having, for example, an external membrane with mixed esters with internal extensions (Table 9.1). The general feeling is that the least likely candidate is a diderm eubacterium because of its double membrane: monoderm bacteria make antibiotics that kill other monoderms but not diderms. It, therefore, seems likely that producing antibiotic resistance was one of the drivers for the later evolution of diderm eubacteria (Gupta, 2011, but see Megrian et al., 2020)).

There is still no accepted view as to the taxonomic status of the LUCA as the three following papers illustrate. Cavalier-Smith (2014) argues for the traditional view that

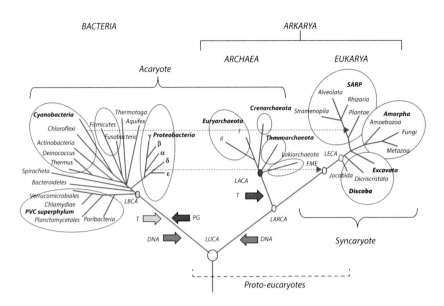

Figure 9.3 The tree of life based on genetic data showing the key prokaryotic phyla and the major eukaryotic phyla groups. The dotted lines show the likely origins of the endosymbioses that gave rise to the mitochondrion (lower) and chloroplast (upper). PG: peptidoglycan; DNA (blue arrows) introduction of DNA; T (pink and red arrows) thermoreduction. LBCA: last bacterial common ancestor, pink circle: thermophilic LBCA; LACA: last archaeal common.

(From Forterre P (2015). *Front Microbiol*; 6:717. Published under a Creative Commons CC-BY, version 4.0 license.)

the LUCA was a very simple monoderm-like eubacterium and suggests that modern Archaebacteria and single-celled eukaryotes are sister taxa. Xue et al. (2003) contend on the basis of an analysis of various contemporary tRNA populations that the LUCA was far closer to the Archaebacteria than to the Eubacteria. Glansdorff et al. (2008) view the LUCA as being a community of varied cells with eukaryotic features so that eubacteria and archaebacteria both represent derivation. The choice depends on which aspects of the limited data available the various authors see as primary and which secondary.

All approaches have to explain how their hypothesised LUCA evolved, and the simplest suggestion is probably that the LUCA had properties common to both Eubacteria and Archaebacteria. One argument in favour of this view is the existence of the contemporary Aquifex family of eubacteria (the Aquificiales in Fig. 9.3). These prokaryotes have a short and limited genome that is only ~30% the size of *E. coli*, they contain many more genes commonly associated with Archaebacteria than do other Eubacteria and are thermophiles that thrive at ~85C and survive at 95C. For all that they have properties associated with the Archaebacteria, they are clearly gram-negative diderm eubacteria because they have the characteristic double cell membrane of this taxon rather than the very different membrane expected of Archaebacteria (Table 9.1). Their mix of properties suggests that their ancestors were evolutionarily upstream of today's Archaebacteria and Eubacteria. This is, of course, the parsimonious view, and contrasts with the alternative hypothesis that today's Aquifex family members represent the downstream simplification of a far more sophisticated ancestor that evolved as a result of either a great deal of horizontal gene transfer or through a eubacterium engulfing an archaebacterium with the two genomes merging in some way.

Readers of the primary literature will soon realise that current discussion about the likely features of the LUCA is wide-ranging and complicated, with various authors appealing to the parsimony, elegance and relative simplicity of their particular view. The difficulty is that prokaryote (some authors prefer the term akaryote, which just means *without a nucleus*) evolution may not have been very efficient and the route from FUCA to LUCA may have explored many blind alleys with indirect, even slow, pathways turning out to be more advantageous in the longer term than an parsimonious and apparently efficient highways. Of one aspect of the history, we can have no doubt: prokaryote evolution from those primitive beginnings to the complexity that enabled the earliest eukaryotes to form was a long and slow business, taking almost two billion years and perhaps trillions of generations.

THE PROKARYOTIC SEAS: EUBACTERIA AND ARCHAEBACTERIA

In the early seas, there were no predators or indeed any obvious brakes on evolutionary experimentation so that, in due course, a considerable range of prokaryotes evolved. Although all were originally grouped together as bacteria, it slowly became clear in the late 1970s that, on the basis of some profound differences, there were two very different types of prokaryotes, the Eubacteria and the Archaebacteria (Table 9.1).

Eubacteria

Eubacteria today are far more common than Archaebacteria and fall into two major groups, those with a single-layered bounding membrane (monoderm) and those with a double layer (diderm). A major diagnostic difference between them is that the cell wall of the former has an external peptidoglycan coat that can usually take up gram stain and become purple (and so are gram-positive) while that of the latter cannot (and are gram-negative). The evolutionary steps that led to the evolution of the major eubacterial phyla are still not well understood, one reason being that the extent of horizontal (or lateral) gene transfer that occurred as early prokaryotes exchanged pieces of genome remains unknown (Weiss et al., 2016). This phenomenon still takes place today as can be seen in *E. coli* conjugation and in the transfer across bacteria of episomes (small circular genomes); these code for proteins that, for example, facilitate antibiotic resistance (Blair et al., 2015), itself a model for evolutionary change.

As time progressed, however, a single-layer outer membrane (monoderm) became two (diderm) and this became the more common choice: monoderm eubacterial species are now found only within the Actinobacteria and subclades of the Firmicutes, which includes the homoacetogens, that may be round (cocci) or rod-like (bacilli). Fig. 9.3 groups today's major Eubacteria taxa, some of whose ancestors had an important role in the evolution of the early eukaryotic cells (see below).

A key evolutionary change within the Eubacteria, in terms of the future of the biosphere, led to the diderm cyanobacterium phylum (sometimes wrongly called blue-green algae–all algae are eukaryotic) that developed the machinery for photosynthesis. This ability to capture light quanta and to use their energy for the synthesis of organic molecules facilitated the production of polysaccharides that could be used structurally for cell walls and coats. On the basis of the detailed structure of later stromatolites in the Tombiana Formation in Western Australia,

TABLE 9.1 SOME PROPERTIES OF CELLS FROM THE THREE DOMAINS OF LIFE

Property	Eubacteria (E)	Archaebacteria (A)	Eukaryotes
Genome	Circular DNA	Circular DNA	Linear DNA
Nucleus	No	No	Yes (A)
Introns	No processing of type I introns	Processing of Introns in RNA-coding sequences	Processing of Introns (A)
Histones	No	Yes	Yes (A)
DNA replication & repair	Simple	Complex	Complex (A)
Transcription machinery	1 RNA polymerase (Pribnow box)	3 RNA polymerases, Complex machinery (TATA box)	3 RNA polymerases, Complex machinery (TATA box) (A)
Transcription initiation code	Formyl-methionine	None	None in nucleus. fMet in chloroplasts & mitochondria [E])
Ribosome	30S + 50S	30S + 50S	40S + 60S (sequences nearer to A)
Citric acid cycle	Elements	Yes	Yes (mitochondria - E)
Membrane	Gram +: single bilayer Gram –: double bilayer Mainly R-glycerol-ester lipids	Single bilayer comprising L-Glycerol-ether lipids	Single bilayer (E) Mainly R-glycerol-ester lipids (E)
Cell wall polysaccharide	Murein	Pseudomurein	A wide range (see Table 10.2)
Photosynthesis	Yes—cyanobacteria	No	Plants & algae—chloroplast (E)

(A) and (E) indicate an archae- or eubacterial origin for a eukaryote property.

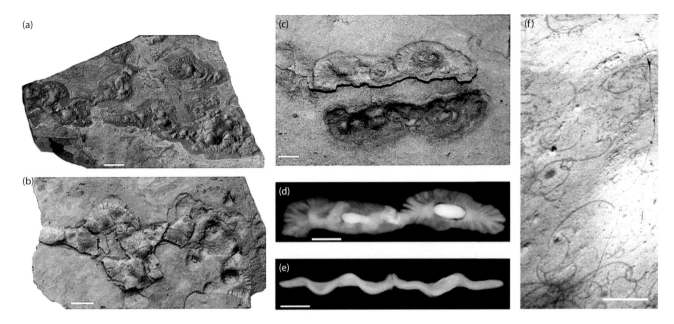

Figure 9.4 a-2: Fossils from ~2.1 Bya deposits in Gabon. (a,b): Two fossils in their matrix. (c): T fossil removed from its context to show its lower side (top) and its impression in the rock (bottom). (d): Micro-CT-based virtual reconstruction of the fossil shown in (c): (volume rendered in semi-transparency), showing radial fabric and two inner pyrite concretions. (e): Longitudinal virtual section running close to the estimated central part of the specimen. (Scale bars: 1.0 cm). (f): Fossils of *Gripania spiralis* (Michigan, 2.1 Bya; Scale bar 2 mm).

(a–e: El Albani A et al., (2010). *Nature* 466:100–104. With permission from Springer Nature.
f: Courtesy of Xvazquez. Published under a GNU Free Documentation License 1.2.)

Cyanobacteria are first thought to have evolved ~2.72 Bya (Flannery & Walter, 2012), although Sleep et al. (2011), on the basis of carbon signatures in early rocks, date the early stages of photosynthesis to almost a billion years earlier.

It was the production of oxygen by cyanobacterium-associated photosynthesis that eventually led to the complete change in the Earth's atmosphere known as the Great Oxygenation Event ~2.4 Bya. Although the availability of free oxygen in the atmosphere eventually enabled animals to evolve, its immediate effects were not only toxic for anaerobic bacteria but resulted in a period of cold (the Huronian glaciation) due to the drop in greenhouse gas concentrations (Kopp et al., 2005). Little of this history is documented in the fossil record, although there are a few examples of what seem to be fossilised bacterial masses from this period and a little later. These include a ~2.4 Bya fungus-like mass (Bengston et al., 2017a; Chapter 12) and the enigmatic and disc-like Gabon fossils from ~2.1 Bya (Fig. 9.4a–e). One example of these was about 12cm wide (El Albani et al., 2014; Aubineau et al., 2018), while another had a spirally coiled ribbon-like structure that resembles *Grypania spiralis*, a similar organism that lived ~1.4-1.3 Bya and was up to 50cm long and 0.2mm wide (Fig. 9.4f; Han & Runnegar, 1992).

Archaebacteria

Examples of this class of prokaryotes were originally found in areas, such as hot springs and salt lakes where little else could survive; they were, therefore, grouped as extremophiles and thought to be relatively uncommon. They are now known as the Archaebacteria and closer study has shown that they are not only present in most habitats but include their own distinctive set of phyla (Eme et al., 2017), albeit that the classification is still in flux. The fossil data on the Archaebacteria is, of course, very limited (Chapter 4) and our knowledge of their evolutionary history comes from computational phylogenetic analysis.

Much is now known about the morphology, structure, biochemistry and genetics of both the Eubacteria and Archaebacteria (Table 9.1), sufficient for it to be clear that they have a common root. They both have circular chromosomes, use the same genetic code, have similar basic transcription machinery and metabolic pathways, together with an outer membrane based on lipid bilayers. These similarities are sufficiently weighty to demonstrate that they both derived from the LUCA that possessed the core molecular apparatuses for reproduction, metabolism and genetic information storage. Molecular analysis shows, however, that there are major differences between the two groups to the extent that it is now accepted, following Woese et al. (1990), that there are three broad domains of life: the Eubacteria, the Archaebacteria, and the Eukaryota (Table 9.1, Fig. 9.3).

These differences are highlighted by their genetic and chemical compositions. The Archaebacteria have more complex genomes than Eubacteria: they possess introns and histones. They also have genetic and protein machinery for transcription and replication that are closer to eukaryotes and more sophisticated than that in Eubacteria. This, together with phylogenetic analysis, shows that the simpler Eubacteria are closer to the LUCA than are Archaebacteria, which clearly evolved later (Cavalier-Smith, 2006). Although their origins still remain obscure, phylogenetic analysis suggests that the first archaebacterium could have evolved from an ancestor of a firmicute, gram-positive eubacteria (Valas & Bourne, 2011).

Perhaps the most noticeable difference between the two is in the outer membranes. Monoderm Eubacteria have a single bilayer membrane mainly composed of R-glycerol-phospholipids held together with ester bonds covered by a coat or wall of murein peptidoglycan; diderm Eubacteria have a double-bilayer membrane with a thin intervening layer of murein. Archaebacterial membranes predominantly have a single bilayer of L-glycerol-isoprene linked together with ether bonds, which are more heat-stable than ester ones (perhaps helping to explain their extremophile behaviour) and are surrounded by a protein coat of pseudomurein. Particularly curious is the fact that Eubacteria use the R-isomer of glycerol for their bilayer whereas Archaebacteria use the L-isomer. These membrane differences are considerable and hard to interpret in evolutionary terms; it is, however, possible that there were ancient parent taxa, now lost, whose membranes were a mix of lipids, with these being more stable than once thought (Shimada & Yamagishi, 2011; Caforio et al., 2018).

FECA, THE FIRST EUKARYOTE COMMON ANCESTOR

Although we have little fossil evidence, phylogenetic analysis suggests that the descendants of the early Archaebacteria and Eubacteria diversified separately for several hundred million years. By then, there seems to have been a rich prokaryotic ecosystem, which included many taxa that were probably little different from those of marine prokaryotes today (Fig. 9.3) and whose population density was sufficient for encounters among them to become frequent. It was from such encounters that the FECA probably arose; this was a cell with a nucleus (for review, see Eme et al., 2017), and its evolution set the scene for the next great stage of life's history.

Many of the mechanistic details of the evolution of the FECA remain obscure, but one fact is clear: standard phylograms, often based on ribosomal gene sequences which give three kingdoms (Eubacteria, Archaebacteria, and Eukaryota) independently spurring off an original root organism (known as the Last Common Eukaryotic ancestor or LECA—see below), are both wrong and misleading. The molecular evidence, summarized in the last column of Table 9.1, together with a great deal of phylogenetic analysis, unequivocally point to the FECA having components that derived from both Eubacteria and Archaebacteria (Figs. 9.3 and 9.6). The evolution of the FECA and the later LECA were not examples of descent with modification from a single species.

It is now clear that the FECA formed through endosymbiosis: this is an extraordinarily rare process by which one cell engulfs another and, instead of the former breaking down the latter, the two come to develop a symbiotic relationship, with the resulting nuclear genome including, as a result of horizontal gene transfer, contributions from both. The idea that cell organelles were acquired through endosymbiosis has a long history: it was originally put forward by Mereschkowski very early in the last century and taken up and developed by Margolis in the 1960s (Katz, 2012), but its proof required comparative DNA-sequence and phylogenetic analysis. It was on the basis of such work that the information given in Table 9.1 was assembled: this shows the separate Eubacterial and Archaebacterial sources of many of the components of eukaryotic cells. This work also allows an evolutionary diagram to be constructed which shows the diversification of the prokaryotes and the origins of eukaryotes and the diversity (Fig. 9.6; Forterre, 2015, and below).

The traditional view has been that the FECA was, on the basis of its membrane structure, likely to have been a eubacterium that acquired its nucleus and most of its genome through endosymbiosis of an archaebacterium (in a genetic context, the FECA should, of course, be seen as an archaebacterium). Subsequent horizontal gene transfer between the two (Dunning-Hotopp, 2011) would have led to the archaebacterial genome, which already had much of the molecular machinery needed for it to become a primitive eukaryote, acquiring the eubacterial genes needed for eukaryotic membrane synthesis and other host requirements though lateral gene transfer (Williams et al., 2013). In this view, the acquisition of a bacterium to form its mitochondrion represented a later and independent endosymbiotic event.

More recently, an alternative possibility for the origin of the FECA has been suggested and this is that the original cell was a methanogenic archaebacterium with a need for hydrogen (for review, see Archibald, 2015). In this scenario, the early eukaryotic cell arose as a result of an archaebacterium engulfing a eubacterium that could satisfy this need. There was then metabolic symbiosis in the composite cell between the hydrogen-requiring archaeabacterium and a cell, such as α-proteobacterium, that could produce hydrogen and carbon dioxide, which became the mitochondrion. In this model, the formation of the eukaryotic nucleus structure arose after the endosymbiosis of a eubacterium. As in the other view, there was then extensive gene transfer to produce a composite genome and the resulting membrane structure reflected an evolutionary preference for eubacterial rather than archaebacterial components. This model does, however, say little about the origin of the internal membranes of the modern eukaryotic cell (see below).

The current view on the origin of eukaryotic cells is now more nuanced. It is that these two mechanisms remain unproven hypotheses, and there could well be alternative, more sophisticated histories involving both endosymbiosis and additional genetic contributions through lateral gene transfer. All we really know on the basis of computational phylogenetics is the nature of the original bacterial cell types from which the genes of modern eukaryotic cells derived. We can make educated guess as to the mechanisms by which the phenotypic and genotypic characters of eu- and archaebacterial cells became integrated to give the eukaryotic cell, but such hypotheses are as yet impossible to test.

The origin of the majority of the FECA's genome has, however, been clarified by the recent discovery of the Lokiarchaeota, a novel archaebacterium phylum from deep marine sediments in the Arctic Mid-Ocean Ridge (Spang et al., 2015). Phylogenetic analysis supports a common genetic ancestry of this organism and the eukaryotes, an argument reinforced by this archaebacterium carrying genes for actin proteins, vesicle trafficking, ubiquitin modification and small GTPase enzymes belonging to the Ras superfamily. All of these genes had previously been considered as being unique to eukaryotes. Here, the evidence supports the traditional view of FECA evolution as the Lokiarchaeota are not methanogenic.

Recently, Imachi et al. (2020) have managed to culture an Asgard archaebacterium (*Candidatrus Prometheoarchaeum syntrophicum*) that is related to the Lokiarchaeota and also found in deep marine sediments. This small (550nm), extremely slow-growing coccus-like cell has no clear internal structure but does have branching protrusions. Its gene signature makes it a good candidate for having been the descendant of an ancient archaebacterium that could have formed the nucleus of the FECA. Imachi et al., (2020) discuss the complex changes needed for such an archaebacterium to adapt to living within a eubacterium and mention that more work on this is in progress.

Although these discoveries highlight the archaebacterial provenance of the eukaryotic genome, they do not explain the origins of the eukaryotic double membranes surrounding the FECA's nucleus, the latter being very different from the single cell membrane of an archaebacterial cell. The obvious suggestion is that the FECA replaced the archaebacterial single-layered L-glycerol-isoprene-based outer membrane with a double-layered D-glycerol-phospholipid membrane. Williams et al. (2013) argue that this may have been simpler than was once thought: this is not only because many of the proteins for both synthetic pathways are common in the eubacteria and archaebacteria, but also because membranes containing both classes of lipid are stable (Shimada & Yamagishi, 2011; Caforio et al., 2018).

An alternative explanation for circumventing the membrane problem is that the original host eubacterial cell was an early member of the Planctomycetes-Verrucomicrobia-Chlamydiae (PVC) eubacterial superphylum; with one possibility being an ancestor of *Gemmata obscuriglobus*. PVC eubacteria today are unusual in having a double external membrane, the inner membrane of which extends into the

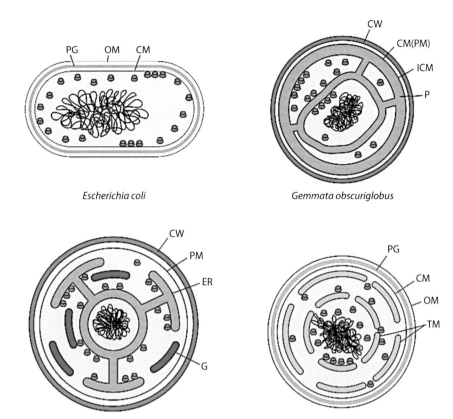

Escherichia coli

Gemmata obscuriglobus

Saccharomyces cerevisiae

Synechococcus

Figure 9.5 Drawings of four contemporary prokaryotic cells with important evolutionary features. *Escherichia coli* is a simple diderm prokaryote showing an absence of nucleus and a diderm outer membrane. *Gemmata obscuriglobus* is a Planctomycetes diderm prokaryote with internal double membranes derived in part from the inner cell membrane layer and a nuclear region enclosed by a double membrane. *Saccharomyces cerevisiae* is a simple yeast eukaryote with a nucleus and limited amounts of endoplasmic reticulum. *Synechococcus* is a diderm prokaryote with thylakoid internal membranes and a candidate endosymbiont for a chloroplast. (CM: cytoplasmic membrane; CW: cell wall; ER: endoplasmic reticulum; G: Golgi apparatus; ICM: intracyoplasmic membrane; OM: outer membrane; P: paryphoplasm; PM: plasma membrane; PG: peptidoglycan; TM: thylakoid membrane.)

(Courtesy of John Fuerst. From Fuerst JA & Sagulenko E (2012) *Front. Microbio.* 3:167. With permission from JA Fuerst.)

cytoplasm and folds to give internal, double-membrane-separated compartments (Sagulenko et al., 2014), one of which is a nucleoid containing the genome (Fig. 9.5). Not only is the double-layered morphology of PVC membranes related to the current eukaryote endomembrane components but so is its protein composition, a fact discovered from an informatics search of the databases (Santarella-Mellwig et al., 2010). All of this implies that an early PVC ancestor of *Gemmata obscuriglobus* is a good candidate for the original host cell, needing only relatively small changes to its cell membranes to achieve full eukaryote membrane morphology. If so, there would have been a ready-made compartment for an archaebacterial genome which would simply have lost its original outer membrane.

If this were the case, the FECA would have formed through endosymbiosis of an archaebacterial cell that donated only its genome, with the remainder of the cell being lost. This possibility has added strength because the Planctomycetes, unlike almost all other prokaryotes, but in common with eukaryotes, have membranes that endocytose external vesicles containing proteins and other constituents (Fuerst & Sagulenko, 2011). This scenario for the origin of the nucleus of the eukaryotic cell is attractive because it is parsimonious, but it should be mentioned that other means of deriving the FECA have been suggested and not yet excluded (e.g., Cavalier-Smith, 2010; Koumandou et al., 2013; Archibald, 2015).

LECA, THE LAST EUKARYOTE COMMON ANCESTOR

The early eukaryotic unicellular organism needed several further advances beyond acquiring a nucleus and an endoplasmic reticulum before it could be characterised as the LECA. These included at the least acquiring (i) a linear, diploid rather than a circular, haploid chromosome; (ii) a complex cytoskeleton that included basal bodies for organizing microtubules for cilial and flagellar movement together with a pair of centriole for organising chromosome movements during the anaphase stage of mitosis; (iii) the ability to synthesise sterols needed for eukaryotic membranes; (iv) a capacity for aerobic respiration; and perhaps (v) an ability to form walled-structures, known as cysts, in times of stress (Porter, 2020). The most obvious and perhaps the most significant of these acquisitions was that of a mitochondrion.

Mitochondria

These intracellular organelles with a double external membrane are present in all but a very few extant eukaryotes and their major role is to generate ATP for their parent cell through the citric acid cycle and the electron-transfer chain. Possession of a mitochondrion by a cell increases the number of ATP molecules available from a molecule of glucose by a factor of six or seven over an anaerobic cell (Weiss et al., 2016). In very basic terms, cells with mitochondria have much more available chemical energy and so have an evolutionary advantage over those lacking them.

Although, it might be thought that mitochondria were much the same in all descendants of the LECA, this is not so, and there has been considerable diversification (Gray, 2012; see Websites). The mitochondria of many invertebrates are anaerobic and less efficient, while those of some eukaryote protists completely lack an electron transport chain. There are even a few amitochondrial species today, although these tend to be parasites that get their energy requirements from their host. Examples include the Parabasalids (these are widely distributed, single-celled parasites within the Excavata phylum; Bui et al., 1996) and the recently discovered *Henneguya salminicola*, a small cnidarian salmon parasite (Yahalomi et al., 2020).

Perhaps more important is that phylogenetic analysis shows that the various classes of mitochondria share a common ancestor and, hence, the endosymbiotic events that led to the LECA only occurred once. Based on sequence homologies and phylogenetic analyses, the most likely eubacterial source for mitochondria was an early relative of the rickettsial group of the α-proteobacteria, all of which are diderms with a double external membrane. Perhaps the most obvious reason for this view is that all eukaryotic mitochondrial genes are homologues of those found in Rickettsia (for review, see Gray et al., 2001; Roger et al., 2017). Apart from being the major aerobic energy source of the cell, mitochondria have other biochemical roles and contain many hundreds of proteins; only a relatively small proportion of which are today coded for by their own DNA (Friedman & Nunnari, 2014). The human mitochondrial genome, for example, has about 16.5kb of DNA and codes for less than forty proteins.

It thus seems clear that the original eubacterium that became a mitochondrion kept much of the original diderm structure but only a small part of its genome; some of the rest was lost and the remainder transferred to the genome of its host cell via horizontal gene transfer. Very little is, however, known about the mechanisms by which this integration happened, or even how the various types of contemporary mitochondria evolved from that ancestor bacterium. It is, however, worth noting that the mitochondrial genome, unlike that of current parent genome, still maintains its original circular form, albeit very much reduced in length. The mitochondrion also maintained its original double membrane, the internal layer of which became the cristae.

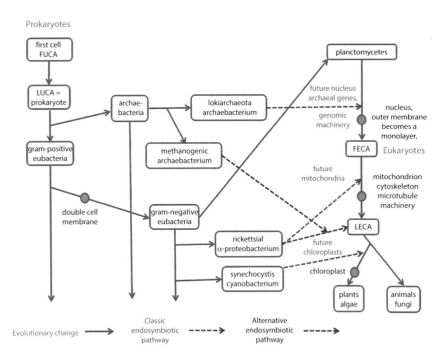

Figure 9.6 Two possible scenarios for the line of evolution from the FUCA to the LECA. These are based on the parent cell of the FECA being an ancestor of contemporary plactomyctes eubacteria with the nucleus coming from either a Lokiarchaeota or a methanogenic archaebacterium. The figure also includes, for completeness, the final endosymbiotic event of the bacterium that became the chloroplast and that gave rise to the last common ancestor of plants and algae.

Membranes and cytoskeleton

A major difference between eukaryotes and almost all prokaryotes is the presence of intracellular membranes, which include the endoplasmic reticulum, a nuclear membrane and the Golgi apparatus. If, however, the original host prokaryote was a member of the Planctomycetes which already had internal membranes, the first steps towards the evolution of an endomembranous system had already been taken. As to acquisition of the cytoskeleton and the microtubule-based mitotic apparatus, the origins of actin and tubulin were clearly eubacterial since proteins, such as MreB and FtsZ, their homologues in contemporary prokaryotes, play a role in organising the eubacterial cell wall (Busiek & Margolin, 2015).

An important cytoskeletal innovation in the LECA were the basal bodies and centrioles that help root flagella and cilia and that organise microtubules, particularly during mitosis (see also Chapter 10). These eukaryotic flagella bore no resemblance to those in eubacterial and archaebacteria prokaryotes: these are simple protein tubes that are not surrounded by cell membrane and that have a distinct morphological organization, molecular constituents and propulsive mechanism. Eukaryote flagella have an outer layer of cell membrane and include nine microtubule doublets surrounding a central microtubule doublet that link to the basal bodies. There is no obvious way that a eu- or archaebacterial flagellum could easily have evolved to become the eukaryote equivalent; Carvalho-Santos et al. (2011) have argued that they were apomorphies that evolved as the FECA developed into the LECA and have suggested possible ways in which this could have happened.

Once the FECA had acquired all these features, it had evolved into the LECA and was thus able to be the common ancestor of fungi and animals and many other phyla, even if some of these plesiomorphies were later lost in some clades. A further step was, however, needed before a descendant of the LECA was able to become the last common ancestor of plants and algae. This was the acquisition of a chloroplast (a plastid), the cell organelle that captures light energy and uses it to energise ATP from ADP and NADPH from NADP. This energy is then used to make sugars and other organic compounds. Chloroplast acquisition is discussed in Chapters 10 and 11.

TIMINGS

It has proven difficult to put times on the acquisition of the various features that mark the evolution of the LECA from the LUCA because the two dating approaches, molecular phylogenetics and palaeontology, do not fully agree. Computational phylogeny, which focuses solely on genomes, has given a range of timings, while the fossil record, which marks the latest stage when a species could have formed, is not always easy to interpret. Porter (2020) has reviewed the molecular clock data and the time windows are wide: the computed dates for the evolution of the LECA and mitochondrion are ~1.7 ± 0.4 Bya and ~1.6 ± 0.2 Bya, respectively. Porter argues that, such was the nongenomic complexity of the final LECA, it took much longer to evolve than would be expected and that the molecular clock data provides an unrealistically ancient time for the appearance of the LECA.

There is, however, recent fossil data that points to an earlier age for the LECA. Yin et al. (2020) have identified acritarchs and cyanobacteria in Chinese rocks that date to ~2 Bya, while. younger spiny, ornamented, organic-walled acritarch microfossils have been found that date from ~1.8 – ~1.4 Bya (e.g., Agić et al., 2017; Knoll, 2015). (*Note*: acritarchs are ill-defined microfossils <1mm in size that have the expected appearance of unicellular eukaryotes with a tough external membrane). Perhaps the key fossils, however, are those that have been found in India by Bengtson et al. (2017b) that date to ~1.6 Bya. On the basis of their morphology, these authors place them in the red algae clade. If these authors are correct, then a fully evolved eukaryotic cell with a nucleus, mitochondrion, plastid, sterol-containing membranes, and other such features must have been in place by then. This date meshes reasonably well with the molecular data (e.g., Parfrey et al., 2011).

Although, future discoveries may prove these figures wrong, it seems sensible for the moment to suggest that the FECA formed ~2 Bya, the LECA with the key properties of a normal eukaryotic cell had evolved by ~1.7 Bya and that a plastid had been endosymbiosed a little before ~1.6 Bya. It thus seems to have taken >2 By (Table 9.2) between the likely appearance of the FUCA, which marked the origin of life, and the FECA, which marked the first eukaryote cell. It then seems to have taken only a few

TABLE 9.2 APPROXIMATE DATES IN THE ORIGINS AND EVOLUTION OF EARLY LIFE

Date	Event	Evidence
~4.2 Bya	Earliest oceans	Zircon radioactive data
~3.7 Bya	First prokaryotes (FUCA)	Carbon isotope balance
~3.5 Bya	Stromatolite microbial mats	Radioactive dating
~2.7 Bya	Early evidence of photosynthesis	Oxide radioactive dating
~2.5 Bya	Increasing amounts of atmospheric oxygen	Oxide radioactive dating
~2.4 Bya	Great oxygenation event	Oxide concentration
~2.0 Bya	FECA	Phylogenetic + radioactive dating (fossils)
~1.7 Bya	LECA—complete eukaryotic cell	Radioactive dating (fossils)
~1.6 Bya	Acquisition of a chloroplast	Radioactive dating (fossils)

hundred million years, a relatively short time by the standards of early evolutionary events, between the initial endosymbiotic events that led to the FECA and the final acquisition of the cell properties that mark the LECA.

It is worth noting that neither the fossil nor the molecular clock data give precise insight into the number of endosymbiotic events between the LUCA and the LECA, and hence on how the mitochondrion was acquired. There does, however, seem to be no doubt that this event was relatively rapid and was soon followed by the independent acquisition of a plastid. Once these two basic cell types were in place, the scene was set for the full richness of eukaryotic diversity to evolve. It should also be pointed out that, if there were other cases of endosymbiosis before ~1.6 Mya, they died out.

Finally, it should be emphasized that, although the essential features of the progress from the FUCA to the LECA are understood, there is still much that is unknown about the first two billion years of life. We have, for example, little idea of how the Eubacteria and the Archaebacteria evolved along different pathways. Similarly, we know very little about the processes by which endosymbiosis occurred and how Archaebacteria and Eubacteria made their respective structural and genetic contributions to the FECA and then the LECA (Gribaldo et al., 2010). And there will always be more to discover the origins of the FUCA.

KEY POINTS

- The original living cell, known as the First Universal Common Ancestor (FUCA), formed almost 4 Bya in the early seas, and probably had an RNA-based genome.

- Its core organic molecules originally came from reactions in the early atmosphere and from meteors which included molecules, such as polycyclic aromatic hydrocarbons.

- The LECA is likely to have formed in the vicinity of deep thermal vents where thermal energy and appropriate catalytic rock was available to facilitate chemical reactions. It slowly evolved to become the LUCA of all subsequent life.

- The LUCA was a very primitive bacterium with a DNA-based genome that probably used today's genetic code, had a single bilayer membrane and possessed sufficient molecular machinery for reproduction and maintenance.

- The stromatolite fossil record shows that mats of prokaryotes were present ~3.5 Bya.

- Over the next ~1 By, prokaryotes diverged widely to give the various groups of monoderm and diderm Eubacteria and Archaebacteria. The latter had membranes and molecular-genetic machinery very different from those of the former. Its genetic machinery was, however, much closer to that seen in eukaryotes.

- Sometime before ~2.4 Bya, cyanobacteria evolved that could use light energy to drive photosynthesis, so turning carbon dioxide into sugars and releasing oxygen.

- By around 2.4 Bya, the resulting oxygen had changed the balance of gases in the atmosphere so that heat-trapping carbon-dioxide levels dropped. This led to the *Great Oxygen Event* that resulted in the Earth becoming very cold (the Huronian glaciation).

- The First Eukaryotic Common Ancestor (FECA) probably acquired a nucleus by endosymbiosis ~2 Bya.

- One possibility is that a eubacterial cell engulfed an archaebacterial cell which became its nucleus, incorporating part of the eubacterial genome into its own through horizontal gene transfer. This was followed by endosymbiosis of an α-proteobacterium cell that carried the molecular apparatus for making ATP. It became a mitochondrion.

- An alternate possibility is that the LECA arose when a methanogenic archaebacterium and a eubacterium entered into a symbiotic relationship eventually forming a cell with a nucleus and a mitochondrion, and subsequently reorganizing its membranes.

- Either way, the resulting cell was the Last Eukaryotic Common Ancestor (LECA) of all subsequent eukaryotic life; it had probably evolved by ~1.7 Bya.

- A later endosymbiotic event was the engulfing of a cell that could use light energy to synthesise sugars (~1.6 Bya). It became a chloroplast, and the resulting cell was the Last Common Ancestor of algae and plants.

FURTHER READINGS

Archibald JM (2015) Endosymbiosis and eukaryotic cell evolution. *Curr Bio*; 25:R911–R921. doi: 9.1016/j.cub.2015.07.055.

Eme L, Spang A, Lombard J, Stairs CW, Ettema TJG (2018) Archaea and the origin of eukaryotes. *Nat Rev Microbiol*; 16:120. doi: 10.1038/nrmicro.2017.154.

Porter SM (2020). Insights into eukaryogenesis from the fossil record. *Interface Focus*; 10(4):20190105. doi:10.1098/rsfs.2019.0105

Pross A, Pascal R (2013) The origin of life: What we know, what we can know and what we will never know. *Open Biol*; 3:120190. doi: 9.1098/rsob.120190.

WEBSITES

- Wikipedia entries on *Archaea, Abiogenesis, Endosymbiont, Flagella, LECA, Mitochondrion*.

- Origin of mitochondria: www.nature.com/scitable/topicpage/the-origin-of-mitochondria-14232356/

The Roots of the Eukaryotic Tree of Life

10

As discussed in the last chapter, the likely date for the formation of the Last Eukaryotic Common Ancestor (LECA) was ~1.7 Bya. This cell probably had a nucleus and an ATP-producing mitochondrion together with a cytoskeleton, intracellular membranes, two flagella, and other innovations. It became the last common ancestor of a wide range of clades that, once it had acquired a plastid, included the Archaeplastida clade that initially contained glaucophytes and in due course red algae, green algae, and plants. Later, through the endosymbiosis of red and green algal protists by unrelated protists, a wide diversity of secondary algae formed (Chapter 11). It is, however, clear that the LECA also gave rise to a second set of organisms. The first key step for this was the loss of the second flagellum and this ancestral cell particularly gave rise to Amoebozoa protists and the Amorphea (initially choanoflagellates and, in due course, the multicellular animals and fungi). Once this had happened, the tree of life had acquired all of its key arborisations.

This chapter considers the details of the evolution of this first protist, events that resulted in the diversity of contemporary eukaryotic phyla (e.g. Fig. 10.1). An early important event occurred when an early protist endosymbiosed a cyanobacterium that became its chloroplast and the last common ancestor of the Archaeplastida. Later changes to protists included clade-specific internal organelles and localisations of membrane functions that included patches of light sensitivity and a wide range of specialisations for engulfing food (for a dramatic example, see Fig. 10.2). Many groups developed extracellular coats and walls, some of which became mineralised. The chapter then considers many of the major protist phyla, and how their social behaviour led to the contemporary multicellular groups of phyla. It ends by reviewing the evolution of fungi, which include both unicellular and multicellular taxa. These are characterized by a cell membrane that includes chitin and glucans and, for multicellular organisms, the possession of hyphae.

Readers should, however, be aware that both the terminology and analysis of these key stages in evolution is complex and subject to change in the light of new data and new perspectives, as are the classifications themselves. The currently accepted taxonomy of early protist evolution is summarized by Burki et al. (2020) and detailed by de Queiroz and Cantino (2020).

We know very little about the early organisms that evolved during the long intervening period of the billion or so years between the evolution of the LECA (~1.7 Bya) and the beginning of the Ediacaran (635 Mya) when large-scale organisms first appeared. This is because such small, soft organisms rarely leave traces in the fossil record. A few enigmatic and primitive fossils that date to before 1.8 Bya have been found in Canada and China (Chapter 9), but their taxonomic status remains unclear (Knoll et al., 2006). More convincing are fossils from India dating to about 1.65 Bya that include multicellular groups embedded in phosphate, some of which resemble today's filamentous and coccoidal cyanobacteria, while others are characteristic of filamentous eukaryotic algae (e.g. Bengtson et al., 2009, 2017b).

DOI: 10.1201/9780429346217-13

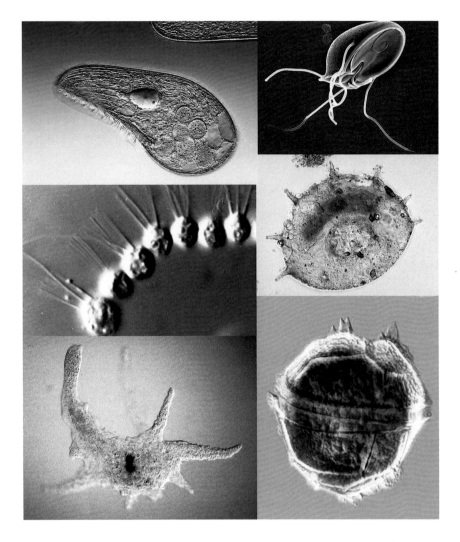

Figure 10.1 Collage of protozoans. Clockwise from top left: *Blepharisma japonicum*, a ciliate (Alveolata); *Giardia muris*, a parasitic flagellate from the Excavata with eight flagella; *Centropyxis aculeata*, a testate (shelled) amoeba; *Peridinium willei*, a dinoflagellate; *Chaos carolinense*, a naked amoebozoan; *Desmerella moniliformis*, a choanoflagellate.

(Courtesy of Frank Fox, Sergey Karpov, Dr. Stan Erlandsen, Thierry Arnet, and Dr. Tsuki Yuuji. Published under a CC Attribution-Share Alike 4.0 International License.)

By 1.2–1.0 Bya, small multicellular cyst-like organisms can be seen in phosphatic nodules that reflect a freshwater environment (Strother et al., 2011). These and other subsequent fossils were probably LECA descendants (the possibility that early examples were late descendants of stem-LECA protists cannot be excluded) and illustrate the early divergence of the major eukaryotic clades (Knoll et al., 2006). The

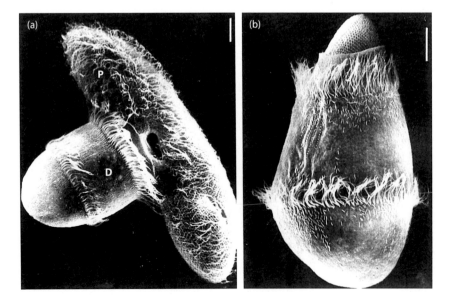

Figure 10.2 **Example of modern eukaryovory** (one eukaryote protist eating another): scanning electron microscopy images of the alveolate ciliate *Didinium nasutum* (D) predating on a *Paramecium* (P). (a): The ciliate attaches to its prey. (b): The ciliate has engulfed almost all of it. (Bar: 10 μm.)

(From Wessenberg H, Antipa G (1970). J. Eukaryote. *Microbiol*; 17:250–270. With permission from the Journal of Eukaryotic Microbiology.)

fossil record for multicellular organisms does however only start to become robust ~1 Bya with the identification of simple red and green algae (Fig. 11.1; Butterfield, 2000; Tang et al., 2020; Chapter 11).

The earliest unicellular eukaryotes (protists) diverged widely to give groups of phyla, the great majority of which now include simple unicellular and colonial organisms (Table 10.1; Fig. 10.1). Subtaxa within a few early phyla evolved to become social organisms and a very few to become stable multicellular organisms (Table 10.2). It is worth noting that, although all animal and plant taxa today are multicellular, the fungi and algae also include unicellular taxa. It should also be emphasized that, while the first three of these clades are monophyletic, the algae are polyphyletic with examples across many bikont phyla (Chapter 11). This is because those not in the Archaeplastida acquired their chloroplast by secondary endosymbiosis.

Although most early eukaryotes lived in aquatic habitats, the fossil record suggest that eukaryotic organisms were to be found on land by ~1 Bya; these include algal mats and small multicellular cyst-like structures with strengthened cell membranes (Strother et al., 2011; Loron et al., 2019). Soon after this, trace fossils of small burrows were associated with these mats and these have been interpreted as being made by "wormlike undermat miners" (Seilacher et al., 1998), albeit that this would place

TABLE 10.1 SOME MAJOR CLADES OF CONTEMPORARY SINGLE-CELL AND COLONIAL PROTISTS

Group (age range Bya*)	Synapomorphy	Representative taxa SC: single-cell protist
Unikont *(single flagellum)* (1.36 ± 0.12)		
Amoebazoa (1.5 ± 0.12)	Tube-like pseudopods	SC: *Chaos carolinensis* Colonial: *Dicyostelium discoideum* (cellular slime mould)
Choanoflagellate	Single flagellum located posteriorly	SC: *Salpingoeca rosetta* Colonial: Proterospongia
Bikont *(two or more flagella)*		
Glaucophyte (Archaeplastida)	Chloroplasts, bikonts	SC algae: *Glaucocystis* Colonial: *Volvox*
Haptista	Haptoneme flagellum Mitochondria with tubular cristae	SC: Coccoliphores Colonial: the Centroheliozoa
Excavata (1.65 ± 0.95)	"Excavated" feeding groove Modified mitochondria	SC: Jakobea Chloroplast: Euglenazoa Amitochondrial: Fornicata Colonial: Acrasidae slime moulds
Cryptistsa	Ejectisome for movement secondary red algae	SC: Crytomonas
CRuMS	A group of orphan unicellular taxa with a range of morphologies	
SAR (1.47 ± 0.16)		
Stramenopila	Thin pseudopods *Heterokonts* Some: secondary chloroplasts	SC: Bicosoecida (no chloroplast) SC: Eustigmatophyceae (chloroplast) Colonial: (without chloroplast) Colonial: diatoms (chloroplasts)
Alveolata	Cortical alveoli (spaces)	SC: ciliates Colonial: Dinoflagellates (chloroplasts)
Rhizaria (1.23 ± 0.72)	Needle-like pseudopods; polyubiquitin insertion	S-C: Foramenifera (shelled) Colonial: radiolarians

* Age from computational analysis. The age of the oldest fossil is much is typically ~40% less. (From Porter, 2020.)

TABLE 10.2 SOME DISTINGUISHING FEATURES OF THE MAJOR MULTICELLULAR CLADES

Clade	Flagella	Centriole	Major cell wall polysaccharides *This list is not comprehensive*
Animals	Single posterior flagellum (opisthakont)	Yes	None (but chitin is the main constituent of invertebrate exoskeletons & fish scales)
Fungi	1 Single posterior flagellum (opisthakont)	Yes	Chitin, β-glucan
Plants	2 Similar posterior flagella (bikonts)	Yes	Cellulose, pectin
Green algae	Bikonts Simple anatomy	Yes	Cellulose + a range of other polysaccharides
Red algae	No flagella	Centroplast	Cellulose, agarose, carrageenan,
Brown algae (stramenopiles)	2 Different posterior flagella (heterokonts)	Yes	Cellulose, algin

the evolution of triploblast bilaterial worms much earlier than other evidence (Chapter 12). We know very little about the lives of these early marine organisms that probably survived on debris, prokaryotes, and one another (Fig. 10.2; Cohen & Riedman, 2018). They existed during the billion or so years between the evolution of the LECA and the start of the Ediacaran Period (635 Mya) slowly evolving and diversifying but leaving little trace. What follows is an analysis of what is known of the properties and systematics of these early eukaryotic phyla.

THE DIVERSIFICATION OF THE LECA

Flagella, centrioles, and basal bodies

One important innovation in the LECA was the presence of flagella that were very different from such organelles in bacteria. Eubacteria have two modes of movement: for swimming, they have a prokaryote flagellum made of flagellin protein fibres that are rotated by an ion gradient. For crawling on surfaces (a process known as twitching!), they use T4 pili, which are based on cell-surface-associated filaments (Daum & Gold, 2018). Archaebacteria use an archaellum for swimming; this type of flagellum includes T4 filaments that are rotated by an ATP-driven motor whose molecular basis is still unclear. Although eu- and archaebacterial flagella look similar they have sufficient differences to make the evolutionary relationships between them opaque.

All eukaryotic flagella, in contrast, are surrounded by an extension of the cell membrane (whereas the filaments of bacterial flagella and archaella are extracellular) and include nine microtubule doublets surrounding a central microtubule doublet and their beating (they do not rotate) is powered by ATP-energised motors (dyneins). They are thus very different from the flagella found in the two prokaryote domains (see Websites). Little is known about the evolution of eukaryote flagella as there is no obvious way that a eubacterial or an archaebacterial flagellum could easily have evolved to become a eukaryote one – in fact, the three "flagellar" types are almost universally acknowledged to be nonhomologous. It is instead thought that eukaryotic flagella were apomorphies that evolved on the eukaryote lineage between FECA and LECA; Carvalho-Santos et al. (2011) have suggested ways in which this could have happened.

These motor-driven whip-like flagella, originally used for cell propulsion in water, are rooted to the rest of the cell cytoskeleton through basal bodies. This latter structure is homologous to the centrioles that help organise microtubules elsewhere in the cell and remains fairly uniform across most eukaryotic taxa. Today, flagella numbers and morphological details are important markers of the lines of evolutionary descent from the LECA.

The key divide in eukaryotes is whether there are one or two flagella. Those taxa with two flagella are the majority and are known as bikonts. Those with a single flagellum are called unikonts, with the subclass of multicellular taxa (the animals and fungi) being classed as opisthokonts (having a rear-facing flagellum). In bikonts, the morphology of the two flagella may be the same (isokonts) or different (heterokonts

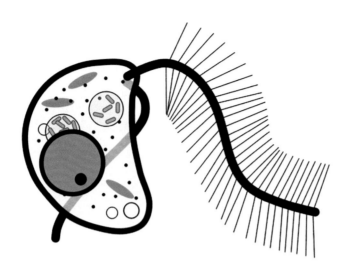

Figure 10.3 Diagram of *Cafeteria roenbergensis*, a heterokont with one normal flagellum and one long flagellum with two rows of Mastigonemes (lateral bristles). As this flagellum moves, it wafts food particles towards the cell.

(Courtesy of Denis Bartel. Published under a CC Attribution-Share Alike 3.0 Unported License.)

or anisokonts). In heterokonts such as the Stramenopila, one of the flagella bears a particular form of hair-like filaments that dramatically alters the hydrodynamic effect of its beating so that it can waft food particles towards the cell (Fig. 10.3). Tables 10.1 and 10.2 list the numbers in the major clades (for further details, see Websites). A few phyla however do include species with more than two flagella: many metamonads have four and some protists, such as the Hypermastigotes, may have hundreds.

In the wide range of clades with two flagella, the presence of flagella is maintained in the multicellular organisms that evolved from protists but usually restricted to motile cells or those with a free surface where they are often seen as groups of cilia that may beat synchronously. The ancestral flagellum form is particularly maintained in cells that require a swimming capability. These include unikonts, such as animal spermatozoa and sponge choanocytes, and bikonts such as the gametophytes (spermatazoa) of green algae and in some fungi and plants (e.g. bryophytes and ferns). The flagellum morphology of sperm thus enables their phylogenetic position to be assigned (Carvalho-Santos et al., 2011). It is worth noting that some groups of eukaryotes completely lack flagella, basal bodies and centrioles; examples include the centrohelid heliozoans and the red algae, whose sperm are moved by local water currents. While both have microtubules, the roles of centrioles are, respectively, taken by the very much simpler centroblasts and polar rings (Dave & Godward, 1982; Cavalier-Smith & Chao, 2003). The evolution of the red algae is considered in Chapter 11.

Cell walls

Another important feature that characterised the various major groups of early multicellular organisms, other than animals and some Amoebozoa, was the inclusion of polysaccharides in their cell walls (Table 10.2). Today, monoderm Eubacteria have an outer coat of murein (peptidoglycan), diderm Eubacteria have a thin murein layer between their two membranes, while some Archaebacteria have an outer coat of pseudomurein. Most eukaryotes have polysaccharides in their cell membrane that differ from those in prokaryotes and are widely distributed within the lipid bilayer (Table 10.2). Cellulose is particularly associated with the cell walls of plants and many algae, while chitin and glucans are in those of fungi. Cellulose is, however, also present in some eubacteria (Ross et al., 1991) and in a few single-celled Amoebozoa as well as being a component of the fruiting-body stalk of the cellular slime moulds, organisms that lie within the Amoebozoa (O'Day, 1979). Genes used for synthesising eukaryotic cell wall polysaccharides, such as cellulose and chitin, have also been identified in Eubacteria (Ross et al., 1991; Steinfeld, 2019; Romeo, 2008) and are thus likely to have been acquired by the LECA through lateral gene transfer.

The chitin and glucan constituents of fungal cell walls are also found elsewhere. Chitin is also the major constituent of invertebrate exoskeletons and fish scales but, until recently, it had seemed to have been dropped from the normal post-fish repertoire of vertebrates. This is because mammals use β-keratin as the key structural proteins for hair, horns, hooves, etc., while reptiles and birds use α-keratin as the structural protein of beaks, shells, nails, scales, shells, and feathers. Chitin has however recently been found in amphibians (Tang et al., 2015; Stern, 2017), while its deposition is an unexpected marker for several human neurodegenerative diseases (Lomiguen et al., 2018). α-glucan is not only found in the fungal cell wall but is also present in some plants (Fesel & Zuccaro, 2016) as are hemicelluloses, polysaccharides related to the glucans. The presence of chitin in the cell membrane is not however restricted to the Amorphea as it has also been found in the fibrils of diatoms, a bikont clade.

It therefore seems that these polysaccharide classes are not restricted to a single phylum, or even a group of phyla. Indeed, many microbial eukaryotes that apparently lack polysaccharides can enter a cyst state of dormancy in which they make polysaccharide-rich walls. There are, however, a few taxa for which cell-wall polysaccharides do seem to be clade-specific. These include the pectins, which are only found in plants, carrageenan and agar, which seem only to be found in red algae, and algin, which is only found in brown algae (see Chapter 11). Taken as a whole, cell-wall polysaccharides are both important and interesting but seem not to be of great taxonomic significance as their presence in a clade is rarely specific and so does not represent a synapomorphy.

Sexual reproduction and diploidy

Two important questions not as yet discussed are whether the LECA was haploid or diploid and whether it reproduced sexually. As all sexual eukaryotes are, for at least part of their life cycle, diploid, there can be little doubt now that the LECA not only had a diploid stage in its life cycle but showed sexual dimorphism (for review, see Goodenough & Heitman, 2014), a view supported by the fact that the essential features of sex chromosomes in both plants and animals are similar, although the details differ (Charlesworth, 2002; Pannell, 2017). This is not surprising as even a prokaryotic diderm like *E. coli* can become partially diploid and show a degree of reproductive dimorphism through its F sex factor, while some Archaebacteria, such as *Haloferax volcanii,* spend part of their life cycle in a post-mating polyploid state (Markov & Kaznacheev, 2016).

The simplest form of reproductive cycle in eukaryotes occurs in those protists that can undergo meiosis to produce haploid spores that meet, fuse and restore diploidy. The best-known examples of such organisms are the fungal microsporidia and the Apicomplexa which are parasitic alveolates. Indeed, all eukaryotic taxa that have been investigated have at least some of the genes necessary for meiosis, even if they reproduce asexually most of the time (Goodenough & Heitman; 2014; Speijer, 2016). There thus seems little doubt that the LECA could reproduce sexually, although this was probably not obligatory.

The drivers for the origin of sexual reproduction still remain unclear. One suggestion is that meiosis evolved as a defense against the production of reactive oxygen species (ROS) after a mitochondrion had been endosymbiosed. It could have acquired the ability on the basis of a set of genes already present in the genome of the archaebacterium that became the FECA's nucleus (Speijer, 2016). Another possibility is that meiosis preceded mitosis. This could have happened if the LECA nucleus continued to divide within its host, so creating a syncytium that would eventually split into separate cells (Garg & Martin, 2016; see Scenedesmus below).

Both suggestions imply that sexual reproduction was not something that happened easily and there are good reasons why this might be so. Sexual reproduction is intrinsically inefficient for several reasons. First, an ability to produce gametes that could mate required the evolution of a set of novel molecular features. Second, there is a cost in both energy and time in making gametes, so further slowing population growth. Third, gametes of the opposite sex (or mating type) have to find one another if a next generation is to be produced, and, in the case of single-celled and colonial organisms in a marine environment, this can have low probability and many gametes will never mate.

There is however a balancing set of reasons that make sexual reproduction an advantageous way of life. In the absence of sexual reproduction, deleterious mutations and any other form of genetic damage can mount within a population to the extent that it dies out (Michod et al., 2008). This phenomenon, known as Müller's ratchet, because it is an irreversible one-way process, is avoided by sexual reproduction which provides a means of losing such mutations. This effect can also be diminished by polyploidy, and this may have been one of the original drivers for diploidy (Maciver, 2016). Equally important is the increase in genetic variability provided by sexual reproduction as this gives genetic flexibility to a cell, so enabling it to survive a wider range of challenges than an asexual organism.

There are many patterns of reproductive activity across the eukaryotic world, some of which are unusual: reproduction in fungi, for example, is pseudo-sexual (Nieuwenhuis & James, 2016), while many plants can reproduce asexually as well as sexually. It is thus clear that the arguments do not come down firmly in favour of one or other mode of procreation. It is however obvious that all large-scale organisms, irrespective of origin, are capable of sexual reproduction and this is almost always the favoured way of producing the next generation.

Plastids

The most important cell organelles that have not so far been mentioned in any detail are plastids. These are the class of intracellular organelles particularly responsible for photosynthesis and starch production (they include chloroplasts but there are several other types). Plastids are found in the Archaeplastida, which include plants, red and green algae and the unicellular, microscopic algae known as glaucophytes, but also, in one or another derivative forms, in many other algal organisms (see Chapter 11 for further details). Much is now known about their origin: the phylogenetic evidence, based on analysing chloroplast genes from a diverse set of organisms (McFadden, 2001), shows that the acquisition of a chloroplast was through the endosymbiosis of a photosynthesising cyanobacterium by an early eukaryote that, as a consequence, became a so-called primary alga. This must have happened before ~1.65 Bya, the likely date of the earliest known red algal microfossils (Chapters 9 and 11; Bengtson et al., 2017b).

The reasons for accepting that a cyanobacterium was the original source of chloroplasts are partly because their thylakoid bimembrane architecture and molecular structure are very similar to that seen in contemporary chloroplasts (Petroutsos et al., 2014) and partly because the core genes on their circular genome show a close phylogenetic relationship to them. It is worth noting that the original genomic content of that original cyanobacterium has been drastically reduced with some genes being transferred to the host genome and some being lost; today, only ~250 genes are present in the largest plastid genomes (those of red algae) and most have ~120 (Jensen & Leister, 2014).

To determine the likely origin of the original plastid, Ponce-Toledo et al. (2017) conducted a very detailed phylogenomic analysis of plastid-encoded proteins and nucleus-encoded proteins likely to have been of plastid origin. This showed that the contemporary cyanobacterium with the most closely related gene set to that in the Archaeplastida is *Gloeomargarita lithophora*, which is normally found in freshwater. They point out that the likelihood of such an origin is reinforced by the fact that contemporary glaucophytes, descendants of the earliest archaeplastid taxon, are exclusively found in freshwater ecosystems.

Most algae that do not lie within the Archaeplastida obtained their plastid through secondary endosymbiosis of an early red or green alga. There is, however, one protist whose plastid chloroplast lies outside the main lines of algal evolution. This is *Paulinella longichromatophora*, a Rhizarian taxon with a chromatophore (a chloroplast equivalent), the only member of its clade with such an organelle. This stand-alone oddity contains some three times as many genes as other chloroplasts. Genetic analysis has now shown that this chromataphore was directly acquired through the endosymbiosis of a cyanobacterium probably related to a contemporary *Synechococcus* cyanobacterium, an event that probably occurred 100–60 Mya (Singer et al., 2017; McFadden et al., 1994; Nowack et al., 2008). This is the only known example of a relatively recent endosymbiotic event that was successful.

PROTIST SYSTEMATICS

Because the fossil record of early protists is so weak, any analysis of early protist evolution has had to depend on morphological and molecular-phylogenetic analysis of the morphology and gene sequences of contemporary taxa to identify the order in which synapomorphies of major organism families first appeared and on comparative genomics and phylogenetics to tease out gene and organism histories.

Until recently, the standard way of partitioning the many clades of protists had been on the basis of cladistic analysis using morphological synapomorphies. Over the past 20 or more years, various biologists have grappled with how to identify those key synapomorphies that would allow the major phyla to be partitioned into natural groups (e.g. Cavalier-Smith, 2014, 2018). As the first part of this chapter has however shown, this is not easy to do. The core reason is that so many properties of the descendants of those early protists have become modified over more than a billion years since the LECA evolved: clades have undergone gene mutation, loss, and duplication, and have acquired new genes through lateral transfer.

The net result has been that as more and more unicellular organisms have been discovered, it has become harder and harder to identify synapomorphies that are unique to particular groups. It is now generally agreed that trying to do this is a fruitless task. The best that can be done is to group organisms in high-level supergroups on the basis of such features as the number and morphology of flagella and their basal bodies, chloroplast type, and the polysaccharides in their cell walls, if present. Showing that these features are clade-specific is however hard as it also requires demonstrating their absence in all other clades and this has proved to be impossible.

The alternative solution that has now been almost universally adopted is to base the taxonomy on phylogenomics (phylogenetic analysis based on whole genomes), or at least a large number of protein-coding sequences (Patané et al., 2018). This is the approach that will be used in this chapter (see also Chapter 7 and Appendix 5). The only limitations here are, first, that the complete set of genomes is not available because many taxa within the more diverse clades are rarely encountered and, second, such analyses do not handle lateral gene transfer easily. Nevertheless, the problems are being overcome and one current phylogram is shown in Fig. 10.4 (there are others that are slight variants on this). It divides the eukaryotic world into seven or eight major clades, for most of which there is a well-established, internal clade structure.

Implicit in Fig. 10.4 are some of the difficulties encountered in establishing this phylogeny. First, some major groups are marked with asterisks: these used to be considered supergroups, mainly on the basis of cladistic analysis, but have been demoted in the light of sequence analysis. Second, there are still some areas of the tree for which the data is still inadequate, particularly the Excavata. This clade is not well established because the data is not yet good enough to establish a robust phylogeny and its internal taxa are connected by multifurcations (unbranched lines). Third, a general problem with sequence-based phylogenies is that they are unrooted (Chapter 7 and Appendix 5) and hence give no direct insight into the morphology of the LECA.

This problem has been addressed in two ways: by comparing the gene sequences of current protists and those bacteria likely whose ancestors may have contributed to the LECA and by examining the details of the morphology of current protists. Phylogenetic analysis suggests that the group that is closest to the LECA is likely to be the Discoba, a taxon within the Excavata with two or more flagella (Sibbald & Archibald, 2017; see also Burki et al., 2020). As to protist morphology, the one feature that is of help here is the presence of basal bodies that facilitate flagella construction. Detailed examination across the many protist phyla (Roger & Simpson, 2009; Yubuki

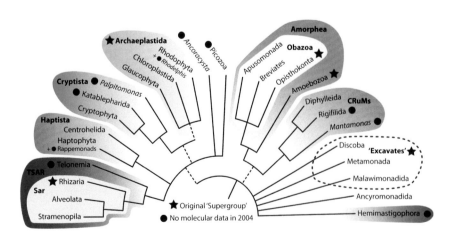

Figure 10.4 The tree of life based on phylogenomics. The coloured groupings correspond to the current "supergroups"; for details, see text.

(Courtesy of Burki et al. (2020) The New Tree of Eukaryotes. *Trends Ecol Evol*; 35:43–55. Published under an International 4.0 Creative Commons License (CC BY 4.00).)

& Leander, 2013) makes it clear that most bikonts together with many unikonts have two basal bodies, and this indicates that the LECA probably had two flagella. Taken together, the observations suggest that the LECA was a bikont likely to be related to the contemporary Discoba taxon.

It should however be emphasised that establishing the basal eukaryotic tree and its likely root organism is an important and active field of research. As a result, the current view may be superseded. Readers should, therefore, check the current literature if they wish to be up to date.

THE ORIGINS OF SOCIAL BEHAVIOUR

The other key set of changes that occurred in the mainly silent period that ended with the Ediacaran were those that, through intragroup signalling and membrane interactions, initiated social behaviour within groups of protists. The data in Table 10.1 clearly show that this happened in many groups of protists, although only a few eventually attained stable multicellularity and differentiation.

The origins of some of this behaviour probably derived from the social behaviour of prokaryotes. Today, many prokaryotes live in colonies or biofilms and there is no reason to suppose that this is a recent innovation (Tolker-Nielsen, 2015). Genome and transcriptome analysis shows that there are proteins present in bacteria that facilitate social behaviour (for reviews, see Mikhailov et al., 2009; Tikhonenkov et al. 2020). Examples include those for intercellular signalling (Visick & Fuqua, 2005), those that facilitate resistance to antibiotics made by other prokaryotes (Peterson & Kau, 2018), adhesion proteins (King et al., 2004) and proteins, such as the *E. coli* F (fertility) factor, possession of which allows a cell to make a tube (pilum) to another cell through which DNA segments can pass (e.g. episomes that carry genes for antibiotic resistance).

Once the adhesion and signaling proteins that facilitated social behaviour had evolved in prokaryotes, they were available for use the earliest single-celled eukaryotes. Indeed, it is quite possible that the LECA itself was a colonial protist. As the data in Table 10.1 demonstrate, colonies and multicellular groups of protists are still a stable form of life in many of the major clades today. Examples include the dictyostelial and plasmodial slime moulds in the Amoebozoa, many choanoflagellates in the Amorphea, chlorophytes and charophytes in the green algae, oomycetes in the Stramenopila, and euglenids in the Discoba. There thus seems little reason to doubt that protist colonies were present across the early phyla. Once such colonies had become stable to the extent that their constituent cells no longer needed to maintain an independent and distinct existence, other than perhaps as spores, multicellular organisms had evolved.

The acquisition of multicellularity

The evolution of large-scale multicellular organisms with distinct cell types seems only to have occurred in the plant, fungal, algal, and particularly animal phyla (for review, see Sebé-Pedrós et al., 2017). The ways in which this happened are not known, but the main focus has been on the Metazoa (the animal phyla). At the end of the 19th century, Ernst Haeckel suggested that the first step towards multicellularity in animals occurred when a group of identical cells came together to form a stable ball (the blastula), and this was followed by a second stage (the gastrea) that included a primitive gut, with cell differentiation then following. Investigating the question of whether Haekel's view (that morphogenesis evolved before differentiation) is correct has provided a framework for considering the evolution of early Metazoa.

Ideally, of course, there would be fossil evidence of these very early life forms, but there is very little of this and those few fossils that we have are hard to interpret, with rare exceptions such as the filamentous red alga *Bangiomorpha pubescens* (Chapter 11). The discussion has therefore focused on small contemporary organisms that show multicellularity, mainly from the Algae and the Metazoa. Today, among the simplest of these colonial protists may be *Scenedesmus*, a genus of colonial, freshwater alga that exists as a linear array of 4 or 8 cells with differentiated end cells (Zhu et al., 2016b). Reproduction involves cells becoming multinucleate and then dividing to form a daughter colony with the same number of cells as the parent (Pickett-Heaps & Staehelin, 1975; Fig. 10.5a). Equally simple,

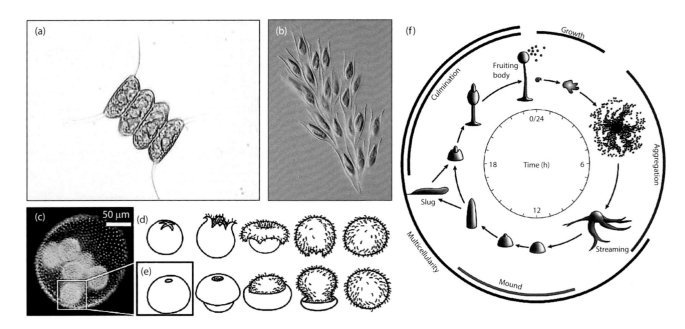

Figure 10.5 Social protists. (a): *Scenedesmus*, a colonial chlorophyte. (b): *Dinobryon*, a heterokont golden colonial alga (a straminopile). (c): A *Volvox* multicolony; in each spherical colony, the flagella are inward facing. (d and e): Two modes of *Volvox* inversion that expose the flagella to the external environment, so completing their development. (f): The life cycle of *D. discoideum*.

(a   From www.microscopy-uk.org.uk/. Courtesy of Mike Morgan. Permission was requested for the third time.
b   Courtesy of Frank Fox from www.mikro-foto.de. Published under a CC Attribution-Share Alike 3.0 Germany License.
c–e Courtesy of Raymond Goldstein, From Haas PA & Goldstein RE (2018). The Flipping and Peeling of Elastic Lips. *Phys Rev E*; 98:052415.
f   From Chisholm RL, Firtel RA (2004). *Nat Rev Mol Cell Biol*; 5:531–41. With permission from Springer Nature.)

but with a different group morphology is *Dinobryon*, a heterokont stramenopile with a chloroplast (Fig. 10.5b).

The Volvox genus shows slightly more sophisticated colonial behaviour. These are relatively recent chlorophyte green algae (the genus seems only to have evolved ~200 Mya - they cannot therefore be seen as early protists). Volvox taxa typically form a hollow ball that contains small spherical colonies of cells with beating flagella and a few spores. These cells initially have their flagella pointing inwards and these need to have their polarity reversed before the colonies break free and become independent. This occurs in a way that is morphogenetically and visually interesting: a space forms in the sphere through which it then inverts so exposing their flagella to the exterior (Fig. 10.5c–e; Höhn & Hallmann, 2011). *Volvox* is however a less simple organism than it might seem as it has both a sexual and an asexual cycle.

Perhaps the best-studied example of a simple organism with complex social behaviour is *Dictyostelium discoideum*, an amoebozoan that alternates between single and multicellular modes of life, with the latter having several different cell types. When bacterial food is plentiful, the mode is of single amoebae that eat and divide. When food runs out for a group of cells, things change: one amoeba becomes a founder cell and secretes cAMP, a signal that diffuses outwards in waves of concentration and instructs nearby cells to migrate up the diffusion gradient towards it. Once all or most of the migrating population has cohered, the composite forms a "slug" that then migrates away over the trail of slime that it produces – hence its name. The slug eventually stops and differentiates: its anterior cells produce a stalk while the posterior cells become spores within a membrane-coated mass at the top to form a fruiting body. This breaks up, so dispersing spores and allowing the cycle to start again (Fig. 10.5f).

D. discoideum development thus involves cell movement, spatial patterning, differential adhesion, and cell differentiation. Because it shows many of the properties expected of a simple multicellular organism; it has been a model organism for the study of social behaviour in eukaryotes for over half a century. As a result of a great deal of work, much of the underlying genetic regulatory system is now known (for review, see Loomis, 2015). The cAMP signaling is secreted in waves and the underlying mechanism for this turns out to have similarities with that involved in regulating the beating of animal hearts, but whether this is homology or homoplasy remains unclear. Some of the molecular details of differentiation are now known, but the molecular basis of the mechanism that divides the slug population into prespore and prestalk cells has yet to be elucidated.

Other examples of very simple organisms with a complex lifestyle are *Syssomonas* and *Pigoraptor,* a pair of near-basal holozoans within the Amorphea (the basal taxon that includes animals): these organisms live in freshwater lakes and have a life cycle that seems to require different cell types (Tikhonenkov et al. 2020). These include flagellates, amoeboflagellates, amoeboid nonflagellar cells, and spherical cysts whose life cycle incorporates aggregate and syncytium-like phases. The complexity of the various cell types in these two examples confirms what has generally been observed across protists, that no simple unikont organism so far studied resembles the blastula or gastrea morphologies formed by the single cell type predicted by Haekel's model. All show evidence of cell differentiation before morphogenetic behaviour.

This evidence suggests that another model is needed for the evolution of the first multicellular organisms. The obvious candidate is the Synzoospore hypothesis put forward by A. A. Zakhvatkin in the 1950s and discussed in detail by Mikhailov et al. (2009). These authors argued on the basis of considerable amounts of early data that morphogenesis-derived structures did not stimulate multicellularity but resulted from the integration of different but already existing cell types. The origin of the Metazoa thus represents the evolutionary transition from independent protists to organisms showing social behaviour, and then to forming stable, multicellular groups in which differentiation occurred. It was only at this stage that morphogenetic events began to occur.

The most likely protist ancestor of the Metazoa is likely to have been an ancient member of the taxon that now includes the choanoflagellates, some of whose species alternate between a social and individual existence and that express both signalling and adhesion molecules (Richter & King, 2013). The first such Metazoan, known as *Urmetazoa*, was probably the first step towards the evolution of a primitive sponge (Müller, 2001). It is, however, turning out that the choanoflagellates are not as closely related to the Metazoa and that sponges are not as simple as once thought (Carr et al., 2020). Little is thus known about the details of the origins of the Metazoa.

FUNGAL EVOLUTION

Among the single-cell protists are the microsporidia, a taxon of parasites thought to infect perhaps as many as a million species that are mainly invertebrates but include some vertebrates (Han & Weiss, 2017). They live within the cells of animals using host metabolism to support their existence; as a result, they have lost large amounts of biochemical capacity and have much diminished mitochondria that are known as mitosomes (Dean et al., 2016). The taxonomic status of microsporidia was unclear for many years, but molecular analysis of their spores shows them to include chitin. Phylogenetic analysis has now confirmed that the Microsporidia are very closely related to the Cryptomycota (see below); they and some other minor unicellular taxa form a distinct clade within the fungus crown group (Spatafora et al., 2017; Bass et al., 2018).

The adaptation that made multicellular fungi so successful was their ability to produce hyphae which collectively make up their vegetative mycelium. Hyphae are long thin filaments that absorb nutrients, breaking down their surroundings to do so if necessary. When hyphae make contact with other sexually compatible hyphae, mating can occur; this ultimately results in a fruiting body that may be above ground from which spores are dispersed into the air (e.g. mushrooms) or buried (e.g. truffles). The latter either disperse spores locally or are eaten by animals that disperse the spores in their droppings.

The ancient fossil record of the fungi is very thin, presumably because of the fragility of the organisms. The oldest example of an organism with apparent hyphal morphology dates to ~2.4 Bya. In this extraordinary and unexpected fossil, filaments branch and anastomose, touching and entangling each other, with mycelium-like structures growing from a basal film attached to the internal rock surface (Bengtson et al., 2017a). If this organism really was a eukaryote with fungus-like morphology rather than a very sophisticated prokaryote colony, it at the least puts the age of the FECA much earlier than the 1.7 Bya suggested by other data (see Chapter 10). Whatever the exact status of this taxon, it is likely to have been a dead end as all contemporary fungi have a mitochondrion and the other features of eukaryotic cells, and this is only known from ~1.7 Bya or perhaps a little later (Shih & Matzke, 2013).

The earliest evidence of eukaryotic fungi on land so far are mats of *Ourasphaira giraldae;* these were found in Canadian shale, then a shallow estuary and date to 1–0.9

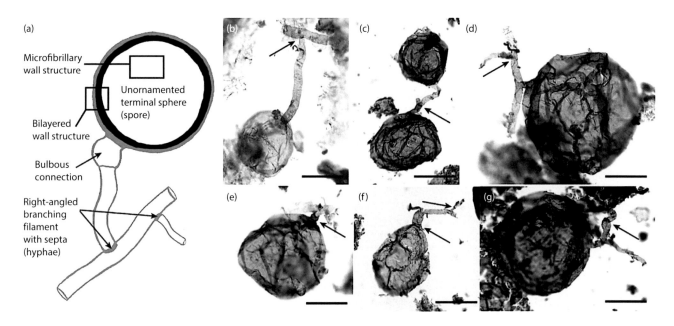

Figure 10.6 *Ourasphaira giraldae,* an early fungus from 1.0–0.9 Bya. (a): Sketch of the microfossil displaying its main features. (b–g): Transmitted light microscopy images showing secondary branching. Arrows show septate connections.

(From Loron, CC. François C, Rainbird RH., Turner EC, Borensztajn S, Javaux EJ (2019) *Nature*, 570:232–23. With permission from Springer Nature.)

Bya, well before plants or animals had evolved (Fig. 10.6; Loron et al., 2019). Another ancient example of fungus (810–515 Mya) was found in Africa by Bonneville et al. (2020) who were able to demonstrate the presence of traces of chitin in their hyphae. Organisms with lichen morphology are seen in the Ediacaran Doushantuou formation (635–551 Mya) suggesting that the symbiosis of fungi and algae is extremely ancient. Fungal fossils are not, however, generally observed until the Ordovician (c. 460 Mya) and only become common in the Lower Devonian (c. 411 Mya), with particularly fine examples being found in the Rhynie chert, a Lagerstätte in Aberdeenshire, Scotland.

By the Pennsylvanian Epoch (318–299 Bya, upper Carboniferous), all the major fungal phyla are represented in the fossil record. Of particular interest here is that, for all their structural fragility, fungi are ecologically robust and seem to have been relatively untouched by the Permian and Cretaceous extinctions (Varga et al., 2019). It was however only sometime after the latter extinction that the classic toadstool morphology seems to have evolved (Varga et al., 2019)

The fungi today are extremely diverse: Tedersoo et al. (2018) grouped them into 18 distinct phyla, which, on the basis of phylogenomic analysis (Choi & Kim, 2017), fall into the unicellular Microsporidia and associated taxa together with three major multicellular groups. These are the Monokarya and the Dikarya that include the Basidiomycota (e.g. puffballs, boletus and chanterelles; Fig. 10.7) and the Ascomycota (e.g. truffles, morels, and penicillium). They can be distinguished on the basis of their hyphae: with that of the Dikarya originally forming from two haploid Monokaryotic hyphae.

The Monokarya includes three major clades, the Zygomycota (e.g. black bread mould), the Chytridomycota, and the Cryptomycota. The Zygomycota, which include chitosan rather than chitin in their cell walls, show sexual reproduction: haploid hyphae of two individuals fuse via a bridge into which a nucleus from each enters. The bridge region separates off and forms a gametangium in which the nuclei fuse. This gametangium then becomes a zygospore, a thick-walled spore which, when it germinates, undergoes meiosis, generating new haploid hyphae, which may then form asexual sporangiospores. These sporangiospores allow the fungus to rapidly disperse and germinate into new genetically identical haploid fungal descendants.

In the Basidiomycota, individual spores are in external basidia, and reproduction is mainly sexual. In the Ascomycota, each hyphal cell has a haploid nucleus which can fuse with another of the opposite mating type, forming diploid zygotes that undergo mainly asexual meiosis. Many Ascomycetes are pathogens with a range of morphologies, with *Fusarium xyrophilum* being a remarkable example. It infects tallow-flowered grasses, producing yellow pseudoflowers that mimic those of the grass, so encouraging insects to spread their spores to (Fig. 10.8; Laraba et al., 2020).

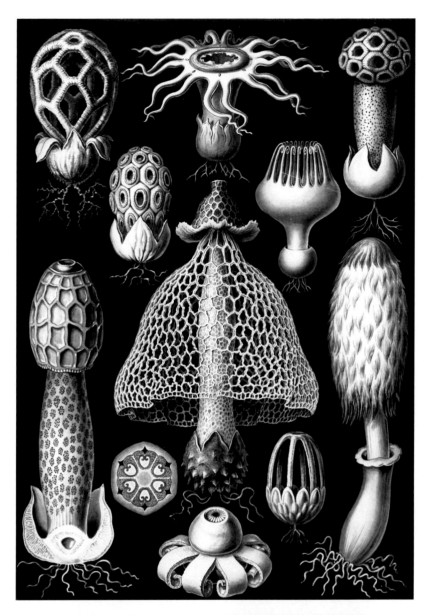

Figure 10.7 **Drawings of mushrooms (Basiomycota).** This classic and beautiful picture is from Ernst Haekel's *Kunstformen der Natur* (1904).

(With permission from Springer Nature)

Figure 10.8 (a): The natural flowers of the grass, *Xyris surinamensis,* on its cone-like spike. (b): The pseudoflowers of *F. xyrophilum*e, a fungus that has infected the grass (scale bar: 5 mm).

(From Laraba I et al. (2020) *F. xyrophilum,* sp. Nov., a member of the *Fusarium fujikuroi* species complex recovered from pseudoflowers on yellow-eyed grass (*Xyris* spp.) from Guyana. *Mycologia*; 112:39–51.)

Chytridomycota are aquatic and can be found in ponds. It is probably for this reason that they have zoospores, motile spores with a flagellum; they are thus the only fungal taxon that maintains the ancestral state (Liu et al., 2006). The Cryptomycota are unusual: the very few taxa are the only fungi that lack chitin in their cell membrane, although there is some in their spores (James & Berbee, 2012). Recent phylogenetic analysis places them very closely related to the microsporidia (Bass et al., 2018).

Multicellular fungi are very widely spread today, with fungal spores being almost ubiquitous: they are present in every land environment and there are even marine taxa (Sun et al., 2019). One reason for their success is that fungi have not only evolved to survive alone in a very wide range of habitats but also to live socially with other fungi (Fig. 10.9). Another reason is that they have also adapted to a shared life with organisms from other phyla as commensals (having little effect on them), as parasites and as symbionts (the classic example here is lichen, where the partner taxon may be an alga or a cyanobacterium - see Chapter 19). Many plants and fungi now have a symbiotic existence and the region of interaction between them is known as a mycorrhiza. As to the relative success of fungi, Bar-On et al. (2018) have estimated that Earth's total fungal mass is more than five times that of all animal species.

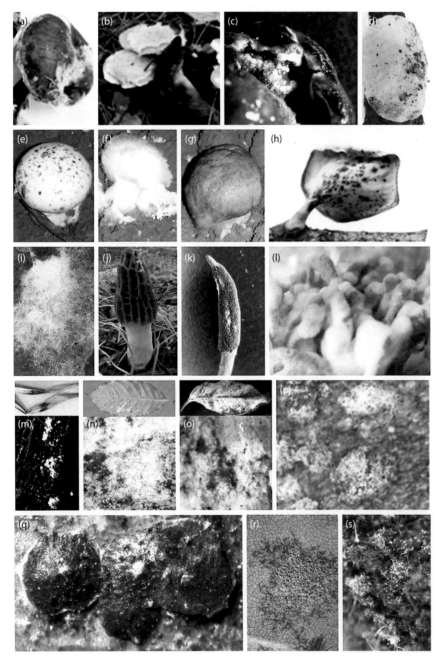

Figure 10.9 Fungal associations. (a): *Hypomyces boletuphus* on *Boletaceae* sp., (b): *Asterophora agaricoides* on *Agaricus* sp., (c): *Hypomyces boletus* on *Boletus* sp., (d): *Hypomyces mycophilus* on Agaricomycetes, (e): *Lecanicillium fungicola* var. *aleophilum* on *Agaricus bisporus*, (f): *Hypomyces rosellus* on *Agaricus bisporus*, (g): *Mycogone perniciosa* on *Agaricus bisporus*, (h): *Hypomyces* sp. on Polyporaceae, (i): *Fusarium solani* on *Tuber* sp., (j): *Diploospora longispora* on *Morchella sextelata*, (k): *Polycephalomyces sinensis* on *Ophiocordyceps sinensis*, (l): *Calcariporium cordycipiticola* on *Cordyceps militaris*, (m): *Clonostachys rosea* on *Alternaria* sp. (n): *Simplicillium lanosoniveum* on rust on a leaf of *Coffea arabica*, (o): *Cladosporium cladosporioides* on rust on a leaf of *Tectona grandis*, (p): *Ramularia uredinicola* on *Melampsora* sp., (q): fungus colonized on stroma of Ascomycetes, (r): fungus associated with *Asteridiella* sp., (s): *Penicillium georgiense* on *Aspergillus* sp.

(From Sun, JZ., Liu, XZ., McKenzie, E.H.C. et al. (2019) *Fungal Diversity* 95, 337–430. With permission from Springer.)

KEY POINTS

- The Last Eukaryotic Common Ancestor (LECA) was formed through a series of endosymbiotic events combined with extensive lateral gene transfer. Suggestions as to how this happened have been put forward, but none are universally accepted.

- As single-celled eukaryotes diversified, a key event was the endosymbiosis of a cyanobacterium by an early such protist that evolved into a chloroplast (>1.6 Bya), so forming a common ancestor of plants, red and green algae, and other more minor phyla (the Archaeplastida).

- Early fossil evidence of eukaryotes is thin but there is some evidence of algal multicellularity and cell differentiation by ~1.65 Bya.

- Over time, protists in other bikont phyla endosymbiosed a red or green algal protist. As a result, contemporary algae are polyphyletic.

- Other diversifications in protists were in cell-wall polysaccharides and in the details of the number and morphology of flagella. Such differences distinguish the major groups of protist phyla.

- Because of the complexity of early protist diversification, it has proved impossible to construct a cladogram of phyla diversification based on synapomorphies. Our current understanding of how the LECA diversified over time is now based on phylogenomics.

- The step to social (e.g. colonial) multicellular lifestyle probably occurred several times in most phyla, although the next step, which was to achieve stable multicellular lifestyle, was less common.

- An early step in this process was sexual differentiation, a facility that probably built on behaviour seen in protists.

- Fungi, characterized by the presence of hyphae and a cell membrane that includes chitin and glucans, originated more than 1 Bya.

- Part of the success of the fungal clade is due to their ability to cooperate with other organisms as commensals or symbionts, or as parasites that exploit their host.

FURTHER READINGS

Burki F, Roger AJ, Brown MW, Simpson AGB (2020) The new tree of eukaryotes. *Trends Ecol Evol*; 35:43–55. doi: 10.1016/j.tree.2019.08.008.

Fenchel T (2003) *The Origin and Early Evolution of Life*. Oxford University Press.

Naranjo-Ortiz MA, Gabaldón T (2019) Fungal evolution: Diversity, taxonomy and phylogeny of the fungi. *Biol Rev Camb*; 94:2101–2137. doi: 10.1111/brv.12550.

Knoll AH (2004) *Life on a Young Planet: The First Three Billion Years of Evolution on Earth*. Princeton University Press.

Sebé-Pedrós A, Degnan BM, Ruiz-Trillo I (2017) The origin of metazoa: A unicellular perspective. *Nat Rev Genet*; 18:498–512. doi: 10.1038/nrg.2017.21.

WEBSITES

- Wikipedia entries for *Flagella Activity, Fungi, LECA, Plastid, Mating in Fungi.*

FUNGI

http://website.nbm-mnb.ca/mycologywebpages/Natural HistoryOfFungi/Home.html

The Evolution of Algae and Plants

This chapter considers the evolution of plants and algae; these are clades characterised by possession of an intracellular plastid, usually a chloroplast, and two flagella. Algae have a very simple morphology, lacking the leaves, shoots, and roots of land plants. Thus, all single-celled and colonial organisms with a chloroplast, together with multicellular organisms that are essentially epithelial are viewed as algae, while more complex organisms with a plastid are plants.

While there are a few early algal fossils, they are not well represented in the fossil record, mainly because such taxa were small with predominantly soft tissues that did not fossilise well. Much of what we know of algal evolution, therefore, comes from cladistic and phylogenetic analysis. The resulting phylogeny is particularly complicated because many taxa with a chloroplast are not placed within the Archaeplastida taxon that includes plants, green algae, and red algae but across the wider Corticata clade. It is now clear that all these apparently anomalous algae originally obtained their chloroplasts by secondary endosymbiosis of a red or green algal protist, as demonstrated by some of them having chloroplasts with three or four layers of membrane.

Plants have a richer anatomy than algae and, as many larger plants are physically robust, particularly those that contain lignin, they have a much better fossil record than fungi and algae. The fossil record is, however, much weaker for those much small, fragile plants that first eked out an existence on land; for these, both cladistic and phylogenetic analysis are generally required to interpret their evolution. The discussion of plant evolution here mainly focuses on the general evolutionary trajectory of the clade and how the major groups evolved. It does not go into the details of the evolution of plant anatomy and clade diversification. This material is covered in the Further Readings at the end of the chapter.

There are four major clades of multicellular organisms: fungi were discussed in the previous chapter and animals will be covered in the rest of this section. This chapter considers algae and plants, which are organisms characterised by possession of a plastid (e.g. a chloroplast) originally endosymbiosed from a cyanobacterium. The apparent simplicity of this statement is illusory for two reasons. First, many algal groups originally obtained their chloroplasts through endosymbiosis of a unicellular red or green alga rather than through direct inheritance. Second, there is a range of plastids with distinct functions (Jarvis & López-Juez, 2013; Sardeli et al., 2019).

Although plastids are informally considered to be chloroplasts, the organelle that captures photons and uses their energy to make ATP and NADH and other compounds, this term should probably only be used for those organelles containing chlorophyll; the most basal of which may actually be the muroplast of glaucophytes that traces its lineage directly back to an endosymbiosed cyanobacterium, although it may have been diminished over time (Chapter 10). In algae and lower plants, the plastid seems to just carry out its original photosynthetic function, but plastids

DOI: 10.1201/9780429346217-14

in higher plants have evolved to fulfil a much wider range of roles. These include fruit ripening, endosperm development, and root gravitropism (Solymosi et al., 2018), while there can even be variation in the plastids associated with different parts of plants (Henry et al., 2020). The range includes, for example, chromoplasts (pigment storage), elaioplsts (lipid production), and amyloplasts (starch synthesis and storage).

The Archaeplastida are very different from the Metazoa, the only other clade that includes large, multicellular organisms with a range of cell types. This should not be unexpected, given that the clades probably separated before ~1.6 Bya at a stage when both were almost certainly unicellular organisms (de Vries & Gould, 2018). In addition to the possession of a plastid enabling them to obtain energy from sunlight, other obvious differences today are that plants have cellulose in their cell walls, and have two rather than one flagellum. In addition, their microtubule cytoskeletons generally lack basal bodies but have acquired other means of handling microtubule morphogenesis (Brown & Lemmon, 2011). Plants and animals have often faced similar evolutionary problems, and it is becoming clear that they have often found different solutions (Durand-Smet et al., 2014).

ALGAE

Algae are defined as simple organisms possessing a plastid. They originated when an early, unknown eukaryotic cell with two flagella (a bikont) endosymbiosed a cyanobacterium that adopted a semi-independent life within the cell, capturing light energy to make energy rich ATP and NADH that could be used by what was now a glaucophyte to make sugars and other compounds. Evidence from phylogenetic and comparative genomics together with fossil data (below) shows that the initial acquisition of a chloroplast was probably a little before 1.6 Bya (Kim et al., 2014). Green and red algae, the two main phyla that are respectively characterised by having a chloroplast (Chlorophyta) and a rhodoplast (Rhodophyta), are thus bikonts within the Archaeplastida. Green plants, the other multicellular member of the Archaeplastida evolved from green algae. There are also several minor taxa of algae in other bikont clades.

Of the three original algal clades in the Archaeplastida (Fig. 11.3), the single-celled glaucophytes are the simplest. They are characterised by being asexual and having murein peptidoglycan associated with the cellulose in their cell wall, a possible relic of the original cyanobacterium that their ancestor endosymbiosed. Although glaucophytes may be the most basal of the Archaeplastida and thus have a key evolutionary role, they seem to be relatively unimportant today as this family of freshwater protists includes only about 15 species. Most of the many multicellular green algal taxa within the Archaeplastida are chlorophytes or charophytes, with the former being mainly protist forms and widely distributed while the latter are mainly freshwater and more complex. The land plants (the Embryophyta) evolved from the charophytes and the two clades together comprise the Streptophyta.

Most algal taxa are aqueous, being found in both salt and fresh-water environments (Fig. 11.1a,b). Some can, however, thrive on land while algal protists and spores can disperse through the air (Tesson et al., 2016). One reason for the success of the algae is that they are capable of forming symbionts with other organisms, such as fungi, sponges, and corals, a topic that will be discussed in more detail in Chapter 19. The majority of algal taxa are small, unicellular protists but three taxa are multicellular: the green algae, the red algae, and the brown algae (also known as kelps); the first two are Archaeplastida bikonts, the last is a stramenopile heterokont.

Green and red algae are very different (Table 11.1). Green algae have well organised chloroplasts that use chlorophylls a and b to capture light, while red algae have rhodoplasts with unstacked thylakoids that use both red phycobilisomes (e.g. phycoerythrin and phycocyanin) and green chlorophyll a to capture light. Chlorophylls in secondary plastids generally reflect their origin but most of those that derive from red algae also produce chlorophyll c. The situation is, however, sometimes more complicated as there is evidence of horizontal gene transfer between red and green algae (Ponce-Toledo et al., 2018). It is also worth noting that red algae have lost their centrioles and flagella; the role of organising microtubules is taken by their very much simpler polar rings (Dave & Godward, 1982; Chapter 10).

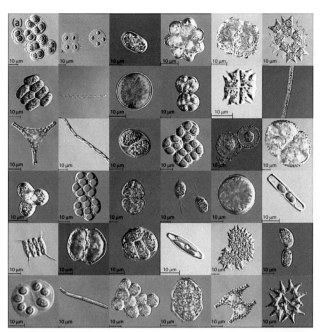

Figure 11.1 **Algal diversity.** (a): The range of microalgae in a Moscow pond; Note that they are mainly colonial. (b): The algal-covered seabed in the national park, NSW, Australia.

(a Courtesy of Alexander Klepnev. Published under a CC Attribution—4.0 International License.
b Courtesy of Toby Hudson. Published under a CC Attribution—Share Alike 3.0 Unported License.)

TABLE 11.1 PLASTID TYPES IN ARCHAEPLASTIDA

Taxon	Plastid	Main chlorophyll types			phycobilisomes	Peptidoglycan in membrane
		a	b	c*		
Cyanobacteria		Y			Y	Y
Glaucophytes	Muroplast	Y				Y
Red algae	Rhodoplast	Y			Y	
Secondary red algae	Secondary Rhodoblast	Y		Y	Y	
Green algae	Chloroplast	Y	Y			
Secondary green algae	Secondary chloroplast					
Paulinella	Chromatophore	Y	Y		Y	Y
Green plants	Chloroplast	Y	Y			

* Chlorophyll c lacks the long tail of the other types.

Algal fossil record

Because so many algal taxa are unicellular, colonial or small, the fossil record is limited but is nevertheless important. The oldest fossil algae are *Rafatazmia chitrakootensis*, *Denaricion mendax,* and *Ramathallus lobatus*; the first two are filamentous and the last is more complex. All are rhodophytes (red algae) embedded in phosphate and come from a site in central India. The claim that they dated to ~1.6 Bya was initially doubted as this seemed too early for a eukaryotic organism, but the dates have been conformed and thus give a minimum time for chloroplast endosymbiosis (Bengtson et al., 2017b).

A rather later algal species showing evidence of having several cell types is *Bangiomorpha pubescens*, a filamentous red alga that lived ~1.05 Bya, with contemporary taxa having very similar morphology (Fig. 11.2a; Butterfield, 2000; Javaux, 2007; Brawley et al., 2017). It had a holdfast for adhering to rocks, with different examples having one of two spore types, both of which are similar to those seen in contemporary *Bangiomorpha*, an observation confirming that sexual dimorphism and reproduction had evolved by then. More recently, Tang et al. (2020) have identified *Proterocladus antiquus*, an early green alga dating to ~1 Bya (Fig. 11.2b), while Vaziri et al. (2018) have discovered more recent algal fossils in Iran that date to the Ediacaran. A further, particularly well documented example from the Cambrian is *Dasyclades*, a small alga that secreted a calcium carbonate shell, a feature which lent itself to fossilisation (Fig. 11.2c,d). This chlorophyte genus still flourishes and is the largest single-celled organism now alive (up to 10cm in length).

Evolution of algae

Although algae are, in one sense, the simplest of the major kingdoms, understanding their evolution is difficult because, in another sense, they are the most complex. The simplicity derives from the fact that the great majority are single cells that are either solitary or colonial (Fig. 11.1a); even those that are multicellular and large, such as the kelps (up to 50 m long) have a very simple, mainly epithelial morphology. Reproduction is also simple and is either through asexual fission or through the production of spores on the epithelial surface. The complexity comes from the fact that detailed morphological and DNA analysis show that algal taxa are distributed across a wide variety of bikont clades.

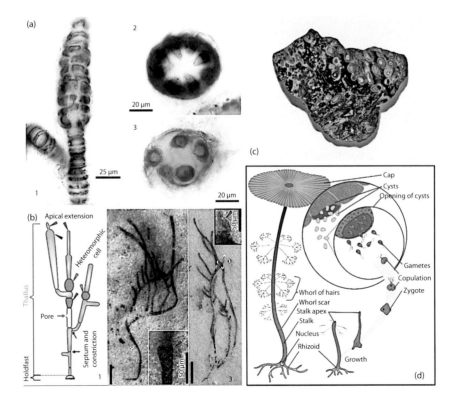

Figure 11.2 **Fossilised algae.** (a): *Bangiomorpha pubescens*, an early red alga dating to ~1.05 bya. 1: mature thallus showing both uniseriate (single-cell width) and multiseriate portions of a filament; 2: transverse cross-section of a filament showing eight radially arranged wedge-shaped cells; 3: transverse cross-section of a filament region in which spheroidal spores have differentiated. (b): *Proterocladus antiquus*, an early chlorophyte (green alga). 1, morphological reconstruction and terminology. 2,3, slender thalli preserved on bedding surfaces showing enlargements of the small white framed areas (inserts). (c): An example of the Dasycladaceae from the Triassic. (d): The life cycle of *Acetabularia*, a contemporary taxon from the Dasycladales. This large single-celled organism has a single nucleus in its holdfast region.

a Courtesy of Nick Butterfield.
b Tang Q, Pang K, Yuan X, Xiao S (2020) *Nat Ecol Evol*; 4:543–549. With permission from Nature Springer.
c Courtesy of "Hectonichus." Published under a CC Attribution—Share Alike 3.0 Unported License.)

It was originally difficult to work out algal systematics because red and green algae are so different. Some while ago, Nozaki et al. (2003) and Cavalier-Smith and Chao (2003) undertook detailed analyses of a range of sequences carried by red algae and a diverse group of other organisms. Their results showed first that the Archaeplastida are monophyletic and second that the red algae are a sister group to the other taxa within the clade, a result confirmed by recent extensive sequencing analysis (One Thousand Plant Transcriptomes Initiative, 2019) and this is reflected in their plastids. As both cyanobacteria and red but not green algae use phycobilisomes to capture light, it is clear that green algae are a derived form (Liu et al., 2019a).

Phylogenetic work has also shown that almost all algae not included within the Archaeplastida had, at one time or another, acquired a chloroplast through secondary endosymbiosis of either a single-celled green alga (characterised by chlorophylls a and b) or a red alga (characterised by chlorophyll a and phycobilisomes). There is a single exception that reflects a second and completely independent primary endosymbiotic event in a very different kingdom. Some 90–40 Mya, *Paulinella*, a Rhizarian protist (a taxon within the SAR group of phyla, Figs. 10.4 and 11.3) endosymbiosed an α-cyanobacterium within the *Prochlorococcus-Synechococcu*s lineage (Gabr et al., 2020). The few contemporary *Paulinella* taxa that descended from that original protist contain a photosynthetic chromatophore, an organelle rather simpler than a chloroplast (Nowack et al., 2008). Its genome, although much diminished, is still much larger than those present in other chloroplasts (Zhang et al., 2017; Gabr et al., 2020).

Further evidence for secondary endosymbiosis comes from the nature of the membranes surrounding the chloroplasts. While those in red and green algae are surrounded by a double membrane, the chloroplasts formed by secondary endosymbiosis can have a triple or quadruple membrane that reflects the original chloroplast membrane together with the bounding membrane of the original cell that was endosymbiosed. The current view of the algal phylogeny given in Fig. 11.3 and is witness to the success of organisms possessing their own ability to use light energy.

Algae have an important role in the world ecosystem as the various taxa provide food for a wide variety of marine life. Red algae that get their colour from phycobilisomes are also commercially important because their cell walls contain agar and carrageenan. Nuri, the seaweed "paper" used for wrapping sushi in Japan is prepared from red algae. Several of the secondary endosymbiont algae are also important for humans. The brown algae, which include the kelps, are sources of food and nitrogen; their colour derives from fucoxanthin, a carotenoid involved

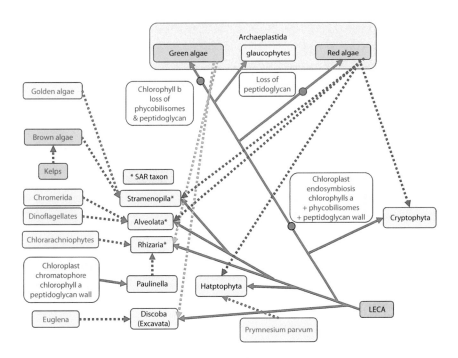

Figure 11.3 A mixed cladogram and phylogram illustrating the evolution of protist phyla that include algal taxa (black type). The blue arrows show clade evolution. The Pale green boxes are apomorphies indicating the evolution of the Archaeplastida. Green and red dotted arrows respectively indicate secondary endosymbiosis of a chloroplast and a rhodoplast. Blocks with red print and grey lines are examples of species from taxa with secondary algae; yellow and green boxes respectively indicate that the clade includes single-cell (or colonial) and predominantly multicellular taxa. Note the anomalous position of *Paulinella* (purple link to newly endosymbiosed cyanobacterium) which formed though a unique and relatively recent endosymbiotic event.

in the photosynthesis pathway. Diatoms, a clade within the Chromista whose cell wall includes silica, are particularly common in the sea and make a particularly important contribution to carbon dioxide absorption and oxygen production. In contrast, *Prymnesium parvum*, a golden alga (e.g. Figs. 10.7b and 11.3), produces a fish-killing toxin. (For details, see Websites.)

PLANTS

The clade of land plants is characterized by organisms with plastids whose tissues are more complex and include many more cell types than those found in algae. The clade is sometimes known as the embryophytes because plants (-phyta) retain their early developing embryos within specialised tissues. Plants seem to have evolved from unicellular green algae that were probably present on Cambrian lake shores (Del-Bem, 2018). Molecular and phylogenetic analysis points to an early member of the Zygnematophyceae, filamentous green algae within the charophyte clade, as being the sister taxon of all land plants and hence that they share a last common ancestor (Szövényi et al., 2019; Wickett et al., 2014). Today, plants generally share two key features: (1) a shoot domain above ground whose roles are to trap light and produce reproductive tissues and (2) a root domain below or at least attached to the ground that stabilises the organism and absorbs nutrients from local material. Plants also encode a circadian clock that has deep evolutionary roots (Linde et al., 2017). In angiosperms, this clock synchronises development, nectar production, fertilization and co-evolution with potential pollinators (Oakenfull & Davis, 2017).

Plants today fall into a clear series of clades of increasing complexity (Table 11.2) that all tend to be widely distributed, other than in the very hot deserts near the equator and the very cold regions near the Poles and on high mountains. Even in those places, a few taxa survive the extremes of heat, cold, and dryness. The angiosperm clade today is by far the most specious and sophisticated; as it includes some 64 orders and 416 families, it is not practical to do justice here to this diversity here (see Further Readings). Taxa extend in length from a few millimeters (e.g. the duckweed Wolffia, Fig. 4.1) to tens of meters (e.g. redwood tress) and in weight from milligrams to tons (Chapter 4). Some have been cultivated for food, particularly the grasses, such as wheat and rice, others for their wood. Some families are taxon rich: the Asteraceae (daisies) include over

TABLE 11.2 THE EVOLUTION OF PLANT COMPLEXITY

Class	Some distinguishing properties	Species no.
Unicellular green algae (e.g. charophyte)	Freshwater organisms containing cellulose and chloroplasts	
Zygnematophyceae (marine)	Colonial filaments of charophyte cells; *Synapomorphies*: spores; green spurs	
Bryophytes (liverworts, hornwarts & mosses—land based)	Land organisms with spores and, non-vascular stems. *Synapomorphies*: cuticles; primitive, non-veined leaves with apical growth	~12K
Clubmoss (lycopodiophytes)	*Synapomorphies*: roots, vascular tissue, microphyll (single-veined) leaves	~400
Ferns (Monolophytes)	Synapomorphies: fronds of megaphyll leaves with multiple veins	~10K
Gymnosperms	*Synapomorphies*: leaves are usually narrow with a single vein; seeds that develop from leaf stems; monoculpate pollen	~1K
Angiosperms	*Synapomorphies*: flowers, seeds that form in an ovary	
Monocotyledons	*Synapomorphies*: One embryonic leaf, narrow leaves with parallel veins, trimeric flowers, 2-8 cotyledons, monoculpate pollen	~60K
Dicotyledons	*Synapomorphies*: angiosperms with broad leaves and branching veins; monocolpate pollen; trimerous flowers	~9K
Eudicotyledons	*Synapomorphies*: Two embryonic leaves, broad leaves with branching veins, tetramerous or pentamerous flowers, tricolpate pollen	~175K

32,500 species; others have only a handful of species: *Welwitschia* and *Ginkgo* are each a single-species gymnosperm.

It should also be noted that the species numbers for the various taxa given in Table 11.2 can only be viewed as approximate. This is because plants are much more tolerant of hybridisation than animals and are also far more capable of displaying tetraploidy, while many taxa even show polyploidy (Alix et al., 2017; Soltis et al., 2018)—it can be hard to tell when a variant becomes a species. Plant genomes have diversified widely and varied extensively over time and their sizes range over a factor of >2000, in part because of the extent of polyploidy (Minelli 2018). Moreover, whereas animals have only one pair of testes or ovaries in which gametes can form, angiosperms can have many thousands of flowers, each with its own reproductive system that develops independently of the others, while gymnosperms can equally have large numbers of gamete-producing cones. There are, therefore, many more opportunities for mutations to become incorporated into gametes in plants than in animals. Variants can thus form more easily and rapidly in plants than in animals

Plant sex chromosomes, on the other hand, are similar in their essential features to those in animals, although the details differ (Pannell, 2017); this provides further evidence that sexual dimorphism was a characteristic of the LECA (Chapter 10). There have also been many cases of Y degeneration so that haploid spores can give rise to both male and female plants (Charlesworth, 2002). In addition, many plants (most ferns, gymnosperms, and angiosperms) can produce both male and female gametes and can often self-fertilize, although there are mechanisms that limit this. Furthermore, many plants can reproduce themselves vegetatively without involving gametes, an observation that emphasises how much more developmentally flexible are plants tissue as compared to animals. Plant reproduction is thus far more varied and complex than in animals as they can reproduce themselves in a far wider range of ways. The conclusions to be drawn from this diversity are that evolutionary plant genetics is rather more complicated than that of animals (One Thousand Plant Transcriptomes Initiative, 2019; Acquaah, 2021; see Further Reading) and that species borders are rather less sharply drawn in plants than in animals (Chapter 22).

Two key features were and are still responsible for the ubiquitous success of plants and, as a result of the food that they provide, of many land organisms. The first is leaves and the second is the association of some species to form symbionts with nitrogen-fixing bacteria. Leaves are particularly specialized for capturing energy from light photons to synthesise ATP and, as a result, release oxygen into the atmosphere (the light reaction). This energy, which is stored in ATP, is then used to produce sugars from carbon dioxide together with a range of other biochemicals that may require the simple nutrients absorbed from soil by plant roots (the dark reactions). Leaves come in a wide variety of taxon-dependent morphologies (Fig. 11.4; Table 11.1). They originated as small growths from moss thallus (undifferentiated tissue) and probably started to enlarge in the early Devonian during a period of high oxygen and low carbon dioxide levels (Beerling et al., 2001; Dahl et al., 2010). The data shown in Table 11.1 give the impression that leaf evolution was a smooth progression marked by increasing complexity, but this was probably not so: cladistic analysis indicates that microphyll and megaphyll leaves evolved separately from thalloid growths (Piazza et al., 2005; Vasco et al., 2013).

Leaves are renowned for the fact that their clockwise and anticlockwise spiral organisations usually reflect two successive terms of Fibonacci series (each successive term is the sum of the two previous ones). This arrangement, which is under a degree of genetic control, can be shown to maximise the amount of light falling on each. This, however, is probably not the explanation of this trait as, on the one hand, there are many exceptions and, on the other, it is also seen in the seeds of a sunflower head, pineapples and pinecones. Kuhlemeier (2017) has suggested that such a leaf arrangement can best be seen as coincidental and an emergent feature of the underlying molecular mechanisms that generate leaves, a conclusion reached by Mitchison (1977) on the basis of theoretical modelling. In this, as in many other aspects of plant growth and development, the plant hormone auxin plays an important role (Swarup & Bhosale, 2019)

The second reason for the success of plants is the ability of the roots of taxa, such as the legumes to attract nitrogen-fixing bacteria, particularly Rhizobia, with which they form symbiotic nodules (Chapter 19; Masson-Boivin & Sachs, 2018). Other plants get their nitrogen from soil which always includes some of a range of bacteria that can

Figure 11.4 The various types of plant leaves. (a): The thalloid-derived leaves of a liverwort. (b): The simple microphyll leaves of a clubmoss. (c): The macrophyll leaves of fern fronds. (d): The needle leaves with a single vein of a gymnosperm fir tree. (e): The parallel veins of monocot iris leaves. (f): The branching veins in the leaves of a young castor oil eudicot plant with two cotyledons together with two smaller normal leaves that are unfolding.

(a Courtesy of "Avenue." Published under a GNU Free Documentation License
b Courtesy of Eric Guinther. Published under a GNU Free Documentation License
c Courtesy of Jamie Dwyer. Published under a CC Attribution—Share Alike 2.0 Generic License.
d Courtesy of Walter Siegmund. Published under a CC Attribution—Share Alike 3.0 Unported License.
e Courtesy of "Drewboy64." Published under a CC Attribution 3.0 License.
f Courtesy of Rick Pelleg. Published under a CC Attribution—Share Alike 2.5 Generic License.)

fix nitrogen. Such prokaryotes are known as diazatrophs and include taxa from both Archaebacteria and Eubacteria (e.g. Cyanobacteria, Azotobacteraceae, and Frankia). Given the importance of nitrogen fixation for plant life and the ability for gene transfer to occur between nodule bacteria and plant roots (Andrews & Andrews, 2017), it is perhaps surprising that there seems to have been no horizontal gene transfer of the nif (nitrogen-fixing nitrogenase) gene between rhizobia and their plant hosts. One reason for this may be that, even if such genes were transferred to a root cell, it would have no way of reaching spore- or seed-producing cells and so being inherited by subsequent generations.

Plant evolution

There is no fossil evidence illustrating the early stages of plant evolution, but phylogenetic evidence (Morris et al., 2018) places it in the Cambrian, perhaps as early as ~510 Mya. The first direct evidence for the presence of plants are spores that are typical of today's liverworts and that date to the lower-middle Ordovician period (~470 Mya; Wellman & Gray, 2000). The earliest fossilised bryophyte plants (liverworts, hornworts, and mosses; Fig. 11.4a) so far discovered date to the early Silurian (~440 Mya). Today, these primitive plant clades have a waxy cuticle that enables them to live out of water and thallus-derived primitive leaves that enable them to photosynthesise (von Konrat et al., 2010; Ligrone et al., 2012). They do, however, lack both roots (any subterranean tissue is for adhesion) and the vascular tissue needed for transporting water; they are, therefore, all small. Computational phylogenetic analysis suggests that bryophytes are monophyletic (Cox, 2018).

The first vascularised plants with leaves and roots were the clubmosses (Lycopodiopsida, Figs. 11.4b and 11.5a,b). These had microphyll leaves that are very primitive, coming directly off the stem, and leaving a characteristic scarring pattern that can be seen in fossils. Clubmosses reproduce through the production of chloroplast-carrying spores from sporangia that form off the base of their stems; these are then dispersed by winds. The earliest examples date to the Upper Ordovician (443–417 Mya), but they became more common during the Silurian (443–419 Mya). Particularly spectacular examples of early plant fossils that date to ~410 Mya (early Devonian; Fig. 11.2) are present in the Rhynie chert Lagerstätte (Scotland) at which early fungi have also been found. The quality of fossils here is due to this Lagerstätte was originally a hot spring whose dissolved silicates led to the fine structure of fossils being preserved Fig. 11.5c,d). A notable example here is *Rhynia gwynne-vaughanii* (Fig. 11.5c.d; also see Websites), a simple branching vascularised sporophyte ~20cm high that may have been related to an early clubmoss.

A further key advance that occurred in clubmosses in the early Devonian was the evolution of lignins, a class of robust polymers that combines with and strengthens cellulose to form wood (Raven & Edwards, 2014). Lignin not only allowed vascularized plants to increase dramatically in size (e.g. some clubmoss trees were ~30m high) but also provided protection against pathogens (Lee et al., 2019). Perhaps unexpectedly, lignin has also been identified in red algae (Martone et al., 2009). It was not initially clear whether lignin is a very ancient molecule or whether its presence in the algae is a homoplasy The answer seems to be the former: lignin-like compounds have also been found in green algae with such organisms having many of the genes whose proteins are required for synthesizing more modern lignin precursors (de Vries et al., 2017).

Soon after wood had appeared, ferns evolved with megaphyll leaves (Fig. 11.6) on which sporangia containing many spores could form. These complex leaves, informally known as fronds, are very much larger than microphyll leaves and the current view is that they represent a distinct and novel evolutionary feature rather being derived from microphyll leaves—they are not homologous. In their detailed

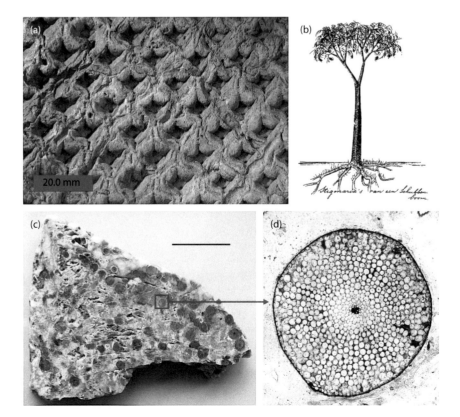

Figure 11.5 **Fossil lycophytes (clubmosses).** (a): surface impression of leaf scars on *Lepidodendron*. (a 30 m clubmoss tree ~300 mya). (b): a 1911 drawing of likely appearance of *Lepidodendron* by Eli Heimans. (c): A piece of rock from the Rhynie chert Lagerstätte (~410 mya) with many fossils of *Rhynia gwynne-vaughanii*, a vascular sporophyte plant related to clubmosses (bar: 1 cm). (d): A thin section of a fossil Rhynia showing the fine detail of the vascular tissue (the veins are about 2 mm in diameter).

(a Courtesy of Peter Coxhead. Published under a CC Attribution—Share Alike 3.0 Unported License.
b Courtesy of "Plantsurfer." Published under a CC Attribution—Share Alike 2.0 England & Wales License.)

Figure 11.6 **Fossilised megaphyll leaves from a pecopteris tree fern ~310 Mya.**

(Courtesy of James St John. Published under a CC Attribution 2.0 Generic License.)

review, Vasco et al. (2013) discuss how megaphyll leaves probably evolved from chloroplast-carrying sporangia through elongation, flattening, and development of laminar tissue, with this apparently happening in several distinct ways. As a result, megaphyll leaves were not only larger and more efficient than microphyll ones but were able to produce spores in large numbers. Their size also helped in harbouring nitrogen-fixing cyanobacteria, collecting plant matter falling from above and, where they met the ground, forming rootlets.

With large leaves trapping energy and wood providing mechanical strength, plants were able to grow much taller and soon forests of high trees dominated the land wherever the climate allowed. These included clubmoss and horsetail trees capable of reaching a height of ~30 m (Fig. 11.5a,b) and tree ferns, such as *Wattieza* (mid-Devonian;) that were up to ~8 m tall. So extensive were the Carboniferous woodlands (354–290 Mya) that they became the source of much of our current stocks of oil and coal. Although contemporary horsetails, such as *Equisetum* are small and have the appearance of clubmosses with microphyll leaves, it seems that these are a megaphyll adaptation and that horsetails are related to ferns (Elgorriaga et al., 2018).

A few so-called seed ferns made seeds, but this line became extinct, and seeds as we know them today instead evolved in gymnosperms. These were descendants of progymnosperms (now extinct), which in turn, evolved from the Trimeropsida (Devonian vascular plants). Progymnosperms were the earliest plants to show such features as cambium (stem cells that enable trees to expand and increase their height) and were thus the first modern trees, although they still produced spores on their leaves. It was not until the Late Carboniferous (~320 Mya) that true gymnosperms that produced seeds from leaves appear in the fossil record. These are mainly conifers whose seeds are made by cones, a derivative of groups of leaves, but the clade today also includes the relatively rare cycads and the single *Ginkgo* species.

The last 250 My have been marked by the consolidation of the early plant clades and the evolution of angiosperms (flowering plants), but the date when this clade evolved has not been easy to establish. This is mainly because the fossil record here is weak, presumably due to the fragile nature of stem and basal angiosperm taxa. Until recently, the earliest angiosperm fossils (e.g. *Leefructus mirus*) (Fig. 11.7b; Sun et al., 2011) had been dated to ~125 Mya, a date compatible with phylogenetic analysis that had placed the evolution of angiosperms in the Cretaceous, at ~130 Mya or a little earlier (Amborella Genome Project 2013; Coiffard et al., 2019; Friis et al., 2016; Hertweck et al., 2015; Hochuli & Feist-Burkhardt, 2013). This orthodoxy was recently challenged by the unexpected discovery by Fu et al. (2018) of *Nanjinganthus dendrostyla*, an early fossil plant that dated to ~174 Mya (the mid-Jurassic; Fig. 11.7a). This plant had many features of an angiosperm including inflorescences (flower clusters) and this interpretation was confirmed by the detailed analysis of Taylor and Li (2018). More recently, Sokoloff et al. (2020) have, however, re-analysed the morphology and suggest that the structures that had been claimed to be petals and carpals in *Nanjinganthus* are more likely to have been small gymnosperm cones. It is not yet clear which interpretation is appropriate. There are as yet no earlier fossils that are clearly monocots; the oldest so far discovered dates to ~105 Bya (Fig. 11.7c,d)

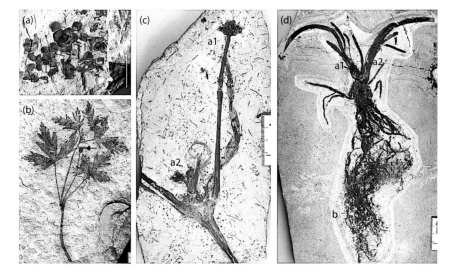

Figure 11.7 Early fossilised angiosperms.
(a): Possible floral inflorescences of *Nanjinganthus dendrostyla*, dated to ~174 mya. (b): *Leefructus mirus*, an early cretaceous dicot with a single Central flower (arrow). (c,d): Two examples of *Cratolirion bognerianum*, a very early monocot angiosperm from the cretaceous showing leaf phyllotaxy, two leaf rosettes (a1 and a2) and root systems (b).

(a From Fu Q et al. Elife. 2018 Dec 18;7:e38827. doi: 10.7554/eLife.38827
b Courtesy of Shizhao. Published under a CC Attribution 2.5 Generic License.
c,d From Coiffard C et al. (2019) *Nat Plant*; 5:691–696. With permission from Nature Springer.)

Angiosperms possess a reproductive system far more sophisticated than anything seen in more primitive plants, even gymnosperms that produce pollen and seeds. The clade, which came to dominate land plants, is characterised by flowers that have a reproductive system surrounded by petals with external sepals (both are leaf modifications). The male reproductive tissues are stamens that are topped by the pollen-producing anthers. The female reproductive system is a central seed-producing ovary, which may be within or below the flower, that is accessed via a style tube, which terminates at the stigma (pollen receptor). Angiosperm ovules are normally fertilised by pollen, but many taxa also show apomixis: this is the production of viable seeds without fertilization—these are, of course, female (Fei et al., 2019).

Dioecious angiosperms, such as hollies, have separate male and female plant types each with its own distinct flowers, and there is thus limited chance of self-fertilisation. The flowers of monoecious angiosperms, such as roses have both stigma and stamens. In these plants, there are often mechanisms ensuring that the seed cannot be fertilised by pollen from its own flowers, so increasing the likelihood of diversification. The steps that led to the evolution of this complex reproductive system from the far simpler one of gymnosperms are not known, but there is a gap between the origin of gymnosperms and that of angiosperms of >150 My. This period is far greater than the ~35 My between the evolution of amphibians whose eggs require an aqueous environment and reptiles that had far more complex amniotic eggs that could develop on land (Chapter 15). Given the subsequent success of angiosperms, this long period suggests that the evolution of flowers with their associated reproductive systems was a particularly complex process about which the little that is so far understood comes from modeling (Sauquet et al., 2017).

Most plants are dependent on wind and water to spread pollen for fertilisation, but the evolution of flowers enabled angiosperms to extend the fertilisation repertoire. Today, the colours and scents of flowers attract animal pollinators: insects and birds drink the nectar fluid produced by nectaries in the cup-like base of the flowers, picking up pollen that they carry to other plants. There is also a very wide variety of animals and birds that disperse seeds, often because they are attracted to the fruit that contains them, but also because some seeds have structures enabling them to adhere to feathers, hair and human clothes. It is probably for these reasons that the angiosperms have diversified so widely to the extent that they are now the dominant plant clade throughout the world (Table 11.1).

The details of early angiosperm evolution and diversification still await clarification from the fossil record, but analysis of the clades has excited considerable interest over the centuries. First Threophrastus (c. 370 BC, Appendix 1) and later John Ray (1682) showed that the flowering plants divide into two obvious classes: 1) those whose embryo develops a single cotyledon (a primitive or seed leaf) and 2) those that develop two cotyledons. The adult plants can easily be distinguished on the basis of their leaves (Fig. 11.4): the dicotyledons have branching veins (most trees and flowering plants), while monocotyledons have parallel veins (e.g. orchids, alliums, grasses, and palms). It is worth noting that the grasses, whose seeds are the major calory source of many modern humans, are monocotyledons that only evolved ~42–35 Mya.

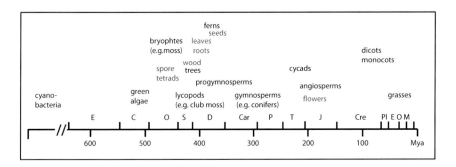

Figure 11.8 This diagram summarises the dates when various plant types first appear in the fossil record. It is clear that most features of plants apart from flowers seem to have evolved around 420–360 Mya, a period known as the Devonian explosion. What the fossil record does not make clear is the detailed sequence of inheritance; this is shown in **Fig. 11.11**.

Putting the fossil record together (Fig. 11.8), It seems that plant anatomical richness developed nonlinearly. It began with a very slow period during which primitive embryophytes evolved; this was followed by a period of increasing organism complexity, but generally little increase in size. It was not until the Silurian, with the evolution of clubmosses with vascularisation, roots and leaves, and a little later, lignin, that land plants started to increase in size and diversity. During the Devonian, the rate of evolution increased rapidly with the evolution of ferns and progymnosperms before slowing during the Carboniferous, the period that was marked by the evolution of seed-bearing gymnosperms. There was thus a rich flora well before the time of the Permian extinction (252 Mya). At an anatomical level, the great majority of the tissues that plants now possess had by then evolved, the exception being the flower-associated structure of angiosperms that evolved during the Cretaceous or a little earlier.

This period since the Devonian has been marked by the major extinctions that ended the Permian (251 Mya) and the Cretaceous (66 Mya). These had devastating effects on animal life, but it is now becoming clear that the effect of these extinctions was much less severe on plant life. Analysis of plant macro- and microfossils across the Permian-Triassic boundary shows little evidence of devastation (Nowak et al., 2019). The K-T extinction similarly had much less effect on plants than on animals: there was a loss of plant species in western North America, relatively close to the impact site in Chicxulub, Mexico (Barreda et al., 2012; Nichols & Johnson, 2008) but much less loss in South America and in other continents. It seems clear that plants, like fungi, are far less sensitive to savage climate shifts than are animals.

Plant systematics

The information from the fossil record is inadequate to indicate the full details of the evolution of the various plant clades because the time points of major changes are too poorly represented. The evidence on morphological differences is, however, sufficient to produce a cladogram of plant evolution on the basis of the main synapomorphies of the major groups of plants given in Table 11.1. Leaves are particularly important here as modifications to their morphology are key markers of evolutionary change (Fig. 11.9; Nikolov et al., 2019). It should be emphasized that the analysis given here is based on a few core synapomorphies and is intended only to demonstrate basic features of plant evolution rather than to provide a full analysis of plant diversity. No one today would construct a detailed cladogram of a set of contemporary organisms on the basis of morphological features alone; it would always be checked, supplemented, and refined by phylogenetic analysis of a range of homologous DNA sequences from these organisms and, in the case of plants with complex chromosome numbers, analysis of their genetic origins (Chapter 8).

The cladogram (Fig. 11.11) shows the specific apomorphies (innovations) that enabled successive taxa to increase in complexity. Green algae have very little structure, but, like all other taxa within the Archaeplastida, contain cellulose (Mikkelsen et al., 2014). Bryophytes have anchoring filaments and very primitive leaves that reflect differentiation of the original stem epithelium and regulate evaporation. Clubmosses were the first clade to possess more complex tissues than bryophytes: they have single-veined leaves, simple roots, and vascularized stems to facilitate water transport that, respectively, derive from primitive leaves, anchoring filaments, and the epithelia of the moss stem. Ferns and all subsequent clades have megaphyll (multiveined) leaves that are modifications of single-veined

Figure 11.9 **Flower symmetry in angiosperms:** (a): Trimerous petals in an open flower of a tulip (monocot). (b): Trimerous petals in the flower of a paw-paw (a magnoliid dicot). (c): Tetramerous petals in the flower of correa alba (a eudicot). (d): Pentamerous petals in the flower of *Crassula ovata* (a eudicot).

(a   Courtesy of Derek Ramsey. Published under a GNU Free Documentation License
b   Courtesy of "Phyzome." Published under a CC Attribution—Share Alike 3.0 Unported License.
c,d Courtesy of JJ Harrison. Published under a GNU Free Documentation License)

leaves. In gymnosperms, the seeds derive from spores but develop from leaves; their characteristic cones are actually modified groups of leaves.

While most of the major plant clades have relatively well-defined synapomorphies (Table 11.1), the angiosperms are more diverse as they include three very distinct clades: monocotyledons, dicotyledons and eudicotyledons (true dicots). Each clade has a distinct pattern of synapomorphies associated with their leaves, flowers, and pollen (Table 11.1, Figs. 11.4 and 11.9). The last of these, an innovation inherited from gymnosperms, is clade-dependent, but all have furrows with apertures that swell and open in the presence of moisture, so enabling the male gamete to escape its coat (Fig. 11.10).

Monocotyledons (e.g. palm trees, alliums, and particularly the grasses) have trimerous flowers, leaves with nonbranching veins and pollen with a single furrow (they are monocolpate). Dicotyledons (e.g. magnolia trees, laurels, water lilies, and the peppers [Piperaceae]) have leaves with branching veins, but also have trimerous flowers and monocolpate pollen. Eudicots (the great majority of angiosperms) also

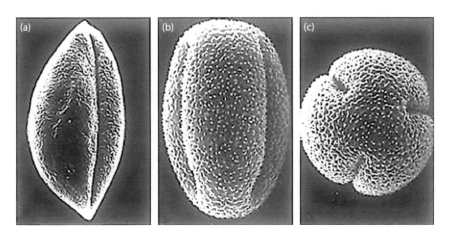

Figure 11.10 (a): Monocolpate pollen of a monocot (lily of the valley—*Convallaria majalis*). (b,c): Tricolpate pollen of a eudicot (*Aspazoma amplectens*). The pollen grains are 40–60 μm in length.

(Courtesy of Peter Sengbusch)

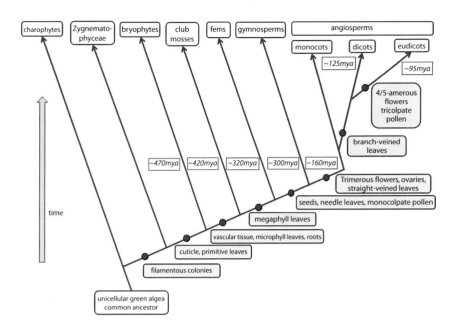

Figure 11.11 **Cladogram showing the evolution of some major plant taxa on the basis of some key synapomorphies (Pale green) that separate the phyla.** The numbers indicate the time of the oldest known fossils of these classes.

have leaves with branching veins but have tetramerous or pentamerous flower and tricolpate pollen (Figs. 11.9 and 11.10).

The evolutionary pathway is thus clear: stem angiosperms diverged off the gymnosperm line and separated into two clades: the monocots and the stem dicots, both of which had monocolpate pollen and trimerous flowers. These stem dicots then developed a pair of cotyledons and modern leaves with branching veins. The clade then branched again to give modern dicots, which are relatively unchanged and eudicots with tricolpate pollen and a more varied range of petals in their flowers (Soltis et al., 2008). It should be emphasized that the simplicity of the analysis for the major clades does not extend to the minor taxa: classifying trees, for example, on the basis of their anatomical characters alone can be particularly difficult because so many species look similar. The solution for grouping living species is, of course, to use molecular phylogenetics to construct phylograms.

KEY POINTS

- The last common ancestor of the Archaeplastida (glaucophytes, green and red algae, and plants) was a bikont protist that endosymbiosed a cyanobacterium >1.6 Bya; this became a chloroplast characterised by the presence of chlorophyll a and phycobiliproteins. Its contemporary unicellular descendants are glaucophytes.

- The red algae evolved from glaucophytes through simplifying the chloroplast and losing flagella.

- Green algae evolved from an early red alga, before it lost its flagella, acquiring chlorophyll b but losing the phycobiliproteins.

- There were several further cases where a bikont protist endosymbiosed a unicellular red or green alga and hence acquired a chloroplast. Organisms with algal features are, therefore, to be found in many bikont phyla.

- All algae are either unicellular or have a simple epithelial morphology that produce spores for reproduction.

- Plants evolved from land-based green algae and have far more complex cells and tissues

- The first plants were bryophytes (e.g. mosses) with no root and only primitive leaves; these were followed by clubmosses with vascularised microphyll leaves and roots. Over time, lignin evolved that combined with cellulose to make wood, so allowing plants to become large.

- Ferns evolved with larger, megaphyll leaves but still reproducethrough spores.

- Gymnosperms evolved during the Carboniferous; they were the first plants to use seeds for reproduction. These were initially made by leaves and only within cones (which derive from groups of leaves),

- Angiosperms evolved from gymnosperms and are characterized by having a flower that has a complex reproductive system comprising that may include both male (a stamen whose anther makes pollen) and female components (an ovary with a stigma and style to receive pollen).

- There are three clades of angiosperms.

  - Monocots have a single seed leaf (cotyledon), trimerous flowers, leaves with straight veins, and monoculpate (single exit) pollen.

- Dicots diverged from monocots and are characterised by two seed leaves, trimerous flowers, leaves with branched veins and monocolpate pollen

- Eudicots diverged from dicots and are characterized by tricolpate (triple exit) pollen and either tetramerous or pentamerous flowers.

FURTHER READINGS

Acquaah G (2021) *Principles of Plant Genetics and Breeding* (3rd ed). Wiley Blackwell.

Minelli A (2018) *Plant Evolutionary Developmental Biology: The Evolvability of the Phenotype.* Cambridge University Press.

Niklas KJ (2016) *Plant Evolution: An Introduction to the History of Life.* University of Chicago Press.

Oborník M (2019). Endosymbiotic evolution of algae, secondary heterotrophy and parasitism. *Biomolecules*; 9(7):266. doi:10.3390/biom9070266.

One Thousand Plant Transcriptomes Initiative (2019) One thousand plant transcriptomes and the phylogenomics of green plants. *Nature*; 574:679–685. doi: 10.1038/s41586-019-1693-2.

Spatafora JW, Aime MC, Grigoriev IV, Martin F, Stajich JE, Blackwell M (2017) The fungal tree of life: From molecular systematics to genome-scale phylogenies. *Microbiol Spectr*; 5:10.1128/microbiolspec.FUNK-0053-2016.

Szövényi P, Waller M, Kirbis A (2019) Evolution of the plant body plan. *Curr Top Dev Biol*; 131:1–34. doi:10.1016/bs.ctdb.2018.11.005.

Willis K, McElwain J (2014) *The Evolution of Plants* (2nd ed). Oxford University Press.

WEBSITES

- The Wikipedia entries on the *Evolutionary history of plant evolution*, *Plastids,* the *Rhynnie chert* and the various *Algae* and plant phyla.
- Hybridization in plant: http://www.plantphysiol.org/content/173/1/65
- Simple introduction to plant physiology: http://www.uwyo.edu/botany4400/lecture%202.htm

The Ediacaran Period and the Early Evolution of the Metazoa

12

This chapter starts by examining the fossil record of the Ediacaran (541–485 Mya). Although there were probably a few small organisms present in the seas before the start of the Period, it was during the Ediacaran that large-scale, multicellular organisms evolved. The early part of the Period is noted for a clade of large flat organisms with a range of morphologies that do not look like anything alive today. They all seem to have been lost during the extinction that ended the Ediacaran.

Towards the end of the Period, however, the fossil record includes small animals, such as molluscs together with microscopic multicellular organisms that, as they were embedded in phosphate, are of remarkable quality. Their morphology makes it clear that they include sponges and the early embryos of multicellular animals. In addition, traces of tracks left by small worm-like organisms from then show that Metazoan evolution was well under way during the Ediacaran, confirming that the roots of the Cambrian explosion lie in that Period or perhaps even earlier.

The second part of the chapter considers the early evolution of the Metazoa in the context of the developmental events that had to have taken place for diploblasts and triploblasts, with two and three germ layers, to evolve. The key triploblast here was *Urbilateria*, the last common ancestor of deuterostomes and protostomes. The chapter ends by considering what can be deduced about *Urbilateria's* anatomy through the analysis of homologous transcription factors in mice and *Drosophila* that are known to play a key role in mediating tissue formation.

During the Cryogenian (750-635 Mya), the Earth went through a period of intense cold known informally as snowball Earth, although slushball may have been a more accurate term. The reasons for this period of cold are not precisely known but, when it ended, the earth's biota seems to have been poised to make the leap from single-cell or colonial life to one where multicellular life forms were present in the seas.

Although there is phylogenetic evidence that metazoan evolution started ~800 Mya during the Tonian (1000–750 Mya; Sperling & Stockey, 2018), the fossil record for this and the Cryogenian (750-635 Mya) is very limited (see below). Nevertheless, there must have been a rich marine biota of protists, both unicellular and social, and perhaps a few small multicellular organisms (Knoll, 2015) well before the start of the Ediacaran (635–541 Mya). This was the period in which the earliest fossil evidence for larger-scale organisms was first found. It was originally called the Vendian but renamed because its first fossils were found in the Ediacara Hills near Flinders, Australia.

During the early Ediacaran, the only well-known form of multicellular life seems to have been a taxon of flat organisms of unknown provenance. They particularly diversified after the Avalon explosion (575 Mya; Shen et al., 2008) and, today, those early marine organisms are mainly preserved as compression fossils. These are impressions of the organism's surface made in sand or mud and maintained when they lithified to become sandstone or shale. The fossil record for the earliest extant

DOI: 10.1201/9780429346217-15

metazoan taxa was laid down during the latter part of the Ediacaran. The full reasons for these bursts of evolutionary activity are not known, but one reason may have been the increase in oxygen level that occurred ~560–550 Mya (Dahl et al., 2010). Associated with this activity was the increase in complexity seen in early Metazoa (Droser & Gehling, 2015).

The fossil record shows that by the end of the Ediacaran, a wide range of organisms were present in the seas. These included primitive sponges, diploblastic coelenterates with radial symmetry (e.g. very early cnidarians), and the first primitive triploblastic organisms with bilateral symmetry. A taxon with this basic morphology became the last common ancestor of all triploblast organisms: it was a small worm-like animal known as *Urbilateria*. Although the original Ediacaran organisms died out following an extinction at the end of the period (Appendix 3), the early metazoan organisms survived, flourished, and evolved to produce all of today's animals.

THE EDIACARAN BIOTA

Early Ediacaran fossils, such as *Dickinsonia costata* and the frond-like *Rangea scheiderhoehni* (Fig. 12.1a,b), are often very beautiful, but remain enigmatic since their exact relationships to both earlier and later organisms remain unclear. A few Ediacaran organisms showed radial symmetry, with some having a holdfast to glue them to rocks (Dzik, 2003). Examples here are *Petalonamae* and *Vendoconularia*; the latter had six-fold radial morphology and is a candidate for an early cnidarian (Ivantsov & Fedonkin, 2002). Some of the later Ediacaran organisms were, however, clearly bilateral and may have been very early examples of the great majority of current animals. Examples of these are *Rangea scheiderhoehni, Spriggina floundensi,* which had aspects of trilobite morphology and *Kimberella quadrata,* which could well have been a primitive mollusc (Fig. 12.1b–d; see below, Fedonkin & Waggoner, 1997).

Ediacaran rocks also include fine burrows that could well have been trace fossils of the primitive worm-like organisms (e.g. *Yilingia spiciformis* [551–539 Mya]. Fig. 12.2; Chen et al., 2019) that had evolved by the end of the period. Supporting evidence for this view comes from the recent discovery of fossils of *Ikaria wariootia* that have recently been found in South Australia and that date to 560–551 Mya. These appear to be very small marine worm-like organisms 3–7 mm in length that burrowed through sand in shallow waters. If this view is correct, such small worm-like organisms are the most primitive members of the Bilateria so far discovered (Evans et al., 2020). It is also

Figure 12.1 Ediacaran fossils.
(a): *Dickinsonia costata*; (b): *Rangea scheiderhoehni*; (c): *Spriggina floundensi*; (d): *Kimberella quadrata*. (Bar: 1 cm.)

(a Courtesy of "Verisimilus." Published under a CC Attribution–Share Alike 3.0 Unported License.
b Courtesy of "Retallack." Published under a CC Attribution–Share Alike 4.0 International License.
c Courtesy of "Merikanto-commonswiki." Published under a CC Attribution–Share Alike 3.0 Unported License.
d Courtesy of Alexei Nagovitsyn. Published under a CC Attribution–Share Alike 4.0 International License.)

Figure 12.2 **Fossil of *Vilingia spiciformi*.** This bilaterian from the late Ediacaran (551–539 Mya) showed chevron-like structure (left), a featureless trail (right) and lateral grooves (white arrows; Bar: 1 cm).

(Chen Z et al. (2019) *Nature*; 573:412–415. With permission from Nature Springer.)

of interest that, while it has generally been thought that all Ediacaran lifeforms were marine, there is evidence that some lived on land (Retallack, 2013).

It seems that, after a period of success for the early sessile Ediacaran organisms, they went into decline (~550–541 Mya) some time before the end of the period (535 Mya). This was accompanied by an increase in the number of sessile eumetazoan taxa (e.g. sponges; Yin et al., 2015), although many of these also died out by the end of the Period. It is not clear whether this was due to environmental pressures or whether they were driven out by increasing populations of early cnidarians, such as jellyfish or by those primitive bilaterians that had by then evolved. The reasons for the end of the Ediacaran (535 Mya) are not known but an important factor may have been the release of the large amounts of hydrogen sulphide that is known to have occurred around then (Wille et al., 2008). By the time that the Cambrian started, the early Ediacaran biota had been lost leaving only small marine organisms; these were the early Metazoa.

Some of the most intriguing Ediacaran fossils of the early Metazoa, remarkable for their quality, are the microfossils (<1 mm) preserved in phosphate that have been found in the Weng'an biota located in the Doushantuo formation in modern-day China (Fig. 12.3; Yin et al., 2007). They date to between 580 and 570 Mya, a time when today's location was in the northern-hemisphere region of the Pannotian supercontinent. The initial suggestion that the fossils included invertebrate embryos was doubted for some time, but Chen et al. (2014) examined further examples of such fossils and were able to establish that these very early organisms show evidence for cell differentiation, germ–soma separation and programmed cell death. Examples of successive developmental stages have also been found (Fig. 12.3a–f) showing that they are likely to be of metazoan origin, although it has proven hard to assign them to a particular taxon. This uniquely rich collection of microfossils also includes examples that were probably sponges and algae as well other small invertebrates (Fig. 12.3g–l; Cunningham et al., 2017)

Until recently, the Doushantuo fossils had been seen as unique, but a further site with similar fossils has now been found that dates to 570–560 Mya and includes a wide variety of small animals and their embryos (Willman et al., 2020). This site is now in North Greenland but, during the Ediacaran, it was part of Laurentia in the middle latitudes of the southern hemisphere (Appendix 3). This new find not only expands our knowledge of early animal morphology but implies that early animal organisms were widely distributed through the Ediacaran world.

One important question is when sponges first originated; here, the fossil and computational work have given slightly different answers. Until recently, the earliest evidence of fossilised sponges had been towards the end of the Ediacaran (Fig. 12.3l; Antcliffe et al., 2014). This contrasted with biomarker and phylogenomic analysis which pushes the time back to perhaps the late Cryogenian (670–635 Mya; Zumberge et al., 2018) or a little earlier (Dohrmann & Wörheide, 2017). One possible reason for the difference in these two dates lies in there being two sorts of sponge: those without and those with biomineral spicules, with the latter being far more likely to become fossils than the former (Tang et al., 2019). The differences between the two lines of evidence would be explained if the softer sponges evolved before the mineralised ones, and it is these latter taxa that date to the late Ediacaran (Chang et al., 2019).

These datings for the origin of the sponges are, however, having to be reviewed in the light of a completely unexpected discovery by Turner (2021). She has found fossilised meshworks of tubules in a reef dating back to ~890 Mya (Neoproterozoic Eon, Tonian Period) that have the structure of contemporary keratose demosponges, a primitive form of soft sponge. This discovery, if confirmed, places the origins of the sponges and hence of the root of the Metazoa >200 My earlier than hitherto thought.

Figure 12.3 SEM images of fossils from the Weng'an biota. (a-f): *Tianzhushania* specimens at various stages of division from a single cell (a) to many hundreds of cells (f) (courtesy of the Swedish Museum of natural history (SMNH) x 6449–SMNH x 6454). (g): *Helicoforamina* SMNH x 6455. (h): *Spiralicellula* (from Tang et al., 2008). (i): *Caveasphaera* SMNH x 6456. (j): *Archaeophycus*, a putative red alga SMNH x 6457. (k): *Mengeosphaera*, an acritarch (unclear species) SMNH x 6458. (l): *Eocyathispongia*, a putative sponge, Nanjing institute of palaeontology and geology (NIGPAS) 16176. Scale bar: (a) 320 µm, (b) 265 µm, (c) 265 µm, (d) 200 µm, (e) 245 µm, (f) 280 µm, (g) 395 µm, (h) 380 µm, (i) 250 µm, (j) 255 µm, (k) 130 µm, (l) 415 µm.

(Courtesy of John Cunningham. h Courtesy of Liu Pengju. l From Yin et al. (2015) *Proc Natl Acad Sci* U S A; 112:E1453–E1460. With permission from PNAS.)

There have now been several published studies that have attempted to make taxonomic sense of the Ediacaran biota. Dzik (2003) showed that they included ancestral forms that could be seen as ancestors of various modern taxa, while Droser and Gehling (2015) concluded that "aspects of the ecology, such as trace fossils, taphonomy, and morphology, reveal that these fossils show characteristics of modern taxa." This view is confirmed by more work (Darroch et al., 2018). Muscente et al., 2019) based on a detailed, network-based analysis of all the fossil data across the major regions of Ediacaran fossils (including Avalon [Canada, 575–565 Mya], the White Sea region [Russia, 560–550 Mya], and Nama [Namibia, 550–542 Mya]), together with a developmental analysis of key Ediacaran organisms (Dunn et al., 2018). This work as a whole clearly indicates that the Metazoa have their ancestral roots in Ediacaran organisms, and hence that their appearance is rather earlier than had previously been thought.

There is an interesting footnote on Ediacaran life. It has long been assumed that all Ediacaran organisms had either been lost or had evolved into forms bearing little resemblance to their original morphology. A recent intriguing observation, however, suggested that one group at least of these Ediacaran organisms might still be extant, hidden in the depths of the ocean. Just et al. (2014) re-examined material that had been dredged up from off south-east Australia in 1986 and fixed in ethanol and formaldehyde. While doing this, they identified two species of small (2–5 mm) mushroom-shaped, diploblastic organisms that they named *Dendrogramma enigmata* and *D. discoides* (Fig. 12.4).

Figure 12.4 **Dendrogramma enigmata and discoides (*) were thought to be surviving organisms from the Ediacaran period.**

(From Just et al. (2014) *PLoS One*; 9:e102976. Published under a CC BY License.)

1 mm

These organisms were superficially like Ctenophorans (comb jellies) or Cnidaria (e.g. jellyfish), but they lacked many of their cell types and tissues, particularly their gonads, although this might have been through immaturity. Just et al. (2014) pointed out that both had a strong morphological similarity to Ediacaran organisms and might have been the relatively unchanged descendants of an early ancestor of the Ctenophera and Cnidaria. Not surprisingly, they ended their paper with the suggestion that a further search for living examples of *Dendrogramma* was needed. This has now been done and phylogenetic analysis of the resulting specimens has sadly shown that the extant organism is actually part of a living cnidarian (a *Rhodaliidaen siphonophore*; Fig. 13.4b) rather than an Ediacaran relic (O'Hara et al., 2016).

THE EVOLUTION OF DIPLOBLASTIC AND TRIPLOBLASTIC EMBRYOS

The first stage of metazoan progression was probably a primitive sponge that had no obvious symmetry, a view supported by the phylogenetic data that puts sponges as a sister taxon to all other Metazoa (Renard, 2018); such a conclusion tallies with expectation. This was soon followed by the evolution of diploblastic organisms with radial symmetry that were composed of epithelia that formed an external ectoderm and an internal endoderm-derived gut; today, these are the coelenterates, which include the Cnidaria and Ctenophora. The last step in these early Ediacaran events was the evolution of the first organism with bilateral symmetry and three germ layers; this evolved into the last common ancestor of all triploblastic animals, a primitive bilaterian known as *Urbilateria*.

Before considering the fossil evidence of the early Bilateria (Chapter 13), we need to consider the embryonic origins of organisms that were far more complex than coelenterates. Little is known of this directly as, although we have early embryos embedded in phosphate (Fig. 12.3; Chen et al., 2014), we as yet have nothing earlier to indicate how they evolved or how they developed. It is, however, possible to use cladistic approaches to probe the order in which layers of complexity were added as early metazoan embryos evolved, and molecular-genetic analysis to identify some of their early anatomical features. It is only from this dual perspective that one can properly appreciate the events that led to the evolution of the Bilateria in the late Ediacaran and their diversification in the Cambrian. The sequence of evolutionary changes discussed below are summarised in Fig. 12.5.

The primary structural feature that categorises the development of an animal is the number of germ layers in its very early embryo. The simplest multicellular organisms, the sponges, have, in effect, a single germ layer that generates primitive epithelia

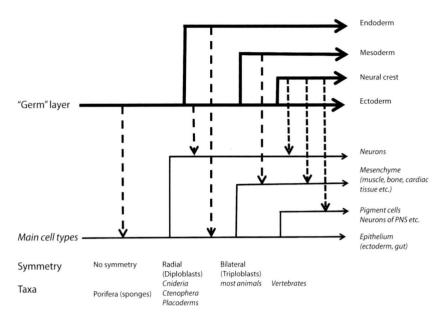

Figure 12.5 Diagram showing the evolution of the germ layers in animals (bold lines) together with some of the cell types cell types associated with each (dotted lines). The lower text indicates the types of organisms that evolved at each stage of the progression.

(Leys et al., 2009) that, in turn, give rise to a range of derived cell types; sponges have little tissue symmetry. Organisms with radial symmetry (the Coelenterata), that include the cnidarians (e.g. jellyfish and sea anemones) and the ctenophores (comb jellies) are diploblasts: they have two epithelial germ layers, ectoderm and endoderm, both of which are single-layer cell sheets. In these animals, the ectoderm forms the external layer while the endoderm forms the internal gut tube. During development, nerve cells and future muscle cells form from the ectoderm (Leclère & Röttinger, 2017).

Protostomes

The very great majority of animals today are, however, triploblast protostomes: their embryos have an epithelial ectoderm that forms the outer cell layer and the nervous system, an epithelial endoderm that forms the gut as it invades a hole in the blastula that becomes its mouth, and a middle layer, the mesoderm. This produces muscles and other mesenchymal structures such as the circulatory system and connective tissue that, in vertebrates, includes the skeleton. (*Note*: Mesenchymal cells, unlike those in epithelia, form 3D masses rather than 2D sheets – see Appendices 6 and 7). The simplest explanation for the origin of the mesoderm is that towards the end of the Ediacaran Period, muscle-forming cells, a late epithelial derivative in coelenterates, separated from the endomesoderm to become a distinct germ layer (Seipel & Schmid, 2005).

There has been considerable discussion over the last 20 years as to whether this view is correct and one possible alternative is that all animals were originally triploblastic, and today's coelenterates reflect a simplification. Recent molecular and bioinformatics work by Wijesena et al. (2017) has, however, reinforced the original view that the Bilateria evolved from a radially symmetric organism. They showed that the β-catenin/Tcf and TGF/BMP signaling responsible for patterning the germ layers in the Bilateria also pattern the endomesoderm of the cnidarian *Nematostella vectensis* as well as having a role in its gastrulation. They also show that some of the DPP/BMP-activated network that drives the simple rhythmic contractions seen in part of the cnidarian muscle system is also used to drive the more complex cardiac muscle activity in the Bilateria. Wijesena et al. (2017), therefore, suggest that it is highly likely that the three germ layers of the Bilateria, together with some of its rhythmic activity, evolved from the two germ layers of the diploblasts. It is hard to disagree with them.

The acquisition of a third germ layer was accompanied by a basic change in early embryonic geometry as radial symmetry became bilateral, mirror-image symmetry; hence, members of the resulting clade are collectively known as the Bilateria. Once in place, of course, mesoderm was available to form a range of nonepithelial cell types other than muscle, and its further differentiation opened up a range of novel anatomical options for new taxa. Natural variation and the absence of predators enabled the resulting novelties to colonise hitherto empty ecological niches.

One possible reason why the Bilateria diversified so much more than the coelenterates comes from the fact that anatomical options available to organisms with mirror symmetry are far greater than to those with radial symmetry because the former has three independent axes along which spatial variation can happen rather than two (there are obvious limitations associated with the radial axis of coelenterates). It is, therefore, unlikely to be by chance that there are only two animal phyla with radial symmetry (Cnidarians and Ctenophera) whereas there are more than thirty in the Bilateria, which can elongate easily (note that echinoderms are initially bilateral and only secondarily show rotational and radial symmetry; see Chapter 14). All of these phyla are protostomes except for the four that are included in the deuterostomes. The most speciate of the former is the arthropods each of whose anterior-posterior segments has provided independent opportunities for a succession of anatomical variants.

Deuterostomes

Although deuterostomes are not properly considered until Chapter 14, it is worth mentioning here the additional evolutionary steps that led to them. The first follows from their mode of gastrulation, during which endodermal cells migrate inwards to form the gut that will extend from the mouth to the anus. The future nature of an embryo depends on the fate of the first opening of this endodermal tube: if it becomes the mouth, the organism is a protostome; if it becomes the anus, the animal is classified as a deuterostome (the chordates, hemichordates and echinoderms). This definition really only reflects early evolutionary forms as the details of gut formation in later protostomes and deuterostomes have been affected by many secondary factors (e.g. yolk morphology or the presence of a placenta and umbilical cord in deuterostomes and the type of egg in protostomes). The second important difference is in the location of the main nerve cord. Protostomes produce their nervous system from neurectodermal cells on the ventral (lower) side of the embryo and the gut is dorsal to the cord. In contrast, the neural tube of deuterostomes, which becomes the brain and spinal cord, forms in the dorsal region of the embryo and the gut lies ventral to it (see Chapter 8). The third is an internal skeleton.

In addition, vertebrates acquired what can be considered as a fourth germ layer, the neural crest. This forms from dorsal ectoderm at the long anterior-posterior boundaries between the lateral regions that will become the external epithelium and the medial region that will become the neural tube. The cells from these two linear domains of neural crest soon migrate ventrally and colonise much of the embryo. There they produce a wide range of cell types and tissues that include neurons whose cell bodies are outside the central nervous system (i.e., sensory nerves, the enteric [gut] nervous system, and part of the autonomic nervous systems), pigment cells and much of the skull (the face, jaws, and parts of the eyes and ears; Steventon et al., 2014). It is, however, likely that the neural crest did not evolve until the Cambrian as it is only then that the fossil record shows chordates with neural-crest-derived tissues, such as teeth (Huysseune, 2009). The earliest known chordates, such as *Haikuella* (Chapter 14) had pharyngeal teeth, although *Pickaia*, a Cambrian cephalochordate and hence a much simpler organism, did not. Today, there is still a class of cephalochordates (the Branchiostomata) that lack neural crest in their embryos and all its subsequent tissues in adults, particularly the sensory tissues of the head region (see Chapter 14, Appendices 6 and 7).

A series of protein networks, signals, receptors and transcription factors underpins all these developmental events (Chapter 8; Appendix 7). Phylogenetic analysis has shown that many of these proteins (e.g. homeodomain transcription factors) have very ancient roots, not all of whose functional origins are yet clear. It is straightforward to see how the networks that drove metazoan cell differentiation and morphogenesis had their origins in social protists, which had several cell types that could express adhesion proteins enabling them to live as individual cells or in groups. It is, however, harder to understand the origins of the pattern-forming (differential-signaling) mechanisms that drove organogenesis in the Metazoa and ensured that originally homogeneous cells differentiated according to their locations on the two and three axes of early coelenterates and bilaterian organisms.

THE EVOLUTION OF *URBILATERIA*

The fossil evidence makes it clear that the evolution of all the core stages in the development of animals, other than that of the neural crest, had taken place well before the end of the Ediacaran. The key organism here is *Urbilateria*, the last common triploblast ancestor of the Bilateria, the clade that includes both protostomes, such as the fruit fly, *Drosophila melanogaster*, and deuterostomes, such as the mouse. Cladistic analysis clearly shows that this organism existed but neither it nor the fossil evidence say much about the likely anatomy of the common ancestor of such profoundly different organisms. There is, however, quite a lot that can be teased out about it from molecular genetics and phylogenetics.

For all of their obvious phenotypic differences, there are some basic morphological similarities between the mouse and *Drosophila*: both are bilateral, triploblastic, have a body plan that includes segments, a nerve cord, a gut and a coelom, as well as having very similar cell types. Although the morphological resemblances break down at the tissue level, both have similar physiological requirements, even though the relevant tissues are based on different anatomical structures. The mouse thus has a four-chambered heart to pump blood to the lungs for oxygenation and then to circulate it round the vascular system. *Drosophila* has a simple heart tube that pumps haemolymph round the body cavity where it acquires oxygen through its diffusion from ectodermal tracheal tubules that are open to the atmosphere and extend inwards into the body cavity. The mouse has a liver that fulfils a wide variety of biochemical functions, many of whose roles in *Drosophila* are fulfilled by the fat body, while excretion is handled by kidneys in the mouse and by the Malpighian tubules in *Drosophila*. Both, of course, have eyes, reproductive systems and associated gender differences, but their respective morphologies are quite different.

What needs to be established is whether pairs of different organs in mouse and *Drosophila* with common functional abilities reflect homologies (common descent) or homoplasies (convergence from different ancestral features). If the former, they were likely to be present in *Urbilateria*. The way that this problem was first approached was through computational analysis to identify possible proteins whose homologues are expressed in such pairs of tissues (De Robertis & Sasai, 1996). Analysing mouse and *Drosophila* genes alone cannot, however, give a complete picture of this ancient organism. This is because individual genes can, over time, be lost from one or another species and their functions replaced by others so that neither the mouse nor *Drosophila* would be expected to include all of the key genes of *Urbilateria* (De Robertis, 2008; Matsui et al., 2009).

A stronger bioinformatics approach to identifying at least part of the *Urbilateria* proteome is to identify the overlap of the genes in a wider range of organisms, a task undertaken by Wiles et al. (2010): they looked for homologues in genomes of some of the major model organisms; these include the mouse, humans, *Drosophila*, *Caenorhabditis elegans* (a nematode worm), and yeast (a unicellular fungus). Their results, best illustrated in a Venn diagram, identified over 2000 genes whose orthologues are common to vertebrates, the nematode worm and *Drosophila*, a number that they viewed as conservative (Fig. 12.6). They also found that many of these genes are also present in yeast, a unicellular fungal taxon whose last common ancestor with animal phyla was probably >1 Bya. These homologues are likely descendants of those genes present very early on in the history of eukaryotic life and are candidate genes for those in *Urbilateria*.

Analysis of these genes showed that they were primarily part of the basic cell biology toolbox, with proteins being involved in mitosis, meiosis, membrane function, cytoskeletal activity, and the biochemical machinery of the cell. This was not unexpected, but also not very helpful for probing the development of *Urbilateria*'s anatomy. One class of protein that one would expect to find in multicellular organisms but not in yeast would have been the signal/receptor pairs that mediate patterning and other developmental events in multicellular organisms but not likely to be found in yeasts. One might also have expected to find components of the various process networks that drive development (e.g. for differentiation and morphogenesis; Chapter 8; Appendix 7). In this context, a particularly interesting network is the β-catenin mechanotransduction pathway that links mechanical stresses to changes in gene expression (Brunet et al., 2013): closely related versions are found in both *Drosophila* and the zebrafish, and a much earlier version was thus probably present in *Urbilateria*. Such genes are not, however, the most helpful for understanding anatomy as they are widely used across a single embryo

There is, however, a further and very different approach to identifying the key genes that might have underlain the development of the anatomy of *Urbilateria*. We know that organogenesis in all triploblasts is determined by high-level transcription factors whose activation leads to the laying down of new tissues. A good test for homologous function, therefore, is to investigate the effect of knocking out a pair of homologous proteins in two very different organisms. If the result is homologous defects in both embryos, it is a

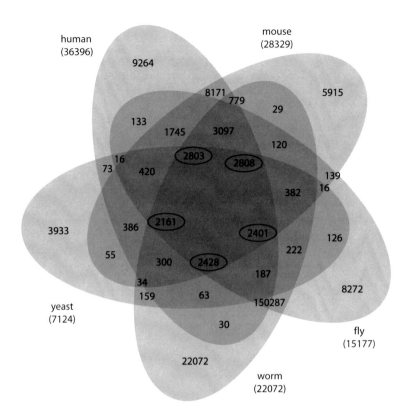

human
(36396)

mouse
(28329)

9264

8171   779

5915

133   29

1745   3097

120

16   2803   2808

73   420

139
16

382

3933   386   2161

2401   126

55

300   2428

222

34

159   63

187

8272

150287

30

22072

yeast
(7124)

fly
(15177)

worm
(22072)

Figure 12.6 Five-way Venn diagram of orthologous genes in humans, mouse, *Drosophila*, *C. elegans*, and yeast. It should be noted that many genes in each species are paralogues. The numbers represent the counts of genes overlapping between one species and the others are circled. All other intersections reflect the genes found in the organism with the largest number in that group.

(Slightly modified from Wiles et al. (2010) *BMC Syst Biol* 4:36. With permission from Nature Springer.)

fair assumption that the last common ancestor of the two species had an ancestral gene whose protein had a similar role (not necessarily an identical one because the ancestor was anatomically different from both of its descendants). If the two species are from either side of the protostome/deuterostome divide (as are *Drosophila* and mouse), then it is reasonable to suppose that the common ancestors of today's homologous genes were present in *Urbilateria,* doing part at least of what they do today.

Some of this knockout work has now been done, and the results have identified several important transcription factors whose homologues are expressed in functionally equivalent organ systems of mouse and *Drosophila*. Key examples are given in Table 12.1, together with the phenotypes of the knockouts. This shows, for example, that the development of the very different eyes and hearts in the two organisms each require the activity of a homologous transcription factor, while the identities of the very different segmented structures both use Hox coding to determine their future. However, the details of the molecular mechanisms by which this coding is first laid down and then translated into morphology still remain unclear and are different in the two organisms.

The current data on transcription-factor knockout observations in mouse and *Drosophila* given in Table 12.1 can be combined with the common features mentioned earlier to give some insight into the likely features of *Urbilateria*. It seems that the fertilized egg of this organism produced a bilaterally symmetrical, triploblastic, protostome embryo whose antero-posterior patterning enabled it to develop into a worm-like organism having a gut, a tail, segmented muscles (which may or may not have derived from overtly segmented mesoderm), a nervous system with perhaps a small ganglionic brain that was probably ventral to the gut (Chapter 8), an olfactory system, small areas of ectoderm sensitive to light, and a reproductive system. It also had a wide range of differentiated cell types that included nerves, muscles, epithelium, and blood cells. Of particular interest is the likely presence of a heart as this suggests that the adult organism was large enough to require a circulation unlike, for example, a nematode worm, such as *C. elegans* that is only ~1 mm long.

The list of early patterning homologies in mouse and *Drosophila* so far identified on the basis of knockouts is unlikely to be complete. For example, homologous genes also seem to be involved in some of the axis and germ layer patterning in the very early development of all animal embryos (Chapter 8; Holland 2015; Barresi & Gilbert, 2019). Actual proof of any role here is, however, very difficult for two reasons. First, the effect of knocking out key genes expressed during the early stages of embryogenesis

TABLE 12.1 SOME HOMOLOGOUS TRANSCRIPTION FACTORS IN MOUSE AND DROSOPHILA* (ADAPTED FROM HELD, 2014)

Drosophila gene	Knockout phenotype affects	Mouse gene	Knockout phenotype affects
Hox system	Segment identity	Hox system	Segment identity
Pax6	Eye formation	Pax6	Eye formation
Tinman	Heart tube	Nkx2-5	Heart muscle
Bagpipe	Gut mesoderm	Bapx1	Gut mesoderm
Byn	Hindgut formation	Brachyury	Tail formation
Ems	Peripheral olfactory neurons	Emx1/	Peripheral olfactory neurons
Otd	Central olfactory neurons	Otx/2	Central olfactory neurons
Dsx	Male sexual development	Dmrt1	Testis & sperm development
GATA-factor (Pannier)	Blood cell maturation	GATA-1,2,3	Erythroid maturation
GATA-ABF (Serpent)	Gut, haematopoeisis, fat body**	GATA-4-5-6	Gut, liver

* Some knockout phenotypes have wider abnormalities than those mentioned here.

** The fat body is the *Drosophila* equivalent of the liver.

is likely to prove lethal, a result that would emphasise the importance of the protein but not its function. Second, it cannot be assumed that the production of a tissue is directed by a single gene alone: the approach will fail if there is redundancy and an alternative transcription factor is available to fulfil the role of the one that had been knocked out (e.g. Ruijtenberg & van den Heuvel, 2015). The absence of an abnormal phenotype would, in this case, be a false negative.

Urbilateria clearly flourished in its marine environment so that, well before the beginning of the Cambrian, its descendants had produced a wide range of species with very different body forms. It is probable that *Urbilateria* had relatively few if any predators and that the selection pressures on its early variants were consequently very low. The net result was that a host of different species rapidly evolved with the range of body forms now seen in the fossil record (Chapter 3). Perhaps it was only towards the end of the Cambrian that populations were sufficiently large and crowded that selection as a result of environmental pressure rather than by functional failure started to become important, but by then phylum diversity was on its way (De Robertis, 2008).

KEY POINTS

- The Ediacaran was the first period during which large organisms evolved.

- Many seem to have been frond-like with holdfasts with which they adhered to the ocean floor. All of these early Ediacaran organisms appear to have been lost, leaving no equivalent such forms today.

- It was during the Ediacaran that modern organisms first evolved, as primitive sponges became more complex and coelenterates developed.

- The fossil record has identified a host of early metazoan organisms that lived towards the end of the period, with microfossils embedded in phosphate showing exquisite detail.

- Worm-like organisms that were bilaterial and triploblastic evolved before the end of the Ediacaran.

- A key example of these was *Urbilatera*, the last common ancestor of all phyla showing mirror symmetry (i.e., protostomes, such as *Drosophila* and deuterostomes, such as the mouse). It lived a little before the beginning of the Cambrian Period.

- Structurally and functionally homologous transcription factors required for organogenesis in contemporary protostomes and deuterostomes have identified some of the likely features of *Urbilateria*.

FURTHER READINGS

Droser ML, Gehling JG (2015) The advent of animals: The view from the Ediacaran. *Proc Natl Acad Sci*; 112:4865–4870. doi:10.1073/pnas.1403669112.

Newman SA, Bhat R (2009) Dynamical patterning modules: A "pattern language" for development and evolution of multicellular form. *Int J Dev Biol*; 53:693–705. doi:10.1387/ijdb.072481sn.

Sperling EA, Stockey RG (2018) The temporal and environmental context of early animal evolution: Considering all the ingredients of an "explosion." *Integr Comp Biol*; 58:605–622. doi:10.1093/icb/icy088.

WEBSITE

- Wikipedia entry on *Ediacaran biota* and the *Avalon explosion*.

The Cambrian Explosion and the Evolution of Protostomes

13

The origins of the Metazoa were in the Ediacaran. Early taxa included sponges (Poriphera), diploblast coelenterates with radial symmetry (Ctenophera and Cnidaria) and triploblast bilaterians with mirror symmetry. It was not, however, until the Cambrian that the Metazoa started to diversify substantially. Most of the new organisms were protostomes (today's invertebrates) but a few were deuterostomes (the Chordata and a few minor taxa).

The first part of this chapter examines invertebrate evolution from the Cambrian onwards, focusing on the cladistic analysis of how the various taxa diverged. The key to understanding the anatomical basis of triploblastic diversification is the realization that, while coelenterate diploblasts such as the Cnidaria have two germ layers and radial symmetry for variation, triploblast bilaterians have three germ layers and tissues that can vary independently over three linear axes. They thus have many more opportunities for morphological variation than diploblasts. The specific features that characterize each phylum within the Bilateria come from early changes to this basic geometry, while diversity within each phylum resulted from secondary tinkering with the patterning of the early tissues of its organisms.

The second part of the chapter considers in more detail the evolution and diversification of the molluscs and arthropods as these are the invertebrate clades showing the most diversity today. The molluscs, which are characterised by a mantle, include organisms with a single or a hinged (bivalve) shell, cephalopods, gastropods, and other marine and terrestrial forms. Arthropods, which are characterised by segmented limbs, include marine crustaceans, myriapods (e.g. centipedes), arachnids (spiders), and hexapods. The last of these is by far the largest phylum, with the most populous clade being the insects, which include beetles, bugs, bees, flies, and butterflies. The chapter ends by considering the changes needed for a marine invertebrate to survive on dry land and how insects acquired the ability to fly.

The term "invertebrate" was originally coined by Lamarck (1809) and today is an umbrella word for the ~30 phyla and more than a million animal species that lack vertebrae. It is important to distinguish between the diploblastic Coelenterata and the triploblastic Bilateria. The latter are all descendants of a last common ancestor that was a primitive worm-like animal, informally known as *Urbilateria*; they originated during the Ediacaran and are the main focus of this chapter. The fossil record now suggests that the various events that define the early stages of triploblast embryogenesis had occurred sometime before the end of the Ediacaran (541 Mya; Chapter 12). It was not, however, until the Cambrian that the fossil record shows the beginnings of the vast diversity of metazoan phyla that we see today (the Cambrian explosion). By the middle of the Period (~497 Mya), there were many distinct triploblast phyla, all being protostomes apart from the very small deuterostome taxon (i.e., the clade that includes the vertebrates).

The fossil record of the early multicellular Metazoa is far better than might be expected, particularly up to the middle Cambrian period. This is because this

DOI: 10.1201/9780429346217-16

record is both extensive and of very high quality, with more of it showing soft-tissue preservation than for any other period. The standard explanation for this is that when these small organisms died they were buried in fine-grained minerals (e.g. mud) under anoxic conditions; they thus decayed slowly enough for the soft tissues to be fossilised. An additional factor may have been that early worm-like phyla that now eat and so remove the soft tissues of dead animals were unlikely to have evolved or at least been widely distributed until after a secondary extinction during the Cambrian (~513 Mya, Zhuravlev & Wood, 2018). This factor may have given earlier Cambrian organisms a better chance of having their soft tissues fossilised than those of later periods (Brasier, 2010).

The Cambrian fossil record

The Ediacaran and Cambrian records of marine organisms suggest that the very great majority of the current animal phyla evolved over ~25 My, a relatively short period in evolutionary terms, although, of course, this estimate is liable to change if new fossils are discovered. The Chengjiang biota in Yunnan, China (Lower Cambrian, 525–520 Mya), for example, includes 16 phyla, with the great majority being triploblastic Bilateria, amongst which are the first chordates and hemi-chordates. The Burgess Shale biota (~505 Mya, Middle Cambrian; Fig. 13.1) in Canada includes at least 15 known phyla, all but one being protostome. The deuterostomes are represented by a single, early cephalochordate known as *Pikaia graciens* (Fig. 13.1a; see Chapter 14).

The phyla that evolved during the Cambrian included many of today's protostomes; they also included some strange organisms that were lost. Two examples that may well have been early arthropods are *Obapinia*, with five eyes and a small proboscis, and *Anomalocaris*, a large predator with a lobed body, two large lobed limbs emerging from the head and a radial mouth (Fig. 13.2b,d). Another

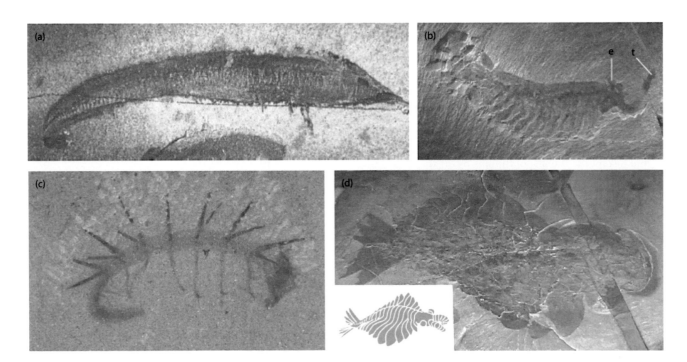

Figure 13.1 **Burgess shale organisms.** (a): *Pikaia graciens*, this was one of the few Cambrian chordates and was several cm long. (b): *Obapinia regalis*. this strange organism had five eyes (e) and a trunk or proboscis which terminated in a claw-like appendage (t). (c): *Hallucigenia*. this organism was probably a lobopodian worm. (d): *Anomalocaris*. this large organism (1–2 m long) was probably an arthropod: it had flexible lobes for swimming, a disk-like mouth made of plates, compound eyes and two large jointed appendages protruding from the head (insert: sketch of *Anomalocaris*).

(a Courtesy of Simon Conway Morris
b Courtesy of Martin Smith.
c Courtesy of Keith Schengili-Roberts. Published under a CC Attribution-Share Alike 3.0 Unported License.)

Figure 13.2 Trilobite morphology.
(a-c): Three trilobites showing unusual morphology. (a): *Acadoparadoxides briareus* (Cambrian). (b): *Cheirurus insignis* (Cambrian). (c): *Walliserops trifurcatus* (Devonian). (d): cast of the compound eye of *Pricyclopyge binodosa* (Ordovician; Bar: 2.5 mm).

(a Courtesy of Mike Peel. Published under a CC Attribution-Share Alike 3.0 Unported License.
b Courtesy of "Vassil." Published under a CC Attribution-Share Alike 3.0 Unported License.
c Courtesy of Kevin Walsh. Published under a CC Attribution Generic 2.0 License.
d From Clarkson et al. (2006) *Arthropod Struct Dev*; 35:247–259. With permission from Elsevier.)

strange organism was *Hallucigenia sparsa* that had a long thin body with limbs and dorsal spikes (Fig. 13.1c). It was originally thought to have represented a phylum that became extinct but is now considered to be either an onychophoran, an early ancestor of today's velvet worms (Smith & Ortega-Hernández, 2014), or a member of a basal taxon within the Ecdysozoa, the clade that today includes arthropods and nematodes (Smith & Caron, 2015).

One feature of the Cambrian organisms that is perhaps surprising given that they were so basal, is that many had eyes (Figs. 13.1a,b and 13.2), with the number ranging from two to at least five (this is not a particularly large number given that most of today's spiders have eight eyes). Metazoans inherited photosensitivity from protist ancestors (Gavelis et al., 2017) and eyes seem to have developed surprisingly rapidly (Chapter 20). Some organisms apparently had simple eyes, but several had compound ones with many ommatidia (Fig. 13.2a–d). Such eyes are particularly seen in trilobites because their lenses were made of calcite derived from dorsal cuticle and so were stable to decay processes (Clarkson et al., 2006). Nothing is known about the underlying ommatidia of trilobites as these would, of course have decayed rapidly after death.

A very early example of an example of a trilobite was *Cindarella eucalla*, a lower Cambrian organism (~520 Mya; Zhao et al., 2013) from the Chengjian Lagerstätte. Each of its eyes seems to have had about 2000 ommatidia, and it is reasonable to assume that integrating and interpreting all these optical inputs would have required relatively sophisticated neural processing by its brain. Trilobites were a species-rich taxon characterised by having a segmented body that was composed of a central and two lateral lobes. These marine arthropods evolved the early Cambrian and died out in the extinction that ended the Permian.

The fossil record demonstrates that, by the end of the Cambrian period, animals that were representative of least 16 protostome phyla had evolved. Such diversity makes it clear that the core molecular mechanisms responsible for cell differentiation, pattern formation, morphogenesis, and physiological function were also in place by then. The Cambrian was thus a period when a host of organisms evolved that were variants of the basic geometry of a simple triploblastic organism and, as it were, tried their luck. The reasons for this burst of diversification are not known, but probably include the range of empty ecological niches, the paucity of predators and perhaps genomes and protein networks that then had fewer regulatory constraints than now. The forms of those taxa that were successful are still reflected in today's protostomes. The rest of this chapter considers the evolution of their major clades.

THE PROTOSTOME WORLD

Porifera

Taxonomically, sponges are metazoan animals because their cells show the key synapomorphies of the clade: their cells, particularly the choanocytes, have a single flagellum, their membranes have no associated carbohydrates, and they lack chloroplasts. Sponges are, cladistically, a sister group to the Eumetazoa (Fig. 13.8) both on molecular phylogenetic grounds and because they lack organs and tissues in any normal sense (Fig. 13.3a,b; Leys et al., 2009). They are amorphous structures that are tethered to the sea floor and are mainly small, although *Xestospongia muta*, the giant barrel sponge, may grow to be more than 1 m high. They can contain ten or more cell types and are often strengthened by biomineralised spicules. Their physiology is very simple: water is drawn in through pores as a result of the beating of choanocyte flagella and passed out through the open osculum at the top, with food particles being phagocytosed on the way. Reproduction is through choanocytes differentiating into germ cells.

The details of the evolutionary origins of the early sponges (Porifera) are unknown. This is partly because the fossil record is weak, although it seems to go back ~800 My (Mah et al., 2014; Turner, 2021; Fig. 12.3l), and partly because the provenance of many examples that were considered to be Cambrian Porifera is now less certain than once thought (Antcliffe et al., 2014). It does, however, seem likely that the first step on the pathway to sponges was the evolution of choanoflagellates, unicellular opisthakonts, to become social colonies. Although the details are not clear, these eventually became choanocytes grouped in simple sponge like structures that eventually became more complex (Brunet & King, 2017). This was not a rapid process: while recent work has confirmed the relationship between the two cell types, it has also shown that choanoflagellates are less closely related to the Metazoa than had previously been thought (Carr et al., 2020). It now seems clear that, although the Porifera may be the simplest taxon in the Metazoa, they are not particularly simple.

Coelenterata

Rather more complex than Porifera are the Coelenterata. (Fig. 13.4) that show radial symmetry and the possession of a coelenteron (gastric cavity). The two contemporary coelenterate phyla are the Ctenophora (comb jellies) and the Cnidaria (e.g. jellyfish, anemones, hydrozoans, and corals) whose synapomorphy is the presence of cnidocysts (small sting-bearing projectiles). Both phyla are diploblasts whose embryos have only ectoderm-derived and endoderm-derived epithelial layers,

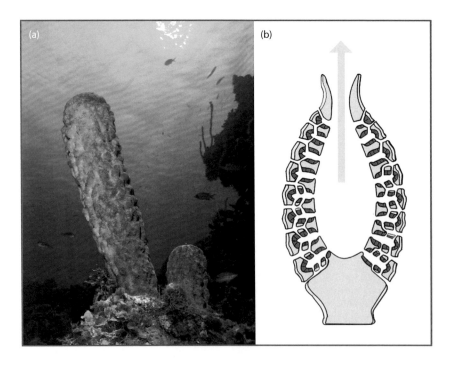

Figure 13.3 **Porifera.** (a): *Aplysina archeri* (the stovepipe sponge). (b): Basic sponge anatomy. Yellow: pinacocytes (epithelial external cells); red: choanocytes; grey: mesohyl (gelatinous matrix); pale blue: water flow.

(a Courtesy of Nick Hopgood. Published under a CC Attribution-Share Alike 3.0 Unported License.
b Courtesy of "Philcha." Published under a CC Attribution-Share Alike 3.0 Unported License.)

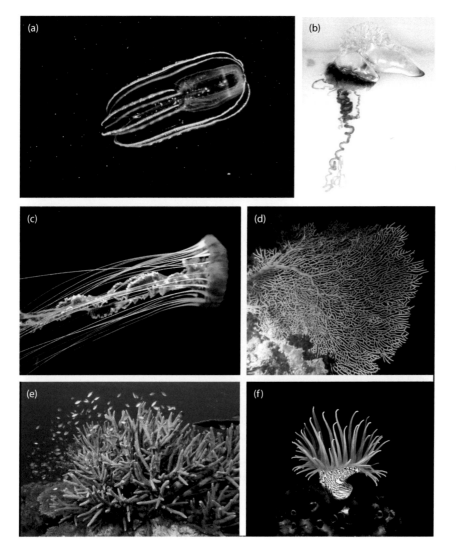

Figure 13.4 **Coelenterata diversity.** (a): *Bolinopsis infundibulum*, a ctenophore. (b-f): Cnidaria. (b): a siphonophore, *Physalia physalis* (Portuguese man-of-war). (c): *Chrysaora melanaster*, a jellyfish. (d): *Annella mollis*, a gorgonian. (e): *Acropora cervicornis*, a stony coral, (f): *Nemanthus annamensis*, a sea anemone.

(a   Courtesy of "Bastique." Published under a CC Attribution-Share Alike 3.0 Unported License.
d,e,f Courtesy of Frédéric Ducarme. Published under a CC Attribution-Share Alike 4.0 International License.)

separated by mesoglea (jelly). Their nerves, cnidoblasts, and muscles derive from the outer epithelium while the gut is derived from endoderm (Leclère & Röttinger, 2017). Ctenophora use cilia for locomotion whereas cnidarians are either sessile or move using their tentacles and through the pumping of their muscular bells.

Apart from their obvious morphological disparities, it is particularly noteworthy that the respective neurons of Ctenophera and Cnidaria are different (Moroz et al., 2014), with those of the Cnidaria being much closer to the rest of the Eumetazoa. It has yet to be established whether the two neuron types evolved independently or whether an early ancestral neuron originally present in a very early primitive Metazoan diverged substantially in Ctenophora (Ryan & Chiodin, 2015). The two taxa are not closely related, and it is not known whether their radial symmetry is a synapomorphy or a homoplasy. Recent work does, however, place the Ctenophora in a basal position as a sister group to other Metazoa and perhaps related to Placozoa (see below; Laumer et al., 2019).

Coelenterata evolved during the Ediacaran, although very few of their fossils have been identified from this or any other period because they are entirely composed of soft tissues that do not lend themselves to preservation. The best-known Ediacaran example is *Haootia quadriformis:* this has been identified as a cnidarian organism and is currently the oldest known organism with muscles (Liu et al., 2014). The occasional Medusa from the Cambrian has also been found (Fig. 13.5; Cartwright et al., 2007), but the early fossil record as a whole includes few coelenterates.

An outlier taxon that was originally hard to classify is the Placozoa. These are small flattened epithelial balls (~1 mm diameter) filled with a matrix and stellate cells; they have a ventral feeding surface rather than a gut, and external cilia that

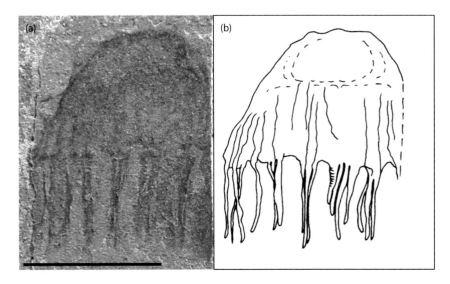

Figure 13.5 (a): A jellyfish (a member of the Medusazoa) from the Cambrian. (b): A drawing of its likely form. (Bar: 5 mm).

(From Cartwright et al. (2007) *PLoS One*; 2:e1121. Published under a CC Attribution License.)

enable them to move slowly. It is not known whether their structure is basal or whether it reflects simplification. They are perhaps the simplest organism to have a true epithelium and so can be considered to lie within the Eumetazoa, as one of its defining synapomorphies is the presence of this tissue structure. The genetic evidence (DuBuc et al., 2019) is beginning to show that the Placozoa are related to Cnidaria and Ctenophera as all express the Chordin/TgfB signaling associated with axial patterning (DuBuc et al., 2019). If so, it is likely that they share a common ancestor, with the morphology of the Placozoa probably simplifying over time.

This picture of the early evolution of basal Metazoa suggests that social choanoflagellates probably gave rise independently to sponges and to coelenterates with the latter line leading to the Cnidaria and Ctenophora, although we know nothing about how this happened. The placozoans may represent the descendants of an early coelenterate taxon that lost most of its tissues while the Bilateria, on the basis of neuron homology, reflect evolution from stem coelenterates (Nielsen, 2019). The cladogram of animal evolution given below in Fig. 13.8 reflects this view.

The Bilateria and protostome diversity

The key innovation that led to the Bilateria was the evolution of a third germ layer, the mesoderm, which became the source tissue for muscles and other mesenchyme-derived cell types. For reasons that are not fully understood, the innovation of triploblasty was accompanied by three modifications to early embryo geometry. First, radial symmetry changed to become bilateral. Second, gastrulation became more complex as it now involved both gut formation and mesoderm reorganisation. Third, antero-posterior patterning evolved and with it the possibility that initially similar regions along this axis could acquire different structures and functions; this facility increased as organisms lengthened and is associated with a directionality of movement..

The last common ancestor of the Bilateria, the root organism of the clades that show these capabilities, was *Urbilateria* (sometimes known as the Urbilaterian, see Chapter 11). A little later, a group of very primitive but still extant marine flatworms known as the Xenacoelomorpha separated off the main line (Cannon et al., 2016). These primitive triploblasts have a gut with an entrance but no exit; they lack a coelom, nephric system and circulatory system. They survive because their flatness allows oxygen to be absorbed through the skin. In due course, the coelomic cavity evolved; this surrounds the gut and contains the major organs. As all taxa in the resulting clade have an excretory system, the clade is known as the Nephrozoa: it includes the protostomes and deuterostomes, with the obscure and simple Xenacoelomorpha (Chapter 14) being a sister taxon.

As taxa diverged, their early development evolved in different ways. Some formed a blastula whose cells were arranged spirally; their descendants include annelid worms and molluscs. Some had cells that proliferated during their early development (e.g. deuterostomes); some prioritized differentiation and morphogenesis early and increased in cell number later (e.g. some arthropods).In other taxa, the fertilised nucleus divided many times to give a plasmodium, with the membranes separating the dividing nuclei not forming the distinct cells of the blastula until this proliferation

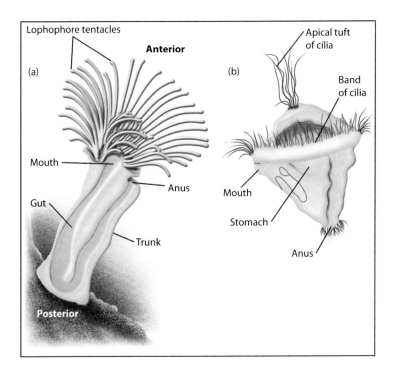

Figure 13.6 **Lophophores and Trochophores.** (a): Drawing of a phoronid (horseshoe) worm (Brachyozoa) that has a pronounced lophophore around its mouth. (b): Drawing of typical trochophore larva showing its core organs.

(Courtesy of Peter Aston. From https://sites.google.com/site/wi16bio124section070group4prol/classification)

stage had finished. Such meroblastic cleavage (Scholtz & Wolff, 2013) is a common mode of early development in the Insecta (e.g. *Drosophila*), which on the basis of the number of its species is probably the most evolutionarily successful of all the eukaryotic phyla.

The development of many taxa includes a larval stage. In some arthropods, the larvae look like small adults; in others, they have very different forms (e.g. butterflies go through a caterpillar stage). Lophophorates, which include bryozoans and brachiopods, have a lophophore as a synapomorphy; this small larva has a circular ridge of tentacles with cilia that can waft food particles into the mouth. (Fig. 13.6a). In contrast, most Trochozoa, which include molluscs, other than cephalopods, together with various members of the Annelida clade, which includes Annelids, Pogonophora (tube worms that are now known as the Siboglinidae) and Sipuncula, produce a trochophore larva (Fig. 13.6b). This is a small transparent sphere composed of an ectodermal epithelium, an endodermal gut, an internal band of mesoderm and circumferential bands of external cilia that are used for feeding and movement (Barresi & Gilbert, 2019; Nielsen, 2018; Wanninger & Wollesen, 2019). Both types of larvae are major constituents of plankton. Although adult lophophorates and trochozoans have no obvious morphological similarities, phylogenetic analysis shows that they share a distant last common ancestor (Philippe et al., 2005). They are, therefore, grouped together in the Lophotrochozoa superphylum.

All these taxa evolved during the Cambrian, and it seems that secondary changes to these basic forms soon led to a wide variety of new groups and species. The resulting organisms were able to occupy hitherto empty ecological niches for which there were few if any competitors and hence unusually low selection pressures. It thus seems that mutation resulted in a much greater degree of phenotypic variation during the Cambrian than was ever to be seen again. The reasons for this are not known, but it may have been because many of the molecular mechanisms that stabilize the protein networks that drive development and metabolism had yet to evolve. If so, molecular systems were more readily changed by mutation then than now, becoming more stable to mutation as time passed.

The earliest evidence of protostome evolution comes from the substantial record of soft-body fossils from the Cambrian (Giribet & Edgecombe, 2019) but there are few of these in Lagerstätten from subsequent Periods. Most of the later evidence for protostome evolution comes from fossils with an exoskeleton (e.g. arthropods) or a shell (e.g. molluscs). This is supplemented by impression fossils (e.g. those of large dragonflies, see below), organisms caught and preserved in amber from tree resin, and trace fossils that record animal activity, such as movement tracks and burrows.

Today there are some 30 protostome phyla, of which the great majority are either worms or small marine animals, almost all of which are either soft-bodied or have a fragile exoskeleton. Their ancestors have left relatively little trace in the fossil record unless, like the Brachiopoda and some Bryozoa, they had robust shells. Most are relatively simple, and their morphology today gives little information as to how their anatomy evolved and became organised, although changes in the morphology of those with shells do mark their diversity and evolution. While a few of the larger protostome taxa are widely represented both in the fossil record and in living fauna, some are much rarer: taxa, such as the Cycliophora, the Loricifera, and the Phoronida each have ten or fewer living species that have so far been discovered.

All protostomes today have an ectoderm and a gut together with nervous, reproductive, excretory and muscular systems. In several common phyla, particularly the annelids and arthropods, the ectoderm is segmented. Although it seems unlikely that even an underlying molecular segmentation is a synapomorphy for protostomes (Clark et al., 2019), it is intriguing to note that homologues of the homeodomain genes, a set of which underlie spatial patterning and segmentation, are found in plants, animals, fungi and algae (Holland, 2013). These genes have very deep ancestral roots: homeodomain proteins have, for example, been found in *Capsaspora owczarzaki,* a unicellular protist that is a sister group to the choanoflagellates (Sebé-Pedrós et al., 2017).

Given the wide distribution of these gene groups, it is not surprising that a range of relatively obscure protostomes outside the core segmented phyla (i.e., annelids and arthropods) also show anterior-to-posterior repetition of tissues. These include the Onychophora (velvet worms), and the tardigrades that have unjointed legs. There are also unsegmented protostomes such as the Monoplacophora, single-shelled, deep-water molluscs whose fossils date back to the Cambrian with early internal anatomy showing evidence of segmentation (Kocot et al., 2020). It is not clear whether segmental patterning evolved once, sometimes being later suppressed and even later reappearing, or several times.

Protostome systematics

Linnaean taxonomy of the core animal phyla is, given their wide range of anatomical features, relatively straightforward. It has, however, proved to be a little harder to untangle their lines of very early evolutionary change using cladistics and molecular phylogenetics. This is because the sheer length of time since the most basal unikont-derived multicellular organism evolved in the Ediacaran (Irwin, 2015) is so long that such phylogenetic analysis of proteins cannot be done with confidence. There is, however, sufficient genetic information in the 18s ribosomal RNA to generate a phylogram based on organisms from across the Metazoa (Fig. 13.7; Giribet et al., 2000); this gives the likely lines of descent from *Urbilateria*, with any recent ambiguities capable of being resolved using molecular phylogenetic analysis. Such a phylogram does not, of course, indicate when during the Paleozoic Era the various anatomical features that mark each phylum evolved.

The relationships within almost all of the contemporary metazoan clades have, as a result of this effort, now been worked out in very great detail. The resulting phylograms are so extensive that they cannot easily be fitted onto a printed page and can really only be integrated and visualised computationally. The reader interested in the details of a particular taxon is directed to the appropriate Wikipedia article for further phylogenetic information.

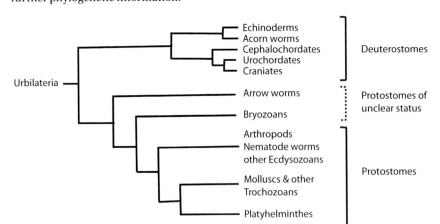

Figure 13.7 A phylogram linking some key modern phyla back to urbilateria, the last common ancestor of deuterostome and protostome triploblasts. Much of the molecular analysis underlying this tree is buttressed by information from developmental observation; This does not, however, apply to arrow worms and Bryozoa as their developmental status is not completely clear. (Tree compiled using data on 18S ribosomal rDNA from Giribet et al., 2000.)

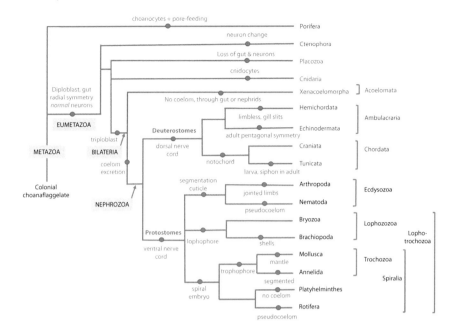

Figure 13.8 **A cladogram showing the evolution of the major metazoan phyla.** Colonial flagellates probably date back more than 1 Bya and the oldest sponges to ~800 Mya (Turner, 2021). Basic organisms for many of the superphyla (right) probably evolved during the Ediacaran and for the individual phyla before the end of the Cambrian. Key synapomorphies are shown in blue type. The position of the placozoa reflects the work of DuBuc et al. (2019).

It is, however, possible to derive a plausible cladogram of the main metazoan phyla based on shared and derived characteristics (Fig. 13.8). Here, it is worth noting that the evolutionary sequence is not yet fully understood since the origins of apomorphies, such as segmentation and embryo spirality are unclear. Ectodermal segmentation may, as already mentioned, have arisen independently several times, but it is also possible that it only arose once, perhaps in *Urbilateria*. If so, it was then lost in most phyla, being maintained in only a few that included the chordates, annelids, arthropods, and some molluscs. The evolution of embryonic morphology is equally unclear: some have a coelom (annelids and molluscs) while others (e.g. Platyhelminthes) do not. In contrast, arthropods and nematodes have a nonspiral embryo with the former having a coelom within the mesoderm and the latter having a pseudocoelom between the mesoderm and endoderm. Understanding the molecular basis of these very basic alternative developmental pathways and their consequent anatomical phenotypes still remains a challenge.

THE MAJOR PROTOSTOME PHYLA

From the Cambrian onwards, the arthropods and then the molluscs together came to dominate the protostome world as measured by size, by complexity, by number of species and numbers of individuals. That said, the ecological importance of the many worm phyla, which have relatively simple anatomy and are mainly hidden or small, should not be underestimated: their many taxa are very widely found in soil, on the seabed and as parasites.

Many types of molluscs are now present in the sea and on land, with all possessing an external mantle and most a foot or its derivative. They may also have shells made of calcium carbonate or chitin that can be internal (e.g. squids have a gladius) or external (e.g. snails, oysters, and the *Nautilus* have a shell; Hohagen & Jackson, 2013). Some 65,000 molluscan species have been identified, and there are probably more as there are many habitats that await exploration. Arthropods can swim, walk and fly and are characterised by the presence of jointed limbs, segmented bodies (in the embryo at least) and chitin exoskeletons, sometimes strengthened by calcium carbonate. Over a million living arthropod species have so far been described and, again, there are doubtless many more to be discovered.

These two phyla are discussed here because, in terms of diversity, size, and fossil record, they are by far the most widespread of the protostome phyla and those about which most is known. Readers interested in the wider evolution of protostome diversity should consult Clarkson et al. (1998), while details of the evolution of insects, the most speciate arthropod subphylum, are given in Grimaldi and Engel (2005), who focus on the phenotypic data, and Misof et al. (2014) who summarise the phylogenomic relationships (for all, see Further Reading).

Molluscs

It is hard to think of another phylum with such a wide range of forms as the molluscs. They include the marine and land-based gastropods (e.g. single-shelled limpets and snails and unshelled slugs), bivalves, such as oysters, scollops and clams, cephalopods, which may have an external shell (nautiloids) or a small internal skeleton (e.g. the squid gladius that is made of chitin) and several taxa of small, marine organisms (Table 13.1). They also show a considerable range of sizes extending from perhaps 2 mm for the smallest snail to ~13 m for the female giant squid, albeit that much of this length is due to its tentacles.

Because of their shells, molluscs have a rich fossil record. The oldest known organism now considered to be a mollusc is the limpet-like *Kimberella quadrata* that was found in the Ediacaran Hills of Australia and dated to 558–555 Mya (Fig. 12.1d; Wanninger & Wollesen, 2019). Larger examples were certainly present in the lower Cambrian and, by the end of the Period, there were many examples of gastropods, cephalopods, and bivalves that are now preserved in the fossil record. Molluscs were solely marine organisms until the Carboniferous (~ 300 Mya) when the earliest land gastropods have been found, although these seem not to have been common until the beginning of the Cretaceous (145 Mya).

The core synapomorphy that unites these very different taxa into a single clade is the presence of an external mantle that is capable of laying down a range of external shells or an internal gladius. Between this and the main part of the body is the mantle cavity that contains the exits from the anus, kidneys (nephridia) and gonads and the gills (or lungs), together in many cases with a pair of osphradia (chemical sensors). All molluscs, other than bivalves, also have a radula, a rasping tongue-like ribbon for scraping food. Many also have a foot which has evolved to fulfil a range of functions. In land gastropods such as the snail, it is the organ of movement; in marine gastropods such as the limpet, it provides adhesion to a substratum while in others it is a swimming organ. In bivalves, it aids burrowing; in cephalopods, its contractions provide propulsion, while its peripheral region is the origin of its tentacles (Shigeno et al., 2008). On the basis of these and other synapomorphies, it is possible to suggest some of the likely anatomical features of an early mollusc (Fig. 13.9).

Molluscan development is characterised by a blastula with a spiral arrangement of cells whose axes are determined by signaling pathways: dorso-ventral polarity and

TABLE 13.1 THE MAJOR GROUPS OF LIVING MOLLUSCS

Lineage group	Taxon	Properties
Aculifera		Molluscs without shells.
	Aplacophora	Shell-less worm-like organisms with a spiny cuticles, a radulae and a small mantle cavity.
	Chaetodermomorpha	Elongated Aplacophoran worms without a foot.
	Neomeniomorpha	Aplacophoran worms with a narrow foot in a central groove.
Polyplacophora		Molluscs with a flexible shell of 8 shell plates and a long foot.
Conchifera		Molluscs with shells.
	Cephalopoda	Cephalopods with a prominent head, and a set of tentacles that evolved from the foot. There are two taxa, those with a shell (the nautilus is the only living example although ammonites were once common) and the Coleoidea in which the shell is internalised (squids) or has been lost (octopuses).
	Monoplacophora	Molluscs with a simple shell, no evidence of body torsion and some segmentation (repeated muscles and nerves).
	Bivalvia	Marine and freshwater molluscs with a shell of two hinged parts that lack a head and radula.
	Scaphopoda	Molluscs with a tusk-shaped shell open at both ends. The wide opening is for the foot and the narrow one for egg and sperm release and for respiration through water flow.
	Gastropoda	Molluscs with either a single shell, which is often coiled (e.g. snails) or with a minimal or no shell (slugs). The body often shows a degree of rotation (torsion) with respect to the foot. Gastropods are found on land, in sea and in fresh water.

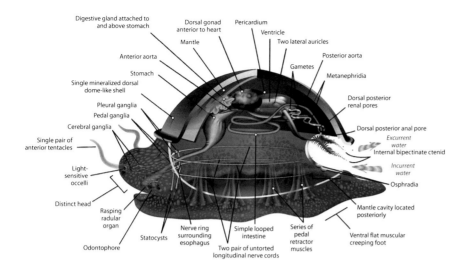

Figure 13.9 The likely anatomical features of an early mollusc. This drawing is based on what is known of fossil architecture (e.g. *Kimberella quadrata*) and clade synapomorphies.

(Courtesy of "KDS444" from Wikimedia Commons. Published under a CC Attribution-Share Alike 3.0 Unported License.)

left-right determination are respectively regulated by dpp/chordin gradients and by the nodal pathway. Once antero-posterior patterning is in place (mainly determined by Hox and ParaHox transcription factors; Biscotti et al., 2014), a trochophore larva then develops and eventually metamorphoses into an adult mollusc. These larvae provide a core constituent of sea plankton. By this stage, most cell fates seem to have been determined and adult anatomical structures reflect this lineage. It is worth noting that the development of cephalopods is somewhat different from that of other molluscs, mainly because of their large eggs have considerable amounts of yolk. They consequently have meroblastic cleavage and do not go through a trochophore stage (Shigeno et al., 2008; Deryckere et al., 2020).

There is now enough fossil data to produce a fairly detailed cladogram of the evolution of the many mollusc taxa (Wanninger & Wollesen, 2019); the key features are shown in Fig. 13.10. This phylogeny divides the phylum into two sets of taxa,

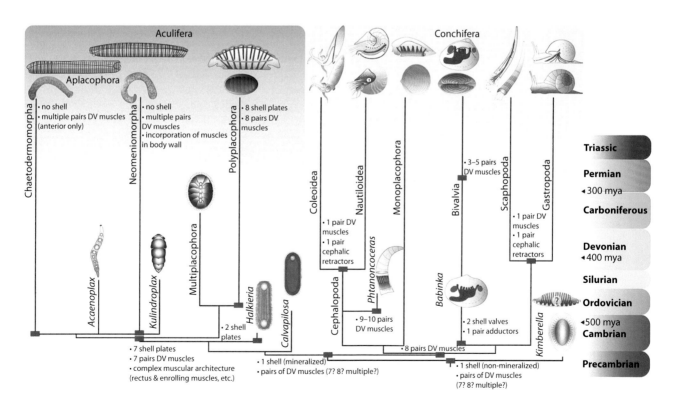

Figure 13.10 A cladogram showing the paths to contemporary mollusc phyla. For more details on extinct phyla.

(From Wanninger and Wollesen (2019) *Biol Rev*; 94: 102–115. Published under a CC by License.)

the Aculifera that lack a shell and the Conchifera that possess a shell or conch. Note that the slugs are included in the latter because they descended from shell-bearing gastropods that lost the ability to make a shell. Analysing the full complexity of the Mollusca phylogenetically has not been straightforward; this is mainly because of the difficulty of getting enough sequence data from some of the rarer taxa to be able to probe the roots of such an ancient and diverse phylum.

Arthropods

The roots of the arthropods lie in the late Ediacaran or possibly at the base of the Cambrian. They are characterised by a linear, segmented body with jointed limbs protected by an exoskeleton. This body plan (or bauplan) proved to be extremely successful and a wide range of body forms evolved rapidly. Many early variants, such as *Anomalocaris* and the trilobites (Figs. 13.1 and 13.2), are now extinct, but the success of today's arthropods is shown by it being the phylum with by far the largest number of recognised species (Fig. 13.11). Indeed, it is hard to identify a marine, land or air-based environment without arthropods. Perhaps unexpectedly, there are even many arthropods in the Antarctic waters, including most of the world's sea spiders (Dietz et al., 2019).

The complexity of a phylum with more than a million species is such that untangling their evolutionary relations and timelines has proven difficult and contentious. However, the combination of morphological, palaeontological, and molecular data is

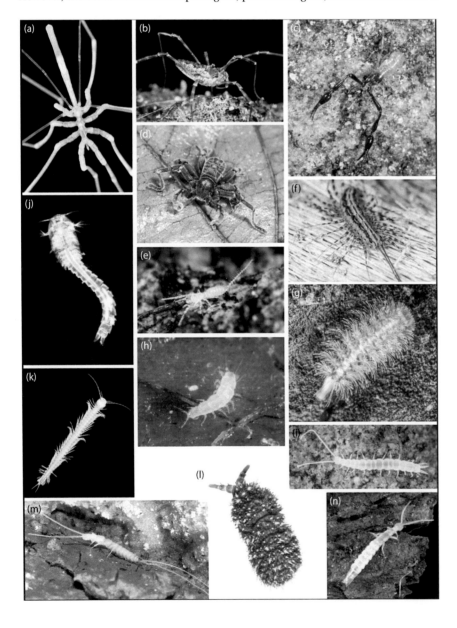

Figure 13.11 Examples of major arthropod taxa today. (a): *Colossendeis* sp. (Pycnogonida); (b): *Odiellus pictus* (Arachnida, Opiliones); (c): *Gymnobisium inukshuk* (Arachnida, Pseudoscorpiones); (d): *Cryptocellus iaci* (Arachnida, Ricinulei); (e): *Eukoenenia* sp. (Arachnida, Palpigradi); (f): *Parascutigera latericia* (Myriapoda, Chilopoda, Scutigeromorpha); (g): *Eudigraphis taiwaniensis* (Myriapoda, Diplopoda, Polyxenida); (h): *Pauropus* sp. (Myriapoda, Pauropoda); (i): *Hanseniella* sp. (Myriapoda, Symphyla), (j): *Lightiella incisa* (cephalocarida); (k): *Xibalbanus tulumensis* (remipedia) (l): *Holacanthella duospinosa* (Hexapoda, Collembola); (m): *Podocampa fragiloides* (Hexapoda, Diplura, Campodeina); (n): *Japyx* sp. (Hexapoda, Diplura, Japygina).

(From Giribet and Edgecombe (2019). *Annu Rev Entomol* 57:167–186. With permission from Elsevier.)

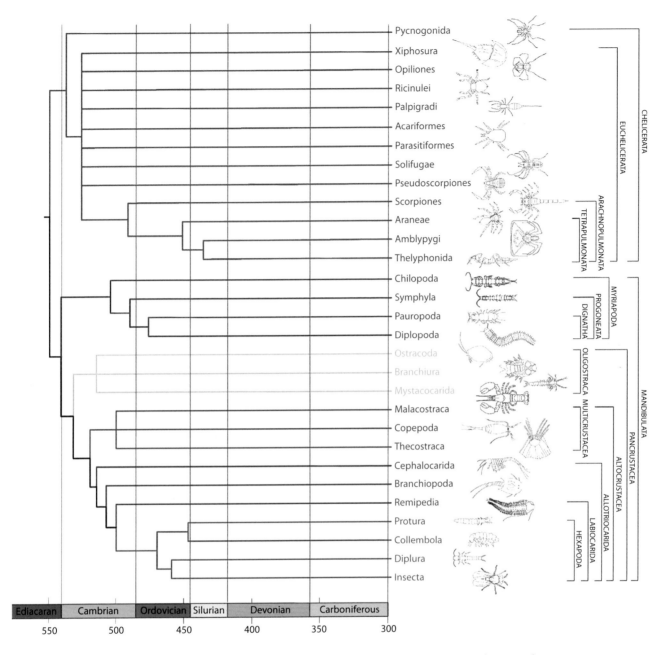

Figure 13.12 **Chronogram and phylogeny of the evolution of contemporary arthropods based on a wide range of sources.**

(From Giribet G & Edgecombe GD (2019) *Curr Biol*; 29:R592-R602. With permission from Elsevier.)

now beginning to give a fairly clear picture of some of the early events in their evolution and to unravel their taxonomic relationships (Fig. 13.12). The rest of this section summarises these for the major arthropod taxa but cannot do full justice to the subject. The reader who wants to go more deeply into this complex subject should consult Giribet and Edgecombe (2019) and explore the Wikipedia entries for the various taxa.

In basal arthropods, the first segment had a pair of antennae that evolved in different ways as the clade diversified into the Mandibulata with jaws, the Artiopoda (trilobites), for which the antennae were sensory, and the Chelicerata (spiders etc.) for which they are pincer-like claws (Sharma et al., 2015). This last modification was highly exaggerated in *Anomolocaris* (Fig. 13.1d) but we do not know whether their large claspers represented a homologue or a homoplasy.

The Mandibulata today includes two major clades: 1) the Pancrustacea (hexapods and crustaceans, whose synapomorphies include details of ommatidia morphology and neuron development) and 2) the Myriapoda (named on the basis of their many legs).

A key clade within the Pancrustacea is the Multicracea which includes marine lobsters and shrimps, crabs, krill, barnacles, copepods, and land-based woodlice; these have four or more pairs of legs and often many swimmerets (Oakley et al., 2013). The hexapods include the insects that have three tagmata: the head, thorax, and abdomen with three pairs of walking limbs on the thorax. The Hexapoda also includes three very minor wingless phyla that each have internal mouthparts, these are the Collembola (springtails), the Protura (coneheads), and the Diplura (wo-pronged bristletails) The Myriapoda includes centipedes and millipedes with up to 750 pairs of walking limbs.

The main taxon within the now-extinct Artiopoda was the trilobites (Fig. 13.2), characterised by three axial lobes (left, right, and pleural) on the dorsal exoskeleton of their multisegmented body (Fig. 13.2). The thousands of identified trilobite taxa were, like all early animals, marine organisms. The clade survived from before the mid-Cambrian (~520 Mya) to the end of the Permian (252 Mya). They had 2-4 pairs of head limbs, while each of their body segments had a pair of double limbs and so were biramous: the inner was a walking limb and the outer a gill for absorbing oxygen from sea water. This morphology is maintained today in, for example, marine arthropods such as crabs and in the larvae of a few insects such as the mayfly.

Today, the Chelicerata (e.g. spiders, mites and ticks, horseshoe crabs, and scorpions) are all characterised by the possession of chelicerae: these are paired feeding parts adjacent to the mouth, which derive from a modification of the first or second segment and, in some taxa, carry venom. It is worth noting that venom production has evolved independently several times across the arthropods and is a homoplasy (Senji Laxme et al., 2018). The Chelicerata have two tagmata (body regions): the anterior prosoma (cephalothorax), which includes the mouth and the body with typically six pairs of appendages (an anterior pair of chelicerae, a more posterior pair of sensory "palps" and four pairs of walking legs) and the posterior opisthosoma which includes the abdomen, heart and respiratory organs, together with a pair of spinnerets for web production.

The approximate number of species in each of today's major arthropod clades is a measure of its success. The hexapods include >1 M named species that are mainly insects. The clade also includes the Crustacea (shellfish, crabs, woodlice, barnacles, etc.) with ~67K species, the Chelicerata whose main taxon is the Arachnida (spiders, ticks, mites, scorpions, etc.) with includes ~60K species, and the Myriapoda (centipedes and millipedes that usually have one pair of legs per segment) with ~15K species. These totals are, however, now declining at a measurable rate (see publications associated with Wagner et al., 2021).

The history of the evolution of arthropod diversity is one of how mutation affected the patterning and fate of individual segments that mark the antero-posterior axis during early embryogenesis. Part of this, as discussed in Chapter 8, is associated with the downstream effects of Hox coding; the morphology of each species within a taxon reflects mutational tinkering with this patterning and other aspects of development (Chapter 8 and Appendix 6).

The transition to land

A further interesting question about the arthropods is how a phylum that was marine for several hundred million years acquired the ability to live and thrive on land and eventually came to be such a presence in the skies. The early, small marine arthropods were able to absorb limited amounts of oxygen from seawater through their thin cuticle, but this limitation constrained growth. The evolution of gills in association with limbs enabled arthropods to absorb more oxygen from sea water and so allowed them to increase in size.

Two further anatomical changes were required for members of the clade to colonise land. The first was a skeleton that was both impervious to water so as to avoid desiccation and strong enough to bear the animal's weight; a hard chitin exoskeleton satisfied both of these criteria (growth occurs through laying down a new soft, inner exoskeleton and sloughing off the old volume-limiting cuticle, so allowing the new one to expand and harden). Nevertheless, as even the largest land-based arthropods today (e.g. the robber crab, *Birgus latro*, which can weigh up to 4.1 kg.) is much smaller than many marine ones (e.g. lobsters can exceed 20 kg), the problems of how the exoskeleton of a land organism might adapt to cope with large body weights were only rarely solved. A most unusual exception was the Carboniferous myriapod *Arthropleura amata*, a land-based arthropod; it was ~2.1 m long and evolved at a time of high oxygen levels (see below). This enormous organism, whose trackways have been well-studied (Briggs et al., 1984), had some 30 segments and about 30 or 60 pairs of limbs (depending on whether its segments were fused) that helped support its considerable weight. Neither the origins nor the physiology of these enormous myriapods are known, although the increased oxygen levels when they lived may have played a facilitatory role (see below).

The largest arthropods were, however, marine organisms. *Arthropleura amata* was smaller than *Jaekelopterus rhenaniae*, an early Devonian eurypterid or sea scorpion (now extinct) that is the longest arthropod known to have existed. This animal was about 2.6 m long and had chelicerae that added about 0.5 m to this length (Braddy et al., 2008). Hibbertopterids, another eurypterid arthropod taxon, were not as long but were much bulkier and probably heavier (Lamsdell & Brady, 2010).

The second anatomical innovation required for life on land was a means of absorbing oxygen from the air and there is evidence from trace fossils that arthropods colonised land as long ago as ~490 Mya (Garwood & Edgewood, 2011; MacNaughton et al., 2002; Pisani et al., 2004). The most common solution, seen in insects, were tracheae; these start as pits (later spiracles) on the ectodermal surface that invaginate to form tubes that spread throughout the body of the animal. Oxygen taken in at the surface can thus reach every cell in the animal, although the mechanism is not particularly efficient and probably limits insect body diameter. Tracheae have evolved several times (Harrison, 2015) and one of the earliest fossilised animals with them was *Pneumodesmus*, a millipede from the Mid-Silurian (428 Mya). It is likely that insects, which are an essentially land-based clade, could only have flourished once this mechanism of oxygen absorption had evolved.

As to the mechanism of trachea formation, Franch-Marro et al. (2006) have shown that the location of the cells from which each trachea develops is part of the region from which limbs develop, with differing signaling pathways leading to the two fates. It thus seems that the original gill phenotype (still seen in crustacea) may well have been switched to a tracheal one. Land crabs, in contrast, have maintained gills but need to keep them moist to enable their surface to absorb oxygen. Curiously, a different solution was found by woodlice: they develop small tracheal-like lungs in their hindlimbs and so no longer need gills.

Book lungs were a further solution to the problem of obtaining oxygen that evolved in land arachnids (spiders and scorpions). These are alternate layers of flattened epithelial pockets: one set is open to the air and the other is filled with haemolymph (the arthropod equivalent of blood) that is continuous with the coelom; as a result, oxygen exchange occurs between them (Kamenz et al., 2008). Book lungs are located in ventral cavities within the abdomen that have an opening to the air. The earliest known organisms with this feature are horseshoe crabs. Despite their name, these are not crustaceans, but a taxon that lies within the Arachnida (Ballesteros & Sharma, 2019).

The transition to flight

The Insecta is the only protostome taxon whose members can fly, an ability that probably evolved during the latter part of the Carboniferous (~320 Mya). It is only recently, however, that the fossil record together with with molecular studies on insect development have shown how wings are likely to have evolved in insects. In the past, there were two candidate mechanisms for their evolution, that wings evolved from gills on the dorsal limb branch of primitive insects such as the mayfly or that they formed from sclerite plates on the dorsal part (the notum) of a thoracic segment.

Molecular analysis now points to a third option, that both gills and sclerite plates came to be involved in wing development. Niwa et al. (2010) examined the development of two insects with dorsal thoracic appendages during their larval stages: mayflies that have gills and bristletails that have styli that form from the notum and are substrate sensors. They studied the expression of two transcription factor genes, *apterous* (*ap*) and *vestigial* (*vg*) together with *wg*, the gene for the wnt signalling protein. They found that *wg* and *vg* were both expressed and active in gill and stylus development, but in different parts of the segment, while *ap* was only expressed in the notal area. In bristletails, the enlarged styli are associated with the expression of all three genes.

Niwa et al. then suggested that wings evolved in apterygote insects through modulation of two regulatory pathways involving the *wg* and *vg* genes that acted in different parts of the segment: one would have induced a motile stick-like dorsal limb (i.e., a branch rather than a gill), the other would have activated a flat outgrowth at the boundary of dorsal and lateral body regions. They then proposed that wings evolved as these two tissues combined and extended; here, the *ap* protein could perhaps have played a role in the enlargement process. This view has been strengthened by the recent work of Clark-Hachtel and Tomoyasu (2020) who showed that a gene network similar to the insect wing gene network operates both in the crustacean terga (the dorsal plates in each segment) and in the proximal leg segments.

Recent fossil evidence has given further weight to this proposal. Prokop et al. (2017) have examined fossils from carboniferous palaeodictyopteran nymphs, such as *Rochdalia parkeri* from the late carboniferous (c. 310 Mya). These are the simplest insects with wings—they actually had three pairs of wing pads on their thorax (Fig. 13.13a). These wings were continuous with notum plates, but their hinges were more ventral, just dorsal to the limb appendages and, hence, located in the region of the original gills.

Once the first insect able to use such primitive wing pads was able to glide, a host of evolutionary niches clearly became available. By the time of the Permian, there were at least ten winged-insect taxa mainly, if not all, with two pairs of wings (like contemporary butterflies, dragonflies, and damselflies). From these evolved a wide variety of flying insects as mutational tinkering modified these structures. In the Diptera (e.g. *Drosophila*), the forewing remained but the hindwing became a balancing organ or haltere; in others, the hindwing remained but the forewings became either halteres, as in the Strepsiptera (an order that are mainly parasites), or wing covers (elytra) as in the coleoptera (beetles), the most speciate arthropod taxon (~350K species have been catalogued).

An interesting aspect of the fossil record is that it not only reflects evolutionary history but can be an important measure of the environment in which ancient organisms flourished. Today, most flying insects are small, with the largest living butterfly, the female Queen Alexandra's birdwing, having a wingspan of ~250 mm and a body ~80 mm long. Early in the evolution of flight (~300 Mya, late Carboniferous),

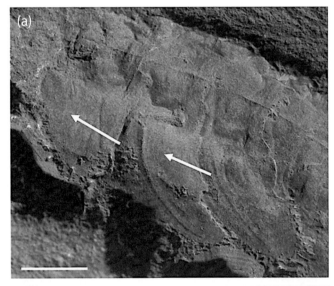

Figure 13.13 (a): A fossil of Rochdalia parker showing the primitive wing forms. (Bar = 3 mm.) (b): Cast of a *meganeura* fossil whose wingspan was about 400 mm. It is on display at the evolution gallery of the Museum des Science Naturelles, Brussels.

(a From Prokop et al. (2017). *Curr Biol*; 27:263–269. With permission from Elsevier.
b Courtesy of "Hcrepin." *Archimollus-en-svg* from Wikimedia Commons. Published under a CC Attribution-Share Alike 3.0 Unported License.)

however, compression fossils show that *Meganeuropsis permiana*, a genus related to today's dragonflies, had wingspans of up to 700 mm, a body length of ~300 mm and an estimated mass of several hundred grams (Fig. 13.13b).

Theoretical analysis suggests that such insects would not be able to take in enough oxygen through their spiracles to fly in today's atmosphere (Gauthier & Peck, 1999), which itself would not have been dense enough to support their flight. Some thought has, therefore, gone into working out how these enormous insects were able to fly. The most likely answer seems to be in two parts: first, they were probably able to increase tracheal efficiency through "breathing" cycles of thoracic compression and relaxation; second, the atmosphere then probably had more oxygen than now (up to 35% as compared to today's 20%) and so would have been denser (Berner, 1999; Dudley, 2000). It is probably because of the high oxygen levels during the Devonian that large predatory fishes and gigantic Myriopoda such as *Arthropleura amata* were able to evolve (Dahl et al., 2010).

The initial aerial radiation was clearly successful as members of many contemporary insect taxa can fly. These include butterflies, moths, dragonflies and damselflies, bees and wasps, beetles, cicada, locusts and even cockroaches, whose modern form appeared in the early Cretaceous (~140 Mya). The success of these insects is just one facet of the overwhelming importance of the arthropod clade for the contemporary biosphere. Bar-On et al. (2018) have estimated the biomass of the core phyla and major groups of organisms in gigatons of carbon (see Table 26.1). As might be expected, the plants as a group are the heaviest (~4.5K GT-C) but, among the animals, the figures for humans, molluscs, fish, and arthropods are 0.06, 0.2, 0.7, and 1.2 GT-C, respectively. It is the arthropods that dominate the animal world in terms of both mass and species numbers.

KEY POINTS

- The earliest Metazoans evolved from social choanoflagellates during the Ediacaran. By the mid-Ediacaran at the latest, one line had led to sponges and another to coelenterates, organisms with radial symmetry and two germ layers, ectoderm and endoderm.

- Later in the Ediacaran, an organism formed that had mirror symmetry and a third germ layer, the mesoderm. One of its descendants was *Urbilateria*, the last common ancestor of the Bilateria.

- The subsequent explosion in the Bilateria during the early Cambrian was mainly due to molecular patterning taking place along the antero-posterior axis of the descendants of *Urbilateria*. Here, a particularly important theme was segmentation.

- All of these early taxa were protostomes and marked by the first opening in the blastula becoming the mouth. Such organisms also have a ventral nerve cord.

- Many of the protostome forms that evolved during the Cambrian provided the basis of today's invertebrate phyla. The dominant phyla today are the arthropods, with jointed limbs and a chitin exoskeleton, and the molluscs, with a foot capable of many roles and an external mantle that can secrete an external shell or an internal gladius.

- Most of the other protostome phyla are either variants on worm morphology or small marine organisms less than a few cm in length, some of which have shells.

- The phylogeny of the protostomes has now been determined with some accuracy on the basis of both molecular and cladistic analysis.

- Major taxa of molluscs have shells and include gastropods (e.g. snails with a single shell), cephalopods (e.g. squid have an internal gladius while the nautilus has an external shell) and bivalves (e.g. oysters and scallops).

- Among the animals, the arthropod phylum is the most successful in terms of total mass and species number—there are now, for example, ~350K ctalogued species of beetles, the largest family in the Insecta.

- The key to this success has been body segmentation that forms early in embryogenesis. Arthropod diversity reflects the many ways in which the development of individual segments has occurred.

- Although they were originally marine (e.g. the Crustacea and the trilobites), many arthropod taxa have flourished on land, while some insect groups have since acquired flight.

- The largest land clade is the Insecta, a subclade of the Mandibulata; these animals have a body with three pairs of legs. The upper parts of two of the leg-bearing segments are able to form wings or wing derivatives such as the wing covers (elytra) of beetles and the balancing organs (halteres) of Diptera and Strepsiptera.

- There are other two important land arthropod clades within the Mandibulata: the Myriapoda that includes centipedes and millipedes and the Chelicerata (marked by chelicera: pincer-like claws that are part of the head) that includes spiders and scorpions.

FURTHER READINGS

Clarkson E, Twitchett R, Smart S (1998) *Clarkson's Invertebrate Palaeontology and Evolution.* John Wiley.

Grimaldi D, Engel MS (2005) *Evolution of Insects.* Cambridge University Press.

Misof B, Liu S, Meusemann KB (2014) Phylogenomics resolves the timing and pattern of insect evolution. *Science* 346:763–767. doi: 10.1126/science.1257570.

Pandian TJ (2018) *Reproduction and Development in Mollusca.* CRC Press.

WEBSITES

The wikis on *arthropods, molluscs* and *Lagerstätte.*

Fossil collections
- The Field Museum, Chicago: www.fieldmuseum.org/science/blog/mazon-creek-fossil-invertebrates
- Sam Noble Museum, Oklahoma: https://samnoblemuseum.ou.edu/common-fossils-of-oklahoma/invertebrate-fossils/

- Smithsonian Natural History Museum: https://naturalhistory.si.edu/research/paleobiology/collections-overview
- Natural History Museum, London: https://www.nhm.ac.uk/our-science/collections/palaeontology-collections.html

Deuterostome Evolution: From the Beginnings to the Amphibians

14

This and the following two chapters summarise evidence on the evolution of the deuterostomes, with the main focus being on the history of the vertebrates. Most of this evidence comes from anatomical analysis of the extensive fossil record and of contemporary organisms: a range of shared and derived anatomical relationships (synapomorphies) among living vertebrates can be traced back through a record of specimens extending back more than 500 Mya. The result of combining these two lines of study, occasionally supplemented by phylogenetic and embryological analysis, is a detailed history of chordate evolution (organisms with a notochord). Such is the extent of the information now available, however, that it is simply not practical to be comprehensive. The aim of these chapters is to highlight the major anatomical transitions that marked the history rather than to provide zoological detail.

This chapter starts by covering some key steps in the evolution of the minor deuterostome clades, but its main concerns are the evolution of the fishes and their transition to stem tetrapods, which in turn, gave rise to amphibians and stem amniotes. The former, like all early marine deuterostomes, are also known as the anamniotes; this is because their eggs lack a protective amniotic sac and so generally require an aqueous, or at least a damp environment in which to develop. Chapter 15 briefly discusses the evolution of the amniote egg that freed animals from depending on water for their embryos to survive and then considers the history of reptiles and birds. Chapter 16 covers the evolution of mammals. Together, these three chapters show that the fossil evidence is sufficient to enable all the vertebrates to be included within a single clade that connects today's rich vertebrate fauna back to the earliest chordates of the lower Cambrian (>529 Mya). A summary of the timeline of Chordate evolution from the base of the Cambrian onwards is given in Fig. 14.1.

Perhaps surprisingly, molecular-phylogenetic analyses of contemporary species play only a secondary role in understanding the evolution of the vertebrate clade. This is because the structural differences between the major branches of the vertebrata are profound, and there are no extant intermediaries. Elucidating the history of chordate evolution has, therefore, depended heavily on its remarkable fossil record which shows that all the major deuterostome phyla had evolved by the early Cambrian (for review, see Satoh, 2016). However, because soft tissues rarely lasted long enough to be fossilised, the post-Cambrian anatomical record for vertebrates consists almost entirely of bones and teeth. The former were preserved by the presence of hydroxyapatite while the latter contain dentin and particularly enamel, which is ~96% mineral and the most robust substance made by chordates. Fossilised skeletal material from across the phylum has not only demonstrated how marine vertebrates evolved into large terrestrial and aerial organisms but also illuminated some of their physiological properties. Such information has enabled generations of anatomists to unravel the evolutionary history of vertebrates in considerable detail.

Deciphering the fossil record means finding anatomical evidence showing that one taxon evolved from another. This, in turn, means closely examining the anatomy

DOI: 10.1201/9780429346217-17

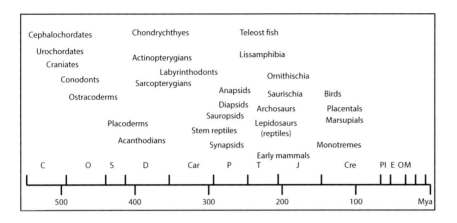

Figure 14.1 The fossil record data for the chordate taxon. The start of a word roughly marks when the group first appears in the fossil record. No formal attempt has been made to link groups, other than through proximity because the times when fossils are first observed in dateable rock do not provides adequate evidence for lines of descent.

of sets of similar fossilised organisms to identify which of their features can be seen as shared and derived (synapomorphies), so implying descent, and which are novel (autapomorphies). Note that the age of the fossilised organism plays no role in these assignments; this is because species can continue to exist long after others have evolved from them. Thus *Haikouichthys*, a relatively sophisticated chordate, was present during the lower Cambrian ~515 Mya (Chapter 4; Shu et al., 2003a), while fossils of *Pickaia*, a much simpler chordate have so far only been found in rock from the middle Cambrian (~505 Mya). Such are the vagaries of the fossil record. The date of an individual fossil indicates only when that particular organism died; it says nothing about when its species originated and when it became extinct. Cumulatively, of course, fossil dates indicate the direction of time, but such dates cannot be part of the criteria for determining evolutionary relationships.

The early chordates were wormlike in their external appearance (Fig. 12.2), but organisms with clear fish features had evolved from them by the start of the Ordovician (485 Mya). The fossil record for early fishes is relatively good, although there are still a few gaps: for example, our knowledge of conodonts, the earliest known fish-like organisms, comes mainly from their dental elements. Of particular importance in the context of this chapter is the set of Devonian fossils that showed how pectoral and pelvic fins evolved to become limbs over a period of ~30 My (~393–360 Mya). The transition from water to land required further major changes to the skeleton that particularly resulted in the limb and girdle bones becoming strong enough to bear the weight of the body on land. The fossil record of these changes is reasonable because they took place in large fishes: sizeable and heavy bones are obviously far more likely to be fossilised than small light ones. The equivalent fossil record covering the evolution of pterosaurs and bats from relatively small ancestors as they developed flight is far weaker because the transition is marked by the thinning and lightening of already small bones. The vertebrate fossil record always has a strong bias toward larger and more recent organisms.

MODERN DEUTEROSTOME ANAMNIOTES

Today, all anamniotes, other than amphibians, are marine organisms and the very great majority are fishes (Table 14.1). Each clade traces its ancestry back to the lower Cambrian or perhaps the late Ediacaran to a deuterostome that evolved from an earlier protostome taxon (Chapter 8; Erwin & Davidson, 2002). In protostomes, the neural cord is ventral to the gut and a key evolutionary change in the early evolution of chordates was the reversal of that morphology to the dorsal nerve cord seen in deuterostomes, a process whose molecular basis is now quite well understood (Chapter 8; Fig. 8.5). It was originally thought that this event happened several times because contemporary deuterostomes have such a wide range of morphologies. This is now know to be wrong as phylogenetic analysis of extant deuterostomes has shown that they all share a common ancestor; the clade is thus monophyletic (Ikuta, 2011). The protostome-to-deuterostome change only happened once and all contemporary deuterostomes evolved from that common ancestor. The early stages of deuterostome evolution are still not universally agreed, but the view expressed here is that held by many workers in the field.

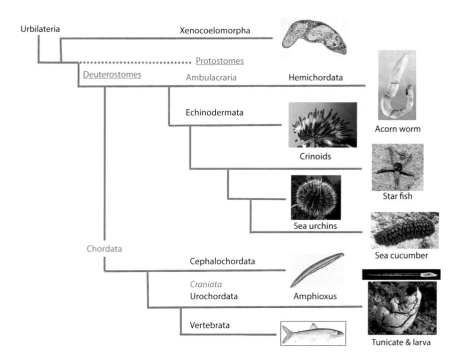

Figure 14.2 Phylogram showing the evolution of the deuterostomes. Note the larval form of the tunicate. For details, see text.

(*Acorn worm:* Courtesy of "Necrophorus." Published under a CC Attribution-Share Alike 3.0 Unported License.
Crinoids: Courtesy of Alexander Vaseniv. Published under a CC Attribution-Share Alike 3.0 Unported License.
Starfish: Courtesy of Nick Hopgood. Published under a CC Attribution-Share Alike 3.0 Unported License.
Sea urchin: Courtesy of Nick Hopgood. Published under a CC Attribution-Share Alike 3.0 Unported License.
Sea cucumber: Courtesy of Alexander Vaseniv. Published under a CC Attribution-Share Alike 3.0 Unported License.
Amphioxus: Courtesy of Giovani Maki. Published under a CC Attribution-Share Alike 3.0 Unported License
Tunicate: Courtesy of Nick Hopgood. Published under a CC Attribution-Share Alike 3.0 Unported License.
Tunicate larva: From Satoh et al. (2003) *Nat Rev Genet*; 4:285–95. With permission from Springer Nature.

Xenocoelomorph: From Curini-Galletti et al. (2012) *PLoS One*; 7(3):e33801. Published under a CC Attribution 2.5 Generic License.)

In addition to the dorsal location of the neural cord, chordates show several other ancient synapomorphies in their embryological development. These include a notochord and branchial (pharyngeal) arches, although some of these features can be lost after the larval stage (e.g. in tunicates). A further core synapomorphy is gut morphogenesis: when the gut forms from endoderm in anamniotes, the first opening becomes the anus and the second becomes the mouth, a form of development particularly apparent today in the gastrulation of sea urchins and amphibians (Fig. 14.2). This basic feature of deuterostome morphology has, however, been secondarily lost in animals with a large yolk or that are nourished by a placenta

TABLE 14.1 SELECTED PROPERTIES OF LIVING ANAMNIOTE CHORDATE TAXA

Urochordates (tunicates)	Typical chordate larvae that metamorphose to give an adult with a sac-like body with syphons
Cephalochordates	Mouth, no eyes or fins, segmented muscles, adult notochord, pharyngeal gills (e.g. *Amphioxus*)

All fishes have an early notochord (replaced or augmented by vertebrae), gills, eyes and neural-crest-derived tissues.

Agnathan fishes	Jawless mouth, midline fins, cartilaginous skeleton, (e.g. lamprey)
Chondrichthyan fishes	Jaws, paired fins, cartilaginous skeleton (e.g. sharks)
Osteichthyan fishes	Jaws, bony skeleton
Actinopterygii (Teleosts)	Fins with bony rays (e.g. salmon)
Sarcoterygii (Dipnoi)	Fleshy fins with bones and terminal rays (e.g. lungfish)

Most amphibians have an aquatic, fish-like larval stage that metamorphoses to give a terrestrial adult tetrapod. Today's amphibian clade, the lissamphibians, evolved from earlier amphibians whose ancestors were stem tetrapods with sarcoperygiian ancestry.

Lissamphibians (Frogs, Newts, etc.)	Normal teeth, larva that are typically aquatic and that metamorphose into terrestrial adults

Note: The Ambulacraria (echinoderms and hemichordates) are deuterostomes but not chordates.

via an umbilical cord. Here, food is presented directly to the midgut of the embryo, which thus has to function early in development. In such cases, the normal mouth and anus develop later.

It is likely that the early deuterostomes soon separated into two clades, the Ambulacraria and the Chordata, which together include five main taxa (Satoh et al., 2014); there is no evidence suggesting that any early chordate clade was lost. The Ambulacraria includes the Echinodermata (~7000 species, e.g. sea urchins) and the Hemichordata (~120 species that are predominantly acorn worms). The Chordata includes the Cephalochordata (~30 species) such as *Branchiostoma floridae* (commonly known as *Amphioxus* or the lancelet), the Urochordata (tunicates; ~2000 species) and the Craniata (animals with a protected brain; ~65,000 species). It was thought for many years that the Xenacoelomorpha, a group of very primitive marine flatworms (~400 species; Chapter 13) that lack a coelom, should be included in the deuterostomes (Philippe et al., 2011), but more recent work (Cannon et al., 2016) suggests that they are a sister group to both the protostomes and the deuterostomes and something of an evolutionary dead end (Chapter 13).

Ambulacraria

The echinoderms are particularly distinct with their characteristic synapomorphy of pentagonal symmetry. Their most populous taxon is the starfish, but the phylum also includes sea urchins, sea lilies (crinoids), brittle stars and sea cucumbers. The fivefold symmetry of starfish and sea urchins is specific to the adult phenotype: the embryos start off as bilateral but reorganise once they have reached the larval stage (Smith et al., 2008). Hemichordates such as the acorn worm have a larva with a stomochord tube in its anterior region that probably plays a structural role but may not be homologous to the notochord (Kawashima et al., 2014). The elongated adult has gill slits in its pharyngeal region. A cladogram of the early stages of deuterostome evolution, illustrated with modern examples, is shown in Fig. 14.2.

The minor chordate clades

These include the Cephalochordata and the Urochordata (tunicates) that probably evolved from an early cephalochordate. The cephalochordates are rare: there are only about thirty known species grouped into three different genera (the *Branchiostoma*, *Epigonichthys*, and *Asymmetron*; Subirana et al., 2020). The single well-studied species is *Amphioxus*, a *Branchiostoma* species: this very primitive fish-like organism has no obvious cranial features other than a mouth and a small brain region. It does, however, have a substantial notochord that persists throughout life providing rigidity for the segmented muscles in the adult. In more advanced vertebrates, this notochord is reduced to a thin, solid cylinder whose main role during early development is to act as a signaling centre.

Tunicates have a larval form similar to a tadpole with a head/body region that includes primitive eyes, a notochord, and a tail with a degree of segmentation in its musculature (Hudson, 2016). These larvae have a more sophisticated neuronal system than *Amphioxus*: tunicate embryos include ectoderm-derived neurons that are characteristic of the neural-crest- and placode-derived (sensory) tissues seen in craniates, although much of this anatomical detail is lost during metamorphosis (Graham & Shimeld, 2013; Stolfi et al., 2015; Horie et al., 2018). This process involves dissolution of the brain, notochord, and tail and the development of an external tunic and an internal siphon, together with a cement gland that they use to glue themselves to rocks (Sasakura & Hozumi, 2018).

The developmental relationship between current Craniata, Cephalochordata, and Urochordata has been clarified by the work of Horie et al. (2018). Craniata have a rich sensory system with organs for seeing, smelling, and hearing that develop in the early embryo from interactions between sensory epithelial placodes and neural crest cells. Cephalochordata have none of these sensory organs, but *Amphioxus* does express genes such as BMP2/4, Pax3/7, Msx, Dll, and Snail that are characteristic of early neural crest during very early development (Holland & Holland, 2001). In vertebrate embryos, this tissue differentiates into migratory neural-crest cells but, in A*mphioxus*, no neural crest or placodes form, although there is transient expression of markers for neural crest derivatives (Jandzik et al., 2015). In contrast, molecular markers for both sensory placodes and neural crest are expressed in the urochordate *Ciona intestinalis*; this tunicate also has sensory neurons. It is thus straightforward to construct a basic cladogram showing the early evolution of deuterostomes (Fig. 14.2).

That said, relatively little is known about morphological and molecular development of any of these species as compared to those of the Craniata, although there are a few similarities in their early phases. All chordates show evidence of A-P body patterning in their early stages and form a dorsal nerve cord (Holland, 2015). Fritzenwanker et al. (2019) have investigated body elongation in *Saccoglossus kowalevskii*, a hemichordate acorn worm, and shown that, as in other bilaterians, it involves cWnt and Notch signaling. This points to a common mechanism of elongation that evolved very early in the history of the Bilateria.

The situation for axial differentiation is more complex. All taxa encode *Hox* genes but, although they show colinearity in the Craniata and Hemichordata and some colinearity in Urochordata, they do not in the Echinodermata (David & Mooi, 2014, Gaunt, 2018). This last situation may help explain the unusual adult echinoderm forms (David & Mooi, 2014). Possession of *Hox* genes does not, of course, prove that they are used to interpret early patterning: both Urochordata and Echinodermata seem to make little use of the *Hox* cluster in their early development (Gaunt, 2018). As of now, the nature of early patterning in these deuterostomes is beginning to be unpicked, but we await the full story.

Craniata

This major clade, a synapomorphy for which is a head with sense organs and a brain covered by a protective layer of bone or hard cartilage, includes the agnathan or jawless fishes (now known as the Cyclostomata, see below) and the Gnathostomata (deuterostomes with jaws). The clade probably evolved in a urochordate ancestor that did not undergo metamorphosis, so retaining a head. Among the steps that led from a primitive chordate to the first fishes (discussed below) was the production of cartilaginous vertebral support elements, an apomorphy of somite differentiation (Appendix 6). The earliest indications of these are seen in *Haikouichthys*, a deuterostome from the Lower Cambrian (Shu et al., 2003a). Vertebrae are now present in all fishes and their descendants, with the exception of the hagfishes, and this is probably because their ancestors lost them.

Further key advances were the evolution of novel tissues that derived from neural crest and the cranial placodes. The neural crest forms from neuroectodermal cells at the boundary between future epidermis and neural tube. The crest cells later migrate throughout the embryo, particularly giving rise to the peripheral and gut nervous systems, sensory ganglia and much of the face region. The cranial placodes are thickened areas of surface ectoderm that will form the epithelial component of paired sense organs (e.g. the future auditory epithelium and nasal sensory epithelium, together with the future lens and cornea of the eye). They also contribute neurons to the cranial sensory nerve ganglia.

Over time, the other major innovation in the Chordata were the production of the various skeletal tissues that include bones, teeth, and scales; these, of course, are the materials that are preserved in fossils. Later Cambrian chordates had three such skeletal elements. These included a notochord containing glycosaminoglycans and dental elements (odontoids) made of dentine and enamel located in the pharynx. The third component was a set of vertebral elements, which seem initially to have contained glycosaminoglycans that was later supplemented with collagen to make cartilage and that were only occasionally fossilised,. These were soon supplemented in early fishes such as the ostracoderms by armoured plates of dermal bone laid down by mesenchymal cells that were subjacent to the external ectodermal layer that was later the source of scales.

Early fish skeletons were probably cartilaginous rather than bony and composed of glycosaminoglycans, glycoproteins, and collagen, laid down by chondrocytes. The skeletal fossil record only becomes substantial after the evolution of bone whose production required the evolution of two distinct osteocyte-based mechanisms. The earlier was the production of dermal bone, which is made directly by mesenchyme that derives from somite dermatomes and from neural crest; this was first used to produce the cranial bones that protected the brain and later for body armour. The second was endochondral bone that derives from somite sclerotome that gives vertebrae and ribs and from lateral plate mesoderm, the source of most long bones. In both cases, a cartilage template is first laid down by chondrocytes and then replaced with bone by osteocytes (for details, see Appendix 7; Barresi & Gilbert, 2019; Rolland-Lagan et al., 2012).

As the vertebrae evolved, the notochord was reduced in size and over time, seems to have lost its skeletal role. Today, the notochord secretes sonic hedgehog during early vertebrate development; this signaling protein initiates the development of neurons in the overlying neural tube and vertebrae in the adjacent somites. In adult vertebrates, other than Dipnoi (lung fishes), the notochord is lost in late development with its remnants becoming part of the intervertebral discs (Holland, 2005).

The major current fish clades include two with cartilaginous skeletons and two with bony skeletons (Fig. 14.3). The first cartilaginous group are the agnathans (lampreys and hagfishes) that lack jawbones but have a circular mouth opening with many small rasping teeth and are hence called the Cyclostomata. Although they lack paired fins, they are clearly an advance on *Amphioxus* since they have eyes, a skeleton, and embryonic neural crest cells. The lampreys have cartilaginous skulls and vertebral elements (arcualia), complex eyes and a dorsal fin. The slime-producing hagfishes are much more primitive: they have cartilaginous skulls with eyespots but neither fins nor vertebrae, although the latter probably represent a secondary loss (Ota et al., 2011; Oisi et al., 2013). Incidentally, the hagfish has a curious eating habit: it bites the flesh of other fish then ties a knot in itself which it slides to its head; it then pushes the knot over its head and pulls its mouth and a morsel of flesh away from the source of its dinner.

The second group of cartilaginous fishes are the Chondrichthyes that include the sharks, rays, and ghostfishes (sometimes known as chimaeras); these have vertebrae, jaws, and paired fins with cartilaginous rays, but lack swim bladders. Some taxa have true scales while others have dermal denticles, but all have gill slits through which water taken in through the mouth exits after flowing over the internal gills. Ghostfishes, unlike sharks, have upper jaws that are fused with their skulls; they also have opercula (gill covers) and separate anal and urogenital openings. Their three pairs of large permanent grinding tooth plates contrast with the many sharp and replaceable teeth of sharks (Stevens et al., 1998). Fig. 14.3 is a cladogram showing the synapomorphies of the various clades.

Most fishes today have a bony skeleton, fins with bony rays and swim-bladders. They are members of the Osteichthyes which has two subtaxa: the great majority

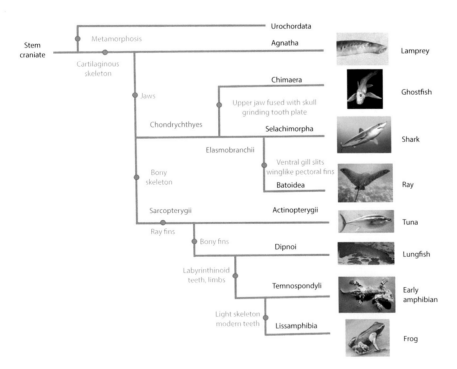

Figure 14.3 A cladogram showing the evolution of the major modern anamniote vertebrate classes together with some of the key synapomorphies.

(*Lamprey:* Courtesy of M Buschmann. Published under a CC Attribution-Share Alike 3.0 Unported License.
Ray: Courtesy of John Norton. Published under a CC Attribution 2.0 Generic License.
Tuna: Courtesy of "aes256." Published under a CC Attribution 2.1 Japan License.
Lungfish: Courtesy of "Tannin." Published under a CC Attribution Share Alike 3.0 Unported License.
Early amphibian: Courtesy of Ryan Soma. Published under a CC Attribution Share Alike 2.0 Generic License.

Courtesy of Charles C Sharp. Published under a CC Attribution-Share Alike 4.0 International License.)

of species have webbed fins with proximal elements and distal rays, both made of bone (the Actinopterygii; *actin* = ray; *pterm* = fin); their most common taxon is the teleosts, which includes almost all current fish species. The second (and much smaller taxon) is the Sarcopterygii which today includes only the Dipnoi or lobe-finned fishes (lungfish and coelacanths). [*Note:* the sarcopterygian clade (sarco = fleshy) includes every animal that evolved from a stem sarcopterygian fish and so includes all tetrapods.] Their lobed fins are notable for including long bones and joints, rather than the small bony elements of the Actinopterygiian fin, although they end in bony rays rather than digits. It is also worth noting that some sarcopterygian teeth are complex with internal foldings, a feature that later characterised the earliest amphibians.

THE EARLY DEUTEROSTOME FOSSIL RECORD

Due to fortuitous fossilisation processes in the early-to-mid Cambrian period (Chapter 3), there are some very early aquatic chordates whose soft tissues have been preserved and that display clear evidence of segmented muscles. This synapomorphy of the Chordata is more often preserved than any skeletal evidence in early vertebrate fossils that predate the evolution of bone (Fig. 14.4). This is probably because muscle is composed of dense protein, while the notochord and any early skeletal support tissues were made of nonmineralised cartilage that was hydrated and so easily lost.

The most primitive of the known chordates in terms of tissue anatomy is *Pikaia gracilens* (Fig. 4.3a). The fossils date to ~505 Mya but their primitive anatomy suggests that the taxon separated off the main deuterostome branch considerably earlier. This species was discovered among soft-tissue fossils in the Middle-Cambrian Burgess Shale in Canada. *Pikaia* had myotomes (segmented muscle blocks), a notochord, a mouth opening and probably gills but no jaws, teeth or eyes. It may thus be close to stem-group chordates. If *Pikaia* has contemporary descendants, they are likely to be cephalochordates as they have similar anatomy. Detailed analysis shows that the *Pikaia* clade became a sister clade to the cephalochordate line early in deuterostome evolution (Morris & Caron, 2012; Zhang et al., 2018).

Chordate fossils that are older than *Pikaia* examples but with more complex anatomy have been found in the Chengjian biota in China. Examples particularly come from the Lower Cambrian Maotianshan shales and include nine species with well-preserved, soft-tissue structures. These include *Haikouichthys* and *Myllokunmingia* (524 Mya), which seems to have had a cartilaginous skull and skeletal structures.

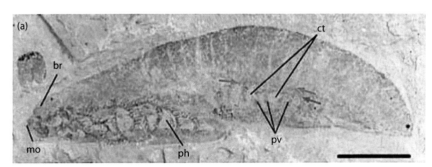

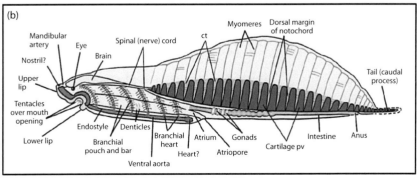

Figure 14.4 *Haikouella*, **an early chordate from Cambrian (~515 Mya).** (a): an example of the fossil. (b): a drawing showing the anatomy. (br, brain; Ct, connecting tissue; Mo, Mouth; Ph, Pharynx; Pv, protovertebrae.) (Bar = 0.2 mm.)

(From Chen J-Y (2009) Int. J. Dev. Biol. 53:733–751. With permission from J-Y Chen.)

Other primitive chordates from these shales include *Yunnanozoon*, which was probably a hemichordate from the same period (~520 Mya, Shu et al., 1996), and *Shankouclava anningense*, a tunicate. For discussion of the relationships amongst these organisms, see Janvier (2015).

Haikouella jianshanensis was another slightly later chordate from this shale (520–515 Mya). Analysis of several hundred examples shows that it was ~2.5 cm in length and had gills, a notochord, chevron-shaped muscle segments, a heart and blood vessels, eyes, a nerve cord and reproductive organs (Shu et al., 2003b; Chen, 2009; Fig. 14.4). It also had tiny odontoids (primitive teeth ~0.1 mm) in its pharynx; these are, so far, the earliest evidence of chordate hard tissue and were clearly used to break up food already engulfed rather than for biting. Such taxa are clearly near to the stem taxon of the vertebrates but were still more like small marine worms than fish in appearance. The lower Cambrian chordate *Haikouichthys ercaicunensis* (520–515 Mya) from the Qiongzhusi formation in Yunnan, China was a little more advanced. As mentioned, its fossils show soft tissues, including a dorsal fin, and are currently the oldest to show traces of cartilaginous vertebral elements (Shu et al., 2003a). Even though it still had segmented gonads, *Haikouichthys* seems more like a very primitive jawless fish than *Haikouella*.

THE FISH FOSSIL RECORD

It is clear that the early chordates of the Cambrian period diversified, with key vertebrate features appearing rapidly (on an evolutionary time scale), soon producing taxa of ever-increasing richness (Fig. 4.2). The earliest known vertebrates that were unequivocally fishes were the Middle-Cambrian conodonts (~515 Mya or possibly a little earlier). These small, segmented eel-like creatures (Donoghue et al., 2000) lacked vertebrae and jaws and so were agnathan, like contemporary lampreys and hagfishes. Instead of teeth, however, they had a complex feeding apparatus of dental elements (Fig. 14.5, Donoghue & Sansom, 2002) that are now used as *index fossils*. Conodonts survived for a very long time, only becoming extinct a little before the end of the Triassic (~200 Mya).

It is with the evolution of bone that the fish fossil record starts to become substantial (for details, see Janvier, 1996). The discussion below makes no attempt to cover the whole extent of their diversity but aims only to show the lines of descent that led to the contemporary fish clades. The earliest known fishes that have been well preserved were the various groups of ostracoderms that evolved during the Ordovician (488–454 Mya; Fig. 14.6a). These agnathan fishes were relatively small organisms characterised by a cartilaginous skeleton and plates of dermal bone that formed in the mesenchyme of the skin and protected their heads and bodies. This broad taxon is subdivided on the basis of the details of head and body armour.

One taxon includes the heterostracans, now extinct, that had head shields of dermal bone, a body covered with trilayered scales and a skeleton made of aspidin, a tissue that has now been shown to be acellular dermal bone (Sansom et al., 2005; Keating et al., 2015). More important were the ostracoderms that seem to have been the first fishes to use gills for respiration. Another group that derived from the ostracoderms were the anaspids that remained jawless and lost their body armour. They were probably the ancestors of today's agnathans. Included in this group was

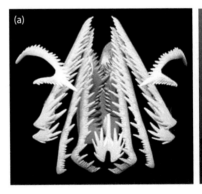

Figure 14.5 (a): Rostral view of a model of the dental elements of an ozarkodinid conodont. Each element is ~1–2 mm long. (b): Reconstruction of a conodont.

(a From Donoghue and Sansom (2002). *Microsc Res Tech* 59:352–372. With permission from John Wiley and Sons.
b Courtesy of Nobu Tamura. Published under a CC Attribution-Share Alike 4.0 International License.)

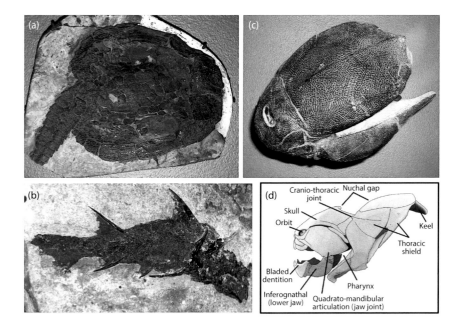

Figure 14.6 **Jaw evolution.** The jaw bones of fish derive from the first branchial arch of agnathan fishes. (a,b): Schematised cranial morphology of an early agnathan fish and of an early jawed fish. (c): The placoderm *Coccosteus decipiens,* an early jawed fish (gnathostome) with dermal head plates.

(a,b From Larsen WJ (1997) Human Embryology. 2nd edtn. W B Saunders Company. With permission from Elsevier.
c   Courtesy of Nobu Tamura. Published under CC B-Y-SA 3.0)

Sacabambaspis janvieri, an arandaspid that possessed a posterior notochordal projection rather than a tail (Pradel et al., 2007). Its fossils are sufficiently well preserved for details of its bone structure to be analysed: its headshield was made of dermal bone whose layered structures included dentine, acellular bone and enamelloid with pulp and spongy spaces; as a result, some tissue appears to be made of compacted teeth. Since no evidence of an axial skeleton has been preserved in any fossils of ostracoderm fishes, any vertebral elements are likely to have been small and cartilaginous.

In the early Silurian (~440 Mya), the agnathan ostracoderms gave rise to the placoderms that were the first fishes known to have jaws with teeth (Figs. 14.6c,d; 14.7c), with *Romundina stellina*, being the oldest known. Later examples were *Qilinyu rostrata* and *Entelognathus primordialis* that had plates protecting their heads and show successive changes to jaw morphology (Rücklin & Donoghue, 2015; Vaškaninová et al., 2020; Zhu et al., 2013; Zhu et al., 2016a). On the basis of modern developmental work, placoderm jaws and teeth formed from the mesoderm- and neural-crest-derived mesenchyme of the first pharyngeal arch, an embryonic tissue that had previously given rise to the anterior skeletal support of the first gill opening (Fig. 14.7a,b). By the mid-Silurian period (~430 Mya), a new taxon, the acanthodians, evolved with protective spikes (Fig. 14.6b). They probably diverged from the placoderms, but their line of descent is not yet precisely known because the fossil record is thin here (Brazeau & Friedman, 2014).

Both placoderms and acanthodians had paired fins and a cartilaginous endoskeleton with a layer of perichondral bone that was, on the basis of contemporary chordate development, made by cells that derived from lateral plate mesoderm (Brazeau et al., 2020). The earliest known fossil showing evidence of vertebral bone, which derives from the sclerotome component of somites (Appendix 7), was an acanthodian; its vertebral cartilage was permeated with bone spicules. The placoderms, in contrast, maintained a dermal skeleton of bony plates. Incidentally, there is some evidence that the young of this taxon developed internally: a fossil placoderm has been found that had an umbilical cord connected to an internal embryo (Long et al., 2008, 2009).

Even though they displayed evidence of endochondral bone, the ancestors of today's cartilaginous fishes, such as rays and sharks, were probably early members of the acanthodian taxon that lost its protective spikes and eventually the ability to make bone. Burrow and Rudkin (2014) report that *Nerepisacanthus denisoni*, a beautifully preserved acanthodian fish from the late Silurian period (~420 Mya) had scales whose detailed structures are typical of those seen in later chondrichthyan organisms. These cartilaginous organisms only appear in the fossil record at ~409

(a)   Jawless fish

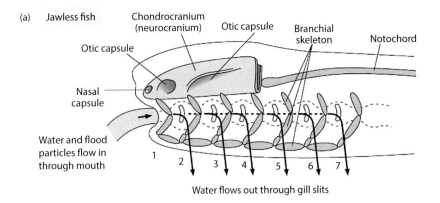

(b)   Primitive jawed fish

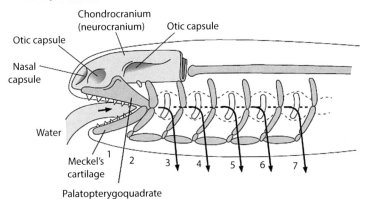

(c)

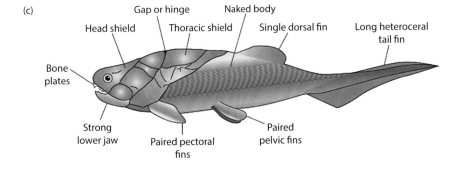

Figure 14.7 Fossil fishes from the Devonian. (a): *Cardipeltis bryanti*, a lower Devonian ostracoderm showing the armour of the ventral surface. (b): *Ischnacanthus gracilis*, a lower Devonian acanthodian (note the spikes). (c): *Bothriolepis canadensis*, a middle-Devonian placoderm. (d): Reconstruction of the skull of *Dunkleosteus terrella*, a late Devonian placoderm. Note the armoured skull case and jaws with only membranes bones; this is because the teeth have been worn away or lost (from Anderson & Westneat, 2007).

(a   Courtesy of James St. John. Published under a CC Attribution 2.0 Generic License.
b   Courtesy of James St. John. Published under a CC Attribution 2.0 Generic License.
c   Courtesy of "Smokeybjb." Published under a CC Attribution-Share Alike 3.0 Unported License.
d   Courtesy of "Steveoc86." Published under a CC Attribution-Share Alike 3.0 Unported License.)

Mya: Miller et al. (2003) have reported an early fossilised shark with a braincase and paired pectoral fin-spines whose basic morphology demonstrated its chondrichthyan provenance.

The details of the early diversification of the fishes are still not entirely clear as it seems that early forerunners of the line that eventually led to the Osteichthyes (bony) and the Chondrychthyes (cartilaginous) fish clades survived for quite a long time. An example is *Janusiscus schultzei*, an early Devonian gnathostome (~415 Mya) with features of both taxa (Giles et al., 2015). By the mid-Devonian, the two distinct gnathostome (jawed) clades had evolved: the Chondrichthyes with cartilaginous skeletons and the Osteichthyes with bony ones, synapomorphies of which are endochondral bone and modern dentition (Rücklin et al., 2012). The Osteichthyes soon gave rise to two distinct subclades, the Actinopterygii and the Sarcopterygii. The former, which always seem to have been the most common fish taxon, had simple rayed fins while the latter had fleshy lobed fins with the pectoral and pelvic ones containing distal articulating, jointed bones and terminal rays. Today, most fish are teleosts (descendants of early Triassic Actinopterygii), while

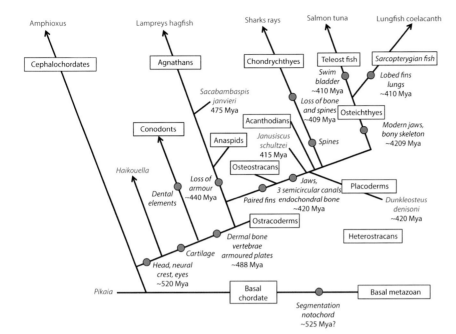

Figure 14.8 Cladogram showing the evolution of the contemporary fish families from a basal chordate. Families are in plain font and species in blue italics. Synapomorphies together with the dates when they are first seen in the fossil record are shown in black italics.

the Sarcopterygii have been reduced to lungfish (Dipnoi) and coelocanths. The importance of the Sarcopterygian branch is that it was within this clade that tetrapods evolved.

The information given in the paragraphs above is sufficient to produce a diagram giving the likely lines that led from Cambrian chordates to the various major taxa of modern fishes (Fig. 14.8). The different relative rates of skeletal change during the middle Devonian period (~410 Mya) emphasise the importance of depending on the anatomical details of the organism rather than the time when it lived for elucidating evolutionary relationships.

One of the reasons why there may be some uncertainty in the lines of descent is that the fossil evidence does not fully explain the evolution of endochondral bone. First, it seems likely that some of the major early fish taxa were paraphyletic and less useful for cladistic purposes than once thought. Second, as the molecular genetics of dermal bone and endochondral bone development are closely related (Wood & Nakamura, 2018); the latter may have evolved or have been lost and restored several times. One reason for suggesting this is that the Chondrichthyes seem to have evolved from the acanthodians, a taxon that included some of the earliest species with endochondral bone, while the Osteichthyes derived from the placoderm line.

Two papers have helped clarify the situation. Eames et al. (2007) have shown that the early stages of endochondral bone are laid down in modern shark embryos (e.g. in the neural arches of the vertebrae). More recently, Venkatesh et al. (2014) have analysed whole-genome sequences of the elephant shark and found that the absence of bone in its endoskeleton is probably due to the lack of genes encoding the secreted calcium-binding phosphoproteins needed for laying down endochondral bone. It thus seems that the presence or absence of bone hangs on whether just a few genes are expressed.

Finally, it is worth mentioning the example of *Tullimonstrum gregarium* from ~300 Mya as it demonstrates that some details of the evolution of marine organisms still remain opaque (Box 14.1; Fig. 16.7). This curious and anomalous fossil has been interpreted as both a mollusc and a chordate. It is mentioned in this chapter because the balance of the evidence currently points to it being the latter.

FROM WATER TO LAND

Two key events define the transition of vertebrates from marine to fully terrestrial organisms: an ability to thrive on land, discussed here, and the much later evolution of the amniote egg, (Chapter 15). The ability to survive out of water demanded a

BOX 14.1 THE TULLY MONSTER

Tullimonstrum gregarium lived during the late carboniferous (~300 Mya) in what is now Illinois (it is the official state fossil) and several thousand examples have been found. It was up about 35 cm long, had a segmented body, clearly pigmented camera eyes at the end of a transverse bar on its head, and a proboscis with what appear to be teeth. It probably had fins but lacked bones. As there are no obvious ancestors with these traits and the organism seems to have left no descendants in the fossil record, this animal has attracted a considerable amount of analytical work.

The anatomy of this strange animal has been hard to classify, and various authors have suggested that it might be a primitive agnathan fish related to lampreys, a mollusc or even an organism whose ancestry could be traced back to *Anomalocaris*, a very early arthropod predator thought to have become extinct during the Cambrian (Chapter 13). Over the past few years, there has been a series of recent studies that argue for and against it being a vertebrate or a mollusc.

- McCoy et al. (2016) characterise *Tullimonstrum* as a vertebrate on the basis of anatomical features such as that its proboscis has teeth, and it seems to have a caudal fin, a notochord and pharyngeal arches (Fig. 14.9).
- Clements et al. (2016) use time-of-flight secondary ion mass spectrometry to show that pigmented microbodies in the eyes of *Tullimonstrum* are melanosomes, a feature of vertebrate eyes.

- Sallan et al. (2017) reanalysed the data, pointing out that features that impressed McCoy et al. (2016) as demonstrating a vertebrate ancestry, were actually less precise than thought and may well have reflected taphonomic artefacts. These authors also point to *Tullimonstrum* features that are more typical of an arthropod than a chordate taxon.
- Rogers et al. (2019) show that the chemical signatures of the eye melanosomes in *Tullimonstrum* are more typical of those in a modern cephalopod eye than a modern vertebrate eye (both taxa, of course, have camera eyes)
- McCoy et al. (2020) show by Raman spectroscopy that the rock that replaced *Tullimonstrum* soft tissue includes material with a spectral signature typical of vertebrate protein rather than that of marine invertebrates, which normally includes chitin.

While the jury is still out, the balance of argument, particularly the likely presence of teeth, now seems to go with *Tullimonstrum* being a chordate, but further evidence may change this view. What, however, is particularly noteworthy in this case study is, first the wide variety of physical and chemical techniques used for studying the chemical nature of fossilised tissues and, second, the fact there are still organisms in the fossil record that have yet to be found a place in the great tree of life.

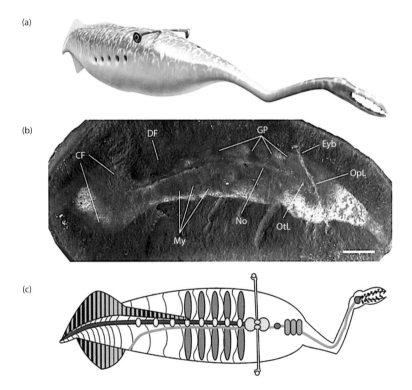

Figure 14.9 ***Tullimonstrum gregarium* seen as a chordate.** (a): Its likely appearance. (b): The annotated fossil (bar: 9 mm). (c): Its internal anatomy on the basis of it being a chordate.

(From McCoy et al. (2016) *Nature* 532:496–499. With permission of Springer Nature.)

major reorganization of fish anatomy, most noticeably the evolution of both fore and hind limbs for movement, and of substantial pelvic and pectoral girdles to anchor these limbs, while a strengthened skeleton was needed to support the weight of the body (Clack, 2012). The reorganization also involved changing the mode of oxygen uptake from a gill system to one based on lungs, which also required modifications to the heart. Finally, given that development took place in water, a mode of novel metamorphosis was required to turn an aqueous embryo and larva into a land-based stem tetrapod. All this took some time to happen.

For several reasons, it is clear that the ancestor of tetrapods was a sarcopterygian rather than an actinopterygian fish (see also Chapter 6). The most obvious of these is that the former has pectoral fins whose proximal skeleton includes bony homologues of the humerus, ulna and radius and whose pelvic fins include homologues of the femur, tibia, and fibula. Actinopterygian fins lack these bones; their fins are strengthened by small bones, mainly homologous with girdle bones. In both taxa; the terminal parts of the fin are rays of dermal bone. Extant sarcopterygian fishes include the *Dipnoi* (lungfish) and the *Actinistia,* whose only living taxon is the coelacanth, *Latimeria*.

There is both anatomical and genetic evidence suggesting that an ancestor of the Dipnoi was the likely ancestor of tetrapods (Brinkman et al., 2004; Venkatesh et al., 2001). The anatomical evidence is particularly convincing: like terrestrial vertebrates, Dipnoi have lungs that develop from pouches off the oesophagus and the early development of modern lungfishes and amphibians is similar, with both differing from that of contemporary teleost fishes (Clack, 2012; Semon, 1901). Today, lungfish fall into two classes: the Australian *Neoceratodus* genus that have working gills and a single lung; the two other genera, *Protopterus* from Africa and *Lepidosiren* from South America, both of which have two lungs and atrophied gills. Dipnoi are clearly the living sister group of the Tetrapoda.

The question of whether the evolutionary events that led to early amphibians took place in salt or fresh water has not been definitively answered (for detailed discussion, see Clack, 2012). Two reasons originally suggested that these events occurred in fresh-water lagoons: first, all modern lungfish live in fresh water whereas coelacanths are marine fish and, second, there were droughts during the Devonian which would have been a selection pressure for organisms able to survive when ponds dried out.

The view today is a little more nuanced as it is known that contemporary lungfish can survive in brackish water; it is quite possible that swampy shallow lagoons would have provided an appropriate environment of marine vegetation in which variations in water levels would have favoured limb development. An added argument that supports a lagoon origin comes from the fossil record: similar organisms have been found in places that were very distant from one another. Such similarity is best explained by migration across bodies of sea water, a conclusion supported by the molecular data (George & Blieck, 2011) and by isotope analysis of fossilised bones, which suggests that the original animals were euryhaline (i.e., tolerant to a wide range of salinity; Goedert et al., 2018).

It is, however, the limited fossil record that has been key to understanding the events that led to the evolution of land-based vertebrates from marine ones. Over the last two decades or so, the anatomy of a series of late-Devonian fossils, some recently discovered and others re-examined, has illuminated the nature of the later stages of the transition from sea to land that took place during Devonian (419–358 Mya (Table 6.1; Fig. 14.10). Most of these fossils fall within the tetrapodomorph total group: this includes all the tetrapods (the crown group) and their sarcopterygian ancestors, which are more closely related to living tetrapods than to living lungfish (this ancestral group is known as stem tetrapods).

The fossil evidence (Chapter 6) makes it clear that primitive limbs had evolved well before the evolution of pectoral and pelvic girdles with sufficient strength to support the weight of the organism on land. Although Clack (2012) considered that the selection pressure that favoured limb evolution was the ability to wade through dense weed, recent work has suggested that it is more likely that limbs evolved from fins that facilitated the ability of partially submerged fishes to walk on the bottom of shallows while their heads were above water and their body weight was supported by water. One reason for this view is that some sarcopterygian fishes had dorsal rather than lateral eyes, which would only have been useful for aerial sight above water. The other reason comes from fossilised trackways dating back to ~390 Mya, well before fish with limb-like fins are seen in the fossil record

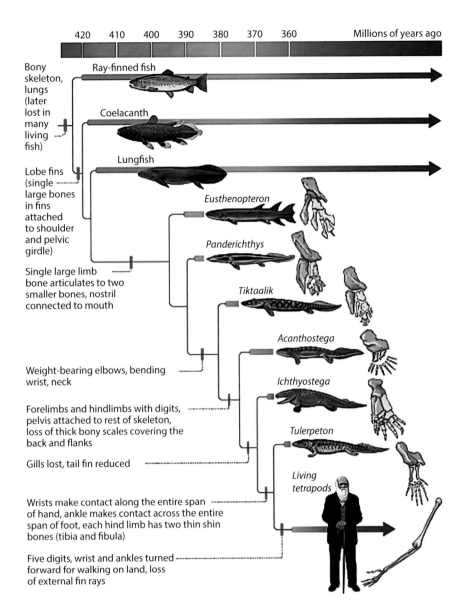

Figure 14.10 A cladogram showing the evolution of the major fish families from a basal chordate. Families are in bold and species in italics font. Synapomorphies, together with the dates when they are first seen in the fossil record, are shown in italic and standard Roman font. Branch points represent unknown last common ancestors.

(From Zimmer C (2013) *The Tangled Bank: An Introduction to Evolution.* Roberts & Co. Permission applied for by me and by AEL.)

(Fig. 14.11; for review and detailed references, see Ahlberg, 2019). Early trackways only reflect walking across the bottom of shallow waters, but later ones (~365 Mya) show signs of body drag on land.

It has recently become clear that, between the sarcopterygian fishes with rayed fins and the later basal tetrapods, there was a group of intermediary organisms known as elpistostegids that is now extinct (Fig. 14.12). The apomorphy that marks this group, which includes *Penderichthys* and *Tiktaalik*, is a primitive autopod that includes unjointed digits (radials) and only some wrist bones as well as other features intermediate between sarcopterygian fishes and tetrapods (for review and detailed references, see Ahlberg, 2019).

It is, however, an earlier sarcopterygian fish, the lobed-fin *Eusthenopteron* (385 Mya. *Note*: dates are for particular fossils, and do not imply a start or end date for a species), that makes a very plausible early ancestor for the tetrapods: apart from its fin bones, it had labyrinthodontoid teeth, nostrils that connected with the oral cavity and a neck joint that allows the head to turn, all of which are seen in the earliest amphibian fossils. Although only hard tissues have been preserved, *Eusthenopteron* would, on the basis of these features, probably have had the primitive paired lungs and limited breathing capacity seen in today's lungfish. This fish also had well-formed fins, and it now seems likely that it could walk on the bottom of shallow water. In this context, fins were probably extending and strengthening—this was an exaptation for the later evolution of limbs that could walk on land.

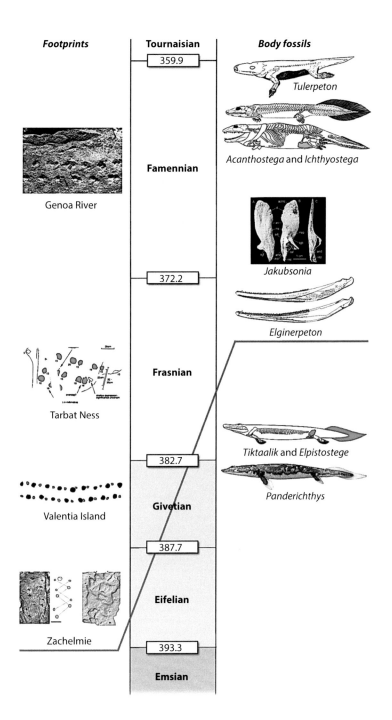

Footprints

Genoa River

Tarbat Ness

Valentia Island

Zachelmie

Tournaisian

359.9

Famennian

372.2

Frasnian

382.7

Givetian

387.7

Eifelian

393.3

Emsian

Body fossils

Tulerpeton

Acanthostega and *Ichthyostega*

Jakubsonia

Elginerpeton

Tiktaalik and *Elpistostege*

Panderichthys

Figure 14.11 **Fossil trackways from Middle and Late Devonian.** These date back to ~385 Mya, far earlier than the fossil dates of the likely organisms that made them (linked by the red line). The boxed dates mark sub-Period boundaries and details of the organisms are shown in Fig. 14.12.

(From Ahlberg, PE (2019) *Trans R Soc Edin* (Earth Env Sci):109, 115–137. With permission from Per Ahlberg and Cambridge University Press.)

Limb evolution

This topic has been discussed in detail in Chapter 6 and only the key points are given here. The sarcopterygian fossil record includes a series of species that lived over the period ~395–348 Mya (most from the Devonian and the early part of the Carboniferous). These demonstrate the steps that led from fish with bony fins to an organism with tetrapod limb morphology; they are detailed in Table 6.2 and illustrated in Fig. 14.10. The long bones of tetrapod limbs were present as homologues in the lobed fins of early lungfish, but functioning on land required them to lengthen and strengthen, a process that required growth plates. This advance was already present in *Eusthenopteron*.

The key innovation in limb development, however, was the evolution of autopods (wrists, ankles, and digits) that are made of endochondral bone and that replaced the fin rays made of dermal bone. Digit formation was a two-stage process with unjointed radials appearing in elpistostegids before jointed digits evolved in basal tetrapods.

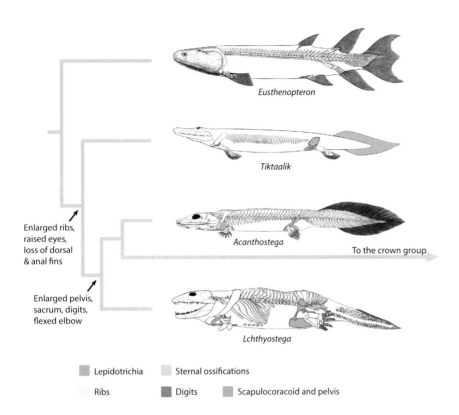

Eusthenopteron

Tiktaalik

Acanthostega

To the crown group

Enlarged ribs,
raised eyes,
loss of dorsal
& anal fins

Enlarged pelvis,
sacrum, digits,
flexed elbow

Ichthyostega

Lepidotrichia ▪ Sternal ossifications

Ribs ▪ Digits ▪ Scapulocoracoid and pelvis

Figure 14.12 A cladogram showing some key nonlimb features that evolved as late sarcopterygian fish evolved into early elpistostegids.

(From Ahlberg, PE (2019) *Trans R Soc Edin* (Earth Env Sci):109, 115–137. With permission from Per Ahlberg and Cambridge University Press.)

An example of the former is *Panderychthys*, an epistostegid with unjointed radials of endochondral bone (~370 Mya), while *Elpistostega* had both unjointed radials of endochondral bone and rays of dermal bone in its fin (Davis et al., 2004; Cloutier et al., 2020). Once tetrapods had evolved, the numbers of jointed digits seen in the various species ranged from three to eight (Table 6.1), and it seems to have taken some millions of years for the number to stabilise at five.

As discussed in Chapter 6, elpistostegids such as *Panderichthys* and *Tiktaalik* had many tetrapod features, including bony limbs adequate to drag them up a beach, a flexible neck and simple girdles, but still lacked jointed digits—they were stem tetrapods rather than basal tetrapods. The synapomorphies that mark the latter clade are autopods with digits homologous to those in later tetrapods, wrist bones (carpals), a flexible elbow, substantial ribs and, particularly, a sacrum. *Note*: fossil datings are unreliable indications of anatomical progress because many of these species survived side-by-side for millions of years.

It was the evolution of the sacrum that enabled early tetrapods to walk on land: this is the set of fused posterior vertebrae that anchors the pelvis, providing it with rigidity and strength. Some of these features are seen in *Ichthyostega*, *Acanthostega*, and *Ventastega*, indicating that these elpistostegids were very close to the last common ancestor of the tetrapods. However, as each possessed lateral line organs for detecting pressure variation in water, they were essentially aquatic animals (Ahlberg, 2019).

The ten species of fossils mentioned in Table 6.1, some of which are shown in Fig. 14.10,12, have been ranked on the basis of their autapomorphies. These do not, however, give a clear line of descent and so the cladogram shown in Fig. 14.10 should be viewed as provisional—further species will be required. Thus, for example, *Panderichthys* (~380 Mya) has bony digits but a rigid neck whereas *Tiktaalik* (~375 Mya) has rayed digits and a mobile neck. Similarly, a comparison of *Ichthyostega* (~374 Mya) and *Acanthostega* (~365 Mya) shows that the former had stronger forelimbs while the latter had stronger hindlimbs. The evidence as a whole suggests that by ~375 Mya there was a broad range of marine taxa and early elpistostegids living near the coastline that were evolving to exploit the opportunities that this niche offered.

Girdle evolution

A key difference between an organism that swims and one that walks on land is that the former does not need its weight to be supported. Walking while supporting the body so that it does not drag on the ground requires that the pectoral and pelvic girdles to which the limbs are attached are rigid. The pectoral girdle of today's tetrapods includes the scapula (shoulder blade), the clavicle, and the coracoid bones. Comparative anatomy has shown that the scapula derived from the scapulocorocoid bone, part of the non-weight-bearing pectoral girdle present in both actinopterygian and sarcopterygian fish that supports the pectoral fin and is attached to the skull base (Holland, 2013). The growth of the pectoral bones from late embryogenesis in contemporary coelacanths shows how its primitive girdle develops (Mansuit et al., 2020). Equally important to the development of the tetrapod pectoral girdle was its change in position from the base of the fish skull to the level of the posterior cervical vertebrae, although it does not form a joint with them (Fig. 14.11). This separation of the girdle from the cranium allowed stem tetrapods to have had a mobile neck.

Evolution of the weight-bearing pelvic girdle with its three bones, the ilium, pubis and ischium, was more complex. Lungfish have a minimal pelvis composed of two pubic bones that are jointed to the pelvic fins but, as they lack an ischium, there is no link to the fused vertebrae of the sacrum. Elpistostegids such as *Tiktaalik* (375 Mya) had a substantial girdle with broad iliac and pubic processes and acetabula (hip sockets) but also lacked the ischium bones (Shubin et al., 2014). Although it may have had a degree of terrestrial competence, the general anatomy of this organism makes it clear that the function of the pelvic girdle was mainly to facilitate enhanced propulsion through water. Boisvert et al. (2013) have analysed the steps needed for the ischium to evolve and suggest that the most important stage was the dorsal migration of some pre-cartilage cells into the early pubis area so as to provide the raw material for the new bones; some repatterning of their differentiation was also necessary to ensure that three bones formed rather than two.

Breathing

A further major step required for life on land was the ability to breathe air rather than to absorb dissolved oxygen from water. As fish nostrils are blind-ending sacs lined with olfactory epithelium whose sole purpose is to provide a sense of smell, a key step in the evolution of breathing was the extension of the nostrils through to the oral cavity. The oldest organism in which this feature has been noted in *Kenichthys* (Table 6.1), a small sarcopterygian fish from the early Devonian (~395 Mya). However, as none of the transitional fossil material includes soft tissues, comparisons between modern lungfish and amphibians are needed to understand the physiological changes associated with the move to land. On this basis, it likely that early lungfish had a cardiovascular system much like that of their modern descendants: this includes primitive lungs in addition to gills, together with a heart with a single ventricle and a partially divided atrium separating most of the pulmonary and systemic flows. This heart is a simplified version of the three-chambered heart of extant amphibians. Note also that the thoracic cavity increased in size as marine organisms became better adapted to a terrestrial existence (Fig. 14.12).

Lungfish today breathe through buccal pumping (using their cheeks) as do today's amphibians (see Websites), although amphibian breathing is strengthened by the additional force provided by the transverse abdominal muscles (Simons & Brainerd, 1999). In addition, amphibians have moist skin that can exchange gases (cutaneous respiration). Neither lungfish nor amphibians have a diaphragm; indeed, the only nonmammalian taxon to have this anatomical feature is the Crocodilia, but this muscular sheet seems neither to play a major role in their normal breathing nor to be homologous with the mammalian diaphragm (Munns et al., 2012). Taken together, the evolutionary steps required to progress from the cardiovascular and pulmonary systems of Dipnoi to those of the early amphibian (Fig. 6.5) do not seem as great as they once did.

Metamorphosis

All stem tetrapods and early tetrapods, like contemporary amphibians, have eggs that develop under aqueous conditions to produce larvae similar to those seen in fishes. The final ability required for the evolution of a land-based tetrapod was a means of metamorphosis. Today, amphibian metamorphosis from the aqueous larval to the terrestrial adult stage is entirely dependent on thyroxin, a hormone made by the tadpole's thyroid gland (for review, see Brown & Cai, 2007). When this hormone is released naturally into the bloodstream of amphibian larvae, a host of changes take place to the anatomy and physiology of the organism. These include losing the tail, remodeling the gut, and replacing the pronephros with the mesonephros and the gills with lungs. It is noteworthy here that experimentally administered thyroxin brings about metamorphosis in axolotls, a species that normally retains larval morphology throughout life, although they do, of course, continue to grow (this retention of a juvenile form is known as paedomorphosis—see Chapter 18). In the presence of the hormone, axolotls acquire a salamander-like appearance and the normal anatomy and physiology for living on land (Demircan et al., 2016).

Little is known about the stages by which these changes evolved, but the importance of thyroxin in larval remodeling was probably already established in fishes. Today, it is the key hormone that mediates the processes required for teleost fishes to attain adulthood (Campinho, 2019). These changes include sexual maturity, vision, and adult anatomy. The most intriguing of these changes are those that convert normal-looking flatfish larvae into adult flatfish, a process that requires the movement of one eye to the opposite side of the head and changes the dorso-ventral to the left-right axis (Campinho, 2019).

AMPHIBIANS

The fossil record shows that, once stem tetrapods had evolved into tetrapods, they diversified widely: this is presumably because there were no competitors when they first colonised land. Over time, one line became amphibians and another became stem amniotes, which very slowly evolved to give the amniote clade (Chapter 15). The amphibians thrived unchallenged for many tens of millions of years, giving rise to a host of species, some of which were very large, before declining during the Permian period, which was when stem amniotes gave rise to reptiles and stem mammals. Many amphibian taxa became extinct during the early Cretaceous, in part perhaps because the requirement of their tadpoles for water limited the habitats that they could occupy.

The main early amphibian taxon was the Temnospondyli (~360–150 Mya), whose defining features were relatively massive triangular skulls and labyrinthoid teeth with internal foldings, a plesiomorphy from sarcopterygian fish (Fig. 14.13a, b). They also had large vertebrae, each of which was in several parts, while many taxa also had dermal armour. The heaviness of the skeleton was presumably to support their weight on land and perhaps for physical strength. Their size range was broad: the largest known is *Prionosuchus* (~270 Mya) and was ~9 m long. Of the other taxa, the best known are the Lepospondyli; these first appeared in the Early Carboniferous ~350 Mya; they were much smaller and lighter than the earliest amphibians, as well as having modern vertebrae. Their taxonomic status is unclear, and it is still not known whether they were amphibians or stem amniotes (see Websites).

The only contemporary amphibian clade is the Lissamphibia. Their early fossil record is unfortunately poor, presumably because they were relatively small and so had fragile bones. The oldest lissamphibian fossil is from the early Triassic (~250 Mya), but dating based on molecular phylogenies has suggested that they probably evolved earlier (Marjanović & Laurin, 2007; San Mauro, 2010). The origin of the lissamphibians is still not fully known: current evidence, however, suggests that their most likely ancestors were the Amphibamiformes, a group of Temnospondyl tetrapods that first appeared in the Late Carboniferous (Atkins et al., 2019).

Today, the relatively few taxa within the Lissamphibia include toads, frogs, newts, salamanders, axolotls and the limbless caecilians. All lissamphibians have larvae with fins, gills and a tail structure, reflecting their evolutionary history. Most larvae develop in water, although some are internally incubated (in the mouth or in

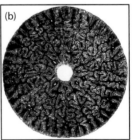

Figure 14.13 **Early amphibian features. a:** Reconstructed fossil of the Temnospondyl amphibian *Eryops megacephalus*. **b**: Drawing of a labyrinthodontoid tooth showing details of the convolutions.

pouches—such behaviour gives clues as to how the amniotic egg might have evolved—see Chapter 15). With the exception of the aquatic axolotl, the larvae of all extant amphibians eventually undergo metamorphosis: they modify their bodies, develop limbs, and replace their gills with lungs, so allowing the adult animals to breathe air and live on land.

KEY POINTS

- The earliest deuterostomes date back to the Cambrian and the quality of the soft-bodied fossil material from these early marine organisms is remarkable.

- The fossil data, combined with cladistic analysis of extant deuterostomes, shows that the earliest taxon within the Chordata was the Cephalochordata; these animals had a notochord but lacked cranial tissues, paired sense organs, and neural crest cells. The earliest known example was *Pickaia elegans* from the Cambrian and the only extant species that has been closely studied are a few members of the *Amphioxus* genus.

- Separate spurs off the *deuterostome* line led to echinoderms and hemichordates (acorn worms) and to the chordates (tunicates, cephalochordates, and craniates.).

- Tunicates have a swimming larval form with a head, a notochord, and a tail. The tunicate adult forms once the larva undergoes metamorphosis, losing most of its characteristic chordate structures.

- The earliest known examples of the Craniata were complex wormlike organisms from the lower Cambrian such as *Haikuella* and *Haikouichthys*.

- Over the next ~200 My, fish evolved with a wide range of morphologies. The earliest were the agnathan (jawless) conodonts that lacked vertebrae and are mainly known through their complex dental elements. Today, the agnathans (the Cyclostomata) include only the hagfish and lamprey that have cartilaginous skeletons.

- Soon, most fish had developed jaws from their first pharyngeal arches (the Gnathostomata) and some developed dermal plates for protection. These formed from dermal mesenchyme.

- In one branch of the early fishes, probably the Acanthodia, skeletal cartilage began to be replaced by endochondral bone that derives from somitic sclerotome and lateral plate mesoderm; these became the Osteichthyes.

- These early bony fishes gave rise to the Actinopterygii (today's teleost fishes) and the Sarcopterygii (today's lungfish and coelacanths). Both had rayed fins but the pectoral and pelvic fins of the Sarcopterygii were fleshy and include proximal bones that are homologues of those of tetrapods.

- The ability to produce skeletal bone was later lost in a taxon that may have derived from the placoderms became the Chondrichthyes, which are characterized by a cartilaginous skeleton. Their contemporary descendants include sharks, ghostfishes and rays.

- Limbs developed from the fins of sarcopterygian fish during the Devonian, probably to facilitate bottom-walking in shallow swamps.

- Stem tetrapods evolved from Sarcopterygii in a series of steps that first led to the elpistostegids; these had primitive limbs terminating with unjointed radials rather than jointed digits.

- Basal tetrapods acquired jointed digits, together with lungs that evolved from pouches off the gut, an enlarged and sophisticated heart and strengthened girdles.

- The final step required for amphibians to evolve was a change to thyroxin-induced metamorphosis. In fishes, this leads to adult maturity, but in amphibians (other than axolotls) it was the hormone that initiated the changes that turned an aquatic tadpole into a land-based adult tetrapod.

- The evolution of tetrapods took ~30 My and was completed before the end of the Devonian. They gave rise to two clades: the amphibians and the stem amniotes.

- Amphibians were the main land animals throughout the Carboniferous (359–299 Mya) diversifying into several families, one of which gave rise to the Lissamphibia, the only surviving amphibian clade. This now includes frogs, toads, salamanders, axolotls and caecilians,

---

FURTHER READINGS

Ahlberg, PE (2019) Follow the footprints and mind the gaps: A new look at the origin of tetrapods. *Trans R Soc Edin (Earth Env Sci)*;109:115-137. doi.org/10.1017/S1755691018000695.

Clack J (2012) *Gaining Ground: The Origin and Evolution of the Tetrapods* (2nd ed). Indiana University Press.

Janvier P (1996) *Early Vertebrates*. Clarendon Press.

Janvier P (2015) Facts and fancies about early fossil chordates and vertebrates. *Nature*: 520:483–489. doi.org/10.1038/nature14437.

Long JA (2011) *The Rise of Fishes: 500 Million Years of Evolution* (2nd ed). Johns Hopkins University Press.

Satoh N (2016) *Chordate Origins and Evolution: The Molecular Evolutionary Road to Vertebrates*. Academic Press.

WEBSITES

- Wikipedia entries on *Buccal Pumping, Deuterostomes, Evolution of Tetrapods, Lepospondyli.*
- Chordate evolution: www.bio.utexas.edu/faculty/sjasper/Bio213/chordates.html
- Deuterostome evolution: www2.gwu.edu/~darwin/BiSc151/Deuterostomes/Deuterostome.html
- From fish to amphibians: www.uky.edu/KGS/education/education-links-Devonian.php
- Evolution Cartoon: From deuterostomes to humans: https://upload.wikimedia.org/wikipedia/commons/5/5c/Ancestors.gif

Vertebrate Evolution: Stem Mammals, Reptiles, and Birds

15

By ~365 Mya (early Carboniferous) and ~30 My after the first tetrapods walked on land, two distinct clades had evolved: amphibians and stem amniotes. Towards the end of the period (~312 Mya), the latter had become amniotes. Much of the reason why this step took so long was due to the major changes needed for eggs that had hitherto developed in water to become able to develop on dry land. This particularly required the development of new embryo-derived membranes to encase the growing embryo in liquid so ensuring that it did not dry out. The resulting amniote egg became a synapomorphy for the amniote clade.

The early amniotes diverged to give two further clades that can be distinguished on the basis of their temporal fenestrae (spaces between the skull bones adjacent to the temporal bone—Chapter 6). The synapsid stem mammals had a single fenestra while the diapsid reptiles had two. The former (pelycosaurs and then therapsids) dominated the Permian world until the Permian extinction, which ended the Paleozoic Era, when all synapsid taxa were eventually lost except the dicynodonts; these then slowly evolved to become mammals. The diapsids radiated widely and dominated the world during the Mesozoic Era. Soon, they branched to give further clades that included euryapsid reptiles (e.g. the ichthyosaur sea lizards that are now extinct) and anapsid reptiles (turtles) characterised by their dorsal and ventral plates. Further diversification within the diapsids led to the archosaurs. A synapomorphy for these is an antorbital fenestra in the skull; the archosaurs included the Crurotarsi (e.g. crocodiles), the pterosaurs and the dinosaurs.

The origin of the various clades of diapsid reptiles are marked by a series of further anatomical innovations. A key example was the development of an antorbital fenestra between the eye and the nostril: this was a synapomorphy for archosaurs, although it was later lost in crocodiles. Other changes to diapsids included alterations to ankle morphology, so separating crocodiles and dinosaurs; subsequent alterations also occurred in forelimbs, enabling them to become wings, first in pterosaurs and then in the very different line that led to birds.

After a very long period of success (~185 My), most members of the wider reptile clade were lost as a result of the K-T extinction (~63 Mya). When the dust of the extinction cleared, the clade was reduced to the now-extant reptiles; these particularly included diapsid lizards and snakes, anapsid turtles and archosaur birds and crocodiles. In the emptied ecological landscape, the synapsid stem mammals radiated widely at the expense of the reptile clade.

This chapter first discusses the anatomical changes that took place as the early amniotes evolved. It then uses the fossil record to discuss the evolution of the synapsids in the Permian and the diapsids in the Mesozoic. The chapter concludes with sections on the K-T extinction and on the evolution of flight in the pterosaurs and birds.

When the earliest amphibians evolved from basal tetrapods towards the end of the Devonian, they thrived because there were only invertebrates and plant taxa on land, the largest of which were progymnosperm trees. As most amphibians today need to live near year-round regions of non-saline water to allow their tadpole larvae to develop, it was probably so then; as a result, large areas of land remained

DOI: 10.1201/9780429346217-18

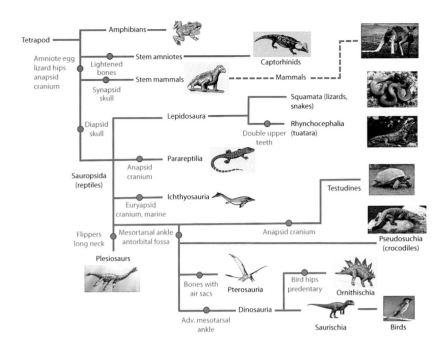

Figure 15.1 Cladogram showing reptilian evolution from a basal tetrapod. The pictures show represent examples of clades, with living taxa on the right.

(Captorhinid
Amphibian: Courtesy of Nobu Tamura. Published under a CC Attribution-Share Alike 3.0 Unported License.
Rhyncocehalia: Courtesy of Sid Mosdell. Published under a CC Attribution Generic 2.0 License.
Squamata: Courtesy of Mark Herr. Published under a CC Attribution-Share Alike 3.0 Unported License.
Parareptilia: Courtesy of Nobu Tamara. Published under a CC Attribution-Share Alike 3.0 Unported License.
Ichthiosauria: Courtesy of Dmitry Bogdanov. Published under a CC Attribution-Share Alike 3.0 Unported License.
Stem mammals: Published under a CC Attribution-Share Alike 3.0 Unported License.
Testudines: Courtesy of "John5199". Published under a CC Attribution Generic 2.0 License.
Pseudosuchia: From Brown et al. (2013) *ZooKeys*; 266:1–120. Published under a CC Attribution-Share Alike 3.0 Unported License.
Pterosauria: Courtesy of Dmitry Bogdanov. Published under a CC Attribution-Share Alike 3.0 Unported License.
Ornithischia: Courtesy of Nobu Tamara. Published under a CC Attribution-Share Alike 3.0 Unported License.
Suarischia: Courtesy of Fred Wierum. Published under a CC Attribution-Share Alike 4.0 International License.
Plesiosaur: Courtesy of Dmitry Bogdanov. Published under a CC Attribution-Share Alike 3.0 Unported License.)

unoccupied by vertebrate life. It took another ~40 My until the late Carboniferous before another line of tetrapods evolved to become early amniotes that were freed from this aquatic requirement and so able to colonise regions away from the sea, rivers, and lakes. What made this possible, of course, was the evolution of the amniote egg, which is impervious to evaporation and includes sufficient water for the embryo to fully develop without the need for a larval stage. When amniotes hatch or are born, they can survive and later reproduce away from stable bodies of water.

Amphibians and stem amniotes, which gave rise to the diapsid reptile clade (the sauropsids) and the synapsid stem mammals (Table 15.1) seem to have coexisted until the end of the Carboniferous (~299 Mya), when there was then a period of global cooling, the forests declined markedly and the larger amphibian taxa were lost. At this point, amniotes, by then divided into a range of taxa distinguishable on the basis of skull morphology (Table 15.1), had no serious vertebrate competition away from bodies of water and rapidly diversified. With the start of the Permian, the synapsid clade became dominant, first through the pelycosaurs (e.g. the sail-backed dimetrodons) and then the therapsids. Following the Permian extinction (259 Mya), however, they went into rapid decline, with the few remaining taxa eventually mainly eking out an existence as small nocturnal animals (Chapter 16).

During the Mesozoic Era (the Triassic, Jurassic, and Cretaceous Periods), the hitherto minor diapsid clade became dominant and rapidly diverged to give the archosaurs (dinosaurs, pterosaurs, crocodiles, and eventually birds), sea lizards (plesiosaurs and ichthyosaurs) together with snakes, turtles, lizards, and a few other minor taxa. Most of these clades became extinct following the K-T extinction (~66 Mya) that ended the Mesozoic Era and led to the beginning of the Cenozoic Era. At this point, the synapsids, which had meanwhile evolved into small mammals, emerged from their burrows and radiated widely to give the mammalian fauna that now populate the land, sea, and sky (Chapter 16).

The broad synapsid taxon (Table 15.1) that included the very great majority of the many thousands of Mesozoic land vertebrates has now been dramatically reduced. There are the two remaining archosaur subclades (the Aves and a few Crurotari taxa that include alligators, caimans, and crocodiles), and three Lepidosaura taxa: the Squamata (lizards and snakes), the Testudinata (turtles and tortoises) and Rhynchocephalia whose solitary member is the tuatara of New Zealand. The data on the anatomical features given in Table 15.1 allow the construction of a cladogram showing the evolutionary history of the main amniotic lines (Fig. 15.1) up to the K-T extinction. This is the major topic of this chapter.

TABLE 15.1 THE MAJOR AMNIOTE CLADES				
Basal amniotes	Skulls lacking temporal fenestrae (e.g. the captorhinids)			
Anapsid reptiles	Skulls secondarily lacking temporal fenestrae (e.g. turtles that evolved from an originally diapsid line)			
Diapsid reptiles	Skulls with two temporal fenestrae			
Archosaura	Diapsid skulls with antorbital and mandibular fenestrae			
	Crurotarsi	Mesotarsal ankle joint	*Lizard-hips* quadrupedal gait (e.g. crocodile)	
	Avemetatarsalia	Advanced mesotarsal ankle joint		
		Theropod dinosaurs	*Lizard hips*, bipedal (today: birds with beaks and 3-digit wings)	
		Sauropod dinosaurs	*Lizard hips*, quadrupedal gait with massive legs with minimal digits (extinct, e.g. brachiosaurus)	
		Ornithiscian dinosaurs	*Bird hips*, predentary bone on lower jaw, mesotarsal ankle joint (extinct e.g. stegosaurus, hadrosaurs)	
	Crurotarsi	Mesotarsal ankle joint	*Lizard-hips* quadrupedal gait (e.g. crocodile)	
	Pterosaurs	Keel, light bones, mesotarsal ankle joint, wing made of finger digits (extinct, e.g. *Pterodactylus antiquus, extinct*)		
Lepidosaura	Sprawling gait		**Squamata**	Lizards & snakes
			Rhynchocephalia	Tuatara
Euryapsids	Skulls with a single, supratemporal bilateral temporal fenestra (e.g. ichthyosaurs)			
Synapsids	Skulls with a single, infratemporal bilateral temporal fenestra			
	Stem mammals	Single lower jawbone (dentary) and differentiated teeth (e.g. pelycosaurs)		
	Mammals	Synapsid-derived skull, warm-blooded, mammary glands, hair, mainly placental (see Chapter 16)		

ANATOMICAL INNOVATIONS IN STEM AMNIOTES

The diversity of diapsid reptiles in the Mesozoic Era was at least as great as that of the mammals today. In some ways, it was even more dramatic as it culminated in land animals that would have dwarfed elephants and included flying pterosaurs almost the size of giraffes (Witton, 2013) as well as very large marine reptiles (the plesiosaurs and ichthyosaurs). Untangling the history of what must have been many thousands of Mesozoic reptilian species has depended on interpreting fossilised skeletal material and trace fossils such as footprints, eggshells and nests. This section covers some of these important defining anatomical features but starts by discussing the evolution of the amniote egg, the key unifying synapomorphy for all reptiles and mammals. It was this innovation that allowed both clades to diversify so widely.

The evolution of the amniote egg

It took ~30 My for tetrapods to evolve from sarcopterygian fishes with all taxa having eggs that developed under aqueous conditions. It then took almost another 50 My for the fossil record to show that animals had evolved that could produce amniotic eggs (sometimes called cleidoic eggs), which could develop away from water (Fig. 15.1). This enormous span of time makes it clear that the evolution of such an egg was no trivial matter. The problem with understanding how the amniote egg evolved is that there are no intermediate forms to be seen today, and there is thus little direct information about the steps by which it happened.

The eggs of contemporary teleost (most bony) fishes have an external chorionic membrane and a vitelline membrane that encloses the zygote and the yolk. The first stage in their development is successive divisions of this zygote with the resulting cells migrating to cover the yolk surface; this process is known as epiboly (see Websites). This layer of cells partially retracts to form a disc within which the embryo forms, leaving an extended cellular membrane lining part or all of the egg (the yolk sac). This membrane becomes vascularised and consumes the yolk, so nourishing the embryo through the midgut region. Contemporary lungfish and most lissamphibians are different: their eggs have no separate yolk or vitelline envelope, just yolk granules within embryonic cells, with the whole egg being directly involved in embryogenesis. In due course, each type of embryo becomes a larva, swims away from the jelly surrounding the egg and becomes independent (del Pino, 2018). This mode of embryogenesis was generally maintained in the Amphibia,

The yolked amniotic eggs of contemporary reptiles and birds are far more complicated: they have a water-impermeable but gas-porous shell and an internal chorionic layer that forms after internal fertilization and also only allows gas transmission (Fig. 15.1b). Soon after fertilization, the embryo develops as a disc on the surface of the yolk-sac-enclosed yolk and, in due course, makes two more extraembryonic water-impermeable membranes that together enable it to develop in the absence of any external water. These are the amniotic membrane that encloses the embryo in amniotic fluid and the allantoic membrane that forms the embryo-attached allantoic sac, a reservoir for nitrogenous waste. The allantoic blood vessels mediate the exchange of carbon dioxide and oxygen through the gas-porous chorion and shell. The various membranes incorporate cells from the ectodermal, mesodermal, and endodermal germ layers of the embryo, with membrane morphogenesis usually commencing soon after the start of gastrulation (for review, see Ferner & Mess, 2011).

Membrane production is thus a major and complex part of the early stages of embryogenesis in amniotes, and it is only once these membranes are in place that the early amniote embryo can continue its development. The amniote egg with its extraembryonic membranes is a synapomorphy of all reptiles and their descendants including the placental mammals, even though their eggs lack both a shell and a substantial amount of yolk. What still remains unclear is the steps that led to the formation of the amnionic and allantoic membranes.

The development of fishes and some unusual frog taxa that lack a larval stage give some insight into the routes that might have led to the evolution of the amniote egg from that of a stem tetrapod. One approach could have been to build on the solution adopted by some early chondrichthyan fishes, that of internal fertilization, simple placental development, and live birth of small fish (Ahlberg et al., 2009). This route is similar to that shown in some Mesozoic marine reptiles (Fig. 15.3) and in today's oviparous reptiles. It was clearly the basis for the later evolution of marsupial and eutherian mammals (Chapter 16). It was not, however, used in reptiles, stem mammals and monotremes that lay shelled eggs.

Other possibilities come from those few amphibians that do not lay eggs in water: an example is the *Cocqui* taxon of tree frogs (e.g. *Eleutherodactylus coqui*); these do not go through aqueous and tadpole stages but hatch as froglets. Another example is the taxon of hemifractid or marsupial frogs whose females carry fertilised eggs in pouches on their back; in some species, they develop to the tadpole stage before being released into water while in others they become froglets. What marks out hemifractid eggs as unusual is that the embryos develop from a disc of cells located on the surface of the yolk, much like those of teleost fishes. Elinson and Beckham (2002) suggest that the reason why such development can take place in both types of amphibian is that their eggs are unusually large. As they provide enough yolk to nourish advanced development, adult development does not require a free-swimming tadpole stage. It seems probable that this last option was the basis for the evolution of the amniote egg, with there being relatively direct routes from the teleost-type fish egg to the hemifractic frog egg and independently to the amniotic egg.

It is thus likely that the early amniote egg was something like a teleost egg with three innovations: (1) a strong outer membrane or shell; (2) a water-impermeable amniotic membrane enclosing both the embryo and the amniotic fluid that surrounds it; and (3) an allantoic membrane for waste (Fig. 15.1b). Here, the outer membrane developed as an alternative to the jelly layer seen in anamniotic eggs, while the allantois may have evolved from a diverticulum off the larval bladder (Needham, 1931). The origins of the amnion itself, however, remain unclear.

Given the complexity of the steps leading to the amniote egg and the indirect link between early membrane production and later hatching, it is not surprising that its evolution took so long. The clade of organisms that reflect the stages in the evolution of the amniote egg are known as stem amniotes (Fig. 15.2c). Today, understanding

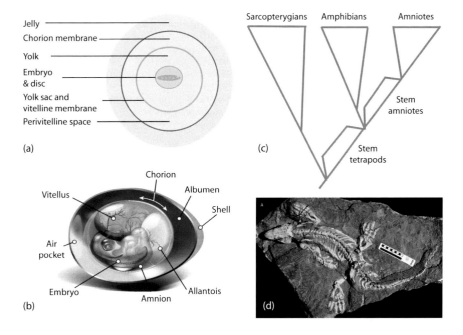

(a)

(b)

(c)

(d)

Figure 15.2 Diagrams of anamniote and amniote eggs. (a): A fertilised anamniote egg showing the embryo lying on top of the yolk. (b): A fertilised chicken egg showing the embryo with its typical internal amniotic membranes. (c): Diagrammatic cladogram showing the core features of the evolution of the amniotes. (d): The fossil of *Orobates pabsti*, a stem amniote.

(b Courtesy of "KDS444". Published under a CC Attribution-Share Alike 4.0 International License.
d From Nyakatura JA et al. (2015) *PLoS One*; 10(9):e0137284.)

how the amniote egg evolved remains a major unsolved problem and, unless there are extant remarkable soft-tissue fossils yet to be found showing some of the membrane detail, it is unlikely that the exact sequence of events that resulted in the evolution of the amniote egg will ever be discovered.

The evolution of the early amniote skeleton

Early amniotes had a series of other anatomical innovations that over time distinguished them from amphibians, the most noticeable of which was a lighter skeleton that enabled them to move more rapidly than amphibians. During their first 320 million or so years of evolution, the amniotes diversified rapidly and, over time, the different clades were marked by various skeletal apomorphies. These were changes to the skull, jaws, teeth, and limbs, some of which had obvious selective advances in facilitating speed and flexibility; an example of such an animal is *Orobates pabsti,* generally considered to be an early reptile from the lower Permian (~260 Mya; Nyakatura et al., 2015). Later differences, such as those that distinguish between the hip bones of ornithischian and saurischian dinosaurs (see below) are harder to explain.

Skull fenestrae

Amphibians and stem amniotes such as the captorhinids were anapsid: they had a solid cranium to which muscles were attached externally. By ~306 Mya, the early anapsid clade split into two branches: the Eureptilia (true reptiles) and the Parareptilia (next to reptiles) with the latter seeming to have died out by the beginning of the Triassic. Soon after the evolution of the Eureptilia, spaces appeared between the postorbital and squamosal bones anterior to the orbit (the space in the skull for the eye); these would particularly have allowed the lengthening and strengthening of the jaw muscles. Such spaces are known as *fenestrae* (an alternative word is *apsid* from the Greek for a loop or an arch). The Eureptilia branch soon separated to give one branch with a single temporal fenestra and another with two such spaces, an

Figure 15.3 The fossil of *Stenopterygius*, an ichthyosaur. It shows the mother and the small fetus (arrows) that she was giving birth to when she died.

(Courtesy of "Steveoc 86". Published under a CC Attribution-Share Alike 4.0 International License.)

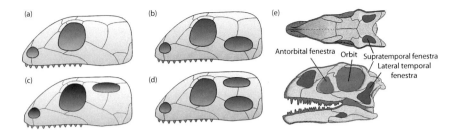

Figure 15.4 **Skull fenestrae of the main reptile clades.** (a): An anapsid skull with an orbit for the eye but no postorbital fenestrae—a characteristic of both early amniotes and later turtles (bones: j: jugal, p: parietal, po: postorbital, q: quadrate, qj: quadratojugal, sq: squamosal). (b): The skull of a synapsid stem mammal with the lower (infratemporal) fenestra. (c): A euryapsid skull with the upper (supratemporal) fenestra of, for example, the ichthyosaurs. (d): A diapsid skull with lower and upper fenestrae characteristic of reptiles. (e): Top and side views of a diapsid skull with the additional antorbital fenestra that was a defining synapomorphy of the archosaur clade; the possession of large fenestrae would have lightened the skull.

upper and a lower. These are respectively known as synapsid stem mammals and the diapsid reptiles (Fig. 15.4). Basal members of the latter clade are sometimes known as the Sauropsids, with a very early example being example *Petrolacosaurus* (304–300 Mya).

The most important taxon in both the shorter and longer but not the intermediate term were the synapsids, a group with only lower fenestrae, whose earliest known species is *Archaeothyris*, a swamp-dweller in what is now Nova Scotia (~306 Mya). The synapsids became the pelycosaurs and therapsids that dominated the early and then later Permian landscapes but, following the Permian extinction, were soon reduced to the cynodonts (i.e. dog tooth) clade with the diapsid reptiles becoming the dominant clade. During the early part of the Mesozoic, the cynodonts slowly evolved to give the three taxa of mammals that would flourish after the K-T extinction (Chapter 16).

During the early Permian, one taxon of diapsids lost its fenestrae (Fig. 15.4) and became secondarily anapsid; these evolved to become the Testudines, a taxon that includes tortoises and turtles (for their embryonic development, see Rice et al., 2016). Those that lost the lower fenestra are known as the Euryapsida; relatively little seems to be known about their diversification, but they included the ichthyosaurs, which were large marine lizards up to 16 m in length. It was, however, the original synapsid clade of reptiles that particularly flourished, particularly after the Permian extinction (~251 Mya) and were to become the dominant taxon of the Mesozoic Era. Their evolution is the main topic of this chapter.

It should be emphasised that subsequent changes to early skull morphology have altered or even led to the loss of temporal fenestrae. For instance, the bridges that separated the upper and lower fenestrae were lost in some dinosaurs, there is only one fenestra in modern lizards, and there are none in snakes. Nevertheless, the details of the fossil record together with the presence of other shared and derived characteristics (synapomorphies) confirm that contemporary lizards and snakes had diapsid ancestors. In mammals, parts of the bones that originally surrounded their fenestrae soon became the bones of the zygomatic arch (Chapter 6). Because the temporal fenestrae are such a key synapomorphy in distinguishing the origins of the major land–animal clades, relatively few other anatomical features are required to establish the basal branches of the cladogram describing the evolution of the major taxa of limbed vertebrates that evolved from early tetrapods.

A further space that appeared in the skull of some diapsids was the antorbital fenestra. This is an opening between the lachrymal and jugal bones that lie between the orbit and the nasal bone. It became a synapomorphy for the archosaurs and is now apparent in birds, although it has been lost in crocodiles. While this opening could have been a diverticulum off the nasal cavity used for a gland or for muscle attachment, homology with modern birds suggests that it may well have housed a sinus space which would have had the dual function of lightening the skull and acting as a sound resonating chamber (Witmer, 1987, 1995).

Today's descendants of the early amniotes now fall into one of three groups: the synapsids led to mammals; the diapsids to reptiles (lizards and snakes) and then to archosaurs (crocodiles and birds), while the anapsids, which derived from diapsids, includes tortoises and turtles. As a result of the K-T extinction, the rich archosaur fauna of the Mesozoic that included thousands of dinosaur, pterosaur and sea-reptile taxa were all, apart from the birds and crocodiles, lost.

Jaws and teeth

Fish and amphibian teeth showed little diversity and were directly attached to jaw bones. Over time, however, there were many changes to teeth in reptiles: in some taxa, they became socketed (a synapomorphy for dinosaurs) and started to range in size. A common feature was the presence of protocanines (lengthened upper teeth) that were used to tear flesh, with some having two such canines on either side of their upper jaws (animals with more than one tooth morphology are known as heterodonts). Carnivores tended to have dagger-shaped teeth for ripping meat. Herbivores had chisel-shaped teeth that could rip vegetation from plants and trees but could not be used for chewing (molars were a later synapsid innovation). For this, both herbivorous and carnivorous dinosaurs adopted the eating strategy still used by birds today: they swallowed small stones that were trapped in their gizzard or stomach where they could grind plant matter. Such gastroliths whose surface has been smoothened by grinding are sometime found associated with dinosaur bones.

A further innovation in an early dinosaur taxon was the evolution of a new bone in the anterior part of the lower jaw; this is the untoothed predentary and became a synapomorphy of the ornithischian dinosaurs. It combined with the premaxilla (or rostral) bone of the upper jaw to produce the beak seen in some ornithischian clades (Nabavizadeh & Weishampel, 2016). (The changes to the skull and jaw bones that took place in the synapsids are discussed in Chapters 6 and 16; synapsid teeth are discussed in Chapter 16.)

Pelvic bones and tail

Stem tetrapods of the Permian had a pelvic girdle composed of the three linked bones: the upper ilium which attaches to the sacrum of the backbone, the backward-facing ischium and the pubis that point downwards and forwards. The acetabulum (the socket for the head of the femur) was a space in the region where the three bones met (Fig. 15.5). This lizard-type hip and pelvic morphology was maintained in all early reptiles, stem mammals (LeBlanc & Reisz, 2014) and saurischian dinosaurs. Later, herbivorous ornithischia, however, evolved a bird-hipped morphology in which the pubis extends posteriorly and ventrally to the ischium (Fig. 15.5). This name is slightly unfortunate as the ornithischians evolved before the birds that descended from saurischians with lizard-hip morphology; the acquisition of the bird-type hip was a secondary change in birds that are a saurischian clade.

Many dinosaurs were large, bipedal, and rapid carnivores. They also had very long tails, and it is now clear that these were not for brushing off insects but played an important role in balance. They did this through being extended and rigid. For this purpose, muscular contraction enabled tendons to lock the tail horizontally. This arrangement changed in the clade that became birds: the feathered tails became much shorter and more aerodynamic, with the tail bone being reduced to a pygostyle of a few fused vertebrae (Pittman et al., 2013).

Limbs

Of particular functional importance were changes in the archosaur ankle. Permian reptiles together with Triassic archosaurs had stiff ankles and inflexible tarsals (Martinez et al., 2011). Early in archosaur evolution, this bone geometry changed with the evolution of a mesotarsal (lower-leg/foot bones) joint; this featured the astragalus and calcaneum bones forming a hinge and so providing flexibility (Thulborn, 1980). This type of mesotarsal joint is a synapomorphy for the Crurotarsi clade (old name: Pseudosuchia) within the Archosauria, which today includes the Crocodilia and the Aves. This joint further evolved to give an ankle in which two tarsal bones (astragalus and calcaneum) form an advanced mesotarsal joint with roller capacity, so further increasing its flexibility (e.g. Nesbitt et al., 2017; Fig. 15.5b). This joint is a synapomorphy for the Avemetatarsalia (bird-foot bones), a clade that includes all the dinosaurs and birds.

Archosaur synapomorphies not only included an antorbital fenestra, hinged ankles and socketed teeth: they also had a ridge on the femur (the fourth trochanter) that was used for the attachment of muscles that probably facilitated bipedality; this feature was lost in some later quadrupedal archosaurs such as the crocodiles and sauropod dinosaurs. It is at first sight odd that developing what seem such minor anatomical variants in the limb bones and the skull were key steps on the evolutionary journey of the dinosaurs as they went on to dominate the Mesozoic world.

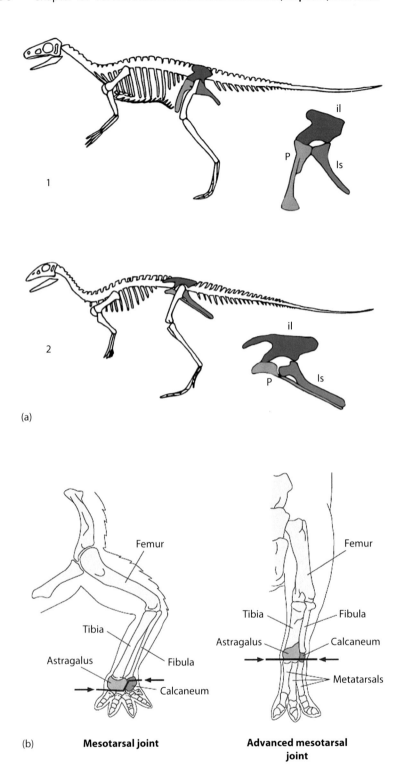

Figure 15.5 Dinosaur skeletal features.
(a): Drawings of the two forms of dinosaur pelvic bones. **1**: *Eoraptor*, a basal Saurischian, with a lizard-hipped, forward-directed pubis. **2**: *Lesothosaurus*, a basal Ornithischian, with its bird-hipped, rear-directed pubis, (P: pubis; il: ilium; Is: ischium). (b): Archosaur ankle joints showing the line of flexion (arrows). The mesotarsal joint shown by all members the Crurotarsi clade, which includes the crocodiles and other related taxa, is between the calcaneum and the astragalus, the two bones of the ankle. b The advanced mesotarsal joint shown by all members of the Avemetatarsalia is between the ankle bones and the metatarsals and is more flexible. Note: The Crurotarsi had a mainly sprawling gait while the Avemetatarsalia had a more upright gait.

(a From Kardong (2014) *Vertebrates: Comparative Anatomy*, Function, Evolution, 7th ed. With permission from McGraw-Hill.)

THE FOSSIL RECORD OF MESOZOIC REPTILES

The fossil data from the late Palaeozoic and Mesozoic amniotes is extensive but incomplete as it focuses on larger animals. This record includes the larger early amphibians, the Permian synapsids and the larger Mesozoic reptiles, with amniote skeletons being distinguished from those of amphibians on the basis of anatomical detail and the relative lightness of their bones. While there are many specimens of large stem mammals from the Permian, the fossil record is inevitably weaker for the smaller taxa of the late Carboniferous and early Permian.

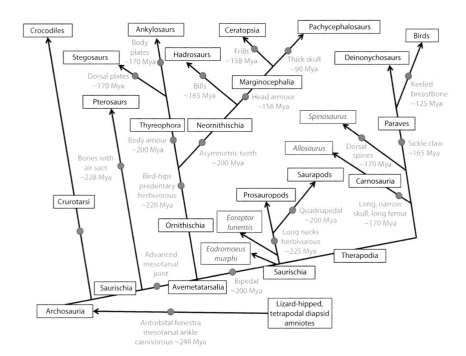

Figure 15.6 **The radiation of the archosaurs.** This cladogram shows the major taxa and some of the synapomorphies that defines the clades, together with the approximate times when they first appear in the fossil record. There is still some ambiguity in the exact details of the dinosaur cladistics and some allocations of early taxa should be viewed as provisional.

The most diverse set of Mesozoic fossil material is from the archosaur radiation; this was so successful that the clade came to dominate the earth throughout the Era (a cladogram of the best-known archosaur taxa is given in Fig. 15.6). Some thousands of species have so far been catalogued, and there are likely to have been many more that have either yet to be found or that have left no fossil record. There is, for example, relatively little fossil data from the Mesozoic for small pterosaurs and birds whose bones were small, light, and fragile. Such was the ubiquity of the archosaur clade that there was barely any ecological niche left vacant for the other major clade of vertebrates, the synapsid stem mammals. During the Triassic, these were essentially reduced to a group of small-bodied, burrow-living animals. The archosaurs thus ruled the land and air for the whole Mesozoic, a period of almost 200 My, an era that was ended by the K-T extinction (66 Mya).

The Palaeozoic Era

The exact timings of when the various groups of amniotes evolved during the later parts of the Palaeozoic and whether the appearance of the various fenestrae were unique events are still contentious, as the early reptile fossil record is thin. The most recent approach (Ford & Benson, 2020) has unpicked the details of amniote evolution by subjecting as complete a list of anatomical autapomorphies as possible to cladistic analysis. The detail provided in this paper is much greater than can be given here and illuminates the richness of Mesozoic life (see Further Readings).

Carboniferous Period (359–299 Mya)

Stem amniotes with some early amphibian features, such as triangular skulls, small legs, and short tails probably emerged by ~355 Mya; these Microsauria may have been a clade of the Lepospondyli, but their exact taxonomic status is still unclear. The late Carboniferous period is marked by the appearance of tetrapods that seem to have had both amphibian and reptilian features. Amongst these was *Casineria* (c 335 Mya; Alibardi, 2008; Paton et al., 1999), a lizard-like organism with pentadactyl clawed limbs adapted to land, but we know very little about it as the sole fossil lacks hind limbs and most of its head.

Cladistic analysis using a wide range of anatomical features places the evolution of amniotes at ~324 Mya (Ford & Benson, 2020), a little earlier than once thought. The earliest fossils of likely amniotes, which are marked by pentadactyl limbs and skeletons much lighter than those of early amphibians, together with clearly characteristic footprint trace fossils, appear in the late Carboniferous (315–310 Mya; Falcon-Lang et al., 2007). Two early taxa from this time with an unequivocally reptile-like body skeleton were the captorhinids and the more amniote-like *Thuringothyris mahlendorffae* (Müller & Reisz, 2005; Müller et al., 2006). Both amphibians and these early amniotes had anapsid skulls. Araeoscelidian reptiles with two temporal fenestrae on either side of their crania do not appear in the fossil record until ~304 Mya; these were among the earliest sauropsids, with an example being *Petrolacosaurus* (~302 Mya; Reisz, 1977). It was not long, however, before the sauropsid clade diverged to give several branches: diapsids maintained two temporal fenestrae on each side of the skull, synapsids had a single lower opening and euryapsids kept only the upper ones.

Towards the end of the Carboniferous, the Earth's temperature dropped markedly. The consequent increase in polar glaciation and the decline in equatorial water levels together with major losses of rain forests seems to have led to a major decline in amphibians while apparently having little effect on early amniote populations (Dunne et al., 2018). It was this taxon that, therefore, dominated the world of land animals from the start of the Permian to the end of the Cretaceous, a period of over 230 My.

Permian period (299–252 Mya)

Soon after the beginning of the Permian, the synapsid clade of stem mammals became dominant. Early examples that have a sprawling gate are known as pelycosaurs and later ones with a more upright gate as therapsids. The earliest-known pelycosaur is *Archaeothyris* (~306 Mya) and the best-known taxon is that of the Dimetrodons (295–272 Mya); these were large carnivorous reptiles famous for their sail of elongated neural spines whose role may have been temperature regulation (Fig. 15.7). Curiously, a similar sail was also present on *Spinosaur*, a later diapsid dinosaur (112–97 Mya)—this is clearly an example of homoplastic evolution.

In contrast, early diapsids such as *Petrolacosaurus* (~302 Mya, a reptile about 40-cm long) seem to have played a minor role in the Permian fauna. Among the few other early diapsids that have been found are *Claudiosaurus*, an iguana-like swimming reptile and the Younginiformes (terrestrial and aqueous lizard-like reptiles). The best-known members of the diapsids, the archosauromorphs, do not appear in the fossil record until the end of the Permian: the earliest known examples date to ~250 Mya (Ezcurra, 2014; Marsicano et al., 2016).

The great Permian extinction (252 Mya) had devastating effects on the therapsids: the great majority were lost and the diminished synapsid taxon was reduced to a small group of rodent-sized animals that seem mainly to have become nocturnal and probably lived in burrows. These started to radiate during late Triassic, evolving to become mammals during the Jurassic period (Chapter 16). For this reason, the Permian synapsids used to be called the mammal-like reptiles or protomammals, although stem mammals is now the standard term.

The Mesozoic Era (252–66 Mya)

Triassic Period (252–201 Mya)

For no obvious reason, it was the diapsids that flourished in the Mesozoic Era, and they soon gave rise to two main groups: the Lepidosauria and the Archosauria. The main synapomorphy of the former is overlapping scales; this reptile clade included such oddities as the late Triassic *Kuehneosaurus latus* (c. 210 Mya), a lizard with winged forelimbs that it probably used to attenuate falling speeds during gliding (McGuire & Dudley, 2011). The clade went into later decline and today has relatively few taxa, just

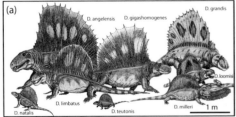

Figure 15.7 **Dimetrodon, a pelycosaur taxon.** (a): Reconstructions of various members of the Dimetrodon taxon. (b): Skeleton of *Dimetrodon incisivum* showing the arrangement of the vertebrae, spines, hip, and limb bones.

the widely distributed squamates (snakes and lizards), and the rhyncocephalians. The latter were once very common but are now reduced to a single species, the tuatara, now found only in New Zealand, which is marked by a double row of upper teeth.

The exact timing for the appearance of archosauromorphs (the most basal members of the Archosauria taxon) is unclear: the earliest known examples date to ~250 Mya (Ezcurra, 2014) and had a mesotarsal ankle (Fig. 15.5). The clade gave rise to the Crurotarsi, the Pterosauria, and the Avemetarsalia. The first of these included the Crocodilia, animals with massive skulls and two sets of dorsal protective plates, together with several minor clades. Other than the Crocodilia, most of these reptiles, together with some amphibian and non-archosaur-reptile taxa, were lost in the extinction at the end of the Triassic (c. 200 Mya). The Pterosauria, which evolved towards the end of the Triassic (Britt et al., 2018), were flying reptiles with membrane wings stretching from a much lengthened fourth finger to the ankle (Fig. 18.2). The clade had mesotarsal ankles, very light air-filled bones and a keeled breastbone for flight-muscle attachment (this is a homoplasy shared with birds). The origins of the pterosaurs are not well understood, but they may have evolved from an early archosaur such as *Scleromochlus taylori* (see below).

The Avemetatarsalia were particularly successful, and the clade includes all archosaurs that are nearer birds than crocodiles; it thus includes the dinosaurs that are characterised by lizard-hip (Saurischia) or bird-hip (Ornithischia) morphology and by more than ten other anatomical features including an advanced mesotarsal ankle (Nesbitt, 2011). Thousands of archosaur species are now known, and it is only possible to mention the major clades here. It should also be emphasised that the full details of the archosaur radiation are not fully known but, as new early species continue to be found, the history of the taxon is becoming clearer (e.g. Nesbitt et al., 2017).

The earliest known dinosaurs were the Saurischian eoraptors, such as the small carnivorous *Eodromaeus murphi* (237–228 Mya), which dates to ~15 My after the Permian extinction (~251 Mya). These reptiles were about 1m long and had heterodont teeth so that they could probably eat both animal and plant matter. A relatively short time later, a line branched from the Saurischia to form a new clade that became the Ornithischia. This diversified widely, particularly in the Cretaceous, to give the large radiation of herbivorous dinosaurs that was characterized by the presence of a predentary bone that extended the lower jaw and facilitated plant eating; indeed, all ornithischians were, on the basis of their teeth, herbivorous (for details on the evolution of herbivory, see Chapter 19). A little later, the Saurischia split again to give the theropods and prosauropods (or suaropodomorphs); the former were active carnivorous predators while the latter were originally omnivorous but later becoming herbivorous. They were initially bipedal but had started to become tetrapedal before the end of the period.

Harder to pin down cladistically were the marine reptiles; these included the Mosasauridae (true reptiles) and the ichthyosaurs and the plesiosaurs, which also evolved during the Triassic. The former seem to have evolved first by diverging off the euryapsid line, while the latter seem to have diverged from the diapsid line before the formation of the crurotarsal joint. The fossil record is opaque on how these two lines left land to become marine reptiles, although the situation may be clarified by the analysis of some recently discovered fossils (Motani et al., 2015; Huang et al., 2019). The ichthyosaurs looked very much like modern dolphins and this shared morphology is a further example of homoplastic or convergent evolution.

The Triassic ended ~201 Mya with the emission of large quantities of greenhouse gases and global warming, perhaps associated with major volcanic eruptions. Irrespective of the cause, the results were dramatic and most of the reptilian radiation including the large marine species and many Archosauromorpha were lost, leaving a much-diminished set of taxa. This major extinction particularly affected the Crurotarsi (although a few at least of the Crocodilia survived) and some dinosaurs, together with amphibian and non-archosaur-reptile taxa.

Jurassic Period (201–145 Mya)

The beginning of the Jurassic was marked by a major expansion of the dinosaur taxon, and they soon came to dominate the landscape. The theropods evolved into a wide variety of mainly carnivorous bipedal dinosaurs ranging in size from the late-Jurassic *Compsognathus* of ~2.5 kg to the semiaquatic *Spinosaurus* that weighed more than ten tons and was larger than the later *Tyrannosaurus rex*. Soon, the originally bipedal prosauropods had evolved to become large, pillar-limbed tetrapodal sauropods that were herbivorous. Examples were *Diplodocus*, *Apatosaurus,* and *Supersaurus*, whose estimated weight was ~34 tons, perhaps five times the weight of an African bush elephant, the largest land animal alive today. Analysing how very large dinosaurs moved is a question that can only be answered through computational modeling (Sellers et al., 2013). At the other end of the size scale were many taxa of small carnivorous theropods, with the Paraves being particularly important as it was from this taxon that the birds evolved (see below). It was also during the Jurassic that pterosaurs diversified widely and some eventually became very large, with a wingspan that could exceed 10 m (see below).

The Ornithischia also radiated during the Jurassic, forming a broad group whose detailed cladistic analysis is still under discussion. The most recent analysis has two major taxa: the Thyreophora or armoured dinosaurs, which included the Stegosauria and the Ankylosauria and the Neornithischia, characterized by having thickened enamel on their lower teeth that facilitated the eating of tough plants. The Neornithischia included the Marginocephalia or fringe-headed dinosaurs (Figs. 15.6 and 15.8d,e). subclades of which were the quadrupedal Ceratopsia

Figure 15.8 **Examples of dinosaur taxa.**
(a): Skeleton of *Stegosaurus stenops*.
(b): Skeleton of Deinonychus (less its tail).
(c): Drawing of the skeleton of *Allosaurus fragilis*, a theropod. (d): The skeleton of Triceratops Trorsus, a ceratopsian theropod.
(e): The skeleton of *Pachycephlosaurus wyomingensis* (inset: skull showing the thickened skull roof). (f): Skeleton of *Scolosaurus cutleri*, an ankylosaur. Note the armoured plates on the body and club-like protrusion on the tail. (g): Drawing of the skeleton of *Edmontosaurus regalis*, a hadrosaur. (h): Skull of *Saurolophus osborni*, a crested hadrosaur. (i): Drawing of the skeleton of *Notocolossus gonzalezparejasi*, a titanosaurian sauropod. Insert: The skull of *Camarasaurus lentus*, another sauropod, showing its peg-like teeth and spacious fenestrae. (Bar 1 m).

and the heavy-skulled bipedal Pachycephalosauria. The Ceratopsia had a horny frill around the bill (e.g. *Triceratops*) whole likely roles were protection and display; the taxon also had an additional rostral bone on the upper jawbone giving the animals a beak-like mouth. The Pachycephalosauria had a very thick skull roof (~4 cm), and it has been suggested that these animals used this for fighting during mate competition. A second major clade in the Neornithischia was the bipedal Ornithopoda, characterised by a horny beak and an elongated pubis but with an unarmoured head. This clade included the iguanodons, the first dinosaurs to be discovered, and the later hadrosaurs, many of which had large hollow cranial crests (e.g. *Parasaurolophus walker*, Fig. 15.6e).

The Jurassic seems to have ended with more of a whimper than a bang: it brought major faunal transitions but no substantial extinctions (Tennant et al., 2017).

Cretaceous Period (145–66 Mya)

This period is named after the major chalk cliffs laid down late in the period. These are composed of coccoliths, which are tiny plates of calcium carbonate produced as a protective coat for coccolithophores (unicellular algal protists). The Cretaceous takes its name from *creta*, the Latin for chalk, and is sometimes abbreviated to K (short for *kreide*, the German for chalk) as in the K-T extinction that ended the Mesozoic. This long (79 My) period of mainly benign weather was marked by the break-up of the single major continent of Pangea into the beginnings of the separate continents seen today. It was also the time when flowering plants evolved and insects diversified.

During the Cretaceous, the dinosaur taxa expanded widely, particularly over the first third of the period (Lloyd et al., 2008). Perhaps the best known of the new species was the saurischian *Tyrannosaurus rex*, a very large bipedal carnivore that dominated the plains of what would become North America towards the end of the Period. This area was also populated by large groups of ornithischian dinosaurs, particularly the many taxa of ceratopsians with bony frills (e.g. *Triceratops*; Gates et al., 2016).

The other ornithischian clade that particularly flourished in the Cretaceous were the hadrosaurs. These were the largest of the ornithischian dinosaurs and were marked by the rostral bone of their upper jaw extending outwards, giving the animals a beak-like mouth; many of the reptiles also had a crest on the top of their heads (e.g. Xing et al., 2017; Figs. 15.6g,h and 15.9). Superb skeletal material has been found in

Figure 15.9 The hadrosaur radiation. This mainly took place during the Cretaceous and included a wide range of taxa that included some very large species up to ~15 m in length.

(Published under a CC Attribution-Share Alike 3.0 Unported License.)

Hadrosauroidea

North America and particularly in China: a wide variety of material from the Late Jurassic and Early Cretaceous (130–122 Mya) has, for example, been found in the Jehol biota that is part of the Liaoning Lagerstätte in NE China. Study of the material there has also revealed dinosaur nests and eggs together with fine detail of early bird morphology (Pan et al., 2013; Tanaka et al., 2018).

The K-T extinction and the beginnings of the Cenozoic Era (66 Mya)

Towards the end of the Cretaceous, there were major volcanic eruptions (Self et al., 2008) which may have had a warming effect on the world's climate and done serious ecological damage. It is, however, now well established that the K-T extinction, which dramatically ended the period, was primarily due to an asteroid falling on Mexico (Schulte et al., 2010; Chiarenza et al., 2020). This produced the Chixulub crater and released enormous amounts of particulate material into the atmosphere, blocking out solar radiation and leading to a period of cold. The result of this is thought to have been the period of extreme and rapid global change during which there were effects on the carbon cycle that lasted for at least 5000 years (Renne et al., 2013).

The net effect of these global changes was the extinction of the dinosaurs, pterosaurs, and most birds, together with all marine reptiles other than some turtles and marine crocodiles. ~75% of species were lost and, when the earth recovered, it was the mammals that particularly diversified (Longrich et al., 2016). Of the reptilian clade, only the birds flourished; by the early Cenozoic they had evolved into two large groups: the Palaeognathae such as ratites (e.g. moas and kiwis) that are mainly ground-living birds and the Neognathae, which includes all the flighted birds (Brusatte et al., 2015). After the extinction, the great majority of the ecological niches that had previously been occupied by the great Mesozoic radiation of reptiles had new mammalian occupants, and the vertebrate world had irreversibly changed.

THE ORIGIN OF FLIGHT

Flight evolved within two very different clades during the Mesozoic: Triassic archosaurs gave rise to the pterosaurs wheras it was not until the late-Jurassic that theropods began to show bird-like features (e.g. *Archaeopteryx*), with birds (e.g. *Yanornis*) evolving during the Cretaceous. The pterosaurs died out during the K-T extinction, but some birds survived and later evolved to give the more than 10,000 avian species seen today.

Pterosaurs

The pterosaurs probably evolved from very early Triassic archosaurs, such as *Scleromochlus taylori*, a small lizard-like animal whose anatomy suggested that it exhibited springing and jumping behaviour but whose phylogenetic provenance is unclear (Nesbitt et al., 2017). As pterosaurs have a mesotarsal ankle, they must have spurred off the main archosaur clade before the evolution of the advanced mesotarsal ankle (Fig. 15.5; see Nesbitt et al., 2017). Little is known of pterosaur evolution as no intermediates have been found but, in addition to archosaur plesiomorphies, the clade displays several synapomorphies that establish their ability to fly and distinguish them from dinosaurs. The most dramatic of these is their extended forelimb morphology with an extremely long fourth digit. This was the attachment point for a broad membrane (the brachiopatagium) that, in some cases at least, extended to the hindlimbs (Figs. 15.10 and 18.2) and is occasionally preserved in fossils (Kellner et al., 2010).

Three key features that facilitated pterosaur flight are homoplasies shared with birds. First, their bones were lightened by the presence of air sacs (Buchmann & Rodriguez, 2019). Second, they were probably warm-blooded, being mainly covered in hair-like filaments that would have maintained internal heat (Witton, 2013). Third, they had a large, keeled breastbone for the attachment of enlarged flight muscles; they also had a strong, rigid shoulder girdle and some fused bones (Wang et al., 2017). Although the details of how these changes were achieved during development are not known, Tokita (2015) has pointed to some likely molecular mechanisms.

The range of sizes is noteworthy: the smallest known pterosaur had a wingspan of ~25 cm while the largest so far identified, *Quetzalcoatlus northropi,* had the extraordinary wingspan of >10 m, weighed as much as 250 kg and had a head-to-tail

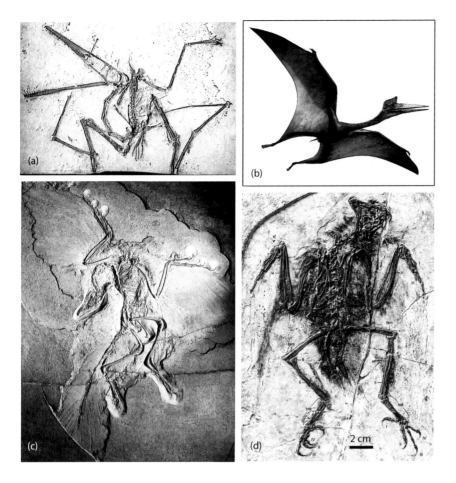

Figure 15.10 **Pterosaurs and early birds.**
(a): Fossil of *Pterodactylus antiquus*, a late Jurassic pterosaur. Note the large, lightened beak. (b): Reconstruction of *Quetzelcoatlus northropi*, a pterosaur with a wing diameter >10 m. (c): The fossil of *Archaeopteryx lithographica* in the Berlin Museum für Naturkunde. The bone structure shows that the animal was a dinosaur with clear evidence of long feathers on its forelimbs and tail. (d): The fossil of *Zhouornis hani*, a cretaceous bird; unlike *Archaeopteryx*, it has lost the forelimb claws.

(a Courtesy of S.U. Vidovoc. D.M. Martill, M. Martyniuk. Published under a CC Attribution-Share Alike 3.0 Unported License.
b Published under a CC Attribution-Share Alike 3.0 Unported License.
c Courtesy of H. Raab. Published under a CC Attribution-Share Alike 3.0 Unported License.
d From Zhang et al. (2014) PeerJ; 2:e407. Published under a CC Attribution-Share Alike 4.0 International License.)

length only a little less than that of a giraffe (Fig. 15.10a,b; Witton, 2013). Pterosaur heads had toothed jaws, although these became more beak-like and less toothed during the Cretaceous. Many had crests that may have served an aerodynamic function. All were, however, lost in the K-T extinction.

Birds

Although birds today have the antorbital and diapsid fenestrae of archosaurs, they do not at first sight fit neatly into any obvious archosaur taxon: they have, for example, the bird-hipped pelvis of ornithiscian dinosaurs, the omnivorous abilities of early saurischian dinosaurs and the breast keel and light, hollowed bones of pterosaurs. They, alone among extant diapsid reptiles, are warm-blooded and feathered. Although Huxley in the 19th century had felt that the skeletons of small carnivorous theropods such as *Compsognathus* were similar to those of birds, it was commonly held for a long time that the clade was a very early spur off the diapsid line that developed independently of other taxa. These views were challenged in the 1960s by John Ostrom and Robert Bakker on the basis that the fossil and other evidence meshed better with Huxley's view than any alternative. Although the light bones of birds do not fossilise well, there is today a clear record of the transition from theropod dinosaur to bird (Xu et al., 2014; see also Websites).

The fossil data show that it was late in the Jurassic that a taxon of small, probably feathered saurischian theropods diverged to become the Paraves, which were characterised by a raptorial claw on a highly modified second digit of the hind limb (Wang et al., 2017). In due course, the birds seem to have evolved from this taxon, although it is becoming clear that there were attempts at flight in other taxa (Wang et al., 2019b). The Cretaceous fossil record for birds is relatively good; in particular, well-preserved specimens have been found in the Jehol biota, part of the Yixian Lagerstätte in Lioning, NE China (Pan et al., 2013).

The first fossils recognisable as having feathers are from ~165 Mya in China, although the best-known example is still *Archaeopteryx* (~150 Mya), first identified in Germany over 150 years ago. (Fig. 15.10c). These species seem to have lived in trees and on the ground and, even though they had feathers, their bone structure suggests that they moved differently from living birds. During the Cretaceous, these newly evolved species shared the air with pterosaurs and, on the basis that both are found in the late Mesozoic fossil record, they co-existed adequately well.

The origin of feathers is now known to have begun rather earlier than was once thought. Simple filamentous structures in the skin have been noted in both theropods and ornithischian heterodontosaurs, and it is thus possible that they were present quite early in dinosaur evolution (~201 Mya; Alibardi et al., 2009; Zheng et al., 2009; Lowe et al., 2015a; Benton et al., 2019). Simple filaments are also known in pterosaurs and their origin may thus be linked to the rapid diversification of archosaurs in the Triassic. Phylogenetic analysis suggests that some of the genes involved in feather development today, such as new kinds of keratins, were present in the common ancestor of crocodiles and birds (Alibardi, 2006. Branched or "true" feathers are, however, much younger, probably dating back to ~165 Mya (Zhang et al., 2008; Lowe et al., 2015a; Eliason et al., 2017). Their presence on many theropods seems initially to have been for insulation, then display, then gliding and only later for flight (Alibardi et al., 2009; Benton et al., 2019; Xu, 2012). The evolution of flight feathers is thus a clear example of an exaptation.

The presence of feathers and filaments, like hair in mammals, is evidence that the dinosaurs with this feature showed a degree of endothermy (warm-blooded physiology) because they always have an insulation effect. This would have been counterproductive in cold-blooded animals as it would inhibit warming up in sunshine, a necessary behaviour shown by contemporary cold-blooded reptiles. It, therefore, seems likely that those dinosaurs with feathers also had the higher metabolic rates associated with a degree of endothermy, although not necessarily to the extent seen in living birds and mammals. This feature was key to the evolution of flight, whose energy demands require a high metabolic rate.

Later, small, fully feathered Jurassic dinosaurs such as those of *Anchiornis huxleyi* (~155 Mya) and *Archaeopteryx* (Fig. 4.8a; ~150 Mya) were clearly transitional between dinosaurs and extant birds. Both are important: the former has sufficient detail in its melanosomes for its feather colours to be identified (red, black, and melanin-free, Li et al., 2010); the latter not only possessed the dinosaur features of lizard-hipped pelvis and jaws with sharp teeth but also had bird-like features, such as a wing-like forelimb with three clawed digits. Although it has generally been thought that they were able to glide but incapable of flight (Longrich et al., 2012), more recent work based on feather dynamics suggests that this view should be reconsidered (Wang et al., 2019). Both have bony features consistent with new forms of moving consistent with early aerial locomotion that was at least homologous with flapping, although they did not have a sternal keel to anchor flight muscles.

Flight, of course, requires more than feathers, and a succession of enabling changes took place during the late Jurassic and early Cretaceous. During the former, the first simple flight/gliding feathers appeared, while the furcula or wishbone that strengthens the shoulder girdle became more prominent (Benton et al., 2019). During the Cretaceous, this clade of theropod stem birds underwent further changes that included a loss of weight through the lightening of bones, and the formation of a deep sternal keel for the attachment of enlarged pectoral flight muscles (e.g. Lee et al., 2014). In addition, the tail was replaced by a much smaller pygostyle of small, fused vertebrae, the jaw bones were encased by a beak and the teeth were lost, their role being taken by the gizzard (Louchart & Viriot, 2011); forelimbs adapted to become wings and the lizard-hipped pelvis became bird-hipped (for review, see Brusatte et al., 2015). This last feature incidentally mirrors the morphology of the ornithischian dinosaur pelvis and hips but, as mentioned earlier, is a homoplasy and not a synapomorphy. An example of an early Cretaceous bird with these features is *Confuciusornis* (~125 Mya) which had huge wings for its body size and asymmetrical feathers. The evolutionary line from saurischian theropod to modern bird is now clear and unambiguous (Brusatte et al., 2015).

The question of whether full homeothermy (constant body temperature maintained, for example, by endothermic metabolisms) rather than just endothermy evolved in birds or whether they inherited it from parent dinosaurs has yet to be answered unequivocally. There is now good evidence that some of the anatomical features associated with endothermy were present in various dinosaurs, and even in very early archosaurs: in addition to keratin structures on their skin, they had the upright posture associated with fast movement and the type of bone morphology associated with endothermy in birds and mammals (Seymour et al., 2004; Lovegrove 2017), while their breathing was more efficient than that of mammals (Claessens et al., 2009; see Lovegrove, 2017, for a discussion of the evolution of homeothermy).

Large dinosaurs would have shown a degree of endothermy because of their small surface:volume ratio, which would have lessened heat loss as compared to that of smaller animals with a larger ratio. It does, however, seem clear that small, fast dinosaurs like the theropod ancestors of the birds could only have maintained the speed appropriate to their morphology if they had some means of keeping their bodies warm; feathers would have helped, but a considerable degree of metabolic warming, would probably also have been required (Grady et al., 2014; Lovegrove 2017). A similar argument holds for pterosaurs: given the large amounts of energy needed for flight, it is hard to imagine that they would have been able to fly if they had had only the ectothermic metabolism of lizards or just a minimal degree of endothermy.

The material given in these last two sections supplements that in Chapter 6 to show that there is an impressive amount of fossil detail to aid our understanding of two major transitions in Mesozoic animals: from fishes to land-based tetrapods and from theropod dinosaurs to bird. Taken as a whole, the evidence clearly shows that from the first chordates of the Cambrian period through to the tetrapods, and from the amniotes to the variety of animals that were present in the late Cretaceous period, the vertebrates form a monophyletic clade. There do, however, remain some important gaps in our knowledge of this period. One is the evolutionary route from a fertilised marine egg and larva to the amniote egg whose membranes enclose sufficient water to enable vertebrate development to take place on dry land, allowing an organism to hatch as a juvenile. Others are the details of the changes that led to the early evolution of pterosaurs, ichthyosaurs, and plesiosaurs.

KEY POINTS

- The key step that enabled stem amniotes to evolve into the line that led to reptiles and stem mammals in the early Permian was the development of the amniote egg.

- The evolution of this egg, which occurred over a period of ~40 My towards the end of the Carboniferous, allowed embryos to develop in a non-aqueous environment. How this happened is still not fully understood.

- The key synapomorphy that distinguishes reptiles from stem mammals is that the former had a double (diapsid) and the latter a single (synapsid) temporal fenestra on either side of the cranium. This division occurred in the late Carboniferous.

- The lines of descent from the earliest amniote to all contemporary vertebrates can be described within a cladogram on the basis of known synapomorphies.

- Synapsid stem mammals (e.g. *Dimetrodon*, a pelycosaur) dominated during the Permian but most were lost in the Permian extinction (252 Mya). The remainder was eventually reduced to small cynodonts that slowly evolved into mammals during the Mesozoic.

- The diapsids were the predominant land vertebrates during the Mesozoic, radiating to give snakes, lizards, turtles, and particularly archosaurs, which included pterosaurs, crocodiles with mesotarsal ankles and dinosaurs with advanced mesotarsal ankles.

- The dinosaur clade originally retained a lizard-hip morphology, and this branch gave rise to carnivorous theropods and vegetarian sauropods, with the former giving rise to the birds that secondarily evolved bird-hip morphology.

- Pterosaurs evolved within an archosaur clade during the early Triassic and diversified widely before being lost in the K-T extinction; the largest, *Quetzalcoatlus northropi*, had a wingspan of >10 m.

- Birds evolved from late-Jurassic theropod saurischians that already had feathers, the evolution of which was initially selected to provide insulation and later for display in dinosaurs that had a degree of endothermy. Their adaptation for gliding and then flight occurred much later and was accompanied by a host of anatomical changes.

- A second dinosaur branch developed so-called bird-hip morphology, and these ornithischians gave rise to the armoured dinosaurs (e.g. *Stegosaurus* and *Ankylosaurus*), the Marginocephalia (e.g. Pachycephalosauria and the Ceratopsia, such as *Protoceratops* and *Triceratops*), and the Hadrisauriformes (e.g. the iguanodons and hadrosaurs).

- All archosaurs, except crocodilians and ground birds, together with most other land and marine reptiles, were lost in the K-T extinction (66 Mya).

- With the coming of the Cenozoic, the small mammals emerged from their burrows and diversified widely. Similarly, the remaining birds regained flight and expanded into many new niches.

FURTHER READINGS

Benton MJ (2000) *Vertebrate Paleontology* (2nd ed). Oxford: Blackwell Science.

Chatterjee S (2015) *The Rise of Birds: 225 Million Years of Evolution.* Johns Hopkins University Press.

Ford DP, Benson, RBJ (2020) The phylogeny of early amniotes and the affinities of Parareptilia and Varanopidae. *Nat Ecol Evol*; 4:57–65. doi.org/10.1038/s41559-019-1047-3.

Sues H-D (2019) *The Rise of Reptiles: 320 Million Years of Evolution.* Johns Hopkins University Press.

Witton MP (2013) *Pterosaurs: Natural History, Evolution, Anatomy.* Princeton University Press.

WEBSITES

- Wikipedia entries: *Paleobiota of the Yixian Formation, Pterosaurs* and the *Archosaur clades*
- The diapsids: http://tolweb.org/Diapsida
- Zebrafish development: https://zfin.org/zf_info/movies/movies.html
- Bird evolution: https://www.biointeractive.org/classroom-resources/great-transitions-origin-birds
- Dinosaur teeth: https://www.fossilera.com/pages/dinosaur-teeth

Vertebrate Evolution: Mammals

16

Mammals evolved within a small taxon of synapsid stem mammals that survived the Permian extinction (252 Mya). As a likely response to the presence of archosaur predators in the Mesozoic Era, this taxon was mainly reduced to small animals that left only limited amounts of fossil material in the early years of the Era. Enough, however, remains to understand how the mammalian skull evolved and to identify a few early members of today's major clades: the Prototheria (the monotremes), Metatheria (marsupials), and Eutheria (placentals). The K-T extinction that ended the Mesozoic Era 66 Mya led to the loss of most diapsid reptiles, leaving only turtles, birds, lizards, and snakes. This freeing-up of ecological space enabled the survivors of the three taxa of small synapsid mammals to radiate early and widely during the early Cenozoic and become the dominant vertebrate clade across the world.

Today's land mammals include a rich variety of animals that extend in size from shrews to elephants. Mammals also evolved to give bats that could fly and squirrels that could glide, together with three groups that became marine: the cetaceans (e.g. whales), the pinnipeds (e.g. seals) and the sirenians (e.g. manatees). As the evolutionary events of the Cenozoic are relatively recent on the evolutionary timescale and as many of the early mammals were large enough to have robust bones, there is a very rich record of fossilised skeletal material. There is also an increasing amount of genetic and embryological information on the origins of some of the soft-tissue anatomical synapomorphies that define the mammalian clade.

This chapter starts by describing the three modern mammalian clades and their Mesozoic fossil record. It then discusses the developmental changes in the skeletal, neuronal, and reproductive system during the Mesozoic that led to the evolution of some key synapomorphies for each of these clades. The last part of the chapter starts by summarising how the different mammalian clades evolved during the Cenozoic Era in the different continents that formed as Pangaea broke up around the time of the K-T extinction. As it is not practical to discuss here the full arborisation of the mammalian clade, the text focuses on the key anatomical changes that led to the radiation of mammals across the continents and on a few taxa with particular features. These include defensive adaptations, the evolution of flight, and the return of mammalian life to the sea, particularly the evolution of whales. Readers wanting to know more about the fossil record and zoological details of the mammalian taxa during the Cenozoic are referred to the books in Further Readings listed at the end of the chapter.

For all their diversity, the ~4000 walking, swimming, and flying mammalian species of today fall into only three high-level clades: the monotremes (Prototheria), the marsupials (Metatheria); and the placentals (Eutheria). Although each has its own features (Table 16.1), there are several cranial skeletal synapomorphies that distinguish them from the early stem mammals and later mammaliaforms that were their ancestors (Chapter 6.) Mammals also have soft-tissue synapomorphies in addition to their clade-defining mammary glands. These include a large brain

DOI: 10.1201/9780429346217-19

TABLE 16.1 SOME FEATURES OF THE THREE MAMMALIAN CLADES

Property	Prototheria (Monotremes)	Metatheria (Marsupials)	Eutherians (Placentals)
Middle ear bones	3	3	3
Molars with numbers per half jawbone	Tricuspid in juveniles	Multicuspid, 4	Multicuspid, 3
Neocortex	Yes, small	Yes	Yes
Inter-hemispheric communication	Anterior commissure	Anterior commissure	Ant. commissure + corpus collosum
Cochlea	Straight	Coiled	Coiled
Epipubic bone	Yes	Yes	No
Hair	Yes	Yes	Yes
Whiskers with touch sensitivity	No	Yes	Yes
Endothermy	Incomplete	Yes	Yes
Mammary glands	Skin patches	With nipple	With nipple
Pouches	Platypus—No Echidna—Yes	In most cases	No
Reproductive, excretion and defecation anatomy	F: Oviduct M, F: Common cloaca M: Penis for reproduction	F: Uterus & vagina M, F: Common cloaca M: Penis for reproduction	F: Uterus, vagina; separate anus & urethral exits M: dual-purpose penis + anus
Embryonic nutrition	Yolk, uterine secretions	Choriovitelline placenta	Chorioallantoic placenta
Egg size	~4 mm	130–350 μm	~70 μm
Post-natal requirement	Altricial	Altricial	Dependent to independent

with a particularly increased neocortex, hair and associated endothermy, and for marsupials and placentals, a coiled cochlea and major clade-specific changes to the reproductive system.

There are also specific synapomorphies associated with each of the three clades. The very few monotremes (the duck-billed platypus and the four species of echidna or spiny anteaters) are now only found in Australia and New Guinea and, as their reproductive system demonstrates (below), they are a sister group to placentals and marsupials (see also Fig. 16.1). Adult monotremes lay eggs, have hair or quills, and only a degree of endothermy. They lack nipples: their mammary glands secrete milk directly onto the skin where their young (known as puggles) can lick it. Echidnas have a pouch for rearing their young, so providing protection for them. Platypuses lack a pouch so that their young have to cling to their mothers; adults have a leathery bill rather than teeth, although teeth were present in extinct species and are present in puggles.

Marsupials (such as the kangaroo) have four molars and up to three premolars on each side of their upper and lower jaws with some milk teeth not being replaced with adult ones (Williamson et al., 2014); they also have a range of pouch types (Tyndale-Biscoe & Renfree, 1987). Pregnant metatherian females develop a simple yolk-sac placenta that nourishes the embryo until soon after its forelimbs have developed functionality. At this point, the embryo detaches from the placenta, enters the birth canal and is born. Further development is then altricial (parent-dependent): the neonate climbs to the pouch situated in the belly area using its prematurely large forelimbs—this is a classic example of heterochrony (Keyte & Smith, 2010; Chapter 18). Within the pouch there are mammary glands with nipples, and the mouth of the neonate is shaped for plugging onto one of them; the resulting milk flow enables the neonate to complete the embryonic and fetal stages

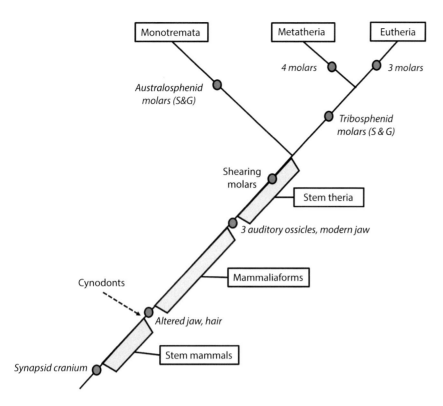

Figure 16.1 **Early mammalian evolution.** A simplified cladogram showing the main lines leading from an early reptile to the modern mammalian taxa. Note the use of tooth morphology to distinguish early fossil material. (Adapted from Davis, 2011.)

of development. After a rich history of evolution and diversification, particularly in Australia and the Americas, (Williamson et al., 2014), metatherian mammals are now mainly found in Australia. In South America, they have been reduced to the many species of opossums and shrew opossums, while only opossums are found in North America.

Placental mammals replace all their original milk teeth and have three or fewer molars on each side of the upper and lower jaws. These teeth have different enamel crown morphologies from those of marsupials (von Koenigswald, 2000; Davis, 2011) and such dental features are useful in identifying fossil material. They also lack the small epipubic bones of the pelvis used for muscle attachment in other mammalian taxa, having lost them during the Cretaceous (Novacek et al., 1997). Other important placental synapomorphies are the advanced brain with its enlarged neocortex and corpus callosum communication system and the sophisticated reproductive systems that allows extensive development *in utero*. Although the conceptus of marsupials has a single trophoblast layer that leads to a short-term choriovitelline placenta, it is only in placentals mammals that the trophoblast layer is capable of forming an intimate relationship with the uterus that results in long-lasting chorioallentoic placenta (and there are a range of these; Morriss, 1075). As a result, some eutherian neonates are capable of an immediate and considerable degree of independence, other than in their need for milk. This is particularly important for herbivores whose survival requires the young to be able to run with the herd soon after birth.

Today, placental mammals have come to dominate all land masses other than Australasia (excluding New Zealand which, until the arrival of humans had only flying mammals) and Antarctica. They occupy almost all the land niches for large animals, competing for these with only a few reptiles that include the bigger snakes, tortoises, and large lizards (e.g. the iguanas and the Komodo dragon, *Varanus komodoensis*). In Australia and, to a lesser extent the Americas, marsupials still do well but eutherian mammals, through migration or through importation for farming, are now common. Eutherian mammals have also been successful in marine environments where the separate clades of whales, dolphins, seals, and sea cows hold their own against large fishes. They seem, however, to have been less successful in the air: bats tend only to fly at night when most birds are asleep, probably to avoid insectivore competitors and predators.

Placental mammals are of immense importance for the world's ecology, and their evolution is also highly relevant to us: we are species within this clade, and the most important one in terms of numbers and influence. By learning about the evolution of related taxa, we learn about our own roots and this chapter can thus be seen as background to Section 5 on human evolution.

THE MAMMALIAN FOSSIL RECORD FROM THE MESOZOIC

The fossil record for the later mammals is substantial and covers all the larger animals in sufficient detail to construct a detailed cladogram (Fig. 16.2). The major gap in the record is the lack of full skeletal details for most of the small, early stem mammals (known as mammaliamorphs) that survived the Permian extinction and the rodent-like mammaliaforms (or mammaliaformes) into which they had evolved by the end of the Triassic. Their most common remains are their skull bones and very small teeth (Chapter 6). The record is much better for the early mammals of the later Mesozoic (Grossnickle et al., 2019) and particularly rich for the taxon of multituberculates. These shrew-like, social mammaliaforms with complex teeth originated in the Jurassic and survived for ~166 Mya until about 40 Mya, longer than for any other mammalian genus (Yuan et al., 2013; Weaver et al., 2021). Other important gaps are for small, light animals, such as the early bats.

The initial radiation of the synapsids started towards the end of the Carboniferous period, and they soon became the dominant vertebrate clade. Although pre-Triassic

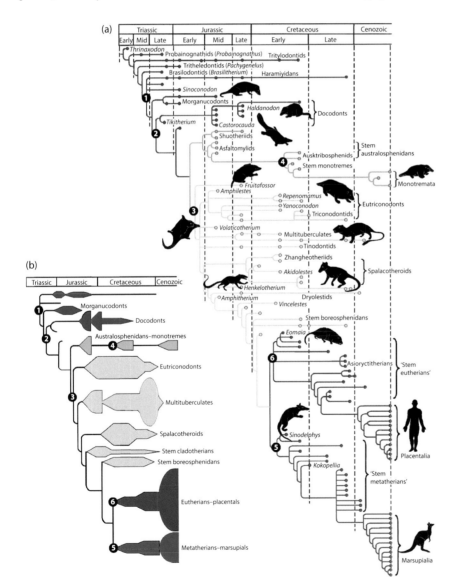

Figure 16.2 Mammalian evolution.
(a): Cladogram showing the evolution of contemporary mammalian clades from their roots in the early Mesozoic Era. The cladogram particularly demonstrates how few of the early clades survived. (b): summary of population diversity and extinction of the many mammaliaform and early mammalian species of the Mesozoic Era.

(From Luo et al., (2007) *Nature*; 450:1011–1019. With permission from Springer Nature.)

synapsids are known as stem mammals, there was little that was then mammalian about them. The dominant taxon then was the Pelycosauria, (e.g. *Dimetrodon* with its spiny dorsal sail, Fig. 15.7), but it was supplanted in the middle Permian (~270 Mya) by the Therapsida, a grade that included stem mammals of all sizes from rat to cow. The most noticeable difference between the two taxa was that the former had a more sprawling posture while the latter were more upright. During the Permian, the synapsids flourished, dominating the great land masses (Kemp, 2005; Kardong, 2014).

Most synapsid taxa were lost during the Permian extinction (~252 Mya), with only three surviving into the Triassic. Of these, the dicynodonts (beaked herbivores; Fig. 16.3a) and therocephalians (carnivores) eventually died out, but the cynodonts survived. This taxon was likely to have been close to the last common ancestor of mammaliaforms. The cynodonts and their immediate descendants initially included both omnivores and herbivores with a range of sizes. The largest known is *Lisowicia bojani* from the late Triassic (length: 4.5m: height 2.6m, and mass: 9 tons); this animal, which had erect limbs, was larger than a bull elephant (6 tons; Sulej & Niedźwiedzki, 2019). An important example is *Psueodotherium argentines*, also from the late Triassic, whose anatomical features suggest that it was an early mammaliaform. One identifying feature here was its Harderian gland, a characteristic of contemporary mammals, whose secretions are used for cleaning fur (Wallace et al., 2019).

By the Jurassic, as archosaur hegemony increased on land, the mammaliaforms seem to have been reduced to a group of small, nocturnal, and shrew-like insectivores

Figure 16.3 Fossils from mesozoic mammaliaform. (a): A dicynodont. This line of stem mammals became extinct during the Triassic. (b-d): *Jeholbaatar kielanae*, an early mammal from the Cretaceous. (b): The exposed fossil; the arrow points to the bones of the inner ear ossicles. (c): An enlarged view of the bones of the inner ear. (d): A drawing of these bones. (Ap, anterior process of malleus; Ct, foramen for chorda tympani nerve; Et, ectotympanic; Fp, footplate of stapes; In, Incus; L, lateral; Ma, body of Malleus; Mb, manubrium of malleus; Sb, surangular boss; St, Stapes.).

(a Courtesy of "Ghedoghedo." Published under a CC Attribution—Share Alike 3.0 Unported License. b,c,d From Wang et al., (2019) *Nature*; 576:102–105. With permission from Springer Nature.)

that probably inhabited burrows from which they emerged at night. This niche diminution led to what is known as the *nocturnal bottleneck* (Gerkema et al., 2013), with the only surviving taxa being those that evolved apomorphies enabling them to cope with these conditions.

There are two reasons for accepting that this is what happened. First, most of the fossilised synapsid material from the late Triassic and early Jurassic periods comes from very small insectivores with proportionately small teeth; there is no skeletal material from larger synapsids that would have been much more likely to have been preserved, had they existed (Ungar, 2010). Second, many common mammalian properties today reflect the life of ancestral animals that were small, nocturnal, and needed to work hard to maintain their body temperature. These include hair, endothermy and the presence of brown fat that can be metabolised rapidly to raise heat and is particularly found in hibernating mammals. Surviving species also strengthened those senses that facilitate living in the dark: these include acute hearing (resulting from an enhanced inner ear), sensitive touch (particularly whiskers), relatively large eyes and a strong sense of smell.

As the fossil evidence of these small animals (usually less than 10cm long, excluding the tail) is often just teeth and only sometimes jaws, other changes have to be inferred on the basis of very limited skeletal data. Three cranial skeletal features are particularly important in species identification (see Chapter 6 and below). First was the presence of the two new middle-ear ossicles, the malleus and the incus, that evolved from two of the reptilian jawbones, the lower articular and the upper quadrate (Anthwal et al., 2012; Fig. 16.3b–d). These ossicles amplify sound beyond the level transmitted by the stapes alone and so indicate enhanced hearing ability, (Gill et al., 2014; Rowe et al., 2011; Wang et al., 2019a). Second was the lower jaw becoming a single bone, the tooth-bearing dentary; third was the enhanced cusp organisation of the molar teeth, which indicate the nature of mastication and hence diet (Gill et al., 2014). The descendants of the cynodont line slowly started to acquire these mammalian features during the early Mesozoic Era. An example of a large mid-Triassic (~230 Mya) mammaliaform is *Trirachodon*, a social animal some 50cm in length that possessed a secondary palate (Groenewald et al., 2001), while the slightly later *Morganucodon* was on the borderline between mammaliaforms and mammals (Table 6.2).

It is impossible to know exactly when the first animal evolved that nourished its offspring with milk and this is, of course, because the mammary glands do not fossilise. The fossil evidence does, however, suggest that the three clades of today's mammals first became distinct in the mid-Jurassic. It seems that the initial division was between the early Prototheria (e.g. *Ambondo* ~167 Mya) that laid eggs and a stem placental group that led to the Eutheria and Metatheria. *Dryolestes* (Late Jurassic, 163-145 Mya) was a probable example of an intermediary as it had therian teeth but an uncoiled cochlea (Table 16.1; Luo et al., 2011a).

From its rodent-sized beginnings, the synapsid clade seems to have diversified over the Mesozoic and some three hundred or so known early mammalian taxa have now been identified (Luo, 2007); they were all relatively small, so avoiding the attention of archosaurs. By the mid-Jurassic (174-164 Mya), these early mammals included such species as *Volaticotherium antiquum* and *Arboroharamiya allinhopsoni* that were gliders and *Castorocauda lutrasimilis* that swam. These species were related to the monotremes and probably evolved before the stem placental split.

Volaticotherium antiquum was an insectivorous mammal with a gliding membrane (known as the patagium) similar to that of a modern-day flying squirrel (Meng et al., 2006). Its teeth were highly specialized for eating insects, and its limbs were adapted to living in trees. The patagium was insulated by a thick covering of fur and was supported by the tail as well as the limbs. The discovery of *Volaticotherium* provided the first record of an early gliding mammal, but at least one other, *Arboroharamiya allinhopsoni*, has now been found in China (Han et al., 2017) with five middle-ear bones (this was probably because the initial components of the malleus remained distinct). It dates to ~160 Mya and both are considerably earlier than other gliding mammals that have hitherto been found.

Castorocauda lutrasimilis, although it lived at about the same time as *Volaticotherium*, was very different: it showed features seen in modern semi-aquatic mammals, such as beavers, otters, and the platypus. Fossilised impressions indicate that some webbing was also present between the toes (Ji et al., 2006), the caudal vertebrae were flattened dorso-ventrally, while the tail was broad with scales interspersed with hairs that were sparser toward the tip. As such a tail is very similar to that of modern beavers, *Castorocauda* presumably used it for swimming in much the same way as its modern counterparts.

Modern mammalian dentition appears in the fossil record ~170–160 Mya. Most mammaliaforms had molars adapted to shearing food but some species, such as the multituberculates, eventually evolved with molars that also had a grinding ability; these are tribosphenic teeth (three cusps) that are now seen only in young monotremes (Fig. 6.8; Chapter 6). The earliest organism with this feature is *Shuotherium* (~167 Mya), identified only on the basis of its jaw and lower teeth (Kielan-Jaworowska et al., 2004). The sister clade to such proto-monotremes diversified widely, with an example of an ancestral mammal being *Dryolestes* (~155 Mya). Although it lacked a coiled cochlea, the internal skull anatomy indicates that it had the auditory innervation seen in the metatheria and eutheria (Luo et al., 2011b). These examples, together with those mentioned below, clearly confirm that much of the diversification that led to the major mammalian clades took place during the Jurassic.

The oldest marsupial fossils so far identified date to ~125 Mya. An early example is *Sinodelphys*, an organism with fur and four molars in each half jaw that was adapted to an arboreal lifestyle (Luo et al., 2003; Williamson et al., 2014). This fossil was found in China region, then part of the Pangaea supercontinent which was in the process of breaking up into Northern Laurasia and Southern Gondwanaland. The details of the dispersal of the marsupials across the various landmasses remains unclear because the fossil record is poor, but their remains have been found in North and South America and eventually in Australia (~50 Mya), which they reached by crossing Gondwanaland before it fully fragmented.

The earliest current evidence of a eutherian mammal is the fossil of *Juramaia sinensis* that lived ~160 Mya, with skeletal analysis showing that it possessed three tribosphenic molars and other more minor placental characteristics (Luo et al., 2011b). Other early placentals are continuing to be found and an example is *Sasayamamylos* from Japan (~125 Mya), identified by the three molars and four pre-molars in its jaw (Kusuhashi, 2013). Such mammals were initially small, mouse-sized animals, and seem not to have exceeded the size of a small dog before the K-T extinction (66 Mya). The skeletal changes that distinguish them from marsupials are relatively minor: they include the replacement of a deciduous premolar (a milk tooth) with an adult premolar, the loss of the epipubic bone and the gain, in some families, of a baculum (penis bone). It does, however, seem odd that the oldest placentals so far identified are so much earlier (~160 Mya) than the oldest marsupials (~125 Mya). It would, therefore, not be surprising if there were older marsupial fossils still to be found.

The fossil evidence as a whole shows that, by the end of the Cretaceous (66 Mya), there was a wide variety of mammalian species (Fig. 16.2), but that much of this rich monotreme, marsupial and placental fauna was lost in the K-T extinction. The much greater loss of all marine reptiles except turtles, and all archosaurs except the crocodiles and birds, emptied a wide variety of ecological niches; the subsequent absence of large synapsid predators facilitated the radiation of the three small, warm-blooded mammalian taxa to fill them.

THE EVOLUTION OF SOME KEY MAMMALIAN FEATURES

Modern mammals are different from stem mammals and mammaliamorphs: they are endothermic, hairy and have brains that are proportionately large as compared to other vertebrates. They also have good hearing, vision, olfaction, and touch. Some, at least, of these properties are, as already mentioned, consistent with their last common ancestor having been a nocturnal, burrowing animal. It seems to have taken most of the Mesozoic Era for the various mammalian skeletal and other apomorphies to have evolved from the more basal properties of the synapsid stem mammals.

It is not immediately clear whether the length of this period reflects the difficulty of making these changes or the relatively low selection pressures on the synapsids. The fact that the clade had been reduced to small mammaliaforms in the early Mesozoic probably reflected the harsh conditions initially prevailing as the planet slowly recovered from the causes of the Permian extinction. By the end of the Triassic, large archosaurs had evolved, and it seems likely that they then exerted sufficient predatory pressure on the mammaliaforms that they had no choice other than to adapt to a nocturnal existence. If so, the fact that many millions of generations were required before they had evolved into recognizable mammals probably reflects the sophistication of the changes. The nature of some of the molecular mechanisms by which genomic variation leads to such phenotypic changes will be examined in Chapter 18. Here, the skeletal and soft-tissue changes that occurred during the synapsid-to-mammal transition are considered.

Changes to the skull

As the evolution of the mammalian skull has been discussed in detail in Chapter 6, only a few summary points are made here. There were four major changes to the synapsid cranium during the early-middle Mesozoic Era that made them recognizably mammalian.

1. Alterations to the jaw hinge that led to the evolution of the malleus and incus; these new middle ear bones increased the aural sensitivity of the already existing columella, which became the stapes.
2. Changes in the bones surrounding the temporal fenestra that led to formation of the zygomatic arch and strengthened jaw musculature.
3. The development of the secondary palate that separated the nasal and oral cavities, so allowing animals to breathe and chew or suckle at the same time.
4. Changes to the teeth and particularly the evolution of grinding molars, which enabled animals to feed on material that would otherwise have been hard to break down in the alimentary canal.

Changes to soft tissues

Metabolism: Homeothermy and hair

Homeothermy or the maintenance of a constant body temperature (usually through endothermy, or metabolic activity) requires more food-supplied energy than does ectothermy, for which body warming mainly derives from the sun's heat. It does, however, have the key advantage that the body's metabolic enzymes always work at their optimal temperature. An endothermic animal can thus maintain a high level of activity that is independent of the external temperature; it is noteworthy that mitochondrial activity in muscle, heart, and lungs is much higher in mammals than reptiles (Hulbert & Else, 1989).

A degree of homeothermy is likely to have evolved before hair, as there is little reason for a cold-blooded animal to develop a pelt: its insulation would slow warming as much as it would help maintain heat. The fossil evidence for hair dates back to at least the middle Jurassic (~165 Mya), with evidence for hairs being seen in *Castorocauda* ~160 Mya (above, Ji et al., 2006), or a little before the earliest evidence of feathers in dinosaurs (~155 Mya, late Jurassic; Rauhut et al., 2012). Both structures are insulators and have keratin as their primary molecular constituent, albeit that hair includes α-keratin and feathers β-keratin (Alibardi, 2006; Dhouailly, 2009); α-keratin, first seen in amphibians, is today a normal constituent of mammalian and bird integument (skin) and horns while the latter is only present in birds (Alibardi, 2003; Wu et al., 2015). Hair evolved to have a further property: it could generate touch sensitivity through its associated mechanoreceptors, an ability particularly enhanced in whiskers.

The evolution of hair required two mechanisms: one for producing hair follicles, the other for regulating interfollicular distance (hairs and whiskers have different separations). Alibardi (2012) has suggested that hair follicles evolved through co-opting groups of cells with keratin-production machinery to produce long strands of protein rather than the matrix seen in skin. As to the spacing, Dhouailly and Sengel (1973) showed that the patterning mechanisms for generating hairs and feathers are essentially interchangeable with those for producing scales in reptiles (and probably in pangolins too). This mechanism involves still-unknown interactions between the surface epidermal epithelium and the underlying dermis of the skin: the epidermis lays down a rough grid of locations in the dermis and a hair-making papilla forms at nodes. The ability to form the keratin-based dermal scales that are found on pangolins provide an interesting example of an amniote plesiomorphy more commonly seen in reptiles, such as snakes (For a comparison of the properties of keratin in feather, scales, and whale baleen, see Wang & Sullivan, 2017).

The brain and sensory systems

An early and key feature of the mammalian line was an increase in brain size as compared with nonmammalian vertebrates of similar body size (Kemp, 2005). Rowe et al. (2011) used X-ray tomography to study the braincases of two early mammals of similar size, *Moganucodon* and *Hadrocodium* (Table 6.2). They showed that, although the latter had a slightly more evolved jaw, the former had a brain that was ~40% larger than the latter after adjusting for size differences and that was within the expected range for a modern mammal. They also found that *Hadrocodium* had a relatively enlarged neocortex, cerebellum, and smell-associated tissues (olfactory bulbs and piriform cortex). Rowe et al. (2011) viewed these increases as being driven by the demands of improved neuromuscular coordination, an enhanced olfactory system that included the turbinate bones of the nose (these facilitate breathing) and an enlarged olfactory epithelium, as well as the tactile receptors associated with body hair, particularly whiskers. Here, it is of note that rodents have a morphologically defined area of the neocortex, the barrel cortex, to which whisker sensors map (Erzurumlu & Gaspar, 2012).

The evidence suggests that a key difference between mammaliaform and earlier stem mammal brains was the development of the neocortex from the earlier and simpler dorsal cortex. This feature is now a key feature of all placental mammals, particularly primates (Medina & Abellán, 2009; Kaas, 2011). The mechanisms by which this happened are not yet known, but studies on comparative transcriptomics in the context of vertebrate brain development are beginning to identify some of the genomic and other changes that may have been responsible for the major changes that led to the modern mammalian brain (Molnár et al., 2014; Briscoe & Ragsdale, 2018; Epifanova et al., 2019; Strzyz, 2021).

One of the most important novel cortical structures, seen only in modern placentals, is the corpus callosum, a large midline region below the cortex whose fibres link the two sides of the brain. This supplements the more limited communication provided by the anterior commissure present in monotremes and marsupials (Suárez et al., 2014). The complex signaling that instructs neurons to cross between the hemispheres during development is now beginning to be worked out, and it has become clear that the FOXG1 transcription factor plays a key role in mediating these events (Fothergill et al., 2014; Cargnin et al., 2018).

The other significant evolutionary improvement was to hearing ability; this was facilitated by the straight cochlea seen in the monotreme inner ear becoming coiled, a feature of marsupials and placentals. This change increased the amplitude sensitivity to and frequency resolution of sound without requiring additional space within the temporal bone. Its evolution was presumably a secondary response to the enhanced aural abilities of the three-boned middle ear. Together, these features required an increased functional capability for the organ of Corti, which converts vibrations in the cochlea to nerve impulses, and an enhanced neuronal processing ability, particularly in the auditory area of the neocortex.

The reproductive and urinary systems

The evolutionary origins of the mammalian reproductive systems are in those of stem amniotes that also gave rise to reptiles. Given that the reproductive system of contemporary Prototheria is similar to that of reptiles, that of stem mammals probably was too. The typical male reptilian reproductive system today includes internal testes, one or two penises, which are used only for procreation, and a cloacal region that incorporates exits for urinary and fecal material. The female has an internal reproductive system of two ovaries and two oviducts, which sometimes meet proximally, before exiting via one or two vaginas into the cloaca which has one or two clitorises (for details of the reptilian external genitalia, see Gredler et al., 2014). Most reptiles lay eggs that develop externally, but a minority are ovoviviparous or viviparous, either retaining the egg in the oviduct or dispensing with the shell. Both modes of live birth have enabled lizards and snakes to expand into environments hostile to the development of a laid egg, such as hot, dry desert areas where there is a risk of desiccation or cold area inhospitable to incubation.

The only adaptations required for ovoviviparity in reptiles are the softening of the hard shell and the increased vascularisation of the lower part of the oviduct, so allowing gas exchange between the allantoic and oviducal vessels. In truly viviparous reptiles, such as the European skink *Chalcides tridactylus*, the diameter of whose eggs are only ~3 mm, a complex placenta forms through interdigitated apposition of the vascularised chorioallantoic membrane and the highly vascularised uterine epithelium (Blackburn & Flemming, 2012). Such reptilian placentas enable the exchange of nutrients and nitrogenous waste as well as gases; they probably indicate the evolutionary basis for the various mammalian placentas (Morriss, 1975; Van Dyke & Griffith, 2018).

Mammals have extended this range of reproductive strategies and there are clear-cut differences between the three taxa (Fig. 16.4). In female monotremes, there are paired internal ovaries and oviducts that terminate in a reptile-like cloacal opening, which includes a vagina as well as an exit for defecation and secretion. In males, the cloaca is just an exit for waste products. In both, the urorectal septum, which in more advanced mammals separates the bladder from the rectum, does not fully descend so that a single exit serves both purposes. In males, the two sperm ducts (vasa deferentia) that carry sperm from each internal testis, meet in the future prostatic area to produce a common duct that penetrates the penis, from which there may be up to four exits. The penis develops from the genital tubercle at the distal end of the cloacal membrane, situated between the hindlimb buds. In females, this tubercle becomes the clitoris.

Reproduction in female monotremes shows features of both egg-laying and viviparous reptiles. They have a small egg (the Echidna egg is ~1.4 mm) with very little yolk, within which only the early stages of development take place (Selwood & Johnson, 2006). As the parchment-like shell is retained, the embryonic membranes do not make direct contact with the uterine wall. The embryonic vasculature extends around the yolk sac to absorb nourishment from it, but it also receives maternal secretions from the oviduct that pass through its parchment-like shell; some of this is stored for post-laying consumption (Cruz, 1997; Renfree, 2010). After the eggs are laid through a cloaca very similar to that of reptiles, the mother keeps them warm for a further ten or so days until they hatch as puggles at about the same stage of development as those of new-born marsupials. Like marsupial newborns, they have enlarged forelimbs, which they use for moving about on the fur-covered maternal abdomen where they lap the milk secretions.

The urinary and reproductive systems of monotremes are thus not very different from those of reptiles apart from the production of milk which substitutes for yolk in providing nutrition for the later stages of devlopment. The evolutionary changes in the urogenital systems of marsupial and placental mammals are far more substantial (Table 16.1 and Fig. 16.4), although there is no fossil evidence as to when and how these happened. The first likely change that distinguished the last common ancestor of marsupial and placental mammals from monotremes was the further descent of the urorectal septum towards the perineum. This tissue would have divided the internal cloacal tube into a dorsal part continuous with the gut and a ventral part that contained the exits from the ureter together with the oviduct in females and the sperm ducts in males. The dorsal part became the rectum while the ventral part became a bladder which terminated in a short urethral tube that voided into the cloaca. The two oviducts or sperm ducts came to lie between the bladder and the rectum, and extended into the cloacal region.

In female marsupials today, the urorectal septum still only descends part of the way down the cloacal cavity so that the vestigial cloaca retains its function as a single exit for the bladder and gut. In these animals, each oviduct evolved to include a uterus and a lateral vagina with an exit in the cloaca that is the entry point for sperm. There then evolved a new third central connecting tube which became a separate birth canal (Fig. 16.4). In parallel with the double vaginal opening, the marsupial penis became bifid in its distal region and so was able to enter both vaginas. It is worth noting that a bifid penis is a very occasional human congenital anomaly, suggesting that the placental line separated after the bifid penis had evolved.

The testes in male marsupials descend into an external scrotal sac that is ventral to the penis, unlike the scrotum of most placental mammals that is dorsal to it (Fig. 16.4). There has been interesting but unresolved discussion about the reasons why marsupial and most placental mammals need an external scrotal sac while all other vertebrates do not (Kleisner et al., 2010). One explanation is that the testes need to be cooler than the body, but this does not apply to organisms with internal testes. These include monotremes, a few eutherian mammals, such as the tapir and the hedgehog, and birds, whose body temperature can be as high as 43C (Prinzinger et al., 1991).

In the development of placental mammals, the urorectal septum descends to, and fuses with, the perineum. In both males and females, this full descent results in separate exits for the anal canal and the urethra. In females, the distal parts of the

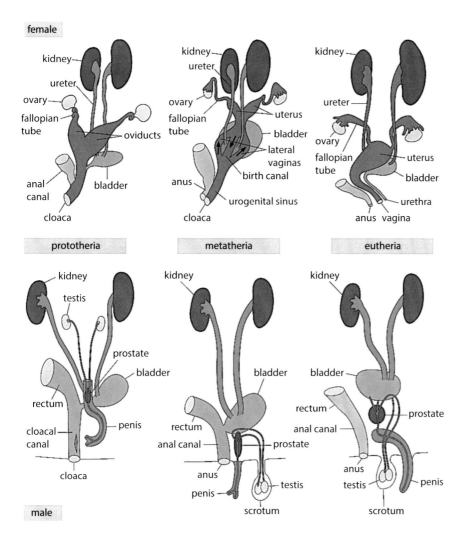

Figure 16.4 **The urinary and reproductive systems of female (upper) and male (lower) monotreme, marsupial, and placental mammals.** These diagrams, which only show core features, are intended to be representative of the different anatomies rather than to feature particular animals.

two reproductive ducts fuse at their proximal ends to give a single vagina between the urethra and the anal canal. The other major external change was the development of two sets of folds on either side of the cloacal (now urogenital) membrane. In the female, these become the inner labia minora and outer labia majora, which collectively protect the clitoris and the urethral and vaginal exits (for developmental and molecular details of these events in the mouse, see Greenfield & Bard, 2015).

In male placentals, the two sets of folds have roles that are very different from those in females. The outer pair extend posteriorly and seal over to form the scrotum into which the testes descend. The inner folds seal to form a tube that extends the urethra to form the shaft of the penis, distal to which is the initially solid and enlarged genital tubercle (phallus) that will become the glans. Sperm leave the testes through the vas deferens tubules, which enter the prostate, and anastomose with the urethra. The terminal part of the urethra is a tube that forms within the glans penis and provides the exit for both sperm and urine.

It is curious that almost all male vertebrates only use their penis for reproductive purposes and urinate through their cloaca; in contrast, a penis evolved in eutherian males that serves both purposes. It is still hard to see how or why such a complex penis evolved as it seems unlikely to matter to four-legged placental male vertebrates whether they urinate through a penis or from an exit within the cloaca.

Embryogenesis

In monotremes, oviductal secretions only help nourish the monotreme eggs for a few days (up to about the 19-somite stage in the case of the platypus); one or two eggs are then laid directly into a temporary pouch on the mother's abdomen (echidna) or into a well-insulated nest (platypus), where they are maintained at a temperature of ~32°C until the fetus-like puggles emerge about ten days later. Milk, secreted onto the abdominal skin surface by two patches of modified sebaceous glands, is lapped by the young (Schneider, 2011). Milk may even be secreted before the fetus hatches, so nourishing it through the porous shell. Thus, although monotremes have an essentially reptilian reproductive system, they are less fully viviparous than some reptiles, while having the typical mammalian characteristics of maternal care through homeothermy and lactation.

Marsupial development is different from that of monotremes in egg size, nutritional mechanism, and developmental detail (Table 16.1), as well as in anatomy. Marsupial eggs are small, having a thin membrane coat and little yolk. The early cleavage stages give a spherical morula, nourished by uterine secretions that pass across the membrane. Following rupture of this membrane, the embryo is initially sustained by yolk via vascularization of the yolk sac, with secretions from the closely apposed uterine epithelium also providing some nutrition until birth. In late pregnancy, yolk is supplemented by a choriovitelline placenta that forms when the yolk sac fuses to the chorion and adheres to the wall of the uterus, with vascularisation leading to the formation of a small umbilical cord (Freyer et al., 2003). Although this placenta is very different from the chorioallantoic placenta of placental mammals, the underlying molecular genetics of the two are similar, so emphasising their common ancestry (Guernsey et al., 2017; Laird et al., 2018).

After ~4-5 weeks, the marsupial joey is born at a stage similar to that of hatching monotremes, and roughly equivalent to that of a 14-day mouse embryo. Unusual features of the tiny marsupial neonate are its precocious clawed forelimbs that enable it to climb to the pouch, its early olfactory abilities that allow it to follow a maternal scent trail to the pouch, and its advanced mouth that enables it to latch on to the teat for lactation. The Tammar Wallaby (*Macropus eugenii*), for instance, is born at 28 days of gestation weighing 350–400 mg; it stays permanently attached to the teat for 14 weeks, by which time it weighs ~100g. It remains in the pouch until about 27 weeks after birth, accessing the teat as required; it then continues to suckle at will for several months after it leaves the pouch (Hickford et al., 2009). One possible driver of the evolution of such development was the benefit of holding the fetus close to the mother and so denying predators access to her offspring. A possible limitation that resulted from this neoteny was that other potential modes of forelimb evolution would have become restricted; if so, this would help to explain why there are no marsupial equivalents of bats or seals (McKenna Kelly & Sears, 2011).

Placental eggs have little if any yolk (the typical egg diameter is ~70μm) and thus require an early-forming placenta. This starts when the external trophoblast cells of the very early embryo become embedded in the uterine wall where they elicit a strong immunological response that results in locally increased vascularization. The subsequent few stages of embryogenesis are primarily concerned with the production of the extraembryonic membranes, as a result of which the chorion, allantois, and umbilical vasculature soon form. The first two of these provide the fetal contribution to the placenta, which forms in close association with the domain of enhanced uterine vasculature; this extends to form the umbilical cord (for review, see Selwood & Johnson, 2006). While the exact details and degree of trophoblastic invasion and placenta formation depend on the taxon, the result is an enhanced supply of fetal nourishment that allows the female to give birth to a neonate that is often relatively self-sufficient. A particular evolutionary advantage of the chorioallantoic placenta is that it enables intrauterine development to proceed for a long time, so providing an opportunity for advanced brain development before birth.

Mammary glands

The ability to produce milk for feeding newborns is not only a synapomorphy for all mammals but was an important and perhaps early step in the series of changes to reproduction and development that characterise the clade. The simplest way of doing this is shown by monotremes that have 100–200 lactatory glands on their abdomen that produce milk, hair, and sebum. Since pups lick the milk that collects on these hairs, it is likely that the glands were modified from those that originally made just hair and sebum, with their origins perhaps dating back to the mammaliaforms of the Mesozoic (Oftedal, 2002). In due course, the sebum and hair-making abilities were lost as the glands coalesced into the branching ducts and terminal nipples of marsupial and placental mammary glands.

Oftedal (2012) has investigated the likely origins of milk phylogenetically: genomic analysis of milk-protein sequences has suggested that they date back to early synapsids ~310 Mya (late Carboniferous) with some of the constituents evolving even earlier. In the Permian, the likely role of milk was to provide fluid and nutrition to developing mammaliaforms through absorption across their porous, parchment-like shells so supplementing the small amounts of yolk. By the late Triassic (~210 Mya), rich milk was probably providing nutrition to the miniscule newborns of small mammaliaforms as in today's monotremes.

For all of the complexity of the placental reproductive system, it is clear that the advantages of an extensive period of internal development have allowed the clade to become the largest, most successful and most sophisticated of all modern land animals. They have had limited success in the air and the numbers of contemporary taxa of gliding mammals and flying bats are quite small as compared to the many bird species. Mammals have, however, had some success in the sea to which they returned several times (below; Uhen, 2007), with perhaps the most dramatic example of such evolutionary opportunism being the evolution of the Cetacea ~50 Mya, or ~320 My after the first stem tetrapod left the sea for dry land.

THE EVOLUTION OF THE CETACEA

The slow break-up of Pangea into northern Laurasia and southern Gondwanaland was well on its way to completion by time of the K-T extinction (Fig. 3.1). Soon after the beginning of the Cenozoic Era, Laurasia had separated into two major land masses, North America and Eurasia, while the supercontinent of Gondwanaland was slowly fragmenting to give the individual continents of Africa and South America, India, Australasia, and Antarctica (Appendix 3 and Book Website). Mammalian diversity today still reflects these events. The Boreoeutherian placentals evolved in Europe, Asia, and North America, the Afrotherian placentals evolved in Africa and the Xenarthran placentals evolved in the South America. South America, Antarctic and Australasia cohered longer than the other major land masses, and it was across this domain that the marsupials flourished. Phylogenetic analysis confirms that mammalian diversity today reflects these events: Tarver et al. (2016) dates the origins of the various eutherian clades (Fig. 16.5) to 100–85 Mya when Pangaea was starting to fragment. This was at least 20 My before the K-T extinction (~66 Mya).

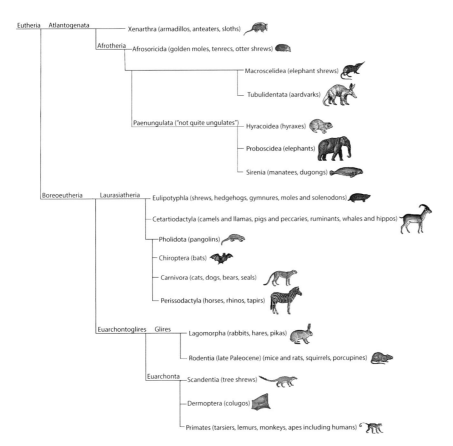

Figure 16.5 Molecular phylogram of the major groups of modern mammals. This clearly demonstrates the relationship between the major clades and their geographic origins.

(From https://en.wikipedia.org/wiki/Evolution_of_mammals. Published under a CC Attribution—Share Alike 3.0 Unported License.)

Figure 16.6 A range of eutherian mammals. (a): Flying fox bat. (b): a colugo. (c): A fossil glyptodon. (d): A pangolin. (e): An armadillo. (f): A spiny anteater. (g): An aardvark. (h): A tapir. (i): An African elephant.

(a   Courtesy of Hasitha Tudugalle. Published under a CC Attribution Generic 2.0 License.
b   Courtesy of Norman Lim.
c   Courtesy of "WolfmanSF." Published under a CC Attribution—Share Alike 3.0 Unported License.
d   Published under a CC Attribution—Share Alike 4.0 International License.
e   From www.birdphotos.com. Published under a CC Attribution—Share Alike 3.0 Unported License.
f   Courtesy of Gunjan Pandey. Published under a CC Attribution—Share Alike 4.0 International License.
g   Courtesy of "warriorwoman531." Published under a CC Attribution Generic 2.0 License.
h   Courtesy of "Rufus46." Published under a CC Attribution—Share Alike 3.0 Unported License.
i   Courtesy of Yathin S. Krishnappa. Published under a CC Attribution—Share Alike 3.0 Unported License.)

Of the three major mammalian clades, the monotremes seem never to have been particularly successful as their fossil record is low; only the platypus and spiny anteaters remain today, restricted to Australia, and New Guinea. The reasons why they have not diversified more widely remain unclear. In contrast, the marsupials (Metatheria) originally spread widely, being particularly successful across Gondwana, being lost from Antarctica as it moved towards the South Pole but surviving in South America and Australasia. There they filled many of the niches now occupied by eutherian taxa and showed similar anatomical adaptations (all are examples of homoplasies). More recently, however, they have been almost completely displaced in South America by eutherians as relatively few metatherian taxa seemed able to compete with these, an exception being the opossum which, once the Panama land connection had formed, migrated north into North America where it is still common. Today, the major habitat of Metatheria is Australia, where the once-wide arborisation of taxa has now been reduced to kangaroos and wallabies, koalas, wombats, bandicoots, and possums.

The eutherian mammals that originated in Gondwanaland diversified in Africa, giving rise to elephant shrews, aardvarks, elephants, and related species, such as the tenrecs of Madagascar (the Afrotheria) and in South America to a range of mammals that included armadillos, anteaters, shrews, and related animals, such as the now extinct glyptodons (the Xenarthra). The limited diversification of the Eutheria in South America may have reflected local competition from the Metatheria whose wide variety of taxa included carnivorous species. The broadest range of Eutherian mammals (the Boreoeutheria), however, evolved in Europe, Asia, and North America, the continents into which Laurasia fragmented. Taxa included ruminants, carnivores, bats, and primates (Fig. 16.5). Many of these species reached Africa in the Cenozoic through the contacts that the continent maintained with Asia (via Arabia) and Europe (via Spain). Large mammals flourished and further diversified in Africa to an extent seen nowhere else.

Little is known of the selection pressures that led to the very distinctive forms of mammals in the different regions, but strong if not complete geographic isolation combined with specific geological features, such as mountains and ecological factors, such as forestry were clearly important factors, as the distinct fauna (and flora) of the isolated island of Madagascar makes clear (Samonds et al., 2013). Although all mammalian taxa are derivatives of early terrestrial tetrapods, there are some important taxa that show specific external differences, which reflect the effect of particular selection factors (Fig. 16.6). These include those that developed unusual features, such as a protective exterior of bone or scales or a dual-purpose proboscis, together with those that became able to glide or fly and those that became marine.

Defensive adaptations

Protection from predators in mammals can be based on speed (e.g. deer), kicking (e.g. zebras), biting (e.g. carnivores), horns and tusks (e.g. many herbivores), thick skins (e.g. cows, rhinoceroses, and elephants), and spines (e.g. hedgehogs, porcupines, and spiney anteaters) but there are some animals that have developed even more formidable defences. These include the pangolins that are in the Boroeutheria clade and have large keratin-based scales that, when the animal curls into a ball, make it invulnerable to predators. A similar well-known example is the armadillo, a South American taxon from the Xenartha clade (Fig. 16.6c–e). The protection of these animals comes from heavy leather plates that also render them impregnable when they roll up into a defensive ball, a behaviour also seen in spined mammals, such as hedgehogs that are also boreoeutherian mammals. Perhaps the most impregnable of all, however, was the *Glyptodon*, a recently extinct relative of the armadillos and another member of the Xenartha. This large animal developed bones (osteoderms, often known as scutes; Fig. 16.6c) in its skin and had no known predators until humans arrived in South America.

Probosces with a second function

All vertebrates have a sense of smell; even fish have blind-ended nostrils whose cavities are lined with sensory epithelium. There are several families of mammals in which the nose or snout has become extended and taken on a second function. These include the Australian spiny anteaters (monotremes) and a range of placentals that includes the South American anteaters (xenarthrans), tapirs (boroeutherian ungulates), and aardvarks and elephants (both afrotherians). As this anatomical feature lengthened independently in the different taxa, it represents homoplastic evolution. Anteaters and aardvarks use their lengthened nose for probing for insects, particularly ants. The South America anteater also has a particularly long tongue and other intriguing features, such as an internal penis and testes (Fig. 16.4) and an unusually low body temperature for a mammal, features that reflect the long isolation of the Xenarthran clade from the other major Eutherian clades.

The largest extant land mammals are elephants and are marked by their ground-reaching trunks; these afrotherian placentals now fall into two major groups: African and Indian. The latter reflects an early migration when a group left Africa, moved East and ended up in India where it evolved to give a distinct taxon whose most noticeable features are smaller ears and an ability to be semi-tamed, a property of great use to humans. Elephants also migrated north and gave rise to the mastodons and other taxa, all of which are now extinct. Embryologically, the trunks of both elephants and tapirs derive from the upper lip and nose and are used prehensively for grasping as well as breathing. Elephant trunks also have touch, manipulation, and sound-production abilities. The extended elongation that led to the elephant's trunk was probably a response to the increasing size of the head and the resultant shortening of the neck to facilitate supporting its weight. As the animal increased in size, it required a means of getting food from the ground and other vegetation sources, and the solution seems to have been a lengthened and dextrous trunk (Shoshani, 1998).

Flight

Bats, the only flying mammals (Fig. 16.6a), clearly evolved from a clade that could glide, an ability that is seen in a few small forest-living mammalian taxa today. The advantages of being able to move from one tree to another at height are obvious. Contemporary examples come from the Metatheria in Australia (the feather-tailed possum, the greater glider, and the flying phalanger) and from the Eutheria, with species that include flying squirrels, anomalures (rodents) and colugos (close to primates, Fig. 16.16.b; Panyutina et al., 2015). The anatomical innovation that makes gliding possible is the patagium, a membrane of skin and hair that extends in its fullest form from the forelimbs to the hindlimbs and even the tail. It is a homoplasy as it evolved independently in at least half a dozen clades.

True flight in mammals is restricted to the bats (Chiroptera) whose wings evolved very differently from those of pterosaurs and birds; it is characterised by hyper-extended digits that maximise the size of the patagium (Figs. 16.6a). Bats have now diversified widely and can be very large: the giant golden-crowned flying fox,

Acerodon jubatus, can weigh 1.6 kg and have a wingspan of 1.7 m. Most bats have poor eyesight but excellent echolocation capacity—they are mainly creatures of the night.

The exact taxonomic status of bats was unclear for a long time because their anatomy is so distinct and their fossil record so poor: no intermediates between early bats and related species have been found, mainly because their bones are light and fragile. The situation has recently been clarified by phylogenetic analysis, which places bats in the very broad Laurasiatheria clade within the Boroeutheria. This includes shrews, ungulates, whales, and carnivores, with the Chiroptera being a sister group to a clade that includes Carnivora, cloven-hooved mammals and pangolins. Computational dating suggests that the ancestors of bats diverged from this clade ~60 Mya, or soon after the K-T extinction (Phillips & Fruciano, 2018).

Back to the sea and the evolution of whales

Three very different groups of mammals have left land for the sea: their descendants include the Cetacea (whales and dolphins), the Pinnipeds (seals, walruses, and sea lions), and the Sirenia (dugongs and manatees). Phylogenetic analysis has confirmed that the Pinnipeds are related to bears, diverging in the early Cenozoic (Springer & Gatesy, 2018) while the Sirenia diverged during the Eocene as a sister group to elephants and hyraxes and, more distantly, the aardvarks (Pardini et al., 2007). The exact origin of the Cetacea initially proved hard to analyse using standard phylogenetic techniques.

Aristotle first recorded some 2000 years ago that dolphins were not fish, but it seems to have been John Ray, an important 17th century biologist, who was the first to realise that whales and dolphins were mammals. The evolutionary adaptations that enabled the Cetacea to cope with marine life involved major changes to the tail, the body shape, and the limbs, all of which required growth and other modifications to pre-existing anatomical features rather than the evolution of novel ones. Equally important are the anatomical and physiological modifications to the respiratory, auditory, and renal systems that enabled land mammals to adapt to a marine environment (Reidenberg, 2007). The fossil record says almost nothing about how these latter changes occurred as they mainly involved soft tissues.

The early fossil evidence shows that a clade of four-legged carnivorous artiodactyls (even-toed ungulates) evolved to become marine (Bajpai et al., 2009): these fossils show a slow extension of the tail, the nasal openings moving cranially and unifying, and the external parts of the hindlimbs being lost, leaving only a rudimentary internal femur and pelvis (below, also see Websites). Curiously, there is some evidence that the diminished pelvis still plays a minor role in male sexual activity (Dines et al., 2014). Until relatively recently, the fossil data was not detailed enough to construct a cladogram back to the last common ancestor of whales and land tetrapods because the anatomical differences among the fossils then were too substantial (the branches of the cladograms are long, Chapter 7) to generate an unequivocal line of descent.

For similar reasons, it is not possible to work out the lines of descent using phylogenetics based on analysing DNA-sequence data: there are too few contemporary artiodactyls for an unambiguous phylogenetic tree to be produced on the basis of inter-species gene or protein homologies (Chapter 7). Fortunately, there was another class of sequence data that turned out to be more helpful: these were retrotransposons or mobile genetic elements that are found in all eukaryotes but are usually not transcribed. These *interspersed nuclear elements* can be short (such SINEs have ~500bp) or long (LINEs have ~6000bp), but the reason why they are useful here is that, once they have integrated into the genome, they tend to remain there and so can be treated as just another heritable feature.

The key early study that clarified the relationships in this clade was that of Nikaido et al., (1999) who set out to identify SINEs and LINEs that were characteristic of the *Artiodactyla* and that were, therefore, potential synapomorphies for the clade. Nikaido et al., identified twenty such sequences that were shared by various current member of the Artiodactyla and also seen in whales. They were then able to use a standard cladistics package to reconstruct a unique tree with twenty insertion events on the basis of which organisms carried particular SINEs and LINEs. The results (Fig. 16.7) show clearly that whales are indeed a member of the Artiodactyla, with the hippopotamus being their closest and the camel their most distant ungulate land relatives today. Such a phylogeny gives no direct clue as to when the lines leading to hippo and whales (grouped taxonomically within the Whippomorpha) separated, but the fossil evidence suggests that it was at least 55 Mya.

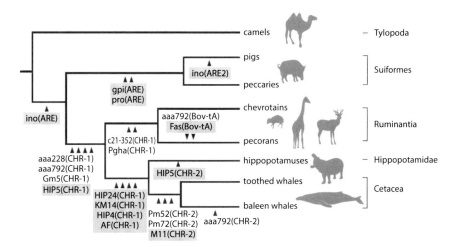

Figure 16.7 Whale phylogeny. A cladogram showing the phylogeny of whales and other even-toed ungulates from an early artiodactylan ancestor using SINEs and LINEs as characters to determine branching points.

(From Gatesy et al., (2013) Mol Phylogenet Evol; 66:479–506. With permission from Elsevier.)

More recent fossil discoveries have further clarified the general lines of descent of the Cetacea from ungulates, the key discovery being the presence of an artiodactyl ankle in early cetaceans (Gingerich et al., 2001). Analysis of further fossils has shown the series of steps by which an unknown species of Artiodactyl gave rise to two clades (Berta, 2017). One eventually led to the Anthracotheriidae, with an example being A*nthracotherium* (~15 Mya), whose terrestrial descendant is likely to be the hippopotamus; the other gave rise to the cetaceans (McKenna, 1975). Fig. 16.8 shows

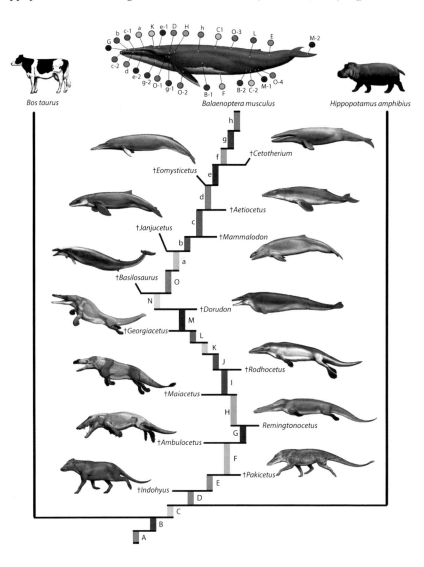

Figure 16.8 Diagram showing the evolution of modern whales from an ungulate ancestor indicating the appearance of anatomical apomorphies (colours) in animals representing major steps on the way.

(From Nikaido et al. (1999). *Proc Natl Acad Sci U S A* 96:10261–10266. © National Academy of Sciences; with permission.)

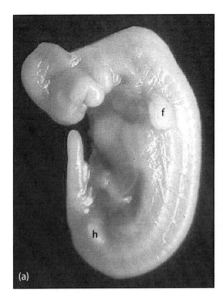 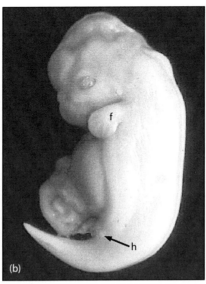

Figure 16.9 Dolphin embryos at 24 d (a) and 48 d (b). Note how the forelimb bud (f) develops and prouces digits while the hindlimb bud (h) almost disappears over this period.

(From Bejder and Hall (2002) *Evol Dev* 4:445–458. With permission from John Wiley & Sons.)

reconstructions of the sequence of some of species that mark the evolutionary path that led to the Cetacea (Gatesy et al., 2013).

A different perspective on how cetaceans evolved from quadruped mammals comes from embryology. Unlike all other mammals, cetaceans lack hind limbs, although all have very rudimentary pelvises, and small internal femurs with no connecting joints in a similar location to the hindlimbs of a normal mammal. Bejder and Hall (2002) have analysed limb formation in dolphin embryos: they found that the 24-day embryo has a hindlimb bud that is very similar to its forelimb bud and essentially indistinguishable from the early hindlimb buds of quadrupedal mammals, such as the mouse (Fig. 16.9). By 48 days of development, however, things are very different: although the forelimb bud of the dolphin has developed by enlarging and forming all normal bones including digits, as in any terrestrial tetrapod, its hindlimb bud has vanished externally, leaving small internal rudiments.

The embryological explanation for this is straightforward: patterning in the early limb-bud is controlled by its distal region, the apical epidermal ridge (AER). This ensures that the bones are laid down in proximal-to-distal order: in the fore-limb bud, the scapula is laid down first, then the humerus, with the digits appearing last, and this is followed by a sustained period of growth (Barresi & Gilbert, 2019). Similar events occur in the hind limb bud: the hip bones are laid down first, then the femur and so on. The presence of a rudimentary pelvis and femur in whales suggests that the patterning and proliferation activity controlled by the AER in the early hind limb ceases immediately after the femur has been initiated; the small amounts of tissue present in adult animals reflect the minimal amount of subsequent differentiation and growth. The reason why the femur is internalized is not known as later stages of dolphin development have not been studied. It is, however, likely that the diminutive early hindlimb simply becomes incorporated into the body as the fetus enlarges. The cetaceans thus carry their evolutionary history as a plesiomorphy (the ability to make a hindlimb) but have their own apomorphy (early cessation of hindlimb development). This story is the exact embodiment of Darwin's *descent with modification*.

This chapter has focused on the evolution of the broad diversity of the mammals over the Cenozoic, to the extent that they dominate the large fauna of every continent. Today, mammals are present on land, in the sea and in the air. The one major mammalian taxon that has not been mentioned in this chapter is the Hominini, a clade that spurred off the primate line ~7 Mya, slowly broadening and then narrowing, so that today there is only a single species left, *H. sapiens*. Our evolution is discussed in Section 5, but it is worth pointing out here that the effect of contemporary humans on the rest of the Mammalia has been dramatic. Over the last few thousand years, many large mammals from across Earth have become extinct as a result of human activity across the world: obvious examples are the giant ground sloths and glyptodons (Xenarthra), mammoths (Afrotheria), aurochs (Boroeutheria) and the Tasmanian tigers (Metatheria), but there are more. Today, many large mammals are under severe threat as humans take over their habitat and hunt them, and it is a fair bet that fewer of these wonderful animals will be here in a few generations' time. The current Epoch is the Anthropocene (Chapter 26), and its fossil record will sadly be marked by the extinction of many, large mammals.

KEY POINTS

- Today's mammals originated from stem mammals, synapsid amniotes that diverged from diapsid reptiles~ 300 Mya.

- Early synapsids can be distinguished from all other amniotes because they have a single temporal foramen on each side of their skull. This and the surrounding bones have since changed to become part of the zygomatic arch.

- Such synapsids (e.g. Dimetrodon, a pelycosaur) were the predominant reptiles Throughout the Permian.

- Only three synapsid taxa survived the extinction that ended the Permian; two of these were later lost to leave only the cynodonts, the ancestral sister taxon of all of today's mammals.

- During the early Mesozoic, the cynodonts radiated and one lineage evolved into a group of animals that eventually became small and lived nocturnally, probably in burrows (this was the nocturnal bottleneck), slowly evolving the mammalian synapomorphies seen today (e.g. endothermy, lactation, hair, and three middle-ear ossicles).

- They also acquired properties that facilitated nocturnal activity, such as relatively large eyes (other than species, such as moles, that now lead mainly subterranean lives and are almost blind) and strong senses of hearing, touch, and smell, properties that remain characteristic of most of today's mammals.

- The fossil record and other data illustrate the evolution of many of the specific anatomical features of contemporary mammals. Particularly important are those of the skull that demonstrate increasing brain size.

- The three extant clades of mammals are the monotremes (Prototheria), marsupials (Metatheria), and placentals (Eutheria). The fossil record and phylogenetic analysis show that these clades separated well before the K-T extinction (66 Mya).

- The three mammalian clades can be particularly distinguished by their reproductive systems and modes of embryogenesis: the oviparous monotremes are closest to stem amniotes, while the viviparous marsupials, and placentals evolved later.

- The extensive adaptive radiation that led to the variety of placental mammals today is almost certainly due to the long gestation period that a full placenta allows. This has particularly enabled the eutherian brain to enlarge and enhance its abilities; it also enables herd species to have young that can move soon after birth.

- The global distribution of the main mammalian taxa (the Metatheria, the Boroeutheria, the Afrotheria and the Xenarthra) today reflects the location of their ancestors at around the time of the K-T extinction when Pangaea was fragmenting into contemporary continents.

- There are now many mammalian taxa that can glide, but only the Chiroptera (bats) that can fly.

- Three clades of land mammals adapted to a marine existence: the Cetacea (e.g. whales), the Pinnipeds (e.g. seals), and the Sirenia (dugongs and manatees). Whales and dolphins are descended from even-toed ungulates and share a last common ancestor with the hippopotamus; this has been established on the basis of anatomical, embryological and genomic evidence.

- In cetacean hindlimbs, the pelvis bones and femurs start to form normally but development of further bones and growth of existing ones are blocked a little later in development.

FURTHER READINGS

Berta A (2017) *The Rise of Marine Mammals: 50 Million Years of Evolution*. Johns Hopkins University Press

Kardong KV (2014) *Vertebrates: Comparative Anatomy, Function, Evolution* (7th ed). McGraw-Hill.

Kemp, TS (2005) *The Origin and Evolution of the Mammals.* Oxford University Press.

WEBSITES

- Wikipedia entries on *Marsupials, Monotremes,* and *Patagium,* together with those for the various fossils and mammalian species mentioned in the text
- Evolution of bats: www.smithsonianmag.com/science-nature/bats-evolution-history-180974610/

- Evolution of early mammals: www.nature.com/articles/d41586-019-03170-7
- Video carton of whale evolution: ocean.si.edu/through-time/ancient-seas/evolution-whales-animation

SECTION FOUR
THE MECHANISMS OF EVOLUTION

This section considers the question of how evolution happens or, more specifically, by which mechanisms do existing species give rise to new ones? This is, of course, the question that Darwin and Wallace posed to themselves in the late 1830s and 1850s, respectively. In both cases, their answer was to suggest that selection of normal variants that were favoured (Darwin's word) in novel environments would lead to the formation of new species. It took, however, almost a century to come up with a detailed explanation of speciation and even today, there are areas where our knowledge is thin. This is partly because change is slow and hence hard to follow and partly because it is complex as it involves all levels of life from the genome to the environment. Today, more than 150 years after publication of *The origin of species*, we can put a great deal of meat onto the sparse bones of Darwin's original and deep perception. Two particularly difficult problems that nevertheless remain hard to understand are the dynamics of speciation and how mutation leads to major changes in phenotypes.

The rediscovery of Mendel's work from 1866 (see Websites) early in the 20th century led to the appreciation of two important points: first, that existing heritable changes could spread through a population as a result of mating and, second, that novel mutations could arise both spontaneously and through their production by X-rays and mutagens, such as N-ethyl-N-nitrosourea (ENU). By the 1920s, geneticists, such as Fisher, Haldane and Sewell Wright had constructed a theory of *evolutionary population genetics* that incorporated the selection of variants. A decade later, embryologists, such as Waddington, were beginning to understand how mutations that affected early embryogenesis could produce variant adults.

The key step on the way to understanding speciation came from the work by Mayr and others by the 1950s. They showed that speciation generally starts with a subpopulation becoming reproductively isolated from its parent population in some novel habitat. As this subpopulation would carry only a random sample of the genetic variance of the parental total because of genetic drift, it would, therefore, have a slightly different phenotype distribution. Whether that subpopulation thrived or

DOI: 10.1201/9780429346217-20

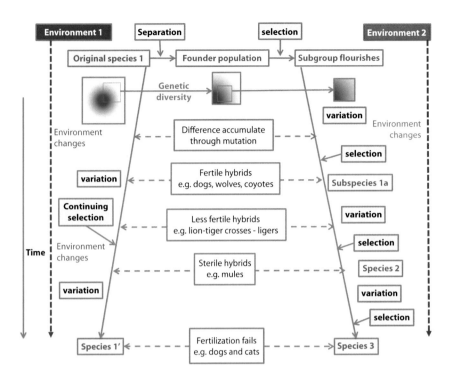

Figure S4.1 **The process of novel speciation.** A small founder population breaks away from its parent population and migrates to a new environment carrying with it only a subset of its genetic diversity. Most if not all of that founder population will fail to flourish in its new environment. It is, however, possible that a subpopulation carrying even less genetic diversity, but a particular spread of phenotypic traits, will flourish and this group will eventually give rise to a new and slightly different population. In their different and changing environments, both parent and sub-founder populations continue to evolve, but along different trajectories. As they do so, interbreeding, should it happen, becomes less, and less successful. Eventually it fails and the successors of the original subpopulation will have given rise to a new species. The full sequence of events may take many thousands, or even millions of generations.

failed in its new environment would depend on whether the blind hand of selection favoured some of its variants so that they left more fertile offspring than others (i.e. they were fitter). If so, they would become the predominant and eventually the only form. At this stage, the population would be recognisable as a subspecies, because it could still interbreed with the parent population to produce hybrids. Over long periods of time, however, further mutations in the two populations would render them increasingly genetically distinct and, in due course unable to interbreed. They would then have become separate species. Today, it seems strange that all of these insights were achieved before anything was known about DNA or molecular genetics.

These events are summarized in Fig. S4.1, and the various aspects of this complex process are discussed in the chapters of this section.

Stage 1: **Variation**. All populations include variants at levels of both the phenotype and the underlying genotype. Chapter 17 looks at variation in contemporary species, the nature of the classes of mutations that underpin them and how they lead to heritable variation. Chapter 18 considers the nature of the variations that led to some of the better-understood changes that characterize evolutionary change and species differences. Here, it is clear that almost all anatomical variants reflect the effect of mutations that affect the complex protein pathways and networks that drive normal embryogenesis.

Stage 2: **Adaptation to a new environment**. When a small sub-population of some species becomes reproductively isolated in a new habitat, it can adapt and thrive in many ways. Chapter 19 first considers some of these adaptations, particularly camouflage, mimicry, and exaptation; it then reviews how organisms can interact with one another to provide what Kropotkin (1902) called *mutual aid*. Important groups here are symbionts where two species come to depend on one another and holobionts where an organism together with the microbiome (microbial community) that it hosts can be considered as a composite organism for physiological and selection purposes.

Stage 3: **Selection.** A subgroup of a population thrives if it has features that enable it to produce more fertile offspring than the rest and Chapter 20 discusses the three major ways by which this can happen. Selection may result from the properties of the organism enabling it to thrive in the environment—this is *natural selection*. Individual males and females will leave more offspring if they have a stronger chance of obtaining mates—this is *sexual selection*. One group may also thrive at the expense of others if its individuals can behave in ways that strengthen the likelihood of group members supporting one another better than do other groups—such altruistic behaviour is known as *kin selection* because members of such groups are usually related.

Chapter 21 considers the quantitative theory of *evolutionary population genetics*. This integrates selection with population genetics and predicts how allele distributions change over time. A particular aim of the chapter is to explain those properties of small groups that allow phenotypic change to occur rapidly. The chapter ends with a discussion of coalescence theory, which integrates gene mutation with population genetics to produce a means of studying population histories. This theory is treated qualitatively as few biologists have the mathematical skills to explore its full richness.

Stage 4: **Speciation**. In separate populations that are reproductively isolated, the genetic differences over time will lead to increasing degrees of breeding incompatibility, and eventually novel speciation. Chapter 22 examines the various definitions of a novel species, how groups can become reproductively isolated, the speed of change and the final changes to genes and chromosomes within a group that are the final and key stage in irreversible species formation.

Variation 1: Mutations and Phenotypes

17

This chapter focuses on how offspring may differ from their parent to increase variation in the population. Because of historical random mutation, new mutations and genetic drift, every trait within a population includes a distribution of phenotypic variants. Sexual reproduction remixes the parental chromosomes so that each offspring has a random 50% of the combined parental genome, together with a few new mutations. The resultant genotype adds to the genetic mix of the population and can occasionally lead to a phenotypic variant outside the normal distribution. Provided that this change is heritable, it can result in evolutionary modifications, provided, of course, that it survives selection.

The first part of this chapter discusses phenotypic variation and the ways in which an offspring's phenotype can differ from that of its parents. It also discusses heritable diseases as the underlying mutations give some insight into how mutation can effect change that may be evolutionarily significant. The second part of the chapter examines some ways in which abnormal phenotypes can be produced by genetic change. These include the effect of mutations on individual proteins, horizontal gene transfer, large-scale genomic reorganisations and transgenerational epigenetic inheritance (non-canonical mechanisms of genetic inheritance). The main thrust of the chapter, however, is on how mutations in individual genes affect the function of the proteins that they encode and their downstream effects on the phenotype. It introduces the following chapter that examines the effect on the phenotype resulting from mutations that alter protein networks and how this can lead to more substantial changes in the phenotype.

In 1859, Darwin began the last paragraph of Chapter 5 of *The Origin of Species* with the sentence: "*Whatever the cause may be of each slight difference in the offspring from their parents—and a cause for each must exist ....*" One can feel Darwin's frustration at not knowing anything about the origins of variation, which he had realised was the basis for all evolutionary change. Indeed, it is no coincidence that the ease of selective breeding in domestic pigeons on the basis of identifying variants occupied a major part of his first Chapter. Pigeon breeding was then a popular hobby and many beautiful variants could (and still can) readily be produced in a dozen or so generations of selective breeding of natural variants (Fig. 17.1). Variation is ubiquitous in the living world and can be observed in all species at all levels from the genome to adult anatomy, although it is most obvious in external features, such as pigmentation and size.

The basic explanation for the origin of variation came some 35 years after *The Origin* with the discovery of mutation by Bateson and de Vries in the context of the re-emergence of Mendel's work on genetics, albeit that a full understanding of mutation had to wait for the discovery of DNA in the 1950s. Little of normal variation across a population is, however, generated by new or even recent mutations in DNA sequences as the likelihood of this happening at a particular site is very low: the estimated rate for humans is ~1.1×10^{-8} per base per generation (Roach et al., 2010). Most variation in sexually reproducing organisms comes from the shuffling of genes through recombination that takes place during zygote formation and from the

DOI: 10.1201/9780429346217-21

Figure 17.1 Pigeon diversity. (a): Capuchin pigeon. (b): Blue bar grizzle frillback pigeon. (c): English trumpeter pigeon. (d): pigmy pouter pigeon. (e): Archangel pigeon. (f): Oriental frill pigeon.

(a Courtesy of Jim Gifford. Published under a CC Attribution—Share Alike Generic 2.0 License.

a Courtesy of Jim Gifford. Published under a CC Attribution—Share Alike Generic License.

b Courtesy of Graham Manning. Published under a CC Attribution—Share Alike 3.0 Unported License.

c Courtesy of Jim Clifford. Published under a CC Attribution—Share Alike Generic License.

d Courtesy of "GyrR." Published under a CC Attribution—Share Alike 3.0 Unported License.

e Courtesy of Jim Gifford. Published under a CC Attribution—Share Alike Generic License.)

effects of random mating, although the latter factor may be modified through sexual selection. This random sampling of genes leads to what is known as *genetic drift* and is a significant part of the reason why offspring can have phenotypes that differ from their parents (Masel, 2011). Other ways of changing gene frequencies within a population (this is the geneticist's view of evolution) are by natural selection and through gene flow, which particularly occurs when two distinct populations meet and interbreed.

As a result of these various modes of genetic change, each offspring has a genome that represents a random 50% of the genomes of each parent (a result of meiosis) and supplemented by a few new mutations that occur during development. Each of its siblings has a different sample, making it distinctive from, but still closer to their parents and siblings than to any other member of the species. Genetic variability within a species is ubiquitous across the living world and is more extensive in organisms that reproduce sexually. It is particularly obvious in humans as no two individuals other than identical twins look the same, and even they slowly diverge in appearance as environmental effects and somatic mutations affect the maturing phenotype. Each animal thus has its own unique genome and, if one looks carefully enough, its own unique phenotype. There is even variation in bacteria, which can change their phenotype if selection is high enough, as the rapid acquisition of antibiotic resistance and an ability to degrade plastic show (Urbanek et al., 2018). The former, incidentally, is a particularly interesting area as it involves both hypermutation for its origin and conjugation (the prokaryote mechanism of gene transfer), for its spread (Culyba et al., 2015). It is this variability that, given a suitable environment, enables new species to evolve.

PHENOTYPIC VARIATION

If one just considers populations of humans and dogs, two species within each of which different strains can easily interbreed, the obvious respective differences are in size, facial detail, skin pigmentation, hair details, and behaviour. The first four of these are relatively straightforward to discuss in the context of events taking place during embryogenesis, but the last is not. This is for two reasons: first, we have little idea of how behaviour is generated by neuronal anatomy and physiology, let alone coded in the genome; second, it is hard to separate out how much of the behavioural phenotype is intrinsic to the organism and how much is the result of environmental influences (the nature/nurture argument). Cultural inheritance is particularly important for humans and also, incidentally, birds (Aplin, 2019), but cannot be considered in the context of this chapter as little is yet known about its genetic basis. It will, however, be considered in Chapter 25, which discusses human evolution, because the increasingly sophisticated creation of artefacts over time provides a measure of mental sophistication and manual dexterity.

Although there is no such thing as "the normal" phenotype for any species, it is reasonable to assume that any individual member of a particular species can be viewed as relatively normal if each heritable aspect (or trait) of its phenotype falls within an expected range and if it can easily breed with other members of the group. Outbred populations in which there is free interbreeding have a considerable degree of genetic variation. Such variation is much less in strains that have been selected on the basis of specific features. Some of these effects have been noted in human founder populations (below and Chapter 24), but limited genetic pools are also seen in inbred dog and mouse strains, while black C57 BL/6 and white BALB/c mice strains have been inbred to minimise within-group genetic variation. This limited pangenome (the entire gene set of the strain) defines what is known as the genetic background of a strain; small pangenomes facilitate, for example, immunological experiments as transplants can be done within the group without rejection.

It is also reasonable to assume that a healthy offspring will mainly possess protein types inherited from its parents, with just a small proportion, because their amino-acid sequences are slightly altered through recombination or mutation, being slightly different in functional effectiveness. Overall, the offspring's phenotype will thus be distinct from that of each parent, allowing for sexual differences. These modifications will appear as natural variants, in the sense that they will not seem remarkable. Our knowledge of the extent of such variation comes partly from observing normal breeding and partly from direct sequencing.

The most important model system for genetics at this level is the human because we have studied ourselves in depth, originally for medical reasons and more recently, because of the easy and cheap availability of genome sequencing, for scientific ones. The 1000 Genomes Project (2015, see below) has found that, given a reference human genome, a typical human has 4–5 million mutations of various types that affect ~20 million base pairs. In more medical terms, about 2% of the human population have a diagnosable, but minor genetic disorder and a further 1% have a serious abnormality, with most of these reflecting major chromosomal defects.

The difficulties with such figures are that, because offspring with serious disorders will be less likely to reproduce, it can be hard to know whether such abnormalities are heritable, reflect environmental effects on the embryo, or are the effect of somatic mutations. More to the point perhaps is that few, if any, offspring or even adults can be identified as having a heritable feature that, on its own, seems likely to give that individual a selective advantage. This is the difficulty in using normal organisms to explore evolutionary change: the extent of the genetic and phenotypic variation seen in the normal population of any organism is usually unimpressive; it is far easier to identify phenotypes that are likely to be detrimental than those that are advantageous. In practice, humans use selective breeding of animals and plants with apparently advantageous traits for commercial purposes but usually on too small and too short a basis and in too constrained an environment for the full evolutionary implications to become clear.

Unusual variation: Sports and anomalies

Although most of an offspring's traits lie within the normal spectrum, parents occasionally produce offspring whose phenotype lies at the periphery of the normal spectrum. In breeding circles, these are known as *sports* and their features are often heritable, particularly those that affect size, and pigmentation. An example is the spotted zebra (Fig. 17.2) whose pattern almost certainly derives from a degenerate version of reaction-diffusion kinetics within a complex protein network (Chapter 18). It is not, however, known whether any offspring of this animal also had spots. The selective mating of sports is the basis for breeding pigeon and dog variants (Fig. 12.1) and is often more successful for minor anatomical features than for physiological abilities, such as speed: the likelihood of two champion horses producing an equally talented offspring is much less than their owners might hope. This is probably because so many complex traits underpin speed that the chances of an offspring inheriting all the protein components of the networks with the appropriate rate constants to finetune them for speed is low.

Rarer are heritable anomalies whose phenotypes affect major anatomical features; the best-known examples are the homeotic mutants in *Drosophila* in which a tissue deriving from a particular segment is replaced by one from a different segment on

Figure 17.2 **This spotted *E. quagga burchelli* zebra is an example of a "sport."**

(From Bard JBL (1981) J. Theor. Biol. 93:363–385. With permission from Elsevier.)

the basis of mutations in Hox genes (e.g. the Antennapedia mutant replaces a head antenna with a limb) (Chapter 8; many other examples are discussed in Held, 2021). Such homeotic mutants can interbreed, and the pattern of their inheritance was a source of study in the middle of the last century, before the molecular basis of the phenotypes was known. These occasional pattern variants are important as they indicate that mutations in the genes that code for network constituents can underpin major anatomical change. These homeotic mutants were perhaps the key stimulus that drove research into the molecular basis of development (evo-devo) during the 1980s and an early triumph was the discovery of Hox genes.

We do not normally think of there being sports in the human phenotype, but humans with extraordinary abilities do arise. There was no reason, for example, to suppose that just one of the six children of Hermann and Pauline Einstein, who were each no more than on the clever side of normal, would be the most brilliant physicist of the 20th century. But this ability was not inherited by either of Albert Einstein's sons. Indeed, it seems to be rare that children inherit extraordinary abilities shown by one or even both talented parents. This is probably because the numbers of protein alleles that underpin such qualities are large and the genetic shuffling that leads to the gene complement of offspring will be different from that of both parents (in genetics, this effect is known as *regression towards the mean*). Inheritance of specific phenotypes cannot be predicted, but it can be tuned under controlled breeding conditions (Chapter 20). Even then, it can take many generations to finetune to the requisite extent even for relatively simple phenotypes like feather patterns and colours.

Variation leading to human disease

Heritable disorders and abnormalities linked to humans are important partly because they help understand normal gene function, and partly because they indicate what can happen if the boundaries of normal variation are stretched. Many heritable genetic syndromes have been identified and published in the literature, but more readily accessible is the Online Mendelian Inheritance in Man (OMIM) database with its extensive and well-referenced entries (see Websites). Many human genetic disorders reflect a recessive single-gene mutation, but others are more complex; very few are advantageous (malaria resistance associated with heterozygous sickle-cell anaemia is an exception—see Chapter 21).

Analysis of heritable disorders has another use: the identification and geographical analysis of group-specific mutations can cast light the historical origins of the disorder. Risch et al. (2003), for example, reviewed the data for the eleven most common diseases specifically associated with Ashkenazi Jews, a distinct group from Eastern Europe, and found that their dates of origin fell into three groups: >100, ~50 and ~12 generations ago. They point out that these periods reflect the dispersion of Jews from Israel by the Romans, their arrival in central Europe, and a founder effect (Chapter 21) associated with their arrival in Lithuania. Similarly, Diaz et al. (2000) were able to show on the basis of linkage disequilibrium analysis that the mutation

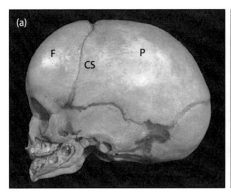  

Figure 17.3 **Craniosynostosis.** (a): The skull of a normal new-born child). (b): A CAT scan of the skull of child with Crouzon's syndrome. Here, proliferation in the coronal suture (CS) between the parietal (p) and frontal (F) bones has ceased and the suture has closed prematurely so fusing the adjacent bones and forcing the growing brain to expand upwards. This distorts normal bone growth on either side the sagittal suture at the top of the skull and the result is a misshapen head. (c): One of the causes of this syndrome is a dominant Cys342Tyr mutation of the extracellular IIIc domain of the FGFR2 signal receptor (the arrow points to the membrane binding domain of the receptor.

(From Morriss-Kay GM & Wilkie AOM (2005) J. Anat. 207:637–653. With permission from John Wiley & Sons.)

in the GBA gene that underlies Gaucher's disease in this population probably arose some 55 generations ago (confidence limits: 100–30).

Other examples of heritable disorders are often spontaneous mutations that are maintained within the population but are not associated with an obvious disease, presumably because negative selection pressures are low and there is a low background rate of spontaneous mutation. A curious example of a rare, but stable phenotype is the dual trait of wet earwax (cerumen) and underarm (axillary) odour. Individuals carrying the rs17822931 SNP (single nucleotide polymorphism) of ABCC11, an ATP transporter protein, have wet earwax and strong axillary odour. The latter can be sufficiently worrying for affected individuals to have their axillary apocrine glands removed from each armpit (Nakano et al., 2009).

Many disorders can be understood on the basis of a protein loss or uncontrolled gain of function, although the former usually requires both proteins to show mutations. Such potential medical conditions include a predisposition to breast cancer through mutations in the BRCA1 or 2 genes, which are involved in repairing DNA breakages, and the various connective tissue diseases, many of which derive from mutations in the collagen proteins.

More important, in an evolutionary context at least, are heritable disorders in which an aspect of normal anatomical development fails as these show how large-scale change can happen. An example here is craniosynostosis, recognized by a misshapen head. Normally, the developing skull enlarges uniformly to accommodate the expanding brain through a balanced sequence of pre-osteoblastic proliferation and osteogenic differentiation at the sutural edge of the growing bones so that these bones progressively increase in size. If this this sequence becomes unbalanced, one or more of the sutures can undergo premature fusion (craniosynostosis); other sutures are then forced to expand more than normal to provide space for the expanding brain - the net result is a misshapen skull. Many craniosynostosis syndromes are due to activating mutations of an FGF receptor gene (Fig. 17.3. Morriss-Kay & Wilkie, 2005; Martínez-Abadías et al., 2013). Premature fusion of one or more cranial sutures in humans and mice can also be caused by mutations in a group of very different genes that, in addition to those encoding the FGFR receptors, includes transcription factors, such as Twist1, MSX2, and signals, such as EPHA4 (Holmes, 2012). The involvement of this range of regulatory proteins clearly points to the involvement of several networks in the growth and differentiation of skull bones.

Some of the most theoretically interesting human congenital abnormalities are those that affect patterning and whose phenotype includes missing or duplicated tissues. One example is the Holt-Oram syndrome (see the OMIM website): its characteristics include a missing thumb and aberrant heart septation, both of which result from mutations in the TBX5 transcription factor gene. The issue is complicated to analyse because this protein, like many others, clearly has different roles in different tissues. (*Note*: it is still difficult to investigate events downstream of transcription factors, particularly during development.)

Genetic disorders that affect developmental anatomy can often be traced back to genes that code for proteins that include signals, kinases (energy-transfer proteins; these often function within protein networks), as well as receptors and transcription factors. A further degree of complexity is seen in heritable disorders that affect behaviour: the molecular reasons for such changes are rarely known even when a candidate gene has been identified. It is, however, important to note that, where a mutant gene has been shown to have a severe impact on some trait, this does not mean that a gene for that trait has been identified, as is sometimes claimed. All that has been identified is a gene whose role in creating that trait is sufficiently important that a mutation in it can have a severe impact on the final phenotype. A car analogy makes this clear: a faulty spark plug will stop a car engine from working, but the plug is not the engine. Consider the Pax6 gene: this codes for the key transcription factor that drives the development of the eye in all animals; mutation in both copies of the gene results in a failure of eye development (Chapter 8). The role of this transcription factor is to set in motion the processes that lead to eye development, it is not the process itself. To carry the car metaphor a little further: there are many ways to stop a motor from working, but it is much harder to improve its performance. Doing this often requires a substantial redesign which, in turn, demands minor changes to multiple components.

The significance of genetic diseases in an evolutionary context is that they show that relatively simple mutations can have two types of effects. First, they can modulate the function of an individual structural protein: mutations in myosin, for example, can lead to muscle disorders (Tajsharghi & Oldfors, 2013); second, they can affect the function of network activators, such as receptors and transcription factors, blocking or modulating all the changes that normally follow from the activity of such proteins. Such deleterious changes give at least a clue as to what might have been the effect of mutations that have had had a positive anatomical effect in times past, a topic discussed in the next chapter.

Secondary variation

The discussion so far has looked at the core features of heritable change, but it is important to realise that a novel change in some specific anatomical feature in a developing organism rarely stands alone. Sometimes the prime cause is clear. If, for example, extra digits form as a result of a familial predisposition to polydactyly, this is due to a mutation in the patterning system. The phenotype is an extra digit, but this not only includes a set of extra phalanges (the key patterning change); it also includes the associated musculature, vascularization, nails, and skin. In this case, the primary effect of the mutation is clear. In other cases, it can be hard to know which aspect of the phenotype reflects the prime effect of the mutation. The Feingold syndrome, for instance, has a phenotype that includes microcephaly, short stature, and numerous digital anomalies; the immediate cause remains obscure. The genetic defects are mutations in the MYCN oncogene, which is located on the short arm of chromosome 2, although it has also been associated with further mutations elsewhere (Marcelis et al., 2008; Burnside et al., 2018). We still have little idea as to how such a mutation could lead to disconnected abnormalities in growth and patterning. Sometimes, it is possible to demonstrate experimentally how the novel phenotype has been generated, as two examples show. First, expression of Pax6 in an abnormal site in the head of *Xenopus* leads to the formation of a small, ectopic eye with all the major tissues expected (Chow et al., 1999). Second, the insertion of FGF-containing beads into the flank of an early chick embryo results in an ectopic limb with all of its normal tissues (Kawakami et al., 2001). In both cases, the patterning cues affect all the surrounding tissues so that they participate in the formation of the new tissue module (Appendix 7).

In other cases, however, the evolutionary success of an innovation requires secondary features to form that are embryologically independent of that innovation and often affect the physiology of the organism. The example of the giraffe makes the point. The advantages of its long neck are that it allows the animal to forage on trees too high for other ruminants, but the success of this change required major adaptations to the cardiovascular system as the heart has to pump blood some 2m higher than is required in other animals. The heart has thus had to change as the neck elongated: its overall size is much as expected for the size of mammal, but its ventricular wall has thickened substantially (Mitchell & Skinner, 2009). The corollary of this is that blood pressure in the limbs is far higher than normal and, to stop edema here, the walls of the lower-limb arteries have also thickened substantially to withstand the pressure (Petersen et al., 2013; see also Fig. 18.9).

Such linked evolutionary changes are known as *co-evolution*. In the case of the giraffe, one can envisage that the extent of neck lengthening was initially limited by the abilities of the cardiovascular system. For the neck to continue extending, the cardiovascular and other systems also had to change, so that there was a slow cooperative, step-by-step change to the animal's phenotype. A similar effect probably occurred in the evolution of long-necked dinosaurs, such as *Diplodocus*, but there is, of course, no evidence here as none of the soft tissues have been fossilised. A particularly interesting example of co-evolution is shown by the set of fossils whose analysis shows how a succession of the descendants of an ancient terrestrial ungulate became whales. As this happened, there were major changes to limb, head morphology, body morphology, lung physiology, and hearing, with the effectiveness of each at any stage depending on the changes made other properties (Fig. 16.8; Bajpai et al., 2009; Gatesy et al., 2013; Geisler, 2019).

GENOTYPIC VARIATION

Phenotypic variation within some trait across a population is underpinned by variation at every underlying level of scale: there are mutations in the DNA, polymorphisms in proteins (e.g. the human blood groups) and, in some cases, downstream changes in the outputs of protein networks. These result in minor differences in anatomical structures, such as in local degrees of growth, the routes that nerves take and the secondary details of the vascular system, and in physiological function, such as, for example, in muscle activity and innate behaviour. Some people are natural athletes and others will never be, no matter how hard they train.

Investigation of high-level functions is difficult because it is hard to work out the extent to which some unusual aspect of an individual's phenotype derives from its genome, from random events during development and from the effects of environmental factors. Indeed, in the original work on population genetics by Fisher (1930), the variance parameter, which is a measure of the spread of some trait in a population, was partitioned into an inherited component (due to genes whose composition could not at that time be specified) and an environmental component (whose effects were hard to articulate and quantify), with each being given equivalent weight in the formal theory.

If, in a breeding experiment, the phenotype of a particular offspring shows a novel feature, one's immediate response is to suggest that the change reflects the effect of a novel germline mutation, and this may be so. Alternatively, it may reflect the effects of either a somatic mutation or a rare combination of parental gene variants already present within the population. Basic breeding work, if it can be done, will show whether the trait reflects a somatic or germ-line mutation and, for the latter case, will indicate whether the mutation is recessive, dominant or sex-linked and whether it breeds true or whether it only sometimes shows its phenotype (this is imperfect penetrance and reflects a degree of genetic buffering).

Normal variation in the sense of discrete change, as for example the petal colours of flowers, is conceptually straightforward to understand in terms of the activity of a single or a few genes and perhaps their polymorphisms. Thus, the various polymorphisms of trichohyalin (TCHH), a protein expressed in hair follicles, are associated in European hair with whether it is straight, waved or curly (Medland et al., 2009). Quantitative changes in a phenotypic character, such as skin-pigment intensity and body growth, height or shape, are much harder to explain on the basis of mutation in a single gene (Fig. 25.5). It is much easier to see such change as the result of the action of several genes, and the proof that this was compatible with Mendelian genetics was an early breakthrough made by Fisher in 1918 (see Fisher, 1930). Today, we are more likely to view continuous change as a result of the sum of the actions of the mutation-induced protein variants in a network. Jacob (1977) suggested that the appropriate way to think about the production of small mutation-induced changes was to consider it as *tinkering* with the genome, but the effects of this tinkering on the resultant phenotype are usually much harder to understand in complex eukaryotic organisms than in prokaryotes.

MUTATION

Long-term heritable mutation underpins all evolutionary change. In the case of animals, this only refers to mutations in eggs and sperm; somatic mutation might affect reproductive ability in an individual, but there are as yet no known mechanisms by which such somatic mutations can be transferred to germ cells. (This was a problem that Darwin tried to solve by the suggestion of panspermia, that sperm was made up of contributions from active tissues—see Appendix 2). The situation is slightly different in plants: in angiosperms, for example, each flower is an independent source of male and/or female gametophytes and hence of variation. One has only to look at

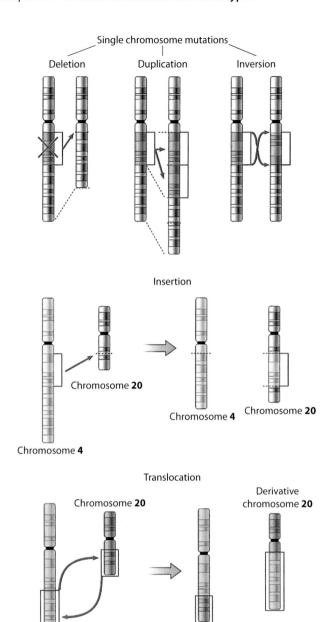

Figure 17.4 **Examples of the many types of mutations that can change chromosomes.**

(Courtesy of Yasssine Mrabet.)

the numbers of olives on a tree to realise how much more rapidly evolutionary change can occur in plants as compared to, say, in mammals.

The last common eukaryotic ancestor (the LECA) must have had a simple genome, probably a simple and single chromosome, that reflected its recent prokaryotic ancestry. Over the past ~1.7 By, this chromosome has changed beyond measure as it has split and diversified to become the genomes of all of today's eukaryotic organisms. Every original sequence will have mutated and perhaps duplicated or inverted, while a few will have been lost (Fig. 17.4). Other sequences may have changed because of cross-over during meiosis or because of other smaller changes. In addition, new sequences may even have been acquired through horizontal gene transfer from other organisms (e.g. parasites or holobionts). These processes occurred and kept occurring as every new species formed. Mutation never ceases, even in *E. coli* (below).

Simple DNA circular genomes have become long chromosomes with the abilities to bind histones and fold, to transpose great lengths to new regions and to accommodate retrotransposons (Shapiro, 2011). At a more macroscopic level, there has been polyploidy and the generation of new chromosomes, each providing an opportunity for future evolutionary change as a result of mutation. The resultant diversity is illustrated by the fact that in the insects, at one extreme, the male jack-jumper ant, *Myrmecia pilosula*, has a single chromosome (the female has two) and at the other the butterfly, *Agrodiaetus shaharami*, has 268 chromosomes. Changes in the genotype over evolutionary timescales have been just as profound as they have been in the phenotype.

Comparison of human genomes makes it clear that the majority of mutations that represent variation within a population are to a single base pair (below; 1000 Genomes Project, 2015). In the case of protein-coding sequences, these often do not change the encoded amino acid because of redundancy in the genetic code. Such synonymous (neutral) mutations can, however, be a first step towards a more substantial change caused by a second mutation to the codon. Mutations that do change an amino acid (nonsynonymous mutations) may have an effect that is advantageous, deleterious or, as often happens, results in no noticeable change to function other than to increase the size of the pangenome. More serious and potentially lethal are those mutations that result in the production of one of the stop codons (UAG, UAA, and UGA).

Some insight into the nature of mutational activity has come from the *E. coli* long-term evolution experiment in which phenotypic and genetic changes in 12 replicates of an initial *E. coli* population are being followed, so far for more than 60,000 generations, with samples being analysed every 500 generations (Good et al., 2017). Here, the only selection pressure is that which the populations of bacteria impose on themselves over time under serial propagation. Perhaps the key result so far is that the populations have never become genetically stable, mutations arise, there are minor changes in phenotypes (e.g. the ability to metabolise citrate), mutation rates change a little, but mutation itself never ceases. One might even think that the population behaves as if it were always striving to become fitter!

It is interesting to explore the extent of the classes of genomic sequences present in contemporary genomes. A simple bacterial genome typically has relatively few types of sequence: those for encoding proteins and regulating their expression together with those that code for the various classes of RNA and perhaps a few other functions (see the GO database, websites). Analysis of eukaryotic genomes has increased the number of sequencing functions to some hundreds of sequence types of which only a small percentage are involved in coding for proteins and regulating their expression. Most of the genome is occupied by sequences with a host of other functions, such as helping fold the long DNA chain into the more compact chromosome. Readers with a bioinformatics background who are interested in the full set of coding classes in DNA should explore the Sequence Ontology (see Websites), a resource that expresses this information in the format of a mathematical graph (see Appendix 1)

Perhaps the most curious of the sequence types in an evolutionary context are transposons and retrotransposons. These are sequences that have or once had the ability to change their chromosomal location: the former are now maintained in their site while the latter are translated into RNA, then reverse-translated to make a new copy of themselves that can insert into a new genome location. Transposons can occupy large amounts of the genome: in maize they comprise ~90% of the genome while in humans this figure is about 44%, of which only a small proportion are active, with their most apparent role being in disease (Mills et al., 2007; Payer & Burns, 2019). One possible role of their evolutionary significance, in plants at least, is their ability to drive variation; another is that they can be a source of sRNA, or RNA sequences less than 200 nucleotides in length, that can regulate gene expression (Lewsey et al., 2016; Niu et al., 2019). Nevertheless, It has always seemed strange that, if the great majority of transposons are silent and have no role, they can occupy so much of the genome without being lost; their presence would seem to impose an unnecessary burden on the replication machinery of every cell, particularly germ cells.

The effects of mutation on the genotype

It has been clear since the iconic work of Watson and Crick in the 1950s that mutation in the genome underpins the genetic component of the variation in phenotypes across a population. It is, however, only recently, with the availability of relatively cheap whole-genomic sequencing, that it has become possible to discover the extent of such genomic variation, with most work being done on humans. The publication of the early results of the *1000 Genome Project* (2010) has given important baseline data on this. If one takes what one can think of as a reference human genome and looks at the likely variants within it, the results have shown that any individual will have in their ~30K protein-coding sequences:

- 10–11K variants that will lead to an amino acid change in a protein
- 10–12K variants that will not give an amino acid change in a protein
- ~200 in-frame deletions and ~ 90 premature stop codons
- 50–100 variants that may well to lead to an inherited disorder

In an evolutionary context, the silent mutations in both protein-coding and nonprotein-coding regions (e.g. for enhancers, repressors and other untranslated sequences) provide a reservoir for rapid potential change by further mutation or by recombination through mating to produce new, functional changes to the genotype (Shapiro, 2011). An example of this will be discussed in Chapter 19 where the work of C.H. Waddington on genetic assimilation is described.

It is also worth noting that the mutation figures given above probably reflect an asymmetry in the contributions from the egg and the sperm. All oocytes are laid down early in ovarian development and, as Haldane noted almost a century ago, mainly include the original mutations from the mother because relatively few cell divisions will have occurred in these germ cells since fertilization. Mature sperm, however, reflect the many mutations that will have occurred in the primary sperm cells (spermatogonia) that keep dividing throughout a potential father's life. Although there are DNA repair mechanisms within cells, one or two mutations get through each division and these accumulate, progressively reducing sperm quality (Sartorius & Nieschlag, 2010). Similarly, somatic mutations can arise in the non-germ cells of adults and these can accumulate in rapidly dividing tissues, such as gut, haematopoietic system and skin. If these mutations affect the growth and apoptosis pathways, they can result in cancer.

Although one sometimes thinks of new mutations as the obvious source of variation, this is not so because the rates of spontaneous mutation are so low. It is worth repeating that most of the novelty in an offspring's genome derives from the fact that each of its chromosomes differs from that of the parent from which it is inherited. This is because each pair of parental chromosomes undergoes cross-over and genetic recombination at some random position during meiosis, so that each daughter chromosome includes parts of both parental chromosomes. Each offspring, therefore, has a gene mix that, of course, reflects that of each parent but is unique. Novel mutations formed after fertilization as opposed to those that are inherited are just another, occasional addition to the mix.

The effect of mutation on individual genes

DNA sequences that underpin any eukaryotic phenotype fall into three classes: 1) those that encode proteins and are both transcribed and translated; 2) those that are only transcribed, such as intron sequences and the various RNA populations (rRNAs, tRNAs, siRNAs, miRNAs, etc.); and 3) those that are untranscribed, such as enhancer and repressor sites. Mutations in enhancer and repressor sites can have important effects on gene expression, although they can be a megabase or more distant from the genes whose expression they regulate (Anderson & Hill, 2014; Schoenfelder & Fraser, 2019). Fortunately, techniques for identifying them have now been developed (Dailey, 2015; Shin, 2018).

Histones play an important role here as the genomic folding that they mediate can bring distant regulatory sequences into close proximity with the promoter sequences that they affect (Harmston & Lenhard, 2013). Enhancer/repressor analysis, together with that of miRNA and siRNA both of which modulate mRNA translation, is a difficult but important area of current research. As to proteins, while the occasional mutation can introduce a stop codon that truncates them, the more common effect of mutation is to modify the protein's shape and/or charge distribution. Each of these affects its ability to interact with other proteins and with substrates: this is partly through the overall shape of the protein and partly through binding interactions, with both leading to changes in rate constants. This is the effect of genetic tinkering, and it can affect all aspects of the phenotype.

Because there are normally two copies of a gene, the phenotypic effect of a mutation on one allele is usually recessive: the normal protein is usually present in sufficient quantities to cover for the mutated one, and the heterozygote shows only a weakened phenotype (this is known as a dosage effect) or no effect at all. In some cases, however, the mutation leads to an anatomical or physiological disorder, many of which have been discovered, and studied in humans on the basis of its phenotype.

Two classic examples of human diseases due to recessive mutations are those that cause cystic fibrosis and sickle-cell anaemia: the former is due to recessive mutations in the CFTR gene, which codes for a cell-surface regulator involved in chloride transport, and the latter to mutations in the β-globin gene which codes for an oxygen-transporting protein. If the homozygote shows a severe phenotype, it is likely to be selected against and the mutation may well be lost; it will only survive if the heterozygote is fertile. The mutation is most likely to spread through the population if the heterozygote phenotype has a selective advantage. The classic example of such a balanced polymorphism is, of course, sickle-cell anaemia, where the heterozygote affords some protection against malaria (Chapter 21).

A particularly common effect of mutation on a protein is to slightly alter its secondary structure; this, in turn, alters its interactions with other proteins and with smaller substrates molecules. An example of the former is the binding of ligands to receptors and here, binding may fail, be too weak or too strong; these changes not only reflect a change in the rate constants associated with the interaction, but, for example, in whether the downstream response to the signaling is blocked or constitutive. For proteins that interact with small molecules (e.g. enzymes and kinases), the kinetics of binding may likewise be increased or decreased. An example here is the Ehlers–Danlos syndrome: the mutations are in the collagen gene and their effect is to weaken interprotein binding; this results in stretchy skin and hyperextensible joints (Malfait & De Paepe, 2014). Particularly important in this context are the effects of mutation on protein networks: any effect on binding constants will change the kinetics of protein-protein interactions and, in turn, the dynamic properties of networks (Chapter 18).

Occasionally, a gene mutation is dominant, and this is usually because either there is only a single copy of the gene in the genome (i.e. it is on the X or Y chromosome in the male or there has been a deletion) or the effect of the mutant protein negates that of the normal one. An example here is a gene that codes for a ligand receptor which the mutation makes constitutive: such a receptor always behaves as if it is in the ligand-activated state and this results in a permanent states change. Thus, an intracellular network normally controlled by such a receptor will always be in the "on" state, irrespective of the behaviour of the normal receptor. Such mutations have been found in the various FGF receptors and result in, for example, dwarfism and craniosynostosis (above).

Although such dominant or single-copy genes can theoretically spread rapidly through a population as the phenotype will be shown by 50% of all offspring of an affected parent, most so far discovered are deleterious, and their reproductive implications are, therefore, likely to be unfavourable. The frequency of many dominant genes that affect development seriously is that of the background mutation rate (this is approximately 1:50,000 per protein site per generation). There is certainly little evidence to suggest that such mutation is a normal route for driving evolutionary change.

Larger-scale genomic changes

The very earliest bacterial genomes included at most a few hundred protein-coding genes and contained perhaps 50,000 base pairs (Chapter 9). The human genome, which is not particularly large, has some three billion base pairs. The massive increase in the genomes of all multicellular animals has not only come about through the inclusion of transposons, but through a long series of gene and chromosome duplications, each of which allowed new sequences to mutate independently of their parent sequences. Such duplications can be studied by identifying similar gene sequences within an organism and tracing them back to common ancestors using standard phylogenetic algorithms (Chapter 8, Appendix 5; for review, see Lallemand et al., 2020).

Particularly interesting is the fact that some of these duplications took place soon after the formation of the Bilateria, often before the protostome-deuterostome split ~520 Mya (McLysaght et al., 2002). Initially, a duplicated gene was likely to have been identical to its original sequence, but differential mutation over time can lead to paralogues, with each acquiring a distinct function, one of which probably retained its original role. The evolution of the Hox gene family from a protist precursor gene provides a classic example of how a gene family can arise from a series of duplications, each of which led to further diversification (Chapter 7; Gehring et al., 2009). There are several mechanisms that lead to duplication (Shapiro, 2011), but more important is the fact that a coding sequence will not be differentially expressed until it has acquired the appropriate noncoding sequences, such as promoter and activator sites appropriate to a new location in the organism. The evolution of novel genes that are functional is not a simple matter.

Not all duplicated sequences will survive. Some lose their regulatory sequences and become pseudogenes. Others may maintain a similar class of function if key parts of their sequence are not affected by mutation. Important examples here are the DNA-binding motifs that characterize families of transcription factors. These include helix-turn-helix (e.g. homeodomains), zinc-finger, leucine-zipper and several other motifs. Slightly less constrained are the membrane-location domains that characterize cell-surface receptors and other such proteins. These domains are regions of non-polar amino acids that are soluble in lipids and form hydrophobic bonds. This feature allows such regions to be identified computationally (Nastou et al., 2016).

Perhaps the most powerful genomic change in generating diversity is polyploidy, when the complete chromosome complement is duplicated. Events here can be followed computationally with a particularly interesting and important example being the two rounds of whole-genome duplication that took place in the genome at the base of the metazoan lineage ~326 Mya. Sacerdot et al. (2018) showed that this happened through a series of steps. First, a prevertebrate genome composed of 17 chromosomes duplicated to 34 chromosomes. This was then reduced to 27 following seven chromosome fusions, before a further complete duplication gave 54 chromosomes. After the separation of the lineage of the jawed vertebrates (Gnathostomata) from the jawless agnathan fishes (the Cyclostomata = circular mouths), four further fusions took place to form the genome of ancestral bony vertebrates) which had 50 chromosomes. A series of further series of clade-specific changes led to the diversity of chromosome number in land vertebrates today.

The situation is even more dramatic in plants where chromosome number can range from 4 in the angiosperms (e.g. the monocot *Colpodium versicolor* and the dicot *Haplopappus gracilis*) up to 1260 in *Ophioglossum reticulatum*, a fern. For reasons that are unclear, plants seem more tolerant of polyploidy than animals. Perhaps more important, Alix et al. (2017) were able to show that polyploidy often occurred prior to or simultaneously with major evolutionary transitions and adaptive radiation of species. This is presumably because the generation of apparently "spare" chromosome results in immediate genetic redundancy and represents an opportunity for the evolution of novel gene function. It seems clear that polyploidy and whole-genome duplication provide key drivers for evolution, divergence, and speciation in both plants and animals.

In the context of speciation, alterations to chromosome structure and number have a further importance: while minor genomic changes have only secondary effects on reproduction, major ones, such as a change in chromosome number, will affect fertilization in subsequent generations through non-disjunction. A classic example here is the mule, which is the offspring of a donkey and a horse, whose last common ancestor lived 4.5–4 Mya and that respectively have 32 and 31 pairs of chromosomes. As a result, mules are almost always infertile, although the very occasional fertile exception has been noted (Ryder et al., 1985). Horses and zebras will also interbreed, and the resulting hybrids have also proved to be infertile. Here, no exceptions would be expected because the three zebra species (*E. grevyi*, *E. quagga*, and *E. zebra*) have 23, 22, and 16 pairs of chromosomes (Chapter 22). Given the differences in the chromosome numbers of horses and zebras, it still seems surprising that interbreeding is even possible and that hybrids will develop and thrive. Such observations imply that alteration of chromosome numbers in some clade is likely to have occurred through a series of small steps rather than through dramatic changes.

HORIZONTAL GENE TRANSFER

One mechanism that allows an organism to acquire new genes is horizontal gene transfer; this is the transfer of genetic material from one organism to another, which then incorporates it in its own genome. It was this mechanism that facilitated the evolution of the eukaryotic cell through endosymbiosis of eu- and archaebacteria (Chapter 9), and its importance in other contexts is now becoming apparent (for review, see Soucy et al., 2015; Gilbert, 2019; see Websites).

Two examples of lateral gene transfer in eukaryotes have been particularly studied. The first is the facility of weevils and some beetles to metabolise plant material, an ability that was acquired through the horizontal transfer of genes that encode cellulose-digesting proteins from an ascomycete fungus some 200 Mya (Kirsch et al., 2014). The second is the behaviour of *Wolbachia*, a parasitic bacterium that is commonly found in arthropods, with the two organisms leading a symbiotic existence. Many examples are now known in which insect genomes have acquired a range of *Wolbachia* sequences through lateral gene transfer (Le et al., 2014; Cordaux & Gilbert, 2017). Of particular interest is the observation that malaria- and dengue-carrying mosquitos infected with *Wolbachia* have a shortened life and reduced infectivity (Jiggins, 2017).

THE ROLE OF THE ENVIRONMENT IN GENERATING VARIATION

It is easy to forget the extent to which the environment affects the phenotype. In humans, pigment intensity in light-skinned people depends on the degree of exposure to the sun, weight depends on food consumption and physical endurance depends on exercise, while much of what humans do depends on what they are taught (cultural inheritance). Moreover, as every gardener knows, the development of plant morphology is highly dependent on the environment (Gilbert & Epel, 2015). A particularly dramatic animal example where the environment plays a deterministic role is in the gender of some reptiles, such as crocodiles. In mammals, gender is fixed at conception by the presence or absence of a Y chromosome; crocodiles do not have a sex-specific chromosome and their gender is determined by the temperature at which the embryo develops its genital system. At 31°C, males and females are equally likely to form; below 30°C, females predominate and above 32°C, males predominate (Crews, 2003). It is also known that temperature-dependent sex reversal can take place in more basal adult reptiles, such as the Australian bearded dragon (Holleley et al., 2015). A similar but opposite effect occurs in turtles (Valenzuela & Adams, 2011); here, the effect of global warming is to feminize some turtle populations (Jensen et al., 2018).

An interesting and long-discussed question is whether the changes to an organism that result from environmental stimulation or physiological activity might be inherited by the offspring. Such ideas can be traced back to Lamarck who was the first serious evolutionary biologist, albeit that he was feeling his way in uncharted territory. Because there was then no other means known for effecting evolutionary change, his views were widely accepted throughout the 19th century. Thus Darwin (1859) explicitly writes about ".... variability from the indirect and direct action of the external conditions of life, and from use and disuse" because he knew of few alternative sources of variation. The idea of acquired inheritance was, however, dismissed by Weismann (1889), who discovered the continuity of the germ plasm, on the grounds that he knew of no mechanism by which it could be implemented.

Today, the question is phrased differently, often in terms of whether developmental plasticity can be replaced by developmental determination. This would mean that a particular facet of the phenotype that was once a reflection of an environmental pressure would instead become part of normal phenotype. This important subject is discussed below and also in Chapters 19 and 20, particularly in the context of developmental plasticity.

Transgenerational epigenetic inheritance (TEI)

One concern about evolutionary change is that it is slow. Lamarck considered that acquired changes in one generation would be passed on to the next, a view accepted by Darwin who proposed the mechanism of panspermia to explain how this would happen. Such views fell into abeyance following the publication of Weissman's cell theory at the end of the 19th and our increasing understanding of mutation over the 20th century, but they are taking on a new resonance today. This is partly because our knowledge of molecular genetics suggests that there are non-canonical mechanisms by which change can be transmitted and partly because there are observations that the standard theory of inheritance still cannot explain. As of now, however, there is no rapid way yet known of passing on the effect of the environment on the phenotype to the next generation in a way that is stable over the longer term.

A slightly more modest question is to ask whether it is possible to inherit changes in a way that does not depend on random mutation, Mendelian inheritance or genetic drift. Such mechanisms are generally referred to as *epigenetic*, a term originally coined by Waddington (1953) to mean *above* genetics in the sense of how genes were used, but this meaning has changed; the word is now used for mechanisms by which gene expression can be modulated in ways that do not reflect normal mutation. Its particular importance in evolution is that such novel modes of changes in gene expression can under a few circumstances be transmitted to the next generation. The general name for such events is *Transgenerational epigenetic inheritance* (TEI) and the subject is now being widely discussed (e.g. Gilbert & Epel, 2015; Bošković & Rando, 2018).

One reason for interest here is that there some inheritance effects that are hard to explain in the standard way. A recent example comes from Yehudah et al. (2015) who found a statistically significant difference in the methylation of the FKBP5 gene between Holocaust survivors and their children: in the former it was high and in the latter it was low (Yehuda et al., 2015). This work did not, however, examine whether the methylation patterns were present in the newborn children or acquired as a result of their development and environment. Other examples come from plants and nematodes (Schmid et al., 2018; Skvortsova et al., 2018).

Over the past 20 of so years, it has been shown that standard gene transcription can be modified in several unorthodox ways that, before our understanding of molecular genetics, would have been considered impossible or at least highly improbable. These include DNA methylation, the activity of small RNA sequences that are present in gametes, histone-mediated chromosome changes and prion activity. Before looking at how these mechanisms work, it needs to be emphasised that, for a proposed evolutionary mechanism to be convincing in an evolutionary context, the changes that it elicits have to be stable over many generations. While some nonmutation mechanisms can have an immediate effect on the next generation, only RNAi and histone effects have yet satisfied a medium-term criterion; the experimental work is still too recent to recent to be certain of very long-term stability.

Methylation

The most studied mode of epigenetic change is through site-specific CpG methylation in the genome: this regulates gene expression, sometimes in an environmentally determined way (Dor & Cedar, 2018; Varriale, 2014). The phenomenon is particularly apparent in animals, but also occurs in other phyla, particularly plants (Rodrigues & Zilberman, 2015; Bartels et al., 2018). Methylation effects are thus obvious candidates for studies on unorthodox heritable change. The molecular basis of DNA methylation in eukaryotic DNA is that DNA methyltransferase (DNMT) enzymes convert cytosine bases in CpG islands to 5-methylcytosine, and this blocks or at least modifies transcription.

The best-known examples of methylation are the imprinted genes that are differentially methylated in human male and female germ cells; they mark the parental origin of the gene expression sequence (Monk, 2015). This methylation is normally initiated after any earlier methylation has been stripped out. In mouse and humans, ~100 genes are expressed in a parent-of-origin-specific manner due to partial or full silencing. This silencing involves DNA methylation that is resistant to later demethylation reprogramming. Such genes not only play a developmental role (e.g. in placental function), but also help regulate physiological behaviour (Peters, 2014).

Methylation has been studied most closely in mice and humans, and it soon became apparent that there are two obvious problems in environmentally directed DNA-methylation effects being incorporated into animal genomes. The first was that there has to be a means by which such changes could be incorporated into the distinct and essentially isolated populations of germ cells that differentiate very early on in animal development—no good mechanism for this has been suggested. The other is that, in the mouse at least, any methylation present is partially removed from primordial germ cells, with another wave of demethylation occurring as they start to differentiate into eggs and sperm (Hajkova & Lane, 2002; for review, see Skvortsova et al., 2018). It used to be thought that this demethylation was complete; it is now, however, clear that it is not entirely so, but the amount remaining in a subsequent generation is not precisely known. It is thus hard to see how any environmentally induced methylation acquired by an adult could be transferred to its eggs or sperm. Iqbal et al. (2015) investigated whether such inheritance could happen by inducing methylation in pregnant mice (so avoiding the first problem) and studying the resulting offspring and grandchildren. They did indeed observe novel methylation in the immediate offspring, but found that any effects were lost in the grandchild generation.

Although there is now a large body of information on transgenerational methylation and its effects (see Skvortsova et al., 2018), it is still hard to point to any significant effects that extend beyond three or four generations (Gilbert & Epel, 2015).

Noncoding RNA

A wide range of noncoding-RNAs are present in both male and female gametes (Conine et al., 2018; da Silveira et al., 2018; Zhang et al., 2019). This range includes micro-RNA (miRNA), gene-silencing, double-stranded RNA (siRNA), long, noncoding RNA (lncRNAs), and piRNA that interacts with piwi subfamily of Argonaute proteins to mediate transposon silencing. Changes in all of these various molecular types could in principle lead to heritable alterations in gene expression (Jodar et al., 2016) and hence could play a role in epigenetic inheritance (Yan, 2014). Such activity would be recognized by its non-Mendelian effects: because the RNAs are present in all gametes and would have some effect on all gene sequences to which they can bind, the presence of one or two SNPs in such sequence is unlikely to affect such binding.

Examples of such RNA inhibition (RNAi) can result in the modification of histones whose effect is to repress transcription in yeast and to initiate methylation in germline DNA. Such RNA silencing has been particularly studied in *C. elegans* (for review, see Woodhouse & Ashe, 2020), and it is worth noting that, of the 23 genes so far shown to exhibit non-protein-associated regulation of gene expression, eight are heritably stable within the germline and more have shown long-term stability. It is thus likely that future research will show that in other organisms too the many RNA types in gametes will heritably regulate gene expression during development in a non-Mendelian way.

Chromosome modification

Perhaps unexpectedly, chromatin organization can regulate gene expression. It has long been known that the folding of DNA brings into the vicinity of transcription complexes distant repressor and enhancer sequences. It is now clear that histone structure can also modify gene expression, while mutations in the various histones within the nucleosome complex can have non-Mendelian effects on gene transcription. Kundaje et al. (2015) have explored these and other epigenomic modifications in 11 human genomes making it clear that such modifications are more common and important than hitherto thought.

Prions

Perhaps the most unexpected candidate mechanism for transgenerational epigenetic inheritance is prion mediation. These are the dual-conformation proteins that lead to the formation of the amyloid strands that are normally associated with diseases, such as scrapie and Creutzfeld-Jakob disease. Prion-like effects have, however, been found in fungi and elsewhere (for review, see Manjrekar & Shah, 2020). Two early examples were identified in the budding yeast *Saccharomyces cerevisiae*: these were the nonmendelian inheritance of the ψ cytoplasmic suppressor and a mutation the *URE2* gene that results in an inability to metabolise ureidosuccinate. Prion-associated effects have now been found in other genes.

The formation of the prionic state seems to occur by mutation at low frequency and would be expected to be lost except in situations where the abnormal form conveys some advantage. The maintenance of prionic proteins that are useful suggests that the few cases where they can result in dreadful diseases are the exception rather than the rule. It is also worth noting that this mode of TEI is genetically stable and, once formed, its presence in an organism reflects positive selection.

Epigenetic inheritance is currently an active area of research, but, before any example of an apparent such change can be accepted, it will have to meet three tests. First, that change has to be maintained over the long term (at least tens of generations). Second, it has to be shown that the change does not depend on normal mutation, except insofar as the proposed mechanism provides a directed way of achieving this. Third, a mechanism needs to be demonstrated that explains how, in cases where the modifying molecules are not expressed ubiquitously, they can become part of the gamete. It will be interesting to see whether it will be possible to build on the current preliminary data to produce a convincing mechanism by which environment-induced changes initiated in parents can be maintained in the long term.

As of now, it does not seem likely that any such changes are of major importance: most of the anatomical and physiological phenotypic traits in both animals and plants derive from events taking place during embryogenesis and so are generally insulated from the effects of the environment. Moreover, there is little evidence of phenotypic change taking place rapidly, except under conditions of extreme selection. One would not, therefore, expect epigenetic inheritance to play a major role in mediating heritable change here. Nevertheless, it is clear that we do not yet understand every aspect of heritable change.

THE WIDER CONTEXT

There is now a very large body of work showing that SNPs and other direct mutational events are responsible for the very great majority of heritable changes, even if we do not always understand how the changes in genotype result in changes to the phenotype (an area discussed further in Chapter 18). However, as the evidence on transgenerational epigenetic inheritance has shown, they cannot fully explain every aspect of heritability. Moreover, we still do not know whether environmental responses can be heritable. Here, it should be said that the most improbable taxon in which this could happen is that of amniotes; this is because germ-cell production and post-fertilisation embryogenesis are generally independent of and insulated from events taking place in the rest of the organism

If there are unorthodox mechanisms of change in animals, they are most likely to be seen in organisms, such as fishes. This is partly because of the sheer number of eggs that are fertilised and partly because the complexity of their phenotype can produce a very wide range of variants. Fish eggs are exposed to the vagaries of the environment as soon as they are fertilised and are thus immediately exposed to external pressures. We still do not fully understand the rapid evolution of the cichlid varieties in Lake Victoria (Chapter 22), but certainly cannot yet exclude the possibility that unorthodox mechanisms of variation played some role there.

Perhaps the most likely clade in which heritable epigenetic change might occur rapidly are plants, and this is for several reasons. First, animals have only a single pair of gamete-producing organs set aside early, and these tend to be insulated from the effects of the environment. Plants, in contrast, can have many independent reproductive organs, particularly gymnosperm cones and angiosperm flowers, all of which that are directly exposed to the environment. This enable them to produce vast numbers of fertilised gametes with greater genetic diversity than would be expected in animals, even fishes. The sheer number of fertilised seeds that they make and distribute means that it is easier for plants than animals for advantageous new variants to flourish. If there are ways in which the environment can affect gamete formation and development in heritable ways, they are more likely to be displayed in plants than in other clades (Bošković & Rando, 2018).

Nevertheless, even if epigenetic inheritance does turn out to play a role in long-term genomic variation, it will just be one member of the rich repertoire of mutational events that provide a steady flow of variants that drive evolutionary change. It is important to bear in mind here that it is inappropriate to consider such change as deriving from a single event in a static background. Any individual mutational modification exists in the context of a constant flux of change across the genome and often in the environment in which its host organism lives. Throughout the population, breeding, mutation and genetic drift are altering every genomic site, albeit very, and very slowly. Mutation and the consequent phenotypic changes never stop.

KEY POINTS

- Phenotypic variation is present in every trait, tissue, and feature across a population and arises from genotypic variation. Most variants are not considered abnormal.

- Variation is particularly noticeable in parameters, such as size, pigmentation and susceptibility to disease.

- Much of our knowledge of mutation and its effects come from the study of human disease, together with genetic engineering and the use of mutagens in model organisms.

- There are occasional organisms whose features, such as pigmentation and size, are outside the expected distribution; these are sometimes called sports. Such features, when heritable, can be the basis for selective breeding of new varieties and strains.

- The primary phenotypic effect of a mutation derives from the effect that it has on the protein that it encodes. If this is to the structure of a protein, it may function abnormally so affecting network outputs and enzyme activity. If the mutation is in a control sequence, abnormal amounts of protein may be produced.

- Major insight into the genetic nature of variation has come from genetic sequencing and from mutations that lead to anatomically abnormal development.

- Genetic differences across a population underpin natural variation. These mainly derive from random mutation, random crossover during meiosis, random chromosome recombination following fertilisation, random breeding, and genetic drift.

- Rarer events leading to genetic variation include transposon movement, horizontal gene transfer and chromosome duplications and rearrangements.

- Transgenerational epigenetic inheritance (inheritance other than through normal mutational change) has long been thought impossible. Mechanisms by which this could happen have now been discovered and are being investigated,

- Such effects are most likely to be identified in plants as their germ cells are widely produced from leaves (gymnosperms) and flowers (angiosperms) that are exposed to the environment.

FURTHER READINGS

Barresi MJF, Gilbert SF (2019) *Developmental Biology* (12th ed). Oxford University Press.

Held LI jr. (2021) *Animal Anomalies*. Cambridge University Press.

Nei M (2014) *Mutation-Driven Evolution*. Oxford University Press.

WEBSITES

- Wikipedia entries on *Horizontal gene transfer, Methylation, Mutation, Temperature-dependent sex determination, Transgenerational epigenetic inheritance.*
- The Online Mendelian Inheritance in Man database of genetic disorders: www.omim.org
- The sequence ontology and the GO ontology are accessible at the OBO Foundry: www.obofoundry.org/ontology/so.html
- Translation of Mendel's original paper from 1866. http://www.esp.org/foundations/genetics/classical/gm-65.pdf

- Mutation and evolution: https://cshperspectives.cshlp.org/content/7/9/a018077.full https://www.nature.com/scitable/knowledge/library/mutations-are-the-raw-materials-of-evolution-17395346/ + associated links
- Methylation: https://www.nature.com/scitable/topicpage/the-role-of-methylation-in-gene-expression-1070/

18 Variation 2: Evolutionary Change

The last chapter considered the nature of variation seen in normal populations of multicellular organisms and the types of effect that mutations in proteins have on their phenotypes, almost all of which were deleterious and hence unlikely to be evolutionarily advantageous. This chapter starts by looking at examples of major anatomical apomorphies that characterise speciation and then considers how they might have been generated by mutations. The approach is to ask questions of the following form: given a pair of species that are known to have evolved from a common ancestor, what is the nature of the variation that would have been necessary in the descendants of that ancestor to achieve the resulting changes? A series of examples is given, with many focusing on limb morphology because change here is not only easy to recognise but often understood at the molecular level. This analysis shows that phenotypic novelties cannot in general be linked directly to the effect of a single mutation in a specific gene but are far more likely to derive from changes that affect the output of the signaling and process networks that drive developmental change.

The second part of the chapter examines what is known about the effect of mutation on the proteins that comprise these systems. Their components as a whole include signals, receptors, and transcription factors, together with the many proteins that comprise the process networks for growth, differentiation, apoptosis, and morphogenesis. The general conclusion is that simple mutations in these various components will indeed be able to effect complex phenotypic changes.

The chapter as a whole reaches two further conclusions. First, because all these mutations are in normal genes, they are readily heritable. Second, because networks have a single output that can be modulated by mutations in several proteins, heritable changes in network outputs can occur more rapidly than changes that depend on mutations that affect a single protein.

Questions of how mutations in DNA protein-coding and control sequences can lead to a change in an organism's anatomical and physiological phenotypes are still not fully answered. It is straightforward to see how small phenotypic variants in mutated proteins can be propagated across a population over time—the theory of population genetics describes how this happens and is discussed in Chapter 21. It is much harder to understand the origins of tissue diversity and to answer questions about how minor mutation in protein sequences produce substantial anatomical changes. In the 19th and early 20th centuries, the prevailing view was that large changes occasionally happened and could be inherited—evolutionary change by such large steps was called *saltation*.

One approach that was put forward by the evolutionary geneticist Richard Goldschmidt in the 1940s was that occasional *hopeful monsters* appeared as a result of mutation that mainly affected events in early embryos. Organisms carrying such major mutations could in principle be fertile, and the novel phenotype thus stood a low but finite chance of becoming part of the surviving population; it could, therefore, help found new species. This view was ridiculed at the time, particularly because

DOI: 10.1201/9780429346217-22

there was no clear way in which such a phenotype could occur or be propagated across a population.

This view was very different from the standard evolutionary perspective in the 1940s and 1950s, which was the Neo-Darwinist Modern Synthesis: this was based on analysing phenotypic traits that were assumed to be underpinned by a single gene or, for complex traits, a few genes, although there was then no clear idea of what a gene was or how it actually worked. Perhaps curiously, there seems to have been relatively little discussion then about the origins of tissue variation during embryogenesis. One reason was that most evolutionary geneticists knew very little about embryology and so didn't give adequate weight to the fact that anatomical change generally starts as embryological or larval change. One of the very few biologists of the premolecular age to integrate evolutionary change with embryology was C.H. Waddington, and some of his work is discussed in Chapter 20.

The question of how changes in DNA protein-coding and control sequences can lead to downstream changes in the anatomical phenotype is still not one that is easy to answer, even after more than a century since the discovery of mutation. This is because, as the chapter on evo-devo has demonstrated, single proteins do not themselves create cellular structures and rarely even structural features, although they may activate or play an important role in their developmental or physiological function. Indeed, the changes that result from mutations in a single protein tend mainly to be deleterious and hence unlikely to give any selective advantage (Chapter 17).

There is a further experimental problem: the pathway that leads from a mutational change to a phenotypic change is generally complicated as it traverses the levels of scale from a DNA change through a protein change to a network change through to changes in cells, tissues, and the organism's total phenotype (Appendix 1). Moreover, most of these events are initiated during early embryogenesis where practical mutation analysis is limited to deleterious genes as there are no assays for beneficial ones. Hence, it is in only in relatively few cases that we can understand the details of how mutation leads to successful phenotypic novelties. This chapter discusses a different and more helpful approach to the problem of understanding large-scale anatomical change.

SOME ORIGINS OF ANATOMICAL CHANGE

It is hard to pin down the origins of past change by studying current populations because rarely, if ever, are there appropriate variants. It is easier to analyse and compare the tissue differences between contemporary organisms that share a common ancestor. In principle this could be done on the basis of genomic differences; this is, however, impractical, partly because there is no case where the full set of genes that contribute to a particular anatomical phenotype is known and partly because, even if they were, it is not yet possible to interpret how the resulting proteins cooperate to produce that anatomical phenotype. Indeed, even if the DNA of some ancient ancestral organism could be sequenced and compared to that of a successor, the information on the differences would still be hard to interpret.

An alternative approach is, therefore, taken here. The first section of the chapter compares the anatomical features in pairs of related extant species and infers the nature of the variation needed for descendants of their last common ancestor to have achieved the different phenotypes of contemporary organisms (Bard, 2018). Although this is not a standard way of approaching variation, it has some qualitative similarities to the far more quantitative approach of computational phylogenetics (Chapter 8); this identifies the most likely common ancestor sequence that could have generated contemporary ones, and the way in which this happened. Indeed, this approach has already been used to infer some of the anatomical detail of *Urbilateria* (Chapter 12).

This section thus provides an evolutionary analysis of some examples of closely related but different organisms that are different in one or more obvious ways (for further examples, see Held, 2014; for summary, see Table 18.1). Several consider vertebrate limb differences, and this is for two reasons; first, anatomical changes are easy to identify and, second, the molecular basis of limb development is now relatively well understood (McQueen & Towers, 2020). The examples as a whole clearly suggest that many variants in limbs and other anatomical features can be

TABLE 18.1 SOME CLADE-DEFINING FEATURES

Phenomenon	Likely basis	Network involvement
Geckos are much smaller than skinks	Growth ceases early	Changes in timing networks
Axolotls do not metamorphose	Inadequate thyroxin production	
Freshwater sticklebacks lose their spines	Loss of appropriate Ptx1 enhancer	Spine production inhibited
The number of neck vertebrae	7 in mammals, variable in birds	Modulation in Hox patterning and growth
Beak size in birds	Size and strength controlled by BMP4 signaling and calmodulin	Growth and bone deposition
Vertebrates and fish have complex hair and scale patterns	Reaction-diffusion kinetics	Patterning and pigmentation
Giraffe neck elongation	Growth in neck vertebrae	Patterning and growth
Melanosome pigmentation in skin	Modulations in patterning	Pigmentation networks
Limb phenomena		
The mouse *doublefoot* mutant has ~8 digits	Mutation in patterning network	Indian hedgehog-induced patterning
Marsupials have premature fore-limb development		Timing/patterning
Cetaceans have miniscule, internalized hind-limb rudiments	Developmental activity ceases very early	Timing/patterning
Leaping tetrapods have elongated feet	Hoxa-11 activity maintained	Timing/patterning
Adult non-swimming birds lose their interdigital web	FGF activity maintained with apoptosis then activated	Timing/patterning Apoptosis
Digit number can increase	Elephants, moles, and pandas have a sesamoid digit	Patterning/growth
Digit number can decrease Birds have 3-wing and 4-foot digits	Repatterning of fore- and hind-limb plates	Patterning/apoptosis
Flying animals have very different digit lengths		Patterning/growth

linked back to heritable changes in the output of the complex networks and pathways that drive embryonic development, particularly those that affect patterning, timing, and growth. This section is, therefore, followed by one that discusses how mutation affects the outputs of these networks.

Organism size

The most obvious difference between related species is in size. The dwarf gecko and the Solomon Islands skink are two lizards that are, apart from pigmentation differences, very similar in morphology (Fig. 18.1). They do, however, have one other important difference in their phenotypes: the adult gecko is ~16 mm long while the adult skink is ~800 mm long (Fleming et al., 2009; Hagen et al., 2012), a difference that reflects 5–6 extra cell divisions averaged across the organism. The cause of this difference is clearly that the gecko stops growing relatively soon after it has developed its final form, while the skink keeps growing for a long time after development has ceased. The likely origin of the change is thus associated with differences in the settings of some internal regulatory clock that regulates systemic proliferation.

Timing changes

A change in the time at which a developmental event occurs in a specific organism is known as *heterochrony* and is a common evolutionary mechanism for creating

Figure 18.1 **Growth difference in reptiles.**
(a): *Sphaerodactylus ariasae*, the dwarf gecko, is 16 mm long, while the Solomon island skink (b): is 830 mm long. The only obvious differences are in pigmentation and size.

(a Copyright S. Blair Hedges. Printed with permission.
b Courtesy Dr. Tim Vickers.)

variants. Paedogenesis reflects a delay of development, sometimes to the extent that juvenile characters are maintained in the adult (neoteny). Precocious development speeds up embryogenesis and an example is the premature development of fore limbs and mouths in marsupial embryos (below, Chapter 16). Work on heterochrony dates back almost a century to when Rowntree et al. (1935) injected thymus extract into pregnant mice and found that the offspring were born with temporally advanced features that included open eyes and ears, with females having a developed vagina and males showing premature testis descent. This example thus demonstrated that changes in the timing of systemic hormone levels could accelerate development (for other examples, see below and Websites).

Perhaps the best-known example of neoteny is the axolotl. Most amphibians develop as tadpoles and, at some point in development, their thyroid is activated and thyroxin is secreted. This, in turn, activates the process of metamorphosis and the larva becomes an adult. Axolotls have lost the ability to produce this thyroxin spike and can only undergo metamorphosis if their thyroid is externally stimulated through, for example, ingesting iodine. In its absence, they grow to become what are essentially enormous larvae up to 300 mm long.

Limb variation

Variation in the development of the vertebrate limb is particularly important as an evolutionary model system. What follows are comparisons of pairs of very similar organisms with different aspects of limb morphology.

Growth changes

Wings are a characteristic of birds, bats, and pterosaurs but their digit and phalange proportions are different. In pterosaurs, only the fifth digit extends; in bats, the four fingers extend; in birds, digits 2 and 3 are lengthened (Fig. 18.2). It is clear that the patterning and timing networks that control the local proliferation networks for phalangeal growth are operating differentially in both species-specific and digit-specific ways to achieve the three modes of wing anatomy that are capable of flight.

Timing variants in vertebrate limb development

Marsupial forelimbs. In most tetrapods, the hindlimb buds start to form very soon after forelimb-bud initiation; development of both is at roughly the same rate and completed by the time of birth, so enabling many newborn animals to walk. Marsupials, such as wallabies and kangaroos are different: they are born at an early developmental stage, having the hind limbs expected for their age, but with exceptionally well-developed clawed fore-limbs that they use to migrate from the exit of the vagina to the pouch (Chapter 16). There is, thus, a heterochronic difference in the formation of marsupial fore-limbs whose early development is much faster than that of their hind-limbs (Keyte & Smith, 2010; Chew et al., 2014). It is clear that this change in the fore-limb timing networks reflects both a major increase in the activity of the growth networks and premature differentiation in the fore-limb tissues.

Hind-limb diminution in cetaceans: Whales and dolphins lack visible rear limbs, each having only a rudimentary pelvis and reduced femurs that are internalised (Fig. 16.9). At 24 days of development, the early dolphin hind-limb bud is much like its fore-limb bud and typical of those seen in other limbed vertebrates (Fig. 16.9; Bejder & Hall, 2002). At 48 days, however, when digits are forming on the developing

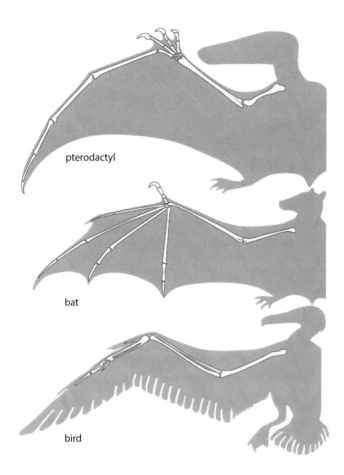

Figure 18.2 **Wing skeletons.** The skeletons of a pterodactyl, bat and bird matched in size, so demonstrating how each, in its own way, extends its digits to enable a wing to form.

John Romanes who (died 1894).

fore-limb, the hind-limb bud has virtually disappeared; its development has stopped and growth of such tissues that have formed (primitive hip bones and femur) is severely diminished so that only their rudiments are left in the adult. Given that whales evolved from normal four-limbed ungulates (Bajpai et al., 2009), the variant that underpinned limb loss was clearly a change in the timing network that led to the premature cessation of its development and growth.

Patterning changes in limb bones

Most limbed vertebrates have in their hind-limbs a proximal femur and a more distal tibia-fibula bone pair that articulates with the talus (or astragalus) and calcaneum which comprise the ankle and heel. The exceptions are amphibians (except for salamanders), the tarsiers (monkeys found only in SE Asia) and galagoes or bushbabies (small African nocturnal monkeys) which all show a similar homoplasy: they have a much-elongated foot with talus and calcaneum elongated to resemble a duplicated tibia-fibula pair (Fig. 18.3; Held, 2014). This extended link in the lower part of their hind-limbs gives them extra leverage in their lift-off, with galagoes, in particular, being renowned for their jumping and bounding abilities (Aerts, 1998). It is worth noting that a wide range of non-jumping mammals (e.g. horses) increase their speed by lengthening their legs; they do this by extending their carpals and tarsals, while ostriches have very long tarso-metarsus bones for this purpose.

The origins of this apparent duplication have only been studied in the *Xenopus* toad: the mechanism by which the pair of elongated bones form derives from a variant in the way that Hox coding normally patterns limbs. In most embryos that have been examined, the Hox expression patterns are approximately the same in fore- and hind-limbs. Blanco et al. (1998) compared the Hoxa-11 expression domains in the developing fore- and hind-limbs of *Xenopus* and noted three differences. The gene is expressed for a longer time, over a larger region and closer to the distal tip in the hind- than in the fore-limb. The view of the authors is that the extra time of expression of Hoxa-11 in the region that would normally become the ankle is the major reason why the bones that would normally become the calcaneum and talus acquire

Figure 18.3 **A drawing of a frog skeleton showing (arrows) the apparently duplicated tibia-fibula pair that have evolved by elongation of two of the ankle bones.**

(Courtesy of Lewis Held. From Held (2014) *How the Snake Lost Its Legs*. Cambridge University Press.)

morphologies similar to those of the tibia and fibula. This anatomical homoplasy is thus due to a change in the timing networks that regulate limb patterning.

Webbed feet

While land-living birds, and indeed all limbed vertebrates, have distinct toes, most birds and mammals that swim (e.g. ducks, gulls, and swans, together with beavers, otters, water opossums, and the platypus) have webbed feet. This latter state actually reflects the way in which autopods (i.e., hands and feet) develop from the early foot- or hand-plate of ectoderm-covered mesenchyme within which the digits form. Webless birds and mammals require an additional step for their development: the epithelium and mesenchyme between the still-forming digits of the hind-limb rudiment are lost through activation of the apoptosis network in the interdigital regions.

It is now known how this loss of tissue happens: in the webless chick, FGF2 signaling in the foot leads to the expression of the Msx2 transcription factor, and the activation of BMP signaling in the interdigital region. These jointly activate the apoptosis network in the mesenchyme and ectoderm between the digits (Gañan et al., 1998), as a result of which these tissues are lost. In the webbed duck foot, FGF2 signaling does not take place, Msx2 is not expressed and apoptosis does not occur, even though BMP is present. All early land vertebrates had distinct digits, indicating that apoptosis activation is part of normal limb development. It is clear that the variation required for webbed feet to form was merely the turning off of a local FGF signaling network so blocking apoptosis.

Variation in digit number

Although most land vertebrates have five digits, birds have three (fore limb) or four (hind limb). Experimentation has shown that digit number is primarily controlled by a gradient of the sonic hedgehog signaling protein across the early hand or foot plate. This is established by a small region at the edge of the handplate known as the zone of polarizing activity. This gradient activates the formation of digit modules (complete digits with nails and joints) in ways only partially understood, although limb-bud width is certainly one component (Towers et al., 2008). The simplest evidence for this is the large number of digits seen in the mouse *doublefoot* mutant whose plate breadth is substantially increased (see below for more detail; Yang et al., 1998; Babbs et al., 2008).

Ectopic digits

There is another way of forming an extra digit, and this is through the increased growth of a sesamoid bone in the normal autopod. Such bones are usually small and do not link to other bones through joints but are formed within tendons. There are many such sesamoid bones in the vertebrate skeleton and several in the limbs, with the best-known being the patella (kneecap). This bone has an enigmatic evolutionary history: it first appeared in amphibia, was absent in most early mammals and dinosaurs then reappeared in birds. Today, it is found in most amphibians, birds, and monotreme and placental mammals, but not generally in marsupial mammals or reptiles (Samuels et al., 2017). Although it is possible that it is frequently lost but

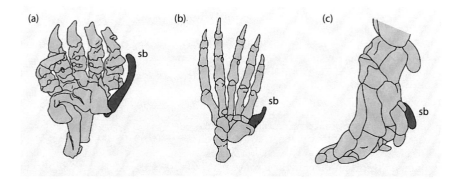

Figure 18.4 Ectopic digits in the forelimbs of (a) a mole, (b) a panda, and (c) an elephant. (sb: sesamoid bone.)

(a From Mitgutsch C, Richardson MK, Jiménez R et al. (2012) Biol. Lett. 8:74–77. With permission from The Royal Society.
c From Miller CE, Basu C, Fritsch T et al. (2008) J. R. Soc. Interface 5:465–475. With permission from The Royal Society.)

has kept re-evolving, it is far more likely that its presence reflects mutations in the signaling system that drives patella morphogenesis

Ectopic digits, particularly an extra (sixth) digit, have independently evolved from sesamoid bones several times in specific members of very different families (Fig. 18.4). Particularly well-known examples are those in pandas that use them to shred bamboo leaves, in moles to improve digging, and in elephants where they presumably help spread weight. Although these ectopic digits have evolved in only one taxon within three very different clades, they share three common features. First, they occur in both the fore and hind autopods, irrespective of whether they have a function in the latter region. Second, they all derive from enhanced growth that extends the existing sesamoid bone, which is adjacent to the first digit. Third, they each lack joints and nails or claws—none is a standard digit module although its external appearance can be superficially normal apart from the lack of a claw (Davis, 1964; Hutchinson et al., 2011; Mitgutsch et al., 2012).

The sesamoid bones in the autopods of most mammals cease growth when they are very small. In the three cases with sesamoid thumbs, the "thumb" clearly derives from the maintenance of the proliferation network that enlarges and extends the early sesamoid bone. It is worth noting that proliferation is still regulated as growth is asymmetric: the proliferation rate (or possibly osteoblast deposition) is preferentially proximal-distal: the bone is elongated rather than spherical. Perhaps the key point here is that the sixth digit does not derive from sophisticated repatterning of the autopod but from the enlargement of an existing feature.

There are also at least two examples of more complex ectopic digits. The first is in the aye-aye, a Madagascan lemur, which has a distinct pseudothumb on its fore-limb consisting of a bony part that is an expanded radial sesamoid and a dense cartilaginous extension (Hartstone-Rose et al., 2019). This pseudothumb receives attachments from muscles that facilitate abduction, adduction, and opposition, which together facilitate dexterity. Harstone-Rose et al. (2019) do not report on whether the aye-aye has a sixth toe. The second example is the frog *Babina subaspera*. Normal frogs have four and five digits on their fore- and hind-limbs, but digit one of its fore-limb is a compound digit that includes a small, cartilaginous thumb element with two jointed parts. In *Babina subaspera*, this pseudothumb extends and becomes distinct, even though it remains cartilaginous (Tokita & Iwai, 2010).

Other evolutionary variants

Stickleback spines

An example in which the basic molecular biology and the evolutionary timescale are both understood comes from anatomical changes in the stickleback fish. Marine sticklebacks have pronounced pelvic spines for protection; freshwater sticklebacks lack these spines, reflecting a variant occasionally present in marine sticklebacks (Cresko et al., 2004). This variant seems to have become a defining feature of freshwater sticklebacks several times, some as recently as 12 Kya.

Molecular analysis has shown that pelvic-spine formation results from local Pitx1 activity; the wide expression of this transcription factor in the fish is controlled by four enhancer regions. In the freshwater sticklebacks, the enhancer region for the pelvic spines has been lost and the gene is not expressed there (Chan et al., 2010; Schluter & Conte, 2009). As a result, the downstream module of signaling and process networks responsible for spine production is not activated. Spine production thus seems to be under the control of a molecular switch that turns on or off the appropriate genetic module for making it.

The number of neck vertebrae

The number of cervical vertebrae in extant tetrapods is highly variable: amphibians have very few, while diapsids, which include birds, crocodiles, and dinosaurs, have numbers ranging from about ten to forty. Synapsids, particularly mammals, have (with very few exceptions) only seven cervical vertebrae; in this respect, there is no difference between mice, giraffes, and whales. The last common ancestor of the synapsids and diapsids was probably a mid-Carboniferous (~315 Mya) sauropsid reptile, with examples, such as *Hylonomus* having had six or seven cervical vertebrae. It thus seems that an early post-separation event was to fix this number in the former and to make it easy to modify in the latter.

The molecular basis for regulating vertebral number is unclear, but Burke et al. (1995) showed that vertebral specification in both groups is controlled by Hox coding: for the cervical region, Hoxa4, Hoxb4, Hoxc4, and Hoxc5 in the mouse (7 vertebrae), together with Hoxd4 in the chick (14 vertebrae) all need to be expressed. The way in which this number is determined remains unknown, but it is clear that the downstream result is the production of a number of vertebral modules, the morphological details of each being determined by its position.

Beak size

The Galapagos finches are the group of bird species whose individual adaptations to particular habitats were a key stimulus for Darwin's thinking about evolution. The birds fall into two broad groups: the ground finches have broad, strong beaks for tearing at cactus roots and eating insect larvae, while the cactus finches have narrow beaks that enable them to punch holes in cactus leaves and eat the pulp (Fig. 4.3). It is now known that beak morphology is controlled by two factors: BMP4 signaling and the levels of calmodulin expression, a protein that regulates calcium concentration within cells (Abzhanov et al., 2006). The strength of BMP4 signaling controls beak breadth and depth, while calmodulin upregulation facilitates beak elongation. It is clear that variation in upstream patterning networks determines the amounts of BMP4 and calmodulin expression in the anterior head and these, in turn, control the growth pathways responsible for beak size (Mallarino et al., 2011).

Vertebrate skin patterns

There are obvious patterning differences in the hair pigmentation of the different taxa of cats, zebras, and giraffes, and in the pigmentation patterns of fish scales. It has long been suspected that reaction–diffusion kinetics underlies these diverse patterns (Bard, 1977, 1981). Such kinetics were invented by the great mathematician, Alan Turing (1952) who, in his only foray into biology, asked the following question: can a group of identical cells with identical biochemical networks linked to allow the diffusion of small molecules (Turing called these morphogens) produce a nonuniform concentration distribution? The intuitive answer is "no" because diffusion would iron out any spatial differences in concentration that might arise.

Turing showed that this intuition was only correct in the case of a single morphogen. If, however, there are two morphogens whose synthesis depends on one another as a result of the internal activity of a protein network, then autocatalytic effects within the network can create spatial patterning with a wavelength that depends on the details of the kinetics. The reason for this is that, under a limited range of parameter values, the uniform concentration distribution becomes unstable and changes to a periodic pattern (this was an early example of chaos theory). This pattern is dynamic because its maintenance requires a continuous source of chemical energy to sustain the concentrations against the dispersive effects of diffusion. Once such a chemical pattern has formed within an array of cells, it is then only a small step to suggest that the peaks and troughs of morphogen concentration represent switch and nonswitch regions for directing pigmentation production.

Turing's paper has excited continuing interest since its publication, and it was shown some time ago that reaction-diffusion kinetics operating over 2-D cellular arrays can generate a wide repertoire of patterns. These are typically arrays of spots but, by varying the kinetics and the boundaries of the patterning domain, the kinetics can can generate stripes that can even bifurcate, with the spatial scale particularly determined by the diffusion constants. Together, these patterns particularly model

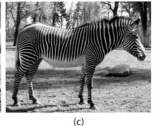

(a)   (b)   (c)

Figure 18.5 **Zebra species.** (a): *E. quagga burchelli* with about 26 stripes. (b): *E. zebra* with about 50 stripes. (c): *E. grevyi* with about 75 stripes.

(a Courtesy of "Gusjr." Published under a CC Attribution Generic 2.0 License.
b Courtesy of Yathin S Krishnappa. Published under a CC Attribution-Share Alike 4.0 International License.
c Courtesy of Thivier. Published under a CC Attribution-Share Alike 3.0 Unported License.)

those seen in the hair of zebras and other mammals, mollusc shells and fish scales (e.g. Bard, 1981; Meinhardt, 1984; Barrio et al., 1999) and may also play a role elsewhere in development. Moreover, because the parameters that control the patterns reflect interaction constants, they are subject to mutation. A particular set of reaction-diffusion kinetics may thus be able to produce a wide range of patterns.

Of particular interest is the effect of heterochrony on the patterns that these kinetics would generate if used during embryogenesis. If the kinetics have a particular wavelength, then the earlier they form in the embryo, the fewer will be the patterning elements that will fit onto the embryo and the larger will each end up. This is because their size reflects their period of growth: the later they form, the greater will be the number of elements that can be fitted onto the early ectoderm and the smaller each element will eventually become because there will be less time for growth after they are laid down.

This effect has been examined for the three different species of zebra: *Equus quagga burchelli* has ~26 stripes and a complicated pattern, *E. zebra* which has ~40 medio-lateral stripes and *E. grevyi* which has ~70 such stripes (Fig. 18.5). Morphological and size details of the early development of horse embryos were measured by Ewart (1897); it is straightforward to show that, if ~400 µm (~20 cells) medio-lateral stripes are spread across 21-, 25-, and 35-day equine embryos, subsequent growth will generate number and geometry of the different body-striping patterns seen in the adults of the three well-known zebra species (Fig. 18.6). Striping variation in zebras can thus be explained on the basis of heterochronic changes in the timing of the activation of reaction-diffusion kinetics.

One further feature of the kinetics that meshes with our knowledge of vertebrate coat patterns is that, while the nature of the patterns is a characteristic of a species, the details are animal-specific: no two zebras have exactly the same striping pattern, and even the two sides of a single animal are different. The solutions to Turing's equations show similar randomness in the details of the patterns that they generate because they depend on instability. Nevertheless, for all of the theoretical importance of Turing's work and the continuous theoretical work that it continues to stimulate (e.g. the work of Bullara & De Decker, 2015) on zebrafish scale patterns, there is still

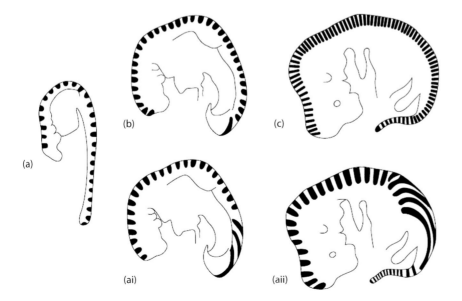

(a)   (b)   (c)

(ai)   (aii)

Figure 18.6 **Likely origins of zebra striping. (a,b,c)** 3-, 3.5- and 5-week horse embryos onto which stripes have been drawn with a 200 µm spacing—more stripes fit onto older embryos. (ai, aii): 3.5- and 5-week embryos showing how stripes laid down at 3 weeks would be expected to look as the embryo enlarges and the posterior region extends.

(From Bard JBL (1977) *J. Zool.* 183:527–539. With permission from John Wiley & Sons.)

no unambiguous chemical evidence that it actually operates in biological systems. After almost half a century of research, however, there are still no alternative models for this class of patterns.

THE EFFECT OF MUTATION ON SIGNALING AND NETWORK SYSTEMS

The last section has shown that many of the complex anatomical changes that occur as species diverge are likely to have derived from mutations affecting the various signaling, patterning, and process networks whose activity drives normal tissue formation in embryos. These include those for patterning, growth, apoptosis, morphogenesis, differentiation, and timing during development, with others that regulate physiological and biochemical processes. The molecular apparatus for such systems involves signaling pathways (signals, receptors, transduction pathways, and transcription factors) together with the resulting new proteins and the process networks e.g. (proliferation, apoptosis, differentiation, and morphogenesis) that they activate (e.g. Figs. 1.1, 8.6, and 25.5; Appendices 1 and 7). This section considers the effect of mutation on these various components.

Signals, receptors, and transcription factors

The sequence of events that leads to phenotypic change is given in Fig. 18.7. Phenotypic change is usually initiated by a signal binding to a receptor that leads to a conformational change in its intracellular component, which, in turn, activates the signal-transduction pathway that then activates the transcription complex. The resulting gene expression can lead to the activation of a process network. The various signaling pathways are each used repeatedly in many different contexts and the effects of mutation on them are independent of the networks that they activate. Mutations in signals and receptors thus act as switches that activate or repress downstream activity and they are, for mutation purposes, typical of the classical Mendelian genes discussed in the last chapter. The situation is not very different in signal transduction pathways, although they may be more complex.

Mutation can, however, modulate the domain over which the signal activation system acts and an interesting example here is the mouse semidominant *Doublefoot* (*Dbf-/+*) mutant (*Dbf-/-* is lethal) mentioned above. The phenotype of this mutant is up to nine digits, all of which are normal (Fig. 18.8; Crick et al., 2003; Malik, 2013). The reason for all these digits is that the mutant limb bud is wider than usual. In the normal limb, the sonic hedgehog gene (*Shh*) is expressed in a discrete region (the ZPA, part of the posterior region of the limb bud); diffusion of SHH protein across the limb bud leads to future digit formation. In the *Dbf/+* mutant, *Indian hedgehog* (*Ihh*) is ectopically expressed along the whole of the growing edge of limb-bud mesenchyme, so broadening the digit-specifying region. The result is an increased number of digits (Yang et al., 1998; Babbs et al., 2008). One importance of the *Dbf* heterozygote is that, because the extra digit modules have normal joints, nails and histology (Malik, 2013),

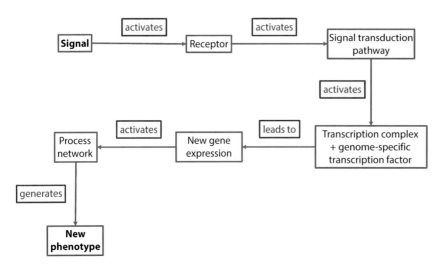

Figure 18.7 The pathway of events by which some a protein signal leads to a change in phenotype. All aspects of this pathways can be altered by mutational tinkering.

(a)

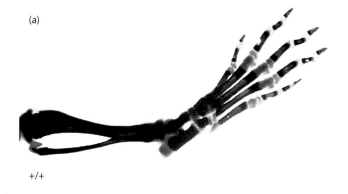

+/+

(b)

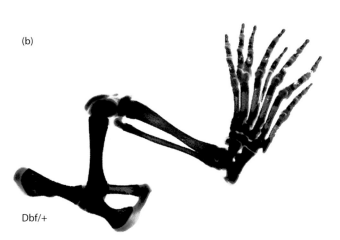

Dbf/+

Figure 18.8 **The *doublefoot* mutant.**
Preparations of the hind-limbs of (a) an E14.5
control and (b) a heterozygote *doublefoot*
mouse mutant at a stage when the early
cartilage of the long bones is being replaced
by bone. (Cartilage is stained blue and bone
red.)

(From Crick et al. (2003) Developmental
mechanisms underlying polydactyly in the
mouse mutant Doublefoot. *J Anat*; 202:21–6.
With permission from Wiley & Sons.)

it helps to explain the range of digit numbers in early land tetrapods (Table 6.2) and perhaps how the basic number came to be five (Chapter 6).

Transcription factors are molecular switches whose activity turns on gene expression and, in this context, can be treated as any other gene although they are particularly complex to analyse genetically. This is because the details of their expression and activity not only depends on signaling but may also be affected by enhancer/repressor activity and by which intron sequences they include, both of which are subject to mutation (Jaruga et al., 2016; Kim et al., 2018; Schoenfelder & Fraser, 2019). As a result, their roles also depend on location and hence on lineage: many transcription factors can have different functions in different locations and at different times, as the example of the Hox A and D genes has shown (Chapter 8).

Much of the evidence on the effects of mutations on developmental pathways and networks comes from the study of mutations that cause human genetic disease and animal variants. The disease effects usually depend on whether mutations are dominant, recessive, or semidominant; in this last case, there is a heterozygote-specific or dosage-dependent phenotype. A few examples are summarised in Table 18.2 and many more will be found in the OMIM dataset (Chapter 17). While most mutations to the components of the signaling/transcription-factor pathways are deleterious, it is not difficult to see that other mutations might modulate the phenotype in a way that could prove advantageous.

Networks

The activity of transcription factors results in the expression of new proteins, many of which have structural and biochemical functions that are well understood. A further and particularly important role of such new proteins in the contexts of development and evolution, however, is to activate the complex protein networks that, as mentioned above, drive developmental change (for examples, see Figs. 1.1. 8.7, 25.5, A1.1, A1.2, A7.2).

TABLE 18.2 GENETIC DISORDERS CAUSED BY MUTATIONS TO THE SIGNALING AND PROCESS PATHWAYS

Signals

Homozygous mutations in the mouse FGF18 lead to major bone abnormalities as the ligand cannot bind to its receptor (Liu et al., 2007).

Receptors

Mice homozygous for BMPR-II receptor mutations die at gastrulation (Beppu et al., 2000).

Humans heterozygous for FGFR2 mutations show premature fusion of skull suture (craniosynostosis; Morriss-Kay & Wilkie, 2005; Chapter 17).

Humans heterozygous for FGFR3 mutations have achondroplasia where limb cartilage fails to form bone and dwarfism results (Ornitz & Legeai-Mallet, 2017).

Signaling pathways

Cancer normally results from mutations that block restraints on proliferation. Affected genes include *ras*, *raf* and the various *MAP kinases*, which are involved in the ERK signal transduction pathway (Dorard et al., 2017).

Mutations in *ras* can also lead to developmental disorders, such as type 1 neurofibromatosis; here abnormal neural-crest-cell differentiation leads to neuronal tumours (Schubbert et al., 2007).

Transcription factors

Pax6-/- homozygote mice lack eyes, although the Pax6 -/+ mouse has small but normal eyes (hence the *small-eye* name).

WT1-/- mice fail to develop kidneys and die, while WT1-/- humans develop WT1 Wilms' tumours associated with the uncontrolled proliferation of kidney peripheral cells that results from a secondary somatic mutation (for review, see Hastie, 2017).

Mutations in the family of *SOX* transcription-factor genes result in skeletal abnormalities (Lefebvre, 2019).

The effect of mutational tinkering with the protein components of process networks is much more complicated than with other proteins because the former work cooperatively rather than sequentially, as the diagrams illustrate. In practice, every protein in a process network, and there may be as many as 100, is subject to such tinkering and there will be a considerable distribution of protein phenotypes within a particular network across a population. There is, however, no reason to doubt that the very great majority of these networks will still produce either the same output or one whose quantitative value is within the fairly narrow limits that define normal variation (such as the extent of the growth in the various features of a human face). Networks are clearly buffered (Nochomovitz & Li, 2006).

However, as minor mutational changes in such networks accumulate within a population over time, it becomes harder for the internal kinetics of the system to buffer their increasing effects. Eventually, a mutated protein, making what might seem an innocuous change to the kinetics of the network, will lead to a significant effect on its output. Should this happen, there are several possible outcomes: first, the network might fail, and this would probably be lethal; second, the network output could be altered to give a phenotype that was outside the normal range; third, an alternative pathway could be activated within a complex network that would generate a qualitatively abnormal phenotype. The latter two options would appear as sports and in principle generate potentially heritable variants in the phenotype. Networks can thus be seen to be amplifiers of variation (Wilkins, 2007).

It is important to distinguish between those networks whose output reflects an on-off state, such as those for differentiation and apoptosis, and those where the response is graded, such those for timing, proliferation, and morphogenetic activity. The differentiation switch seems to be particularly sensitive to major mutation: this is because any mutation that had the effect of blocking or changing a differentiation network other than those affecting minor features, such as pigmentation would probably be lethal for functional reasons. An example of this is the runx2 protein, a key transcription factor involved in bone ossification: if one copy is mutated, humans show cleidocranial dysplasia, if both copies are mutated, ossification fails and the embryo dies (Jaruga et al., 2016; Takarada et al., 2013). Of interest here is the suggestion that secondary mutation in the human *RUNX2* gene or in its regulatory system is one of the reasons why the morphology of bones, such as the clavicle differ between anatomically modern humans and Neanderthals (Green et al., 2010). The situation is similar for apoptosis where mutations that result in abnormal local cell death can change development (e.g. the loss of webbed feet mentioned above), result in abnormal development in embryos (Baehrecke, 2000) or lead to cancers in adults where cells that would be expected to die for one or another reason do not.

In contrast, mutations in the timing, proliferation, and morphogenetic networks can change their outputs qualitatively. Changes in timing networks can result in heterochrony, albeit that knowledge of the details of most developmental clocks is limited, with that underpinning somite formation being best understood (Riedel-Kruse et al., 2007). Much more is known about the networks that govern the heartbeat and circadian rhythms in adults (Noble, 2011; Anderson et al., 2013).

Perhaps the most common of all evolutionary changes, however, are those that affect growth; these mainly derive from the properties of both the signaling and the proliferation pathways and are the most obvious markers of local variation in a population. There is direct evidence that the rates of proliferation can be under both systemic and tissue control. Thus, Hornbruch and Wolpert (1970) found that local mitotic rates within the developing chick limb can vary over a factor of five, with specific rates probably being driven by differential signaling from the apical ectodermal ridge at the distal edge of the limb. It thus seems likely that the proliferation networks are context-tunable, and it is the ability that probably drives a great deal of evolutionary change.

As to the networks that control morphogenesis, mutations that change such properties as movement, adhesion, folding, and tube formation seem to be rare. The best-known is probably spina bifida: in this disease, which shows strong statistical heritability (Jorde et al., 1983), the neural tube either fails to close leaving an open spinal cord or closes late so that the developing vertebrae fail to fully enclose the cord. Although the molecular basis of this heritable change is not yet fully understood, the mutated phenotype is associated with mutations in planar-cell polarity genes that are involved in the control of cell movement within epithelial sheets (Tissir & Goffinet, 2013).

The reader will note that this discussion of networks is mainly based on abnormal phenotypes rather than on how their internal dynamics are affected by mutation. This is because so little is known about how these networks operate. Elucidating this remains a major challenge to biomathematicians. A major difficulty here is that we do not know how much of network complexity reflects alternative function pathways and how much it has the role of ensuring robustness to mutational change, a natural consequence of selection. It certainly seems likely that, in the early days of animal evolution, the process networks were both simpler and less robust than now. They were thus capable of creating greater variety under mutation, albeit at the risk of greater functional fragility.

DEVELOPMENTAL CONSTRAINTS ON VARIATION

There is a joke university exam question from the 1970s, a time when nothing was known of the molecular basis of embryogenesis, that built on the doggerel of Lewis Carroll in *Alice through the looking glass.*

> "The time has come," the Walrus said, "to talk of many things:
> Of shoes and ships and sealing-wax, of cabbages and kings
> And why the sea is boiling hot, and whether pigs have wings."
> *Question*: Recode the pig's DNA so that it does have wings.

This joke implicitly poses a further, more serious question: are there limits on the extent to which mutation can produce novel phenotypes? This question is not one that can be easily answered, but it is immediately clear that the possibilities for phenotypic variation here and more generally are intrinsically limited by various types of constraint. There are physical constraints: is it possible to produce a wing structure capable of producing enough lift to raise the weight of the pig given reasonable estimates of muscle strength? There are developmental constraints: variation can only build on what is already part of the phenotype. Here, one has to ask whether it is possible to duplicate the basic patterning of the region of the pig's future anterior limb system so that the lateral mesenchyme can produce both a superior wing like that of a bird and the original inferior limb? Then there are evolutionary, exaptation constraints: what might be the selective use of a small wing that was physically inadequate to lift the animal?

The most obvious constraints are those provided by the laws of physics. Sea invertebrates, such as arthropods and particularly cephalopod molluscs can grow almost indefinitely because their mass is supported by water; squid can thus be >10 m long and weigh many kilograms; while the blue whale is, at ~200 tons, the largest organism that seems ever to have existed. Land invertebrates are, however, limited in size by the strength of their chitin exoskeleton and by the respiratory efficiency of tracheal tubes that provide oxygen. Goliath beetles are ~10 cm long and their larvae weigh ~100 gm, while the Chaco golden knee spider has a span is 26 cm and weighs about 150 gm. The size limits on land vertebrates are, however, much less because of the strength of their internal bone skeleton: *Supersaurus*, a Jurassic sauropod, was ~30 m long and weighed 35–40 tons.

In the context of a flying pig, the largest flying birds, such as the albatross weigh ~8 kg but need a wingspan of >3 m. A better model is the pterosaur (Chapter 15): *Quetzalcoatlus northropi* weighed as much as 250 kg but had a head-to-tail length only a little less than that of a giraffe and a wingspan of >10 m (Fig. 15.10 a,b; Witton, 2013). If a typical pig weighs ~150 kg, flight would require the evolution of major modifications to its body and the presence of very large wings to produce the necessary lift. The original question is not yet answerable — even in principle.

The local need for oxygen in ATP production can also impose chemical limitations on size: the largest unvascularised tissues in vertebrates are the lens and the cornea; both are sustained through the pumping activity of the adjacent ciliary body and this is only likely to be possible because both are essentially passive structures with minimal energy requirements. The size of land invertebrates is limited not only by exoskeletal robustness, but by the capabilities of tracheal tubes to provide oxygen through diffusion. It is probably for this reason that wheels have not evolved in multicellular organisms: their energy requirements are too great to be met through diffusion, even when aided by a protein pump, while wheels cannot be linked to a blood supply. The only known example of a rapid, freely rotating system is the flagella of bacteria: these are sustained by hydrogen or sodium ion gradients that drive the necessary ATP production, but these only operate over nanometer distances (Oster & Wang, 2003).

Particularly interesting are the constraints imposed on developmental change by the extant geometry of the embryo. A classic example here is the different routes taken by the short superior and very long inferior laryngeal nerves that supply the larynx in the adult giraffe (see Box 18.1 for details). On both sides, the laryngeal nerves originate from the hindbrain as part of the 10th cranial (vagus) nerve and the superior ones migrate straight to the larynx. The branches that become the inferior laryngeal nerves run under the left (sixth) and right (fourth) aortic arches on their way to the larynx. These routes are posterior to the ductus arteriosus on the left, an early heart vessel that becomes the ligamentum arteriosum, and the subclavian artery that supplies the right forelimb. During development, the neck extends so that the nerves also have to lengthen if they are not to cut through the blood vessels they loop round. The reasons for the long nerves thus derive from the geometry of the developing embryo (Box 18.1, Fig. 18.9). Unlike stem cells, embryonic tissues cannot discard their developmental history.

Within these developmental constraints, there is still flexibility for innovation as demonstrated by the fossil record, the anatomical changes documented here and in the wider literature. Perhaps the major point made in this chapter and exemplified by the example of the amphibian limb with its remodeled calcaneum and talus (Fig. 18.3), is that anatomical alterations can generated by modulating network activity and this can happen through relatively modest mutational changes that operate within the constraints of extant anatomical architecture..

Finally, there are the niche constraints. The biosphere is now so occupied that it is hard to envisage a niche that is unoccupied and in which an organism with substantial structural or physiological innovation would flourish. That flying pig would have to compete with eagles for food, and it doesn't take an expert to see which would survive that battle! Indeed, the major bursts of anatomical innovation have always come after major extinctions, when whole arrays of niches became

BOX 18.1.   THE INFERIOR LARYNGEAL NERVE OF THE GIRAFFE

In all tetrapods, laryngeal activity is controlled by the superior and inferior branches of the left and right laryngeal nerves. The superior laryngeal nerves extend from the vagus directly to the larynx. The inferior nerves are the terminal part of the left and right recurrent laryngeal nerves (RLNs). Their routes descend down the neck and into the body of the giraffe (Fig. 18.9). The left RLN loops beneath the ligamentum arteriosum, which links the pulmonary artery to the descending aorta, while the right RLN loops around the subclavian artery that feeds the shoulder and forelimb; both RLNs then ascend the neck to the larynx. The effect of this detour on length is spectacular, particularly on the left side: while the superior nerve is ~20 cm in length, the inferior is ~400 cm.

The apparently bizarre routes of the inferior laryngeal nerves make complete sense in the context of their development. In the early embryo, the heart lies ventral to the early head, before the neck forms. As a result, the embryonic hindbrain, from which the vagus nerve

originates, is directly dorsal to the presumptive larynx and caudal to the aortic (pharyngeal) arches. The routes of the inferior laryngeal nerves are thus caudal to the fourth arch artery on the right (the future subclavian artery that feeds the right forelimb) and the sixth arch artery on the left (this becomes that ductus arteriosus that brings oxygenated blood from the placenta to the descending aorta during development and forms the ligamentum arteriosum when it becomes blocked after birth). As the neck extends, the nerves fibres have to lengthen if they are not to cut through these blood vessels (Fig. 18.9).

As a result, the left recurrent laryngeal nerve of the giraffe is the longest nerve in all contemporary animals. It is, however, shorter than that of the long-necked sauropod, *Diplodocus*: its inferior pharyngeal nerve would probably have been about 28 m long, but even this would have been shorter than the 40–50 m nerve that controlled the movement of the tip of its tail (Wedel, 2012).

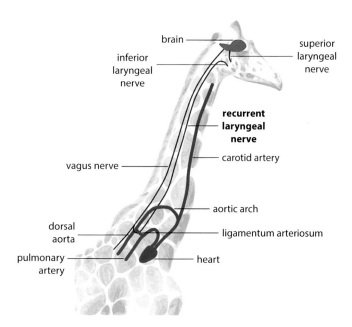

Figure 18.9 The routes of left laryngeal nerves of the giraffe. The route of the inferior nerve is much longer than that of the superior nerve because it gets trapped behind the ductus arteriosus artery early in development. When the neck extends, the nerve has to lengthen if it is not to cut through the artery. (*Note*: After birth, the ductus arteriosus becomes the ligamentum arteriosum.)

unoccupied. Perhaps, given enough time, a flying vertebrate would evolve that had a porcine ancestor, but it would probably change to look more birdlike as bats did when they diverged from their land-based ancestors. Here, it is worth noting that, as that hippopotamus ancestor took to the sea and slowly evolved into a whale, it became more fishlike in form (Fig. 16.7).

Taken together, these constraints make it clear that there are regions of what Maynard Smith et al. (1985) referred to as morphospace that remain unoccupied, and for good reasons. In 1985, it might have been surprising that so much of this space was empty, but this was because so little was then known about molecular genetics and the ways in which tissues form. The work of the last 20–30 years, particularly on protein networks and signaling, has now clarified the principles that underlie tissue generation during development. Similarly, the work on genetic engineering has shown how major changes can be generated through sequence manipulation that affects network outputs, particularly those that affecting patterning and growth. In the limits, however, the potential for anatomical change is not only limited by the constraints discussed here but by the invisible and unpredictable nature of selection that is discussed in the next two chapters.

THE MUTATIONAL BASIS OF TRAIT CHANGE—THE BROADER VIEW

The main conclusion of this chapter is that normal phenotypic change derives from the integrated activity of protein pathways and networks, with most evolutionary change resulting from mutations that affect their output. Because there are so many cooperating proteins in a single network, the effects on the phenotype of a mutation in just one of them is inevitably indirect and consequently hard to predict. In most cases, one would expect the effect of any mutation to be buffered and morphological result to be within the parameters of normal variation. Nevertheless, because the internal properties of complex networks can amplify change (Wilkins, 2007), a single mutation in one of a range of genes or the cumulative effect of several will occasionally lead to substantial change, as the example of craniosynostosis has shown.

This conclusion has two important corollaries. First, because the mutations are essentially minor in their effects on individual proteins within networks, the resulting phenotypic changes can be generated, inherited, and maintained through subsequent generations in the normal way. Second, as mutations in each of several genes within a network can lead to the same or a similar variant, sustainable phenotypic change can be generated more rapidly in these cases than where it depends on a single allele.

The net result is that the spread through a small population of a novel and substantial mutant trait based on changes to process networks can be faster than one might expect if the change in phenotype has a strong selective value. Evidence to support this view comes from the rapid and wide-ranging adaptive radiation events in cichlid fish (Danley et al., 2012; Chapter 22), and the changes in fresh-water stickleback dorsal spines (above) ~14 Kya (Cresko et al., 2004). This chapter has, moreover, shown how relatively minor mutations can have substantial effects on the phenotypes in, for example, digit morphology and number and other examples of patterning. In this context, it seems worth returning to Goldschmidt's suggestion from 1940 that major changes were initiated by mutations producing large-scale phenotypic changes that he called hopeful monsters. This approach, albeit that the results are not monsters, doesn't seem quite as fanciful now as it did then.

KEY POINTS

- Most anatomical and physiological characteristics of an organism are underpinned by the activity of systems of protein networks rather than by the activity of a few individual proteins.

- Evolutionary innovations reflect mutations that affect the activity of these networks.

- Major development changes can be effected through mutations that alter the timing of network activation, thus generating heterochrony.

- Mutations that affect proteins within networks can alter the internal dynamics of that network and hence its output properties.

- Variation in networks is buffered so that the output of the network is generally held within the bounds that are seen as normal. In many cases, mutation will have little or no effect. Very occasionally, the cumulative effects of protein variation can lead to an abnormal network output that results in a change in the phenotype.

- Because these changes derive from mutations whose direct effect is on single proteins, the resulting phenotypic changes are readily heritable.

- When a phenotype depends on the activity of a complex network, a particular change would be expected to be generated independently following mutations in several of its genes.

- Heritable change of the resulting phenotype can hence be achieved far more rapidly than when a change depends on mutation to a single gene.

FURTHER READINGS

Bard J (2013) Driving developmental and evolutionary change: A systems biology view. *Prog Biophys Mol Biol*; 11: 83–91.

Barresi MJF, Gilbert SF (2019) *Developmental Biology* (12th ed). Sunderland, MA: Sinauer Press.

Held LI (2014) *How the Snake Lost Its Legs: Curious Tales from the Frontier of Evo-Devo*. Cambridge University Press.

Wilkins AS (2007) *The Evolution of Developmental Pathways*. Sunderland, MA: Sinauer Press.

WEBSITES

- Wikipedia articles on *Gene regulatory networks* and *Heterochrony*
- Lists of pathways: https://reactome.org/PathwayBrowser/#/

- A set of key protein networks can be downloaded from the "Qiagen pathway map reference guide" site.

Adaptation, Symbionts, and Holobionts

19

The two previous chapters discussed the mechanisms by which a novel variant in the phenotype of some species may arise and how every aspect of that phenotype can vary across a population. If a subgroup within a population of an organism has some novel feature, it will eventually be lost unless it increases fitness within the local environment. This means that organisms with that feature will likely to produce more fertile offspring than organisms lacking it. If they do, that feature is a positive adaptation to the local environment and likely to become a characteristic of the population as a whole. In due course, it may even become a defining characteristic for a new subspecies.

This chapter discusses examples of ways in which a range of species have been successful in adapting to novel environments and niches. Most examples of adaptation reflect minor changes to an organism's anatomy and physiology but others reflect relationships between pairs of organisms. Examples here are the many types of mimicry and camouflage as well as adaptions that facilitate the fertilisation of plants by animals, that optimise offspring numbers so as to maximise the chance of their survival, and that enable organisms to live, in particular, habitats. It is the range of adaptations seen today that defines the glorious breadth of species across the living world.

Particularly interesting are those adaptations that enable organisms to survive through cooperation. Many organisms have co-adapted so that they provide mutual aid to one another as symbionts (e.g. fungi and algae in lichens). Most recently, the importance of holobionts has been recognised: this is a composite organism of a host and all that live on and within it, particularly the microbiome of prokaryotes and sometimes small eukaryotes. Often, the degree of mutualism is so strong that the one cannot survive without the other. In an evolutionary context, holobionts are important because it is the composite organism that is the unit of selection rather than the host, and their collective phenotype may open up new evolutionary opportunities. A particularly important example of this is herbivory.

In the 18th century, it was generally thought that the demonstrable suitability of each organism for the niche in which it lived, the so-called *perfection of nature*, was proof that God created a world in which every organism had its proper place (Appendix 8). The major difficulty in countering this argument is that variation in a particular population is unexceptional, selection is generally very weak and the acquisition of an adaptation so slow that only rarely can the process of adaptation be observed. It was in this context that the examples like the peppered moth, *Biston betularia*, which became black towards the end of the 19th century as city atmospheres were polluted by soot were so important (see below). Equally important are the experimental models discussed in the next chapter because they demonstrate that change can happen under selection. The argument against the *perfection of nature* is that it cannot handle environmental change.

It took almost a century of research to demonstrate that organisms only achieve an apparently perfect fit to their niche after a complex and slow period of adaptation.

DOI: 10.1201/9780429346217-23

Figure 19.1 **Adaptations.** (a): Clownfish (*Amphiprion ocellaris*) resident in purple anemone (*Heteractis magnifica*). (b): The female angler fish accompanied by her very small male mate (arrow). (c): The remarkable camouflage of the leaf insect, a member of the Phylliidae family.

(a Courtesy of Nick Hopgood. Published under a CC Attribution-Share Alike 3.0 Unported License.
b Courtesy of Andrew Butko. Published under a CC Attribution-Share Alike 3.0 Unported License.
c Courtesy of Nandini Velho Published under a CC Attribution Generic 2.5 License.)

If we envisage a small population of an organism finding itself in a new habitat, those that survive will flourish, and new mutations will allow some of their offspring to do better than others in adapting to the new habitat. In due course, the surviving group will seem perfectly matched to this habitat, but it may take a very large number of generations (Fig. S4.1). An element of the process is summed up in the words of Samuel Beckett in his novella *Worstward Hoe* (1983): "Ever tried. Ever failed. No matter. Try again. Fail again. Fail better."

While the previous two chapters focused on how mutation led to phenotypic change, this chapter looks at the types of adaptive options that are available to organisms, together with examples of how they have been realised. The next chapter considers the selection pressures that habitats exert on possible phenotypes to determine which will be successful, as defined by the fitness criterion. In considering these pressures, it is important to keep in mind that fact that, although the examples generally focus on a particular change within a single aspect of the phenotype, every trait in every species is continually subject to mutational change and subsequent selection under environmental pressures. As a result, the resulting organism may represent a compromise in which the whole is optimised for fitness rather than any particular trait (see Websites, Chapter 21). If, however, any facet of an organism's niche changes, so will the selection pressures on the phenotype of that species. Mutation, variation, and selection never cease.

What makes the study of adaptation so fascinating is the sheer variety of ways in which organisms have evolved to mesh with their environment and the means by which they have met the challenges of other organisms that want to eat them or the pressures of the physical world in which they find themselves, be it heat or cold, drought or darkness. The wide range of modes of adaptation discussed in this chapter illuminates the theme of the survival of the fittest. This is the idea that, in a population of organisms showing variation, the group that will eventually become dominant is that whose members adapt best to the niche in which they find themselves by leaving the greatest number of fertile offspring.

Sometimes, adaptation just means that an organism has evolved so that it is camouflaged and is essentially invisible to predators or that some anatomical feature has altered to facilitate access to a particular type of food (e.g. a bird's beak may change, Fig. 4.3). A different type of adaptation occurs when two organisms come to depend on one another, such as the symbiotic relationship between the clownfish, *Amphiprion ocellaris*, and sea anemones (Fig. 19.1a). Here, the fish protects the anemone from parasites, while the fish, in turn, is protected from its predators by the stinging tentacles of the anemone. As part of its adaptation, the clownfish secretes a surface mucus that protects it from those tentacles.

In this case, the two symbionts are physically distinct, but there are many examples where the symbiosis is between an organism and the prokaryote community that it hosts (it's microbiome). In animals, these are mainly located in the gut and on the exterior, while plants have different microbiota above and below ground. In animals, the microbiome associated with their gut facilitates digestion and probably their immune system while the animal provides a safe niche for the constituents of the microbiome (see below). In plants, the microbiome can facilitate growth (e.g. nitrogen fixation, see below) and help suppress pathogens (Compant et al., 2019) while the host can provide a stable environment for the microbes. The composite unit of a eukaryote with its microbiome which may also include unicellular eukaryotes is known as a holobiont and the extent of its significance in an evolutionary context has been recognised only relatively recently (McFall-Ngai et al., 2013; Morris et al., 2018).

The importance of symbionts and holobionts forces us to redefine what, in an evolutionary context, is meant by an individual. Traditionally, it was a member of a species and hence that this was the unit of selection. It is, however, now clear that this definition is clouded by the existence of the various sorts of symbiotic relationships and in such cases, selection acts collectively and in complex ways on mixed communities.

TYPES OF ADAPTATION

Adaptation is an umbrella term that covers any stable feature that helps a species mesh with its environment and enhance its fitness. One has only to look at any book on animals and plants to see the enormous range of anatomical and behavioural traits that have evolved to ensure that a particular strain or species becomes successful in a particular habitat. That said, most have not been subject to experimental investigation to confirm the adaptive importance of their particular phenotypes. Examples that have been studied include topics as diverse as changes to a bacterium's biochemistry that allows it to survive on new food sources, such as plastic (Urbanek et al., 2018), the various heritable physiological adaptions that allow humans to live at high altitude (Beall, 2013), and the replacement in dipteran flies of the pair of rear wings seen in early winged insects with halteres (balancing organs) that increase flight control (Kathman & Fox, 2018).

It is not practical even to begin to list here the vast range of adaptations seen across the biosphere. Each species bears witness in its anatomical and physiological features to the habitats in which it has successfully survived in the past and in which it survives now. Most are unexceptional and reflect colour, size, and behaviour, although a proportion are so bizarre that they merit a mention. Thus, for example, the male angler fish has been reduced in size compared to its female partner to the extent that it lives as almost an appendage on its partner's underside, having adapted to a role where its sole function is to provide sperm (Fig. 19.1b). Another wonderful example comes from the Phylliidae or leaf insects that have evolved to become invisible in a pile of leaves (Fig. 19.1c).

Adaptations can of course be classified and what follows is a discussion of the ways in which organisms change to mesh with their environment contrasting *adaptive* and *convergent radiation*. An instance of the former occurs when a small population of an organism reaches a new and bounded territory, such as a lake or an island: subgroups soon diversify and adapt to distinct ecological micro-niches with the result that, in due course, there are a host of subspecies. Adaptive radiation accounts for the cichlid species flock seen in Lake Victoria (Chapter 22), the diversification of lorises in Madagascar and, of course, the finches in the Galápagos islands that set Darwin off on his interest in evolution.

Convergent radiation or homoplasy is essentially the opposite and occurs when, as a result of living in similar habitats, different organisms evolve to have very similar anatomical features. Well-known examples here are the different types of vertebrate wings, the sabre teeth found in a wide variety of animals (Chapter 5), the camera eye of vertebrates and molluscan cephalopods, and the spines and general morphology of hedgehogs, a member of the Erinaceidae, a boroeutherian mammal, and Madagascan tenrecs, an afrotherian mammal (Chapter 16). Examples of mimicry, which also reflects homoplasy, are discussed below.

It should not, however, be assumed that all features of an organism reflect direct selection and adaptation. There are often anatomical features in an organism that cannot be linked in any obvious way to selection pressures from the environment and an example is the morphology of the liver. The large human liver has four lobes whereas the small mouse liver has five. It is hard to see how such minor species differences can be regarded as an adaptation as there are no obvious selection pressures on liver morphology. They seem more likely to derive from the effects of genetic drift or linkage than on adaptation in response to some physiological pressure (Chapter 21).

Camouflage and mimicry

Many adaptions can be viewed as attempts to change the balance between predator and prey: the former may become bigger, stronger, and more vicious in line with the well-known evolutionary trend for taxa to become larger over time (this known as Cope's rule, and examples are seen in both vertebrates and invertebrates; Chown &

Gaston, 2010; Lamsdell & Brady, 2010). As a result, the latter may become faster and more nimble, and perhaps develop group protection or camouflage. In the latter case, an organism produces an external morphology and colour pattern that mimic its background, rendering it almost invisible to potential predators and so more likely to live longer and breed (Fig. 19.1c). In the most extreme of cases, animals, such as chameleons and cephalopods can change their skin colour and patterning to match a range of environments through sophisticated modulation of their chromatophores and iridophores (Williams et al., 2019).

Perhaps the most famous example of camouflage evolution, probably because the dominant variant changed so rapidly, is the pigmentation of *Biston betularia*, a moth particularly found in the north of England (Fig. 19.2a,b). Originally, the predominant form of this moth had a highly mottled, but light wing and body with this pigmentation pattern camouflaging it when it rested on lichen-rich barks (its caterpillar is camouflaged as a twig). In the 19th century, however, as the industrial revolution proceeded and smoke filled the air of the British Midlands, tree lichens died as carbon was deposited on bark. As this happened, what had been an occasional black morph of the butterfly became predominant and the frequency of the mottled variety dropped.

Now, with improved environmental standards leading to lowered pollution levels, the mottled variety is becoming more common again. The core selection pressure driving change is the extent to which the pigmentation pattern camouflages the moth against predators, although the detailed reasons for this reversion may be complex (Cook & Saccheri, 2013). Moths mainly reproduce annually so, over a period of ~150 years and the same number of generations, the predominant morph has flipped from variegated to black and back again. This is unusually fast for natural selection and reflects a particularly strong selection pressure.

More visually dramatic results arise when organisms take an opposite mimicry strategy: instead of becoming invisible, the organism (the mimic) evolves to look like another organism (the model) that is poisonous or dangerous in some other way and whose dramatic patterning acts as a warning to predators (Mokkonen & Lindstedt, 2015). Mimicry through deception is known as Batesian mimicry after Henry Bates,

Figure 19.2 Camouflage and mimicry.
(a,b): The normal and melanised subspecies of Biston betularia, the normal peppered moth. (c,d): An example of Batesian mimicry: the palatable hoverfly *Syrphus* © has mimicked the morphology of unpalatable wasp *Polistes* (d). (e,f): An example of Müllerian mimicry in two North American unpalatable butterflies: the viceroy butterfly, *Limenitis archippus* (e) and the monarch butterfly, *Danaus plexippus* (f).

(ab Courtesy of Olaf Leillinger published under CC BY-SA 3.0.
cd From Chittka L, Osorio D (2007) *PLoS Biol*; 5(12):e339. Published under a CC Attribution Generic 2.5 License.
e Courtesy of Benny Mazur. Published under a CC Attribution Generic 2.0 License.
f Courtesy of Kenneth Dwain Harrieson. Published under a CC Attribution-Share Alike 3.0 Unported License.)

a 19th century English naturalist, and there are many examples, some of which exist in the absence of obvious current models (Pfennig & Mullen, 2010). A well-known example is the similarity between the harmless hoverfly and the stinging wasp (Chittka & Osorio, 2007; Fig. 19.2c,d;). There is also a further form of mimicry in which two poisonous organisms have evolved to mimic one another. This is Müllerian mimicry, named after the 19th century German naturalist Fritz Müller, and it gives both species a selective advantage if they have a common predator (Fig. 19.2e,f).

Exaptation

It cannot be assumed that an anatomical feature seen today, which clearly plays an adaptive role, originally evolved to fulfil its current function. There is now no doubt, for example, that birds evolved from theropod dinosaurs (cladistically, they are still dinosaurs), but it now clear that feathers did not evolve in the context of flight. Their original function was insulation and hence the maintenance of body temperature (Prum & Brush, 2002). It was this function that helped ground-based birds, whose feathering was more extensive than that of dinosaurs, survive the period of severe cold that followed the meteorite catastrophe responsible for the extinction of all other dinosaur taxa at the end of the Cretaceous period (Fig. 19.3a). Once feathers had evolved, however, they were able to vary in shape and size and so be subject to selection for their properties in other contexts, such as gliding and then flying, as well as for sexual display, as in the peacock (Fig. 19.3b). A character that is selected for its advantages in one context and turns out to have another role in a future and different context in known as an *exaptation*.

Niche construction

An ecological niche is essentially a specific habitat that provides a physical environment, food and adequate protection from predators for particular organisms. In many cases, that niche either drives adaptation or is changed by the species that occupies it. An interesting example here is the beaver, a large semiaquatic rodent with webbed hind-feet, features that reflect adaptation to a riverine mode of life. Following their arrival in a landscape with streams, they build homes (lodges) with underwater entrances and dams that alter both the nature of the landscape and the way in which water is used in it. The wider effect of these dams on their surroundings is impressive: flow is controlled, water quality is improved and sediments sink. These

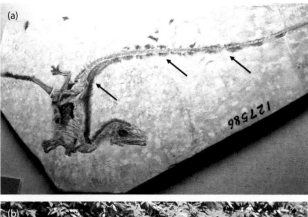

Figure 19.3 (a): **Fossil of *Sinosauropteryx prima*, an early Cretaceous therapod dinosaur with feathers (arrows) for insulations but not for flight.** (b): The display of the male Indian peacock. The prime role of these feathers is for sexual display, although the peacock can fly for short distances.

(a Courtesy of Olai Ose. Published under a CC Attribution Generic 2.0 License.
b Courtesy of Dinesh Kannnambadi. Published under a CC Attribution-Share Alike 3.0 Unported License.)

improvements also change the nature of the fauna and flora that can live in the area. In the short term this is through migration of existing species that benefit from the new environment; in the longer term, this new niche affords opportunities for other species to adapt.

An early and key example of macroscopic niche formation was the change in the atmosphere of the earth that resulted from the evolution of Cyanobacteria ~2.5 Bya or a little earlier; these use carbon dioxide as an energy source, releasing oxygen as part of the process. One effect of this "Great Oxygen Event" (~ 2.4 Bya) was to change the earth's atmosphere by lowering of the levels of carbon dioxide, a greenhouse gas. This, in turn, led to the period of extreme cold known as the Huronion glaciation and the extinction of any prokaryotes that lived near the surface or that might have managed to live on land. In due course, however, the availability of free oxygen facilitated the evolution of aerobic prokaryotes and, perhaps 600 My later, led to the endosymbiosis of aerobic eubacteria that could use oxygen to make ATP. Eventually, one of these bacteria became the mitochondrion of a eukaryotic cell able to provide it with far more energy than had been previously available (Chapter 10). It was probably this event, more than any other, that enabled early eukaryotes to flourish and diversify.

Human evolution is an interesting model in this context. The change in lifestyle from arboreal to land-based and from a quadrupedal to a bipedal mode of walking brought a host of consequences. First, as our ancestors adapted to their new niche, they lost the ability to flourish in the old one, so leaving more space for monkeys and apes. Second, as they adapted their lifestyle to that of hunter-gatherers and then to agricultural farmers, they slowly changed the landscape. Further changes occurred as groups became larger and started to live first in in villages and then in large towns, while also changing the landscape to make it appropriate for providing food. The effects of this on the world's flora and fauna has been dramatic.

At the other end of the scale is the behaviour of the cuckoo, which uses the nest of another bird for its own egg. This has led to three adaptations: first, the cuckoo egg has a short incubation time so that the hatchling emerges early and can push other eggs out of the nest; second, the cuckoo hatchling can mimic the sound of several young birds so encouraging the parents to bring enough food for a brood; third some cuckoo chicks now mimic the appearance of the host chicks (Langmore et al., 2011), so reassuring the host adults.

Parasites

Parasites and commensals are organisms whose niche is the larger organisms that they exploit. The better-known parasites (e.g. tapeworms) have been identified on the basis of the damage that they do, while commensals are harmless. Humans are, for example, hosts to a wide variety of both prokaryotes and invertebrates. The former are part of the holobiont and are necessary for normal life (e.g. the microbiome is needed for full human gut development and function [Hooper & Gordon, 2001]), while some, like the fungus *Aspergillus*, are normally harmless commensals. Human parasites are only problematic when they cause disease; this is either because they damage their hosts directly (e.g. worms) or because they can carry a wide variety of infectious protists, bacteria, and viruses. Humans are not special here: most larger organisms harbour parasites that have adapted to their host. We know most about human parasites, however, because details about the diseases that they cause and the treatment needed to cure them have been far more closely studied than those of any other organisms, even farmed animals.

Some parasites become so adapted to the niche that their host provides that they cannot survive elsewhere becoming specific to that species. Indeed, one of the ways of defining a species is by the uniqueness of its habitat: some apparently similar parasites (e.g. nematodes) can, for example, only be distinguished on the basis of their unique host. The successful parasite does not, however, kill its host but allows it to survive, perhaps in a damaged form (e.g. *Plasmodium falciparum*, the parasite responsible for malaria), so providing a long-term niche for the parasite, and its offspring. Larger parasites may well have their own smaller parasites and will certainly have a microbiome. As Dean Jonathan Swift wrote in 1733:

> So, naturalists observe, a flea
> Hath smaller fleas that on him prey;
> And these have smaller still to bite 'em;
> And so proceed ad infinitum.

Facilitating reproduction

An interesting aspect of animal adaptation is how males and females choose mates (sexual selection, see Chapter 20). Choice is not, however, an option for plants where pollen from the stamens of one individual have to land on the stigma of the pistils of the same or a close member of the species for fertilisation to take place. In the latter case, pollen may be carried by the wind, through water or by animals. The male pollen has to be readily dispersible while the female plant can do no more that have a morphology that facilitates its capture. Many plants have evolved strategies to ensure that a particular insect or group of insects is attracted to male flowers where it picks up pollen and then to flies to female ones where it can deposit it. Examples of successful invitation strategies for plants are odour, morphology, colour, and flower shape.

These events have been particularly studied in *Ophrys aoifera*, an orchid whose flowers have evolved to look very much like bees in shape and colour (Fig. 19.4a). Although these orchids can self-pollinate, their beautifully coloured and bee-shaped flowers also attract male bees that try to copulate with them. While this pseudocopulation is being attempted, the male bee's head comes into contact with the stamens and pollen grains stick to its hairs. When that bee next tries to copulate with another such orchid, the pollen gets transferred to its stigma and fertilization is achieved. There are many other similar adaptations that facilitate reproduction in plants, although there is as yet no direct evidence on how such morphologies evolved. It is, however, possible that they arose as a minor, lower-key variant with a small selective advantage (Vereecken et al., 2012). It then probably required many successive changes, each with a slightly enhanced fitness, to produce the species seen today in which such a phenotype has become the wild type.

Once a seed has been fertilised, it needs to be dispersed and the simplest way is through gravity (the seed drops to the ground). Sometimes, the wind will widen this dispersal, and there are adaptations, such as the helicopter-like spokes of the dandelion, that will facilitate this (Fig. 19.4b). A particularly important angiosperm adaptation was the evolution of fruit as a structure to surround and initially protect

Figure 19.4 (a): **Plant fertilization strategies.** The appearance of the bee orchid *Ophrys apifera* encourages bees to land and pseudocopulate, so picking up pollen. (b): The morphology of the dandelion seed allows it to be dispersed over long distances by winds.

(a Courtesy of Hans Hillewaert. Published under a CC Attribution-Share Alike 3.0 Unported License.
b Courtesy of Piccolo Namek. Published under a CC Attribution-Share Alike 3.0 Unported License.)

seeds as they ripen. Ripe fruit can then fall to the ground or be eaten by animals and birds so dispersing the seeds either because it was not being eaten (e.g. stones) or through being eaten and then excreted in dung which can itself act as a fertiliser. A particularly useful species in this context is *H. sapiens* who requires and consumes vast amounts of plants and fruit and often discard the seeds well away from the original plant.

Increasing offspring numbers

For a species to increase in number over the longer term, each organism in a population needs on average to leave more than one fertile offspring. In practice, this requires producing enough fertile offspring to survive predation, and the way of doing this is yet another form of adaptation. In sexually reproducing species, males always generate large numbers of spermatozoa; females can, in principle at least, produce any number of female gametes, from a single ovum upwards. The theory of population dynamics is helpful here as the growth rate in any population **N** can often be described by the growth equation

$$dN/dt = rN(1 - N/K)$$

Here, **r** is the growth rate while **K** is a parameter that reflects the carrying capacity (e.g. the extent of parental care) of the totality of the environment for offspring. In this equation, the growth rate will be high if **r** is high but much less if **r** is low. The **K** parameter modulates the effect of **r**: the lower is **K**, the higher is the effective value of **r** while high values of **K** have minimal effects on **r**. The value of these parameters for a population reflect an adaptation to its own traits and to the environment in which it lives.

There are two extreme reproduction strategies on this spectrum that help understand population behaviour; they are known as **r** and **K**. Species that display the **r** strategy (**r** is high and **K** tends to be high) have a high potential growth rate and, inevitably, limited parental care; in such cases, relatively few offspring will survive. An example is cod where adult males and females may produce millions of eggs and sperm but take no interest in their future development (Secor, 2015). Plants likewise generally follow the **r** strategy, as demonstrated by the thousands of olives seen on a single tree or the numbers of seeds in each of the many cones produced by gymnosperm trees. The strategy of low **K** is common in mammals where females suckle their young, and in birds where the female may produce one or very few eggs that require being kept warm. Animals that adopt the **K** strategy often invest large amounts of parental time, care and protection of their eggs and young hatchlings.

The choice of where on the **K-r** spectrum an organism comes depends mainly on its evolutionary history. The **K** strategy is associated with a high offspring-survival rate and the **r** strategy a very low one; the latter does, however, have the advantage that a small population finding itself in a novel and beneficial environment can increase its numbers far more rapidly than the latter and so have a greater likelihood of first surviving and then diversifying to give, for example, species flocks. It should be noted that the traditional **r/K** approach is now more used for discussion than for quantitative analysis; population geneticists today employ more sophisticated modelling techniques for analysing population behaviours (e.g. Plard, 2019).

SYMBIOSIS

Every species survives in a niche that allows it to flourish, and this niche will have several components: ideally, it will provide food, a physical home, adequate protection from predators and opportunities for mating, together with a good chance of the resulting offspring living long enough to reproduce. Part of an organism's ability to survive in that niche reflects its adaptation to what it has to offer. Symbiosis (i.e., living interdependently) is a particular form of niche adaptation in which two organisms come to depend on one another and, as a result, become part of one another's niche. This dependency does not mean that their respective adaptive requirements are the same: one symbiont might provide food for both, the other a protective home for the other.

Today, vast numbers of symbiotic relationships between pairs of organisms have been documented; a few have already been mentioned and more are given below. Many show how two organisms can not only depend on one another but go further in their mutual relationship: they jointly change their phenotype to the benefit of both, to the extent that the products of their gene pools can be seen to interact. This phenomenon is known as *coevolution* and can be defined as reciprocal evolutionary change between interacting species. This can sometimes happen to the extent that the two organisms become mutually dependent, as in the case of lichens that are typically composed of fungi and algae (Thompson, 1989; Fig. 19.5d). Here, the adaptation is particularly robust to the extent that, in some cases, the two participating species have undergone new, joint speciation. The older and perhaps more graphic term for this was Darwin's *coadaptation*, as the process involves two species adapting their phenotype for their mutual benefit. Coevolution is, however, a more accurate term as it reflects the deeper biological context within which such change happens.

The importance of coevolution, that two species can do better together than separately, was known long before the earliest thoughts on evolution; indeed, the idea of crop rotations can be seen to depend on it. The early scientist with whom these ideas are particularly associated is Prince Pyotr Kropotkin, a Russian biologist and socialist who lived and worked around 1900. He argued for the importance of mutual aid among plants and animals as the basis for evolutionary change. He held this view, which contrasted to the then prevailing one of evolutionary success depending on the basis of the fight for survival ("Nature red in tooth and claw"), partly on socialist and anarchist grounds, but mainly on the basis of his geographical and other field research in Siberia during the 1860s. His book on Mutual Aid (1902) is an early classic in evolutionary biology (see Websites).

The theoretical context in which symbiotic relationships are best discussed is *fitness*. Here, a mutual relationship with another organism that enhances the ability of each to leave more offspring increases the fitness of both and will be favoured. Parasitism occurs when one organism of the pair increases its fitness at the expense of the other—successful parasites, such as pathogens, tend not to decrease this too much as it risks reducing host numbers over time so lowering habitat opportunities. If the fitness of neither organism particularly benefits from the relationship, the pair are known as commensals.

Even if we just restrict the use of the word symbiosis to eukaryotes, there are a host of examples of mutual symbiotic support in every ecosystem. Leaf-cutter ants cultivate fungi for food, so allowing the fungi to flourish securely; hummingbirds obtain nectar from bird-pollinated flowers that have, in turn, adapted their flower morphology to fit the bird's bill and so ensure pollination (Fig. 19.5c); and acacia trees provide nourishment to acacia ants which, in turn, protect the trees from other insects.

Lichens are, as mentioned, particularly interesting examples of symbiosis (Fig. 19.5d): each is a composite organism of a fungus and an alga or cyanobacterium:

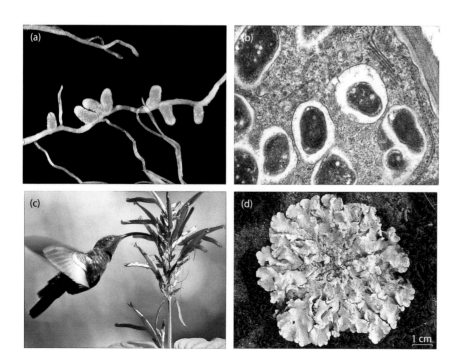

Figure 19.5 **Examples of symbiosis.**
(a): Rhizobia nodules growing on the roots of *Medicago italica*. (b): TEM micrograph of a nodule cell in a soybean root. Individual *Bradyrhizobium japonicum* cyanobacteria are located within symbiosome vesicles. (c): The elongated and slimmed beak of the hummingbird that evolved to drink the nectar of elongated flowers, incidentally facilitating pollination. (d): *Flavoparmelia caperata*, the common greenshield lichen.

(a Courtesy of "Ninjatacoshell." Published under a CC Attribution-Share Alike 3.0 Unported License.
b Courtesy of Charles J Sharp. Published under a CC Attribution-Share Alike 3.0 Unported License.
c Courtesy of Jason Hollinger. Published under a CC Attribution-Share Alike 3.0 Unported License.)

photosynthesis from the algal or cyanobacterial component provides food that sustains the fungal component and this, in turn, provides physical protection and water to the alga or cyanobacteria (Grube & Wedin, 2016). In the many species of lichen, the two components have become integrated so that the resulting structure appears to be a single organism, with most of the structural component being adherent fungal hyphae. The net result is a combined life form that can survive extraordinarily well in many inhospitable environments—indeed, it is claimed that lichens are some of the longest-lived organisms on earth.

HOLOBIOSIS

Of particular evolutionary interest are the symbiotic relationships between animals and the microbial communities that they host, with the composite organism being known as a holobiont. Holobionts often have the interesting property that their genomes can be seen as being almost joint (Bordenstein & Theis, 2015). This is partly because the microbiome is often passed directly from mother to offspring but also because there can be genetic activity and phenotypic properties that reflect interactions between host and microbiome. The holobiont thus has what can be considered as a collective genome that is mutually beneficial to both. In an evolutionary context, this means that the holobiont as a whole is the unit of selection (Guerrero et al., 2013). This section discusses a few of the better-known examples of holobionts across the plant and Animal Kingdoms (Table 19.1).

Although the concept was introduced by Adolf Meyer-Abich in the 1940s (Baedke et al., 2020), it has only started to attract much attention in the last decade as research showed that microbiome-host support interactions were more important and more frequent than had previously been thought. McFall-Ngai et al. (2013) document important examples where the host provides a habitat for the microorganisms which, in turn, increase the fitness of their host. The microbiome can improve and strengthen gut function, activate and maintain the immune system, and can facilitate the development of the host's offspring, particularly in the maturation of invertebrate larvae.

TABLE 19.1 EXAMPLES OF HOLOBIONT ACTIVITY BETWEEN HOST AND MICROBIOME (IF NOT OTHERWISE GIVEN, SEE TEXT FOR REFERENCES)

Eukaryote	Prokaryote	Benefits
Bos taurus (cows)	*R. flavefaciens*	Herbivory and downstream anatomical changes.
Mice gut villi	*B. thetaiotaomicron* (gut bacteria)	Paneth cells in the gut respond to these bacteria by making angogenin-4. This induces capillaries in the adjacent villi. It also kills bacterial competitors.
Humans	A range of gut bacteria	These facilitate digestion, immune health, pain tolerance etc.
Euprymna scolopes (a squid)	*V. fischeri* (luminescent)	The squid produces a light-emitting organ (Rader et al., 2019).
Acyrthosiphon pisum (pea aphid)	*Hamiltonella defensa* *Rickettsiella* bacteria	Protection against parasitoid wasps. Protection against beetle predation: aphid colour changes from red to green sap digestion.
Insects	Various fungi & bacteria	Herbivory.
Insects	*Burkholderia*	Protect stinkbugs against insecticides and defend *Lagria* beetle eggs against pathogenic fungi (Kaltenpoth & Flórez, 2020).
Coral	Microbiome as a whole *Symbiodinium* (alga)	Coral and reef health. Reef-building.
L. luymesi (siboglinid tube worms)	Gammaproteobacteria	Metabolise sulphated hydrocarbons and act as the digestive system.
Rough-skinned newts	Several bacterial types	Provide neurotoxin defense against snakes and other predators

This importance of holobiotic relationships is not restricted to animals: the enzymes for nitrogen fixation are only found in a few bacterial families but they are instrumental in the success of plants and the production, either primarily or secondarily, of food for all land-based animals. The best-known example of this is the nodules of nitrogen-fixing bacteria that form on the roots of legumes (Fig. 19.5a,b). The roots provide a stable environment and organic food for rhizobia bacteria which metabolise atmospheric nitrogen into the nitrogen-containing molecules that they provide to the legumes. There is a wide variety of other plant and nitrogen-fixing bacterial pairings that fulfil this mutual-support role. Plants that do not form such relationships just get their nitrogen compounds from soil bacteria, a less efficient process, and one that commercially often requires nitrogen-based fertilizers.

An important example of holobiosis is the digestion of plant material by herbivores: as only prokaryotes and a few eukaryotic protists express cellulase, this ability depends on microbiome-host interactions in almost all vertebrates and invertebrates, (Zheng et al., 2019). Its evolutionary importance is that, once in place, herbivory opened up major new opportunities for organisms to live in new habitats (Box 19.1); indeed, major dinosaur and mammalian clades (e.g. the ornithischians and the ungulates) have evolved as a result of such holobiosis. Other examples demonstrate that the microbiome of animals can have important roles beyond facilitating digestion. In humans, for example, it is now clear that microbiome activity helps modulate pain (Guo et al., 2019) and enables the immune system to function properly (Postler & Ghosh, 2017). The assay here is simple: if the microbiome is affected, so is the health of the host and this mutual strengthening is a defining property of the holobiont.

BOX 19.1 THE EVOLUTION OF HERBIVORY

The key role of the holobiont in an evolutionary context is that microbiome-host interdependence can open the door to otherwise inaccessible niches, and this can have profound effects. The best-known example here is herbivory (for review see Gilbert, 2019): many vertebrates require plant food as their sole source of nutrition, and, in every case, the ability of that animal to break down cellulose derives from its microbiome as the requisite enzymes (e.g. cellulase and pectinase) evolved only in prokaryotes.

All early land-based animals were, on the basis of their dentition and body features, carnivores, with herbivory evolving long after the evolution of terrestrial animal life. The earliest tetrapod that, on the basis of its fossilised skeleton and teeth, exhibited herbivory was *Eocasea*, an early synapsid pelycosaur that lived during the late Carboniferous ~300 Mya: it had the bulky digestive tract required for herbivory of later animals known to be herbivores (Reisz & Fröbisch, 2014). These authors point to the presence very soon after *Eocasea* of synapsids with the barrel-shaped bodies, spade-shaped teeth, jaw morphology, and other features that characterises herbivory.

It was, of course, only after the evolution of an animal gut that could digest plants would there was any reason for the evolution of the secondary anatomical features associated with herbivory, and they would have been accompanied by the evolution of associated browsing and grazing behaviours. Even today, the genes for the proteins needed to break down plant tissues are still not part of any vertebrate's genome but are synthesised by bacteria and so are part of the holobiome. It seems that every herbivore has developed a community of plant-digesting microbes in its gut. The fact that, after ~300 My of herbivory, no horizontal gene transfer of these enzymes from bacteria to a mammal has been found is not only surprising but emphasises the difficulty of horizontal gene transfer from the microbiome to germ cells in vertebrates.

Of all the anatomical changes associated with herbivory in vertebrates, perhaps the most interesting is the evolution of the cow's rumen, which is not only a stomach chamber in which the assorted microbiota digest plant material but an intriguing example of developmental plasticity. The initially small rumen is first seeded with microbes as a result of the new-born's passage through the birth canal but does not become active until the calf is weaned. This is because the newborn calf has an oesophageal groove, which is opened by the sucking reflex, so allowing the milk to bypass the rumen. Once the calf starts to eat grass, however, the groove closes and food enters the rumen where the microbiome there starts to break it down. As the bacteria proliferate, they produce, inter alia, volatile fatty acids that include butyrate; these not only cause the rumen to enlarge but direct the associated muscle and ruminal papillae to differentiate. Hence, not only have the gut bacteria helped form their own niche, but they have expanded the developmental repertoire of the animal (Gilbert, 2019).

The situation in herbivorous insects is different as their genomes now contain many of the genes that encode plant-degrading enzymes. Kirsch et al. (2014) have phylogenetically analysed the pectin-depolymerase gene in a wide variety of plant-eating beetles and found that they all derived from Ascomycota fungi some 200 Mya via horizontal gene transfer (HGT), which was followed by a series of complex duplications and other HGT events. A crustacean that synthesizes a cellulase has also now been identified (Besser et al., 2018).

Six further examples illustrate the richness and the complexity of further ways in which the microbiome returns with interest the benefits that it derives from the provision of a habitat by its host.

- *Developing normal physiological function in embryonic mice.* Microbiome bacteria produce small fatty acids that are released into the maternal blood, cross the placenta and are recognised by receptors in the developing neuronal, digestive, and pancreatic systems of the fetus. Kimura et al. (2020) show that, as compared to normal mice, germ-free mothers with little or no gut microbiome produce offspring liable to obesity and metabolic syndrome, a liver disorder.
- *The health of coral colonies.* These require the unicellular algal symbiont, *Symbiodinium*, to take up residence in their ectoderm if they are to flourish. If this microbiome component deteriorates, their ability to make reefs is severely diminished. This is because *Symbiodinium* donates to its host most of the organic material it makes as a result of photosynthesis, and it is this material that is used to facilitate reef building (Muscatine et al., 1984; Decelle et al., 2018).
- *Protection against snake bites.* The rough-skinned newts (*Taricha granulosa*) and some other organisms protect themselves against predators, such as snakes through the production of tetrodotoxin (TTX), a neurotoxin that blocks voltage-gated sodium (Nav) channels. Vaelli et al. (2020) have now shown that this toxin is not actually made by the newts, but by several bacterial holobionts that live on the newts' skin. In parallel, the Nav receptors of the newts have become modified so that they are impervious to the effects of TTX.
- *The digestive system of a tube worm. Lamellibrachia luymesi luymesi* are siboglinid tube worms that lack a digestive system. This is provided by Gammaproteobacteria that live within the worms' specialised trophosomes, small organs within the coelomic cavity. The worms survive on areas of the deep-sea floor in areas where chimneys at hydrothermal vents release the sulphur-laden "smoke" of heated water. The worms absorb this water and transfer the sulphates to their trophosomes, where the microbiome metabolises it for its own benefit, simultaneously providing nutrition for the worm (Cordes et al., 2005).
- *The digestive system of termites.* There is an unusual holobiont/symbiont interaction between termites and *Termitomyces* (basidiomycete) fungi where the former cultivate gardens of the latter. This situation is different from the behaviour of leaf-cutter ants, mentioned earlier in the chapter, that colonise fungi for food. In this case, the termites provide plant material to feed the fungi, which break down the cellulose to a form that the termites can then eat. The fungus thus provides part of the digestive system of the holobiont (Aanen et al., 2002).
- *Commercial mushroom cultivation.* This fails if its associated microbiome deteriorates (Carrasco & Preston, 2020).

Holobiont interactions can in some cases lead to gene sharing, and an example of such gene flow that has already been mentioned is that from *Wolbachia* to *Drosophila* (Chapter 17). A further example is in the way that the pea aphid, *Acyrthosiphon pisum*, has acquired the ability to eat sap. ~150 Mya, a gammaproteobacterium, *Buchnera aphidicola*, colonised an aphid ancestor and the two entered into a close holobiotic relationship with the bacterium living within the aphid's gut epithelium. As a result, the *B. aphidicola* genome has now shrunk (Bennet & Moran, 2015), although it can still make the amino acids and peptides that aphids require to exploit sap. The inheritance of the diminished bacterium is now very complex: the aphid gut epithelium exocytoses the bacteria which eventually transfer to the early embryo (Koga et al., 2012). Inheritance is thus matrilineal. The example of *A. pisum* and *B. aphidicola* are two cases where the holobiont organisms come to depend completely on one another—both are now new species defined by their requirement for one another.

Bennett and Moran (2015) point out some of the evolutionary risks that a strong degree of co-dependence brings. First, the degree of specialisation in both partners means that a deleterious mutation in one organism may be hard for the other to compensate. Second, the initial survival requirements of the two organisms may diverge if the nature of the environment of the holobiont changes. Third, natural mutation in the two organisms may lead to genetic incompatibilities in the long-term. It should, however, be said that, given the large number of symbionts and holobionts that are observed in nature, these risks seem likely to be low.

The wider importance of holobionts is that their communal activity creates new variants and opens a host of new evolutionary niches, as the example of herbivory has demonstrated. As Gilbert (2019) has pointed out, symbionts within the holobiont have extended the range of possible phenotypes, facilitated niche construction and helped direct the expansion by the environment of the phenotype of the holobiont.

DEVELOPMENTAL PLASTICITY AND ADAPTIVE CHANGE

The discussion so far has focused mainly on events rather than mechanisms and here the guiding rule was pointed out by Van Valen (1973) who wrote: "A plausible argument could be made that evolution is the control of development by ecology." Evolutionary success for a possible new species requires either that it has already adapted by chance or that it has the ability to adapt to a new environment, as the example of the rumen (Box 19.1) clearly demonstrates. Either way, this means that the developmental trajectory of the new species will be different from that of its parent species. This process of change will inevitably be speeded up if there is some plasticity in the developmental processes that underpin these differences. It is only recently, some 50 years after Van Valen's perception, that ideas of developmental plasticity and what Gilbert et al. (2015) have termed eco-devo-evo are attracting attention (Minelli & Fusco, 2010; Gilbert, 2019; Levis & Pfennig, 2019). The molecular-genetic basis of some examples of developmental plasticity are now being investigated and have been reviewed by Lafuente and Beldade (2019).

Developmental plasticity is the ability of organisms to react to their surroundings by modifying their development. Well-known examples include temperature-dependent gender determination that is shown by a few reptiles (Kohno et al., 2020) and the ability of organisms to respond to seasonal changes. In this case, the maintenance of plasticity is an important adaptive ability. Sometimes, however, groups of organisms may inhabit a range of environments, each with its own feature. In such cases, plasticity will give rise to a range of phenotypes that, if the environment is stable over the long term, may well become fixed in each organism as the plasticity is no longer needed. This phenomenon is known as "plasticity first" adaptation. An experimental example is Waddington's "genetic assimilation" in which an environmentally determined phenotype becomes the normal phenotype (Chapter 20).

Gilbert et al. (2015) cite several examples of developmental plasticity that are known to be a response to environmental pressures. Two, in particular, are visually impressive and so worth mentioning; these are horn production in male Onthophagus dung beetles and context-dependent development in Senegal bichir fish (*Polypterus senegalus*). Onthophagus beetles develop head horns, an ability controlled by expression of the doublesex (dsx) gene (Beckers et al., 2017; Rohner et al., 2020). They will, however, only do this if food supplies are good, an ability that is thus context-dependent and reflects developmental plasticity (Fig. 19.6a).

The Senegal bichir (*Polypterus senegalus*), a bony fish with ray fins (a member of the Actinopterygii), is one of the most extraordinary fishes known. Bichirs have both gills for oxygen uptake in water and small lungs for breathing when on land, which they do through a pair of dorsal spiracles (Graham et al., 2014). In the former environment, they swim normally, but in the latter, they can use their fins as walking limbs (Fig. 19.6b; see Websites). The developmental plasticity of the fish is shown by the observation that fish that develop on land not only have better walking ability but also have strengthened, modified fin bones (Molnar et al., 2017; Standen et al., 2014). As these authors point out, this behaviour emulates some of the early behaviour of stem tetrapods.

Perhaps the key point here is that developmental plasticity encourages evolutionary diversification since those species showing it will be more easily able to colonise new habitats and so generate new species than those whose developmental flexibility is minimal. Adaptation is a first step on the way to full speciation and

Figure 19.6 Two animals that show developmental plasticity. (a): The size of the horns of *Onthophagus taurus* depends on the amount of food available. (b): The size and shape of the bones in the fins of the bichir fish depend on the extent to which they walk on land (see Websites).

(a Courtesy of Jansuk Kim.
b Courtesy of "Royal Ace." Published under a CC Attribution-Share Alike 4.0 International License.)

normally starts with a group displaying a variant property that enables it to flourish in a novel (**Fig. S4.1**) niche. The obvious variant properties for land animals likely to facilitate the adoption of a new niche are those that modify size and pigmentation, although physiological and behavioural variations can also be important. Almost all adult adaptations have their root in embryonic development and many reflect later developmental plasticity.

KEY POINTS

- There is an ever-shifting spread of phenotypes within a population due to genetic drift and mutation and each is subject to selection on the basis of its fitness.

- Successful adaptations are those that enable organisms to improve their chances of reproducing successfully.

- Modes of adaptation include camouflage and mimicry.

- Organisms adapt to produce a breeding strategy that optimises offspring numbers. The strategies extend from producing few offspring that are well looked after (the **K** strategy; e.g. land vertebrates) to laying large numbers of fertile eggs that require little or no parental care (the **r** strategy; e.g. plants and teleost fish).

- Parasites are organisms that have adapted to a niche that is another organism, giving nothing in return.

- Symbionts are pairs or groups of organisms that have adapted to sustain one another.

- Holobionts are composite organisms each of which includes a host organism together with prokaryotic and unicellular eukaryotic organisms for which they provide a niche. In many cases, the mix of species have evolved to provide necessary support for one another.

- Important examples of microbiome contributions to the success of the host are (i) the cellulase-degrading prokaryotes that enable herbivory in a wide variety of organisms and (ii) the human microbiome that facilitates normal digestion and immune activity.

FURTHER READINGS

Guerrero R, Margulis L, Berlanga M (2013) Symbiogenesis: The holobiont as a unit of evolution. *Int Microbiol*; 16:133–143. doi:10.2436/20.1501.01.188.

Herron JC, Freeman S (2014) *Evolutionary Analysis*. Pearson. England.

McFall-Ngai M, Hadfield MG, Bosch TC et al. (2013) Animals in a bacterial world, a new imperative for the life sciences. *Proc Natl Acad Sci U S A*; 110:3229–36. doi: 10.1073/pnas.1218525110.

Saxena RK, Saxena S (2015) *Animal Adaptations: Evolution of Forms and Functions*. MV learning.

WEBSITES

- The wiki article on *Adaptation* (and its associated links), *Holobionts, Symbiosis*
- Video showing Bichir fish walking: https://www.nature.com/news/how-fish-can-learn-to-walk-1.15778

Selection

20

The focus of this chapter is on how advantageous traits and their underlying genes are passed on to the next generation at the expense of less advantageous ones. It starts with a discussion of fitness, a measure of the likelihood that a particular individual in a group will contribute either a particular trait or a gene allele to the next generation. It then discusses the three major modes by which traits with a higher fitness are more likely to be found in the next generation. The first is natural selection: this is essentially the effect of the environment in its widest sense on a population and the example of the camera eye is considered; this is because its evolution now been shown to be more straightforward than used to be thought. The second is sexual selection, which enhances traits that increase the chances of reproduction. There are many examples of animal and plants phenotypes improve an individual's likelihood of mating with an organism of the opposite sex. There are also good reasons why sexual mating is so common when hermaphroditic or parthenogenetic reproduction is both simpler and more reliable. The third is kin selection which acts on traits that enhance the future of the group at the expense of the individuals displaying such altruistic behaviour. In this last case, it is not easy to understand how a trait that is not beneficial to an individual can be selected. Thanks, however, to the work of Hamilton, Price, and others, we now have some insight into how such counterintuitive events can happen on the basis of inclusive fitness, which includes the care of orphans.

The last part of this chapter considers the speed of evolutionary change under controlled selection. The best-known examples involve selective breeding on the basis of a heritable trait that is rare. Producing a small population in which such a trait is part of the normal phenotype typically takes less than ten generations. This is followed by a discussion of the speed of change under artificial conditions when the phenotype is not obviously present in a wild-type population. A key example here is Waddington's work on generating a bithorax (four-winged) phenotype in a population of wild-type *Drosophila*, which normally has only two wings and two halteres (balancing organs). Under extreme selection, Waddington was able to effect this change in anatomy in about 20 generations. Part of the significance of this work today is that it demonstrates the extent and significance of the genetic variation within a normal population.

Selection, as a word, has the ring of an active process but it actually reflects something far more passive in the domain of evolution. Every niche is composed of a host of populations of organisms, each a spectrum of phenotypes and all competing for food and attempting to reproduce. This competition is not only between species but within them. Some populations are independent, some are part of a community of similar organisms while others may depend on support, which may be mutual, from other species. Over time, some species carry on much as they always did, some become more prevalent, others fail while a few change more or less rapidly. In due course, the end results of these competitive interactions can be seen as the

DOI: 10.1201/9780429346217-24

results of the process of selection (Fig. S4.1). Describing these changes over the long term is made more complicated by the fact that aspects of that niche may well have changed over this time, be it organic (animal, plant, fungus, protist, or prokaryote) or inorganic (e.g. climate). Because of the unpredictable nature of these events, selection is, under natural conditions at least, essentially a stochastic and complicated process.

This chapter considers selection within, rather than between species, to consider how allele frequencies can change, an early step on the way to speciation. A key parameter here is fitness: while selection is an end result, fitness is a measure of the immediacy of the process of change. The simplest definition of the fitness of a phenotype in a particular environment at a particular time reflects the likelihood that it will produce offspring that will go on to reproduce in the next generation, although there are other definitions as quantifying this parameter is not easy (Orr, 2009; Akçay & Van Cleve, 2016). One variant has a greater fitness than another in some environments if it leaves more fertile offspring. In this case, it will take relatively few generations on an evolutionary timescale for that variant to become dominant within the population. If, for example, a new heritable trait gives a 1% selective advantage in a sizeable population, it can be shown using the theory of population genetics and reasonable assumptions about genetic heritability, selection pressures, and population size that it can only take ~1000 generations for the trait to have spread through the great majority of the population (Nilsson & Pelger, 1994). On the scale of evolutionary time, this period is almost instantaneous, but for practical purposes, it is a very long time indeed: even for *Drosophila* which has a breeding cycle of about two weeks, a thousand generations represent some forty years.

The implication of this slow rate of uptake of a trait means that it is only rarely possible to follow evolutionary change outside of selective-breeding conditions, where the niche, population size, environmental conditions, and strength of selection can all be controlled. In practice, this mainly means commercial breeding and laboratory-based experiments in which heritable change can be regulated and followed; some examples are discussed later in this chapter. In general, however, research on selection has to be followed in the field and, on the basis of a large amount of such work, it is now clear that the process generally falls into three categories: *natural selection, sexual selection* and *kin selection.*

The other important factor that leads to phenotypic change is *neutral selection*. It has long been known that many minor traits are subject to little or no selection, either positive or negative. An example is those genes that make make human faces different from one another. It might be thought that the frequency of these alleles would be fixed as their presence leads to no associated selective advantage or disadvantage. In 1968, Kimura put forward the neutral theory of evolution which suggests that such neutral alleles are, as a result of random breeding, continually being moved within a population through genetic drift; as a result of this, asymmetric allele populations can arise in small populations. Such alleles may even become a part of the normal phenotype and so may wrongly appear to have been subject to positive selection (Kimura, 1968, 1991). The neutral theory of genetic change is considered in Chapter 21.

NATURAL SELECTION

Natural selection is the umbrella term that covers the influences from the environment in the widest sense that affect an organism's ability to reproduce; it is sometimes called *ecological selection* to distinguish it from sexual or kin selection. Natural selection occurs when events both outside and within a group of individuals encourage one variant in a population to leave more reproductive offspring than another. Natural selection may occur because one animal variant has a better chance of surviving against predators or of catching prey than another from the same species, or can better survive changes in climate, or can access a novel form of food. A plant variant may be more tolerant of temperature tor drought than others, root better in the soil, or have a leaf display that allows it to access more light and so improve its photosynthetic ability, so producing more seeds. In other cases, an organism may be immune to a particular disease or produce zygotes that are more robust (e.g. seeds may become tolerant of particular soils or be able to be quiescent under suboptimal conditions and only germinate when better conditions return).

The list of possibilities is endless and the situation is made more complex for two reasons. First, it is not the trait that is the unit of selection but the complete organism, or for larger organisms, the holobiont (the organism with its microbiome; Chapter 19). Second, different selection pressures generally apply to each of the different traits that together comprise an organism's phenotype. In this context, a mix of minor traits may prove to be more beneficial to a specific variant in a particular environment than a single apparently more important one. It is only under laboratory conditions that the selection pressure for a particular phenotype is unaffected by secondary interactions.

One factor that clearly affects selection is coat colouration, as the butterfly examples mentioned in Chapter 19 have illustrated (Fig. 19.2). Similar arguments apply to vertebrates: thus, polar bears have a much lighter pelt than do brown bears, presumably because the former offers camouflage against ice whereas the latter are less visible in woods (Edwards et al., 2011), while ermine have a dark coat in summer that changes to white in winter) An intriguing example in this context is the stripe pattern on zebras and a range of possible uses have been suggested over the years for this feature which makes them highly visible during the day. Given that almost all herbivores are rather drab in colour and pattern, animals with bright striping should be particularly visible to predators and so subject to negative selection. Possible advantages of bright striping include group recognition, camouflage at dusk when the animals drink, a dazzle effect that might confuse predators both large and small (e.g. flies; Caro et al., 2019), visual cues for mates, and the differential heat absorption of black and white hairs leading to a cooling effect (Cobb & Cobb, 2019). It is quite possible that all, or at least some of these factors ensured the selection of the patterning, but it is hard to work out the precise selective value of striping when experimentation is so difficult. This difficulty is exacerbated by the fact that different species of zebra display different stripe numbers and morphologies, features as likely to have arisen through drift as selection (Fig. 13.5).

For such reasons, the exact explanations for one variant of any species rather than another having a selective advantage are hard to prove in situations where experimental investigation is impractical, and this is the very great majority of cases. The reasons are generally inferred on the basis of comparisons among extant and extinct organisms and their particular environmental context. (*Note*: Readers who do not know them may enjoy Kipling's *Just so* stories about trait acquisition; these were written for children in memory of a daughter who died young.) An improvement on this "because it would have worked" argument is to pose the question of evolutionary stability: if the current adaptation were much weaker, would there be any selection pressure for the frequency of that trait to increase now?

It seems obvious, for example, that because almost all vertebrates have eyes, those with limited or no vision, such as moles, descended from animals who lost vision when they went underground, although eye development in such organisms starts off normally in the embryo (Carmona et al., 2010). Under these circumstances, it was clearly far more important that such animals improved their touch and digging ability than that they maintained good eyesight—hence selection for the extra sesamoid digits that they now have (Chapter 19). For all that the explanation makes sense, it cannot be tested, although the mole phenotype does suggest that, without active selection, facilities deteriorate over long periods (the *use it or lose it* hypothesis).

Sometimes, however, the only convincing explanation for some aspects of the phenotype seems to be one based on the neutral theory rather than selection: examples include why the right and left lungs of humans should have three and two lobes while those of mice should four and one lobes or, as mentioned earlier, or why different zebra species should have different numbers of stripes (Chapter 18). Nevertheless, there seems little reason to doubt that, unless other forms of selection or neutral effects are responsible, it is the blind power of natural selection that decides which of the available variants within a population that, for example, finds itself in a new environment, will eventually become the norm.

Selection as a subject has the advantage that it can be studied quantitatively. Chapter 21 considers the mathematical theory of evolution which integrates population genetics with selection to analyse the future of allele distributions for a particular trait within a population of some organism. It is worth pointing out that the theory is generally used to study a single, large population in a fixed environment. The reality is more complicated for at least three reasons. First, many populations are sufficiently small for both drift and selection effects to be important; second, traits are frequently underpinned by a range of alleles; third, the population being examined is generally competing for environmental resources (food and shelter) with other organisms. In cases where selection within a small population depends on several parameters, it becomes impossible to predict which aspects of the phenotype will be favoured (see Websites for Chapter 21).

The evolution of the camera eye under selection

The easiest means of assaying change over long periods is through those few hard tissues that fossilise. Equally important changes also occur in soft tissues because they reflect physiological function, albeit that they are generally lost from the fossil record. Perhaps the most interesting example of a soft tissue in which anatomy and physiology are closely linked is the camera eye. This has a special place in the history of evolution for several reasons, the most important of which is that it is an extremely complex organ that only seems to work if all of its components are already in place.

Even well into the 20th century, there was some doubt as to whether it was actually possible for a camera eye to form through a series of small changes, with each change being advantageous to the organism. Darwin thus saw that providing an explanation for the evolution of the eye was a necessary test of any theory of evolution, as he wrote in *The Origin of Species*:

> *To suppose that the eye, with all its inimitable contrivances for adjusting the focus to different distances, for admitting different amounts of light, and for the correction of spherical and chromatic aberration, could have been formed by natural selection, seems, I freely confess, absurd in the highest possible degree. Yet reason tells me, that if numerous gradations from a perfect and complex eye to one very imperfect and simple, each grade being useful to its possessor, can be shown to exist; if further, the eye does vary ever so slightly, and the variations be inherited, which is certainly the case; and if any variation or modification in the organ be ever useful to an animal under changing conditions of life, then the difficulty of believing that a perfect and complex eye could be formed by natural selection, though insuperable by our imagination, can hardly be considered real.*

The selection pressures driving improvements in vision are obvious: any organism that can see food, identify a potential mate or avoid a predator has an immediate advantage over one that does not, and the greater the visual acuity, or ability to resolve detail, the stronger this advantage will be. Basic light sensitivity is very ancient and is found in many unicellular eukaryotes. Members of the light-sensitive rhodopsin family of proteins are even found across the prokaryotic kingdom (Böhm et al., 2019; Pushkarev et al., 2018) and they underpin the activity of both camera and compound eyes. There is even a molecular unity underpinning all animal eyes: each has a Pax6 transcription factor whose early expression drives the subsequent development of its eyes (Chapter 8). It is, however, important to note that the visual apparatus is worth nothing if it is not supported by a neuronal image-analysis system linked to muscular activity.

Camera eyes evolved separately in vertebrates and molluscs and, although the geometries are slightly different (Chapter 4), develop similarly. In both cases, the Pax6 evidence shows that each clearly derived from a patch of light-sensitive cells on the head of an early member of the Bilateria that made neurological connections to the brain. The evolutionary route from this primitive light sensitivity to the sophistication of the camera eye is not immediately obvious, although there are a set of extant mollusc eyes that indicate the stages by which this might have happened (Fig. 20.1a; Salvini-Plawen & Mayr, 1977).

Nilsson and Pelger (1994) took an unusual and original approach to explore how long it took for the camera eye to evolve from a small, flat plate of light-sensitive cells backed by a pigmented epithelial layer. They worked on the assumption that the driving force for change was an increase in visual acuity, with a 1% heritable increase giving an organism a selective advantage. They then explored how visual acuity could be increased by series of changes to the geometry of the eye, each of which would increase its acuity by 1%, while maintaining its diameter. They showed that the necessary geometric changes to produce the camera eye from a flat, photosensitive region of ectoderm involved three substantial morphological steps: rounding, pinholing, and then cornea and lens production, each stage of which is seen in the adult eyes of different mollusc species (Fig. 20.1a,b). In practical terms, Nilsson and

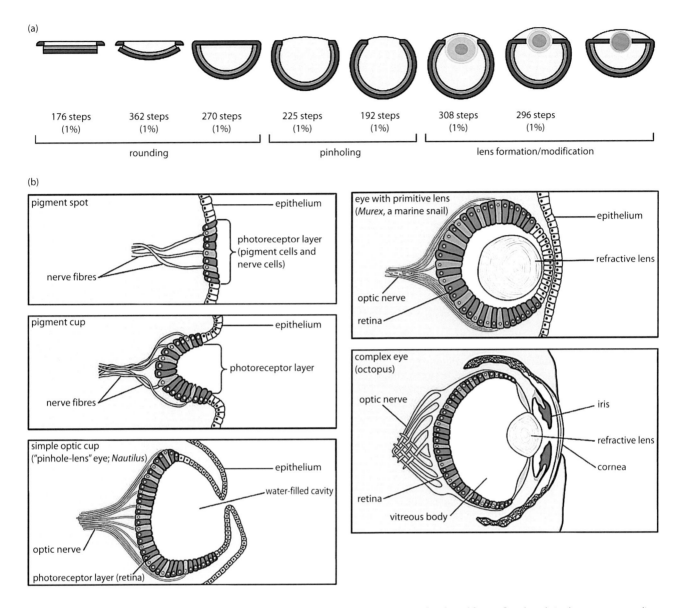

Figure 20.1 The evolution of the camera eye. (a): The likely way by which the camera eye developed from a flat placode in three steps: rounding, pinholing, and lens formation. The number under each eye is the number of modification steps, each increasing acuity by 1%, required to achieve the next eye in the set. (b): Although this set of eyes looks like a developmental sequence, it is actually drawings of the eyes of five adult molluscs with increasing visual acuity.

(a From Nilsson DE & Pelger S (1994). A pessimistic estimate of the time required for an eye to evolve. Proc. Roy. Soc. 256B:53–8. With permission from The Royal Society.
b Courtesy of the Encyclopedia Britannica.)

Pelger showed that, while the flat sheet of cells could only distinguish between light and dark, the final structure could resolve ~100 discrete objects.

Their first step was to analyse the optical properties of these intermediate eyes and calculate the number of steps required to achieve them under the constraints that each increased acuity by 1%, and that eye diameter remained constant. Nilsson and Pelger then used the framework of the population genetics theory (Chapter 21) to calculate the number of generations that it would take for these small improvements to eventually become established. Using the appropriate equations from the theory, they showed that, under the most conservative of assumptions, the evolution of the complete camera eye formation would take about 360,000 generations, or the same number of years for an organism that reproduces annually. This would represent ~0.5% of the Cambrian period (541–485.4 Mya), a surprisingly short period of time for the evolution of such a sophisticated organ, albeit that improvements in visual acuity would need to be accompanied by advances in neural image processing and interpretation.

This analysis, together with the range of morphologies seen in extant mollusc eyes, suggests that the camera eye evolved sequentially from a simple light-sensitive placode during a relatively short time, probably during the late Ediacaran Period as even some jellyfish have lensed eyes (Garm & Ekström, 2010). The significance of Nilsson and Pelger's work is that it shows that strong selection pressures can affect major changes to anatomical detail relatively fast on an evolutionary timescale.

SEXUAL SELECTION

An important aspect of selection is ensuring that organisms breed successfully: if a variant cannot pass on its genes, that variant phenotype will be lost. Considerable effort has, therefore, gone into understanding *sexual selection* nd the evolution of traits that encourage individuals in a population to increase their fitness by increasing the likelihood of mating and producing offspring (possible reproductive strategies have already been considered in Chapter 19). Darwin was the first person to explore these ideas, and they are an important part of his second major book on evolution, *The Descent of Man, and Selection in Relation to Sex* (1871). The reader requiring more detail on this topic than can be given here should consult the texts at the end of this chapter.

The advantages and disadvantages of sexual reproduction

Because sexual dimorphism is almost ubiquitous in the multicellular world, it is easy to think of it as inevitable, but the situation is not as simple as it seems. There are many organisms that are hermaphrodites and so able to produce both male and female gametes. Jarne and Auld (2006) estimated that ~30% of noninsect animals are hermaphroditic, with most being able to act as males or females, albeit with only a small proportion being able to self-fertilise. Although a few fish, such as the clownfish, sea bass, and groupers, are hermaphroditic, most hermaphrodite animals are invertebrates and particularly include snails and slugs, worms, and echinoderms. As is better known, many plants have both types of sex organs in the same flower, but only a proportion of them can self-fertilise or produce unfertilised seeds that develop normally to give female plants (such asexual reproduction is known as apomixis).

There are also many animal organisms that reproduce asexually, although in this clade it is known as parthenogenesis. Many phyla have species that sometimes or even always reproduce parthenogenetically; common examples include nematodes, scorpions, aphids, and some bees and wasps. Honeybees are interesting here: only the queen bee lays eggs and those that are not fertilised become males (or drones) and these are thus produced parthenogenetically. The sole role of drones is to mate with queens and those fertilised eggs that are female become workers and very occasionally queens.

These examples illustrate the fact that although sexual reproduction is an option seen in most multicellular phyla, it is not obligatory. Given that sexual reproduction is intrinsically inefficient, this is not surprising: for a sexually reproducing organism to maintain its population, male gametes not only have to meet up with female ones (by direct or indirect sexual contact) but the female has to produce at least two offspring on average that will in due course reproduce if the population is to least maintain itself. To sustain a population through asexual reproduction, each organism merely has to leave a single offspring; if it leaves more than one on average, the population will expand. If all other things are equal, populations that produce asexually will usually expand more rapidly than those that produce sexually.

What then is the advantage of sexual reproduction? The obvious answer is that the genetic complements of offspring produced sexually, based as they are on recombination and dual parental contributions, will show far greater variability than those deriving from asexual reproduction. In spite of meiotic crossover, there will be much less variation in cases where egg and sperm come from the same adult and even less in organisms that are the result of parthenogenesis. As the previous chapters have made clear, the extra genetic variability that results from sexual reproduction is the key both to buffering against environmental change and to generating the phenotypic variability that confers evolutionary adaptability. In summary, sexual reproduction is more robust than asexual. The evidence that most animals, and certainly all the larger ones, reproduce sexually confirms that the advantages of sexual reproduction outweigh its cost in inefficiency and in the demands that it makes on the phenotype and its social behaviour.

The sexual phenotype

Sexual reproduction requires traits that encourage sexual attraction and mating, and a considerable amount of adaptation and selection has underpinned the evolution of secondary sexual characteristics that help an individual attract the opposite sex or give it a copulatory advantage. The most dramatic examples are those where there is male–male competition, and this can sometimes lead to one male treating all the females in a group as his personal harem, a situation that encourages the evolution of alpha males. Here, there are many examples where males compete for sexual dominance in a group, usually on the basis of strength, and this is one of the drivers of male characteristics. A well-known example of sexual dimorphism is the antlers that male deer alone develop and use to establish dominance during the rutting season (Fig. 20.2a). It is clear that sexual selection favours phenotypic properties that enhance an individual's sexual activity and so increase the frequencies of that individual's gene variants within the next generation.

In other cases, the male chases and the female chooses; to make themselves eye-catching here, males have evolved display characters that make them particularly attractive to females. Two well-known examples of such sexually dimorphic traits are the peacock's feather train (Fig. 19.3b) and the lion's mane. Other examples include the startling display of the male peacock spider, the bright feather pattern of mandarin ducks and the apparently attractive proboscis of the male elephant seal (Fig. 20.2b–d). The peacock train is particularly interesting as its advantage in attracting female peacocks and so improving a male's sexual fitness overcomes its obvious disadvantage in rendering the bird less mobile against predators. Particularly, dramatic examples of sexual dimorphism are known as Fisherian runaway selection (Fisher, 1930); this is because these extreme phenotypes seem to arise through positive feedback, although the reasons why such dramatic phenotypes are so rare remains unclear.

Rather less common across the animal world are cases where it is the females who display phenotypic characters that make them attractive to visually undistinguished males. In some cases, both sexes develop secondary sexual characteristics, and an example here is *Homo sapiens*: during puberty, males develop beards and prominent musculature under the hormonal influence of testosterone while women, unlike female monkeys, develop prominent breasts under the hormonal influence of

Figure 20.2 **Sexual selection.** (a): Whitetail deer (*Odocoileus virginianus*) locking horns during the rutting season. (b): A male peacock spider (*Maratus volans*) in the courtship position. (c): Male (left) and female mandarin ducks (*Aix riculata*). (d): The male elephant seal (*Mirounga angustirostris*) has a much more pronounced proboscis than the female.

(a Courtesy of Brian Stansberry. Published under a CC Attribution-Share Alike 3.0 Unported License.
b Courtesy of "KDS444." Published under a CC Attribution-Share Alike 3.0 Unported License.
c Courtesy of Francis C Franklin. Published under a CC Attribution-Share Alike 3.0 Unported License.
d Courtesy of Broke Inaglory. Published under a CC Attribution-Share Alike 4.0 International License.)

oestrogen. Mating behaviour has thus evolved on the basis of mutual attraction that includes these and other features, although in human populations, there is also a range of strong and varied cultural constraints that regulate sexual behaviour across many human groups.

It is important to note that the possession of secondary sexual characteristics is not always enough to ensure an interest in copulation. Indeed, mating is not the usual primary interest of either sex, but often requires physiological stimulation that may involve a strong time element. In many species, one or other sex can, at particular times of the year, secrete behaviour-changing odours (pheromones) and some of these act as sexual attractants. While female secretion of pheromones to attract males seems more common, the reverse also happens in taxa as diverse as wasps, lizards, and birds. There are also internal biochemical events, such as monthly or annual hormonal production, whose effect is to increase sexual desire. Examples here are the monthly change in the colour of female genitalia in some simians that attract males, and the roughly annual season of musth in male elephants, when their high testosterone levels enhance their interest in mating.

The sex ratio

In almost every species where sexual reproduction takes place, there is a roughly 1:1 ratio of males to females. This primarily derives from the fact that there is an asymmetry in the chromosome complement of germ cells. In many vertebrates and other organisms, sperm have an equal chance of carrying a male (Y) or female sex (X) chromosome, while the female gamete (egg) carries two X chromosomes. Other organisms have the reverse asymmetry: birds, some fish, and other invertebrates have a ZW sex-determination system. Here, male gametes have a ZZ chromosome pair while eggs have a Z and a W chromosome, with the Z chromosome being more gene-rich (for further details on the evolution of the ZW system, see Ellegren, 2011).

Humans are interesting in this context because the implications of the Y chromosome being slightly smaller than the X chromosome have been investigated. This difference enables Y-chromosome sperm to swim slightly faster than X-chromosome sperm (Sarkar et al., 1984) and, as a result, the balance of fertilisation is slightly altered in favour of males (the primary sex ratio). The result is that slightly fewer females than males are borne (the secondary sex ratio). As females have a slightly lower likelihood of early childhood mortality than males (Verbrugge, 1982), the ratio of the two sexes at reproductive age (the tertiary sex ratio) is restored to 1:1. The secondary and tertiary ratios are, however, further altered in some societies because, for example, female foetuses are selectively aborted for cultural reasons.

Fisher (1930) and Hamilton (1967) have provided an explanation as to why the tertiary sex ratio in sexually mature organisms should be 1:1. Their argument was that if producing offspring required an equal contribution from each parent, this figure would provide a stable equilibrium value, alternatively known as an *evolutionarily stable state* or ESS. (*Note*: ESS also stands for evolutionary stable strategy which refers to a behaviour strategy that is more stable than alternatives, at least near equilibrium.) Deviations from this equilibrium would self-correct in time because, were males to be less common at birth, they would have better breeding prospects than females and so produce more offspring. Thus, parents predisposed to produce males would produce more grandchildren than other parents and male births would become more common, restoring the original situation.

This argument fails occasionally in cases where the standard conditions for an ESS are not met. An entertaining example is provided by pyemotic mites (*Pyemotes ventricosus*) that are only ~0.2-mm long: these tiny organisms develop internally, with males inseminating their sisters before they are born. Of the ~250 mites born to a mother, some 92% are female and only about 8% are male (Bruce & Wrensch, 1990). The reason why this figure is stable seems to be either that the mother has some way of reducing the sex ratio or that there is dramatic and lethal competition among the very young males. The phenomenon is referred to as local mate competition.

KIN SELECTION AND ALTRUISM

The discussion in this chapter so far has focused on selection associated with anatomical or physiological variation, in terms of what might be advantageous to the individual. Kin selection, which reflects what is sometimes called *inclusive fitness*, is very different as it analyses heritable behaviour in the context of an extended family group, and particularly considers why an individual might do something that benefits the reproductive success of its relatives and the family group at the expense of its own – this is altruism. How such counterintuitive behaviour might evolve is not an easy question to answer, and the ideas, the theory, and the modelling of how this might occur have intrigued some of the greatest evolutionary theorists, including Darwin, Fisher, Haldane, Hamilton, Maynard Smith, and Price.

There are numerous examples of behaviour that can be viewed as deriving from kin selection (Bourke, 2014). While some come from insect societies (e.g. Wilson, 2000), two well-studied examples involve squirrels. First, a ground squirrel that senses predators will give an alarm signal that allows its group to escape, even though this behaviour risks drawing the predator's attention to itself, and its subsequent death. Of particular interest here is that such behaviour is particularly pronounced if the group includes close family members (Sherman, 1977; Mateo, 2002). Second, red squirrels are much more likely to adopt orphan pups that are related to them than those that are not (Gorrell et al., 2010). Other examples of altruistic behaviour are of the rearing of children by groups that include nonfamily members; these effects have been studied in foxes (Macdonald & Moehlman, 1982) and particularly in the cichlid fish *Neolamprologus pulcher* (Wong & Balshine, 2011; Box 20.1). In both cases, there is a dominant breeding pair whose activities are helped by other, nonbreeding members of the group who seem to derive no reproductive benefit from their behaviour and whose genes are, therefore, lost.

Kin selection would be easy to understand if natural selection were only concerned with increasing the frequencies of genes that are both widely distributed within and beneficial to a population in the sense that possession of the gene increases group fitness. From a population standpoint, it doesn't matter which individuals have these genes, and this is most easily seen through Haldane's joke "I would lay down my life for two brothers or eight cousins," which he later corrected to "I would truly die

---

BOX 20.1 CASE STUDY: ALTRUISTIC BEHAVIOUR IN BREEDING CICHLIDS

A particularly well-studied example of altruistic behaviour is the cooperative breeding shown by *Neolamprologus pulcher*, cichlid fish that live in groups of about 20 in Lake Tanganyika (for review, see Wong & Balshine, 2011). Typical groups include the male breeder (the largest male) and the female breeder (the largest female), with the remaining fishes forming a hierarchy, each with a role in helping with territory defence, territory maintenance, and brood care. As only the dominant pair is sexually active, it has been an open question as to how the helper group evolved. As these fish can be kept in tanks and experimentally manipulated, they provide a good system for studying the details of altruistic behaviour.

Work over the past 30 years has mainly focused on three questions: (1) why do subordinates stay rather than disperse to breed independently elsewhere; (2) why do subordinates forego breeding within the group; and (3) why do subordinates help rear the offspring of breeders? In trying to answer these questions, four hypotheses have been put forward.

1. *Kin selection*: Experimentation supports kin discrimination and positive correlation between relatedness and territory defence, although there seems to be no difference in the degree of help between fishes that were and were not related to the breeding female. There is some support for kin selection explaining that part of the altruistic behaviour that benefits relatives, but it fails to explain altruistic behaviour for nonrelatives, unless the fish cannot tell the difference.
2. *Pay-to-stay*: Here, helpers provide help as a form of rent that allows them to remain part of the group. The expectation here is that failure to help would

result in some sort of punishment and, in the limits, banishment. Although there can be a fair amount of aggression among group members, particularly in the presence of nearby neighbours, no evidence has been found to support this being associated with helping responsibilities (Hellmann & Hamilton, 2019).

3. *Signals of prestige*. This approach justifies altruism on the basis that the more a fish helps, the greater is its prestige within the group so that, for example, if one of the breeding fish dies, the next fish down the hierarchy is more likely to take on that role. The mechanism is supported by modelling but seems as yet not to have been experimentally tested in this system although it has been seen in groups of foxes (Macdonald & Moehlman, 1982).
4. *Group augmentation:* This mechanism of altruism focuses on the mutual benefits that the group obtains from the altruistic behaviour of a proportion of its members. The benefits, in this case, are obvious as they ensure the long-term maintenance of fish with size and vigour. It is, however, hard to see a mechanism that would drive a group of individuals each behaving as if it wished to increase its own fitness to become a group where a few behaved in this way with the remainder never reproducing. Not surprisingly, this hypothesis seems to have been more discussed than tested.

Cichlid altruistic breeding behaviour has now been studied for more than forty years but the mechanisms that drive it are still not completely understood. As of now, kin selection seems to be the major mechanism that underpins social behaviour in cichlids.

to save more than a single identical twin or more than two full siblings." From the point of view of the animal displaying altruistic behaviour, however, it can be hard to identify the direct mechanism (the immediate or proximate cause) by which the appropriate alleles are maintained. This is because the individual who suffers needs to have some physiological trait that makes it behave in a way that is detrimental to itself and may even be lethal, a trait that is not heritable in any obvious way, and is certainly unlikely to be dominant. Indeed, it is not obvious how such a trait might ever emerge under selection.

In the case of the squirrel adoption, the recognition of family is likely to be by smell (Mateo, 2002), and it seems that the altruistic behaviour is thus innate rather than reflecting an active choice. In the particular case of "viscous" populations where there is little dispersal of family members, an organism can safely behave as if all or most of nearby species members are kin. In terms of maintaining the altruistic phenotype, the adoption of orphan red squirrels by other family members does not depend on *personal fitness* but on *inclusive fitness*. The former is defined as the number of offspring that an individual produces that survive to reproduce. The latter counts the number of offspring that an individual rears and that eventually reproduce. For this count, related offspring are scaled according to the degree of their relatedness to the carer, with a sibling's and a cousin's child being, respectively, 1/2 and 1/16 of an offspring equivalent (hence Haldane's joke). Altruism reflects this increase in fitness. *Note*: These definitions of fitness differ from those used for discussions of natural and sexual selection (Chapter 21).

The first theoretical analysis for altruistic behaviour was that of Hamilton (1964; Box 20.2): he investigated the criterion for a gene increasing in frequency within a population when an *actor* performs some act for the benefit of a *recipient* rather than itself. He showed that, provided that the gene was already present within the population and that the number of group members who benefited was larger than the number that were killed, the criterion for the gene increasing in frequency within that population was

$$\mathbf{rB} > \mathbf{C}$$

Here, $\mathbf{r}$ represents the genetic relatedness of the recipient to the actor, $\mathbf{B}$ is the additional reproductive benefit gained by the recipient, and $\mathbf{C}$ is the reproductive cost to the actor. Of the parameters in Hamilton's rule, neither $\mathbf{B}$ nor $\mathbf{C}$ can be predicted. This criterion has, however, been shown to work for the case of squirrel adoption

BOX 20.2 THE DERIVATION OF HAMILTON'S RULE

The following simple derivation of Hamilton's rule gives a flavour of how altruism can be explained. Suppose a trait for altruism is present in a population with a probability $\mathbf{p_i}$ of it being shown by an individual (i.e. it fails to reproduce) and a probability $\mathbf{p_j}$ of that individual benefitting from another's altruism. In the special case where the relationships are linear, the inclusive fitness function $\mathbf{w_i}$ for an individual is given by

$$\mathbf{w_i} = \mathbf{w_0} - \mathbf{Cp_j} + \mathbf{Bp_i}$$

where $\mathbf{w_0}$ is the general fitness associated with nontrait associated genes, $\mathbf{C}$ is the reproductive cost (decreased likelihood of producing offspring) to the individual showing altruism, and $\mathbf{B}$ is the reproductive benefit (increased likelihood of producing offspring) to an individual as a result of the altruistic act of others.

Suppose that the altruistic trait is relatively novel and not yet at equilibrium in the community. The altruism trait will be encouraged to spread if $\mathbf{w_i}$ increases as $\mathbf{p_i}$ increases, and the criterion for this is that

$$\mathbf{dw_i}/\mathbf{dp_i} = -\mathbf{C} + \mathbf{B}\mathbf{dp_j}/\mathbf{dp_i} > 0$$

where $\mathbf{dp_j}/\mathbf{dp_i}$ is a measure of the relatedness between the two probabilities, a constant known as r. This has the value

of 0.5 for siblings and decreases as relatedness decreases (see text). The criterion thus reduces to:

$$\mathbf{rB} - \mathbf{C} > 0 \text{ or } \mathbf{rB} > \mathbf{C}$$

This derivation reflects the simplest possible case and the mathematical derivation of the altruism criterion for nonlinear and multigene conditions is considerably more complicated. For a fuller discussion of the basis of the rule, see van Veelen et al. (2017).

There are two additional comments about Hamilton's rule that should be mentioned. First, the theory gives no indication as to the values of $\mathbf{B}$ and $\mathbf{C}$, and, as it is impossible to predict them, they have to be worked out empirically. Second, Nowak et al. (2017) made the important point that the same data that are used to establish altruism are those that then have to be used to calculate these constants. Hence, the only way to test the theory is to split the data using one part to calculate the constants and the other to explore the validity of the rule (i.e. much like bootstrapping or jacknifing in phylogenetics, Appendix 5). For such an analysis to be valid, this splitting would need to be done many times stochastically and the results analysed statistically.

(Gorrell et al., 2010). More recently, Kay et al. (2020) have considered 200 papers that include quantitative data on altruism. Their analysis shows that, in every case, relatedness among the participants was a key aspect of the phenomenon, even in the 43 cases in which the original authors doubted that this was so.

One obvious concern with the rule was the original suggestion that the molecular complexity of the altruistic trait which involves recognition, neural processing, and physiological behaviour, that can be reduced to a single gene—it obviously cannot. Fortunately, the genetic complexity underlying this complex behaviour is not explicitly included in Hamilton's rule which only refers to fitness, costs, and degree of relationship. The relationship between trait-associated genes and DNA-based genes is considered in Chapter 21.

Kramer and Meunier (2016) have pointed out that the significance of Hamilton's rule is not restricted to altruistic behaviours: as **rB** and **C** represent, respectively, the indirect and the direct fitness consequences of any character of interest, they can be positive, negative, or zero. If both terms are positive, the trait is mutually beneficial; if both are negative, the trait can be seen as spiteful; if the direct component positive and indirect component zero or negative, then the behaviour will be selfish.

There is also a second way (see Kramer & Meunier, 2016) of looking at such social behaviour: this is to view selection as acting both on the individual and on the group of which that individual is a member. This is multilevel (or group) selection theory and in it selection can be deleterious for the individual and beneficial to the group. The underlying idea is that a member of a group will display altruistic behaviour if its sacrifice of the chance to breed ensures that other family members will propagate its genes. Here, selection effectively acts on the group rather than its members. This is seen as altruism, and it is driven by the selection of group benefits. It seems, however, that there is as yet no generally accepted formal theory for these ideas which allows them to be tested.

The theory of kin selection today provides the basis for genetic analysis of the evolution of social behaviour in groups of organisms. A key figure in developing this theory was George Price, who strengthened Hamilton's rule to produce what is known as the Price equation; this describes how gene frequencies change over time as a result of selection (Frank, 2012; Gardner, 2020). It is not discussed in further detail here partly because the derivation and formulation of the equation are mathematically complex and partly because it is not straightforward to use. The essential idea is to separate the evolutionary change in some population traits from one generation to the next into a part that derives from selection and a part that derives from other factors, such as genetic drift. This formalism can be used to describe how a trait or gene frequency changes over time, but not easily.

Price's work has been applied to the development of evolutionarily stable strategies for the behaviour of groups through the use of game theory, an area further strengthened by Maynard Smith (see Frank, 1995). There is now a considerable body of work, both theoretical and experimental, in this area, which is often known as sociobiology, with much of the experimental basis coming from genetic studies of behaviour in insect societies (e.g. Wilson, 2000). Unfortunately, the mathematics required to understand this theory is not simple, while the methodology for partitioning the cause of behaviour between heritable and environmental factors (sometimes called the balance between nature and nurture) is still contentious.

All of this theoretical work was recently thrown into doubt by the simulations of Nowak et al. (2010) on the evolution of eusociality, or social behaviour in groups. Their claim was that Hamilton's rule was inappropriate, particularly because eusocial behaviour evolved in groups of organisms that were not closely related. The authors then showed by simulation that any result that could be obtained by using Hamilton's rule could be obtained under normal Darwinian evolution. This paper in *Nature* was not liked by the evolutionary biology community and the response was a paper in the same journal the following year signed by more than 100 eminent evolutionary biologists (Abbot et al., 2011). This paper claimed that Nowack et al. had misunderstood eusociality, the forces that drove it, the assumptions that lay behind it, and its successes in explaining data. Insofar as the simulations of Nowak et al. were correct, Abbot et al. suggested that they did no more than show that, "if the fitness benefits are great enough, then altruism is favoured between relatives." These two papers were followed by a many others.

A particularly interesting contribution was that of Bourke (2015a) who carefully analysed the data on sex balances and behaviour in groups of Hymenoptera (bees, wasps, and ants). He showed that the predictions on numbers were not met by the ideas of Nowak et al. (2010) but were by standard kin selection theory. In a second paper (Bourke, 2015b), he discussed the odd fact that workers in insect societies sometimes kill their queen. He showed how kin selection theory explains the complex internal dynamics of bee colonies and shows that the closer workers in a wasp species are related to potential worker offspring, the more likely they are to kill the queen, having already reared replacements.

It should be said that kin selection remains a difficult subject both in theory and in practice but its mode of action was clarified by this academic argument. Analysis of the data showed the strength of Popper' view (Popper, 1959) that science advances by testing the predictions of hypotheses and seeing if they can be falsified.

THE SPEED OF CHANGE UNDER SELECTION

A few examples are known in which heritable change is rapid under conditions of strong selection and an example discussed in the last chapter was the alternation of mottled and black morphs of the peppered moth (Chapter 19). Vertebrates can also evolve relatively rapidly, particularly when predation rates are low and there are few if any constraints on variants. This is demonstrated by the novel species that evolved rapidly (on an evolutionary timescale) after the Permian and Cretaceous extinctions. At a more modest level, it also occurs when a founder population invades a novel environment in which there are few predators and selection pressures are hence weak. Perhaps the best-known example of this is the current flock of the many hundreds of species of cichlid fish seen in Lake Victoria whose evolution will be discussed in Chapter 22 (Hajkova & Lane, 2002). In general, however, full speciation in the wild seems to take many tens of thousands of generations if substantial anatomical modifications are involved. That said, it can be very much faster if the phenotypic changes needed require just a few single-nucleotide polymorphisms: the evolution of flounders that lay eggs on the sea floor from an original population that lays them at sea must have occurred within the last ~2400 generations (Schedel et al., 2019; Chapter 22).

Further evidence that anatomic change is generally slow is demonstrated by the rate at which differences in sets of transitional fossils accumulate (Chapter 6). A good example is the series of fossils that bridges the fin of a late sarcopterygian fish, such as *Tiktaalik*, which terminated in rays, and the limb of an early tetrapodomorph (or stem tetrapod), such as *Icthyostega* which had seven digits (Table 6.2). The known fossils date to ~380 Mya and ~374 Mya, respectively, although the precise time ranges for any of the intermediate species are not known. These changes seem to have resulted from fish being able to look for food in shallows with dense seaweed, with there being a selection pressure to push through the vegetation, probably while walking on the sea floor (Chapter 6). Although exact lines of descent are not known and the fossil record is not complete, such data as we have suggested that it took a few million years and perhaps a similar number of generations for autopods with digits to evolve and rays to be lost under selection.

Evolutionary change can be much faster (Chapter 25) but cases where this happens often reflect unusual situations. At one end of the spectrum, rapid change can happen when competition and selection pressures are low so that variants can readily find an unoccupied niche in which they can flourish (e.g. the cichlid species flock, Chapter 22). At the other end, change can occur rapidly under conditions of strong selection, such as selective breeding in which organisms with unfavoured traits are stopped from reproducing. Examples of this are given below and in Chapter 22.

One reason why we cannot be certain as to timings is that they depend on the size of the population undergoing change. If the population is large, it will take a long time before a favourable mutation spreads throughout it; the time will be much less if the founder population is small, and this is particularly so for genetic drift (Chapter 22). In the case of cichlids, the reason why evolutionary change was so fast is almost certainly because the initial cichlid population was small and dispersed across Lake Victoria into even smaller subgroups. Unfortunately, there is no fossil record here, partly because the fish were small, partly because any anatomical changes would be hard to recognise in such anatomically similar populations, but also because numbers in individual groups were so small. Indeed, in any situation where group numbers were small, the fossil record would be expected to be weak.

CONTROLLING SELECTION

Selection pressures under normal conditions are very low and change can take a long time. It is worth recalling that the evolution of the membranes required for the amniote egg to survive under dry conditions took many tens of millions of generations (Chapter 15). Even the switch in the wing patterns from the mottled to the dark morphs of *Biston betularia*, the peppered moth (Fig. 19.2a,b), took scores of generations. If one wants to study the effects of selection, there is no option other than to impose it experimentally. This has been done for the two reasons that are examined here: to improve natural variants for commercial or sporting reasons and to explore natural variation and the range of possibilities that it offers. The latter topic is not easy to investigate and requires carefully designed, long-term experimentation. The example given here is Waddington's work on genetic assimilation, but the classic work of Rice and Salt (1990) on novel speciation, which is discussed in Chapter 22, could also have been included in this chapter (for other examples, see the websites for Chapter 22).

Selective breeding

From time immemorial, humans have undertaken selective breeding to produce new variants that improve domestic animals and plants both for food and for visual pleasure (Van Grouw, 2018). Indeed, the whole horseracing industry is predicated on horses being fast because they were bred from parents chosen for their race-winning abilities. In Chapter 1 of the *Origin of Species*, Darwin focused on pigeons and the ease with which exotic varieties could be selected on the basis of serial interbreeding from minor variants (Fig. 12.1). Today, some of the chance element in breeding can be eliminated because, where a specific gene that confers some trait is known (such as resistance to a selective weed killer in a plant), that gene can be genetically engineered into a host genome and a population bred with the required gene and hence trait in only a few generations.

All breeding starts with breeders recognizing what appear to be favourable heritable features or traits in an individual, then doing their best through selective matings to strengthen and optimise the phenotypic trait within a small breeding population. Ideally, this is done through pairing males and females that both show the trait on the basis that this will maximise the chances of the offspring also showing that trait. However, as discussed in the earlier chapters, many anatomical features depend on the inheritance of complex networks of proteins and, because of the random nature of meiotic crossover, it is not always easy to predict the extent of that trait in offspring (horseracing success in parents is not as reliable a predictor of speed in the offspring as owners might hope).

Selective breeding can be fast and efficient because the selection pressures are so high. Thus, for example, Hershberger et al. (1990) were able to increase the mean weight in a line of coho salmon by 60% in four generations of breeding (see Websites for other examples). Selective breeding has a very long history and is reasonably fast: a rare but interesting feature can usually be selectively bred to give a population that all show this feature, perhaps in a stronger way than in the original, with this often taking less than ten generations (see website). Readers interested in the details of this aspect of variation and selection are recommended to explore Van Grouw (2018), with its spectacular drawings.

Generating anatomical novelties

It is difficult to investigate experimentally the generation of novel anatomical features, but it is worth noting the classic work of C. H. Waddington in the early 1950s. Waddington was a major developmental and evolutionary geneticist who predicted the essential features of molecular genetics in the 1940s on the basis of developmental mutants in *Drosophila* whose phenotypes showed abnormal embryogenesis. He was also the first geneticist to realise that the study of mutations responsible for abnormal embryos held the key to investigating the normal behaviour of genes during embryogenesis; work that he did in the late 1930s, long before anyone had any idea that a gene was represented by a DNA sequence.

Waddington's other main interest was the genetic origins of evolutionary change under selection and, needing a model system to work with, he chose wing development in *Drosophila*, mainly because he had already used mutants to explore its underlying genetics. Originally, flying insects had a pair of wings in each of the second and third thoracic segments (T2 and T3) but the hind wings of ancestral flies were reduced to small balancer organs called halteres, probably during the Cretaceous. Waddington (1953) examined whether it was possible to start with a normal outbred population of wild-type *Drosophila* with two wings and produce a population with four wings in a way that was heritable. He knew that it should, in principle be possible to do this because there was a rare mutant called *bithorax*, that could be induced by X-ray treatment, in which the T3 segment was replaced by the T2 segment, so giving the fly two pairs of wings. It was also known that if fertilised *Drosophila* eggs were treated with ether, a very small proportion also showed the *bithorax* phenotype (such an organism is known as the phenocopy of a mutant).

Waddington investigated whether selective interbreeding of successive generations of ether-induced *Bithorax* flies would lead to a stable population of four-winged flies. He, therefore, treated *Drosophila* eggs with ether and then interbred the few with a *Bithorax* phenotype, again treating the resulting eggs with ether. It took only about 20 rounds of ether selection and subsequent inbreeding for Waddington to produce a population of four-winged flies whose *Bithorax* phenotype bred true without further ether treatment of the eggs and so became a self-sustaining population. (Fig. 20.3a–c).

Waddington called the phenomenon by which such "acquired" characters might be converted into inherited ones *genetic assimilation*, and showed that there were several other such characteristics for which this phenomenon could occur (Waddington, 1953, 1961). There was, however, one prior condition that had to be met for the experiments to succeed: the flies had to come from an outbred rather than an inbred population, so ensuring that there was plenty of genetic variation for selection to act on. Waddington noted that such was the speed of genetic assimilation that the *Bithorax* phenotype was highly unlikely to have arisen through new mutations. This view is supported by more recent work (Nachman & Crowell, 2000).

There has been considerable discussion over the years as to the mechanisms by which genetic assimilation occurs as, at first sight, the phenomenon looks like Lamarckian inheritance, although it is not. Today, the explanation is reasonably clear: the *bithorax* phenotype normally results from a complex set of mutations in the *ultrabithorax* gene group, and its regulatory sequences that result in the

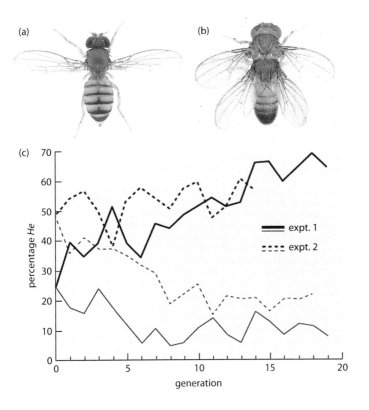

Figure 20.3 Waddington's experiments on genetic assimilation. (a): A normal *Drosophila* fly with two halteres (third thoracic segment) behind the wings (second thoracic segment). (b): A *Drosophila* bithorax mutant whose normal third thoracic region has been replaced by a duplication of the second (broader) region that carries wings. (c): Two sets of experimental results showing the progress of selection over 20 generations for (upper data lines) or against (lower data lines) an initial *bithorax*-like response to ether treatment in two wild-type populations that reacted with rather different frequencies (from Waddington, 1953).

(a Courtesy of Nicolas Gompel ©.
b Courtesy of EB Lewis.
c From Waddington CH (1953). *Evolution* 10:1–13. With permission from John Wiley and Sons.)

repatterning of the third thoracic segment by the second (Pavlopoulos & Akam, 2011). This is because in normal wild-type populations there are neutral mutations within the genes that encode the complex that normally have no effect on the phenotype. Ether treatment selects for combinations of such mutations that together lead to the bithorax phenotype, although the reasons for this are not clear. Further successive rounds of selective breeding capture further mutations already within the genotype of the selected population that strengthen the likelihood of a *bithorax* phenotype. Waddington's experiments with their extreme selection pressures had the effect of concentrating far more *bithorax*-phenotype-related mutations in a fly population in a much shorter time than could ever happen under normal circumstances.

There are implications of this work for selection and for how mutation works. First, it demonstrates that, provided selection pressures are strong enough, unexpected novel structural phenotypes and behaviours can be generated surprisingly rapidly from the genetic variants that exist within a normal, outbred population. In one sense, the importance of Waddington's experiments is that they demonstrate the extent and the selection potential of the genetic variability in a normal wild-type population.

Second, the work highlights the way in which mutation can affect the phenotype. The mutations selected as a result of ether treatment were in those genes that affect the protein activity of the ultrabithorax gene complex, not the genes responsible for either haltere or wing anatomy, which remained unaffected. The resulting change in the activity of the ultrabithorax complex repatterns a whole segment of the *Drosophila* thorax. The effect of this is the suppression of the set of metathorax networks that, among other tissues, produce the haltere and the activation of a set of mesothorax networks that produce wings. This work is a further example of how major evolutionary changes are due to changes in underlying developmental control systems, rather than to changes in the genes that affect the overt phenotype (see Chapter 19).

SELECTION PRESSURES IMPOSED BY HUMANS

The major alterations to global selection pressures today are those imposed by *H. sapiens*, mainly as a result of population increases over the last century. These have changed the biosphere and the climate irreversibly: our increasing demands for land have led to the exclusion of many animals from their natural territory and, as a result, their populations have declined. Our need for fossil-fuel-supplied energy has led to the warming of the world's climate and some fish populations are moving to cooler waters.

The list of extinctions resulting from these changes is depressing beyond measure. The most obvious examples were large marsupials, such as the ground sloths and glyptodonts of South America, large birds, such as the elephant bird of Madagascar and the moas of New Zealand, the mammoths of the northern latitudes and the aurochs of southern Europe and mid Asia. As of now, other large mammals, such as elephants, large cats, and rhinoceroses, are all on the verge of extinction as their natural habitats become part of human territory. These pressures are not only on land mammals; many bird and insect populations are also being diminished. In the seas, whales were only saved from extinction by political pressure, while fish stocks of, for example, cod and tuna are at risk of falling below sustainable levels. The effects of humans are also felt on many smaller organisms and the rate of species extinction has increased by some three orders of magnitude since the beginning of the Anthropocene (Otto, 2018).

Which species are being positively selected as a result of our behaviour? Humans, of course, together with the relatively few animals and plants species that we grow for food and other uses. The animals include mammals, such as dogs, cats, horses, cattle, and chickens together with marine organisms that are now being farmed (e.g. salmon and prawns). As to plants, vast areas of land have now been cleared to grow a very restricted range of food plants, particularly seed-bearing grasses (wheat, barley, etc.) and soybean, while ancient, slow-growing hardwoods are being replaced with fast-growing gymnosperms. The effect of this human-imposed selection is a major loss of natural diversity, not as deliberate policy but as an inevitable consequence of increasing human populations and their demands.

KEY POINTS

- Fitness is a composite measure of how suited an organism is to its environment and so likely to contribute fertile offspring to the next generation. The measure of the success of a mutation is the degree to which it increases fitness.

- Different types of selective pressure affect different aspects of the phenotype.

- Natural selection reflects the effect of all aspects of the environment, such as climate, habitat, predators, and food availability.

- Sexual selection drives changes that increase an individual's likelihood of breeding and encourages, for example, traits that make one sex attractive to the other.

- Although sexual reproduction that enhances the likelihood of reproduction is intrinsically inefficient, it is often favoured over parthenogenesis or hermaphroditism. This is because the mixing of genetic material that results from sexual reproduction increases diversity and hence robustness.

- Kin selection occurs when an individual's behaviour benefits the group at the expense of that individual's likelihood of reproducing and often occurs within the family. The origins of its evolution are still not easy to explain but Hamilton's rule shows that a necessary genetic criterion is that other family members rear orphans.

- Selection in the wild is generally very weak and the acquisition of an adaptation so slow (change may take thousands of generations) that it can rarely be followed during the career of a single human.

- Selection can usually only be studied under artificial or laboratory conditions.

- Artificial breeding based on a rare phenotype in a population can, through strong selection, often lead to a population with that phenotype in less than ten generations.

- Under conditions of very strong, artificial selection, however, the acquisition of novel traits within a wild-type population can be fast (~25 generations). For this to happen there has to be considerable genetic diversity in that population.

FURTHER READINGS

Abbot P, Abe J, Alcock J et al. (2011) Inclusive fitness theory and eusociality. *Nature*; 471:E1–E10. doi:10.1038/nature09831.

Akçay E, Van Cleve J (2016) There is no fitness but fitness, and the lineage is its bearer. *Phil Trans Roy Soc*; 371B, 20150085. https://doi.org/10.1098/rstb.2015.0085.

Bourke AF (2014) Hamilton's rule and the causes of social evolution. *Phil Trans R Soc Lond*; 369B: 20130362. doi:10.1098/rstb.2013.0362.

Nowak M, Tarnita, C, Wilson E (2010) The evolution of eusociality. *Nature*; 466:1057–1062. https://doi.org/10.1038/nature09205.

Van Grouw K (2018) *Unnatural Selection*. Princeton University Press.

WEBSITES

- Wikipedia entries for *Inclusive fitness, Laboratory experiments of speciation, Selective breeding*.
- A simulation of the effects of selection on a simple organism with several traits: www.youtube.com/watch?v=0ZGbIKd0XrM&t=86s

- Neutral theory: www.nature.com/scitable/topicpage/neutral-theory-the-null-hypothesis-of-molecular-839/

Evolutionary Population Genetics

21

Evolutionary population genetics is the mathematical formulation of classical population genetics that has been expanded to include the effects of selection. It thus integrates the modern evolutionary synthesis with Mendelian population genetics. This chapter starts with Hardy–Weinberg equilibrium; this shows that, if a gene has two alleles and breeding is completely random, the allele frequency remains constant from generation to generation. Any change in allele distributions requires that this equilibrium breaks down; this will happen if, for example, genes are linked (nearby on the same chromosome), breeding is nonrandom or if the breeding population is small. In the last case, gene distributions are asymmetric because of stochastic effects known as genetic drift. *Evolutionary population genetics* adds selection to the mix.

This chapter first discusses these various ways of changing allele distributions in a population and then considers the relative importance of selection and drift in driving evolutionary change. Selection is an active process whereby "fit" traits do well, while drift reflects the passive movement of mutations through a population as a result of random breeding; both lead to an increase in asymmetric allele distributions, particularly for neutral mutations not subject to selection. We now know that selection is more important in large and drift in small groups, and that small groups have a set of evolutionarily important genetic properties. These include unpredictable gene distributions, loss of heterozygosity, and the ability for allele distributions to move rapidly through a population. It is these properties that facilitate evolutionary change. Analysis of the theory shows that, while using the theory is relatively straightforward for one or two Mendelian genes whose alleles have direct effects on the phenotype, it is much harder in more complicated cases. This is partly because of the complexity of the equations but also because of the difficulty of estimating fitness and other constants on the basis of the inevitably limited amounts of population data.

The last part of this chapter considers coalescence theory. This builds on Fisher's realisation that variant gene sequences traced backwards in time coalesce onto an ancestral sequence. It is now possible to combine DNA-sequence variation, phylogenetic analysis and population genetics to work out how contemporary sequences are likely to have evolved from common ancestor sequences and how long ago the latter existed (measured in generations). If enough sequences are used for the analysis (typical simulations use whole-chromosome sequences), a population history of the species can be generated. When applied to human data, the results show bottlenecks when the population has been severely reduced. These were often the founder groups that left parent populations to explore new territories.

In 1859, Mendel formulated two key laws of genetic inheritance on the basis of his experimental work with pea plants:

1. The *law of segregation*: Individuals have two alleles for each trait that segregate independently during meiosis.
2. The *law of independent assortment*: The alleles for different traits assort independently of one another.

DOI: 10.1201/9780429346217-25

Mendel's papers were lost for about 40 years and only rediscovered at the beginning of the 20th century. In 1908, Hardy and Weinberg independently showed that, in the absence of any interfering factors, the proportions of each allele in a population would stay constant from one generation to another. Some of these factors were introduced into a much wider model of population genetics that was soon published by Ronald Fisher (1918a,b), a British mathematician. These two papers not only included the effects of selection and laid out mathematically the theory of *evolutionary population genetics* but also put forward the basis of much of today's statistical methodology. (*Note*: Fisher's deserved reputation as a brilliant scientist is marred today by his beliefs in eugenics, the idea that populations would be improved if people with undesirable traits were inhibited from breeding.)

Soon, J.B.S. (Jack) Haldane in England and Sewall Wright in the United States were making major additions to the theory of population genetics and the essential theory was in place within a decade. This theory uses Mendel's laws to investigate how allele frequencies in a population are distributed through breeding and how they can be altered by various factors that include linkage (genes being close to one another on a chromosome and so not being randomly distributed), nonrandom breeding and genetic drift (the effects on gene distribution of random breeding in small populations). For both mathematical and practical reasons, the theory is most useful where variation in a trait can be seen to be underpinned by a genotype based on the alleles of one or at most a very few genes. It is only under these circumstances that it is possible to mesh the theory with the limited amount of numerical information available from phenotype distributions across a population to obtain statistically significant results. Its most important use is in providing a framework within which gene distributions under a wide range of circumstances can be worked out analytically.

Over the next thirty or so years, the theory was further developed by mathematical geneticists, including Hamilton, Mayr, Price, and Kimura, who introduced concepts, such as kin selection, and modernised the treatment of genetic drift. A full theory of quantitative *evolutionary population genetics* was in place by the 1950s and incorporated random and other breeding conditions, linkage between alleles, genetic drift in small population, gene transfer from outside the population (immigration), and selection for the phenotypes of these alleles. Work over the last half century has added the effects of kin selection (altruism), means of exploring genetic history on the basis of DNA variants (coalescence theory) and the analysis of gene polymorphisms across large populations (known as genome-wide association studies or GWAS). The net result is that we now have a rich and complex body of theory for investigating how genes move through a population over time.

Formalising all these ideas quantitively requires a great deal of mathematics that includes probability and statistical theory, population dynamics, stochastic analysis and Markov modeling. Under simple conditions, the resulting equations can be solved analytically, but contemporary models are now so complex that solutions normally require sophisticated computational approaches (see Further Readings). As most biologists lack the mathematical background to handle the theory, the treatment given in this chapter is mathematically light, with the focus being on the evolutionary implications of the underlying genetics. The aim is to explain rather than prove how change in allele distributions can arise and be propagated, and how the data can be used to make inferences about evolutionary events.

Readers who would like a very different perspective on *evolutionary population genetics* might enjoy the Primer simulations (see Websites) that use stochastic rather than analytical methods to simulate evolutionary change. While analytical approaches can usually only give equilibrium results, computational simulations show how that equilibrium is achieved. Although the Primer models are inevitably simplified, they demonstrate the unpredictable ways in which selection alters gene frequencies during the process of change in a small population. The advantage of such models is that they can follow over time the population genetics of a group of individuals with an arbitrary number of genes, the phenotype associated with each being selection. They particularly show that once a group has alleles for several key traits, each with its own selection coefficient, the resultant allele distribution in a population cannot be predicted.

CLASSICAL POPULATION GENETICS

The basic theory of population genetics is based on Mendelian traits that are assumed to be the result of the activity of a gene with two alleles that are distributed across a population. The result will be two or three phenotypes, depending on whether the effect of the pairs of alleles is dominant, semidominant, or recessive. Both Hardy and Weinberg showed that the equilibrium distribution of traits is stable over time (Box 21.1). As this result denies the possibility of evolution, it was clear that one or more of the underlying assumptions of Hardy–Weinberg equilibrium had to fail if change were to occur. These assumptions were that genes were assorted independently (no linkage), mating was random, the population was stable (no immigration or gene transfer), no selection for or against particular phenotypes, and no novel mutation. The modern theory explores the implication of the breakdown of each of these assumptions.

The one factor that is, however, unlikely to be of any importance, in the short term at least, is novel mutation. Nachman and Crowell (2000) estimated the mutation rate for humans and found a value of ~2.5×10^{-8} mutations/site/generation which suggested that only about 175 new mutations became established per generation and these were, of course, distributed across the whole genome. The chances of a gene under investigation altering over the time of any study are thus very low for humans, and this figure is unlikely to be very different for any other diploid organism.

Linkage

Mendel's second law states that gene alleles in one generation are, under standard conditions of random mating, distributed randomly across the next. This, we now know, is because recombination and crossover during meiosis has the effect of shuffling gene alleles. This effect is compounded by a gamete having separate sets of shuffled chromosomes from each parent. However, only a limited number of crossovers between pairs of chromosomes takes place during meiosis so that genes on the same chromosome may assort together or independently with the likelihood of the former (linkage) increasing with the closeness of the genes on a particular chromosome. This effect can be exacerbated by genetic drift as random sorting of one allele may be accompanied by that of a nearby allele on the same chromosome.

Linkage disequilibrium occurs when two genes are sufficiently close that there is a higher probability of their assorting together than distributing randomly and this disequilibrium may affect the phenotype of the offspring. Linkage disequilibrium in sexually reproducing organisms is, however, unstable because further recombination over the generations will eventually lead to the alleles assorting independently. The corollary of this is that if the linkage disequilibrium for an allele combination is stable or increasing over the long term, the accompanying phenotypic effect will reflect positive selection. (Sabeti et al., 2002). *Note*: The appearance of linkage disequilibrium can, of course, also occur if there is strong selection on a particular allele combination.

BOX 21.1 HARDY–WEINBERG EQUILIBRIUM

The first question of *evolutionary population genetics* is how a gene with two alleles, each present in one generation of a large population, is distributed in the next generation if breeding is random. Suppose that the two alleles are named **A** (frequency **p**) and **a** (frequency **q** = **1 − p**). It is obvious that in the absence of selection, the probabilities of the allele frequencies in the next generation are (**p** + **q**)(**p** + **q**). This gives the probabilities for the **AA, Aa,** and **aa** allele combinations as $\mathbf{p}^2$, **2pq,** and $\mathbf{q}^2$ which, of course, sum to 1 (in the case where p = q = 0.5, this gives the classic Mendelian ratios of 1:2:1).

The frequency of **A** in this second generation is given, as expected, by

$$\mathrm{F}(\mathbf{A}) = \mathbf{p}^2 + \tfrac{1}{2}(2\mathbf{pq}) = \mathbf{p}^2 + \tfrac{1}{2}(2\mathbf{p}[1-\mathbf{p}]) = \mathbf{p}$$

Similarly, the frequency of **a** [F(**a**)] in this next generation is **q**. In other words, if all other things are equal, gene frequencies do not change over time, and neither, of course, do the associated phenotypes, whether or not the alleles are dominant or recessive. This situation of stability is known as **Hardy–Weinberg equilibrium**. Evolutionary change thus requires the presence of factors that break down this idealised situation.

Linkage effects can be measured and analysed for their evolutionary implications. Ellis et al. (1994) have, for example, examined the incidence of Bloom's syndrome (BS) in Ashkenazi Jews, a population that was reproductively isolated for many tens of generations and in which the syndrome is particularly common: ~1% of this population carry the mutation whose locus maps to chromosome sub-band 15q26.1 This site is tightly linked to that of the proto-oncogene FES and possession of the pair of mutations is very common in the Ashkenazi Jews. The result is that the population shows a higher frequency of individuals with both diseases than would be predicted This linkage disequilibrium constitutes strong support for a founder-effect hypothesis: the chromosome in the hypothetical founder who carried BS also carried the C3 allele at FES (see also Chapter 17 and Risch et al., 2003). Of the various ways in which allele frequencies can change, however, linkage does appear to be relatively insignificant in affecting evolutionary trajectories (Barton et al., 2007) and will not be considered further.

Nonrandom breeding

Nonrandom breeding clearly leads to asymmetric allele distributions. An obvious example is seen in the mating behaviour of deer: males with the largest antlers reproduce with all the females in their group and exclude males with smaller antlers from reproducing and passing on their genetic complement to the next generation of animals (Malo et al., 2005). It is relatively straightforward to calculate this effect, but it should be noted that the number of breeding individuals for the purpose of describing allele distributions here is not that of the full population, N, but is N_e the *effective population number*. This is the number of reproducing adults and is always less than N as it excludes sexually immature, infertile, and postfertile members of the group as well as allowing for the breeding habits of a population in which most males never have the opportunity to breed.

Herron and Freeman (2014) gave an estimate of how N_e depends on the numbers of sexually active males (N_m) and females (N_f) in a breeding group:

$$N_e = 4N_m N_f / (N_m + N_f)$$

Thus, if a male has a harem of 10 females, then N_e has the value of 3.6 rather than the population number of 11, a number that itself is probably very much smaller than the size of the full group. In this case, the genes of the dominant male will be overrepresented in the subsequent offspring and any different alleles in the remaining 90% of males will be lost. *Note*: Today, values of N_e are often estimated by coalescence analysis (see below and Wang et al., 2016).

Nonrandom breeding can also occur through incestuous mating, which is more likely to happen in small than in large populations (the odd example of male pyemotic mites that breed with their sisters before birth was considered in the last chapter). In terms of evolutionary change, the effect of nonrandom breeding is to lower the effective population size. This diminution also has the effect of increasing the rate of movement of gene variants or new mutations across the population, albeit at the expense of limiting genetic diversity.

Migration

A very different form of nonrandom breeding arises when members from one group of a species join another related group through migration and then interbreed with them (see ring species, Chapter 22). It seems to be almost a rule of nature that, if organisms can interbreed outside their immediate group, they will, and the resulting offspring will have gene alleles from both groups. This transfer of gene alleles from the immigrating to the host population, known as *gene flow*, can be advantageous: if the incomers carry alleles that are beneficial in other environments, the future population may include offspring with widened adaptability as compared to their parents. A particularly well-studied example of this, because we have DNA from the respective populations, is the interbreeding that happened ~50 Kya when *Homo sapiens* migrated out of Africa and into regions habited by Neanderthals and Denisovans (Chapter 24). There are many examples where population dispersals have led to gene flow; the reader interested in this area should consult Saastamoinen et al. (2018).

GENETIC DRIFT

A key criterion for Hardy–Weinberg equilibrium to occur is that the population is sufficiently large for statistical fluctuations in breeding to be unimportant. Genetic drift occurs when the population is not large enough for this criterion to be met; in such cases, allele distributions are unpredictable. The importance of group number in population genetics was realised early on in the development of the theory with a key figure being Sewell Wright (see Wright, 1955, for a historical perspective). It is possible to calculate the effects on allele distributions of low population numbers, but understanding the detail requires some theoretical background (Ewens, 1972).

Genetic drift reflects the effect of random mating on the allele distribution for genes already present in a population. If that population is large, its allele distribution will closely approximate that expected on the basis of an infinitely large population. The smaller the population, however, the more atypical that allele distribution is likely to be simply because random choice can give unexpected results. Coin distributions provide a helpful example. Consider a box filled with coins of different values, with the same number for each value. If one handful is pulled out at random, the coin frequencies in this group will probably be unequal but, as more and more handfuls are added to the group, numbers will approach equality. The mathematics of this is illustrated by the classic coin-tossing experiment. In principle, the likelihood of heads or tails in a run of tosses is 50:50. If the number of tosses is small, the likelihood of any head:tail ratio is given by the appropriate term of the binomial distribution, with more asymmetric results having lower but still realistic probabilities.

The effects of genetic drift are clearly more important in small groups than large ones, as demonstrated by an experiment on the red flour beetle (*Tribolium castaneum*). This study capitalised on the fact that there is a mutant form whose homozygotes are black and heterozygotes are brown. Rich et al. (1979; Fig. 21.1) followed the colours of 24 populations of beetles, each of which started with equal numbers of red and black beetles. Half of these populations started with a small number of beetles (5M, 5F) and half with a larger number (50M, 50F). Over 20 generations, the mean frequency of red beetles in the group with large numbers increased from 0.5 to 0.76 with fluctuation in a given generation of about ±0.25. The mean frequency in the small group also increased to about 0.76 over 20 generations. Here, however, the spread of frequencies

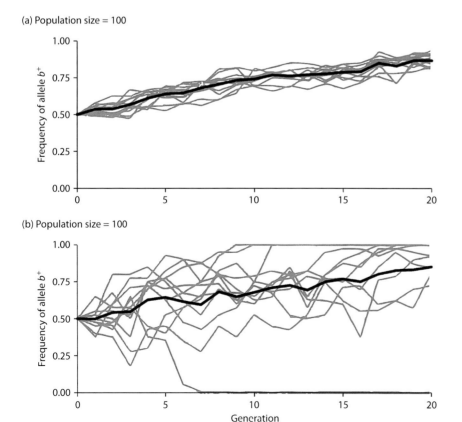

Figure 21.1 The b+ red-colour frequencies in 12 large (**a**, 50M + 50F) and 12 small (**b**, 5M + 5F) populations of red flour beetles. Initially, there were equal proportions of the dominant b$^+$/b$^+$ (red, wild-type) and recessive b$^-$/b$^-$ (black) morphs. The large population slowly loses the b$^-$ mutation across the groups with only minor fluctuations in b$^+$ frequency. In the smaller population, the mean effects are much the same, but the effects of drift lead to a wide spread of results, including the loss of the wild-type b$^+$ allele in one population after seven generations and the loss of the b$^-$ allele in another after nine generations.

(From Rich et al. (1979) Evolution; 33(2): 579–584. With permission from the Society for the Study of Evolution. With thanks to Jon Herron for the coloured and redrawn image.)

was very much greater: one group achieved 100% redness after nine generations (the allele became fixed in the population), while another lost the red allele completely after seven generations.

This classic experiment illustrates the swings in gene frequencies that can occur in small populations under the effect of drift, even in the absence of selection. The speed of the effects in these results cannot, however, be precisely extrapolated to the wild as selection acts on a wide variety of traits. Although the implication so far has been that the genes underpinning selectable traits are randomly distributed under drift, it also randomly distributes alleles for traits that are neutral and not subject to selection. This experiment shows that a low incidence of any allele may rapidly increase by chance or may be lost. The importance of neutral effects is considered below.

The other effect of genetic drift in small populations is that the resulting gene-frequency asymmetries can create more homozygotes than heterozygotes. This can easily be demonstrated: suppose, as a result of random drift in a small subpopulation, two genes which each had a frequency 0.5 in the original population now have frequencies x and 1-x. It is easy to show that the proportion of homozygotes to heterozygotes in each population is

$$x^2 + (1)^2 : 2x(1-x)$$

and that the ratio has a minimum value of 1:1 when x = 0.5, as in the original population. If, however, the gene distributions change so that, for example, x = 0.9, the ratio of homozygotes to heterozygotes increases to 0.82:0.18, or about 4.5:1 (Fig. 21.2). Hence, there are more homozygotes in small genetically unbalanced populations than in large genetically balanced ones and a consequent loss of heterozygosity. In general, the distribution of the alleles will not be balanced, but any allele distribution which increases allele asymmetry will also increase the number of homozygotes. This *loss of heterozygosity* is known as the *Wahlund effect*, and its evolutionary importance is that phenotypes due to recessive genes will become more common in small groups as opposed to large ones.

Such effects of drift give the appearance of selection but, in fact, the changes are just due to the random nature of gene distributions as a result of breeding. Once in place, however, the different phenotypes are accessible to selective pressures with the result that the distribution will continue to change. Details of the important evolutionary implications of drift in the context of founder groups are considered below.

THE EFFECTS OF SELECTION

The key feature that distinguishes *evolutionary population genetics* from classical population genetics is the addition of selection. This is assumed to act on the fitness of a phenotype associated with an allele; this parameter is known as **w** and is a measure of reproductive success. It is, however, important to emphasise that fitness is not an easy parameter to pin down and different definitions are needed for discussions of

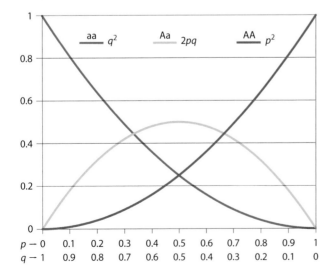

Figure 21.2 **Allele asymmetry and loss of heterozygosity.** The green line shows how the number of heterozygotes that form in a daughter generation depends on the relative frequencies (red and blue) of a pair of alleles.

(Courtesy of "Johnuniq." Published under a CC Attribution-Share Alike 3.0 Unported License.)

TABLE 21.1 DEFINITIONS OF FITNESS FOR EVOLUTIONARY SELECTION

Natural and sexual selection

Absolute fitness	The likelihood that all the genotypes of some trait will appear in the next generation
Relative fitness	The proportion of each of these genotypes compared to that most likely to appear in the next generation

Kin selection

Personal fitness	The number of offspring that an individual produces that survive to reproduce
Inclusive fitness	The number of offspring (their own or related) that an individual rears and that will eventually reproduce, weighted for the degree of genetic relatedness (a sibling's and a cousin's child are 1/2 and 1/16, respectively)

natural and sexual selection as compared with that for kin selection (Table 21.1; Chapter 20). In that case, the emphasis was on the number of offspring reared; for natural and sexual selection, fitness is a measure of the transmission of traits and their underlying genotypes to the gene pool of the next generation. It is also worth pointing out that this constant cannot be measured directly in the wild but has to be estimated on the basis of analysing offspring traits and genotypes.

The selection coefficient **s** is connected to **w** by the simple formula

$$\mathbf{w} = \mathbf{1} - \mathbf{s}$$

where **s** represents the relative disadvantage of the genotype for that trait. Hence, a value of **s = 1** is lethal, while a value of 0.2 means that 80% of the offspring carry that allele. This equation is easiest to use under conditions of controlled breeding in which the values of **w** and **s** can be calculated accurately. Analysing the effects of selection in the wild is much harder as the parameters reflect many aspects of breeding and survival in a particular environment and the value of **s** for a particular aspect of the phenotype can be small. The classic model for selection here is malaria resistance in humans; this is because there is so much data (Box 21.2). In domestic breeding, where selection parameters can be controlled, population genetics gives accurate results for Mendelian traits, but less accurate predictions for non-Mendelian traits (e.g. the speed of racehorses).

In the context of major evolutionary change, a further difficulty in estimating selection pressures in the wild is that they cannot be assumed to remain unaltered over very long periods as conditions frequently change. In addition, one can never

BOX 21.2 MALARIA RESISTANCE AND SELECTION FOR A BALANCED POLYMORPHISM

The classic example of the effect of fitness for a pair of alleles is the mutation in the haemoglobin gene that confers malaria resistance (Jones, 1997). Most people bitten by a mosquito carrying *Plasmodium falciparum* will get malaria. Individuals carrying one copy of a mutation (the S allele) that converts glutamic acid to valine in the haemoglobin β chain have up to 90% protection against infection. This probably because the mutated protein can carry less oxygen than the normal A allele, so that the blood becomes a less favourable environment for the plasmodium. One might suppose that this mutation would become the norm in areas where malaria is very common, but this is not the case: in addition to individuals with a single copy of the gene allele (heterozygotes), there are two homozygotes populations: those with no copy of the protective allele and those with two. The latter population unfortunately have red blood cells that are predominantly rigid, sickle-shaped, and carry diminished amounts of oxygen. As a result, homozygotes for the allele suffer from a series of disorders and usually die young, which is, of course, associated with a very low likelihood of

reproduction. This homozygous disadvantage leads to the recessive allele continually being depleted.

Ridley (2004) has worked through the complete calculations to show how the frequencies of the malaria-insensitive heterozygote, the sickle-cell homozygote, and the malaria-sensitive homozygote for a population in Nigeria can be used to calculate the allele frequencies and fitness values. His analysis shows that the allele frequency ratio (A:S) is 0.877:0.123, and that the fitness coefficients for the AA, AS, and SS phenotypes are 0.14, 1.0, and 0.88, respectively. The difference in fitness values reflects the fact that someone with malaria is less likely to die before reproducing than someone with sickle-cell anaemia, while both are less reproductively successful than the heterozygote.

The sickle-cell mutation is described as a balanced polymorphism because, in regions where malaria is prevalent, the advantages of some individuals having a single copy of the mutation are balanced by the disadvantage of others having two copies. It should be noted that such balanced polymorphisms are rare.

be certain that the phenotypic effect under consideration stands alone and has no interactions with other aspects of the phenotype (see the Primer website). An animal, for example, might benefit from being heavier and stronger in that it might get more mates and compete better for food, but this increase in size might make it more obvious to predators and slower in fleeing from them in a changing environment. It also needs to be emphasised that selection acts on the whole phenotype rather than on a particular trait and one can never be sure, except under conditions of selective breeding, what this will mean in practice (Anderson et al., 2014; see Websites).

One advantage of the use of selection coefficient is in predicting whether a novel allele (deriving, say, from novel mutation) with some selective advantage can survive the vagaries of genetic drift and become established within a population. Haldane (1927) showed that, given some simplifying assumptions, this could happen provided that its selection coefficient $s > 1/N_e$, with the likelihood of this happening being ~$2s$ (this is known as Haldane's sieve; for a modern view, see Wahl (2011) and associated papers). He also showed that the smaller is the effective population size, the more likely it is that a novel mutation will become part of the population's genotype (see Orr, 2010). Kimura and Ohta (1969) then proved that the expected number of generations that it took for such a novel gene to become fixed within a population was $4N_e$. Thus, the smaller the size of the breeding population, the faster and more likely it is that an advantageous new allele will spread through it. Even then, however, this is not a rapid process.

Analysis of selection is further complicated by the need to distinguish between environmental and genetic effects on the phenotype variance. Thus, for example, animal breeding rates will depend on food and mate availability as well as on anatomical features, while plant pollination rates often depend on insect numbers and flower morphology, both of which may depend on weather fluctuations. A supportive environment will favour reproduction while a poor environment will effectively act as a brake on it. There are also plasticity effects in organisms as their development responds to favourable and unfavourable environments (Chapter 19). In practice, many things are going on at the same time, and it can be hard in field studies to keep some factors constant while studying variation in others. All of these environmental effects are, of course, superimposed on the underlying genetic distribution. The early solution to this problem was to partition phenotypic variance into contributions from the effects of heritability and the environment, but this was easier to do in theory than in practice.

PRACTICAL CONSIDERATIONS

The use of *evolutionary population genetics* to predict the future distribution of observed genotypes and phenotypes in a particular population is not simple. In addition to understanding the formal mathematical theory, it requires knowledge of allele frequencies, estimates of the selection, linkage, and other coefficients, some way of distinguishing between environmental and genetic effects on phenotypes, and a model of breeding behaviour. Very few of these can be derived theoretically and most have to be abstracted from the limited amounts of experimental data. Estimates of allele frequencies are straightforward if the trait under consideration depends on, say, protein variants that can be obtained and measured. It can be much harder if it does not, even in cases where there are only two alleles. Under these conditions, underlying gene frequencies and fitness coefficients have to be estimated on the basis of observed phenotypes (Box 21.1), while possible linkage and migration effects may have to be allowed for.

Modelling the evolutionary future of the alleles of two genes each of which may have several alleles adds a layer of complexity. These come from the added number of equations and the further fitness and other constants to be extracted from the population data. Obtaining constants with any accuracy is normally difficult because of the limited amount of such data, and each often has a standard error associated with it. Incorporating all this information into a set of equations is possible, but interpreting their solutions is not simple, because of the intrinsic uncertainties.

Quantitative traits that display continuous rather than discrete variation (e.g. height and longevity) are not obviously explained on the basis of alleles that have discrete phenotypes. This problem was solved by Fisher (1930) who showed that continuous variation in some phenotypic characters could be achieved if alleles of several genes contributed to the phenotype. In such cases, it is also likely that there would be interactions between the gene products (this is known as *epistasis*).

Handling this degree of complexity in practice is particularly demanding because each gene needs its own fitness and coefficients, while covariance effects may also have to be included (Hill, 2010). In most cases, the experimental data is usually inadequate to provide a complete model and can only account for a fraction of the variance in some traits. We now know that such situations are even more complicated because continuous variation often reflects the output of complex proteins networks (Appendix 1) rather than being the effect of a few alleles. In these cases, it is still too difficult to produce a full mathematical model of the situation and the only feasible approach is to model the behaviour of the network as variants of a few artificial *trait genes*.

Any study that involves reproduction also requires a model of breeding behaviour for the population and this includes details of how many individuals in a population contribute to the gene distributions of the next generation (N_e), their reproductive behaviours and age distributions. For any realistic model, the mathematics is generally intractable but there are two simplified models that have been useful here. In the Wright–Fisher model, populations have a constant large size, mate freely but do not have overlapping generations; in the Moran model, populations do overlap, but when one organism is born, another is assumed to die. Such idealised populations have the advantage that they can be formulated mathematically, and this was an important consideration in the 1940s and 1950s, before the availability of computers. Today, the implications of much more realistic population behaviour can, of course, be computationally simulated to model long-term genetic distributions.

NEUTRAL THEORY VERSUS SELECTION

Until the 1960s, genetic change was generally assumed to be driven by one or another form of active selection on new or recent mutations that generated phenotypic change and that would move through a population. Fisher (1918a,b) modelled this on the basis that the movement could be seen as resulting from the effect of allele diffusion or as a result of a branching process and these ideas on drift were explored by other geneticists.

In 1968, Kimura suggested that genetic drift was of greater significance than had previously been thought. This was because it could, in the absence of selection, generate phenotypic change, even in traits for which there were no selection pressures (the neutral theory of evolutionary change). This happened because the gradual accumulation of neutral mutations would eventually lead to phenotypic differences. Such mutations would often be synonymous (replacing one DNA base with another that does not change the amino acid coding) and so silent or have at the most very minor effects on protein structure. For practical purposes, this means that the selection coefficient averaged over evolutionary time |s| had a value of <0.001 (Nei et al., 2010). The predictions resulting from this did not exactly tally with observations but did so once the theory was strengthened by Ohta (1973) who introduced the concept of *nearly neutral mutations* whose phenotypes had a small selection coefficient (|s| < 0.01).

Following the publication of Kimura's and then Ohta's papers, there was heated discussion among evolutionary geneticists as to the significance of neutral drift. Further work showed that the smaller the population, the more significant would be the effect of neutral mutations and hence the more important was the drift of neutral mutations in generating phenotypic variation. This work has now been quantified, and it is possible to calculate the size of the effective population for which selection becomes more important than drift in the fixation of a novel mutation (Neher & Shraiman, 2011).

Perhaps of greater immediate importance was the fact that Kimura's neutral selection made predictions about variation within a population of DNA sequences that would not be expected on the basis of classical theory. One was that variation should be found across the genome, but that its levels should be less in sequences that had a functional role. A second prediction was that the number of mutations in orthologous protein sequences (those that share a recent common ancestral sequence) should increase linearly with divergence time. Both predictions have been borne out, and the key role of drift is now generally appreciated, although quantification of its predictions is often complex.

Two reports emphasise these points. First, a study on the small populations of three-spined stickleback in nineteen weakly connected ponds illustrate the respective roles of selection and drift (Seymour et al., 2019). Each population had its own distinguishing features and detailed phenotypic analysis showed that the major causes of most differences were genetic drift. There were, however, a few features, such as defensive behaviour, that resulted from environmental effects, such as selection and developmental plasticity. Second, in their experiments on the genetic behaviour of the red flour beetle (*T. castaneum*), Weiss-Lehman et al. (2019) showed that as a population extends its territory, it is those members on the expanding edge start that first show the effects of drift. The general view today is that both selection and drift are key drivers of evolutionary change with the former dominating in large populations and the latter in small ones (for review, see Nei et al., 2010; Casillas & Barbadilla, 2017).

SMALL POPULATIONS AND FOUNDER GROUPS

Under normal circumstances, evolutionary change in large populations is very slow because the time taken for a new mutation to move through a population under selection depends on its size. As discussed above, allele-distribution changes are much faster in small, isolated groups in which Hardy–Weinberg stability breaks down because numbers are small. The work discussed in this chapter shows that there are a number of ways in which the genetic future of a small population differs from that of a large one.

1. The most important is the enhanced role of genetic drift, although its effects are unpredictable. This phenomenon reflects the effect of the random frequency of gene alleles that a founder population takes with it when it leaves the parent population. The smaller the founder population the more atypical that allele distribution is likely to be; the larger that population, the closer the distribution is likely to be to that of the parent population.

2. Small groups show a loss of heterozygosity (the Wahlund effect) because there are more homozygotes in genetically unbalanced populations than in genetically balanced ones. The effects of recessive genes thus become more accessible to selection in small populations; again, this may be beneficial or deleterious.

3. In small groups, which often include families, breeding may well not be random and may even be incestuous. The effect of this will be further loss of heterozygosity and reduced variability in the next generation together the enhanced likelihood of rare traits occurring.

4. Any favourable mutations, existing or novel, will spread much more rapidly in a small than in a large group with, as discussed earlier, the expected number of generations that it takes for a favourable mutation to become fixed within a population being $4N_e$ (the effective breeding number in the population; Kimura & Ohta, 1969). This number may well be smaller for a founder population than might be expected. This is partly because an existing favourable allele may be disproportionately represented in the group and partly because family groupings will be overrepresented. The latter will not only bias the genetic character of the group but may well encourage asymmetric and incestuous inbreeding, so decreasing N_e even further.

The future of a small population with a specific allele distribution depends on its habitat. If it is the same as its parent population, it is likely to do less well because its gene distribution would be expected to be less appropriate. If the new habitat is different and has novel selection pressures, then its future depends on its allele distribution. If advantageous, the population will thrive; if disadvantageous, the population will die out. If that subpopulation does thrive in its new environment, it may become first a subspecies with a distinct phenotype and eventually a full species in the sense that the former can breed with the parent line and the latter cannot (Fig. S4.1). The phenomena that drive this progression are variation, selection, further mutation, and speciation. The selection of fit variants is, however, the key step, as Darwin realised, although he had no sense of the importance of population size. The lower the number of breeders in a small population, the more rapidly it will become clear whether that population dies out or thrives.

Such small-group behaviour is the key to novel speciation and this topic is discussed in detail in Chapter 22. The evidence to support this contention comes partly from the theoretical analysis given here, partly from experiments of the type mentioned earlier and partly from population genetics based on coalescence analysis discussed below. This last approach, which is based on DNA genome-wide mutation analysis and population genetics, allows population histories to be reconstructed and particularly identifies bottlenecks where population size has been drastically but temporarily reduced.

QUANTITATIVE AND COMPLEX TRAITS

The discussion so far has focused on simple traits underpinned by the alleles of a single gene; it is clear that *evolutionary population genetics* provides a good explanation for change here and helps explain trait variation in terms of allele number. Many traits important in an evolutionary context cannot, however, be explained easily on this basis and there are two obvious classes: those that vary quantitatively (e.g. height and longevity) and those that are complex because they are known to depend on the cooperative activity of many proteins (e.g. physiological and developmental networks). Such traits may even require the integrated activity of a hundred or more proteins and this complexity poses problems that could not have been foreseen by the theoretical geneticists who devised the theory of evolutionary population some decades before the discovery of DNA in the 1950s and who viewed the genotype as being defined by the phenotypic traits (Boyle et al., 2017).

The original approach to the problem of traits showing quantitative variation was to assume that they are generated by the activities of several key genes, each with several alleles. Such a model can, as Fisher showed, generate continuous variation. Doing so is, however, much easier in theory than in practice and an alternative approach is to handle examples statistically (Roff, 2017). An example is the work of Charmantier et al. (2006) who investigated the relationship between age at first and last reproduction over 36 years in a population of swans (*Cygnus olor*). Using multivariate analyses on the longitudinal data, they showed that both traits were strongly selected, but in opposite directions. Hence, although both traits display heritable variation and are under opposing directional selection, their evolution is constrained by a strong evolutionary tradeoff. These results were consistent with an increase in early-life performance being paid for with faster senescence but said nothing about the underlying genetics. It is still very difficult to use classical genetic approaches to untangle the genetics of complex traits.

It is sometime possible to sidestep all this complexity and one approach for handling unknown genetic information is to use the breeder's equation (see Websites). This simple formulation states that, for some trait Z in one generation, the expected change in its mean value for the next generation is given by:

$$\Delta Z = h^2 S$$

Here, the selection differential **S** is a measure of association between trait values and fitness and so incorporates selection effects, while **h²** is the proportion of trait variation statistically attributed to additive genetic effects. **h²** summarises all the genetic knowledge and can be inferred by measuring the phenotypic similarity of parents to their offspring (Kelly, 2011). The use of this equation sidesteps the need to consider a great deal of genetic complexity.

Gene identification

In terms of experimental molecular genetics, the first step in approaching an understanding of the molecular basis of any trait is to discover the genes that contribute to its phenotype. For this, both molecular and genetic approaches can be used. At the molecular level, there is a host of techniques for analysing mutant organisms to identify mutant genes and to use protein-association discovery methods, such as yeast-two-hybrid and FRET (fluorescence resonance energy transfer) to explore the interactions between normal and mutant proteins (for review, see Mohammed & Carroll, 2013).

The key population genetics technique for identifying unknown genes that may be associated with some trait is GWAS (genome-wide-association studies). This statistical approach identifies genetic variants that are associated with a trait (particularly single-nucleotide polymorphisms, SNPs) in large populations. This methodology is particularly used to study human genetic disease because there is now so much information about them that can be accessed computationally (Bulik-Sullivan, et al., 2015; Sella & Barton, 2019; Uitterlinden, 2016). GWAS studies can analyse data from many thousands of people, albeit that the mathematics of data identification and analysis are inevitably complex. In such studies, many hundreds of genes are often identified as contributing to the trait and even then, there are still components of the variance that remain to be identified (Manolio et al., 2009)

In general, molecular techniques mainly focus on the roles of proteins in specific pathways and networks, while GWAS studies identify the wide range of genes that may affect traits and indicate the extent to which particular SNPs contribute to trait variance (in many cases, very little). In the context of evolution, the former approach has so far probably been more helpful in understanding the molecular basis of a quantitative or complex trait. GWAS techniques are only now beginning to be used in evolutionary studies.

Analysing such enormous amounts of information is not within the capabilities of the classical theory of *evolutionary population genetics* (Cannon & Mohlke, 2018; Hayes, 2013). This is partly because of the number of genes involved in generating complex and quantitative traits but also because the theory assumes that there is a direct link between gene function and trait, with alleles of that gene directly accounting for trait variation. This clearly does not hold for substantial biological features, particularly those involved in developmental, neurological, and physiological traits, where the link between gene and trait is often indirect.

These modern approaches, which accept genetic complexity, contrast dramatically with the classical theory which assumed that complex traits were the result of the activity of one or two genes. A well-known example here was Hamilton's (1964) suggestion that there is a gene for regulating altruistic behaviour, a trait that benefits the group at the expense of the person displaying the phenotype. In the case of the squirrel discussed in Chapter 20, altruism involves an animal first recognizing a predator and then, instead of taking evasive action, behaving so as to raise the alarm for the group with, in addition, group members being prepared to adopt orphans. The suggestion that two complex and interrelated neurophysiological events could be controlled by a single gene, allele or protein seems a gross oversimplification today and was probably not even appropriate as a model in 1964, many years after genes had been identified as being DNA sequences.

Selfish genes

Linking gene alleles with trait variants in this way has an interesting theoretical correlate: in terms of selection, *evolutionary population genetics* can be formulated in terms of alleles or traits. In the former case, selection can theoretically be seen as acting on the alleles, although in practice it, of course, only operates on phenotypes. It was such work that led Dawkins (1976) to argue for the idea that it is genes rather than phenotypes that are selected and that such selfish genes are those that eventually dominate in a population. The sorts of genes that Dawkins was writing about particularly included those that were involved in social and altruistic behaviours (Chapter 20), and were thus involved in group cohesivity, but the argument can be used in much wider contexts.

In the light of modern knowledge, Dawkin's views generally only hold for those genes for which the trait can be directly linked back to the activity of a single allele. It is, however, hard to find an example of a gene that, on its own, is responsible for the evolutionarily important traits that Dawkins had in mind. In practice, traits are high-level phenomena that typically represent a specific variant of the output of a set of proteins and protein networks (Table 14.1). The link between gene and phenotype is indirect because there are several layers of scale separating them, and their effects are diluted in each layer (Table 21.2; Appendix 1).

TABLE 21.2 COMPLEX TRAITS AND THEIR UNDERLYING LEVELS

Level 5: Complex Traits	The observed phenotype for a feature, such as an anatomical, physiological or behavioural property.
Level 4: Network Groups	Each is the immediate generators of complex and quantitative traits (these can often be seen as trait genes, whose variants reflect the range of possible phenotypes).
Level 3: Protein Networks	Each drives an aspect of the phenotypic trait (e.g. *development*: patterning, growth, differentiation; *physiology*: neuronal networks, muscle action, clocks).
Level 2: Proteins	Each has a specific functional role either alone or within a network.
Level 1: Genes	Sets of protein-coding DNA sequences and associated control sequences (these include cis-acting sequences and methylation effects).

An example is provided by the features that make each human face unique, which mainly reflect local growth and pigmentation. The genes underpinning these traits reflect minor mutations in the many proteins that control overall pigmentation and the finely controlled networks that regulate local growth (Level 4 objects in Table 21.1).

It is, however, important to understand why the summary effects of a few artificial trait genes can sometimes provide a good way of summarising the effects of large networks of proteins. The key reason seems to be that, for all the internal complexity of a network that control growth, specific behaviours, and physiological properties, they have a very limited range of outputs. Hence, all that is required for modelling purposes is the minimal of artificial *trait* genes whose collective activity can generate the phenotypic behaviour of a group; nothing needs to be said or understood about the underlying molecular genetics.

COALESCENCE APPROACHES

Standard *evolutionary population genetics* aims to predict the *future* effects of genetic drift, linkage, and selection on trait and allele distributions in a population that carries several genes in the light of its breeding behaviour. It does, however, say nothing about the *past* history of some population. Molecular phylogenetics reconstructs sequence histories on the basis of sequence data but says nothing about time or population behaviour except insofar as it can use mutation rates as a clock. This is because phylogenetics ignores an important aspect of the available data, the detail of sequence variation within a population.

It is possible but not simple to reconstruct genetic and population history through the use of current sequence variants and coalescence theory. The approach derives from an original insight by Fisher: this was that (in today's terminology), if the various polymorphisms in a particular gene sequence within a population are traced backwards in time, they coalesce back onto a single sequence, known as the coalescent, possessed by the *most recent common ancestor* (MRCA) for that sequence. The number of generations back to the coalescent is known as the *coalescent time*.

Coalescence theory (Kingman, 1982, 2000; also see Websites) provides a methodology for modeling the reversal of mutation distributions in a set of DNA sequences over the history of a population back to an MRCA that possessed the original sequence from which later ones diverged. In essence, it reverses population growth, mutation and genetic drift, all of whose details have to be included in the model. Application of coalescence theory to a set of genes not only gives estimates of phylogenies and original sequences but also of the number of generations between nodes and how population size changed over time (Rosenberg & Nordborg, 2002; Li & Durbin, 2011; Wall & Slatkin, 2012). The basis of coalescence analysis is covered below, and its use in analysing human evolution is discussed further in Chapter 24; references to the full theory are given under *Further Readings and Websites*.

Coalescence analysis of a diploid population integrates a model of population behaviour with the phylogenetic analysis of variant sequences to work backwards in time. It starts with a set of **J** genes, each with the spread of sequences associated with a population of **N** individuals, an assumed mutation rate and a model of population behaviour that will include recombination, drift, etc. It then uses stochastic methods in which breeding choices are random (see Appendix 5) to run mutation and genetic drift backwards in time for as many generations as is needed for all the **J** coalescents to have been identified. The process is repeated many, many times with different random mating choices and the results together generate a statistically significant function called the *coalescent*. This is a quantitative model of the genealogy of the **N** individuals on the basis of sequence variation back to a single ancestor sequence.

Two points should immediately be clear. First, a group of polymorphisms for a particular sequence will coalesce at some point in the past that we recognise as the node representing that sequence in their MRCA. If the simulation is then continued for all the **J** sequences, we will eventually reach an overall MRCA for them all, on the way to generating a phylogenetic tree or coalescent for the set of **J** sequences (Liu et al., 2009; Wakeley, 2010). Second, the larger the population of breeding individuals between MRCAs, the more generations will be needed for two sequences to coalesce.

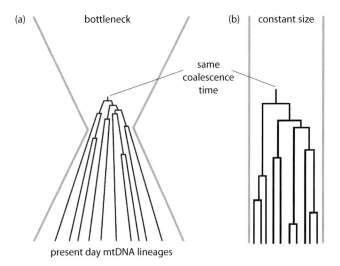

present day mtDNA lineages

Figure 21.3 **Coalesce analysis and population numbers.** A drawing illustrating two scenarios for the coalescence of a set of 10 mitochondrial DNA sequences back to an ancestral sequence, with both taking the same amount of time. (a): There is an early bottleneck, detected by the short lengths of the node branches, followed by a period of exponential growth (long branch arms) and diverging population (broad outer lines). (b): The population remains constant and there is no obvious pattern to the lengths of the branches.

(From Weaver TD (2012). *J. Hum. Evol.* 63:121–6. With permission from Elsevier.)

(Fig. 21.3). Here, it can be shown theoretically that the expected number of generations for this to happen is 2 N_e (Wang et al., 2016). While coalescent theory can be used for constructing phylogenetic trees over relatively short periods (typically ~10 My), its more important use is in exploring the history of the size of the population back to that most recent common ancestor whose original sequences gave rise to the spreads of sequences seen today.

It should be said that the mathematical basis of coalescence theory is complex, and analytical solutions for obtaining the coalescent are only possible for the most simple of conditions (e.g. for the idealised Fisher–Wright and Moran populations mentioned earlier; see Hein et al., 2005). In practice, therefore, such work today is always done computationally. Although the mathematical modelling and computational methods used for this are quite sophisticated, the ideas behind the process are relatively straightforward.

The methodology for extracting the coalescent starts with the realisation that random breeding in a cohort is formally equivalent to each polymorphism of a gene randomly selecting its parent, with a node being established if two polymorphisms choose the same parent. If, in a particular situation, the breeding is not random, the model imposes constraints on selection choices. This process can be repeated for a large set of genes for each member of the population to produce a set of possible nodes that covers all breeding possibilities in the previous generation. The process can be repeated back in time until each gene history coalesces onto a single most recent common ancestor.

The weights assigned to each node depend on the extent of the coding differences in the two polymorphisms: single codon differences are more likely than many-codon differences, with few differences being given more weight than many. This process is then repeated for the previous generation for which, of course, there will be fewer individuals if the population has increased with time. In due course, and perhaps after many generations, there will be a set of histories covering all breeding possibilities, each with its own likelihood. At the end of the set of simulations, one can analyse the resultant spread of genealogies statistically and obtain a set of parameters that will include the most likely genealogy linking the MRCAs, together with estimates and confidence limits for the coalescent time for each and the most likely sequence at each node.

One of the most interesting types of number that emerges from the analysis gives the estimates for branch lengths between nodes, which is measured by the number of generations, itself a measure of time measured in generations. As mentioned above, this figure is an estimate of twice the effective number of breeding individuals. Coalescence analysis thus gives is the population number at each generation node, and so provides a means of tracking population sizes backwards over time.

Perhaps unexpectedly, it turns out that the accuracy of the estimates from a coalescence simulation, as seen in the breadth of the confidence limits, depends far more on the numbers and lengths of genes used than on the number of individual alleles analysed for a particular gene. In practice, significant results about the evolutionary history of a species can be obtained with whole-genome sequence data from relatively few individuals. This is because the sequences on each pair of homologous chromosomes are different and this pair of similar but significantly different strands of DNA, each of length ~150 megabases and containing many thousands of coding and other sequences, provides the basic material for the analysis. Using coalescence analysis, Li and Durbin (2011) were, for example, able to reconstruct the history of African, Chinese, and European human histories over the past 200 Kya on the basis of the complete diploid sequences of just seven individuals.

In particular, they inferred that the African population went through a bottleneck (population decline) from around 16,100 breeding individuals 150–100 Kya to about 5700 at ~50 Kya, while the European and Chinese populations each experienced a severe bottleneck some 40–20 Kya when the population was reduced from about 13,500 to about 1200. (Chapter 24 has more on this subject.) These reduced populations, which acted as founder groups for new populations, had genetic profiles that were a subset of those in the original population because of genetic drift. This effect is the key reason why contemporary humans show characteristics that reflect their geographic origins.

Coalescence approaches can be expanded through the choice of model of population genetics used in the analysis to incorporate other aspects of population dynamics, such as the effects of migration and breeding links between two groups. Today, coalescence theory is a standard technique used by computational evolutionary biologists who wish to explore population genetics over evolutionary history and many other aspects of the history of genetically diverse populations. It is thus a particularly exciting area of evolutionary biology. The major limitation on the use of coalescent theory is the breadth of the confidence limits on estimates of the various parameters that emerge from the statistical analysis: these can be quite wide, to the extent that, unless very large amounts of information are included in the analysis, the information may have to be viewed as indicative rather than quantitative.

CONCLUSION

Evolutionary population genetics is unique: no other area of biology has such a powerful, rich, and integrated body of mathematically robust theory. The great strength of this evolutionary synthesis is that it provides a formal framework for analysing the gene allele and phenotype distributions in populations whose breeding behaviour can be explicitly defined in the context of selection. A century of theoretical work has shown how genes and their alleles flow through populations under a range of mechanisms that include linkage, flow, selection, and drift. The theory allows us to track back populations and sequences in time, identify genes associated with rare phenotypes and show how small populations are the natural breeding ground for novel speciation. The mathematics is not easy and the modelling is complicated, but the rewards have been great.

The modelling can also be used to describe experimental investigations, but only under slightly restricted conditions. Where the focus is on simple traits defined as those whose genotype is the alleles of one or two Mendelian genes that change under pressures that remain stable, the full quantitative theory is coherent and perceptive. It is key to conducting and analysing population studies and in running short-term breeding experiments where selection is strict. The theory is, however, much harder to apply under less controlled conditions, over the long term and in the wild. There are two main reasons for this. First, it can be difficult to get accurate estimates of the parameters needed to solve the equations. This is because the only data that it has to work with are distributions of phenotypes across a population. Second, the theory has to assume that selection and perhaps drift are affecting the dynamics of change of just the aspect of the phenotype under investigation. While this may be so under controlled conditions, it is rather less so in the wild where every trait of an organism is subject to variation and selection, even those that appear stable (Good et al., 2017). Selection rates reflect the effects of the environment as a whole on every trait in the phenotype, not just that being studied.

As a result, the best that can often be done under such circumstances is to estimate the percentage of the variance for some phenotypic change that can be accounted for. The practicalities of research also make it very difficult to study the dynamics of evolutionary change because, except in the rarest of cases, such changes take so long to be generated, even for *Drosophila* whose generation time is about two weeks. The few laboratory studies of evolutionary change are not relevant here because they use extreme selection pressures and exclude organisms that fail (Chapters 20 and 22).

The theory is far less helpful in cases in which the phenotype depends on non-Mendelian genes whose contribution to that phenotype are embedded within complex biochemistry (e.g. the protein networks that regulate growth). These genes are particularly important in the context of evolution as it is their mutated states that eventually result in anatomical change. It is still impossible to model these networks as we know very little about the internal interactions between the proteins and nothing about the rate and other constants needed for numerical solutions. The best that can be done is to hope that the main features of such networks can be modelled by the interactions of a few artificial trait alleles.

In summary, the mathematical analysis of the theory has enhanced beyond measure our understanding of gene dynamics in a population under a wide range of conditions. The theory is useful in investigating the equilibrium behaviour of phenotypes that result from the activity of a few Mendelian alleles under controlled conditions. It is less successful in modelling the dynamics of phenotype and genotype behaviour under conditions in the wild and how change happens there. The theory can only really handle the complexity of phenotypes underlain by sets of interacting genes or networks by grossly oversimplifying the nature of their genotype. This is because the original theory assumed that genes were simple and that phenotypes were underlain by the alleles of a few genes. It is nevertheless remarkable how far those very basic ideas have taken us.

KEY POINTS

- *Evolutionary population genetics* (EPG) is a mathematical model of the future behaviour of a population that describes how fitness, population size and behaviour, selection, and other such parameters can together lead to changes in the distribution of gene alleles over time.

- EPG starts with Hardy–Weinberg equilibrium: this shows that, in the absence of confounding factors, allele distributions within a population do not alter from one generation to the next.

- These factors include genetic linkage, migration (gene transfer), genetic drift, and selection. EPG includes the mathematical description of these effects.

- Genetic drift (the random distribution of parental genes to offspring) is particularly important because the results in small groups can be a skewed subgroup of the parent population.

- The effects of genetic drift can be similar to those of selection, but the environment has no role in generating those effects whose origins are stochastic.

- EPG shows that such small groups have particular genetic properties that encourage rapid change in phenotypes (e.g. the loss of heterozygosity and a rapid distribution of alleles).

- If such groups move away from the parent population to a new environment with different selection pressures, they may die out or flourish.

- Such founder groups represent the early stages of novel speciation.

- An important and recent advance in EPG has been the development of coalescence theory, which is a mathematical reconstruction of genetic history based on analysing DNA-sequence variants (e.g. the differences in the pairs of an individual's chromosomes)

- Such analysis can give an estimate of the numbers in a founder group that led to a particular modern population and of the time (in generations) when it formed.

- EPG has explained many of the principles of evolutionary change but is mathematically very sophisticated and often hard to use in practice, partly because the constants needed to solve its equations are hard to estimate.

- EPG is at its most useful in handling Mendelian genes but less useful in practice in handling complex genetics (e.g. quantitative phenotypes such as height, and complex phenotypes such as behaviour).

FURTHER READINGS

Hamilton MB (2009) *Population Genetics*. Oxford: Wiley Blackwell.

Hedrick PW (2010) *Genetics of Populations*. Sudbury, MA: Jones & Bartlett.

Herron JC, Freeman S (2014) *Evolutionary Analysis* (5th ed). Pearson: England.

Jobling M, Hollox E, Hurles M et al. (2013) *Human Evolutionary Genetics*. Garland Press.

Rosenberg NA, Nordborg M (2002) Genealogical trees, coalescent theory and the analysis of genetic polymorphisms. *Nat Rev Genet*; 3:380–390. doi:10.1038/nrg795.

Sætre G-P, Ravinet M (2019) *Evolutionary Genetics: Concepts, Analysis, and Practice*. Oxford University Press.

Uitterlinden AG (2016) An introduction to genome-wide association studies: GWAS for dummies. *Semin Reprod Med*; 34:196–204. doi: 10.1055/s-0036-1585406.

Wahl LM (2011) Fixation when N and s vary: Classic approaches give elegant new results. *Genetics*; 188:783–785. doi: 10.1534/genetics.111.131748.

WEBSITES

- Wikipedia entries for *Coalescent theory, Fitness, Linkage, Population genetics*
- Breeder's equation: https://www.nature.com/scitable/knowledge/library/the-breeder-s-equation-24204828/
- The neutral theory of evolution: www.nature.com/scitable/topicpage/neutral-theory-the-null-hypothesis-of-molecular-839/
- A simple introduction to the mathematics of coalesce theory: http://cme.h-its.org/exelixis/web/teaching/seminar2016/Example1.pdf

- Simulations of various aspects of selection (the Primer evolution site) on YouTube. Particularly note the link to *Natural selection.*

https://www.youtube.com/channel/UCKzJFdi57J53Vr_BkTfN3uQ

http://www.sfu.ca/biology/courses/bisc869/869_lectures/MHP_Coalescent.pdf

Speciation

22

Speciation not only reflects a temporary end point on an evolutionary path and an irreversible change in the local ecosystem but it also represents an opportunity for future diversity. It is thus the key step in evolutionary change. In living species that reproduce sexually, novel speciation can be recognised by reproductive isolation: individuals from some new species can at best produce sterile hybrids when mating with individuals from its parent and other closely related species. However, for a separated daughter subpopulation to be unable to breed with its parent population, major genetic changes will need to have accumulated, particularly at the level of the chromosome. In most cases, however, this breeding test is impractical and less stringent criteria are used.

This chapter discusses the various ways in which speciation can be recognised, and then considers the mechanisms that lead to the evolution of a new species. This process always starts with a population subgroup (or founder group) becoming reproductively separated from its parent population and finding itself in a novel habitat in which some at least of the subgroup can adapt and flourish. This isolation can, as the evidence shows, happen in a range of ways that depend on the nature of the separation. Important factors in this isolation are geographical and ecological differences between the habitats of the parent and daughter groups together with factors that affect mating choice.

Genetic separation is particularly likely to occur if the founder group is small because the speed with which mutational differences (new and old) spread through a population as a result of random drift and selection increases as its number decreases. Over time, there is a slow accumulation of phenotypic and genetic differences between parent and daughter groups that culminate in an inability to interbreed. The process is particularly demonstrated by ring species, such as the taxon of greenish warbler birds that encircles the Himalayas. These occur when a population splits and eventually meets up again; neighbouring groups can interbreed at all parts of the ring, except at the meet-up point where the accumulated genetic differences preclude it. Such ring species do, however, demonstrate that the definitions of a species are not as sharp as one might imagine.

There are usually three stages in the process of speciation (Fig. S4.1): the first and second are relatively fast and possibly overlapping, while the third often takes much longer (up to millions of generations). The first affects the phenotype and results in individuals in one population becoming unwilling to mate with those in the other (prezygotic isolation). The second is the increase over time in the infertility of any hybrid that does result from such a mating. The third stage is achieved when all the hybrids that form are infertile. At this stage, the two populations have finally become irreversibly separate species. This third stage of speciation usually derives from the accumulation of major genetic changes, such as chromosomal rearrangements to the genome so that any cross-fertilization from a parent species fails through chromosome nondisjunction or asynapsis.

DOI: 10.1201/9780429346217-26

New, distinct species mark evolutionary change—they are the culmination of the processes of variation and selection. The separation of a new species from its parent species on the basis of reproductive incompatibility represents a one-way ticket into the unknown and an opportunity for further new species to form in the future. If such separation only gives a subspecies, whose members are capable of mating with individuals from the parent species, then there is no incentive for novel adaptations to form; indeed, that subspecies could well be absorbed back into the parent species. It is clear on the basis of its title alone that, when Darwin published *On the Origin of Species* in 1859, he had already realised that the key to explaining evolution was understanding how new species formed and had thought through some of the ways in which it might happen.

It is also worth noting that Darwin was the first person to focus on the significance of speciation in this context, with Wallace being the second. There were then further discussions in the late 19th and early 20th century about how speciation might happen. Our current understanding of the process can, however, really be traced back to the work on the modern evolutionary synthesis in the 1930s and 1940s. Mayr, in particular, emphasised that the first step on the way to new speciation was the separation of small populations of reproductively isolated organisms in distinct environments. The research initiated by Fisher. Wright, Haldane on the population genetics of such groups (Chapter 21) and developed by others, particularly Mayr, laid the foundations for our current understanding of how new species form.

DEFINING A SPECIES

The problem of deciding whether a group of organisms represent a distinct species has turned out to be far more complex (and messy) than the early workers in the area ever suspected. For the traditional Linnaean taxonomy, an approach to classification laid down before evolution was accepted, the species is the key level for describing organisms, with each reflecting a set of morphological criteria. Higher levels of the taxonomy are groupings of increasing generality up to the phylum level (Chapter 4), which is defined in terms of the basic tissue geometry possessed by each of its each lower level taxa (e.g. each species within the arthropod phylum has an exoskeleton, a segmented body and jointed limbs). Taxa below the level of the species are seen as variants of the parent species that are recognised partly by their morphology and partly by their capacity to interbreed with the core species. Thus, for example, all tigers are member of the Pantherinae, which are a subfamily of the Felidae family; the classic Asian tiger; *Panthera tiger tiger* is the type species and the Indonesian tiger, *P. tiger sondaica*, is a subspecies. Here, it should be said that the criterion for a type species is not always clear and is often accepted on the basis that it was the first to be formally described.

The apparently clear definition of a species on the basis of its inability to breed with other species is usually inappropriate: this is partly because the criterion is rarely testable and partly because interbreeding can occur between groups that are traditionally recognised as different species. These difficulties are seen in the Canidae: wolves and dogs are clearly separate species but they can interbreed because they have the same number of chromosomes. The definition is also hard to apply to organisms that appear similar but have different habitats (e.g. parasitic worms) and impossible to apply to parthenogenetic and extinct taxa.

Coyne and Orr (2004) discuss the various types of definition of speciation that there are in the literature and just how hard it is to produce one that matches all conditions. This is because each definition has its own set of recognition criteria intended for a particular mode of speciation. Today, the concept of a species is seen as a slippery idea that tries to push groups of organisms which always show variation into well-defined categories. It is not correct, for example, to assume that the definition of a species that is appropriate for vertebrate fossils matches that appropriate for dogs or for worms.

What follows is a discussion of the major types of definition for a species: it shows that none handles all situations. It is, however, important to realise that although we need several criteria to distinguish distinct species, the perspective from evolving organisms is much simpler. A daughter population behaves as if it is a new species if, when given the chance to breed with the parent population, either refuses or produces only offspring likely to be sterile. The problem is that we can only rarely check this experimentally.

The breeding criterion

- A species is a group of organisms with the property that even if one of its members can successfully mate with an organism outside that group, any hybrid offspring will die or be sterile.

This definition is simple and clear and represents the mechanism of speciation, but it cannot usually be widely applied for two reasons. First, extinct species don't breed so the criterion can only be used for extant ones. Second, the practicalities of finding organisms for which the experiments are even worth attempting can be formidable. Where such experiments can be performed, it may be possible to measure a decrease in fertility rates and this may point to two populations that are diverging. This can be for many reasons because successful interbreeding requires a series of steps (i.e. mating, fertilization, embryogenesis, and the production of fertile offspring) and inadequacy in any one of these can represent a step on the way to novel speciation. In practice, the mating test is often too strong a criterion because many species will not breed with other related ones even though the offspring might be viable. Reasons include inappropriate behavioural cues (e.g. wrong courtship rituals), visual cues (e.g. humans would not even think of mating with chimpanzees or bonobos), and physiological cues (e.g. frogs and toads come into heat at different times).

The morphological criterion

- A species is a group of organisms that can be distinguished from other groups on the basis that they look different.

This is the original and, in practice, most used means of distinguishing one species from another, and usually the only definition for extinct taxa. It is not, however, always easy to decide whether, for example, a newly discovered fossil is a representative of a known species or whether it should be considered as a member of a newly discovered species. Detailed analysis of both morphology and other factors, such as when and where it lived, is needed before determining the taxon into which that organism should be placed and whether it should be viewed as a member or subspecies of an existing species, or whether it should be assigned to a new species. This choice is made harder because there is considerable variation within a species, some of which may reflect gender differences that may not be recognised as such in fossilised material.

One gets some sense of the difficulty of the problem when considering with no prior knowledge whether the skeleton of a small Chinese pug that is ~30-cm high and has a short muzzle comes from the same species as a terrier with a normal muzzle or a Great Dane ~75-cm high (Figure 4.2). It would be easy to suggest on morphological and size grounds that pugs were a different species. A similar question could be asked about the relationship between a modern Maasi adult ~2-m high and a Baka pygmy ~1.5-m high. Given the major size and minor morphological differences, it would be easy to suggest that they reflect different species, but one would first have to exclude the possibility that the differences reflected age, gender, strain variation, or diet. In fact, of course, they are simply variants of *Homo sapiens*.

The habitat criterion

- A species is a group of organisms that can be distinguished from other similar groups because it is the only one that can flourish in a particular environment.

This definition is logical, but it is hard to study experimentally species that are defined by their ecosystem; these include host-parasite pairs and communities of species that have become ecologically dependent on one another through co-evolution. In the latter case, the number of species in the group may be large and the interactions among them quite sophisticated (Gu et al., 2015, and below). Decisions on such ecologically defined species is difficult as only a limited number of habitats can be tested.

The molecular criterion

- A species is a group of organisms that can be distinguished from other similar groups because it has a unique molecular signature.

This criterion may be the only one that in practice can be used for bacteria and very simple asexual eukaryotes where there are no obvious morphological or habitat differences to distinguish one group from another. It has, of course, the problem that one needs to produce distinction criteria based on sequence differences, and such criteria are inevitably arbitrary. Here, it is sometimes helpful to include a phylogenetic assay where a species includes some characters that can be seen as shared and derived within a monophyletic context and others that are unique in some functional context. Sometimes, such a phylogenetic assay is the only one available, particularly for organisms that do not reproduce sexually and look identical (Baum, 1992).

The timing criterion

- A species is a group of organisms that can be distinguished from other similar groups because it lived at a different time from them.

A particularly difficult problem is posed by a group of organisms that slowly change over time—this is phyletic gradualism (Gould & Eldredge, 1972). An example is a chalk cliff made of coccoliths (calcite plates that at one time encased coccolithophore microalgae) where those at the bottom of the cliff were laid down long before those near the top. There will be clear differences between two coccolith skeletons at well-separated levels but, the closer they are, the more similar they will be. As there are few if any major morphological step points to identify the start of a new species, it is often impossible to distinguish between nearby organisms in any way other than by the time that a particular organism died, which is reflected in its height level within the cliff (Saez et al., 2003). Under these circumstances, one can only identify groups on the basis of where their morphological features come on a spectrum of chronologically defined features (Donoghue & Yang, 2016).

Although the concept of a species is convenient, these various definitions show that none is generally applicable, and all have fuzzy edges. Even the breeding criterion is dependent on the vagaries of chromosome reorganisation as the very occasional mule that was expected to be sterile has turned out to be fertile (Ryder et al., 1985; Yang et al., 2004). In contrast, all of the Canidae (dogs, wolves, coyotes, etc.) can each of be considered to be a variant of the same taxon as any interbreeding can produce fertile offspring. In contrast, horses and zebras, which are morphologically more similar than, say, a wolf and a poodle whose ferile offspring are known as eskipoos(!), have to be seen as separate taxa because their hybrid offspring are almost all sterile.

In evolutionary terms, a sterile hybrid is of no importance as it cannot breed. Fertile hybrids, on the other hand, are taxonomically problematic because they span two clades. They can be seen as being a first step towards a new species if they interbreed with each other and, if they breed with members of either parent species, provide a means of substantially increasing diversity through gene flow. Perhaps, a deeper reason for the difficulty of defining species, however, is that each is in a state of very slow flux as it replaces another species, evolves further, and then becomes extinct as it, in turn, is replaced by another that derives from it or another closely related species. While we see today's biosphere as a large number of stable species, this is a single snapshot of a very slowly changing species landscape. The fossil record shows that the biosphere looked very different ~15 Mya from how it is today, and we can have no doubt that it will be hard to recognise the species that inhabit the world ~15 My into the future.

HOW NEW SPECIES FORM

Speciation starts in small groups that become reproductively isolated from their parent population (Fig. S4.1). Indeed, any niche that becomes empty or is only partially occupied will soon be invaded by small groups of individual from species in neigbouring niches (Nosil, 2012). The fate of such founder groups depends on how its spectrum of phenotypes handles its new selection pressures and this cannot be predicted; it is, however, likely that many founder populations will be lost. Those fortunate enough to do well will include members whose trait variants enable them to capitalise on the character of their new environment. If local predation rates are low and the habitat range is large, small groups can expand rapidly both in numbers and in range of phenotypes (the *founder flush* effect that is seen, for example, in the cichlid fish of Lake Victoria—below). In due course, the new populations will differ sufficiently in appearance from the original one that they can be assigned to new species on morphological grounds (Box 22.1; Templeton, 2008). This scenario is much more likely if the parent species follows an **r** rather than a **K** scenario, so producing

The best and most direct evidence that speciation is due to the accumulated build-up of minor genetic differences in populations derived from **founder populations** comes from the very few examples of **ring species**. This is a taxon of related species that derives from a single founder population, subgroups of which extend clockwise and anticlockwise around an inhospitable central domain. This geography ensures that interbreeding is limited to neighbouring groups, each of which can be seen as a subspecies. One key feature of a ring species is that the two groups at the limits of the clockwise and anticlockwise migrations eventually meet up. It sometimes happens that, at this point, the accumulated but different mutations in the two groups are sufficient for mating to fail so that the two groups meet the criterion of being different species (Irwin et al., 2001). Three well-known such examples are the populations of Laurus gulls surrounding the North Pole, song sparrows encircling the Sierra Nevada in California, and the greenish warblers (*Phylloscopus trochiloides*) populations around the Himalayas (Fig. 22.1).

The small greenish warbler birds form a taxon of anatomically distinct populations that started perhaps a few million years ago when a single population of birds in the southern Himalayas started to spread east and west. Groups formed a succession of minor variants as they colonised new habitats, each of which had its own ecological features. The genetic differences that arose between neighbouring variants are, however, sufficiently small that fertile hybrids can form at the borders of adjacent territories.

When, however, the east-migrating and west-migrating populations eventually met up in central Siberia, north of the Himalayas, those genetic differences had built up to the extent that the new neighbours could no longer produce offspring. There is thus a discontinuity in breeding ability between the two adjacent groups that derive from the easterly and westerly migrations. These neighbours, therefore, have to be considered as a different species (for more details on these birds, see Websites).

Genetic analysis of the greenish warblers has shown differences between groups that increase with distance of separation, although not linearly. There are, however, no major chromosomal distinctions between the subspecies, a result indicating the recent nature of the separations (Alcaide et al., 2014). It is not yet known, however, whether this interbreeding failure reflects an unwillingness to interbreed or a full genetic incompatibility. It is also not known whether this set of variants is stable. Martins et al. (2013) expect that within 10–50 Ky, each variant subspecies will become a distinct species, and that what is seen

now represents only an intermediate stage on the way to dispersed speciation.

The current family of greenish warbler variants not only highlights the difficulty in being precise about where the border lies between variants and species but it also demonstrates how small differences, none of which alone block the formation of normal intergroup hybrids, can accumulate and lead to speciation.

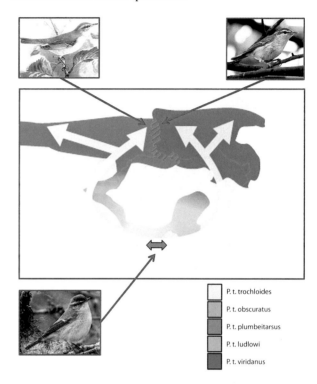

▢	P. t. trochloides
▢	P. t. obscuratus
▢	P. t. plumbeitarsus
▢	P. t. ludlowi
▢	P. t. viridanus

Figure 22.1 Ring species. The taxon of greenish warbler (*Phylloscopus trochiloides*) subspecies that originally spread out from the South to surround the Himalayas and meet up in Siberia. All neighbouring groups can interbreed except in Siberia where the east- and west-migrating populations eventually met.

(Courtesy of Jeff Schoenebeck.
Courtesy of G. Ambrus.

Insert of original *P. trochiloides*: Courtesy of P. Jaganathan. Published under a CC Attribution-share Alike 3.0 Unported License.

Insert of *P. plumbeitarsus*: Ayuwat Jearwattanakanok.

Insert: *P. veridianus*. Courtesy of Dibenu Ash. Published under a CC Attribution-Share Alike 3.0 Unported License.)

very large numbers of offspring (the former emphasises offspring number, the latter parental care—Chapter 20).

Two genetic factors then enhance any existing genetic differences and initiate new ones: drift and novel mutation, with the latter being particularly slow. In due course, the parent and daughter populations will become more and more genetically and phenotypically distinct, particularly if further gene flow between the two fails to occur (Via, 2009). Initially, of course, fertile hybrids will form if the groups are near each other (parapatric separation—below) and separation is incomplete. Over time, however, hybrids tend not to thrive: one reason for this is that, because each population would

TABLE 22.1 THE EFFECTS OF ISOLATION ON HYBRIDS

Extrinsic isolation

Ecological unviability: Normal hybrids cannot find a suitable niche

Behavioural sterility: Hybrids cannot find mates

Intrinsic isolation

Hybrid unviability: Hybrids suffer developmental defects

Hybrid sterility

Physiological sterility	Hybrids have reproductive system defects
Behavioural sterility	Hybrids show inadequate courtship behaviour

have become adapted to its own habitat, the hybrids would be less adapted to both. This phenomenon is known as *reinforcement* or the Wallace effect (Table 22.1).

It is the later genetic changes that cumulatively drive genomic speciation (e.g. the greenish warblers—Box 22.1). This amplification of differences between the two populations diminishes their tendency to interbreed and eventually leads to an inability of the two groups to produce fertile hybrid offspring. Timings cannot be predicted but, as a general rule, hybrids become infertile once there are differences in the chromosome numbers of the two groups, and the larger the difference, the less likely the hybrid is to survive. Horses and donkeys have 64 and 62 chromosomes, respectively, and mules, their hybrid offspring, are viable but generally sterile. Goats and sheep have 60 and 54 chromosomes, respectively: their sperm will fertilise eggs, but the embryos die. The core reason is *asynapsis*, the failure of homologous chromosomes to pair during meiosis, as was shown by Bhattacharyya et al. (2013) when they analysed the infertility of hybrids between two mouse species. It is still surprising that horse–zebra hybrids survive even though there are such major differences in their chromosome numbers (below).

MODES OF REPRODUCTIVE ISOLATION

The first step on the path to speciation is population separation leading to reproductive isolation and, over the past 80 or so years, it has become apparent that such isolation can occur in a host of ways. Coyne and Orr (2004) have listed the main types of separation that lead to groups of organisms lowering their chance of mating with or cross-pollinating their parent populations (Table 22.2). They fall into three main classes: geographic, ecological, and behavioural separation.

Ecological separation factors include local changes in the environment, such as changes in the mineral and organic content of the local ground, which lead to different plants flourishing in different, but adjacent areas. The resultant changes in these niches can, in turn, provide new domains for invertebrate subpopulations to colonise. Behavioural changes are those that affect reproductive activity: as a result of changes to mating calls (sound signals), pattern changes (visual cues), or hormonal cues (periods of fertility), members of adjacent groups may no longer recognise one another as mating partners, a process that can be relatively fast. Geographic isolation, the most important, is slightly more complex, and its categories are considered in a little more detail.

Allopatric separation

The clearest reason for group separation is habitat separation: a subgroup of a population becomes physically separated from its parent population and finds itself in a new environment where it is able to flourish. This geographic speciation has several subtypes (Table 22.2). The most common form seems to be allopatric speciation in which a population becomes separated into two or more subgroups that become located in different territories. The classic example is the evolution of the

TABLE 22.2 TYPES OF SPECIATION (FROM COYNE & ORR, 2004)

Geographic speciation

Allopatric speciation		A subpopulation finds a new niche outside the parent territory (*allo* = other).
Peripatric speciation		A small group becomes separated from its parent population (*peri* = around).
Vicariant speciation		A population is split in two by a major geographic feature (*vicarious* = substitute)
Parapatric speciation		Subpopulation forms at border between two parent species (*pari* = beside).
Sympatric speciation		Subpopulation finds a new niche within the parent territory (sym = together).

Ecological speciation

Barrier isolation		Subpopulations become separated by ecological barriers.
Habitat isolation		Habitat preferences lower the probability of mating between individuals.
Pollinator isolation		e.g. Pollen-carrying insects distinguish flower variants.

Behavioural and reproductive isolation

Mating system isolation		e.g. Self-fertilization.
Behavioural isolation		e.g. Changes in appearance that affect courtship.
Mechanical isolation		e.g. Incompatible reproductive apparatuses.
Timing (allochronic) isolation		e.g. Different breeding or spawning times.
Gametic (postmating, prezygotic) isolation		e.g. Sperm or pollen fails to fertilize.

Note that some of these categories can be seen as overlapping.

monophyletic tribe of 15 or so species that comprise Darwin's finches. It now seems likely that the parent population for this was a group of grassquit birds, *Tiaris obscura*, which originated in the Caribbean islands, then spread to central and South America, eventually reaching the Galapagos Islands ~2.3 Mya (Sato et al., 2001), with separate groups of birds ending up in each island. Once in their new habitats, each evolved separately, mainly as a result of food availability, and eventually formed distinct species (Fig. 4.3a). Ground finches, for example, have broad, strong beaks for tearing at cactus roots and eating insect larvae, while cactus finches have narrow beaks for punching holes in cactus leaves to access the pulp. Even after this long period (>2M generations) of separation, the full process of speciation is not yet complete as fertile hybrids can be bred between some species, provided that the strong breed-specific mating preferences based on song can be altered through imprinting (Grant & Grant, 2008). It is, however, clear that the original driver of adaptive radiation was food type (de León et al., 2014; Chapter 10).

Peripatric speciation

This variant of allopatric speciation occurs when a small founder group breaks away from a parent population and finds itself in a nearby but distinct habitat. This can, for reasons given earlier, lead to particularly rapid change. It is typified by the polar bear that evolved from an ancient Irish brown bear species after a small population became isolated at the northern periphery of a large population (Edwards et al., 2011). The descendants of this group adapted to the icy conditions by losing melanin from their hair, so acquiring camouflage.

Parapatric speciation

If a new niche becomes available that is adjacent to the niche already occupied by a species, it can be colonised by a subgroup of that species. Each group may acquire differences but opportunities for hybridization between the groups remain easy. Whether a new species evolves under these peripatric conditions really depends on the viability of these hybrids. The ring of greenish warblers described in Box 22.1 is an example of parapatric speciation.

Sympatric separation

The other well-known class of geographic separation is sympatric speciation: this occurs where a localised niche within a population's environment alters in some way that enables it to be colonised by a variant subgroup. This is a likely source of speciation in asexual organisms where, for example, a subgroup can benefit from a food source distinct from that of its parent group. Sympatric speciation can occur under laboratory conditions of strong selection and, of course, happens in the wild if two overlapping groups evolve to separately exploit the two niches. Thus, one group of Darwin's finches with broad beaks can survive next to another with narrow beaks, with the two groups being sufficiently distinct that they choose not even to try to breed with one another (Huber et al., 2007). In this case, beak size not only has a role in food choice but affects the sounds and calls that each group makes. As these are part of the courting rituals, beak size has a dual role and is an example of a *magic trait*, one that both drives selection and facilitates mating choice. Although such examples can be quite dramatic, sympatric speciation does appear to be considerably less common than allopatric speciation.

A special case of sympatric speciation is co-evolution, where two organisms can evolve to become mutually interdependent to the extent that the one cannot exist without the other — they are one another's prime environment or habitat (e.g. lichens, Chapter 20). Here, each organism defines the immediate component of the other's environment for the purposes of speciation; what started as mutual support has become co-evolution (Chapter 20). This phenomenon is not restricted to pairs of organisms but can involve whole ecosystems: groups of organisms, both plants and animals, may become so interdependent that loss of a single member of the system has wide and deleterious ramifications (Gu et al., 2015).

THE RATE OF SPECIATION

The evidence from the fossil record suggests that natural speciation is a very slow and gradual process (this is *phyletic gradualism*) with individual species surviving for ~1–10 My before becoming extinct and their niche, if it still exists, becoming occupied by new ones (Coyne & Orr, 2004). This continuous loss of species and replacement implies that there may a *natural extinction rate*, or more accurately, a typical *species lifespan*. Lawton and May (2005) have measured these lifespans and the figures seem to depend on the class of organism: allowing for the vagaries of the fossil record, these are: 4–5 My for marine animals, ~10 My for bivalves, ~2 My for graptolites (early colonial organisms that are withing the Hemichordata and are used as index fossils; Nanglu et al., 2020), and 1–2 My for mammals (for organisms that breed annually, these figures also indicate the numbers of generations needed). This last figure does, however, now seem rather low, given contemporary evidence on the evolution of the Equidae and Canidae (below). However, the pressures now being imposed by humans on a broad range of species during the Anthropocene may be leading to a shortening of the lifespans of many contemporary species, particularly large ones.

The major counterargument to the idea that speciation was always slow came from Gould and Eldridge (1972, 1993) who pointed out that *punctuated equilibrium* provided a better description of evolutionary change in the fossil record. Here, the long periods of slow evolutionary change could be interspersed with, for a variety of reasons, periods of relatively rapid speciation. Such fast speciation certainly occurred in the periods following major extinctions. The fossil record clearly shows that, over the few million years that followed the extinctions that ended the Permian and Cretaceous Periods, there were major and rapid radiations as emptied habitats were recolonised by new taxa that evolved from previously minor taxa that had survived extinction. After the former, diapsid reptiles were replaced by synapsids and, after the latter, these were replaced by mammals.

As predation rates were low in those early empty habitats, speciation was initially rapid but then seems to have settled down and became more gradual. Eldridge and Gould were also able to find examples in the fossil record where long periods of slow change were followed by bursts of rapid change that were not associated with major extinctions but were in accordance with the predictions of punctuated equilibrium. Other longitudinal studies, however, found organisms that showed both types of change: these included trilobites, snails, bryozoans, and dinosaurs (Eldredge & Gould, 1972; Horner et al., 1992). In many cases, however, the fossil record is not rich enough to establish with any certainty how long the process of speciation has usually taken.

More recent work has demonstrated that the early events of speciation can move much faster than is indicated by the fossil record, particularly when a group colonises a new niche in which the novel selection pressures are very strong. In salmon, for example, genetic and phenotypic differences can be seen in as little as 14 generations (Hendry et al., 2007). In terms of further separation, there are population-genetics reasons based on analysing multilocus models of sympatric speciation for believing that the early stages of speciation can take place in ~100 generations (Fry, 2003). Under such circumstances, phenotypic differences can thus arise rapidly in evolutionary terms, even though the sorts of chromosomal changes that lead to an irreversible failure of breeding will take much longer.

Hybrids

Examples in which the morphological and behavioural changes due to selection are relatively fast on an evolutionary timescale say little about how long it takes under unexceptional circumstances for two groups that had separated to become unable to breed. There is, however, a considerable amount of evidence from hybrids, particularly from the *Caninae*, that include wolves, dogs, dingoes, foxes, and jackals. They appear to have separated from a common ancestor species almost 2 Mya, and about the same number of generations, but there is no evidence of hybrid infertility among this group, presumably because all taxa have 78 chromosomes.

Similarly, there seems to be little inhibition to the formation of cat hybrids: lions and tigers that would never normally cohabit will mate under zoo conditions even though their last common ancestor lived 10–15 Mya (Johnson & O'Brien, 1997). The offspring of males and females are known as ligers and tigons and are larger than their parents, with characteristics of both. They retain a limited degree of fertility, mainly because almost all members of the cat family have 38 chromosomes. The exceptions are those in South America, such as the ocelots: these have 36 because two small chromosomes fused to give one larger one (Wurster-Hill & Gray, 1973).

A further, well-studied example is the *Equidae* family for which there is good data on speciation, location, and chromosomal differences (Table 22.3; Chapter 6). In the past, the broad taxon of *Equidae* was widely dispersed, but then went into decline. It seems that the last common ancestor of today's Equidae lived 4–4.5 Mya, with its offspring becoming geographically separated into the groups that are seen today; the most recent split led to the separation of the domestic and Przewalski's horses, an event that happened only 72–38 Kya (Lau et al., 2009; Orlando et al., 2013).

TABLE 22.3 MODERN EQUIDAE SPECIES

Species	Common name	Origin	Chromosome No.
E. ferus przewalksi	Przewalksi horse	Mongolia	66
E. caballus	Horse	N America	64
E. africanus asinus	Donkey (African wild ass)	E Africa	62
E. hemious	Onager (Asian wild ass)	Pakistan, India, Mongolia	56
E. kiang	Tibetan wild ass	Tibet	52
E. grevyi	Grevy's zebra	Kenya	46
E. quagga burchelli	Plains zebra	Southern Africa	44
E. zebra	Mountain zebra	SW Africa	32

Hybrids between *Equus grevyi* and *Equus quagga burchelli* are fertile (Cordingley et al. 2009), presumably because their chromosome numbers are so similar (Table 22.3). Mules, which are hybrids between horses and donkeys, are easy to breed, presumably because the parents also have very similar chromosome numbers and mating behaviour, but only the very occasional mule is fertile (Ryder et al., 1985; Yang et al., 2004). Given that mares first breed when they are 1–2 years old, it seems that hybrid infertility here has required perhaps 3–4 million years of species separation and perhaps half as many generations. What is more unexpected, given that the parents have very different chromosome numbers, is that matings between the nonzebra Equidae and the various zebra species can also produce hybrids (e.g. zeebroids, zeedonks, etc; see Websites), although females have low fertility and males are sterile (in accordance with Haldane's rule—see below). It seems that their genomes share sufficient genetic compatibility for meiosis and normal development to proceed, but that their very different chromosome numbers and assignments preclude either meiosis or early development in the subsequent generation. One curiosity in horse–zebra hybrids is that they have more stripes than their zebra parent and this probably reflects heterochrony: there is postponement of time during embryogenesis when their stripe patterning is laid down in the hair follicles within the embryonic epidermis (Fig. 22.2; Chapter 18).

Speciation speeds under allopatric conditions

The evidence from vertebrate hybrids thus suggests that full speciation often takes a few million generations, and this also seems to be so for invertebrates. Ridley (2004) suggested that novel speciation in *Drosophila* takes some 150–350 Ky. However, as the *Drosophila* reproductive cycle takes about three weeks in the wild, these figures imply that it takes a minimum of several million generations for speciation to occur after separation.

The period required for humans to have evolved from a last common Homininae ancestor seems short for full speciation to have occurred. The complex species route that led to *H. sapiens* began after an ancient ancestor group split away from the clade that now includes chimpanzees and bonobos ~7 Mya, with our taxon appearing ~200 Kya (Chapter 23). Given a reproductive cycle of about 15–20 years for hominids, these timings represent only some 500,000 generations. Today, this separation is marked by a single difference in chromosome number: humans have 23 pairs while great apes have 24. This difference is probably less significant than it might seem: human chromosome 2 results from the fusing of two ancestral ape chromosomes at their telomeres, an event that was followed by the inactivation of a centromere 6–5 Mya (Chiatante et al., 2017).

There is also evidence of human hybrids. When anatomically modern humans reached the Near East and then Europe some 50–40 Kya, they encountered groups of Neanderthals, a taxon that had left Africa some ~350 Ky earlier. Those who continued further east also encountered Denisovans, representatives of a group that probably left Africa a little earlier (Chapter 24). On the basis of the Neanderthal and Denisovan contributions to the contemporary human genome (e.g. Sankaraman et al., 2014), neither of these groups had any difficulty in breeding with humans; not even the behavioural bar to reproduction had been reached after ~350 Ky of separation. Even

Figure 22.2 An *E. quagga burchelli–E. caballus* hybrid bred by Cossar Ewart in about 1895. The zebra father (Fig.18.5) had about 25 stripes, but the hybrid offspring has about 70 stripes with a striping morphology more typical of *E. zebra* than its *E. quagga burchelli* father.

(From Ewart JC (1989) The Pennycuik Experiments. Adam & Charles Black.)

with a slightly shorter reproductive cycle than now, these figures represent ~20,000 generations. This is, however, a relatively small period on the evolutionary time scale.

The period for speciation can, however, be much shorter if the selection pressures imposed by the new environments on newcomers are strong enough. An example of unusually rapid speciation on the basis of allopatric/ecological separation is seen in two closely related groups of flounders living in the low-saline Baltic Sea. Most lay their eggs in the open sea (pelagic spawning); a second, reproductively isolated population live near the coast and spawn son the sea floor (demersal spawning) where salinity is particularly low. Sequence analysis has identified a genetic difference of 2051 biallelic single-nucleotide polymorphisms (SNPs) between the populations (Momigliano et al., 2017). One SNP is particularly important as it affects sperm motility. This mutation allows the sperm of the demersal males to remain motile in low-saline concentrations that would substantially slow the movement of pelagic males. Other adaptations include smaller, more robust, and nonfloating eggs being laid by the demersal fish and larger, less dense eggs by the pelagic fish. This example not only reflects strong divergence but shows that egg buoyancy and sperm motility can act as magic traits driving rapid speciation via reproductive segregation based on physiological conditions.

There is good timing evidence for when the flounder populations separated. The geology of the region suggests that it was only ~8.5 Kya that the originally freshwater Baltic Sea became connected to the North Sea, the origin of the Baltic flounders. This sets an upper date on when the demersal population started to segregate from the original pelagic one, and is matched by phylogenetic simulations undertaken by Momigliano et al. (2017): these also show that the populations shared a common ancestor ~2400 generations ago. As flounders breed only once when they are three years old, this generation number confirms that separation of the two populations only started once the Baltic Sea opened up to the North Sea. The reason why selection was so fast here was that, unusually, the phenotypic changes required for adaptation and segregation only needed a few SNPs, rather than any more substantial genetic changes, to change the reproductive phenotype.

The record for the speed of achieving the isolation stage of speciation is, however, probably held by finches from two Galapagos islands. Lamichhaney et al. (2018) have documented the case of a hybrid finch that followed the breeding of an immigrant male *Geospiza conirostris* from Española with a resident finch (*Geospiza fortis*) on the island of Daphne Major. The hybrid bird bred with a resident *G. fortis* finch, but all subsequent breedings seem to have been endogamous (within the group), even though there was no obvious reproductive reason for this. The likely novel feature that stopped intergroup mating was a change in the beak size in the hybrid that affected both feeding habits and mating song; this is another example of a "magic trait." In this unusual case, the first and key isolation step on the road to speciation has taken only three generations. It is hard to see that this record will ever to be broken as it reflects a theoretical minimum.

SPECIES FLOCKS

The process of speciation has been particularly illuminated by the study of species flocks. These are groups of geographically related but anatomically distinct species, each of which probably evolved from an original founder population that colonised a new area in which there was a variety of distinct niches that encouraged immediate allopatric speciation. The best-known case is again the Galapagos finches, but a far more recent example is the flock of cichlid species in Lake Victoria (Danley et al., 2012) that includes >300 species that can be morphologically distinguished (a sample is shown in Fig. 22.3) and whose genomes are now being investigated (Brawand et al., 2014). The anatomical variants seen in the cichlids of Lake Victoria today particularly include those tissues involved in food choice, such as tooth number and shape, pharyngeal muscle size, and lower jaw shape, as well as those likely to be involved in sexual selection, particularly male scale colouration and superficial ectodermal adornments (Miyagi & Terai, 2013).

It is usually impossible to know with any precision when a process of speciation starts but, in this case, the geological evidence implies that Lake Victoria only filled ~14.6 Kya (Seehausen, 2002). It has been suggested that at this point, the lake was seeded by a small founder population of cichlids from adjacent streams with enough genetic variation to produce the considerable spectrum of anatomical and behavioural

Figure 22.3 Examples of cichlid fishes from Lake Victoria selected on the basis of the predominant component of their diet. (a): zooplankton; (b): algae; (c): larvae; (d): fish scales; (e): snail: (f): small algae; (g): insects; (h): fish; (i): anancestral river-dweller also found in the Lake.

(From Brawand et al. (2014) *Nature* 513, 375–381. With permission of Springer Nature.)

traits seen today (Schluter & Conte, 2009). As to the degree of speciation now achieved, it is simply impractical to classify all these variants on the basis of interbreeding potential and the more traditional morphological assay has had to suffice.

It is reasonable to assume that the original population, wherever its origins, found itself in an environment in which there were few if any predators and was able to expand rapidly with few constraints on phenotypic and genetic variants, many of which found their own niches in the new lake. The assumption that these fish breed annually gives an upper bound of just under 15,000 generations for the full radiation with its novel anatomical features to become established. This is very much faster than the example of sticklebacks where it took little more than 10 Ky and the same number of generations for the change in morphology that followed the evolution of freshwater variants from their marine forebears. In that case, however, the change only involved a minor variant becoming the dominant form (Chan et al., 2010; Chapter 18).

Because Lake Victoria initially had few fish and included a wide range of habitats and niches, the conditions were appropriate for both peripatric and sympatric speciation. There would, therefore, have been little to inhibit the evolution of a very large number of habitat-defined groups. Goldschmidt (1996) has documented how anatomical adaptations, mainly to the oral cavity and teeth, have arisen that have permitted distinct species, located in geographically distinct locations, to feed on specific organisms, such as prawns, insects, other fish, algae, and snails. Among the diversity of cichlid taxa, there are 109 fish eaters, 29 insect eaters, and 21 zooplankton eaters (Fig. 22.4). As to the long-term stability of this diversity, it would be surprising if, as populations increase and move around the lake, these numbers do not slowly decrease as a result of competition between species currently occupying similar but distinct niches.

The evolution of >300 distinct cichlid species with anatomical differences from a small founder population over a mere 14 Ky of evolution gives a speciation time of a few thousand years at most. This period is so much shorter than almost all other examples of speciation that it demands some explanation. Part at least comes from

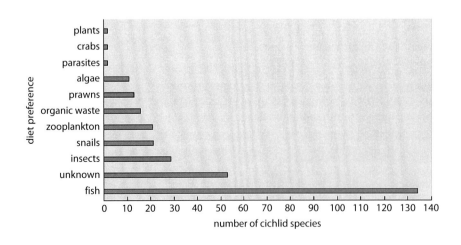

Figure 22.4 The numbers of cichlid species with particular diet preferences in a sample from Lake Victoria. (Redrawn from Goldschmidt, 1996.)

the nature of the **r** selection strategy used by bony fish where thousands of eggs are produced, rather than the very few produced by organisms using the **K** strategy (Chapter 20). Normally, the many embryos and young fish (known as fry) produced as a result of the **r** strategy have a very low survival rate because they are eaten by predators. In a new habitat with few predators, however, **r** selection has the advantage that it allows very rapid growth in a population and, through normal recombination and mutation, the creation of a great number of variants. A second argument is that anatomical change only takes a few thousand generations if it is based on a few single-nucleotide polymorphisms (Momigliano et al., 2017) (Chapter 20).

One argument against claiming that these reasons were sufficient to achieve speciation in a few thousand years (and probably generations) is that, while this figure could perhaps have been adequate for some novel anatomical features to form, it is too short to allow sufficient chromosomal alterations to accumulate for true speciation to occur. It would, therefore, be surprising were not most of these species still able to interbreed and, indeed, laboratory experiments on a few species have shown that hybrids can form (Keller et al., 2013).

An alternate explanation for the extent of cichlid diversity is that Lake Victoria is, for evolutionary purposes, much older than previously thought. The argument for its age being 14,600 years is that it was then that the deepest part was dry. Fryer (2001) has, however, pointed out that, while there is no reason to doubt this geological evidence, nothing is known about the state of the various ponds and rivers that helped fill the new lake or about the various fish and other organisms that colonised it. While it seems that the cichlids in Lake Victoria originally came from one or two founder species, it is quite possible that the populations of fish that colonised the newly formed Lake Victoria were far more diverse than used to be thought and that these last common ancestors lived a very long time before the whole area coalesced as a single lake. If this is so, then speciation will have been rather slower and more typical of other habitats than has been claimed.

The situation has been clarified by Schedel et al. (2019) who have conducted a very detailed phylogenetic analysis of cichlids from Lake Victoria and the nearby Lake Tanganyika and Lake Malawi, which are a major part of the East African radiation (EAR) of cichlids that dates back to ~29 Mya. They analysed ten mitochondrial genes and other sequences from 180 cichlids from across the world, of which about 120 were from the EAR radiation. The analysis dates the origin of the cichlid radiation to a most recent common ancestor that lived 7~4 Mya, Mya. This led to a rapid divergence of the species into several tribes. The Lake Victoria region superflock was found to have an initiation date to ~310 Kya. It thus seems that there was a considerable range of cichlids in the streams and ponds of the region that would become Lake Victoria long before the Lake appeared.

If there are >300 distinct species in Lake Victoria, this reflects >8 sets of binary splittings during the 310 Kya back to a last common ancestor. This gives a minimum speciation time of ~40 Ky, although it could be longer if the initial spread of fish in the lake gave a large number of small groups that each went on their own evolutionary trajectories. Given the size of the lake and the relative absence of predators until recently when Nile perch were introduced, two points can be made. First, the main selection pressures in the Lake were food types and similar-species competition (Fig. 22.4). Second, given that cichlids start to breed when they are about six months old, the speciation time could well have been as little as ~60,000 generations. Even this number may be considered to be low for the generation of significant anatomical change when compared to other organisms. The speed here clearly reflects an **r** breeding strategy in which multiple offspring rather than parental care is emphasised, although cichlids exhibit more such care than most fish (Wong & Balshine, 2011; Box 20.1).

EXPERIMENTAL SPECIATION

The work of breeders over the centuries to produce new variants has shown that, given the presence of a rare but interesting heritable trait in an individual, it is usually relatively straightforward to enhance its presence in a group through selective breeding, so producing a variant. In this sense, selective breeding just speeds up natural selection. Enhancing a trait does, however, fall a long way short of producing a new species from an existing one. Doing this normally requires both generating a new trait and undertaking selective breeding until the two populations at least meet the threshold of being unwilling or unable to mate. Such an effort goes further than the work of Waddington (Chapter 20) who generated a novel trait in *Drosophila* (flies with four wings) but this change did not inhibit breeding between them and wildtype flies. Indeed, Waddington later used this ability for genetic analysis of what had taken place when the lower thoracic regions with halteres were replaced by the middle thoracic region with wings (Waddington, 1961).

There have now been more than a dozen series of experiments that have set out to produce new species on the basis of their failure to breed with the parent population (Fry, 2009; see Websites). The great majority have used *Drosophila* as the model organism. This is because it breeds rapidly, its genetics are well understood, and it is cheap and easy to work with. The key feature of all of these experiments is that a single population is split or splits itself through habitat choice and each group is then grown for many generations in a distinct habitat. Experimental breeding is conducted within each group, usually on the basis of the strength of the novel trait. Intergroup breeding is also undertaken at regular intervals to see if the differential selection to date has been enough to isolate the groups reproductively (e.g. Rice & Hostert, 1993). The experiment continues until reproductive isolation is confirmed or the work has been deemed a failure.

A good example of an early success is the work of Rice and Salt (1990) on modelling sympatric speciation. They took a random population of wild-type *Drosophila* pupae, subjected them to a series of tests as they developed and then selected two populations on the basis of their very different behaviours. Successive sets of offspring from each group were bred separately, with examples of each allowed to attempt to breed with the other population to assay the strength of the separation.

For this work, Rice and Salt (1990) placed the pupae in a maze with choices for accessing food. These were a preference for light or dark (phototaxis), a tendency to prefer flying upwards or downwards (geotaxis), and a liking for acetaldehyde or acetone vapours (smell). They selected two groups of flies on the basis of the behaviour in the maze: the first were those who emerged early from pupae and chose the dark, upwards, acetaldehyde routes to food, while the second were those that emerged late and chose the light, downward, acetone route. After 25 generations of selective breeding within their own cohort, the two groups became unwilling to interbreed. They could thus be considered at least on the way to becoming two distinct species formed under sympatric conditions. Such was the strength of the selection pressure here that this figure is probably near the absolute minimum number of generations for the early, reproductive isolation stages of speciation to occur on the basis of behaviour and in the absence of a magic trait.

It is worth noting that this work did not depend on the presence of any particular trait within the wildtype population. Instead, the original group was chosen for their phenotypic diversity. Rice and Salt were then able to capitalise on further opportunities for diversification that came from genetic drift in the small populations that showed the behavioural options—the number of generations involved in such experiments was almost certainly too low for any mutation to play a role here (Chapter 20).

THE GENETICS OF SPECIATION

As daughter and parent populations slowly become more different, their ability to interbreed declines and there are several irreversible stages in this process. (Figure S4.1) The first is that the two separated populations become less willing to breed with one another, as shown in the experiments just discussed. This process, known as *reinforcement*, can reflect adaptation (above) or mating behaviour, and is shown by the need for specific mating-associated calls of the Galapagos finches. The second step occurs as hybrids become increasing less fertile over time and eventually become sterile. The progression is shown by lion–tiger hybrids (ligers) that have a viability of about 25%, and horse–donkey hybrids (mules) that are now at the fertility/infertility border, with only the very occasional example being fertile (Ryder et al., 1985; Yang et al., 2004). Finally, breeding becomes impossible because, even if organisms from the two groups do mate, fertilisation fails or the embryo dies, usually because of chromosome mismatch (Figure S4.1). Thus sheep have 54 chromosomes and goats have 60 chromosomes and any hybrids that develop are generally stillborn, although a very few survive (see Websites).

The precise genetic reasons for the slow decrease in hybrid fertility over time are not entirely clear but possibilities include increasing genetic incompatibility and decreasing efficiency of meiosis. In vertebrates, it is usually the male that becomes infertile first; this in accordance with Haldane's Rule, which is based on his observation that "when in the offspring of two different animal races one sex is absent, rare, or sterile, that sex is more likely to be the heterozygous (heterogametic) sex" (see Schilthuizen, 2011). This essentially means that, where the sex chromosomes are different (usually males), any genes on such a chromosome that acquire mutations will produce an abnormal phenotype as the effect of the mutant gene cannot be shielded by that of the wild-type one. Further evidence on the nature of hybrid infertility comes from interstrain crosses of various species, with a particular focus on *Drosophila*: these demonstrate that hybrid sterility is most commonly associated with genes on sex chromosomes that particularly affect fertility (Coyne & Orr, 1989; for review, see Rice & Chippindale, 2002).

There are also speciation genes, so-called because they have been genetically identified as playing a role in reproductive isolation; in some cases, this role as been confirmed by genetic engineering, usually in *Drosophila* (Nosil & Feder, 2012). Examples of such genes here are *Odysseus* (*Odsh*), *hybrid male rescue* (*Hmr*), *lethal hybrid rescue* (*Lhr*), and nucleoporins 96 and 160 (*Nup96, Nup160*; Nosil & Schluter, 2011), but the reasons why they facilitate isolation still remain unclear. There are also speciation genes in mice, with six hybrid sterility (Hst) gene loci now identified. Two of these are involved in sperm flagellar assembly (Pilder et al., 1993). A possible reason for lowered fertility here is that weakened interactions between the two distinct flagellar proteins of the hybrid sperm may lead to reduced motility.

In due course, a daughter population isolated from its parent population will become increasingly distinct as each population acquires its own mutations that will affect various aspects of interbreeding. Such changes can affect the geometry of the reproductive systems, the immune responses to sperm, and behavioural differences that further exacerbate mating preferences. In such cases, total reproductive separation can be achieved long before there are significant differences in chromosome numbers.

The key role of chromosomes

It is worth noting that even after >2 My of species separation, lions and tigers can still interbreed to produce fertile liger offspring whereas horses and donkeys produce infertile mules. The reason is that lions and tigers each have 19 pairs of chromosomes whereas horses and donkeys have 32 and 31 pairs, respectively. The key to irreversible species separation, in general, is the evolution of differences in chromosome organisation and number between the two. The initial formation and subsequent spread of such changes through a population are, as the examples given above demonstrate, rare, slow, and unpredictable.

What happens in practice is that as chromosomal rearrangements (e.g. inversions, duplications, joinings and splittings) accumulate, meiosis becomes progressively unlikely because of the increased likelihood of nondisjunction (Chapter 17). Eventually, the chromosomes from the two populations become unable to pair or separate properly, first during meiosis (this implies infertility) and later perhaps in mitosis (normal cell division becomes impossible). At this point, the original subpopulation has given rise to a new species, one that cannot interbreed with its parent population to produce fertile offspring.

Evidence on chromosome damage comes partly from irradiation studies, mainly on *Drosophila*, but also from studies on humans. Jacobs et al. (1992), for example, showed that ~1% of normal females aged 35 or over carry some sort of translocation, deletion, or rearrangement in their somatic cells. A more recent study of couples having spontaneous abortions in China (Fan et al., 2016) showed that ~2% of females and 1% of males had chromosomal abnormalities. In neither of these studies was there evidence of an external cause of the abnormalities, and it thus seems clear that there is, in human populations and presumably those of other organisms, a reservoir of normally unobserved but minor chromosomal defects that form spontaneously (McFadden et al., 1997). It is presumably some of these changes that will eventually drive speciation emerge, with that change possibly associated with linkage to mutations under positive selection. There is, however, no clear evidence on how or when these happen; it seems that their occurrence is rare and stochastic.

Changes to chromosomal organization clearly underpin the breeding definition of speciation but they say little about species defined in any other way. In practice, there is no satisfactory definition of a species that applies under all conditions, but it probably doesn't matter a great deal. This is because, on an evolutionary time scale, species change is normal: as environments alter, small groups of organisms move to them and adapt, eventually producing new species that outcompete and replace older ones there and may also then spread. This is not a rapid event: the fossil record suggests that, as mentioned earlier, species typically last some millions of years before becoming extinct.

KEY POINTS

- A species is best defined as a group of organisms that is reproductively isolated from all other species. The evolutionary implication is that such a species cannot be absorbed back into its parent species but can generate further diversity.

- Species derive from the splitting of lineages: a new one generally starts when a group, becoming isolated in some way from its parent population, includes variants that thrive in their new environment.

- Such speciation generally requires long separation time (> 1M generations) as it requires chromosomal differences to have arisen between the daughter and parent populations.

- In general, however, a species is usually defined as a group of organisms that can readily be distinguished from other groups, usually on the basis of morphological or habitat criteria.

- Isolation can occur for a range of geographic, ecological and behavioural reasons, with the most important being geographic separation (allopatric speciation). This typically occurs when a small subpopulation becomes separated from its parent population (the founder effect) and flourishes in its new territory.

- Population genetics shows that founder populations have low numbers because, the smaller the population, the more rapidly a novel mutant under selection can become established.

- Isolation can also follow the generation of a novel niche in an already occupied region whose ecology is sufficiently different to allow a subpopulation to flourish (sympatric speciation).

- The initial cause of variation in a new population is genetic drift. This is partly because of their asymmetric gene-frequency distributions, which also lead to the emergence of recessive homozygotes, and partly because the speed with which gene variants spread across a group slows as numbers increase.

- As new mutations affect the phenotype, the daughter population increases its genetic distance from its parent population.

- As this happens, any hybrid offspring become increasingly less fertile and, in due course, all interbreeding between the two groups becomes impossible.

- Once a population has become reproductively isolated from its parents, it can continue to produce further subpopulations that can evolve independently before it becomes a new species.

- Selection pressures in the wild are normally low and speciation is very slow. If, however, major adaptive change can result from only a few minor single-nuclear polymorphisms, effective speciation can occur very much faster.

- Full genetic speciation of a sexually reproducing eukaryote based on chromosome structural differences usually takes more than a million generations.

FURTHER READINGS

Barton NH, Briggs DEG, Eisen JA, Goldstein DB, Patel NH (2007) *Evolution*. Cold Spring Harbor Laboratory Press.
Coyne JA, Orr HA (2004) *Speciation*. Sinauer Press.
Shapiro BJ, Leducq JB, Mallet J (2016) What is speciation? *PLoS Genet*; 12:e1005860. doi:10.1371/journal.pgen.1005860.

Via S (2009) Natural selection in action during speciation. *Proc Natl Acad Sci*; 106 (Suppl 1): 9939–9946. doi:10.1073/pnas.0901397106.

WEBSITES

- Wikipedia entries on *Darwin's Finches, Laboratory Experiments of Speciation, Sheep-Goat Hybrids, Speciation*.

- *The greenish warblers and their songs, Zebroids*: www.zoology.ubc.ca/~irwin/GreenishWarblers.html

SECTION FIVE
HUMAN EVOLUTION

No question in evolution has attracted as much attention as that of how humans evolved for the obvious reason that we humans are unique: no other animal possesses our mental, speech, and creative capacities. The topic is important for two linked reasons: first, it is about "us" and we have a natural interest in our biological history; second, there is now more evidence on the evolution of *H. sapiens* from fossils, from genomic analysis and, uniquely, from artefacts than for any other species – humans are a model organism for studying the evolution of all other large organisms.

It was Darwin who initiated serious thought on human evolution through his book *The descent of man, and selection in relation to sex* that was published in 1871, at a time when there was very little evolutionary evidence on the subject. Since then, a very large amount of information about human fossil, genetics, and artefacts has been discovered that has clarified much of the details of how a spur off the primate line in Africa ~7 Mya led to the evolution of *Homo sapiens*, of the various times that humans left Africa for Asia and why human phenotypes across the world are so varied. The three chapters of this section discuss that evidence.

The main aim of the first chapter is to summarise the fossil evidence on the evolutionary steps that resulted in an early group of hominins that descended from the African great apes, eventually giving rise to anatomically modern humans (AMH). All of the early fossils have been found in Africa and mark evolutionary progress there. A few of the later fossils, which date to after the earliest *Homo* taxa had formed (~2.8 Mya), have also been found in Europe and Asia and these provide insight into the early migrations of the various *Homo* taxa.

The second chapter particularly focuses on the historical and contemporary sequence data of DNA from the recent *Homo* taxa and from primates across the world. Of particular importance here is mitochondrial DNA: analysis of its various haplotypes illuminates details of the recent migrations of *H. sapiens* both within Africa and out of Africa to the rest of the world. Sequence analysis also highlights the implications of the interbreeding between AMH and other earlier human taxa (Neanderthals and Denisovans) already in Eurasia as a result of earlier migrations from Africa.

The third chapter looks at the evolution of modern humans. The first part considers what the evidence from early human artefacts says about how and when they evolved cognitive abilities, such as those for tool making, speech and art. This area is, of course, related to the evolution of the human cerebral cortex

DOI: 10.1201/9780429346217-27

with its formidable range of abilities that far exceed those of chimpanzees, even though these animals have sufficient mental capacity to be taught the elements of language (Beran & Heimbauer, 2015). The latter part of the chapter considers the question of how, if all humans trace their ancestry back to relatively recent migrations from East Africa, the various populations across the world have come to look so different today.

Human Evolution 1: The Fossil Evidence

This chapter summarises what is known about the fossilised remains of early human taxa that shared a last common ancestor with the *Pan* clade (bonobos and chimpanzees). The chapter starts with a discussion of the geographic context in which humans evolved in Southern and Eastern Africa and the nature and limited quantity of the associated fossil evidence. It then considers some of the differences that have evolved in human taxa as they diverged from primates, abandoning any trace of an arboreal existence to become completely terrestrial.

The central part of the chapter reviews the main hominin taxa that are now seen as falling into six main groups: possible early hominins (e.g. *Sahelanthropus*), archaic hominins (e.g. the Australopithecus taxa), megadont archaic hominins (e.g. the Paranthropus taxa), transitional hominins (e.g. *Homo habilis*), pre-modern humans (e.g. *Homo heidelbergensis*), and anatomically modern humans (i.e. *Homo sapiens*). The last part discusses the isolated groups of pre-modern humans that have been found across the world, with some of them surviving until relatively recently (e.g. those on the Island of Flores). Of particular interest here is the Jebel Irhoud site in Morocco, which includes almost-modern skulls that date back ~300 Kya. They suggest that the evolution of anatomically modern humans was earlier and that the migration of early humans was more widespread across Africa than had generally been thought.

It is now clear from the location of fossil and associated artefacts that all the essential features of early human evolution took place in what are now the savannah areas of Central, Eastern, and Southern Africa. There is however a well-known saying among human evolutionary biologists that "absence of evidence is not evidence of absence." The further question of whether some very early humans also lived in the wetter and more forested areas of Africa is still unanswered. This is because evidence in such environments would mainly have been degraded and any remnants hard to find. It is also likely that there are human fossils to be found in more desert-like areas of Africa and possibly elsewhere that have not yet been fully explored. Some of the later material recently found in Morocco and Greece is mentioned below.

The context in which early human evolution took place in Africa is now known. The climate and general faunal evidence suggest that late Miocene and Pliocene Africa (7–2.6 Mya) started off as cool but, by ~4 Mya, was ~2–3C warmer than today with oscillating periods of wet and dry conditions, while sea levels were perhaps 25 m higher (De Menocal, 2004). As a result, much of the land was forested; this was the natural habitat of the Hominidae: this is the clade that now includes great ape, chimpanzee, bonono, and Hominini taxa. In due course, first the Homininae separated from the rest of the great apes and eventually the Hominini clade became distinct with this last event occurring ~7 Mya (see Table 23.1 for explanations of terminology).

With the coming of the Pleistocene Epoch there were three ice ages (2.6, 1.7, and 1.0 Mya) that particularly affected more northerly and southerly regions and

DOI: 10.1201/9780429346217-28

TABLE 23.1 TERMINOLOGY ASSOCIATED WITH HUMAN EVOLUTION

Hominoidea		Extant lesser and great apes and extinct taxa more closely related to them than to any other taxon.
Hominidae		All modern great apes (humans, chimpanzees, gorillas, and orangutans) and extinct taxa more closely related to them than to any other taxon.
Homininae		The Hominini and the Panini (chimpanzees and bononos) clades.
Hominini		All *Homo* taxa together with all fossil taxa more closely related to anatomically modern humans than to any other living taxon (Wood & Lonergan, 2008.)
Ardipithecus	5.8–4.4 Mya	A very early hominini genus adapted to life in trees. Other such early taxa are *Sahelanthropus* and *Orrorin* (Grabowski et al., 2018).
Australopithecus	4–2.0 Mya	A genus of early hominins that were probably bipedal and included mainly gracile, but also robust members.
Paranthropus	2.6–1.3 Mya	A genus of early hominins with a robust anatomy and large post-canine teeth (megadontia) that was probably related to *Australopithecus*.
Homo	2.8 Mya to present	A genus of bipedal hominins characterised by increasing cranial capacity.
Geological time periods		
Neogene Period Miocene epoch	23–5.33 Mya	
Pliocene epoch	5.33–2.588 Mya	Start of glacial periods
Quaternary Period Pleistocene epoch	2.588 Mya–11.7 Kya	End of glacial period
Holocene epoch	11.7 Kya–present	

caused sea levels to drop markedly; these also led to the equatorial climate became much more arid and cooler. Over this period, there was a gradual change over non-desert Africa as forested areas became more open. It was this change that seems to have been the driving force for the early Hominini to adapt to life in grasslands; such fossilised limb bone evidence that we have of early *Homo* taxa reflects this adaptation.

Early fossil and geological evidence show that, by ~16–15 Mya, the Hominoidea, the clade that now includes the lesser and great apes and their descendants, mainly lived in forests and woodlands, spending some time on the ground. Over the next ~8 My (middle to late Miocene), subgroups of great apes slowly acquired some of the anatomical characteristics that enabled them to be more upright and better adapted to a ground-based than an arboreal environment. (For a detailed analysis of the great ape fossil data over the period 20–7 Mya, see Andrews, 2020.) Detailed analysis of the late Miocene fossil material from early humans and apes has enabled Andrews to identify likely skeletal features of the last common ancestor of the *Homo* and *Pan* clades (Table 23.2).

Some information on the anatomy of the descendants of the LCA of the hominins is given by the skeletal details of the earliest hominin fossils (below). One should however be aware that many aspects of hominin evolution are not provided by such data, although some may be illuminated by it. While we can, for example, infer that early hominins were bipedal on the basis of their general anatomy and omnivorous on the basis of their teeth, but it is not easy to use such data to probe their physiology and behaviour. Nevertheless, a detailed study of the fossil material has provided some insight into the life history and behaviour of these early ancestors and highlighted the differences between humans and chimpanzees (Box 23.1; Robson & Wood, 2008).

TABLE 23.2 LIKELY ANATOMICAL FEATURES OF THE LAST COMMON ANCESTOR OF THE *HOMO* AND *PAN* CLADES (FROM ANDREWS, 2020)

Projecting canines, P3 honing

Enlarged molars, thick enamel

Deep palate, parallel tooth rows

Zygomatic shifted forward

Moderate mid-facial prognathism, low alveolar prognathism

Little facial buttressing, robust mandible with inferior torus

Divergent hallux, hand not elongated

Long thumb, precision grip, short robust phalanges High angle of femur neck

Elongated ilium, elongated ischium

Primitive tool use, hunting vertebrate prey

Palmigrade, quadrupedal terrestrial locomotion

Environment seasonal woodland

Chromosomes 2n = 48, Y chromosome differences (Chapter 24)

BOX 23.1 SOME *HOMO-PAN*/CHIMPANZEE DIFFERENCES: 7 My OF EVOLUTIONARY CHANGE

GENETIC DIFFERENCES

- Chimpanzees have 48 and humans have 46 chromosomes (human chromosome 2 includes most of chimpanzee chromosomes 2A and 2B; Miga, 2017)
- Y chromosome difference (see Chapter 24)

ANATOMICAL FEATURES IN HUMANS

Changes   Flattened face, defined chin, and enlarged nose

        Diminution of sexual dimorphism, particularly in size

        Teeth and jaw changes associated with a more varied diet

                Loss of the canine-P3 honing complex (canines are no longer sharpened)

                Strengthened enamel

                Shortened and raised tongue and a lowered larynx (associated with speech)

        Much enlarged brain (~1350 as opposed to ~380 gm)

        Much enlarged cerebral cortex

        Various changes to the hand including shortened fingers

        Divergent hallux (big toe) becomes non-divergent

        Various modifications associated with bipedalism

        Coarse body hair has become fine and less dense

        Five rather than 3–4 lumbar vertebrae

PHYSIOLOGICAL FEATURES IN HUMANS

Gains   Sweating

        Obligate bipeds able to run long distances

        A diet that included tubers and increased amounts of meat

        Early weaning but later first childbirth* (~18 rather than ~13 years)

        Enhanced life span*

BEHAVIOURAL AND CULTURAL DIFFERENCES

Gains   Enhanced ability to hunt

        Mental capacity and speech (with associated changes to the oral cavity and larynx)

        Dexterity and a greatly enhanced facility for tool making

        Art: body decorating, sculpture, painting

        Changes in social and sexual behaviour

* In human foraging populations.

Sources: From Andrews, 2020; Duda & Zrzavý, 2013; Lieberman, 2013; Robson & Wood, 2008

Further understanding has come from comparative studies that use computational techniques to analyse a range of behaviours shown by both humans and apes and so by inference by their last common ancestor (e.g. Duda & Zrzavý, 2013; Andrews, 2020). These analyses show that many human behaviours can be traced back to older clades, but others are only shown in *H. sapiens* (Table 23.2). Some, such as tool use, are well known but others that emerge from the analysis include a reduction in sexual dimorphism: in terms of size, human males and females are far more similar in size than are great ape males and females (Kikuchi et al., 2018), a feature that slowly appeared in taxa after the time of the *Australopithecus* taxon (Larsen, 2003). In addition, humans are weaned earlier and go through puberty and finalise their secondary dentition later. They also have behaviours surrounding group activity and copulation that differ from those of chimps and bonobos.

HUMAN FOSSIL MATERIAL

Eugene Dubois was the first to discover an early hominin fossil that he named *H. erectus* because its skeleton was the earliest then known to show an erect posture. This discovery was made in 1891 in Java and led to the idea, now known to be wrong, that the origins of *H. sapiens* were in Eastern Asia rather than Africa. It was not until 1924 that Raymond Dart discovered the first hominin skeletal material in South Africa and named it *Australopithecus africanus* (African southern ape) as it was both older and far more chimpanzee-like than *H. erectus*.

Since then, the bones and teeth of many other ancient taxa have been discovered in Asia and Europe but mainly in Africa. It should, however, be emphasized that the fossilised skeletal material for many taxa is rare. We generally have little idea of either the range of morphologies or of the extent of sexual dimorphism for a particular taxon. This is because such early human fossilised skeletal material as we have is usually incomplete: robust and dense tissues, such as mandibles, teeth, and crania are more common than more fragile ones, such as digits, vertebrae, and ribs; moreover, bones from larger individuals (males) are more common than those of smaller ones (females and children).

In assessing the status of the skeletal material, evolutionary anthropologists need to decide whether it is immediately recognisable as a member of a known taxon, an outlier in a known population, reflects a sex difference or may need to be assigned to a new taxon. Very detailed analysis is needed to justify that the material typifies a new taxon and, if so, to work out how close it is to other taxa known to live at around the same time. Key pointers here are the capacity of the cranium (brain volume), the shape of the face and the detailed anatomy of the limbs (Fig. 23.1).

Analysis of a series of fossil crania has now clearly shown that as hominins diverged from their hominid ancestors (Table 23.2; Figs. 23.1 and 23.2) ~7 Mya and became

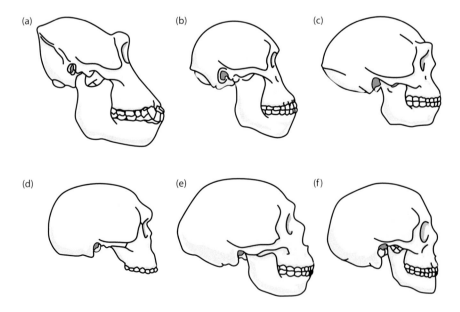

Figure 23.1 Skull shape and cranial capacity. (a): Gorilla (~550 cc). (b): *Australopithecus* (~450 cc). (c): *Homo erectus* (~1000 cc). (d): Pre-modern human ~0.3 Mya (e.g. *H. heidelbergensis*, ~1100 cc). (e): *H. neanderthalensis* (~1500 cc). (f): *H. sapiens* (~1400 cc). Note that the cranial capacities have not been adjusted to reflect body size so that *Autralopithecus* had, for its height (1.2–1.4 m), a larger brain than might have been expected.

(Courtesy of "Vladen666". Published under a CC Attribution-Share Alike 3.0 Unported License.)

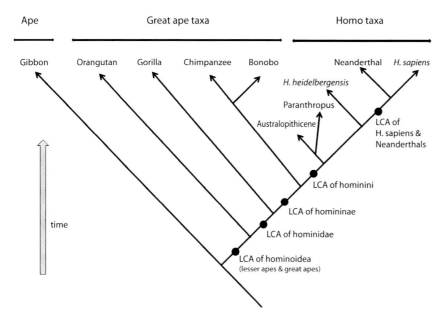

Figure 23.2 A cladogram showing the core line of descent of humans from early primates over the last 8–5 My. (Note: time is not scaled; LCA: last common ancestor.)

increasingly adapted to life on land as opposed to in trees, their cranial volume gradually tripled. Associated with this terrestrial habitat was an increase in cognitive ability and dexterity as assayed by tool production (Chapter 25), probably driven by the availability of improved food, particularly protein, as a result of hunting (Lieberman, 2013).

This enlargement clearly reflects an increased brain size as brain growth drives skull formation (Chapter 17). As the mammalian brain develops, superficial mesenchyme (some deriving from mesoderm, some from the neural crest [see Appendix 7]) forms the cranial bones (e.g. the frontal, parietal, and temporal bones). Sutures between these bones allow them to grow as the brain expands, and these do not finally ossify until well after the brain has completed growing. The final size of the cranium thus reflects the size of the brain. The importance of this mechanism is shown by mutations that force sutures to close prematurely: other, still open sutures have to accommodate the expansion of the brain and this results in a misshapen skull (Figure 17.3; Morriss-Kay & Wilkie, 2005).

The molecular basis of such brain growth is not yet known but the subject has naturally interested evolutionary molecular geneticists and a large amount of work

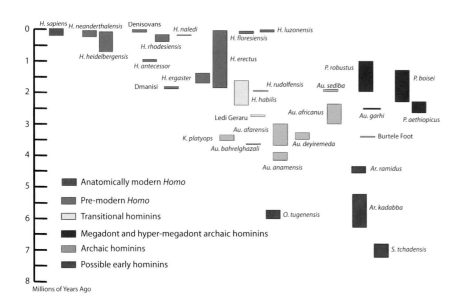

Figure 23.3 **A broad ("speciose") taxonomy of human speciation.** The columns reflect evidence of the earliest and latest identified fossils of each taxon but may be underestimates of the lifespan of each taxon.

(Courtesy of Bernard Wood.)

is currently being published. It is not, however, possible to do more here than give a flavour of the different ways in which this important topic is now being investigated. At the molecular level, Berto & Nowick (2018) are comparing gene-expression in the prefrontal cortex of chimpanzees and humans to identify ways in which they are different. In addition, Fiddes et al. (2019) have shown that genes associated with Notch signalling have been amplified in human brain development while Suzuki et al. (2018) have shown that these genes have an important role in neuronal growth and differentiation, particularly in the cortex. At a more experimental level, Benito-Kwiecinski et al. (2021) have shown that Zeb2, a transcription factor whose roles include epithelial morphogenesis and the epithelia-neural transition in the brain, has differential effects in cerebral organoids made from early apes and human tissue, allowing human organoids to become larger.

In a slightly wider context, Jeong et al. (2021) have shown the extent to which methylation patterns have changed in the developing *Homo* brain as compared to those in the great ape clade, and how these changes affect early differentiation patterns in the developing brain. (The effect of genetic methylation in controlling facial and laryngeal anatomy in Denisovans is discussed below.) Wei et al. (2019) have studied some of the key genes in cognitive networks and pointed to the enhanced role that the human accelerator region (HAR) genes play in human as opposed to chimpanzee brains (see also Doan et al., 2016). The question of how the unique properties of the human brain evolved is particularly difficult to answer, but progress is certainly being made.

THE MAJOR HOMININ GROUPS

As the fossil material has slowly accumulated, the picture of how *H. sapiens* evolved has become clear. This material clearly shows that there was no direct line of descent from an early pre-human ancestor to anatomically modern humans (AMH). Instead, it demonstrates that there was a broad bush of taxa distributed across a few clades, although detailing links between them will need more fossils with intermediate phenotypes than we now have. The picture is much like that for the Equidae (**Figures 6.10** and **6.11**), except that today, only one human species remains. The most recent grouping of human fossils based on their anatomy is that of Wood and his associates (**Fig. 23.3**; Wood et al., 2020; Wood & Boyle, 2016). This is given here.

Possible early hominins (7–4.5 Mya)

The very limited amount of very early skeletal material comes from central and sub-Saharan Africa and include *Sahelanthropus tchadensis*, *Orrorin tugenensis,* and the later *Ardipithecus* hominins (5.77–5.54 Mya; *Ardipithecus* means ground or floor ape). These included *Ardipithecus Kadabba, Ardipithecus ramidus,* which had a reduced tooth honing complex (below), and *Ardipithecus Sahelanthropus*. The last of these had the massive brow ridge seen in today's gorillas and a brain-case volume of 350–400 cc, much like a chimpanzee; it also, however, had features that mark it out as being different from chimpanzees. These include more generalized teeth and a change in the position of the foramen magnum (the oval space at the base of the skull through which the spinal cord passes) from that associated with a four-legged gait to one also capable of a bipedal walking gait. *A. ramidus* also had a grasping hallux (big toe) and the limb morphology typical of arboreal chimpanzees—it was clearly capable of both an arboreal and a ground-based existence (White et al., 2009).

One notable feature of *Ar. ramidus* is the diminution of the canine-P3 honing complex (Suwa et al., 2009). In apes, the tooth arrangement is such that the large upper canines are sharpened by movement against the lower premolars and canines. As hominins evolved, the canines decreased in size and the ability of other teeth to hone them was lost (Delezene, 2015); the geometry of early hominin teeth is important for classifying fossilised skulls.

Archaic hominins (4.5–2.5 Mya)

The *Australopithecus* ("southern ape") taxa included *Au. anamensis* (~4 Mya), *Au. afarensis* (3–2.1 Mya), *Au. sediba, Au. bahrelghazali,* and *Au. deyiremeda*, with *Au. garhi* being on the border of the archaic and transitional hominins (Spoor, 2015). Although a few taxa were robust, these hominins were mainly small, slender (gracile) and 1.2–1.4 m in stature, with long upper limbs suitable for a life that was at least partly arboreal. They showed several features that represent advances on the earliest hominins: the cranial base was shorter, their heads had become more *Homo*-like in orientation indicating an upright posture, their hands were more human-like (e.g. the fingers had shortened relative to the thumb, so increasing manual dexterity) and they had foot anatomy that indicated bipedalism (Kimbel & Rak, 2010; Ward et al., 2012). Dennis et al. (2012) have argued that this was the first hominin to have a copy of SRGAP2, a gene whose protein plays an important role in neuronal development, particularly in synapse formation (Ho et al., 2017). This might account for the fact that the size of the *Australopithecus* brain (390–515 cc) is relatively large for its height, as compared to that of modern chimpanzees.

Australopithecus garhi, found in an area that is now Ethiopia, seems to have been a late member of the clade. The taxon seems to have been relatively short-lived and had an appearance much like other australopithecines, although its teeth were noticeably large, and it is this feature that seems to have been characteristic of megadonts,. This hominin, which may have been transitional between the *Australopithecus* and the megadont clades, may also have been the earliest known user of stone tools (the Lomekwian industry; Harmand et al., 2015) as animal bones located near *Au. garhi* remains have stone-inflicted scars (de Heinzelin et al., 1999).

Megadont archaic hominins (2.5–1 Mya)

These included the *Paranthropus* ("near human") taxa, such as *P. boisei* (Wood & Constantino, 2007). These were similar to, and probably descended from, the archaic australopithecines, but distinguished from them by their robust skeletons, large jaws, and the details of their postcanine and anterior teeth which were large (hence their name). Like all earlier hominins, their fossils are found in the arid areas of the eastern and southern parts of Africa—the chances of finding the fossilised material in forested areas is low. They had a braincase, whose capacity was ~450 cc. The *Paranthropus* taxa, which overlapped with the transitional hominins, appear to have died out leaving no successors.

Transitional hominins (~2.5–1.4 Mya)

This group, which includes *Homo habilis* (2.3–1.4 Mya) and *H. rudolfensis*, marks the beginnings of the *Homo* group (see Table 23.3).

H. habilis and *H. rudolfensis* were similar, but the latter had a flatter face and larger teeth than the former, features that may indicate that the two taxa had different diets. *H. habilis* had short femurs, long upper limbs, and other features showing that it had a mixed arboreal and savannah-based existence. *H. gautengensis*, for which there is only cranial data, may also have been a distinct transitional hominin (Curnoe, 2010): it was found near Pretoria, some 3000-km south of the Olduvai Gorge and has similarities with *H. rudolfensis*, although its exact status is still not agreed. The volumes of the braincases of these transitional hominins had by now increased to 500–800 cc.

The long phalanges of the limbs associated with *H. habilis* suggest that it had the sort of grasping ability associated with at least a partly arboreal life, and this ability may have opened up other possibilities. While some chimps use stones to crack nuts, *H. habilis* made hand choppers and flaked stones to give them sharp edges that were useful, for example, in cutting up meat; hence its name which means "handy man." It thus seems to have been the first hominin to shape stones into tools (Plummer, 2004), an ability that may have been facilitated by its fully opposable thumb. Lieberman (2013) has discussed the various ways in which such tools freed up members of the *Homo* taxa from spending almost all their time foraging and eating, as do modern chimpanzees.

In terms of progress towards humans, the major novel feature of the transitional Hominins relative to the archaic ones was the increased cranial capacity. There were other minor changes to the skeleton: the head acquired a diminished snout (i.e. a less-protruding mouth) and a reduction in tooth size that, together with molar specializations, suggest an increasingly broad diet. There were also further changes to the limbs that would have facilitated bipedalism and dexterity.

Pre-modern *Homo* (1.9 Mya–30 Kya)

This group includes two early and several later taxa (Table 23.2). The earliest is *H. ergaster* ("workman"); when discovered, this species was the earliest to be associated with stone tools, and was named on this basis. Its not-very-different successor seems

TABLE 23.3 THE FOSSIL RECORD OF THE VARIOUS *HOMO* "SPECIES"

Species	Known existence Kya	Location	Adult height cm	Adult weight kg	Cranial volume cm^3	Fossil record
H. sapiens (modern *Homo*)	200–present	Worldwide	150–190	50–100	950–1,800	Still living
Pre-modern humans						
Red Deer Cave people	145–115	China				Very few
Denisovan Hominins	~300-30	Russia (Siberia) Tibet				A few bones; 3 individuals
H. luzonensis	67–50	Luzon, Phillipines				Bones from 3 individuals
H. floresiensis	120–100	Indonesia	~100	25	400	7 individuals
H. sapiens idaltu	160–150	Ethiopia			1,450	3 crania
Early *H. sapiens*	~300–200	Jebel Irhoud Morocco) (Apidima, Greece				
H. naledi	325–236	S Africa	~150		465–600	Many individuals
H. neanderthalensis	350–30	Europe, Western Asia	~170	55–70 (heavily built)	1,200–1,900	Many
H. rhodesiensis	300–120 (perhaps earlier)	Zambia			1,300	Very few
H. heidelbergensis	600–350	Europe, Africa, China	~180	90	1,100–1,400	Very limited
H. cepranensis	900–350	Italy			1,000	1 skull cap
H. antecessor	1200–800	Spain	1~75	90	1,000	2 sites
H. georgicus	1850	Georgia			~560	5 skulls
H. ergaster	1900–1400	Eastern & Southern Africa	~190		700–850	Many
H. erectus*	1900–200	Africa, Eurasia (Java, China, India, Caucasus)	~180	60	850 (early)–1,100 (late)	Many
Transitional Hominins						
H. rudolfensis	~1900	Kenya			700	2 sites
H. habilis*	2200–1400	Tanzania	~150	33–55	510–660	Many

* The key species for "lumpers" looking for taxon parsimony (Wood, 2019).

Sources: From a wide range of sources.

to have been *H. erectus* (1.9–0.1 Mya), examples of which have been found in Africa, Java, and many other places. *H. erectus* was followed by a series of taxa that include *H. antecessor, H. heidelbergensis, H. rhodesiensis, H. neanderthalensis, H. antecessor,* and *H. naledi,* the most recently discovered taxon.

This last taxon was recently discovered in the Dinaledi cave near Johannesburg in South Africa, and the remains have now been dated to ~335-236 Kya (Berger et al., 2015; Dirks et al., 2017). A large number of individuals were found in what seems to have been a burial chamber and a few more in a nearby cave. They were bipedal,

~1.5-m high, had cranial capacities of ~550–600 cc in males and ~465 cc in females. Some features were typical of australopithecines (e.g. curved fingers, ribcage, shoulder, and skull volume), while others were more typical of *Homo* taxa (thumb, wrist, and smallish teeth, together with skull, brain morphology, and a relatively modern face with a nose). The phylogenetic status of this taxon is unclear, but its morphology suggests that it stands outside the main line of human evolution and left no successors.

There seems little doubt that even early pre-modern *Homo* taxa were obligate bipeds, while their enlarged cranial capacity (600–1250 cc and up to ~1700 cc in Neanderthals) certainly suggests an increase in intelligence. Moreover, the changes to their face (they had less protruding lower faces and enhanced noses) would have facilitated the increased amount of breathing required for running, as opposed to walking. If so, and as running requires the ability for the body to lose heat easily, this was probably the period when body hair thinned to a light fuzz and sweat glands spread over the body. Lieberman (2013) has discussed how this ability would have made hunting much easier: chasing animals that cannot sweat causes them to become exhausted more rapidly than those than can, partly because they cannot pant enough to cool down. Moreover, as time progressed, these hominins made increasingly sophisticated stone tools (Chapter 25) that would have facilitated food procurement and preparation. In short, the new abilities shown by the pre-modern Hominins would have enabled them to eat better than their ancestors and so to flourish.

These pre-modern hominins also seem to have been the first to leave Africa in large numbers: evidence of their existence has been found across Asia (see below) dating back to about 1.85 Mya. It is now, for example, clear that the bones of the original Java man, now dated to 1.0–0.7 Mya, were typical of *H. erectus*. The earliest *Homo* skeletal material currently known from outside Africa, however, is probably from the Dmanisi cave in the Caucasus; while it is clearly related to one of the early hominin taxa, it is hard to be sure which because its brain volume is smaller than expected (546 rather than the 850–1100 cc typical of *H. erectus*). On this basis, Lordkipanidze et al. (2013) considered that the morphology of the few examples discovered reflected within-species variation of *H. erectus* and the taxon was actually a subspecies that should be named *H. erectus georgicus*. Schwartz et al. (2014), however, reaffirmed the older view that the morphology of the skull, particularly the mandible, is sufficiently distinct to justify this hominin being a distinct species that should be called *H. georgicus*.

One unusual early group that has been much studied is *H. floresiensis*, a taxon whose skeletal material was recently found on the island of Flores in Indonesia; it had undergone dwarfism, a characteristic trend in isolated island populations (Gibbons, 2018). Its ancestry is unclear but may date back to the earliest emigrations of *H. erectus* from Africa (Aiello, 2010; Tucci et al., 2018). An interesting example here is the material found in the Callao cave on Luzon, in the most northern part of the most northern island of the Philippine cluster (Détroit et al., 2019). Their small bones and teeth have both *Australopithicus* and early *Homo* characteristics so that their exact provenance remains unclear. More bones from pre-modern Hominins have recently been found in Chinese, Indian and Israeli sites, but not yet in sufficient quantities for unambiguous species assignments to be made (Bellwood, 2018; Li et al., 2017; Hershkovitz et al., 2021). There are probably more bones of unknown taxa to be found, and it is to be hoped that together they will help clarify human phylogeny.

The current view is that the last common ancestor of *H sapiens* and *H. neanderthalensis* was *H. heidelbergensis,* or a very near relative, such as *H. rhodesiensis* that has only been found in Africa (~300 Kya; Stringer, 2016; Grün et al., 2020; Table 23.2). The exact status of *H. heidelbergensis* is still unclear: the original (type) *H. heidelbergensis* fossil is just the Mauer mandible that was found in 1907 and dated to 609 ± 40 Kya, but the other material in Africa and Eurasia has since been found (see Websites). It now seems that *H. heidelbergensis* and *H. rhodesiensis* are either the same taxon or are both closely related descendants of *H. erectus*. What does, however, seem clear is that *H. heidelbergensis*-like individuals left Africa perhaps as early as ~600 Kya and migrated first through the Middle East and later spread out, eventually inhabiting much of continental Europe and northern Asia (Grün, 2020). It is from their descendants that *H. neanderthalensis* and the Denisovans evolved ~350 Kya (Chapter 24).

By ~400 Kya, the fossil material shows that hominins of this lineage had settled in Europe, eventually going as far north as Wales, or just south of the extent of the glaciation associated with the then ice age. Of the sites with fossils that have now been found across Europe, a particularly important example is the cave in Sierra de Atapuerca, near Burgos in Northern Spain, that includes skulls and other bones dating to at least 400 Kya (Arsuaga et al., 2014; Jacobs et al., 2019). The fossils were identified as being from *H. antecessor* individuals, but the recent extraction of DNA from teeth has allowed further clarification of the lineage through comparison with the genomes of more recent taxa. As a result, Welker et al. (2020) have now shown that these individuals were very early examples of Neanderthals.

A second related group left Africa and migrated eastwards, eventually reaching Tibet (Chen et al., 2019a). The most important site in which fossilised material from this group has been found is the Denisovan cave in Eastern Siberia, which was first inhabited >200 Kya (Douka et al., 2019; Jacobs et al., 2019). The material from this cave reflects a long period of habitation and includes bone fragments and teeth from 16 individuals. DNA analysis suggests that this comes from both Denisovan and Neanderthal (140–80 Kya) individuals, with the youngest Denisovan dating to perhaps 52 Kya. Of particular interest is a female whose DNA showed mixed heritage and whose mitochondrial DNA was characteristic of Neanderthals, an observation that confirms that the two taxa could interbreed (Slon et al., 2018). No Neanderthal or Denisovan bones have been found dating more recently than ~35 Kya.

An interesting perspective on Denisovan anatomy comes from a recent study by Gokhman et al. (2019, 2020) who analysed methylation patterns in Denisovan and Neanderthal DNA on the basis of such patterns in humans and chimpanzees. These patterns can be linked to link to anatomical features because the methylation of protein-coding sequences is associated with suppressing their expression. Gokhman and coworkers were thus able to predict that Denisovans had a relatively large cranium and a high dental arch, an elongated face and a wide pelvis (the first two were confirmed by a Denisovan skull). Of perhaps greater interest, however, was the extensive hypermethylation of a network of face-associated and voice-associated genes (SOX9, ACAN, COL2A1, NFIX, and XYLT1). The authors conclude that it was the lessened expression activity of these genes that led to Denisovans having a flattened face and a vocal tract closer to that of modern humans than previously thought.

Among the most interesting of the recent finds of early *H. sapiens* fossils are those discovered at the Jebel Irhoud site in Morocco (Hublin et al., 2017). These have been dated to ~300 Kya on the basis of thermoluminescence on fire-heated flints found near the bones and of uranium radioactivity of a fossilised tooth. The authors note that while the skulls have facial, mandibular, and dental morphology that aligns them with early anatomically modern humans, their neurocranial and endocranial features are morphologically more primitive. In addition, some skulls have the brow ridges typical of more ancient hominins and others do not, a feature that may reflect sexual dimorphism. While their exact taxonomic status is still unclear, the skulls (Fig. 23.4) as a whole are much closer to those of modern humans than those

(a) (b)

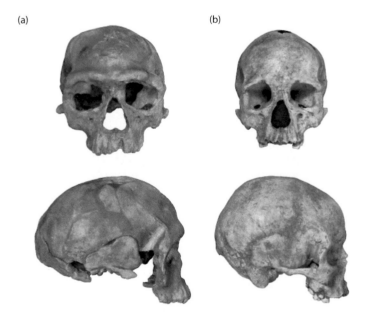

Figure 23.4 Comparison of early and modern skulls of *Homo sapiens*. (a): Frontal and lateral views of a skull from Jebel Irhoud (~300 Mya) shows its elongation. (b): Frontal and lateral views of a modern human skull.

(From Callaway E (2017) *Nature News,* June 7. With permission from Springer Nature.)

of other pre-modern *Homo* fossils. More recently, a similar skull has been found at the Apidima cave in Greece that has been dated to ~210 Kya (Harvati et al., 2019). Such data indicate that the extent of the migration of early modern humans was more extensive than had previously been appreciated.

Anatomically modern *Homo* (from 0.2 Mya onwards)

It used to be thought, on the basis of bones found in several sites, that *Homo sapiens* (also known as Anatomically Modern Humans or AMH), evolved ~180 Kya from a population of *H. heidelbergensis* or *H. rhodiensis* in in eastern and southern Africa (McDougall et al., 2005; Cyran & Kimmel, 2010; Stringer, 2016), and spread across the continent. The Jebel Irhoud fossils found in Morocco are clearly early examples of *H. sapiens* that date to ~300 Kya. They are forcing us to revise the original view and the early stages of the evolution of *H. sapiens* in Africa are now a little more opaque than they once seemed. The current view, based on the likely dating of Mitochondrial Eve and Y-chromosome Adam is that *H. sapiens* dates to ~200 Kya (Chapter 24).

Early human migrations can be followed on the basis of settlement sites. Those outside of Africa include Skhul (~120–90 Kya) and Qafzeh (~100–90 Kya) in Northern Israel (see also Chapter 24). The earliest *H. sapiens* sites in the far east are in Laos (65–45 Kya) and in the Philippines (~67 Kya; Groucutt et al., 2015) but skeletal finding are still thin, and it seems that all who lived in them were lost. There is as yet relatively little fossil or site evidence of early successful migrations of *H. sapiens* out of Africa other than that which resulted in humans arriving in Australia. This is generally thought to have happened ~50 Kya, but this date may be too recent because stone tools found at the Madjedbebe site in Arnhem Land have been dated to 65–50 Kya, although this dating is controversial (Chapter 24). The second major successful migration occurred ~55 Kya and is currently thought to have populated whole world. It is worth noting that the humans who settled in Europe ~40 Kya used to be known as Cro-Magnons or stone-age men, but these terns have fallen out of use. The details of these migrations are discussed in Chapter 24.

For all of the quantitative differences in size and pigmentation amongst AMH subgroups today, it is hard to pinpoint any absolute, qualitative differences amongst them; there is certainly no difficulty in their interbreeding. This ability also applies to humans and Neanderthals, at least to produce female offspring: Mendez et al. (2016) show that there is considerable divergence between the AMH and Neanderthal Y chromosomes. There is thus no reason to doubt that the taxic diversity seen in earlier hominins (Grün et al., 2020) has now been reduced to a single species.

SELECTION AND SPECIATION

The fossil evidence on human evolution, for all the discoveries over the last 25 years, remains frustratingly incomplete. Producing a convincing cladogram of hominin evolution on the basis of anatomy requires the identification of enough shared and derived characteristics among the twenty or so taxa so far identified to produce hypotheses about relationships, but the data needed to do this is still inadequate. The reasons are partly because we don't yet have enough fossil evidence and partly because the differences are qualitative (particularly in size and shape) rather than quantitative. The fossil record does, however, make two clear points. First, there is a clear evolutionary trend between the last common ancestor of chimpanzees and anatomically modern humans, defined by a range of properties, particularly in cranial and limb features (Tables 23.2 and 23.3); second, there was, for most of the past four million years, a high degree of taxic diversity (i.e. evidence of more than one contemporary taxon). There was no single obvious path of descent to modern humans; we can merely follow a rather diffuse track.

The discussion so far has avoided the question of how many species of hominins there were by using the more neutral term of taxon. Wood & Lonergan (2008) have discussed the different views of the "lumpers" who would like to have fewer species and more subspecies and the "splitters" who would favour most anatomical variants having species status. The two views really reflect on how one defines a species. One classic definition is that it is a taxon with which other taxa cannot breed; another is that it has sufficient anatomical characteristics that are sufficiently different from those of other taxa to mark it as more than just a variant (see Chapter 22). Different authors take different views on whether an apparently new feature (e.g. a distinct nose) reflects a genuine novelty or is merely a variant that lies within the expected range of another taxon.

Given the genetic evidence that successful breeding took place between *H. sapiens* and *H. neanderthalensis* and between *H. sapiens* and the Denisovans some 50–40 Kya when they met up in Europe and Asia (Chapter 24), it is clear that all descendants of their likely common ancestor, *H. heidelbergensis* or a near relative, who lived ~0.6 Mya are the same species on the first criterion. Indeed, given the amount of time needed for a distinct daughter population to be unable to breed successfully with a descendant of a parent population (around a million generations, Chapter 22), there is support for the extreme position of there only ever having been a single hominin species, albeit one that has included a great but not an exceptional degree of variation (consider the extent of the range of canine morphs). This was the view reached on molecular grounds by Curnoe and Thorne (2003).

This view is also supported by the chromosome evidence as the genomes of *H. sapiens* and chimpanzees are very close (Wildman et al., 2003; Table 23.1): the current human and chimpanzee chromosome numbers are 46 and 48, respectively, with human chromosome 2 resulting from the fusion of chromosomes 2 and much of 4 in an early Homininae ancestor (Ijdo et al., 1991). Assuming that the hominin line separated from the panins ~7 Mya and the reproductive cycle is around 20 years, the separation occurred at most some 400,000 generations ago, a number that seems rather low to produce separate speciation on the failure-of-interbreeding criterion. It would not, therefore, be surprising if it were possible for human–chimpanzee hybrids to be produced, although the two species would be unlikely to wish to copulate. Such predictions are, of course, untestable as the necessary experiments would never be approved on ethical grounds.

The question of how distinct speciation is recognised in the context of hominin evolution, therefore, depends on one's view of whether the anatomical differences between any two groups are sufficiently substantial for them to be viewed as being different species. As experimentation is impossible here, this will always be an open question, albeit one whose answer is not particularly important. What matters is not whether one or another taxon has species or subspecies status but the details of the changes that led from one taxon to another and why they were successful. For this, the key criterion is that the definition of a taxon has to be broad enough to include, for example, minor differences in size, and narrow enough to exclude individuals with important differences. An example of the lattter is probably the location of the foramen magnum as it reflects on whether members of a taxon walked upright or not.

Given the paucity and fragmentation of the hominin fossil material, even meeting this criterion is not always straightforward (see Lordkipanidze et al., 2013). This is because the possibility cannot be excluded that similar features in different groups may reflect homoplasy (convergent change from different origins) rather than synapomorphy (shared, derived characteristics; see Wood & Harrison, 2011). In principle, therefore, it is probably most helpful at the moment to side with the splitters' view of hominin evolution as this emphasises the secondary anatomical detail. We can hope that future discoveries will provide further hard data that will be sufficient to produce a cladogram of hominin taxa in which one can have confidence. In the meantime, it may be more appropriate to talk about the various hominin taxa rather than the various hominin species or subspecies, and so sidestep the problem.

KEY POINTS

- The line that led to *H. sapiens* (the taxon of anatomically modern humans or AMHs) ~200 Kya, originally split from the chimpanzees ~7 Mya, most probably in sub-Saharan Africa.

- The limited amounts of fossil data indicate that the hominin clade includes a complex bush of taxa. It is, however, not yet possible to produce an unambiguous phylogeny showing the line of descent from the LCA of apes and hominins that leads to *H. sapiens*.

- There were several major and many minor migrations of *Homo* taxa that left Africa for much of what is now Asia and Europe. The earliest was *H. erectus* ~1.5 Mya.

- An African descendent of *H. erectus* was *H heidelbergensis* who lived ~*600 Kya* and who is a likely ancestor taxon for *Homo sapiens*.

- Groups of *H. heidelbergensis* also left Africa ~600 Kya and their descendants were Neanderthals in Europe and Denisovans in Asia. These and earlier migration were lost by ~35 Kya.

- The recent emigrations of *Homo sapiens* from Africa were ~70 Kya, whose descendants are now indigenous Australians, and ~55 Kya that was the source of the other populations across the world.

- When members of migrating groups of *H. sapiens* met up with Denisovans and Neanderthals, successful interbreeding took place as marked by the maintenance of their genes in contemporary non-African human populations.

- The many morphologically distinct hominin taxa can be assembled into six major groups. The question of whether to give species status to sub groups or to consider some to be variants in widely defined species of major groups is still a matter of choice.

- As full speciation leading to reproductive isolation typically takes at least a million generations, all hominin taxa, given their slow reproductive cycle, are probably part of the same species based on the criterion of an ability to interbreed.

- It is, however, more helpful to view the many hominin taxa as distinct in order to build up a detailed picture of how the AMH phenotype evolved.

FURTHER READINGS

Andrews P (2015) *An Ape's View of Human Evolution*. Cambridge: Cambridge University Press.

Jobling M, Hollox E, Hurles M et al. (2013) *Human Evolutionary Genetics* (2nd ed). Garland Press.

Lieberman D (2011) *Evolution of the Human Head*. Cambridge, MA: Harvard University Press.

Lieberman D (2013) *The Story of the Human Body: Evolution, Health and Disease*. New York: Pantheon Press.

Robson SL, Wood B (2008) Hominin life history: Reconstruction and evolution. *J Anat*; 212:394–425. doi: 10.1111/j.1469-7580.2008.00867.x.

Smith TM, Alemseged Z (2013) Reconstructing hominin life history. *Nature Education Knowledge* 4(4):2.

Stringer C (2011) *The Origin of Our Species*. London: Penguin Books.

Wood B (2019) *Human Evolution: A Very Short Introduction* (2nd ed): Oxford University Press.

Wood B, Elton S (eds) (2008) *Human evolution: Ancestors and relatives*. *J Anat*; 4:335–562. Symposium papers.

WEBSITES

- Wikiedia entries for *Homo* (this covers the evolution of Homo taxa), all the other fossil taxa, and the *List of human evolution fossils*

- Reconstructing hominin life history: https://www.nature.com/scitable/knowledge/library/reconstructing-hominin-life-history-96635644/

Human Evolution 2: Genes and Migrations

24

This chapter discusses the information on human evolution and dispersion that can be obtained from phylogenetic analysis of sequence information from three sources of DNA: primates, relatively ancient bones from around the world, and the many contemporary groups of anatomically modern humans (AMHs). This work has provided a genetic history of humans and how they diverged from primates.

The most important information has come from analysing mitochondrial DNA (mtDNA), which is particularly well preserved and carries many group-specific polymorphisms on its haploid genome. Further insight has been provided by coalescence analysis: this integrates population genetics with sequence phylogenetics to provide the population history of a group. Together, these studies have shown that current human populations derive from two emigrations from East Africa, each of ~1500 African AMHs. The first left Ethiopia for the Yemen ~70 Kya, travelled eastwards around the coast, eventually becoming the founder group of indigenous Australians 5–10 Ky later. This journey was much easier then than it would be today as, due to the ice age, vast amounts of sea water were trapped as ice in the polar regions, so lowering sea levels.

The second migration left Africa ~55 Kya, probably through the Sinai Peninsula, with many moving East. A further small founder group eventually reached Siberia, with a subgroup later crossing the then Bering land bridge into America. They then travelled south, eventually occupying North then South America. Some descendants of the original group, however, remained in Turkey and the Levant until the end of the ice age (~45 Kya). Some of these then travelled north then west up the Danube valley into eastern Europe, then spread west across the Mediterranean littoral and the lowlands of the Pyrenees before moving into more northern areas of western Europe.

The western migrating AMH groups interbred with local Neanderthals while the eastern migrating AMH groups interbred with Neanderthals and Denisovans, both of which were descendants of *H. heidelbergensis*-like groups that had left Africa 5–400 My earlier. These now-lost populations have left a genetic legacy of small amounts of their DNA in the contemporary human genome. Details of the migrations are also supported by archaeological evidence from early AMH sites across Europe.

The information on the evolution of AMH that comes from skeletal material is highly biased by the fact that so much of the evidence comes from central, eastern and southern Africa. It thus says little about the dispersal of AMH both within and out of Africa. Deeper insight into the makings and migrations of modern humans comes from analysing and comparing genomes of contemporary *H. sapiens* from across the world and comparing them with those of chimpanzees and with ancient DNA extracted from early human material (Oppenheimer, 2012; Cappellini et al., 2018) and even from sediments in caves that had been occupied by early humans (Slon et al., 2018). Fortunately, there is more of this than one might expect, since DNA decays slowly: analysis shows that it can last >300 Kya under optimal conditions.

DOI: 10.1201/9780429346217-29

This is particularly so for mitochondrial DNA, whose location within these organelles provides additional protection against degradation.

Such comparisons not only tell us about the degree of genetic variation today but, when combined with the fossil and artefactual evidence, provide insight into the various migrations out of Africa that eventually populated the world. This chapter aims to explore that genetic data and provide baseline information for the following chapter that examines the origins of contemporary human diversity. It is, however, important to emphasise that, for all that we know about the fossil, genomic, computational, archaeological, and climate data, details of the many secondary migrations that led to the distribution of AMH across Earth are still incomplete (Groucutt et al., 2015).

The genomes of mitochondrial DNA (mtDNA) and the Y chromosome are particularly helpful in analysing human descent because each only exists in haploid form. This means that sequence changes over time reflect mutation but not meiotic crossover. As mtDNA is present in eggs but almost entirely absent in sperm (Luo et al., 2013), mtDNA is only passed down the female line and all of today's human variants can be traced back to an LCA known as *Mitochondrial Eve* who lived ~176 Kya (see below). Similarly, the Y chromosome is inherited solely from the father and all human variants can be traced back to a last common ancestor known as *Y-chromosome Adam* who, on the basis of coalescence analysis lived ~254 Kya (Karmin et al., 2015; Chapter 21). These authors also provide weight to the existence of a major migration 52–47 Kya because they show that there was at that time a cluster of small founder groups that descended from this male.

INFORMATION FROM CHIMPANZEE GENOMES

The differences between today's human and chimpanzee genomes are minor at the chromosomal level: chimpanzees have 48 chromosomes while humans have 46, the difference being that humans have a single chromosome 2 whereas chimps have two, known as 2A and 2B. These fused with minimal loss of DNA material in an unknown distant ancestor of *H. sapiens* and *H. neanderthalensis* (Miga, 2017). The genome of *H. sapiens* does, however, display a very large number of deletions, duplications, inversions and mutations when compared to that of the chimpanzee. Sequence analysis shows that the ~4% of mismatches include ~35 million single-nucleotide differences (~1.23%; CSaAC, 2005) and ~90 Mb of insertions and deletions (Catacchio et al., 2018; Varki & Altheide, 2005). These are a measure of the extent to which large genomic changes can occur between two groups after ~7 Mya of divergence from an LCA.

If we assume that the average generation time for each species was ~15 years until relatively recently and that the lines separated ~7 Mya, then these differences reflect about ~4.7×10^5 generations of mutation and selection since the separation, or about 40 mutations per generation. Comparisons of the differences between the human and chimpanzee genomes carry a wide range of evolutionary and biomedical implications (e.g. CSaAC, 2005; Varki & Altheide, 2005; Kehrer-Sawatzki & Cooper, 2007; Catacchio et al., 2018), although most of the differences are not yet fully understood.

Particular attention is currently being paid to those differences in gene expression seen in the prefrontal cortex, the key region of the human brain that differs from that of the chimpanzees in size and function (Berto & Nowick, 2018). Differences in gene details and expression may give insights into how the human brain evolved after the lines separated: it is now known, for example, that the alternative splicing patterns are different in the two species (Calarco et al., 2007) and that the enhancers and other regulatory sequence of genes, such as FOXP2, which is involved in human speech production, show considerable evolutionary change (Caporale et al., 2019).

The genes located on the Y chromosomes of *H. sapiens* and chimpanzees have also diverged since the clades separated (Hughes & Rozen, 2012). Although the genetic complements of human and chimpanzee genes are very similar on their respective Y chromosomes and even to that of the rhesus monkey (LCA = 25 Mya), the details of their respective gene organisations are surprisingly different (Hughes & Rozen, 2012). The proportion of the Y chromosome that is occupied by ampliconic sequences (those specifically involved in testis development) is not only larger in chimpanzees (~57%) than in humans (~45%) but more complex. In humans, the majority of repeat structures are palindromes with a copy number of two. In chimpanzees, most of the repeats are present in 4 or more copies, and the most abundant repeat has 13 copies.

The reasons for these differences are still not known but they do emphasise that what seems to be neutral genomic variation and mutation is an active process, even in a region of a haploid genome that one might have expected to be highly conserved over a few hundreds of thousands of generations.

It is interesting to ask why, if the genomes of extant chimpanzees and humans are so close, the phenotypes are so distinctive with respect to surface anatomy, growth patterns and mental capacity? Minor differences have been observed in kinase domains in their respective genomes (Anamika et al., 2008) and in the patterns of gene expression in the two brains (Bauernfeind et al., 2015), but such observations do not really seem to get to the root of the problem. The question really reduces to how relatively minor changes in sequencing produce major changes in phenotypes. From the systems perspective taken in this book, the answer is that the sequence differences affect the protein networks that are particularly involved in growth, pigmentation, and brain development. Here, small changes in the kinetics of these networks would be expected to have important downstream effects on both gene expression and aspects of the phenotype. In this context, the differences in kinases observed by Anamika et al. (2008) may be particularly important as these proteins are known to play a key role in growth pathways (Kosmidis et al., 2014; Chapter 14). This question of brain evolution is considered in more detail in Chapter 18.

ANCIENT *HOMO* DNA

Direct evidence on human evolution comes from analysing DNA sequences, both genomic and mitochondrial, from the various early *Homo* taxa (Oppenheimer, 2012). The most important of these are the Neanderthals and the Denisovans, both of whom were probably descendants of *H. heidelbergensis* or a closely related taxon. These descendants left Africa ~450 Mya and spread out across the Levant and Europe, with a few groups eventually reaching as far east as Siberia. The line that led to the Denisovans seems to have emigrated from Africa before the Neanderthal ancestors (Reich et al., 2010), moving first to the Levant, then north to the Caucasus and then eastwards into Siberia and as far as Tibet. This view is inevitably speculative as the Denisovan fossil record is very much poorer than that of the Neanderthals.

The oldest known example of DNA from these early taxa is that from the mitochondria of Neanderthals buried in the Sima de los Huesos (Pit of Bones) at Atapuerca (near Burgos in northern Spain); this dates to ~400 Kya (Arsuaga et al., 2014). In addition, large amounts of sequenceable genomic and mitochondrial DNA have been retrieved from later Neanderthal bones in southern Europe, Denisovan bones in Siberia and, of course, the many subgroups of contemporary *H. sapiens* (Veeramah & Hammer, 2014).

The oldest full hominin genome so far sequenced comes from a Neanderthal woman who lived ~50 Kya in Siberia (Prüfer K. et al., 2014). Unfortunately, it has so far proven impossible to extract sequenceable DNA from the bones of pre-AMH hominins found on the eastern Indonesian Island of Flores, who were probably descendants of an early migration of *H. erectus* that may well have included the group that included Java Man (Chapter 23). There are, however, good chances that other fossils will be discovered whose DNA will clarify our understanding of the lines of hominid descent. This, together with improved techniques for sequencing and reading ancient DNA, is likely to produce important new insights in the coming years.

Comparisons of AMH mtDNA (16,569 bp) with chimpanzee, Denisovan and Neanderthal mtDNA give mutational differences of 1462, 385, and 202 bases, respectively, with phylogenetic analysis suggesting that the line leading to AMH separated off that leading to Neanderthals >500 Kya (Reich et al., 2010). This is earlier than analysis based on genomic DNA, which has given a date of ~370 Kya (Noonan et al., 2006). One particularly interesting insight has come from comparative DNA analysis of their respective genomes: the protein sequence, if not the regulatory aspects, of the speech-related gene FOXP2 in some Neanderthal DNA was essentially identical to that in modern humans. Speech, of course, requires not only neural function but the requisite anatomical structures for controlling sound, particularly an appropriate larynx and hyoid bone. As Neanderthals had a hyoid bone very much like that of *H. sapiens*, they may well have had some basic language capabilities (D'Anastasio et al., 2013; Johansson, 2014; see also Chapters 23 and 25).

Timing data on these populations comes from three sources: palaeontological analysis of locations, the age (based on radioactive decay evidence) of the ancient skeletal material for each taxon and computational work on genomic and mitochondrial DNA sequences. In the last case, the timings depend on two parameters: the generation time, which is generally taken to be ~20 years or perhaps a little less for very old cases (the generation time for chimpanzees is ~15 years), and the mutation rate, which is estimated to be between 1.1 and 2.5×10^{-8} per base pair per generation (Chapter 7). In this context, analysis of the sequence changes among homologous genes in related organisms gives a figure of ~1.3–1.8×10^{-8} per base pair per generation (Veeramah & Hammer, 2014).

Sequence comparisons of the DNA from *H. sapiens*, the Denisovans and *H. neanderthalensis* has now established that they all descended from a common ancestor, probably a descendant of *H. erectus*, with DNA-methylation analysis even giving details of minor morphological differences (Gokhman et al., 2019). This work as a whole also provides clear evidence of interbreeding between *H. sapiens* and the Neanderthals as they migrated across Europe and Asia (Sankararaman et al., 2014; the arrows in Fig. 24.1). Analysis of this data shows that ~2% of the DNA of modern humans is Neanderthal-derived, while the DNA of the indigenous populations of south-eastern Asia typically includes ~4% Denisovan DNA, but as much as ~6% in Papua New Guinea (Lowery et al., 2013; Akkuratov et al., 2018; Bücking et al., 2019). It is thus obvious that the mutations and other genetic changes resulting from 400 Kya and >20K generations of separation were not enough even to send the two groups on the path to distinct speciation under the criterion of a failure to interbreed.

There is now a body of evidence on the gene variants that *H. sapiens* particularly acquired from *H. neanderthalensis* (for review, see Dolgova & Lao, 2018, and websites). These are associated with keratin filaments, metabolism, muscle contraction, body fat distribution, enamel thickness, oocyte meiosis, brain size, face morphology, and vision. In terms of the level of *H. sapiens* diversity, there are

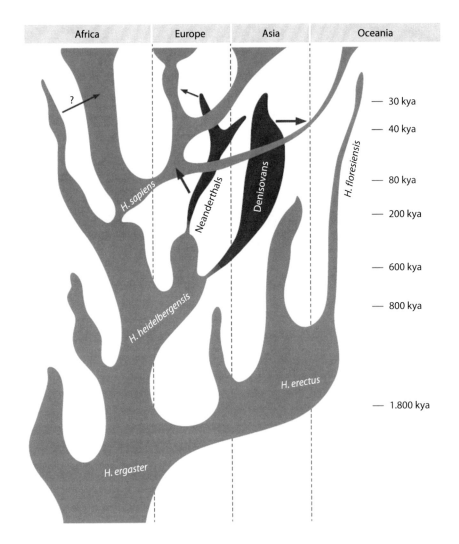

Figure 24.1 Diagram of likely migrations and evolutionary relationships within the later *Homo* taxa based on the fossil and genomic data. Arrows show where known interbreeding occurred, while the arrows with a query show where possible interbreeding occurred. The width of a taxon at a given age is a measure of its estimated population size.

(From Lalueza-Fox C & Gilbert MTP (2011) *Curr. Biol.* 21:R1002–R1009. With permission from John Wiley and Sons.)

also genes involved in the variation of skin and hair pigmentation (see Chapter 25). The Neanderthal contribution to the immune system is of particular importance as it seems that *H. sapiens* acquired most of the major histocompatibility-complex genes from them (Quach et al., 2016). (A Denisovan contribution to the human genome is discussed below.)

CONTEMPORARY HUMAN GENOMES

A further important area that DNA-sequence analysis is clarifying is the extent of genomic variation in contemporary human populations and its likely implications. Analysis of ~650K loci in 938 individuals from 53 populations distributed across Africa, Asia, and Europe has shown that the degree of genomic diversity declines with the distance from sub-Saharan Africa (Li et al., 2008a). This provides very strong confirmatory evidence that AMH originated in sub-Saharan Africa, where there is now far more genetic diversity than anywhere else in the world. This data combined with coalescence analysis shows that populations in the rest of Africa and elsewhere derived from small founder groups that left SE Africa. The descendants of those who reached Asia produced a further succession of such groups that spread across Earth. This, in turn, suggests that genetic drift in successive small founder groups deriving from this loss of diversity (Chapter 23) formed the basis of the differences among the various strains of AMH (see Chapter 25). More recently, similar analyses of Neanderthal DNA in Europe shows that they too had little genetic diversity and so too probably lived in small and perhaps isolated groups (Castellano et al., 2014).

Superimposed on the historical effects of genetic drift on ancient AMH populations are, of course, the effects of interbreeding with Neanderthals and Denisovans and those more recent mutations that have occurred in local populations. Nachman and Crowell (2000) estimated that ~175 such mutations accumulate per diploid genome per generation. Most of these will, of course, be neutral, having no effect on phenotypes and hence on selection. A few will, however, have minor effects on the phenotype (e.g. changes in the curliness of hair which are due to mutations in the LIPH gene; Mizukami et al., 2018) or will lead to recessive diseases. An example of these is Tay-Sachs, a neurodegenerative disorder that is caused by genetic mutations in the HEXA gene on chromosome 15. It is particularly associated with Jewish populations that originated in Russia and Poland (see Chapter 17), but the mutations also arose independently in the Cajun population of Louisiana and in two French–Canadian communities in Canada (Sutton, 2002).

MITOCHONDRIAL DNA

The clearest evidence on the origins of human populations across the world comes from analysing mitochondrial DNA (mtDNA), which, as mentioned above, is only transmitted maternally. mtDNA is circular, a historical relic of its bacterial origins (Chapter 10), and the haploid genotype (or haplotype) has ~16.5K nucleotide pairs with a heavy (guanine-rich) and a light strand. The heavy strand encodes 12 protein subunits, two ribosomal RNAs (12S and 16S), and 14 transfer-RNAs, while the light strand encodes one protein subunit (for oxidative phosphorylation), and transfer-RNAs. Because mtDNA includes no sequences whose proteins would correct copying errors, neutral mutations slowly accumulate, although mutations with a seriously deleterious effect will, of course, be lost. It is also worth noting that there is a range of human genetic disorders that derive from mutations in mitochondrial DNA (Molnar & Kovacs, 2017).

mtDNA from many of the indigenous populations across the world has now been sequenced and the differences closely analysed. The rate at which mtDNA mutates is known as the mitochondrial molecular clock, but this has turned out to be a complicated figure to measure. However, an estimate by Fu et al. (2013), based on a Bayesian analysis of securely dated mtDNA samples spanning 40 Kya, gives figures of 1.57×10^{-8} substitutions per site per year for the coding region and 2.67×10^{-8} substitutions per site per year for the whole sequence. This figure is similar to that of the rest of the human genome (Nachman & Crowell, 2000).

Such sequencing analysis has also shown on the basis of mutation that there are 26 major haplogroups (groups of alleles that are associated with particular populations), with further mutations subdividing them into a range of minor population-associated haplogroups. The key mtDNA haplotype marker for the AMH migrations out of Africa

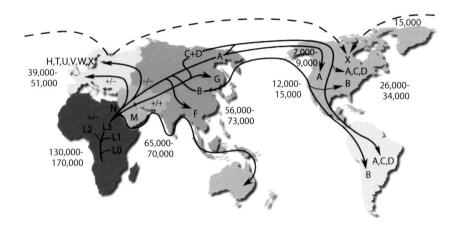

Figure 24.2 The patterns of the migrations of *H. sapiens* based on the main and secondary mtDNA haplotypes (for details, see key. Note: the date for reaching the Americas was ~30 Kya and the N migration was through Sinai.)

(Courtesy of Maulicioni. Published under a CC Attribution-Share Alike 3.0 Unported License.)

was L3, which seems to have arisen in eastern Africa more than 70 Kya, although the details of how this happened remain contentious (Cabrera et al., 2018). This haplotype later subdivided into the N and M groups to which the G and A haplotypes were later added. As successive generations of humans migrated across the world, these mutations built up so that each indigenous population carries the genetic history of its travels. The resulting map (Fig. 24.2) shows the major routes by which *H. sapiens* colonised the world over the period ~70–20 Kya.

MIGRATIONS WITHIN AFRICA

The locations of the pre-AMH fossil material (see Websites) demonstrate that even though the earliest hominins seem mainly to have lived across eastern Africa, they migrated westwards across Africa early, reaching at least as far as Chad where *Sahelanthropus* material dated to ~7 Mya has been found. Although few ancient hominin bones have yet been discovered in north and west Africa, early AMH fossils (Fig. 23.4) have been found at the Jebel Irhoud site in Morocco. As these fossils date to ~300 Kya, they clearly demonstrate that human migration had reached the northeast coast of Africa by then. It is highly likely that more such sites will be found as palaeoanthropological research groups extend their explorations.

Good evidence on the migrations of AMH within Africa has comes from analysis of mtDNA sequences from groups of contemporary Africans; this has, for example, been undertaken by Cyran and Kimmel (2010) using a range of models for the population dynamics. Their results show that it matters relatively little as to which population model is chosen as all give an expected coalescence time within the range 176 ± 15 Kya, although the 95% confidence limits are somewhat wider. This figure provides an approximate date for mitochondrial Eve, the last common ancestor of all anatomically modern females, and for when *H. sapiens* became a distinct species. Coalescence analysis also suggests that the L haplotype is the oldest mtDNA variant and was probably already carried by an ancestor of Mitochondrial Eve ~200 Kya.

Analysis of 2,200 contemporary mt-DNAs (Heinz et al., 2017) confirms that the original L0 haplotype dates to ~201–136 Kya. Today, this haplotype is most commonly found in 74% of the hunter-gather/herder Khoisan (or Khoe-San) in southern Africa but is represented at lower levels elsewhere. Heinz et al., also identified a host of secondary mutations in the various mt-DNA haplotypes that could be used for a finer-grained analysis of descent. Rito et al. (2019) deduced from this data that AMH probably evolved in southern Africa and that Khoisan groups migrated north and east very early (Fig. 24.2).

The next genetic divide is between the L0 haplotype of the hunter-gather/herder Khoisan groups and the L1 haplogroup branch now present in central and eastern Africa (from which non-African populations descend, later acquiring an L3 mutation). There was then a further split (Rito et al., 2019) in the latter group to give the L1'6 haplotype present in western and central Africa and the L2'6 haplotype in eastern Africa. The situation is then complicated by the many subsequent movements of people within Africa. Thus, for example, Skoglund et al. (2017) have analysed ancient African DNA and shown that there was considerable gene flow between Khoisan ancestors and the ancestry of both Malawian hunter-gatherers (~8.1–2.5 Kya) and of Tanzanian hunter-gatherers (~1,400 Kya). More recently, Rito et al. (2019) have shown that there was likely to have been a further Khoisan migration to eastern Africa ~70 Kya, or around the time that a group of AMH left Africa on a journey that eventually led to Australia.

Today, there are four major ethnolinguistic divisions among Africans: Nilo-Saharan speakers from north and northeast Africa; Afro-Asiatic speakers also from across the northern part of the continent; Niger-Congo speakers, the predominant group in sub-Saharan Africa; and the smaller hunter-gatherer (Khoisan) populations in southern and central Africa. This considerable ethnolinguistic diversity is largely paralleled by genetic diversity, as population genetic studies and whole-genome sequence data have shown. These are now allowing computational investigations to be done that will identify genetic differences and population histories across these groups (Choudhury et al., 2018).

MIGRATIONS OUT OF AFRICA

The skeletal and genetic evidence now suggest that there were probably four major migrations of *Homo* populations out of Africa to various parts of the world and probably several minor ones, with members and descendants of some of these groups returning (Hodgson et al., 2014). The most recent migrations can be analysed on the basis of mtRNA haplotypes (Fig. 24.2), but evidence for earlier ones mainly depends on fossils and stone artefacts. The oldest current evidence of a non-African hominin is the ~1.85 Mya *H. erectus georgicus* skeletal material from the Dmanisi Cave in Georgia, which is about 5,000 km from Africa. It is likely that the main reason for this migration was the search of a growing population for new sources of food, although we cannot exclude curiosity and wanderlust playing a secondary role. Reaching Georgia may well have taken many hundreds or even thousands of years, and it is likely that other exoduses from the same period led to other early hominin groups moving to the Middle East and then on to Pakistan, Java, and China (Zhu et al., 2003), with a late-surviving taxon being *H. floresiensis* in Indonesia that only died out some 20 Kya. It does, however, seem that all the successors of this early migration and, indeed, any others initiated before ~700 Kya were lost.

The second migration for which there is dated skeletal evidence occurred 700–500 Kya and was probably when distinct tribes of the ancestors of *H. heidelbergensis* migrated out through the Levant, eventually dispersing across much of both Europe and Asia. Here, mitochondrial DNA suggests that the last common ancestor of these tribes lived ~800 Kya (Lalueza-Fox & Gilbert, 2011). The fossil evidence indicates that the European branch evolved into the Neanderthals and the Asian branch into the Denisovans. They did not, however, remain separate: there is evidence from specimen 11 in the Denisovan cave that the two taxa interbred (Slon et al., 2018; Chapter 23). Both taxa had, however, died out by ~35 Kya, after having further interbred with members of *H. sapiens* (Higham et al., 2014).

The earliest known migration of early AMH out of Africa reached the Apidima Cave in Greece ~210 Mya (Harvati et al., 2019), where what seems to be a primitive AMH skull, together with a later Neanderthal skull (~170 Kya) have been found. A slightly later AMH maxilla (194–177 Kya) has been identified in the Misleya Cave on Mount Carmel in Israel (Hershkovitz et al., 2018). Also found in this cave were stone tools displaying Levallois technology (an advanced method of shaping flints that also marks material found in Jebel Irhoud, see Chapter 25) and, more recently, necklace shells dating to about 120 Kya (Bar-Yosef et al., 2020). Early human bones and artefacts reflecting habitation have been found in several other caves in the Mount Carmel and Galilee regions of northern Israel; these include those at Qesem (~200-100 Kya), Qafzeh (~120 Kya) and Skhul (~90 Kya). Given the low chances of early material being found, it is likely that there were many early human groups in the area. It now, however, seems that all members of early non-African AMH groups were lost ~80 Kya, leaving only Neanderthals in the near east (Wallace & Shea, 2006). Oppenheimer (2003) has suggested that the reasons for this may have included a change in the climate.

The third major migration out of Africa (the black line in Fig. 24.2) was ~70 Kya and was from Ethiopia to Yemen. At this time, the archeological evidence (Fig. 24.3) shows that *H. sapiens* had already extended its territory across Africa, while large areas of Europe and Asia were sparsely inhabited by Neanderthals and Denisovans. This migration resulted in a group of *H. sapiens* eventually reaching Australasia; their descendants are today's indigenous Australians alone; there seems to be no evidence of their survival in any other location that they must have passed through, such as

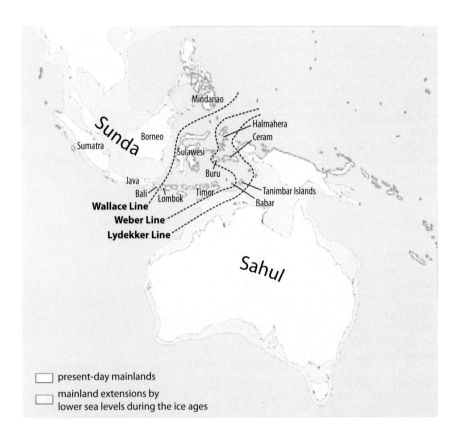

Figure 24.3 **The SE Asia and Australia landmasses (dark) at the time of the last glacial maximum compared with those today (light).** The Wallace, Weber, and Lydekker lines reflect boundaries separating various types of flora and fauna. (See Websites and Oppenheimer, 2009, 2012.)

(Courtesy of Maximilian Dörrbecker. Published under a CC Attribution-Share Alike 3.0 Unported License.)

Indonesia (Macaulay et al., 2005; Nagle et al., 2017; Nielsen et al., 2017). As to their origins, mtDNA analysis shows that today's indigenous Australians have the N and M haplotypes (the first division of the original L3 haplotype carried by the population who left Africa, Nagle et al., 2017), except in so far as there has been interbreeding with recent European immigrants.

The date when humans reached Australia is still not clear: it was long thought to be ~50 Kya. Recent stratigraphic evidence, together with the presence of tools at the Madjedbebe site in Northern Australia has suggested that humans were active in Australia ~65 Kya (Clarkson et al., 2017). This view has been refuted by re-analysis of the archaeological data by Nagle et al. (2017) and genetic data by O'Connell et al. (2018); this reaffirmed the traditional view that AMH probably reached Australia ~50 Kya.

If undertaken today, these migrations from Eastern Africa and onwards across the rest of the world would involve major sea crossings. Then, however, such crossings were much shorter (Fig. 24.3) as the sea levels were considerably lower due to the effects of the last ice age which started ~110 Kya and only ended ~12 Kya. This increase in the amount of glaciation in the polar regions locked enormous amounts of water as ice. As a result, sea levels were much lower, with today's Red Sea, for example, being reduced to an easily crossable mud flat.

The consequent migration path out of Africa for this early group seems to have followed the coastline, traversing the southern Arabian Peninsula, across the mouth of the Red Sea to southern Iran and then the western seaboard of India to Thailand, Myanmar, and Indonesia. The group then crossed the sea to Australia, with the whole journey taking perhaps several thousand years and some hundreds of generations. The last step, reaching Australia, would have been far easier than now because of the low sea levels. In the area at that time, there were three distinct regions connected by relatively short distances of water (Fig. 24.3; Oppenheimer, 1998). The Sundaland land mass incorporated the Indonesian islands into the Asian continental mass, the Sahul area included Australia, Tasmania and New Guinea, while Wallacea was the group of intervening islands. It thus seems as if travel across the seas linking SE Asia with Australia could then be undertaken almost without losing sight of land (Irwin, 1994).

The fourth major migration, and the second of AMH from Africa, occurred about ~55 Kya. This migration resulted in groups of AMH moving east into Asia and then north and west into Europe (Tassi et al., 2015). Coalescence analysis shows that the population that crossed into Asia was ~1500 adults (Li & Durbin, 2011), but it is now clear that not all of them or their descendants stayed there - some returned to Africa (Hervella et al., 2016). It was originally assumed that these migrations were, like that of ~15 Kya earlier, via the southern route from what is now Ethiopia across the Bab el Mandab Straight at the mouth of the Red Sea to the Arabian Peninsula, still narrowed as a result of the ice age and at a time when the weather was relatively cool and moist (Frumkin et al., 2011). Recent comparisons of DNA profiles across Egyptian, Ethiopian, sub-Saharan, West African and non-African populations have, however, changed that view (Pagani et al., 2015). The analysis shows that splits occurred between non-African populations and West African, Ethiopian and Egyptian populations ~75 Kya, ~65 Kya, and ~55 Kya, respectively, (Pagani et al., 2015). Moreover, the Egyptian population is, on the basis of both haplotype and coalescent analysis, the closest to non-African populations. This and other evidence (Tassi et al., 2015) suggest that, while the emigration whose descendants first reached Australia was from Ethiopia via the then narrows to Arabia, the later migration to Eurasia was from Egypt, across the Sinai Peninsula and into the Levant.

This latter population initially moved eastwards, interbreeding as they went first with Neanderthals and then with Denisovans whom they met and eventually supplanted (Dolgova & Lao, 2018; Yew et al., 2018; Fig. 24.4). The evidence for this comes from genome analysis showing that Asian populations include not only Neanderthal polymorphisms but ~4% Denisovan-associated DNA mutations and sequences (Higham et al., 2014; Sankararaman et al., 2016; Bücking et al., 2019). Many of these mutations are associated with metabolism and perhaps the most interesting is the EPAS1 variant found in Tibetans, who are able to live in high altitudes where oxygen levels are low (Huerta-Sánchez et al., 2014). As the EPAS1 protein is a transcription factor that activates the hypoxia pathway when oxygen levels are low, the Tibetan allele improves oxygen transport in people living at high altitudes. It is worth noting that EPAS1 activation in indigenous people living in the high Andes is handled differently: as they age, their EPAS1 methylation levels decline, allowing more of the protein to be expressed (Childebayeva et al., 2019).

As part of the expansion eastwards, different groups separated, with some moving south through New Guinea and on to the neighbouring islands (Pugach et al., 2013; Summerhayes et al., 2010), but probably not as far as Australia: sea levels were then rising as the ice age ended so increasing distance of the sea crossing. Other groups moved northwards towards central Asia and beyond, through China to Eastern Russia (Fig. 24.2, green line). From there, a few people seem to have traversed the

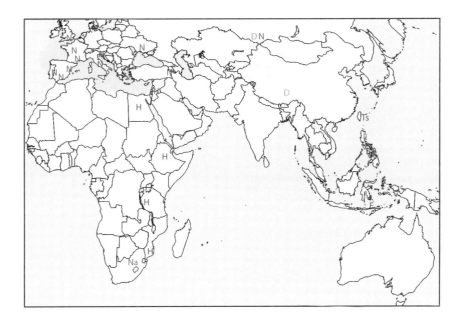

Figure 24.4 Locations of hominin sites across Eurasia just before the group of *H. sapiens* left Africa ~70 Kya and would eventually reach Australia. Key: **D**: Denisovans; **F**: *H. florensis*; **L**: *H. luzonensis*; **H**: *H. sapiens*; **N**: *H. neanderthalensis*. **Na**: *H. naledi*; Rd: red deer people; Ts: *H. tsaichengesis*. The blue and green areas roughly indicate the approximate regions inhabited by the Neanderthals and Denisovans. (Data from variety of sources.)

(Own picture based on map courtesy of "Lofo7." Published under a CC Attribution-Share Alike 4.0 International License.)

Bering land bridge (now the Bering Strait) into America ~30 Kya, much earlier than originally thought, with their descendants moving south and across the continent. Human footprints have now been identified in New Mexico and stone tools found in Mexico, both dating to ~25–20 Kya (Ardelean et al., 2020; Bennett et al., 2021), while there is rock art in Columbia and Central Brazil dating to ~12 Kya (Goebel et al., 2008; Potter et al., 2018; Becerra-Valdivia & Higham, 2020; Neves et al., 2012).

While it is clear that the great majority of the indigenous American population derives from this early founder group, it is possible that there was a later and minor Polynesian contribution to the South American gene pool. There is fossil evidence in Brazil suggesting that a small number of Polynesians crossed the southern Pacific from the Sahul (Australasia and the islands of SE Asia). If so, there could well have been interbreeding between them and descendants of the original population that migrated from Asia and moved south (Neves et al., 2007; Goldberg et al., 2016).

Migration into Europe took place later. As the southern extent of the last ice age reached beyond the Caucasus, the descendants of the group that crossed Sinai ~55 Kya could not migrate northwards until the ice had retreated and this seems to have taken a further ~10 Ky. During this period, the AMH population lived in what is now Turkey, where they met and probably interbred with local Neanderthals. Sometime after ~45 Kya, the ice had withdrawn sufficiently for groups of *H. sapiens* to move north through the valleys of the Danube, and its tributaries towards Germany, then west across the Mediterranean towards Spain. Descendants of such groups eventually spread out across all of Europe, breeding with but soon supplanting the indigenous Neanderthals (Lowery et al., 2013). The reasons for their disappearance are not clear but there are several nonexclusive possibilities. One is intergroup conflict, although this does not mesh with the observations on interbreeding, unless that was through rape. Another could have been breeding success: if the groups of migrating humans were larger than those of the small Neanderthal groups that they encountered and also produced more offspring that bred, the supplanting may have been the consequence of numerical superiority. It is also possible that the enhanced mental abilities of humans as compared to Neanderthals (Chapter 25) enabled them to survive longer, with less infant mortality.

Knowledge of the founder groups that crossed from Africa into the Levant and from Siberia into America again comes from the use of coalescence analysis (e.g. Li & Durbin, 2011). On the basis of reasonable estimates of generation time (20 years) and mutation rates (1.25×10^{-8} per site per generation), this suggests that the modern human gene pool was essentially in place ~100 Kya, with gene flow occurring until 20–40 Kya. More recent improvements in the technique have given estimates of how the population size varied over time (Sheehan et al., 2013). The analysis shows that the effective population of early *H. sapiens*, a subgroup of which migrated into Europe, underwent a severe decline to only a few hundred around 70–100 Kya, before recovering to about 12,500 by ~16 Kya. There was also a similar but much less pronounced bottleneck in the wider African population as well as in animal populations at around the same time. As to their cause, a possible reason is the Toba supervolcanic eruption (Indonesia) that occurred ~74 Kya; this led to a global volcanic winter that lasted up to a decade in some parts of the world (Lane et al., 2013).

THE WIDER PICTURE

The general picture of AMH dispersal across the world based on mtDNA and other data discussed above is confirmed by groups who have looked at whole AMH genomes from a wide range of contemporary and a few ancient groups. The genetic data clearly demonstrate that, the further a human subpopulation is from sub-Saharan Africa, the lower is its degree of genetic diversity. Li et al. (2008a), for example, analysed 650,000 single-nucleotide polymorphisms (SNPs) in 938 unrelated individuals from 51 diverse populations. Their results showed that the relationship between haplotype heterozygosity (roughly, the number of sequence alternatives for a particular gene) and distance from Africa was as expected for a population that started in sub-Saharan Africa, with each successive step in this migration being characterised by the majority of the local population remaining where they were, and only a minority moving on.

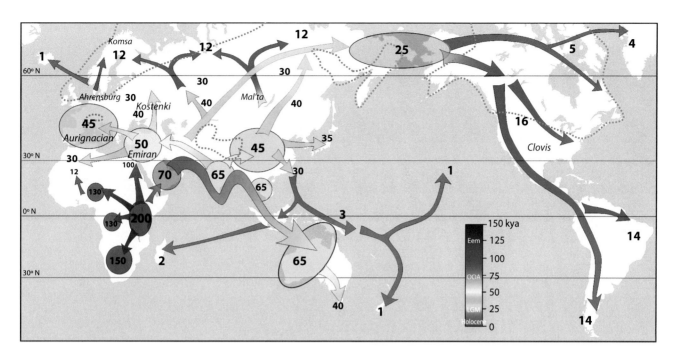

Figure 24.5 Early human habitation sites. Overview map of the peopling of the world by anatomically modern humans based on the archaeological data.

(Courtesy of D. Bachman. Published under a CC Attribution-Share-Alike 4.0 International License.)

Although the discussion in this chapter has focused on the genetic data underpinning the dispersions of *H. sapiens* as they left Africa, there is supplementary evidence on early humans that is completely independent of such data. This comes partly from evidence of tools and artefacts (see Chapter 25) and particularly from ageing the oldest habitations that archaeopalaeontologists have discovered across the world (Metspalu et al., 2004). The locations of these are shown in Fig. 24.5: the similarity between the archaeological data and the genetic data is striking.

KEY POINTS

- There were several important migrations of *Homo* taxa from Africa to colonise much of Asia and Europe; the earliest was ~1.5 Mya while the key one was ~55 Kya.

- Most of our knowledge of recent migrations comes from analysing group-specific genetic markers and how mutation led to these differences. A group's genetic history can thus be traced through these markers.

- The key markers are those in mitochondrial DNA as these are transmitted through the female line alone. All can be traced back to *Mitochondrial Eve*, an early *H. sapiens* woman who lived ~176 Kya.

- Coalescence analysis on DNA sequences across the world (particularly from mtDNA) shows that a succession of small founder groups from two original groups that left East Africa populated the rest of today's world.

- The descendants of the group that left Ethiopia for the Yemen ~70 Kya are today's indigenous Australians. Migration then was easier than now because sea levels were lower as a result of the Ice Age.

- The descendants of the group that left Egypt through the Sinai Peninsula ~55 Kya are mainly responsible for populating the rest of the world.

- Many in this group initially migrated East across Asia with some going south, others going north.

- On their way, they first met Neanderthals and later Denisovans, early groups that derived from ancestors of *H. heidelbergensis*-like. *H. sapiens* initially bred with and eventually supplanted both groups.

- The descendants of those who migrated east eventually reached Siberia. Later members of this group crossed the Bering land bridge and colonized the Americas.

- Others from the original group lived in Turkey until ~45 Kya when the ice of the last glacial period receded. They slowly migrated north up the Danube Valley and then west and north across Europe. On their way, the met, bred, with and eventually supplanted the Neanderthals.

FURTHER READINGS

Jobling M, Hollox E, Hurles M et al. (2013) *Human Evolutionary Genetics* (2nd ed). Garland Press.

Oppenheimer S (2012) Out-of-Africa, the peopling of continents and islands: Tracing uniparental gene trees across the map. *Phil Trans R Soc* 367B:770–784. doi: 10.1098/rstb.2011.0306.

Wood B, Elton S (eds) (2008) Human evolution: Ancestors and relatives. *J Anat;* 4: A collection of symposium papers.

WEBSITES

- Wikipedia entries for *Early human migrations, Human Mitochondrial DNA haplogroups, List of first human settlements, Neanderthals, Mitochondrial Eve*, and *Wallacea*.

Human Evolution 3: The Origins of Modern Humans

25

This chapter discusses two questions. First, how did anatomically modern humans (AMH) acquire their contemporary phenotype? Second, if all AMH groups derive from a common ancestor population that lived ~70 Kya in North-east Africa, why do groups from across the world appear so different today? The molecular data gives little information about the genetic basis of the anatomical and behavioural changes that characterizes the evolution of AMH as there are relatively few genetic differences between them, Neaderthals and Denisovans. The artefactual evidence, particularly stone tools, does, however, provide an alternative measure of intellectual progress. Those from the early Palaeolithic that date back ~3.3 Mya were crude crushers but later tools were increasingly sophisticated. By the Mesolithic (~15–5 Kya), AMH was knapping stone to produce flakes for knives and finer tools for a range of tasks, such as butchering and scraping animal skins. Both Neanderthals and Denisovans made early stone tools, with Neanderthals only seeming only to make sophisticated tools after coming in contact with AMH.

Early human groups used fire ~1 Mya and some Neanderthals certainly buried their dead ~300 Kya. Body painting and bead making were features of both populations and, while Neanderthals and AMH each produced art, that of the former was simpler than that of the latter. The anatomical evidence makes it likely that Neanderthals had some capability for language; if so, this implies that their last common ancestor may also have had that capacity. It does, however, seem clear seems clear that not only was AMH more intellectually capable than Neanderthals, but that they had cognitive abilities not dissimilar from those of today's humans at least ~100 Kya, well before they left Africa to disperse across the world. Together, the artefactual data mark the increasing abilities of *Homo* taxa over the last million years,

As to the question of why, if all extant humans originally descended from East Africans, we are so diverse today, the answer has several parts. The key insight here comes from coalescence analysis, which shows that today's non-African humans derive from a succession of small founder groups that slowly migrated across the world (Chapter 24). The physical differences that exist between the descendants of these groups derive from allele asymmetries mainly resulting from genetic drift and selection. Two other effects have supplemented these genetic differences: novel mutations and the effects of interbreeding with Denisovans and Neanderthals that the migrating groups of AMH encountered. The most obvious of the novel selection pressures encountered by migrating humans were the lowered UV levels in northerly regions which led to lighter skin tones. This trend was reversed when a northern founder group crossed the Behring Strait and moved southward through the Americas. These and other differences reflect minor changes: there are no obvious neuronal or intellectual differences among today's humans: we can all read and write, and we all produce music and art.

DOI: 10.1201/9780429346217-30

TABLE 25.1 HUMAN CHARACTERISTICS

Defining characteristics of *H. sapiens*	Variable characteristics of *H. sapiens*
Short, fine body hair	Skin pigmentation
Bipedal gait	Eye pigmentation
Ability to sweat through skin	Size
Manual dexterity & tool making	Hair colour and curliness
Digestive changes for a terrestrial diet	Minor facial and body parameters
Speech	
Cultural inheritance	
Artistic creativity	
Burying the dead	

Anatomically Modern Humans differ from their ancestor species in many ways (Table 25.1). Some of these changes are physiological and anatomical, with those in the limbs, for example, reflecting the movement from an arboreal to a savannah-associated habitat; others reflect increases in cognitive abilities. For the former, Lieberman (2013) has discussed in detail how the driving force for hominid natural selection was ease of access to food: variants able to feed off the land (ground plants and animals) rather than from trees (fruit, nuts, and the occasional arboreal animal) were able to move to a new and rich ecological niche. The ability to achieve success here required a digestive system appropriate for a ground-associated diet and hands that were suited to picking; the latter was provided by a bipedal gait, albeit that this less well adapted to tree life. In addition, more advanced taxa would have hunted, and this would have required an ability to run over long distances; with this came a loss of body hair and the ability to sweat, together with a flattened face and an extended nose which together facilitated breathing.

These changes, which took several million years to achieve, were accompanied by intellectual advances facilitated by an increased brain size as measured by cranial capacity. This chapter particularly considers the two questions of when a descendent of *H. heidelbergensis* first became intellectually indistinguishable from modern humans and, second, given that AMH probably evolved in around East Africa, what led to the heritable differences that characterise the diversity of humans seen today. The former question particularly focuses on the evolution of cognitive abilities, the latter on the minor heritable differences that distinguish the various groups present across the world today.

The difficulty in exploring the evolution of cognitive abilities is that all the evidence is indirect. Some inferences can be made on the size of fossil brains and such features of their surface anatomy as can be obtained from the interior surface of skulls, but they are limited in the information that they provide. More helpful are behavioural and molecular comparisons that indicate differences in brain function in apes and humans (Wei et al., 2019). This is an active and important area of current human evolutionary biology but, because of the extent of the gap and the lack of intermediate taxa, progress is slow. Perhaps, however, the most direct and interesting information comes from analysing the historical progress of the many human artefacts that have been found as these reflect increases in both cultural inheritance and mental ability.

THE RISE OF *H. SAPIENS*

Cultural inheritance

A key characteristic of the *Homo* family is *cumulative cultural inheritance*, the ability for one generation to learn some skill from its parent generation, use, and perhaps improve it and then teach it and any improvements to the next. This particularly applies to tool making: a wide range of stone tools has been discovered across the world dating back almost 3 Mya. The ability to make tools and pass on tool-making skills requires manual dexterity, intellectual competence and an ability for adults to teach and the young to learn. In terms of human evolution, if similar artefacts can be associated with both *H. neanderthalensis* and early *H. sapiens*, taxa whose ancestors separated ~500 Kya, the basic ability to make such artefacts was one likely to have been associated with their last common ancestor, probably a descendant of *H. erectus,* such as *H. heidelbergensis* or a near relative. If only one of these two more recent species has a particular ability, it probably arose after the groups separated — it is unlikely that a cognitive ability would have been lost.

Tool-making is not a uniquely hominid facility as a wide range of animals use available material for a variety of purposes (see Websites). Examples include chimpanzees, and even capuchin monkeys, that use stones to crack nuts and twigs to access insects, while birds use twigs and other matter particularly to make nests and beavers use logs to make dams. Even some marine animals use tools (Mann & Patterson, 2013). Many of these abilities can be seen as part of direct genetic inheritance, but there are many examples in which animals have been shown to learn particular behaviours from their parents. Whitehead et al. (2019) give the examples of food choice and location, how to recognise and escape predators, and which migratory pathways to take through their environments. Many animals display such teaching including cetaceans, primates, birds, and of course, bees. Often this behaviour requires both genetic and cultural inheritance; it has been analysed, for example, in the upbringing of wild mongooses (Sheppard et al., 2018).

The inheritance of human abilities goes beyond this simple cultural inheritance because each generation can pass on new knowledge and skills to the next not just through visual demonstration but also through speech and very recently through writing. It is obviously important to know whether such cumulative cultural evolution (CCE) is unique to us or is also seen in other animals. Mesoudi and Thornton (2018) have reviewed this subject in detail and their supplementary material gives 15 examples where CCE has been clearly observed. A few are in birds, one is in a fish, one is in a whales and the rest are in apes.

The example of whales is interesting as they can communicate through clicks and song (see Websites) for sexual selection (Smith et al., 2008a) and other social behaviours. Most of the work, however, has been conducted on chimpanzees and a particularly interesting example is the relationship between Washoe, a chimpanzee that was captured as an infant and taught sign language, and Loulis, a baby chimpanzee that she adopted after her own babies had died. It is clear that Washoe taught sign language to Loulis when she was very young and the two communicated through signs (Fouts et al., 1989). The evidence as a whole clearly shows that human cumulative cultural evolution derives from the more limited abilities seen in the great ape clade.

Tool-making

A good measure of the facility for cultural inheritance and increased ability to carry out complex tasks comes from the record of toolmaking. Although chimpanzees, and even capuchin monkeys, can use stones to crack nuts and twigs to access insects, there is no evidence of either species being able to make more sophisticated tools (Mangalam & Fragaszy, 2015). The use of stone (lithic) tools by hominin taxa goes back >3 Mya to the Palaeolithic (the old stone age) and only ended some 8–2 Kya (depending on the location) with the manufacture and use of bronze, an alloy of copper and other metals and sometimes other elements. Large numbers of stone tools have been discovered over the last two centuries, with many coming from Africa and Europe, particularly France (for examples, see Fig. 25.1; for reviews, see Conard & Delagnes (2010) and Websites).

The importance of the artefacts associated with the various hominin taxa is that they provide clues as to when mental skills were acquired. The increase in sophistication of these artefacts over time illustrates the implications of brain enlargement as hominins slowly evolved. It should, however, be emphasized that, as skills were acquired in different regions at different times, they are not useful for detailed evolutionary timelines. This is probably because skill acquisition by a specific group depended on its location and the extent of its isolation. In practice, these lithic artefacts are grouped according to the degree of skill needed to make them rather than when they were first used, with the individual styles being characterized as *industries.*

The earliest evidence for tool making comes from the grooved and fractured bones from Dikika and Gona (in modern Ethiopia) and date to ~3.4 Mya, the Australopithecine period (McPherron et al., 2010). The stone tools made then and a little later are simple choppers (so-called *Mode 1*) that were created by hitting one stone with another. This technique had improved by ~2.6 Mya when the predominant

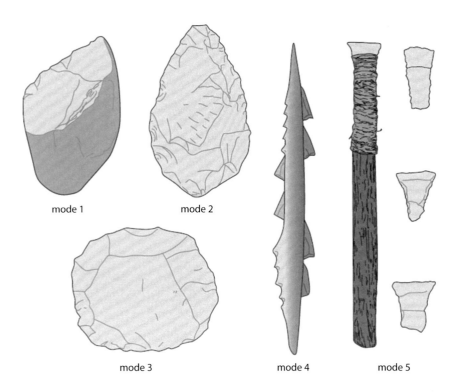

mode 1   mode 2

mode 3   mode 4   mode 5

Figure 25.1 **Paleolithic stone tools.** Mode 1: Oldowan chopping tool (~5 cm across, 3.4–1.6 Mya, *H. habilis*). Mode 2: Acheulean handaxe (~5 cm across, 1.7–300 Mya; *H. erectus*). Mode 3 Mousterian scraper made using the levallois technique (~8 cm across, 300–30 Kya; *H. neanderthalensis*). Mode 4: Aurignacian harpoon (~15 cm long, 50–10 Kya; *H, sapiens*). Mode 5: microliths, with one tipping and strengthening an arrow (~3 cm long, ~40–10 Kya; *H. sapiens*).

(Modes 1-4: courtesy of José-Manuel Benito and published under CC BY-SA 2.5.

Mode 5: from Clark JGD (1936) *The Mesolithic Settlement of Northern Europe.* With permisson from Cambridge University Press.)

hominin was *H. habilis:* sharpened edges were created from river stones by banging them with other stones to removes flakes (the process known as knapping). This style of tool, which could be used for butchery and for woodworking, is known as the *Olduvai industry.* This is because so many were first found in the Olduvai Gorge, the site in Tanzania at which many early *Homo* taxa were found, but the production of such tools occurred widely across Africa.

The next level of tool sophistication dates to ~1.7 Mya, around the time when pre-modern humans (probably *H. ergaster)* first left Africa. This was marked by the production of hand axes in which a stone had been carefully shaped and flaked to produce an ovoid object with a blade and point at one end and a smoother hand-holding area at the other. These *mode 2* tools are known as the Acheulean industry and are named after the site at St. Acheul, France, where many were found.

Progress then seems to have been slow until ~300 Kya, when Neanderthals in Europe (Monnier, 2012) and Denisovans in Siberia (Jacobs et al., 2019) developed the use of knapping to remove fine flakes from larger stones, particularly flint, and so were able to make slim knives, and scrapers (*Mode 3*, the Levallois technique). Such techniques continued to be used and are known as the as the Mousterian industry because examples were first found at Le Moustier, near the Dordogne River in southwest France. In Africa, hand axes held in hafts for greater leverage were being made ~200 Kya (Rots & Van Peer, 2006), and this may be the first innovation that can be associated with the early stages of AMH evolution (McBrearty & Brooks, 2000; Henshilwood et al., 2001). By ~70 Kya, stones were being heated over fires to facilitate knapping and the production of small tools.

When the later dispersals of AMH from Africa reached Western Europe ~40 Kya, there was an overlap of AMH with Neanderthals; various improvements in the tool design can be dated to this event. The final lithic work by Neanderthals was marked by toothed stone tools (the Châtelperronian industry, 35–29 Kya) and these may reflect interactions between the two taxa, with the latter date possibly marking the final demise of the Neanderthals. AMH stonework in France was marked by the very careful knapping required to give long blades (*mode 4*, the Aurignacian industry, and 40–28 Kya), and by the polishing required to give smooth-surfaced stone artefacts (although the oldest tools of this type were found in Japan—see Websites). *Mode 5*, the last in the staged series, covers the production of small stone tools that were used for delicate work and for arrow and spear tips (microliths). They are characteristic of the Gravettian (29–22 Kya) and Magdalenian (17–12 Kya, early Mesolithic) cultures, which were first identified in La Gravette and La Madelaine, sites near streams off the Dordogne River in SW France. Later improvements in stone working during the Neolithic led to the production of better knives, finer axe heads, and grain-grinding tools.

Social living

There are other tantalising clues about the increasing mental abilities of *Homo* groups over the last million or so years. The oldest is probably the use of fire: there is now clear evidence from the Wonderwerk Cave in S Africa of burnt plants and bones that date to ~1 Mya and are associated with *H. erectus* (Berna et al., 2012). Evidence of early hearths dated to 350–200 Kya comes from the Qesem caves in Israel, which were probably then inhabited by Neanderthals. Some 130–100 Kya, *H. sapiens* was using tools, fishing, and making fires on the South African coast (Grine et al., 2017).

Burying dead bodies seems to be a rather more recent innovation than fire: the earliest evidence comes from a collection of Neanderthal bones dropped into a deep pit in the Sima del los Huesos (the pit of bones) at Atapuerca in Spain >400 Kya (Arsuaga et al., 2014). There is also evidence that *H. naledi*, an early pygmy group in S Africa, similarly disposed of their dead. At least fifteen Naledi skeletons were found in the hard-to-access Dinaledi cave that lies at the end of the Rising Star cave complex that extends ~80 m from its entrance and dates to 335–236 Kya (Dirks et al., 2015, 2017). The earliest known *H. sapiens* burial sites date to 135–100 Kya (Vandermeersch & Bar-Yosef, 2019) in an area of the Levant where *H sapiens* and Neanderthals co-existed for some time before that population of *H. sapiens* disappears from the record.

Further evidence of early hominin activity comes from the non-stone material associated with sites of domestication. The oldest examples so far known are the wooden throwing spears found in what is now a mine in Schöningen, Germany, where preservation conditions were remarkable. These date back to ~400 Kya and are associated with *H. heidelbergensis* activity (Thieme, 1997). By ~77 Kya, *H. sapiens* was using bones to make tools, such as the awls and points that have been found in the Blombos cave near Cape Town in South Africa (Henshilwood et al., 2002). The earliest evidence of sophisticated Neanderthal bone technology is of lissoirs, which are bones (e.g. ribs) shaped for softening leather, from ~52 Kya in Mousterian camps (Soressi et al., 2013).

Speech

The greatest gap in our knowledge of early hominins is, of course, when speech evolved. It is this facility, more than any other, that is the key to a sophisticated degree of cumulative cultural evolution. Apart from any mental abilities, making the range of sounds needed for speech required a set of anatomical changes to the soft-tissue structures of chimpanzees and were probably present in the last common ancestor of the *Homo* taxon. Lieberman (2011) has discussed the nature of these changes: they include the lowering of the larynx with its vocal cords, the loss of laryngopharyngeal air sacs, the shortening and rounding of the tongue, and the shortening of the oral cavity. Most of these features reflect changes to non-bony tissues that are lost soon after death; it is, therefore, hard to know the precise details of the ways in which these evolutionary changes happened, but they certainly took place in Africa.

Linguistic analysis of modern African languages shows that there are four major groupings: Hunter-gatherer populations in southern and central Africa, Nilo-Saharan speakers from north and northeast Africa; Afro-Asiatic speakers from across north and east Africa; and Niger-Congo speakers who are the predominant group in sub-Saharan Africa. These divisions mesh with genetic variation (Choudhury et al., 2018) and clearly show that, before early humans spread out from SE Africa to colonise the rest of the continent, they were able to talk. This puts a minimum date of ~200 Kya for the evolution of basic language. It may, however, be considerably older as recent evidence has suggested that both Neanderthals and Denisovans may well have had at least some of the anatomical morphology required for speech (Chapter 24).

A key tissue here is the hyoid bone, an unjointed bone in the neck situated immediately above thyroid cartilage of the larynx; it has muscular attachments to both the larynx and the tongue and plays a role in the tongue movements that facilitate speech in modern humans. A well-preserved hyoid bone from the skeleton of a Neanderthal discovered in a cave in Israel and dating to ~60 Kya has now been analysed (D'Anastasio et al., 2013). It is similar in both internal architecture and biomechanical behaviour to those of the hyoid bones in modern humans.

As these details of hyoid anatomy almost certainly reflect use, it is quite possible that there was a capacity for speech in Neanderthals, although whether this reflects common descent or homoplasy is not known. If the former, it seems likely that a descendant of *H. erectus,* such as *H. heidelbergensis* also had the appropriate physiology for speech: this taxon certainly had a brain capacity little different from that of modern humans, and rather larger than its antecedents. It is thus possible that it was with *H. heidelbergensis* that the *Homo* taxon began to develop the ability to pass on immediately acquired knowledge to the next generation in nonvisual ways, with primitive language enhancing visual teaching. This ability would have enabled group expertise to increase dramatically, with effects on selection that were almost certainly far faster than those of genetic drift and mutation. An interesting perception on such advances comes from Hrdy (2009) who has discussed the extent to which social behaviour drives improvements in cognition.

Art

While there is no evidence to show that chimpanzees left to themselves will paint, they can certainly be taught to do so and apparently to enjoy the results (Fig. 25.2a). The evidence that early hominids had artistic interests is much weaker, and its strength depends on the view that one takes of some very early stones that have human and animal forms (see Websites). An example of these, that was found in Morocco and

Figure 25.2 **Chimpanzee and African art.**
(a): A painting by Congo, a chimpanzee.
(b): The Tan-Tan venus (500–300 Kya; Morocco). (c): A carved block of ochre from the blombos cave; Insert: a shell bead from the cave (~70 Kya; South Africa). (d): The painting of an animal from the Apollo 11 Cave (~25 Kya; Namibia).

(a Published under a CC Attribution—Share Alike 4.0 International License.
b Courtesy of Christopher Henshilwood.)

dated to 500–300 Kya, is known as the Tan-Tan Venus. It originally seems to have been a stone with some features of a female body that may have been accentuated by scratching with another stone and has traces of ochre on its surface (Fig. 25.2a; Morriss-Kay, 2010). Its status, however, is still contentious.

A little later, 250–200 Kya, there is evidence suggesting that, as groups moved around, they carried pieces of ochre, a soft, pigmented stone with a colour range of yellow to red. The most detailed evidence of how early AMH lived in Africa has been found in the Blombos cave that was inhabited ~100Kya. Artefacts found there include evidence of paint-making tools, engraved pieces of ochre, awls and points, together with body adornments, such as beads made from shells (Fig. 25.2c; Henshilwood et al., 2002, 2011). Body-colouring material and perforated shell beads dating to ~82 Kya have also been found in Morocco (Bouzouggar et al., 2007).

In Europe, pieces of ochre and black pigments (charcoal and manganese dioxide) have similarly been found in Neanderthal sites (Roebroeks et al., 2012). Apart from its role in body painting, ochre has long been used for tanning, colouring human bones, and engraving. Neanderthal body adornments have also been found in Spain and France and these may have carried symbolic implications (Caron et al., 2011; Zilhão et al., 2010). Body adornment thus seems to have been a pleasure of both Neanderthals and early AMH.

All early representative art, as we understand it today, was, until recently, thought to have come only from AMH sites across the world (see below), with Neanderthals' artistic abilities being restricted to decorating bodies, shells, and beads. Recently, however, evidence has come to light showing that Neanderthals had some painting ability too: a few relatively simple pictures have been discovered on cave walls in Spain that date to ~65 Kya, or 25 Kya before the arrival of *H. sapiens* (Fig. 25.3a;

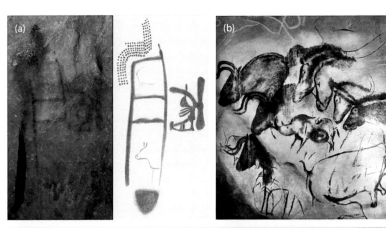

Figure 25.3 Early European art.
(a): A Neanderthal painting dated to ~65 kya together with an artist's view of the paintings underneath the surface salt layer. (b): Bulls from the paintings at the chauvet cave (~30 Kya). (c): Aurochs from the paintings at Lascaux (!7.3 Kya).

(a From Hoffman et al. (2018) *Science*; 359:912–915. With permission from the American Society for the Advancement of Science.
b Courtesy of "Thomas T." Published under a CC Attribution Share-Alike 2.0 Generic License.
c Courtesy of Prof Saxx. Published under a CC Attribution—Share Alike 3.0 Unported License.)

Hoffmann et al., 2018). Thus far, the oldest known examples of AMH representational art in Africa are the bovid pictures near the Apollo 11 cave in Namibia that date to ~25 Kya or a little earlier (Fig. 25.2d; Mason, 2006). It is unfortunate that there are no other early African art remains: any wooden artefacts would by now have decayed, with those from wetter regions being lost first. Once again, absence of evidence is not evidence of absence. It has always been the case that study of any aspect of early hominin evolution needs more data.

Most examples of early representational art date to after the arrival of AMH in Europe. Perhaps the oldest is a statuette of a man with a lion head that is made of mammoth ivory and that survived in pieces. It was found in a cave at Hohlenstein-Stadel in Germany and dates to 40–35 Kya, or relatively soon after *H. sapiens* reached Europe (Chapter 24). It was found and assembled in 1939 and, after further fragments were found, reassembled in 2013 (Fig. 25.4a; see Websites). Another very early piece of European figurative art that dates to ~40 Kya is the venus (the generic name for a small female sculpture) from Hohle Fels. This is a small, headless woman carved from a mammoth tusk that was found in the Hohle Fels cave in southern Germany. Its meaning is unclear but many of the features suggest that it may have been a self-portrait of an AMH woman who had recently given birth (Fig. 25.4b; Morriss-Kay, 2011). An important site in eastern Europe that has an abundance of early painted and carved art is Dolní Věstonice in the Czech Republic, which includes work dated to ~30–20 Kya. It is also the earliest site in which there is evidence of ceramics production, an ability soon found in other sites in eastern Europe (an example of a venus from this site is shown in Fig. 25.4c; Farbstein et al., 2012; Vandiver et al., 1989).

The greatest examples of early representational art so far found are without doubt the paintings in French caves that were produced by early European humans. These include those of Chauvet, which was originally dated to 32–30 Kya but may be a little more recent and Lascaux, which dates to 17.3 Kya (Fig. 25.3d,e). The latter are particularly wonderful and, after seeing the Lascaux paintings, Picasso is reported as having said "we have learnt nothing." Apart from the great skill and artistic ability shown by these early AMH artists, the paintings obviously carry important cultural significance. As the caves had no natural light, those early artists would have had to work using light from fires, torches, and perhaps primitive tallow candles. Moreover, the pictures would have had to be painted on the basis of sketches and memories. The difficulties must have been formidable and the pictures, sometimes located far from the cave entrance in hard to access sites, would not have been painted simply for pleasure. Efforts are now being made to interpret the meaning behind some of these paintings and so to understand something of the cultural and spiritual life of those early Europeans (Gabora & Steel, 2020; Lewis-Williams, 2002; Morriss-Kay, 2010, 2011).

The artistic creations of *H. neanderthalensis* and *H. sapiens* in Europe highlight the effects of their ~400 Ky of separate evolution. The anatomical and genetic differences between the two taxa were small enough to permit interbreeding, but, on the basis of their respective artefacts, the mental differences were large enough for AMH to replace Neanderthals. It seems clear that their different abilities in making tools and works of art and possibly in their respective powers of speech and hence cultural inheritance were among the features that enabled AMH to supplant Neanderthals. It should be noted that evidence of early human artistic abilities were not just to be found in Europe: there are traditions of representational art in the cultures of indigenous Australians (~16 Kya, Ross et al., 2016), early Europeans, Indonesians (~40 Kya, Aubert et al., 2014), and Meso-Americans (~9 Kya, Neves et al., 2012). It is,

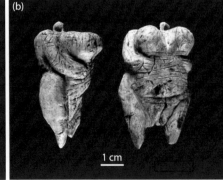
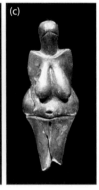

Figure 25.4 (a): The lion man from the Hohlenstein–Stadel cave; this restored wooden carving is of a human figure with a lion's head (Germany, 40–35 Kya). (b): The Hohle Fels venus (~40 Kya). (c): The ceramic venus from the cave at Dolní Věstonice (30–20 Kya; Czech Republic).

(a Courtesy of Dagmar Hollmann, Wikipedia. Published under a CC Attribution—Share Alike 4.0 International License.

b Courtesy of Nicholas Conard, University of Tübingen.

c Courtesy of "Danny B." Published under a CC Attribution—Share Alike 2.5 Generic License.)

therefore, highly likely that these abilities were present in their ancestors who left Africa ~70 and ~55 Kya (Oppenheimer, 2003; Morriss-Kay, 2010).

The artistic abilities of *H. sapiens* as seen in Europe ~40–30 Kya were little different from our own; they are more sophisticated in their visual representation than the very few examples seen in Africa ~77 Kya and in Neanderthal sites in France (Fig. 25.3a), although these may not reflect the full repertoire of what they created. Today, it is clear that the earliest members of *H. sapiens* to reach Europe possessed a fully developed sense of artistic creativity, with any advances on that of their African ancestors the result of cumulative cultural inheritance. It is slightly sad that we are unable to interpret the inherent symbolism in early human paintings (Morriss-Kay, 2010). Just as creation stories are an inherent component of Australian art that looks like decorative patterns to the untrained eye, the linear patterns made by Neanderthals (Fig. 25.3a) and those drawn alongside painted animals in European AMH sites are not random doodles and must have had clear meanings when they were created. There is a fascination in looking at the world through the eyes of our ancient ancestors and perhaps further discoveries will help explain their perceptions of the world.

The modern brain

All of this artifactual data immediately raises the question of what was special about the AMH brain that distinguished it from those of earlier hominins and particularly of Neanderthals. There was certainly no obvious difference in gross size; if anything, the Neanderthal brain was slightly larger than that of AMH. Although it is normally difficult to infer brain detail from skull features, such as indentations and capacity, two distinctions between AMH and Neanderthals have been noted. First, there are slight differences to the superficial cerebral vasculature: that of AMH is richer than that of Neanderthals, and this affords better protection for the brain against the cold than just hair (Rangel de Lázaro et al., 2015). Second, as the *Homo* genus evolved, the parietal region of the brain broadened, with that of AMH being noticeably larger in both absolute and relative size than that of Neanderthals (Bruner et al., 2011).

The importance of the parietal region is its association with mental self-representation, speech, visuo-spatial integration, body image, memory of experiences and goal-related activity (Bruner et al., 2014). It is certainly likely that an increase in the relative size of this region would reflect an increase in such functions. If so, it gives some of the reasons why AMH were more successful than Neanderthals in Europe, once they were sharing the same territory. It is, however, unlikely to be the full answer as AMH were less successful in earlier mixed groups in Israel (~100–90 Kya): Neanderthals survived there for some time after that population of AMH seems to have been lost (Chapter 24).

Finally, it is worth asking whether the data gives any clue as to when modern members of *H. sapiens* with their full mental faculties evolved. The existence of pictures on the walls of caves implies that the artists then were able to remember what they had seen and so had the ability to draw images from memory or imagination. This ability, which has been called the "mind's eye" (see Morriss-Kay, 2010), must have evolved long before AMH left Africa and probably even before they dispersed across Africa. It is not possible to be precise about dates here, but it is hard to think of core faculties beyond language and a visual imagination that are needed to distinguish us from our premodern-human ancestors. There is certainly no reason to doubt the hypothesis that anatomically modern humans, with all the faculties that characterize this taxon today, were present in south-eastern Africa well over 100 Kya.

THE ORIGINS OF HUMAN DIFFERENCES

Early anatomically modern humans expanded across Africa and there were several dispersals of *H. sapiens* out of Africa. The first of the latter group to succeed in the long term left Africa from Ethiopia ~70 Kya: their descendants are today's indigenous Australians. It was, however, a further emigration ~55 Kya that left through Sinai that eventually populated the rest of the World. As both groups were small and presumably not dissimilar, the particularly interesting question arises as to why their descendants are so varied in appearance today, even within Africa (Choudhury et al., 2018). What has happened over this relatively short period to achieve such differences in human appearance and genetic makeup?

Before approaching such questions, it is important to point out that these differences are essentially superficial. In terms of the general properties that make us human, such as basic anatomy and core cognitive abilities, such as speech and cultural inheritance, love of music, tool-making, and artistry, we are all much the same. The differences only apply to such features as skin and eye pigmentation, body build, hair, minor facial features, and a few disease susceptibilities (Balaresque & King, 2016).

As these differences are all heritable, they must reflect the effects of early mutations that were either recessive or occurred after each migration out of Africa. Two areas of genetic investigation have proved particularly useful in unpicking this history: coalescence analysis and group-specific mutation. Coalescence analysis combines DNA phylogenetic approaches with a population growth model to explore genetic and population history (Chapter 24). The study of group-specific mutations, identified by their differential effects on the phenotype, has proven helpful in identifying group identity as well as likely selection pressures. Coalescence work has been strengthened by two other approaches: first, such information can be combined with phylogenetic analysis to calculate when particular mutations first became established in a population; second, animal studies can be helpful in identifying the likely effects of these mutations on physiological properties and anatomical development.

Genome-wide association studies (GWAS; Chapter 21) have also given some information on human history. GWAS involves the analysis of single-nucleotide polymorphisms (SNPs) and other mutations in whole genomes across populations on a statistical basis, to see if any can be associated with specific phenotypic features (Peterson et al., 2019). GWAS generates a host of genes correlated with some phenomenon, albeit that not all turn out to play an important role in them. GWAS analysis is an important first step in understanding the genetic basis of variation.

Untangling information gained from anatomical and genetic analysis is further complicated because those primary migrations out of Africa and then onwards across the world were followed by secondary migrations, some of which were back to Africa (Hodgson et al., 2014). Over the subsequent thousands of years, there have been explorations, colonisations and wars that were followed by further migrations, interbreeding and consequent gene flow. As a result, there are few if any areas today whose populations are the sole descendants of those AMHs that first colonized them. Investigating this complex history on the basis of genetic analysis alone has not been easy. Coalescence analysis is, however, beginning to unpick some of the detail (Li & Durbin, 2011, Chapter 24) and has shown that the early migrations occurred as a succession of relatively small founder groups of perhaps one or two thousand breeding individuals left their home over the last 70 Kya to explore unknown territory. The technique has also revealed details of the later migrations between populations.

It is the loss of genetic variation resulting from the increasingly restricted genetic makeup of these successive groups that has mainly been responsible for the changes in general appearance of their descendants (Box 25.1). As has already been mentioned, genetic variation is highest in African populations, while the further a population is from Africa, the lower is the degree of genetic variation (Henn et al., 2012). Each time a founder group broke away from its home group to explore new lands the amount of genetic variation in the new population diminished further and

BOX 25.1 THE GENETICS OF SMALL POPULATIONS

The key factor that makes the genetics of small populations different from that of the larger ones from which they originate is the effect of genetic drift (Chapter 21). Drift represents the unbalanced sampling of the genetic distribution of the parent group that is found in small daughter groups. Such small groups have genetic properties that make them different from parent groups.

- The smaller the group, the more likely it is to have fewer gene variants as compared to the parent group with these variants being atypical and less balanced.
- There is a loss of heterozygosity and more recessive homozygotes in the daughter group.

- New, successful mutations spread much more rapidly through small than through larger ones.
- Incestuous breeding is more likely in small than in large groups and this reduces the effective breeding number N_e of the population (Chapter 21).
- As a result of their distinct genetic profiles, selection affects the two groups differently.

The effects of a succession of small founder populations are the main reasons why there is much less genetic and hence less phenotypic diversity in, for example, Japanese and indigenous South American populations than in African populations.

its phenotype changed as the distribution of alleles within that population changed (Fig. S4.1).

The small group that left Africa ~55 Kya produced a succession of secondary founder groups: one moved south-east towards Indonesia; another moved north-east through Russia towards Siberia with a subgroup of these crossing the Bering Strait to Alaska (~30 Kya). The group that remained in Turkey led to groups that moved northwards as the ice receded (~40 Kya). The most recent founder groups are those that resulted in the successive colonisations of Polynesia, New Zealand, Madagascar, and other islands in the Indian and Pacific Oceans (Figs. 24.2 and 24.5; Friedlaender et al., 2008; Ioannidis et al., 2021).

The details of all these migrations are still being studied often through whole-genome sequencing of both ancient and modern DNA. Analysis of the genomes of individuals who lived 3–4 Kya in Ireland, for example, has revealed that there were migrations from the fertile crescent (this includes Egypt, much of the northern part of the Middle East and the region extending south to the Persian Gulf) to Southern Russia and then to Ireland (Cassidy et al., 2016). More recent examples of forced migrations are the mass movements of slaves from Africa to the West Indies and UNITED STATES from the 17th to 19th centuries. The current emigrations from Syria and Afghanistan towards Europe, and the trans-Mediterranean migration of African groups, are just the latest example of the very long history of humans leaving their homeland, mostly in search of a better life.

Two further factors increased the genetic differences in the descendants of the early groups of *H. sapiens* that left Africa. First was the genetic enrichment that resulted from interbreeding with local *Homo* groups (Neanderthals and Denisovans) that migrating AMHs met. Second were the mutations in existing genes that arose after the departure from Africa. (*Note*: It is hard to find even a single new gene that has evolved since then). In both cases, novel alleles would have spread rapidly, on an evolutionary timescale, across these small groups (Box 25.1; Chapter 21). Any phenotypic properties that resulted from these genetic effects would, of course, have been subject to selection and the most obvious of these is the increasing loss of pigmentation in skin and hair as early groups migrated north (see below).

It is, of course, much easier to identify and analyse the phenotypes of genetic alleles that are deleterious than of those that are beneficial. The genes accessible in the OMIM database (see the *Online Mendelian Inheritance in Man* website) include many examples of group-specific heritable disorders. A well-known instance is the Ellis-van Creveld syndrome found in the isolated Amish population of Pennsylvania (Muensterer et al., 2013). This is an autosomal recessive disorder whose symptoms include polydactyly and congenital heart disease; it derives from mutations that effect the EVC1 and EVC2 proteins. These transmembrane proteins are in cilia and play a role in Hedgehog signaling. The disease is extremely rare except in the Pennsylvanian Amish, and its origins have been tracked back to a couple in an early, small founder group that settled in that state. More examples of such group-specific mutations associated with disease are given below.

An interesting but unanswered question is why the early AMH settlements proved unstable in the sense that small groups found it necessary to leave what were probably secure, but relatively isolated settlements. Reasons may have been an insufficiency of local animals for hunting and of plants for gathering or there could have been intergroup conflict. Another possibility, however, may have been an innate wanderlust: the urge to explore seems deep-rooted in the human psyche as is still shown by, for example, the *walkabout* of indigenous Australians (Drumm, 2018). Here, young men go off into the bush, surviving alone for some months as a rite of puberty, something that many until recently repeated at a later stage in their life. This urge to travel into empty territory clearly satisfies some innate urge in males at least, but it is not one that we understand yet.

Skin pigmentation

The most obvious example of an area-specific selection pressure is the local amount of UV radiation. The human response to its deleterious effects is to increase the degree of skin melanisation (Meredith & Riesz, 2004), a trait that reflects an inherent degree of melanosome production that can be modified by developmental plasticity (tanning). The details of this characteristic strongly depend on latitude and altitude and amounts of melanin are much higher in equatorial Africa than elsewhere. The indigenous population of Australia (15–30°C south of the equator), for example, that almost certainly descended from a group from Ethiopia (10°C north of the equator) are considerably paler than modern Ethiopians. All other non-Africans populations probably descended from a group that lived near the Sinai Peninsula (30°C north of the equator) whose likely skin hue is best matched by today's northern Egyptians, who are themselves much paler than Africans from nearer the equator. The skin hues and associated local UV levels are shown in Fig. 25.5.

One constraint on any discussion here is that our knowledge of when early AMH first covered themselves and so protected their bodies from UV is limited. Recent

Figure 25.5 **The relationship between human skin pigmentation and local UV intensity.** (a): The distribution of skin colour in native populations across the world in 1940 as graded on von Luschan's chromatic scale. This map is adapted from data collected by R. Biasutti on the "distribution of the varying intensity of skin colour" in indigenous populations (see Websites for further details). (b): UV intensity across the world in erythermal units, which are essentially a measure of the damage that UV does to skin.

a From Jablonsky & Chaplin (2000) Published under a CC Attribution-Share Alike 3.0 Unported license.
b From Box et al. (2018). *J Transl Med;* 16:252. Published under a CC Attribution-Share Alike 4.0 International License.)

discoveries in Morocco (Hallett et al., 2021) do however suggest that furs were worn ~100 Kya. Nevertheless, the migration and pigmentation maps (Figs. 24.4 and 25.5) make two clear points about climate and hue, First, the gradient of pigmentation differences is most striking in northern Europe and Asia where it ranges from light brown in southern regions to a virtual absence of skin pigmentation in northern of countries, such as Finland (50–60°C north of the equator). Second, the migration map shows that, while populations moving northwards towards Siberia lost colour (today's Inuits are also pale), they regained some of the lost hue once they crossed the Bering Strait, as they moved south through the Americas over the last ~10 Ky.

A similar effect marks the SE Asian migration into Australia: here skin hues lightened a little as humans moved south. The small region of particularly dark hue in NE India reflects the altitude effect of high UV levels in the Himalayas, while that in Sri Lanka and South India particularly reflects the hue of the Tamil population who live near the equator (possible genetic reasons for melanin modulation are discussed below). It is, however, clear that, in general, skin hue reflects the selection pressure of environmental UV levels, and an obvious expectation is that the indigenous inhabitants of equatorial South America, should they survive, will darken a little over the coming millennia.

The genetic basis of pigmentation

Optimal levels of pigmentation in a particular environment reflect the balance of having enough UVB reaching the skin where it is necessary for the photosynthesis of vitamin D_3, but not so much that it results in deleterious sunburn or skin cancer (Jablonski & Chaplin, 2010). Skin hue depends partly on the ratio of brown/black eumelanin and red pheomelanin and partly on the number, size and density of melanosomes in skin epithelium; all of these are under genetic control. The details of how this balance is achieved are not, however, easy to discover as >100 genes have been associated with pigmentation (Sturm, 2009). Serre et al. (2018) point to networks of ~30 proteins involved in the biosynthesis and transport of the key pigment proteins, while D'Mello et al. (2016) also emphasise the complexities of the signaling pathways that regulate melanin production (Fig. 25.6).

Many mutations have been discovered that modulate the balance of melanins produced in melanosomes (for review, see Jeong & Di Rienzo, 2014). An important example is the melanin-associated gene OCA2, loss of which leads to oculocutaneous albinism (lack of melanins in the eyes and skin). A mutation in this gene is associated with skin lightening, probably because the mutant leads to diminished tyrosine uptake (a key amino acid component of melanins). This mutation is particularly found at high frequency in East Asian populations. Simulations by Smith et al. (2018) using coalescence methodology show that the time back to the last common ancestor possessing the mutant OCA2 alleles was 20–15 Kya.

Some of the other key genes involved in pigmentation synthesis are SLC24A5 (also known as NCKX5), SLC45A2, GRM5, TYR27, MATP, and OCA2. SLC24A5, which is involved in lowering calcium levels in melanosomes, has a variant that arose 28–22 Kya and is associated with decreases in pigmentation levels. This seems to have been under positive selection in white Europeans and is now common. Variants in other genes are also associated with a loss in pigmentation. A mutation in SLC24A5 is also present across a central swathe of Asia but is not found in South India, an area some 10° north of the equator where the dark-skinned Tamil population is located (Mallick et al., 2013). As the latter mutation occurred well after the separation of east- and west-migrating humans, this, as well as other pigmentation-affecting mutations, is likely to have arisen independently in the two populations (Jeong & Di Rienzo, 2014). A second pigmentation-lessening variant of SLC45A2 seems to have been introduced to Europe at around the same time, but at lower frequency. In addition, the *MATP* gene, which may have a role in tyrosinase trafficking in melanocytes, has an L374F variant that is found in areas of Europe where lighter skin is common. In contrast, the H615R mutation in the *OCA2* gene is only present in East Asians. In short, the genetics of pigmentation is complicated!

A different perspective on the evolution of pigmentation differences comes from the work of Mathieson et al. (2015), who analysed the DNA of 230 ancient (8.5–2.3 Kya) western Eurasians. These analyses show that the mutations in *SLC24A5* and *SLC45A2* that are associated with weak pigmentation seem to have originated in what is now Turkey and were brought north by migrating Anatolian tribes—it was probably these tribes who introduced agriculture to Europe. It was also during this relatively recent period that the TYR and GRM5 variants associated with low pigmentation

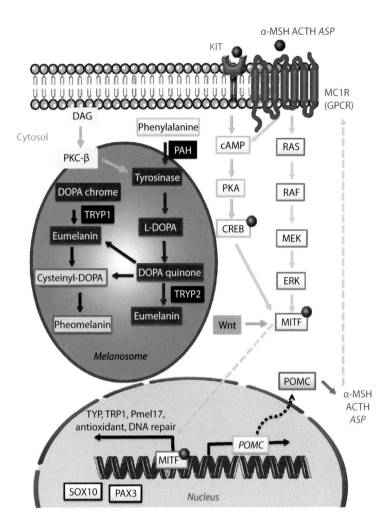

Figure 25.6 Eumelanin and pheomelanin are catalysed by specific melanogenic enzymes (black) whose production is under the control of the MITF transcription factor. MITF is activated by signaling pathways that include PKC (brown), cAMP (blue), MEK (purple), and WNT (orange). MITF expression also drives the expression of a number of genes including SOX10 and PAX3 and plays a role in a host of further activities.

(From D'Mello et al. (2016). *Int J Mol Sci*; 17:1144 Available from MDPI under an open-access license.)

also entered northern populations. It is thus likely that the specific combination of pigmentation alleles that is responsible for low pigmentation levels in western and northern Europe today is relatively recent, probably only reaching its current high level of frequency within the past 5000 years.

One interesting aspect of human pigmentation changes in the wider evolutionary context is the light that they cast on the speed of selection. Based on a generation time of twenty years, it is estimated that humans left Egypt to colonise the world ~250 generations ago. If that population had the same degree of pigmentation as that of contemporary Egyptians (noticeable but not deep), this was sufficient time for the migrating population to lighten as it approached the Arctic (in Scandinavia today, there is little trace of skin pigmentation) and the Bering Strait and to darken again as it continued into South America. This gives a period of ~100 generations for the particular change to spread across a small population under a selection that optimised the body's need for UV light while producing enough melanin to afford protection.

Eye colour

The other pigmentation feature that varies across the world is eye colour, with those in Africans and Asians being predominantly dark brown while those in Europeans having a spectrum of colours that include blue, grey, hazel, and even green. Eye colour derives partly from the nature of the melanin laid down in the iris and partly from light-scattering effects. Brown eyes have high melanosome levels in the iris epithelium and overlying stroma, while blue eyes have less melanin in the stroma, with the colour being primarily the result of light scattering. Green eyes are essentially blue eyes, but with small amounts of amber-coloured pheomelanin in the stroma. These add some yellow to the blueness, so resulting in a green hue. Other colours reflect a different balance between these constituents (see Websites).

The genetics of eye colour are complicated: at least 16 genes are known to be involved in networks that are, of course, closely related to those for skin melanosomes (Lin et al., 2016). Key examples are OCA2 and HERC2 in which an intron regulates the expression of OCA2 (White & Rabago-Smith, 2011), and whose rs12913832 mutation is strongly associated with blue eye color. This latter mutation can be tracked back to hunter–gatherers from present-day Italy and Georgia 14–13 Kya and appears only to have become almost fixed in parts of northern and western Europe by ~8 Kya. This rapid increase in frequency, particularly at high altitude, seems likely to have been driven by selection, although the reasons are not yet clear. One possibility is that there is some selective advantage to scattering rather than absorbing light in regions where the intensity of sunlight is diminished.

Genetic disorders

Reports in the medical literature of disease susceptibilities across the world make it straightforward to identify group-specific disorders with the many hundreds so far found being documented in OMIM (see Websites). Some, such as the Ellis-van Creveld syndrome found in the isolated Amish population of Pennsylvania (Muensterer et al., 2013) have already been mentioned. Others that are the result of specific mutations and whose genetic inheritance pattern follows simple Mendelian laws include cystic fibrosis (in northern Europeans), Tay-Sachs disease (Jews from Eastern Europe and Cajuns in South Louisiana), thalassaemia (a haemoglobin abnormality found in people who live around the Mediterranean Sea), lactose intolerance (native Americans, East Asians, and some Europeans) and alcohol intolerance due to an alcohol dehydrogenase mutation (common in East Asians). The group restrictions reflect the origins of the mutations, while their degrees of penetrance are a measure of the time since each first occurred. The reason why these deleterious mutations have not been lost due to selection is that most are recessive and have limited or no effect on a heterozygote's reproductive ability. Even those that are homozygous may have little effect on inheritance if the associated disorder develops well after puberty.

Body differences

One feature that characterizes distinct groups is body size. It is generally found, for instance, that people living in hot climates have a thin body shape and long limbs which facilitate heat loss, while people in cold climates are more likely to be stocky and short-limbed, features that preserve body heat (Ruff, 2002). This is not, however, the full story because, superimposed on the genetics of anatomy are the environmental effects of diet and health care. As Bentham et al. (2016) have found, many populations across the world have increased significantly in height over the last century with the greatest increase being in South Korean women and Iranian men who are ~20 cm higher now than then. The key reasons here are improvements to diet and health as a result of increased prosperity.

GWAS studies have identified many genes associated with variation in height, body size and facial features in humans (Balaresque & King, 2016; Richmond et al., 2018), all of which reflect local modifications of growth. It is, however, extremely hard to work out the specific function of individual proteins. The mouse is the only practical model for experimental work here and the phenotypes of genes that are mutated or knocked out in it are sometimes hard to translate to humans, because height, size and facial features are so different in the two organisms. It thus seems unlikely that we will be able to discover in the near future the mutations in the individual proteins and their regulatory sequences that are responsible for the wide range of human features.

Even such a clear-cut question as to the genetic reasons why the Dutch are the tallest nation in Europe or why there is such a spectrum of heights in Africa (Maasai men are ~1.9 m and men from pygmy tribes are ~1.5 m) have yet to be answered. In the case of the Dutch, it should be noted that, even thousands of years ago, there was a height gradient across Europe with northerners being taller (Mathieson et al., 2015). One obvious possibility is that the quality of the Dutch diet was and continues to be good. While this may be so, it does not seem to have been significantly better than that of neighbouring countries. The genetic reasons for why there is a peak of height in the Netherlands still remain unclear, but genetic drift would certainly be expected to be part of the answer.

Even a phenotype as simple as the degree of hair curliness has proven hard to understand. The trichohyalin (TCHH) protein is involved in determining the secondary structure of keratin filaments (the key protein in hair) and GWAS analysis has identified TCHH variants that are associated with straight hair as opposed to curly. A GWAS investigation in Han Chinese found, for example, that the rs3827760 mutation in the EDAR gene was the predominant cause of straightness (Wu et al., 2016). These gene variants together, however, only accounted for a small proportion of the variance in the straightness phenotype and these authors concluded by suggesting that hair straightness may be affected by a large number of genes, each having subtle effects on the phenotype.

Cognitive abilities

This is an area that still excites more heat than light and an interesting text on why societies evolved at different speeds is Jared Diamond's *Guns, Germs, and Steel* (Diamond, 1997). This starts with the report of a conversation between the author and a politician from New Guinea who raises the question of why people from Europe had so much more "cargo" (i.e. material goods) than his own countrymen, who were clearly just as smart as Europeans. The rest of the book looks at the possible answers and makes it clear that key to those early societies that prospered was not innate ability but the presence of large animals that could be domesticated. These were not only an easy source of high-quality protein, but allowed ploughing and agriculture to become efficient, so liberating people from a hunter-gatherer subsistence life. This, in turn, enabled such privileged groups to live better, so giving them more time for building better homes and spending more time on what we would now see as cultural activities. The result was the beginnings of modern society.

One key message of Diamond's book is that the number of large mammals that are readily domesticable is very few; they include cattle, sheep, horses, goats, pigs, camels and llamas. The single area in which the majority of these were originally native was the Fertile Crescent (Egypt and the near East); many fewer are found in other areas of the world. It is thus no coincidence that it was in the countries of the Fertile Crescent that skills, such as writing were first developed, although writing schemes were invented independently in five early, but prosperous societies, all of which were sophisticated for their times. These were Iraqi cuneiform (~5.2 Kya), Egyptian hieroglyphics (~5.1 Kya), picture writing in China (4-3 Kya), the Indus scripts in India (~3 Kya), and the Mayan culture's glyphs (~2.5 Kya) in South America (Robinson, 2007). All are based on logogram characters that represent both syllables and concepts. The first useful alphabet was developed later in the Phoenician region of the Levant (~ 3 Kya); technically it was an abjad as it lacked vowels (as do contemporary Arabic and Hebrew abjads). These were, however, introduced for Greek ~2.8 Kya. Alphabet-based writing then spread rapidly to give the script diversity across the world that we see today.

These early intellectual successes were, however, as much due to opportunity as ability: those societies that had lagged behind showed that they had sufficient intellectual ability to read and write as soon as they were exposed to education. As every *H. sapiens* group has these abilities, it is clear that the necessary brain structures and neuronal abilities were established in all African populations before the start of migrations across and out of Africa. In more recent times, Africans have flourished outside of their original continent for many centuries; there were, for examples, black Africans who prospered in Tudor England (16th century; Kaufman, 2017).

An important insight into the abilities of people in different cultures comes from museums of comparative anthropology. A particularly interesting example of this is the Pitt Rivers Museum of Oxford, U.K., as it displays exhibits by theme rather than by distinct cultures. Among its many cases are those for tools, money, pottery, musical instruments, weapons, decorative clothing and jewellery, with each containing artefacts from across the world. Together, these cases make a single important point: the abilities and interests of every culture from around the world are similar and the differences in what they make and how they dress derive from the materials to hand and the local needs. We have no evidence to suggest that there are any obvious dissimilarities in the innate mental abilities of the many groups that live on earth. There are, of course, easily observable differences among us, but they are skin-deep and cultural.

ARE HUMANS STILL EVOLVING?

The simple answer to this question is most likely to be *no* in terms of genetic novelty; this is because humans no longer meet the criteria for novel genetic change to occur. There is, of course, a degree of evolutionary change occurring today, but this is due to migration and consequent interbreeding between members of groups from different continents rather than as a result of new mutations. The novel gene combinations resulting from such interbreeding may, of course, lead to slightly altered traits, but full evolutionary change will depend on whether such novel traits that might be seen as beneficial can benefit from selection to become common or even dominant within their populations, or even survive the effects of drift. Here, the likely number of generations for a novel gene to survive genetic drift and become fixed in a population is ~$4N_e$ (the size of the breeding population; Chapter 21). In today's cities, where most people live and breed, N_e is a very large number. The net result of travel and inter-group breeding is that populations across the world are slowly becoming more homogeneous, local interbreeding populations are becoming larger and the extent of group-specific features is becoming diminished.

It is worth asking whether, even if a new mutation led to a novel phenotype with what might seem to be beneficial features, would it be subjected to positive selection in a way that would change genetic profiles in a large population. The answer is unlikely to be yes because women now control the extent of their reproduction far more than in the past. The decreased number of offspring to which each now gives birth all have a far greater likelihood of surviving than in the past. As the measure of a successful mutation is the extent to which it enhances the production of offspring that carry the associated allele, any change in its frequency within the population is likely to take place very slowly. Hence, the power of positive selection in human populations is much less than it was.

The situation is different for new deleterious mutations that, in the past, would have been fatal or unlikely to have been passed on to further generations. Such is the power of modern medicine that many individuals showing such mutant phenotypes would previously have died before reaching reproductive age; today, however, they can enter the breeding pool. Deleterious mutations are hence far less likely to be lost now than in the past. There is no doubt that modern medicine has blunted the blind power of selection for humans.

The one environment in which genetic change would be likely to occur is in isolated populations for which $\mathbf{N_e}$ is small and selection is hence relatively fast. These, however, now exist in only a few places, such as the Amazon rain forest (Salgado, 2021) and, even were $\mathbf{N_e}$ as low as 50 for a group, it would still require many hundreds of years for a novel mutation to become widely distributed within it. Given the speed with which small indigenous populations are being lost as a result of social and commercial pressures, all are likely to have been dispersed long before any new alleles could become established. The corollary of this is that those alleles unique to any such small group will be brought into the wider populations at the expense of the individuality of those small groups being lost. The net result is that we are becoming more, not less genetically similar; this is not a recipe for evolutionary change.

KEY POINTS

- The best evidence on the evolution of the mental capacity of AMH comes from analysing and comparing the artefacts that they and Neanderthals made.

- Over several million years, stone tools became increasingly sophisticated, but it seems that the most complex tools made by Neanderthals followed their meeting up with AMH.

- Both groups decorated their bodies with pigments and created art, but the painting and other artwork of AMH was considerably more sophisticated than that of Neanderthals, albeit that very little of the latter's art has been found.

- The evidence on whether Neanderthals could speak mainly comes from analysing their hyoid bones, which are attached to both the larynx and the tongue, and so facilitate speaking. As their hyoids and those of AMH are very similar, it is likely that they, like AMH, had language. It thus seems that the basics of speech were present in their last common ancestor who lived >500 Kya and may have been *H. heidelbergensis*.

- Across the world, the different AMH groups have similar abilities: all can read, write and do arithmetic; they also have comparable artistic interests (music, body decoration, painting, etc.). There is, therefore, no reason to doubt the conclusion that all of these abilities were present in AMH well before the emigrations from Africa.

- Different racial groups today look distinct from each other because the world was populated through a succession of founder groups, each with less genetic diversity than its parent group and an asymmetric sampling of its gene pool due to genetic drift.

- Further differences arose for three main reasons. First, as each group migrated into unknown territory, it was subjected to new selection pressures. Second, it acquired new mutations, some of which are seen today as group-specific diseases. Third, their gene pool was enlarged by interbreeding with Neanderthal (particularly those who migrated into Europe) and Denisovans (those that went East).

A particularly instructive marker here is skin pigmentation, which was diminished in AMH groups who reached the far north but regained when those that migrated across the Bering Strait from Russia to northern America moved south. The selection pressure here is a balance between the need for enough UV to be absorbed by skin to make vitamin D and the need for UV-absorbing melanin to protect against the deleterious effects of high UV levels.

It is likely that because of the interbreeding of long-distance migrations of populations with their new host populations, traits and genes that were group-specific are now becoming part of the greater gene pool. This means that there is a trend towards greater human similarity and that novel variants are unlikely to have the opportunity to become prominent.

Today, the small, isolated groups in which diversity flourishes are being lost and human evolutionary change seems to be coming to an end.

FURTHER READINGS

Aimé C, Laval G, Patin E et al. (2013) Human genetic data reveal contrasting demographic patterns between sedentary and nomadic populations that predate the emergence of farming. *Mol Biol Evol*; 30:2629–2644. doi:10.1093/molbev/mst156.

Colonna V, Pagani L, Xue Y, Tyler-Smith C (2011) A world in a grain of sand: Human history from genetic data. *Genome Biol*; 12:234. doi:10.1186/gb-2011-12-11-234.

Diamond J (1997) *Guns, Germs and Steel*. Vintage Books, London.

Jobling M, Hollox E, Hurles M et al. (2013) *Human Evolutionary Genetics* (2nd ed). Garland Press.

Kaufman M (2017) *Black Tudors: The Untold Story*. Oneworld Publications.

Lieberman D (2011) *Evolution of the Human Head*. Harvard University Press.

Lieberman D (2013) *The Story of the Human Body: Evolution Health and Disease*. Pantheon Press.

Narasimhan VM, Patterson N, Moorjani P et al. (2019) The formation of human populations in South and Central Asia. *Science*; 365: eaat7487. doi:10.1126/science.aat7487.

Pugach I, Stoneking M (2015) Genome-wide insights into the genetic history of human populations. *Investig Genet*; 6:6. doi:10.1186/s13323-015-0024-0.

WEBSITES

- Wiki entry for the history of writing, *Stone Tools, Tool Use by Animals, Eye Colour*, and *Whale Vocalizations*, together with the various art sites, such as *Dolní Věstonice*.
- Japanese stone tools: https://core.ac.uk/download/pdf/5105122.pdf
- Map of human skin colours in the 1940s: https://commons.wikimedia.org/wiki/File:Unlabeled_Renatto_Luschan_Skin_color_map.svg
- Online Mendelian Inheritance in Man: https://www.ncbi.nlm.nih.gov/omim
- Stone age tools: http://humanorigins.si.edu/evidence/behavior/stone-tools
- Examples of what may have been the earliest stone art: https://davidneat.wordpress.com/tag/venus-of-tan-tan/

Conclusions

26

This last chapter starts by briefly reviewing the progress made over the last few years in understanding the details of the tree of life and the mechanisms of variation, selection, and speciation. It then considers the challenges facing evolutionary biologists today. While there remain details to be filled in on evolutionary history, they mainly concern expanding our knowledge of the mechanisms of evolutionary change. These particularly involve understanding how mutation leads to changes in the anatomical phenotype; we also need richer models for selection and the molecular basis of speciation that reflect the complexity of these processes. The chapter goes on to ask what is missing from our core picture of evolution; three obvious topics are 1) whether there are faster ways of mediating heritable change than mutation; 2) the evolutionary basis of human behavioural and cognitive abilities; and 3) the origins of life. The chapter ends by considering the future of humans; it concludes that the combination of population pressure and climate change does not suggest that it will generally be successful.

In 1859, Darwin ended *The Origins of Species* by comparing the evolution of the living world to that of a 19th century English riverbank.

It is interesting to contemplate an entangled bank, clothed with many plants of many kinds, with birds singing on the bushes, with various insects flitting about, and with worms crawling through the damp earth, and to reflect that these elaborately constructed forms, so different from each other, and dependent on each other in so complex a manner, have all been produced by laws acting around us. These laws, taken in the largest sense, being Growth with Reproduction; Inheritance which is almost implied by reproduction; Variability from the indirect and direct action of the external conditions of life, and from use and disuse; a Ratio of Increase so high as to lead to a Struggle for Life, and as a consequence to Natural Selection, entailing Divergence of Character and the Extinction of less-improved forms. Thus, from the war of nature, from famine and death, the most exalted object which we are capable of conceiving, namely, the production of the higher animals, directly follows. There is grandeur in this view of life, with its several powers, having been originally breathed into a few forms or into one; and that, whilst this planet has gone cycling on according to the fixed law of gravity, from so simple a beginning endless forms most beautiful and most wonderful have been, and are being, evolved.

Apart from its lovely imagery, this paragraph is remarkable for three reasons. First is the comment in the last sentence, that evolution never ends but is a natural part of life; at the time, this may have been the most revolutionary point in the whole book. Second is the middle part, which lays out the basic skeleton of the mechanism by which evolution works. Third is the almost contemporary emphasis on the complexity of variation and selection. More than 150 years of research has shown that Darwin was right in almost everything that he wrote, other than the comment on use and disuse and the emphasis on higher animals. What generations of evolutionary biologists have done is to add substance to the bare bones of Darwin's mechanism of evolutionary change.

DOI: 10.1201/9780429346217-31

For Darwin, perhaps the most surprising of our new insights would probably be the extent to which the dynamic of evolution is unhurried. Indeed, it is far slower than Darwin would ever have suspected, given his knowledge of selective breeding and his belief in pangenesis, a mechanism based on Lamarck's view, that abilities acquired by parents could be passed on to offspring. Today, it is clear that novel variants can take hundreds or even thousands of generations to predominate, while the process of speciation, in the sense of full reproductive isolation, may well not be complete after a million generations. Ligers, the hybrid offspring of lions and tigers are still fertile, as is the very occasional mule, the offspring of a horse and a donkey, even though each pair of parents shared a last common ancestor several million years ago. It is over this long period that the mutations responsible for speciation accumulate as a result of selection and drift.

This book has considered much of the evidence for this view and its implication for life on Earth over almost four billion years. The purpose of this final chapter is to review the current state of evolutionary biology and to point to areas where our knowledge is incomplete. The chapter ends by considering the future of human evolution.

THE ACHIEVEMENTS OF EVOLUTIONARY SCIENCE

Every branch of biology has participated in producing the modern understanding of evolution, as the previous chapters have demonstrated, but perhaps a few contributions from biologists of the last generation stand out.

Palaeontology: Increases in the detail of the fossil record have clarified the past of every aspect of the biosphere since the Ediacaran. This record and its gaps continue to set problems that the rest of evolutionary biology has to solve, such as how the amniote egg and the human brain evolved.

Population genetics: Darwin asked how new species originate on the basis of variation and selection. Population genetics has produced the answer by showing that small groups of organisms that break away from a large parental group have a set of genetic properties that, on an evolutionary time scale, facilitates rapid change and speciation.

Informatics: The production of the programs that allow DNA sequences to be compared has allowed, for example, phylogenies to be constructed for clades where there is no fossil record, shown how the eukaryotic cell resulted from the endosymbiosis of several bacteria and enabled population histories to be estimated. The importance of these programs is incalculable.

Developmental biology: The tools of molecular genetics and experimental developmental biology have been applied to a wide range of animal embryos. The results have shown how organisms with very different morphologies each have an essentially common underlying molecular toolkit based on homologous signaling systems and process networks that is used to make new tissues. This work has not only strengthened our understanding of evolutionary diversification but has cast light on the early evolution of multicellular organisms.

CONTEMPORARY CHALLENGES

These achievements, together with all the other work already discussed in the book, make it clear that we now have a solid and robust understanding of the two major arms of evolutionary biology: the history of life, which is known in all its essentials, and the mechanisms of evolutionary change whose principles at least are now clear. The history can now be traced back to a very simple bacterium, the first universal common ancestor (FUCA). The knowledge that supports this statement comes from a vast amount of phylogenetic, fossils, and developmental data. The evidence to support our understanding of the various mechanisms that drive evolutionary change does however sit on a smaller experimental base than does evolutionary history. That is because change is so slow in the wild that, with the exception of a few fortuitous cases, we can only probe it under experimental conditions. This is not to say that we don't understand it or that its experimental basis is thin, but that we need detail to be filled in. If, a century ago, we were surrounded by a dense fog of ignorance, today that fog has generally cleared and we live in a generally light mist: in some areas, it is only the lightest of clouds but there are others where the fog is lifting more slowly than we would like. The following few pages point to areas where we need to see a little better.

The history of life

The animal fossil record

Knowledge about extinct organisms is always provisional and likely to be changed in the light of new fossils. Such fossils are, of course, always needed to provide more information on the speed of evolutionary change and, in particular, whether there is evidence of saltatory or rapid morphological modifications to phenotypes. Significant remaining gaps in animal evolutionary history are the details of the early history of eukaryotic existence, the origins of the large organisms of the Ediacaran and Cambrian periods, the evolution of early arthropods and molluscs and the events that led to mammalian diversity. Future understanding of early life forms will depend on fortuitous discoveries of soft-bodied fossils suitably preserved in the rocks from these ancient periods.

Plant evolution

An area that is discussed comparatively lightly in this book is evolutionary botany. Unfortunately, early, unwooded plants do not fossilise well so the details of their early evolution are not easy to follow. There is also a gap in our knowledge of early angiosperm evolution: the evidence on how these early flowering plants evolved is thin and what there is remains ambiguous; more fossils are needed to clarify the situation (Chapter 11). There is also the difficulty of plants being far more tolerant of interbreeding than animals and it is therefore much harder to discover genetic constraints on breeding and evolutionary change. It is obvious that the mechanisms of plant physiology, biochemistry, and development are very different from those of animals (Meyerowitz, 2002; Edel et al., 2017); as more information accrues in this important area, it will be interesting to see if there are lessons on evolutionary constraints to be drawn from further comparisons between the two clades.

Phylogenetics and computational evolutionary biology

Much of the phylogenetic analysis so far carried out has focused on the comparative analysis of protein-coding sequences. Understanding the ways in which these proteins are mobilised will come from computational analysis of the sequences that regulate their activity. This work will involve analysis of the various classes of regulatory RNAs (microRNA, siRNA etc.) and cis-acting sequences. Another important gap that needs to be filled is the population history of species other than humans; it is likely that interesting results will emerge as coalescent analysis is applied more widely. In the wider context, we still lack an integrative computational model that captures the full richness of evolutionary change, but this may be too difficult to achieve, partly because of the complexity of the description, partly because of the different scales of activity involved (Fig. 26.1) and partly because details of the rate and other constants involved in modifying protein networks are so hard to obtain.

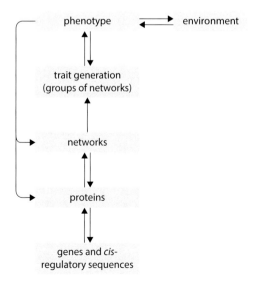

Figure 26.1 **Diagram illustrating some of the key levels of scale and functional interactions.** Note that, as there are both upwards and downwards feedback interactions in this system, causality is distributed.

Evo-Devo

The last forty years of research have shown how different types of embryos use similar molecular processes in generating structures and functions and so help us to understand how phyla became distinct. Much less is, however, known about the origins of the different cell types that are present in most of the Bilateria, some of which were also present in early coelenterates and sponges. Work here will probably depend on integrating evo-devo and computational phylogenetics, perhaps by studying simple organisms that reflect the early stages of animal life. Of particular importance will be information on the evolution of the nervous system. The behavior of neurons is so complex and their use so diverse that it is hard to believe that they evolved more than once or changed much once they had formed – the very ancient type of neuron seen in the Ctenophora is the only atypical form known (Moroz et al., 2014).

Three other areas that need further investigation are developmental plasticity, the ways in which protein networks generate outputs and the evolution of the amniote egg and early amniote embryogenesis. The first reflects the innate ability of an organism to change its phenotype as a result of changes in the environment (Chapter 19). We know little of the mechanisms by which this can happen, but they are important: the fastest way to achieve a new phenotype is to have it, or something close to it, as a developmental option that can become fixed. As to the second, the internal dynamics of networks that may include up to a hundred proteins are currently not understood and there is no obvious approach to unravelling these dynamics. We need this information to understand development and to understand how mutation leads to anatomical diversity, but we currently lack the techniques to unpick network activity. The third awaits the chance discovery of very ancient embryos whose soft tissues have been perfectly preserved. We can only hope that there are Lagerstätte from the Carboniferous and Permian awaiting discovery.

The mechanisms of speciation

There are two reasons for our limited understanding of these mechanisms. The first derives from the slowness of change in the wild. Unpicking the evidence for how change happens has mainly depended on laboratory studies in which selection pressures can be speeded up under controlled conditions and on those occasional situations where natural phenomena have illuminated facets of evolutionary behaviour. Examples here are ring species, species flocks, hybrids, and extinctions. A second reason is that we do not fully understand how mutational change leads to phenotypic change except in cases where there is a direct link between them (i.e., through the activity of Mendelian genes). In most cases, that link is indirect as the mutations affect proteins that cooperate with others in networks. This gap is particularly wide in the areas of developmental anatomy, neurobiology, and behaviour. Our understanding of developmental molecular genetics is helping with the first of these areas, but it is not clear that we even have the tools to approach the other two.

Mutation and variation

The major unsolved problems here are how mutational changes to proteins lead to substantial changes in an organism's phenotype, particularly its anatomical details. This book has argued that a key step will be understanding the ways in which mutation affects the output of protein networks involved in signaling, timing, and driving development change. The hardest problems to be solved here are, of course, in neurobiology: we know very little about the cellular basis of behaviour. The best model system for this is probably *Caenorhabditis elegans* for which mutants have been identified with altered behaviour (Goodman & Sengupta, 2019).

Symbionts and holobionts

The interest and importance of symbionts have long been known, but the realization of the significance of host-microbiome interactions in holobionts, particularly as a unit of selection, is relatively recent. They are particularly interesting for understanding the mechanisms of interspecies support, their role in local ecosystems and the opportunities that they provide for colonizing niches that would not have been open to the host species alone. Their study may also indicate lines that are of medical importance. There is clearly a lot to do in this new area of biology.

Selection

Under controlled laboratory conditions, selection can be restricted to a single parameter, be it a physical or behavioural demand, a mating opportunity or even just food access. It is thus relatively straightforward to work out what is going on. Under natural conditions, where there is interdependency, it is much harder to analyse selection pressures. The various pressures that determine an organism's fitness in a complex environment have to be subsumed within one or two umbrella parameters whose values cannot be predicted but have to be measured. It will be interesting to see whether it is even possible to provide a means of estimating the full range of selection pressures for organisms living in the wild.

Speciation

The problems of predicting when the extent of the genomic differences between a parent and a daughter population are sufficient to block interbreeding are very hard to solve. This is because novel speciation is a slow and random process: it starts with small-scale mutations that affect mating behaviour and eventually leads to major chromosomal changes that result in a failure of hybrid fertility. While the first step can be very rapid if populations are subject to very strong selection on the basis of novel behaviours (Rice & Salt, 1990), it is still hard to identify the anatomical and physiological phenotypes responsible for such changes, let alone understand their molecular basis. A further difficulty is that the occasional mutations that drive speciation need to spread through the population via random drift, only sometimes modified by selection. In all, the whole process of speciation is so slow and multifaceted that it is hard to be optimistic that there will ever be a quantitative, predictive theory.

Evolutionary population genetics

This theory has provided the quantitative framework for discussing the effects of gene alleles in the context of breeding, mutation, and population numbers. Its successes are obvious and well documented. There are, however, a series of problems that limit its use. The most important is that of linking alleles and traits. The theory defines a gene on the basis of its associated phenotype. This has been wrong since the 1960s, when it became clear that a gene is correctly defined as a sequence of DNA with a particular function. Analysis of craniosynostosis, for example, has clearly shown (Chapter 17) that the same phenotype can be obtained by mutations in several very different genes while many traits are now known to be underpinned by complex networks of proteins. As the classical theory cannot incorporate this level of complexity, a much better method of linking the reality of molecular genetics to the quantitative theory of population genetics is needed.

A further problem arises when one tries to use the quantitative theory to study organisms in their natural habitat in order to understand how allele differences lead to phenotypic ones. The experimental data are rarely rich enough to obtain the constants needed by the equations and so make further testable predictions. An example of this is, as just mentioned, understanding the dynamics of selection in the wild when there are several selection pressures and variation is multifaceted. Simulation makes it clear that the future behaviour of a population is so unpredictable under realistic conditions that it can only be explored using stochastic modelling.

IS ANYTHING MISSING?

The most difficult question to answer for the theory of evolution or indeed any substantial theory that works well is whether there are parts of the model that are missing. This is not a trivial question, and there are evolutionary biologists who feel that the standard view of evolutionary biology that forms the core of this book is inadequate (e.g. Laland et al., 2014). Their view is that the standard theory ignores important components of evolution, particularly the evolution of the phenotype, a facet of evolution that this book has aimed to explore.

The speed of evolutionary change and noncanonical heritability

Most evolutionary change seems to be slow and is generally measured in many thousands of generations. Some evolutionary biologists therefore feel that the canonical mechanisms of evolution are inadequate to explain evolutionary change.

The context for looking for non-canonical change comes from Lamarck's view (1809) that anatomical adaptations made by a parent in response to the environment would be passed on to the next generation. That approach to evolutionary change was believed by Darwin and all other biologists until the late 19th century when it was seen to be incompatible with Weissman's theory on the continuity of the germplasm. Any example of acquired adaptation would of course require one or more mechanisms by which the acquired changes that underpinned a phenotypic change in an organism would be transferred to its reproductive organs and gametes (hence Darwin's theory of pangenesis). The difficulty is that these organs are isolated from the other tissues in animals, although this is less so in plants, algae, fungi, and other phyla. As of now, no mechanism is known by which this can happen through orthodox molecular genetics. Nevertheless, non-canonical mechanisms of heritability, originally known as soft inheritance and now called *transgenerational epigenetic inheritance* are now beginning to be explored and interesting evidence is emerging. What these mechanisms have in common is that they modify gene expression in unusual ways.

The fastest of these is post-fertilisation methylation of CpG sites in the genome, which can occur during development and later in life (Chapter 17). The difficulties here are that, in animals at least, developing gametes are shielded from the rest of the organisms and almost all methylation is removed during their development. A few cases of methylation-associated heritable changes have been noted, but they have been lost after a few generations. Other mechanisms of non-Mendelian transgenerational epigenetic mechanisms include the effects of chromatin remodeling, prion inheritance, and RNA regulators of gene expression, many of which are found in gametes and so can affect development (Chapter 17). Gilbert and Epel (2015) reviewed some 30 examples of epigenetic inheritance across 13 organisms and found that very few persisted beyond a few generations unless they reflected mutation, indirectly at least.

The most convincing examples of epigenetic inheritance are those in which small molecules of RNA present in gametes modulate gene expression so affecting, for example, histone structure. The importance of this phenomenon is that RNA sequences can bind to complimentary DNA sequences even those that are slightly different but a little shorter. Such binding can regulate expression in a dominant way, while mutations in the RNA can have effects that are not obvious or canonical. Much of the work in this area has been conducted on *C. elegans* (for review, see Woodhouse & Ashe, 2020) and it will be interesting to see if such effects are common in other organisms. It will also be important to explore the details of how such inhibitory RNAs work and the effects of mutation and other changes to the system. The key test for this and for all other mechanisms of non-canonical mechanism of evolutionary change will of course be providing a mechanism whereby, in those cases where the modifying molecules are not expressed ubiquitously, the changes that they mediate in the genome can become intrinsic and so acquire long-term stability.

It should, however, be emphasised that relatively rapid evolutionary change can occur as a result of well-established, canonical mechanisms (Chapter 22). Possibilities include

1. When population numbers are small and selection pressures are low so that the effects of genetic drift are amplified. In such cases, variants rapidly flourish and an example is the formation of the flock of cichlid species (Fig. 22.1).
2. When selection is artificially high under laboratory conditions and only specific variants are permitted to reproduce (Waddington, 1953, 1961; Rice & Salt, 1990).
3. When so-called magic traits evolve: these increase both adaptation to an environment and reproductive likelihood and so show extreme fitness (Huber et al., 2007; Momigliano et al., 2017; Lamichhaney et al., 2018; Chapter 22).
4. When change builds on favourable mutations that are already present in the organism, particularly if selection pressures are strong. This is most clearly seen both in selective breeding and in developmental plasticity when a possible option becomes the normal phenotype.
5. When mutated genes are part of a complex network: because network outputs depend on many genes, it is reasonable to suppose that the same change in output can be driven by mutations in several of the genes within the network.

There is, as of now, little reason to doubt that the standard evolutionary theory is adequate to account for the very great majority of the changes that have taken place over evolutionary time and that are continuing to occur. It has, however, never proven sensible in science to suggest that anything in the biological world cannot happen and this is particularly true at the molecular and genomic level where our knowledge is still incomplete. It will be interesting to follow work in the area of transgenerational epigenetic inheritance and acquired adaptation and see whether it turns out to be important, an occasional evolutionary option or whether it is eventually to be one of those many scientific ideas that were inadequate, in this case because the changes turned out not to be stable enough to survive the slowness of selection under normal circumstances or the effects of genetic drift.

The origins of novel phenotypes

Perhaps the weakest aspect of our understanding of evolution today is the origin of phenotypic variation, which can affect anatomy, physiology, neurology, and behaviour in animals. Molecular genetics has clearly shown that such events are far too complex to explain on the basis of simple Mendelian genetics. Most traits are graded and are the output of a hierarchy of events, each taking place at a different level of scale (Fig. 26.1); these extend from activities taking place in the genome to selection based on environmental pressures acting on a particular population. Some levels are easier than others to understand: our knowledge of genomes and proteins, for example, is now quite deep but that of networks and their effects is still inadequate, mainly because of their internal sophistication. The situation is made even less tractable by the fact that there is a feedback between these levels. The full degree of the complexity of all these events has only been realized in the last twenty or so years.

The main difficulty in understanding how mutation leads to trait alteration is that there is no good theory of how normal traits form, events that mainly take place during embryogenesis (Appendix 7). If we do not understand the details of normal development, it can be hard to analyse change. A major problem here is causality which, because of feedback, is hard to follow. Upwards causality is relatively straightforward in that genes define proteins, proteins assemble in networks, networks produce functional outputs, and so on. Downwards causality is, however, equally important: the environment can change the phenotype through, for example, food availability, UV exposure, increasing levels of carbon dioxide and altered levels of predation: changes here affect lower levels of scale through feedback to organism behaviour and even to the activities of networks and the proteins that regulate gene expression. The net result is that causality is distributed and the relative significance of its various components is unpredictable.

The effect of mutation on the proteins that make up these rich systems is thus hard to fathom, unless its effect is to block an event (e.g. *Pax6* incapacitation) or point it in another direction. In almost all cases known, the effect of mutation is to decrease rather than increase fitness (e.g. the result is a diseased state). We do not know how mutational change leads to phenotypic change; this is because we do not understand the dynamics of protein networks. Advances in this area will probably come from developmental anatomy rather than studies of neurology or behaviour, if only because the mutant phenotype is easier to recognise. It would, however, reflect a triumph of hope over experience to expect solutions to these problems in the near future.

Neurobiology and behaviour

Our lack of understanding of the initiation and subsequent evolution in the neural control of physiology, of innate and acquired behaviour and of cognition are the largest gaps in our knowledge of evolution. This is because the subject is so complex and regulatory control so dispersed.

Origin of life

The elephant in the room for all discussions of evolution is, of course, the origin of life. We still do not really understand the details of the evolution of a primitive bacterium known as the first universal common ancestor (FUCA) that was able to reproduce itself, the key criterion for life. As discussed in Chapter 9, there are indications that this bacterium arose from an RNA-based world, that energy for its reactions came from electrical energy in the atmosphere and thermal energy from hot vents on the sea floor, that some at least of its basic organic constituents initially came from rocks originating from outside the Earth and that it formed in the vicinity of rock surfaces that acted as a catalyst for chemical reactions. The difficulty is that, for cells to evolve that are capable of division, many different capabilities each have to be in place and able to interact with one another. There is clearly much to come in this area.

THE EVOLUTIONARY FUTURE OF *HOMO SAPIENS*

More is known about the evolution of our species than of any other; the fossil, genetic, and artefact data from populations across the world have enabled us follow hominin anatomy and early artefacts from its beginnings ~7 Mya to now in ever more detail as we approach the recent past. The bush-like picture of human evolution provides a model system for the evolution of other species. Nevertheless, the early fossil evidence is still too thin to produce a cladogram of human evolution. More early fossil data and ancient DNA samples are needed to clarify human phylogeny and to give some insight into the molecular basis of the minor growth differences between human groups. Equally interesting would be the discovery of further artefacts, particularly very early ones as they would provide insight into one of the most interesting aspect of AMH evolution, that of how mental capacity increased.

Any discussion of the evolutionary future of *Homo sapiens* does however pose questions that no other species has ever faced. This is because humans dominate the world's ecosystem to an extent that is unique in the long history of Earth. Our population, currently around eight billion and estimated to reach a billion or so more towards the end of this century, runs the planet almost as its private garden, paying minimal attention to its other inhabitants and their needs except insofar as they can be beneficial or detrimental to us.

One gets a sense of this dominance from the estimates of Bar-On et al. (2018) on the biomass of the core phyla and major groups of organisms in gigatons of carbon (Table 26.1). The millions of arthropod species together provide just under 50% of the total animal mass – they are far and away the commonest animals. The single human species does, however, overshadow every other species: it alone provides ~2.5% of the total animal mass and requires almost twice as much weight in livestock for food. A measure of this imbalance among the vertebrates is that the total human mass is three times as much as all the other non-livestock mammals combined. This dominance raises two obvious evolutionary questions: are humans still evolving and are they destroying their own habitat?

Is the human species still evolving

This question was discussed in some detail at the end of the last chapter, and the conclusions were that, while some genetic mixing will occur as a result of interbreeding between groups, there is little chance of novel features accumulating as a result of

TABLE 26.1 GLOBAL ORGANISM MASSES IN GIGATONNES OF CARBON (GT C)

Phylum group	GT C	Animal taxa	GT C
Plants	4500	Arthropods	1.2
Bacteria	70	Fish	0.7
Fungi	12	Molluscs	0.2
Protists	4	Annelids	0.2
Animals	2.5	Cnidaria	0.1
Algae	<1	Livestock mammals	0.1
		Humans	0.06
		Nematodes	0.02
		Non-human and non-livestock mammals	0.007
		Birds	0.002

Source: From Bar-On et al. (2018)

mutation. Given the success of human reproduction today, it is hard to detect or even envisage a positive selection pressure that would lead to any reproductive advantage for a novel phenotype that occurred by chance. It is equally so for negative selection at the population level: even climate effects, such as high levels of UV, that might once have been a selection pressure on humans are now substantially diminished because we have buildings and clothes to protect us from any deleterious effects.

Genetic drift is equally unlikely to lead to the emergence of novel alleles and phenotypes. *Evolutionary population genetics* shows that the typical number of generations for a novel gene to survive the vagaries of genetic drift and become fixed in a population is ~$4N_e$ (the size of the breeding population; Chapter 21). Even for the very few small, isolated populations whose breeding population is perhaps 50 such as those that live in the Amazon Rain Forest (Salgado, 2021), this will take ~4000 years. The very great majority of humans live in far larger breeding pools and those few human populations that do meet such conditions are now either being lost or being absorbed into larger populations, so contributing to the wider gene pool. It is hard to draw any conclusion for humans other than that, for the first time in the history of life, the effects of evolution have been suspended for them.

The effect of humans on the planet

Modern medical science is helping the world's population to increase; this is partly because fewer people are dying due to disease, trauma, or childbirth and partly because people in richer countries are living longer. The resulting rise in the world's population is only slightly ameliorated by the fact that, as humans begin to control their own reproduction, they have fewer offspring. The world's population has tripled over the past 70 years and the current growth rate of ~1% a year is predicted only to halve by 2050 (see Websites). Even the COVID pandemic, for all its disastrous effects in many counties, has had only a trivial effect on the total number of humans. At the level of the individual, this diminution in selection pressures is wonderful: it enables many, many more people to lead long healthy reproductive lives than has ever before been possible. It is only at the level of the global population that the downside of the medical and other advances can be seen.

The demands of this population for more and better food and increasing amounts of energy are putting heavy pressures on the planet. For energy, much will depend on its source. Solar and wind energy will be greenhouse-gas neutral, but fossil-fuel-derived energy will continue to increase atmospheric carbon dioxide. Given the need for global transport and for heat in winter, it is simply unrealistic to hope that solar energy will make more than a secondary contribution to world energy consumption in the near future. It seems that further increase in greenhouse gas levels is now inevitable. This will be supplemented by the demand for meat, which is a symbol of prosperity in many cultures. Livestock also produce greenhouse gases, particularly methane and carbon dioxide, through flatulence, belching, and manure degradation. Further production of cattle will not only amplify the amounts of greenhouse gases and increase world temperatures but result in further loss of forest habitats for other animals. We are already seeing the devastating effects of climate change as the planet warms; it is not of course possible to predict what evolutionary changes will result in the long term under the new set of selection pressures.

The rises in temperature will be self-amplifying for three reasons. First, there will be an increased demand in summer for energy-consuming air conditioning. Second, as Antarctic ice melts, global warming will increase because less solar energy will be reflected back into space and more absorbed by sea and land (Wunderling et al., 2020). Third and equally serious, the resultant melting of the Sibderian permafrost could well lead to the release of the vast amounts of methane trapped under it, a gas whose greenhouse effects are far more serious than those of carbon dioxide (Walter et al., 2018). It is hard to see how these effects can be reversed.

Evolutionary biology brings a long-term perspective to the eventual consequences of climate change. Many extinctions that have occurred over the last ~2.5 By have mainly come about as, for one or another reason, world temperature altered. This has most frequently been due to changes to atmospheric CO_2 levels (Glikson, 2016). The earliest of these that we know about was the great oxygen extinction that occurred ~2.4 Bya: as a result of early cyanobacteria activity, oxygen levels rose and carbon dioxide levels dropped; this led to an extreme ice age that fortunately had little long-term effect on evolution because the lower levels of the seas were unaffected. Much later, cooling also occurred as a result of the asteroid collision with the earth

~63 Mya: the resulting atmospheric debris blocked sunshine, raised pollution levels dramatically and resulted in the K-T extinction.

Extinctions that have arisen because of global warming are more significant in the context of the current conditions, with the Permian extinction having been the most devastating in its effect on life. This extinction was almost certainly caused by massive volcanic eruptions that led to the burning of major coal beds; together, these events resulted in the release of some three trillion tons of carbon into the atmosphere (Grasby et al., 2011). It was not long before temperatures rose and these heating effects, probably compounded by the noxious gases in the atmosphere, led to the loss of much of marine life.

A more recent model for what is happening now is the Paleocene–Eocene Thermal Maximum (PETM), a period of ~200 Ky that occurred ~55.5 Mya following a period of perhaps 20 Ky during which CO_2 levels slowly rose by about 0.2 billion tons a year, for reasons that are still not clear. As a result, average global temperature rose by 5–8 C (Glikson, 2016), all polar ice melted, sea levels rose and there were major changes to animal life. Adaptation of life-forms to those changes was facilitated by the facts that the rate of temperature increase was slow and the Earth was not heavily populated by animals. A further ameliorating factor was that those animals unable to tolerate the heat had time to migrate to cooler regions. The thermal maximum was, however, marked by extinction of some marine organisms and changes in speciation, with many modern mammals first appearing soon after this period.

Almost all these changes took place slowly over many thousands of years. In contrast, events today are proceeding far more rapidly as the rate of increase in the amounts of CO_2 being pumped into the atmosphere now is far greater than then (Blunden & Arndt, 2018; World Meteorological Organization, 2020), with the current figure being ~3.5 billion tons a year (see Websites). The amount of carbon dioxide released into the biosphere since the beginning of the industrial revolution ~200 years ago is already causing world temperatures to slowly increase. As a result, global weather on land is becoming less stable as seen in stronger hurricanes, unpredictable droughts and terrifying numbers of major forest fires. The sea is slowly warming and its pH declining as it absorbs carbon dioxide. The joint effects of the expansion of water with temperature and the increasing volume as the Antarctic ice melts is leading to flooding and the loss of low-lying land. Already, fishes are migrating from central latitudes to cooler water (see Websites).

If carbon dioxide levels, human populations and the extent of environmental destruction all continue to rise, it is hard to come to any conclusion other than that human life will rapidly become much harder and less stable. We already have evidence of this as poverty, exacerbated by population pressures, wars and climate change, are today driving migration from impoverished hot countries to rich, cooler ones. The massive increase in world numbers over the last century, an increase that continues with everyone rightly wanting a better life for themselves and their children, is beginning to place impossible demands on world resources for energy, food, and land. In short, although the genetic future of *Homo sapiens* as a species is not yet under threat, one cannot be optimistic about its future happiness or stability.

FINALLY ...

When one looks at how well adapted any species is to its environment, it is easy to feel that they were made for one another (the so-call "perfection of nature"), but a far better metaphor is Dawkins' *Blind watchmaker* (Dawkins, 1996) where the predominant driver of success is trial and error and eventual success, with each species and all its variants continually being tested by selection, to destruction if necessary. This goes on until the original species eventually becomes extinct, its niche becoming occupied by fitter descendants or incoming organisms.

Although our knowledge of how species evolve is clear in principle, the specific details remain unknown for most of them. In contrast, we have a good macro-scale picture of what happened over the last ~3.8 By; this is the history of life. The distinct contributions made by cladistic analysis of the fossil record, phylogenetic, and coalescence analysis of homologous sequences and the application of molecular genetics to understanding common mechanisms underpinning the development of very different organisms have each illuminated every aspect of the history of life. It is a wonderful achievement and human society as a whole is richer for knowing both this broad history and its own ancient roots.

This book has explored the highways and many of the byways of evolutionary biology, making it clear that coming to grips with the full breadth of the subject is unfinished business. This is because the task of discovering the history of life and the mechanisms of change builds on the whole breadth of biological knowledge: evolution is both the pinnacle and the base of biology. We know a great deal, but not enough. The general picture of how evolution works has been clear for more than 70 years. Two generations ago, the new problems were about genetics, DNA, and mutation; for the last generation, the new areas were computational phylogenetics, coalescent theory, and evo-devo, tasks that are still keeping us busy.

For the next generation of evolutionary biologists, one can hazard a guess that, apart from the discovery of new fossils and the light that they cast, the focus will be on the effect of mutation on protein networks, perhaps the role of epigenetics on mutation and about what coalescence theory has to say about ancient populations. Questions about evolutionary neurobiology and the origin of life are likely to be for the longer term. The former will probably require a far better understanding of neurophysiology than we have today, while the latter will require experimental genius. All of this is based on what we know now. Who can foresee what perceptions and approaches future research successes in biology and the other sciences will bring to the study of evolution? There is more to be done – there always is.

WEBSITES

- Wikipedia articles on *Climate Change, Climate Change, Fisheries.*
- World population data: www.worldometers.info/world-population/#table-historical
- World CO_2 emissions: www.worldometers.info/co2-emissions/co2-emissions-by-year/

APPENDICES

The following eight appendices serve several purposes. Some are to provide important background to readers whose knowledge may be thin in one or another area (those on systems biology, model systems, and developmental biology). Some are intended to fill in detail for interested readers (those on rocks and fossils, on the history of the Earth and on phylogenetics). The last two (on the history of evolutionary thinking and on creationist approaches) are there because of their general interest. The appendix on history shows how the subject slowly arose through attempts to make sense of the world on the basis of evidence rather than theology. The appendix on creationism is in two parts. The first gives the creationist viewpoint and seeks to provide answers to questions about evolution. The second gives the scientific viewpoint and replies to questions from a creation viewpoint.

Systems Biology

Systems biology is an approach to biology that focuses on two difficult areas. The first is understanding and modeling protein networks. These are interacting sets of proteins whose communal outputs are responsible for much of development (e.g. growth, patterning, and differentiation) and physiology (e.g. cardiac rhythms and neural activity). Mutations that affect these networks and their outputs are major causes of variation within populations and hence can be drivers of evolutionary change. The second aspect focuses on biological complexity. Evolutionary change results from the interaction that extend in scale from mutations in genes to selection pressures from the environment. To achieve a variant, the effect of a DNA mutation has to work its way up several levels of scale, through proteins, networks, tissues, and organs up to the organism. Working through the details can be complicated beyond measure. Fortunately, such complexity can at least be made explicit through the use of mathematical graphs, the natural language of systems biology.

Physicists have always appreciated that their subject involves two very different sorts of approach. First, events at a particular level of scale have to be understood and formalized; and second, the interactions between the different levels have to be clarified. The understanding of the effect of heat and energy on gases provides a good example: at the everyday level, there are laws, such as those of Boyle and Charles that describe the effect of heat and pressure on gas volume; underneath these are the laws of thermodynamics that use the macroscopic parameters of energy, temperature, and entropy to provide a theoretical framework for the behavior of gases. Underpinning all of this is the thermal motion of gas molecules, which is described by the laws of statistical mechanics that, in turn, provide a theoretical basis for thermodynamics and hence for Boyle's and Charles' laws.

Physicists are fortunate here. First, they often deal with systems that turn out to obey relatively simple laws, provided that the distances being considered are greater than those for which quantum mechanics is needed. Second, there is a large body of applied mathematics that can be used to help work out the numerical implications of these laws under specific conditions. Biologists are less privileged. Mathematical descriptions are restricted to particular models that operate in very few areas (e.g. population genetics, biomechanics, and some areas of physiology), and even then, the treatments are rarely simple. Moreover, there is only one single overarching theory in biology and that is the theory of evolutionary change, a subject that does not lend itself to mathematical formalization. This is partly because it represents the results of independent events at many levels and is thus intrinsically complex, and partly because evolutionary change depends on random events, such as mutation at the genomic and unpredictable effects on selection that reflect events in an organism's environment.

Over the last decade or so, biology has had another layer of complexity added to an already complicated picture: it has become clear that many proteins do not operate alone but are part of complex networks that have a functional output, such as initiating growth, that could never have been predicted on the basis of the basis of the properties of individual protein constituents or the structure of the controlling networks. Systems biology is partly an attempt to understand how these protein networks operate (the narrow view) and partly an attempt to provide a framework for integrating the complexity of events that take place at different levels of scale (the broader view).

THE NARROW VIEW

This aspect of systems biology is primarily interested in understanding how molecular networks and pathways work. Over the last couple of decades, it has become clear that as many as a hundred or so proteins can cooperate within a single network, often activated by a signal protein binding to a receptor, which produces a functional output. Well-known examples include signal transduction pathways: these ensure that an external signal protein that binds to a membrane-bound receptor eventually results in a downstream protein entering the nucleus and activating a transcription complex. Such pathways can be surprisingly complex and can include up to a hundred or so proteins (e.g. the epidermal growth factor (EGF) pathway, Fig. A1.1). Other examples of networks include the rho-GTPase protein network that controls cytoskeletal behaviour and hence much of morphogenesis

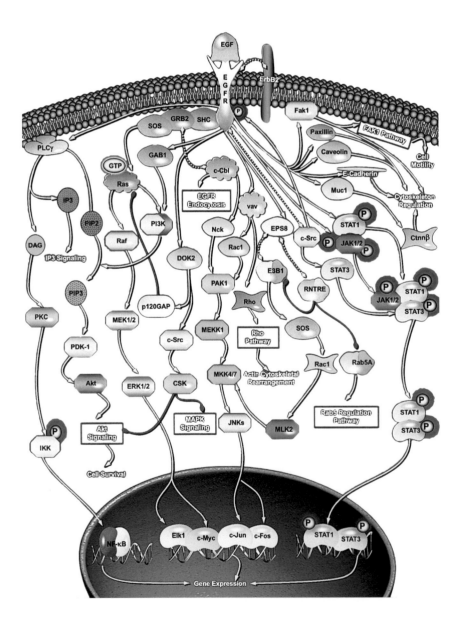

Figure A1.1 **Key features of the mouse endothelial growth factor (EGF) signaling network.** EGF signaling is widely used across mammalian embryos, particularly in activating both proliferation pathways and context-dependent differentiation. The whole network contains >80 proteins arranged in many subnetworks, the function of most of which are unknown.

(From https://proteinlounge.com/pathway. php. Courtesy of ProteinLounge.)

(Fig. 1.1), the vascular endodermal growth factor pathway that regulates the formation of new endothelial blood capillaries (see below), and the Krebs cycle that generates adenosine triphosphate (see Websites).

What is immediately noticeable about these pathways and networks is that it only possible to elucidate the specific roles of individual proteins in relatively simple cases, such as that for Notch-Delta signaling (Fig. 8.6; McIntyre et al., 2020). It is generally impossible to deduce both a particular protein's activity from its immediate interactions with other proteins, unless its core function is already known (e.g. it is a kinase), and its precise contribution to the functional output of the network. This is because the output activity is the end result of the interactions among the many proteins that cooperate within a single network or pathway.

Understanding how such complex pathways work qualitatively and quantitatively underpins what can be seen as the *narrow view of systems biology*. Making numerical sense of these protein networks is particularly difficult for several reasons. First, their normal mathematical formulations as sets of differential equations can only be used in cases where the law of mass action applies. Unfortunately, the concentrations of the participating proteins and substrates in cells are often so low that the law of mass action has to be replaced using stochastic methods (Kirkilionis, 2010). Second, such equations can only be readily solved near equilibrium, a condition that may not be obeyed. Third, the rate constants for the equations are rarely, if ever, known. Such networks are thus very hard to model computationally.

The situation is made more complicated when one considers the effect of mutation on the outputs of networks, unless its activity as a whole has been modified by a mutation that rendered the signal or receptor incompetent. Some insight here can be obtained by studying the effects of targeted mutations on their constituent proteins, but there is a long way to go before we can elucidate the underlying principles of network dynamics or predict the effect of mutations on specific networks. One does not, however, need to understand the fine detail of their roles in development and physiology to see the evolutionary significance of mutations that affect network outputs.

THE BROADER VIEW

This second aspect of systems biology comes from the fact that nothing in an organism is isolated; any activity involves other components at the same level of scale and usually at levels above and below it; these may extend from the genome to the environment (Table A1.1). Indeed, it is impossible to understand any rich aspect of

TABLE A1.1 EVOLUTIONARY EFFECTS OF DNA MUTATION AT DIFFERENT SCALE LEVELS

Population levels	Scientific domain	Evolutionary significance
Ecological environment	Ecology and geography	Affects the general context for selection
Social behavior	Ethology	Affects sexual selection
Phenotype frequencies	Population genetics	Changes in frequencies
Individual levels		
Whole-body structures	Anatomy	Creates behavioral, functional, and anatomical variants
Organ function	Physiology	Variants available for selection
Embryogenesis	Development	Variants available for selection
Cell and organelle function	Cell biology	Generates anatomical and other variants
Protein networks	Systems biology	Generates changes to network outputs that work upwards to change the phenotype
Proteins and other molecules	Molecular biology and biochemistry	Changes in protein function
DNA mutation	Genomics	Leads to DNA sequence variants

biology without an appreciation of this. The situation is made even more complex because there are often direct feedback interactions between levels that can be both downwards and upwards. The following three key themes underlie this broader view of systems biology.

Events within each level are complex

Interaction within each level are rarely simple (Table A1.1). Gene expression, for example, is modulated by activators, repressors and methylation, while short- and long-term weather changes can affect the local ecological system in arbitrary ways. There is a further problem: while the activities at one level depend on events at a lower level and must, of course, be compatible with them, they cannot be predicted from them. Thus, the rate constants that govern protein dynamics cannot be derived from the DNA coding sequence nor can the substrates that interact with enzymes be predicted on the basis of their primary sequences. Similarly, the output of a network cannot be predicted on the basis of its constituent proteins.

There are interactions between levels

There are always upwards and downwards feedback effects between levels. The former occurs when, for example, a simple signaling molecule activates a complex protein network that leads to changes in expression and subsequent change in, for example, morphogenesis. Other examples of upwards control are the effects of neuronal networks on behavior and population numbers on the greater environment. An example of the downwards control is the determination of alligator sex: this depends on the ambient temperature when the egg is developing—males have a higher chance of developing if the temperature is above 32°C while females will develop if the temperature is below. What needs to be remembered is that, when analysing any complicated event, interactions within and between several levels may need to be taken into consideration.

Causality is distributed

The combination of complexity within levels and feedback both within and across them means that it can be difficult to work out exactly where causality lies. In the case of alligator sex determination, the immediate or *proximate* cause is ambient temperature, but there are other causes. The temperature-sensitive system presumably evolved because there was a selective advantage in having more males at higher temperatures, while the development of the early reproductive system depends on protein activity and novel gene expression. In general, events at many levels contribute to change, and the complexity of the feedback systems means that causality is distributed within and between levels; as Noble (2012) put it: "no level is privileged." It is almost always true that, where a complex event is claimed to result from a single cause, that explanation is an oversimplification.

Within the context of evolution, one needs to know how mutational change affects the whole organism; in particular, what are the immediate and long-term implications of heritable mutations on organisms as they work their way up from the genome, through protein and networks, to affect the phenotype. It should be emphasized that the immediate effects of most mutations are often very minor or even unnoticeable — the most common results of knocking out a gene from the mouse is that the loss has no effect on the phenotype. In the longer term, however, it is the cumulative effect of mutations that is responsible for the variation seen across a population of organisms. Very little is, however, known about how all this happens, and this point is discussed further in Chapter 18.

In short, complexity lies at the heart of life, and understanding the biosphere requires us to grapple with and find ways of unpicking that complexity. Many areas of biology have always known this to be so; important examples from traditional areas include the complex physiological models of heart activity and nerve conduction. Although systems biology is considered to be a very modern area, it actually has deep and old roots.

A History of Evolutionary Thought

There is a view that history is irrelevant to science because today's knowledge makes yesterday's approaches redundant. It is, however, difficult to appreciate the complexity of modern views of evolution without a sense of the historical context in which they developed. It is also important to pay tribute to the many serious scientists who fought their contemporary orthodoxy to try to make sense of the story of life, and on whose shoulders we all stand. This appendix provides a brief summary of the history of evolutionary science from the creation myths to the molecular era. The reader who wants to know more of the early work should explore the references at the end of the appendix, particularly Mayr (1982) and Browne (1995).

Until relatively recently, the world of life seemed static: the Sun rose in the east every morning, and the stars, moon, and planets appeared to circulate around us in a predictable way. Even when Galileo showed that the Earth rotated around the Sun, there was no reason to let these new views of the solar system affect ideas of natural history. The animals and plants of one's grandchildren were the same as the animals and plants of one's grandparents, as far back as history went. The very idea of evolution seemed wrong for two reasons, both of which are articulated in the first chapters of Genesis. First, animals were each perfectly suited to their environments, all of which were populated, so there was no need for new creatures. Second, animals could only breed with members of their own kind, so new species could only be formed by *de novo* creation. The few who did think about such things realized that either the living world had been there forever or that, in the dim and distant past, it had been created. If the former, then there was nothing to be said; if the latter, then some great creative force had done it and the problems were who or what and how. The thoughts here gave rise to the many wonderful creation myths (see Websites at the end of the appendix).

The best-known of these myths, in the Western world at least, are the two at the beginning of Genesis. The first (Genesis Chapter 1 and Chapter 2, verses 1–4) has God creating the universe in 6 days, starting with the physical world, then the living world, and, on the last day, Adam and Eve. All this work was followed by a day of well-deserved rest, the Sabbath. Modern textual analysis suggests that this story was not actually written by Moses in the 13th century BCE, but by one or more Hebrew priests around the 6th century BCE; it is still hard not to be overwhelmed by its wonderful sonorous phrases ("And it was the evening and the morning of the first day...").

The second version of the creation (the rest of Genesis Chapter 2) is much more mundane in tone: first, the world and man were created, then the other living organisms, male and female, and, finally, and because there was no creature for man to mate with, Eve was created through God taking one of Adam's ribs and refashioning it as a woman. This version was probably written some 300 years earlier in Judea and includes the story of the Garden of Eden and the recognition of good and evil (Friedman, 1987, and website on the documentary hypothesis). Some rabbis contend that the two versions are a single story, with the former concerned with the physical and the latter with the spiritual world. The tones and the language of the two stories, not to mention the timings, are so very different, however, that one can only feel that this has to be seen as a sterling attempt to reconcile the irreconcilable—unless, of course, both are regarded as fables.

In an odd way, we can feel that those who invented these and the many other creation stories were among the first scientists, looking for explanations of the complexities of the world. They were not, of course, real scientists because they never thought it necessary to test their ideas or to work through their implications. Getting to this stage was to take some thousands of years, and it is only in the last 200 or so that we have begun to use an evidence-based approach to questions about where and how life arose, whether each of the species had always been as it is now or whether change had occurred and, particularly, how ideas might be tested.

What follows is a summary of how we reached today's understanding of evolution, and it is a history of the work of many biologists and philosophers, some of whom had to argue against those who believed that their own traditional myths were literal accounts of creation, irrespective of any facts to the contrary. Even today, when the evidence for evolution is overwhelming, there are still those who would prefer to believe ancient religious narratives. Appendix 8 considers the arguments for these two views.

THE EARLY DAYS

Until about the 18th century, the very great majority of the people in the Western world accepted the biblical view that God created the world. Over the preceding centuries, a few people had tried to make sense of the living world, mainly within this theistic framework. The earliest biologist about whom much is known is Aristotle (died 322 BCE). However, while he collected a great deal of biological knowledge and tried to systematize it and make sense of it, it seems unlikely that he ever considered that species were anything other than fixed. Similar views were held by his successor, Theophrastus (died 287 BCE) who wrote the *Historia Plantarum* (a 10 volumes of the *History of Plants*) that was used as a resource for many centuries. The one early thinker to suggest that the biological world could change was Titus Lucretius Carus (died 55 BCE): in a six-part poem entitled *De Rerum Natura* (*On the Nature of Things*). He suggested (part five) that living things had a naturalistic rather than a divine, origin. Thus, in Leonard's translation (2008):

> How merited is that adopted name
> Of earth – "The Mother!" – since from out the earth
> Are all begotten.

Although various people interested in biology made comments on the possibility of evolution over the early years, the first substantial writings seem to have been by the Arabic writer Abu Al-Jahiz (known as "the goggle-eyed" because of an eye malformation) who lived in late 8th and 9th century Baghdad and who was one of the very few who read the living world correctly (Bayrakdar, 1983). In his *Book of Animals*, it seems that he wrote something along the lines of:

> Animals engage in a struggle for existence for resources, to avoid being eaten and to breed. Environmental factors influence organisms to develop new characteristics to ensure survival, thus transforming into new species. Animals that survive to breed can pass on their successful characteristics to offspring.

The accuracy of this translation and the extent to which Al-Jahiz had such evolutionary ideas has recently been doubted (see Websites at the end of this appendix and Stott, 2012), while it is clear that Al-Jahiz thought that all changes happened within a theistic context. Nevertheless, his ideas certainly became known in the following centuries, as did the suggestion of Ibn Khaldun (14th century) that man derived from monkeys.

For the next five centuries, and in order to avoid religious controversies, such ideas about evolution were rarely discussed outside of the very small domain of academic medicine. Even George-Louis Buffon (1707–1788), the first major systematic biologist and director of the King's Garden (Jardin du Roi) in Paris, was careful not to publish his real views until he was very old (1778). His *Histoire Naturelle, générale et particulière*

(39 volumes, 1749– 1789) was the first comprehensive discussion of biology and environment, and much of it was attacked by the church. Buffon's response, although he himself was unsure about evolution, was to apologize but to retract not a word! The next few generations of biologists were brought up on these texts.

It was perhaps the unlikely figure of a Swiss lawyer, who, in the 1770s, set the scene for modern thinking about evolution. Charles Bonnet (1720–1793) was, however, not only a lawyer, but also a considerable experimental biologist who discovered parthenogenesis and plant and insect respiration, while also studying regeneration and mental disorders. His most interesting evolutionary idea was that man was the culmination of a progression of organisms up the 52 steps of the ladder of being. This long-established picture of life's complexity started with matter and progressed through plants, worms, insects, shellfish, fish, birds, and so on to its ultimate triumph, man. The idea that there was such a hierarchy of living organisms is one that went back to Plato and perhaps to the first version of creation in the bible, but it seems to have been Bonnet who suggested that progression was possible. He also believed in preformationism, the idea that the sperm contained a small homunculus that could develop, once it met the fertile environment of the egg, much as seed developed when planted in Earth. Bonnet got a lot wrong, but he made evolution a concept that serious people could consider.

In the last decades of the 18th century, ideas of biological evolution were clearly in the air, for all that the church tried to stop them. Notable authors who discussed evolution were the philosopher Denis Diderot (1713–84), who published an encyclopedia in France that included chapters on geology and evolution, James Burnett (Lord Monboddo, 1714–1799) in Scotland, and Erasmus Darwin (1731–1802) in England. The last, a country doctor and Charles Darwin's grandfather, was also a serious naturalist who held original evolutionary views, especially on sexual selection. However, he remained publicly silent for many years until 1796, when he completed *Zoonomia*. This included the substantive sentence:

> *Would it be too bold to imagine, that in the great length of time since the earth began to exist, perhaps millions of ages before the commencement of the history of mankind, would it be too bold to imagine, that all warm-blooded animals have arisen from one living filament, which THE GREAT FIRST CAUSE endued with animality, with the power of acquiring new parts, attended with new propensities, directed by irritations, sensations, volitions, and associations; and thus possessing the faculty of continuing to improve by its own inherent activity, and of delivering down those improvements by generation to its posterity, world without end!*

His final work on evolution, a poem entitled *The Origin of Society*, was published posthumously.

Although these authors all came to the conclusion that evolution had happened and each is historically important and interesting, none of them has anything scientific to say to us today; this is mainly because they never tested or properly worked through their ideas. Their significance today is that they were at least prepared to think about rational as opposed to theistic explanations of life.

THE MOVE TO EVOLUTIONARY THINKING

The turn of the 19th century was a pivotal time in the history of evolution and, before discussing what then happened, it is worth standing back and looking at the wider context. Four factors were pushing progress, two biological, one geological, and one philosophical. The first of the biological drivers was fossils. These had always seemed anomalous, particularly the presence of the shells of marine organisms a long way from the sea. They had been commented on as far back as Xenophanes (c. 570–478 BCE) and Herodotus (484–425 BCE) in Greece, and Pliny the Elder (23–79 CE) in Rome. The standard explanation, promulgated by theologians for many centuries, had always been that they had been organisms that died out, particularly during Noah's flood; fossils were just their remains.

Stott (2012) discusses how the problem of such fossils had worried Leonardo da Vinci (1452–1519), and it was in 1508 that he noted fossil oyster shells present on the mountains around Milan and started to analyze them. He showed that the groups of shells included both young and old individuals and that they were arranged much as they were found when alive. He concluded that they could neither have been washed there nor moved there on their own and correctly deduced that, at some time in the dim and distant past, the Apennine Mountain range had been under water. It was not only shells that were problematic; fossilised bones had been noted as curiosities as early as 2000 years ago in China. Examples were being found in Europe, and, by 1665, the subject of fossils was of sufficient interest for Robert Hooke to give a talk on them to the Royal Society in London. By the 18th century, bones much larger than those of living animals had been discovered in the Ohio valley and fossil collecting was becoming not only fashionable but of serious geological interest.

The second biological driver was the appreciation of just how well-adapted organisms were to the habitats in which they lived. Here, the 39 volumes of Buffon's *Histoire Naturelle, générale et particulière* published over the period 1748–89 were important because they included so much detail about so many animals. Carl Linnaeus (1707–1778), his almost exact contemporary, was equally significant, because his systemization of some thousands of plants and animals was becoming a standard part of education. The question of why closely related animals were so well adapted to their niches was one that could sensibly be asked for the first time. The idea that God designed each animal for its correct place (the perfection of nature) was beginning to seem less credible.

The third was geology: the subject, as we know it today, started at the end of the 16th century with the geological investigation of rock strata and the religious impetus to find evidence for Noah's flood. The former proved more useful to science and, by the middle of the 18th century, was being popularized by Buffon in his *Histoire* and by Diderot in his *Encyclopédie*. The key person, however, was James Hutton (1726–1797), who worked out a great deal about layering and geological forces. He particularly suggested that not only was the Earth very old, far older than Bishop Ussher's then-accepted biblical creation date of 4004 BCE, but also in a continuous state of flux. One reason for his taking this view was his discovery of ancient fish bones buried in the Salisbury Crags cliff in the south of Edinburgh. By the 1790s, geologists, particularly William Smith in England, had realized that series of rock strata could be identified on the basis of the fossils that they contained, and a sense of the progression of organism complexity was becoming apparent. It seems odd today that the early geologists seem not to have taken a major part in the many discussions over the next century as to how this progression occurred.

The final driver was philosophical: the late seventeenth and eighteenth centuries had been the Age of Enlightenment when intellectuals set out to reform society on the basis of reason. A key theme was to throw out ideas grounded in faith and tradition and start again. This was the environment in which the great physicists Isaac Newton and Gottfried Leibniz worked; it was impossible to be what we would now call a scientist and be untouched by these wider ideas. It was in this general context of inquiry that Bonnet, Erasmus Darwin, and a few others were expanding the boundaries of biological thinking. Not everyone, however, was at ease with such changes and the tensions were particularly apparent in the Museum of Natural History in Paris around 1800 and the following few years, when three major figures of biology fought out their ideas about whether evolution had happened.

THE EARLY 19TH CENTURY

In 1793, the revolutionary government in Paris enlarged the Jardin du Roi to be the Museum of Natural History and appointed 12 professors, included among whom were Étienne Geoffroy Saint-Hilaire, George Cuvier, and Jean-Baptiste Lamarck. The fine detail of the natural world had only recently begun to be studied academically and each of these three focused, almost obsessively, on a particular area. Étienne Geoffroy Saint-Hilaire (1772–1844) was the professor of zoology and an embryologist, paleontologist, and anatomist. His work on the diversity of vertebrates led him to the view that there was an underlying structural plan to animals. He was the first person to think seriously about anatomical similarities among unrelated organisms, such as the wings of bats and birds. He explored these ideas in the context of ontogeny (a term used in various ways but here meaning *the development of form*) and came to the view that evolution and species change (he used Lamarck's term, *transmutation*) probably occurred but involved major jumps. He was also of the view that structure determined function, unlike Cuvier, who took the view that it was the needs of function that determined structure.

Georges Cuvier (1769–1832) was the professor of natural history, and perhaps the greatest comparative anatomist and paleontologist of his day; his wonderful collection of tissues and fossils is maintained virtually unchanged today as part of the Gallery of Paleontology and Comparative Anatomy at the Jardin des Plantes in Paris, a place that remains a joy for any biologist to visit. Cuvier is known today partly for his work on taxonomy, but particularly for his discovery of extinctions. For this, he carefully analyzed the fossilised organisms in adjacent layers of rocks and found that the organisms in one layer often bore no resemblance to those in the layers above or below it (for instance, land animals could be covered by sea animals). He deduced that extinctions had actually occurred and that these had been followed by new bouts of creation, and hence that there was no evidence to support the idea of evolution. Although Cuvier was religious, he saw himself as a serious man of science whose views were grounded in knowledge; he, therefore, did not argue against evolution on grounds of faith, but on grounds of data; the trouble was that he had too little data and what he had, he misunderstood.

Lamarck: The first evolutionary scientist

Jean-Baptiste Lamarck (1744–1829), the third of the trio, was originally a soldier and then a botanist, but was made the professor of insects, worms, and microscopic animals (or invertebrates—a term that he coined) at the Museum (Burkhardt, 2013). Lamarck's fame comes from his work on insect classifications and on evolution, with his views here being shaped over the period 1795–1809 and finally published in his *Philosophie zoologique* (1809). Gould (2000), in an article that is required reading for anyone interested in the history of evolutionary thought, has pointed out that the early chapters of this book see evolution as organisms climbing Bonnet's ladder of life. The picture in the appendix of the book is, however, very different: Lamarck had by then realized that, on the basis of anatomical geometry, earthworms and parasitic worms could not be in the same class since the anatomy of the former not only had a coelomic cavity but was very much more complex than that of the latter. It became clear to him that the only acceptable explanation of this problem, and, by extension, other types of anatomical differences, was that evolution must have occurred not by climbing a ladder, but by descent and branching (Fig. A2.1). Lamarck was thus the first person to put forward the idea of branching descent on the basis of evidence, and this is the foundation of all modern thinking on evolution.

Lamarck also grappled with the mechanisms by which anatomical change might happen. This was a problem that Darwin was not able to solve, merely commenting in Chapter 5 of *On the Origin of Species* that "Whatever the cause may be of each slight difference in the offspring from their parents – and a cause for each must exist…" Lamarck, as Gould (2000) has pointed out, came to the view that organisms had two abilities, to complexify and to adapt, and that the resulting changes were heritable, although he did not express his ideas very clearly. Lamarck's views were accepted by everyone, including Darwin (Chapter 12), until August Weissman showed in the 1880s that they were incompatible with his theory of the continuity of the germplasm.

Lamarck today has a general reputation completely at odds with the importance of his work and the reason is that he antagonized Cuvier, who believed in change through successive rounds of extinction and creation rather than heritable adaptation. After Lamarck died, Cuvier wrote an excoriating obituary in which he accused Lamarck of being a theoretician who went far beyond the data in his views on evolution (see Websites). Gould (2000) quotes a particularly damning paragraph from this obituary.

> *These [evolutionary] principles once admitted, it will easily be perceived that nothing is wanting but time and circumstances to enable a monad [single-celled organism] or a polypus [sea anemone] gradually and indifferently to transform themselves into a frog, a stork, or an elephant…A system established on such foundations may amuse the imagination of a poet: a metaphysician may derive from it an entirely new series of systems; but it cannot for a moment bear the examination of anyone who has dissected a hand…or even a feather.*

Even at the time, the obituary was regarded as unfair because Lamarck was well known as a practical and distinguished biologist, but Cuvier's views have prevailed

Lamarck's last linear order of nature

from *Philosphie zoologiques* (1809)

1. Les Mammifères ⎤
2. Les Oiseaux   ⎟ Animaux
3. Les Reptiles   ⎟ vertébrés
4. Les Poissons   ⎦

5. Les Mollusques ⎤
6. Les Cirrhepedes ⎟
7. Les Annelides   ⎟
8. Les Crustacés   ⎟ Animaux
9. Les Insectes   ⎟ invertébrés
10. Les Vers   ⎟
11. Les Polypes   ⎟
12. Les Raiaires   ⎟
13. Les Infusoires ⎦

Lamarck's first depiction of a branching model for the history of life

from the appendix to *Philosphie zoologiques* (1809)

Additions

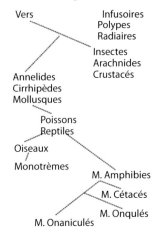

TABLEAU

Servent *à montrer l'origine des differents animaux*

Vers   Infusoires
              Polypes
              Radiaires

              Insectes
              Arachnides
              Crustacés

Annelides
Cirrhipèdes
Mollusques

      Poissons
      Reptiles

Oiseaux

Monotrèmes

            M. Amphibies

              M. Cétacés

                M. Onqulés

    M. Onaniculés

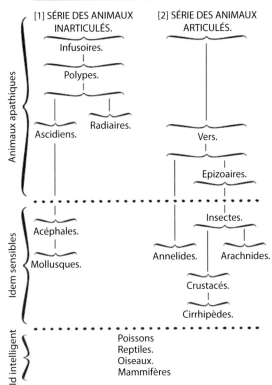

ORDRE *présumé de la formation des Animaux, offrant 2 séries séparées, subrameuses* (1815)

[1] SÉRIE DES ANIMAUX INARTICULÉS.

[2] SÉRIE DES ANIMAUX ARTICULÉS.

Infusoires.

Polypes.

Ascidiens.   Radiaires.

                        Vers.

                        Epizoaires.

Acéphales.

Mollusques.   Annelides.   Arachnides.

                        Insectes.

                        Crustacés.

                        Cirrhipèdes.

Poissons
Reptiles.
Oiseaux.
Mammifères

Animaux apathiques

Idem sensibles

Id intelligent

Figure A2.1 **Lamarck's three successive views of phylogeny.** The first two are from the text and the appendix of his Philosophie zoologique (1809). The third was published in 1815 (See Gould, 2000).

in spite of the writings of everyone who has looked into the history more carefully. Today, after some 200 years, the work of Gould and others is restoring Lamarck's position as the first serious evolutionary biologist and as the key forerunner of Darwin, as Darwin himself eventually realized.

The other person in the first half of the 19th century who laid the groundwork for modern ideas of evolution was the Estonian, Karl Ernst von Baer (1792–1876; Fig. A2.2), who was a major figure in early 19th-century biology, and one for whom Darwin had considerable respect. Although his work ranged across the sciences, it is

Figure A2.2 **A picture of Karl Ernst von Baer on an old Estonian 2 kroon banknote.** von Baer is probably the only embryologist to have been honored in this way.

as an embryologist that he is remembered today. Most of his work was technical: he discovered the blastula, the notochord, the human egg and, with Christian Pander, the triploid structure of the gastrula. von Baer is important here because his laws, based on the examination of a great variety of embryos, provided the groundwork for integrating embryology and evolution, in spite of the fact that he never properly accepted evolution. His two key ideas were: (i) that the general features of embryos within a group of species emerge before specific ones and (ii) that embryos of "higher" forms do not resemble the adults of "lower" ones, only their embryos.

THE ERA OF DARWIN

Although evolutionary thinking progressed little in the 50 or so years between Lamarck and the Darwin/Wallace era, it is clear that the idea of evolution was being widely discussed during this period. Robert Grant in England explored homologies, while Robert Chambers, a self-educated publisher, wrote *Vestiges of the Natural History of Creation,* which he published anonymously in 1844. This book, which proposed that the solar system and life had both evolved, was immediately controversial. While the public bought it in droves, churchmen and academics damned the book on grounds of religion and science, the latter because it went far beyond the evidence then available without producing any mechanisms as to how evolution had happened. This risk of such public notoriety was the key reason why Darwin held back from publishing his ideas on evolution for so long. It is interesting to note that, when Darwin first read *Vestiges*, he was stunned to discover that Chambers' analysis of evolutionary change matched his own (Browne, 1995), although Chambers' contribution to evolutionary thought seems to have been mainly forgotten. It is noteworthy that, in the first edition of *The Origin of species*, Darwin only once mentions the author of *Vestiges* and then to mock him, while he does not cite Lamarck at all.

The crucial publication that eventually provided the key to understanding evolutionary change was Thomas Malthus' *An Essay on the Principle of Population,* which had actually been published some decades earlier (the most important version was printed in 1803). This argued that human populations tended to increase exponentially and were kept in check by natural constraints, such as sickness, famine, and war. The inverse of this idea was taken up by both Wallace and Darwin: that a subpopulation of natural variants could escape these population constraints and flourish as a new species if it found an environment for which it was suited.

Alfred Russell Wallace (1823–1913) was a self-taught naturalist from a very modest background who became the first person to explore the distribution of animal species within their geographic environment while supporting himself by collecting tropical animals to be sold to collectors and museums in England. Wallace essentially invented the science of biogeography and is still remembered for identifying what is now known as Wallace's line, the boundary marking the western limit of Australasian fauna. Already convinced that evolution was a reality, Wallace became interested in why related species lived in different environments. The crucial insight came to him while he was recovering from a nasty bout of fever (probably malaria) on Ternate, a remote eastern Indonesian island adjacent to the larger island of Halmahera: he realized that Malthus' essay held the key to understanding how fitness determined why one species would flourish in a particular environment while another would not.

He wrote up his ideas in 1858 and sent the paper to Charles Darwin (1809–1882), the most distinguished biologist of the time, who had been quietly working on evolution for more than twenty years and was shocked to think that he might lose the credit for all this effort. On the advice of his friends Charles Lyell and Joseph Hooker, Darwin abstracted a manuscript on the same theme that he had originally written in 1847 but not published and gave both articles to Lyell and Hooker, who sent them to the Linnaean Society, and they were published together (Darwin, 1858 Wallace, 1858). In one of the great miss-readings of history, Thomas Bell, then the president of the society, mentioned in his 1859 presidential address that the previous year had not been marked by any of those striking discoveries that at once revolutionize the department of science on which they bear (Guerrero, 2008).

Although Wallace was a remarkable scientist and deserves some of the credit for discovering natural selection, he rightly has a secondary role in the story. One reason is on grounds of priority: although Wallace had been thinking about speciation for some years, his insight came in around 1858; Darwin had started thinking about speciation towards the end of his 5-year voyage around the world in the *Beagle* in 1836; his notes include comments on the island-specific features of giant tortoises (Fig. 4.3b; Browne, 1995). Darwin started his first notebook on evolution in 1837 (according to his autobiography), a year or so after he had returned. Well before 1844, when *Vestiges* was published, Darwin had reached Wallace's (1858) conclusion; indeed, all of Darwin's colleagues knew his views on evolution well before Wallace wrote his letter.

The second reason is because of the depth of their respective research. Wallace's analysis derived from his knowledge of natural history and his academic interest in speciation. Darwin, however, had explored the evidence for evolution across the whole of biology because he realized that to make the public case for evolution, he had to produce broad-based and convincing arguments. In particular, he had to show that variation was innate and commonplace. It was no coincidence that the first chapter of *On the Origin of Species* (1859) included information about breeding pigeon varieties (book cover; Fig. 4.3a).

The book itself focuses on a single question: what are the implications of the variation that any observer can see within a species? The answer, buttressed by a wide range of examples from across the then biological spectrum, starts with the fact that, left to themselves, populations grow exponentially but that this growth is constrained by the environment. Where variants arise, these may disappear or flourish (and may even replace their parent population) as selection pressures from that environment dictate (the *struggle for existence*). Such Malthusian competition (*natural selection* or *survival of the fittest*) acting on the nonuniform population leads to *descent with modification* (which is, of course, evolution), and these two ideas, which underpin the book, are the key to understanding the origin of new species.

It is also worth noting what is not in the book: although it is obviously about evolution, the term is barely mentioned in any explicit way. One reason was that the word then had an embryological meaning, while a second was that he did not want to upset his wife, who held traditional Christian views. Rather than discuss issues that were obviously contentious, Darwin focused on the ideas and implications of branching descent and hence on the mechanisms by which new species could arise from existing ones. The details of these, as Mayr (1982) points out, had not previously been discussed in any substantive way, other than by Lamarck who had not really produced anything that could be viewed as a mechanism.

Perhaps surprisingly, Darwin's relatively brief discussion of the past in *On the Origin of Species* is postponed until Chapter 10 (on the imperfections in the geological record) and Chapter 11 (on the fossil record). Here, the information is used within the context of the theory rather than to assert that evolution had been acting since life had originated. Darwin clearly understood the evolutionary implications of his work but, at this stage, chose neither to emphasize them nor to consider the place of man. This was mainly because he realized the likely public opprobrium to which he would be subjected – he did not want to be treated like the anonymous author of *Vestiges of the Natural History of Creation*.

For all his hopes to the contrary, the book was extremely controversial, exciting public discussion and religious fury from some but not clerics. Thus, the future Cardinal Newman (1801–1890) wrote to a colleague in 1868 that he was not really worried about how and when God created matter and laws nor was he disturbed by the idea of evolution, ending his letter by saying "I do not [see] that 'the accidental evolution of organic beings' is inconsistent with divine design—It is accidental to us, not to

God" (see Websites). Others were, however, less tolerant. Darwin took little part in the controversy, but his views were stoutly defended by Thomas Huxley (1825–1895) who held a famous debate on evolution with "Soapy Sam" Wilberforce, Bishop of Oxford, in 1860 at the Natural History Museum in Oxford that seems to have ended in a draw.

The result of all this controversy, however, was that the subject of evolution was now firmly in the public domain. Over the next few years, Darwin brought out further editions of *Origins* in English, and it was widely translated—it was perhaps the first international bestseller about science. He then worked on various other areas of biology, so that it was not until 1871 that he produced his second book on evolution, *The Descent of Man, and Selection in Relation to Sex.*

The *Origin* had carefully avoided any discussion on the evolution of man apart from a cryptic comment towards the end saying that, in the distant future, "Light will be thrown on the origin of man and his history." In *The Descent of Man*, Darwin explored the implications of beauty and courtship behavior in driving evolutionary change (sexual selection). In his next book, *The Expression of the Emotions in Man and Animals* (1872), Darwin set out to show that emotions and psychology that might have been thought to be uniquely human could have evolved from animal behavior. His real conclusion was that man was just another animal and the rules that governed evolution elsewhere in the living world also applied to humans. While this work was inevitably controversial, the British public came to realize that Darwin, for all that he was reclusive and controversial, was a man of very great distinction. When he died in 1882, he was buried in Westminster Abbey, the only biologist to lie alongside royalty and the great and the good of England.

It is sometimes claimed that, because Darwin was an upper-middle class and well-connected member of society, it was much easier for him to be a scientist than someone like Wallace, who was relatively poor and had no background advantages. This is true but only a small part of the story: Darwin was very rich because his father and wife were rich and, indeed, never had to do a day's work in his life. The reality, however, was that he had a restless mind and unstoppable energy for work (in spite of a long-term debilitating illness picked up during the *Beagle* voyage). Over a period of 40 years, he worked unceasingly and wrote a dozen books and many papers describing his experiments and theories on just about the whole range of biology as it then was. He was perhaps the most remarkable biologist of all time, and one of those few people about whom the more one reads, the more impressive that person seems.

This is not to say that Darwin got everything right or that he explained every aspect of evolution. In his search for causes of variation, for instance, he was sometimes less than critical in his reading of the data. One example is his acceptance of telegony, the belief that, when a female who has already had offspring is impregnated by a second male, influences of the first male remain to affect subsequent offspring. Likewise, his theory of pangenesis to explain how variation in one generation became incorporated into the next was the crudest Lamarckism: he suggested that each domain of the adult body produced pangenes that reflected its anatomical details, and that these then travelled to the testis or ovary where they became incorporated into sperm and eggs. In this way, environment-induced change in all parts of the adult would be transmitted to the next generation. In an early example of testing the predictions of a theory experimentally, Francis Galton (1822–1911), who was Darwin's cousin, disproved the predictions made by the pangenesis hypothesis through a series of blood transfusion experiments on rabbits (for details, see Bulmer, 2003). Darwin also had little sense of quantitative biology or how population dynamics worked – indeed, the word *population* doesn't occur in *The Origins*.

Such minor criticism is irrelevant to the importance of Darwin: before him, the idea of evolution was barely science; after him, no one who had read his writings with an open mind could doubt its reality or the way in which it happened. Through his ideas and the massive body of scientific data that he marshalled to support them, Darwin opened wide the door to acceptance of the ideas of evolution, and his successors flooded through it – his basic insights were right and he changed biology forever.

THE 19TH CENTURY AFTER DARWIN

The half century after Darwin saw a very large amount of work in all aspects of evolutionary biology across the scientific world (Mayr, 1982). For various reasons, England focused more on descent with modification than on natural selection; France and the United States remained uncomfortable with the idea of evolution for the next 20 or 30 years, and it was only in Russia and Germany that Darwin's views were completely accepted. Evolutionary research was very strong in Russia until the 1930s; at that point, Trofim Lysenko (1898–1976), an agronomist, denounced Darwinian analysis as being incompatible with Marxist-Leninist thinking and claimed that evolution proceeded along simple Lamarckian lines which could be seen as properly in line with Stalinist views of communism. As a consequence of this flattery, Stalin gave Lysenko power to implement his views and the result was a disaster for Russia: there was a serious famine when his apparently evolved crops failed, and this was accompanied by the disgrace of many fine biologists and the blocking of serious evolutionary work in Russia for several decades.

The main reason for Germany initially accepting evolution was the effort put in by Ernst Haeckel (1834–1919), but he was so aggressive and so anti-Christian in his views that there was widespread rejection of the theory of evolution across Germany towards the end of the 19th century. Ernst Haeckel is better known today for his biogenetic law "Ontogeny recapitulates phylogeny" or, in more modern language, development repeats evolution. This idea, developed around 1870 and built on older ideas, holds that the development of later-evolved animals builds on the evolution of earlier ones so that adult features of early evolving animals are seen in the embryos of later ones.

The classic exemplar of ontogeny recapitulating phylogeny was Haekel's claim that the pharyngeal (branchial) arches of all early vertebrate embryos reflect the presence of the gill slits of adult fish. Haeckel, as has been widely pointed out, was wrong: in this specific case, the branchial arches of fish embryos are actually primitive stem tissues. In fish, they give rise to external gill tissues, but in reptiles and their descendants they form the tissues of the jaw and neck – there is no recapitulation of adult tissues. In fact, Haeckel should have known better because von Baer had already published the correct view when he wrote that embryos of higher forms do not resemble the adults of lower examples, only their embryos (Gould, 1977).

For all that it was wrong, however, Haeckel's biogenetic law provided a framework for comparative embryology and stimulated a great deal of research in the latter part of the 19th century, particularly on comparative developmental anatomy (Gould, 1977). Important discoveries that resulted from this work were the relationships among the chordates and the distinction between protostomes and deuterostomes (the mouth forms from the first cavity in the blastula in the former, and from the second in the latter). By the end of the 19th century, however, the biogenetic law had rightly been abandoned and comparative embryology had been superseded by experimental embryology under the stimulus of Hans Driesch and Wilhelm Roux, who had been Haeckel's student.

By this time, the key evolutionary arguments had turned to the relative importance of hard inheritance, the effect of intrinsic variation, and of soft, sometimes called Lamarckian, inheritance, the result of environmentally determined change that became intrinsic, and whether variation proceeded by slow continuous change or substantial discontinuous jumps (saltations). The key figure here was August Weismann (1834–1914), who discovered the germ theory of inheritance (Weismann, 1889). This says that inheritance is carried by sperm and eggs, and that the crucial factor in generating variation is crossing-over of chromosomes during meiosis. Although Weisman was initially tolerant of soft inheritance, he decided in 1883 that it simply could not happen and such was his authority that, as Mayr (1982) points out, Weismann's neo-Darwinism set the scene for all discussion on evolution until well into the 20th century.

It is interesting to see the extent to which the ideas of evolution became the stuff of popular culture, to the extent that the poet Alfred, Lord Tennyson's 1850 phrase "Nature, red in tooth and claw" came to be seen as representing natural selection, perhaps because it tallied so well with Victorian ideas of capitalism. An important counterbalance, albeit initially ignored, came from the Russian scientist Pyotr Kropotkin who was born a prince but became a socialist and anarchist. His view, based on the work of the geographical survey that he ran in Siberia during the 1860s, was that much of evolutionary change depended on *Mutual Aid,* the title of his 1902 book, an early evolutionary classic that is still interesting to read. In this, he argued that evolutionary advances depended far more on what organisms could do for one another than on the struggle for existence with other organisms. Today, we would say that Kropotkin's focus was more on the importance of ecological niches, altruism and kin selection than on competition.

THE EARLY 20TH CENTURY

It was in the context of trying to discover whether inheritance was discontinuous in around 1900 that Corren, Tschermak, and particularly de Vries independently realized the importance of Mendel's work on genes, which had been published in 1865 and mainly forgotten. What Mendel showed was that inheritance was particulate and achieved through discrete genes, and hence that the blending of characteristics could not happen. This idea, that mutations in genes were responsible for variation, led to a split in the field. Those who took Mendel's work to heart were either mathematicians or experimentalists. The former worked on *evolutionary population genetics* and how mutations would spread through a population, The latter, who became the first evolutionary geneticists, were interested in the mechanisms of gene activity and mutation, with many being originally trained in embryology (e.g. Hugo de Vries, Thomas Hunt Morgan and, later, Conrad Waddington, universally known as Wad). The other school was composed essentially of naturalists and their focus was on diversity, paleontology, evolutionary taxonomy, and, in particular, mechanisms of speciation.

The most immediately profitable work was done by the mathematicians. In 1908, Hardy and Weinberg independently showed that the distributions of the alleles of a gene in some population would remain stable over generations in the absence of selection and other confounding factors. (It is claimed that Hardy proved this using a pencil on a dinner napkin.) In 1918, Ronald Fisher published two ground-breaking publications that showed how novel mutations would spread through a population under selection and laid out the basis of the theory of population genetics (at the same time, he produced much of modern statistical theory).

His brilliant work was then widened by two further mathematicians, J.B.S. (Jack) Haldane and Sewall Wright, who developed ideas on the effects of linkage, genetic drift, and population. They also showed that the effects of a complex phenotype could be partitioned into an inherited (genetic) component and an environmentally determined component. The result was the production of the body of work known as *Evolutionary Population Genetics* (EPG). Over the next thirty or so years, the theory was further integrated with Darwinian selection by theoretical geneticists, such as Hamilton, Mayr, Price, and Kimura, who introduced the concepts, such as kin selection, and modernised the treatment of genetic drift. All of this work was done at a time when genes were defined in terms of the phenotype that they underpinned as nothing was known about DNA.

The schools of evolutionary biologists interested in genetic mechanisms (experimentalists and mathematicians) and the naturalists who studied organisms moved so far apart in their interests that they barely communicated. It was not until 1937 when Theodosius Dobzhansky (1900-1975) published *Genetics and the Origin of Species,* which integrated the theoretical work on genes and mutations with experimental work on animal populations that they started to come back together. Once this had happened, the biological community as a whole realized that there was a series of common problems that had to be solved collectively if evolution was to be properly understood. Different groups then focused on specific areas. Naturalists asked questions about diversity, behavior, and selection within populations of organisms and, in particular, looked at the relationship between species and environment. Paleontologists and comparative anatomists worked on evolutionary trends in anatomical structures. Taxonomists focused on species and their interrelationships, while geneticists were interested in genes and characters (we would now say *genotypes* and the *phenotypes*) and how change occurred.

A particular problem for more theoretically inclined biologists was the nature of variation. A key early scientist here was Sergei Chetverikov (1880-1959), a Russian scientist who worked on *Drosophila* flies that he collected in the wild in the 1920s. He was able to show on the basis of genetic analysis that there was a great deal of hidden variation in a population. One interesting question was whether genetic change through mutation occurred smoothly or in discrete jumps; the organism data suggested that this was smooth, while Mendelian genetics suggested it was discrete. This problem was solved by Fisher, who showed that the cumulative effect of sets of genes that each operated discretely would be able to generate continuous change.

This work did not, however, explain how substantial anatomical variants could form during embryogenesis and whether they could be explained by Mendelian genetics. Goldschmidt (1940) felt that small changes, such as mutations in the sorts of genes then known, were inadequate and suggested that major anatomical changes, such as were seen in the fossil record arose through very occasional mutations that had major effects on the phenotype. This was the saltation hypothesis, and the resulting organism would be what Richard Goldschmidt called a "hopeful monster." Most would be lost, but the occasional mutant might breed and lead to a population of new variants that could survive and lead to new species. This suggestion was not met with much approval; indeed, it was widely ridiculed.

The only supporting evidence came from the work of C.H. Waddington (1905–1975), an early *Drosophila* geneticist and embryologist, as well being perhaps the major theoretical developmental biologist of the 20th century. He showed in 1953 that, provided selection was strong enough, the large amounts of variation seen in a normal *Drosophila* population was sufficient to produce genetically stable flies with four wings rather than the normal phenotype of two wings and two halteres (balancing organs); this was an apparent example of a saltation, albeit one that reflected an earlier evolutionary state. He thus demonstrated that major anatomical changes could occur within a normal population, although it is only relatively recently that molecular explanations for such observations have become available (see Chapter 20 for details).

These important ideas led to the formulation of the modern evolutionary synthesis of the 1940s and 1950s that came from a group of biologists that included Theodosius Dobzhansky, Julian Huxley, and Ernst Mayr. This synthesis was strengthened in the 1960s by Motoo Kimura, who realized that genetic drift, or how genetic variants were distributed throughout a population as a result of random interbreeding, played an important role in distributing mutations across a population that were not subject to selection (so-called neutral mutations). In small populations, these could lead to the appearance of new stable phenotypes, giving the impression that their appearance reflected selection when it was, in fact, due to the stochastic effects of random breeding. Together, these geneticists produced a framework that integrated evolutionary thinking with a mathematical model of evolution based on population genetics, genes, variation, and mutation. The key points of what was now the *Modern Evolutionary Synthesis* were that:

- Within a population, there is slow accumulation of mutations that have small effects and these lead to genetic variation and a resultant expansion of phenotype variants.
- Through random interbreeding and genetic drift, mutations and allelic alternatives are distributed across the population, and this is the basis of normal variation on which selection acts.

This synthesis could not only be used to analyze data but was able to provide a genetic basis for the answer to Darwin's original question about how a new species can arise from an existing one. The additional important insight from Mayr and others in the 1950s was that novel speciation would happen most easily if a small subpopulation became reproductively isolated from the parent population in a new environment with novel selection pressures. This was because the synthesis had shown that, the smaller the population, the more rapidly gene frequencies would change under selection and the more different the populations would become.

The separated group would, as a result of genetic sampling, inevitably have a phenotype and genotype distribution that was different from that of the parent population. If some in this group were fitter than the others in their new environment (more of their offspring survive to reproduce in their new environment), natural selection (Malthusian competition) would eventually lead to their descendants dominating the population. Eventually and over a very long time, further mutation would lead to those in the new group being unable to breed with members of the original, parent population, and the descendants of the variant would have become a new species.

It is also worth noting that the modern evolutionary synthesis excludes or downplays several alternative ideas that had been previously thought important; among these were soft (or Lamarckian) inheritance and saltation (mutations that have major effects on the phenotype). In the early 1970s, however, Eldridge and Gould suggested that the second of these points might not have been correctly interpreted and suggested that, rather than through the slow accumulation of small changes (or phyletic gradualism), evolution might proceed through periods of stasis alternating with short periods of rapid change, a process that they called punctuated equilibrium. This was obviously so after a major extinction (see Chapter 4), but there was a considerable argument as to whether it might occur under more stable conditions.

The evidence that Eldridge and Gould cited to support their ideas came from longitudinal studies on trilobites and snails, while others have identified a range of organisms from bryozoans to dinosaurs whose evolutionary changes show punctuated equilibrium (Gould & Eldredge, 1972; Horner et al., 1992). This was not to say that change occurred through a single mutation that had a major effect (saltation) but that the weight of accumulated mutation became enough to reach a tipping point and so produce a rapid change. Analysis of the fossil record also showed that evolution in many species did not depend on saltation but could be adequately described by the idea of phyletic gradualism. It is probably correct to say that the two views reflect the opposite ends of a continuous spectrum of evolutionary change, but the reasons why this spectrum exists are unclear. They will probably remain so until we have a deeper idea of how variation builds up in a population, and how and why specific phenotypes are selected.

TAXONOMY

The other, more theoretical area that changed thinking on evolution in the 20th century was the work of a line of biologists who had been concerned with how to produce a classification of organisms that reflected their evolutionary history. The standard taxonomy, based on Linnaeus' 18th-century naming scheme, grouped living organisms on the basis of shared and distinct features, usually anatomical. Such a taxonomy is essentially based on the formal rule

$$< Organism\,A >< is\,a\,member\,of\,the\,set\,of >< Group\,B >$$

The result may well be to link groups with a common evolutionary history, but that requirement is not a necessary part of this type of classification. Such relationships can be linked to form a hierarchy, and this is an example of a mathematical graph (Appendix 1).

In the 1950s, on the basis of earlier thinking, Willi Hennig put forward an alternative classification system that he called phylogenetic systematics (Hennig, 1966), and is now also known as cladistics (the word *clade* derives from the Greek word for branch). This classification was based on Darwinian thinking whereby organisms were related through a common ancestor if they shared derived characteristics, and that the more of these that they shared, the closer was the relationship. This classification was thus between individual species, rather than groupings of species, with the relationship between them being not *<is a member of the set of>* but *<derives with modification from>*. The information associated with satisfying this latter criterion allowed two species with a particular shared and derived character to trace their ancestry back to a last common ancestor that was the first to show such a character. This information could be represented as a tree known as a cladogram (Chapter 5).

It took some time before cladistics became accepted as the evolutionarily appropriate way of grouping species, one reason being that the cladistics terminology, while very precise, contained many new terms that many still find complicated (Chapter 5). For a few decades, the two taxonomic approaches lived uneasily alongside one another until it became clear that the higher levels of Linnaean taxonomy had to be made evolutionarily coherent and thus consistent with cladistics taxonomy. Thus, for example, the fossil evidence showed that today's hippopotamuses and cetaceans shared a common ancestor, and this relationship had not been included in Linnaean taxonomy; a new group, the Whippomorpha, was therefore invented to include the two groups (Chapter 16). The net result of such work is that cladistic and Linnaean taxonomies are, by and large, consistent in the way that they group contemporary organisms, for all that they have different roles and structures.

A formalism based on inheritance (a phylogeny) of features turned out to be particularly suitable for the new molecular age. Although cladistics views two organisms as being related if they share derived, anatomical characteristics identified in the fossil record (so producing a cladogram hierarchy), the methodology for grouping organisms equally applies to shared mutations in a DNA sequence (this gives a molecular-phylogram hierarchy). Once sequencing became possible, informaticians were able to produce programs for analyzing sequence similarities and differences, and this work led to the new subject of molecular phylogenetics.

This provided a particularly robust means of classifying and grouping species within an evolutionary hierarchy. Because the analysis uses only contemporary sequences, its obvious roles are in examining the natural groupings of extant species, and in identifying ancient relationships where there are no fossil data. The methodology can thus, for example, identify those DNA sequences that were shared by a last common ancestor of mouse and *Drosophila* and others that arose alter the clades separated. Of equal importance is that the method can be used for all living organisms, irrespective of whether there is a fossil record, and can thus be used across the biosphere.

THE MOLECULAR ERA

The Modern Evolutionary Synthesis dominated all thinking about the subject until the molecular era, which started late in the 1950s, and is still an important component of evolutionary sciences. What seems remarkable today is the fact that the founders of this synthesis constructed a whole theory on the genetic basis of life without having much idea about what a gene was, how mutation worked, or what variation actually meant at the molecular level. Even more remarkable is how, given what we now know, that theory is still being used and has remained essentially unchanged. This is because the synthesis is a theory of phenotypes, organisms, and populations for which molecular details are essentially irrelevant and cannot be incorporated in any detail (Saylo et al., 2011). It was for this reason that the discovery of the structure of DNA and molecular biology in the 1950s initially had little effect on evolutionary thinking beyond explaining how DNA mutations altered proteins, although this insight did provide the first mechanistic insight into the links between genotypic and phenotypic change.

Three subsequent molecular advances have immeasurably increased our understanding of the molecular basis of evolution and provide the context for much of evolutionary work today. The first comes from the insight discussed above that it is possible to use computer algorithms to analyze the DNA sequences of similar genes in different organisms (see Chapter 8). The second derives from the early work of Waddington on how evolutionary change reflects anatomical changes during embryogenesis and that this, in turn, reflects mutation-derived variation (see Chapter 20); this area of work is now part of evo-devo. The third comes from the more recent realization that many proteins not only make their own contribution to building anatomical structures in embryos, but also work cooperatively in complex networks; understanding how mutation affects such network is the key to explaining how change happens in embryos and is an important aspect of systems biology (Chapter 18, Appendix 1). The evolutionary implications of systems biology for variation have yet to be worked through fully, but a preliminary analysis is considered in Chapter 26.

DNA sequence analysis

While cross-species comparisons of DNA sequences have not changed our basic ideas of eukaryotic evolution, DNA sequencing has, since the 1970s, revolutionized our knowledge of the genotype and helped fill in many gaps in our understanding of the history of life, particularly where there is no fossil evidence. An important example is explaining how the eukaryote cell evolved. It is now clear on the basis of sequence comparison that it occurred as a result of endosymbiosis, an extremely rare process by which one prokaryote engulfs another with the two maintaining a long-term symbiotic relationship. Although the details are still being discussed, it is clear that archaebacteria and eubacteria both contributed to the eukaryotic cell which formed ~1.7 Bya (Chapter 10). Some while later, such a cell endosymbiosed a cyanobacterium which became a chloroplast, so creating the last common ancestor of algae and plants. These early eukaryotes diversified to produce the major clades (the phyla) present today, be they unicellular, social of multicellular.

Sequence analysis has been a key to understanding eukaryote evolution, particularly, in animals. Experimental work using molecular-genetic technology has enabled us to assign anatomical features to the earliest bilaterian organism (Chapter 12) and to understand how an early protostome invertebrate gave rise to the first deuterostome. It has also allowed clades of organism families to be reconstructed right across the biosphere that are far more accurate than is possible using the limited phenotypic characters of fossilised and living animals. Sequencing has also shown the extent of genetic variation within populations: the is demonstrated by the data generated by the "1000 Genomes Project" (2012). Such variation can be used to track

Rocks, Dates, and Fossils

The history of evolution is to be found in the fossil record, remnants of organisms that have died and become embedded in rock. This appendix examines how rocks are made, dated and analysed for clues as to the climate when they were laid down. It also considers how dead organisms are fossilised and how unfossilised organic material is dated. It, thus, provides the background for Chapter 6, which discusses the fossil record, and Section 3, which summarises the history of life.

ROCK TYPES

The most common rock (~65%) in the 5–70 Km-thick outer crust of our planet is igneous (e.g. basalt and pumice). It emerged from the earth's core by volcanic activity and was forced toward the surface as magma; this slowly cooled to form the great masses (e.g. granite) that are known as plutons. The other basic type of rock is sedimentary (~8%); this mainly forms from the deposition or sedimentation of particles that eventually become compressed into large masses. These particles are derived from other rocks eroded by weathering, tectonic forces and the effects of vegetation, together with organism-derived detritus, such as the calcium-rich skeletons of corals and invertebrates. Occasionally, sediments that are salts formed when mineral-rich solutions, such as sea water, dry out; these are known as evaporates. Sediments subject to strong compressive forces form rocks that are often stratified, having layers that reflect the successive sedimentation deposits from which they have formed; these are classified by the size of the deposited particles and the mode of deposition. The third major class is metamorphosed rock (~27%), which is formed by extant rock that has been subjected to heat and pressure: calcareous sediments, for example, produce marble, while sedimentary shales and mudstones form slates and gneiss.

If, however, one looks only at surface rock, the proportions change and ~66% is sedimentary. This is where fossils are mainly to be found: they represent the structures of dead organisms immortalised as stone. Occasionally, the fossilised material in original sedimentary rock (e.g. shale and limestone) can still be seen after the rock has metamorphosed (e.g. to slate or marble), but it will usually have been severely distorted. Rock type, on its own, however, tells us nothing about evolution as rock is continually being made and changed. Making sense of the fossil record requires knowledge of the temporal and the spatial relationships among the fossil-bearing rock types. It should be emphasised that the current fossil record represents only a small sample of what has been laid down. This is because the land surface of the world is mainly covered with vegetation and relatively little, other than desert and areas subject to weathering, such as coastal cliffs, has fossil-bearing rock that is easily accessible.

Nicholas Sterno (1638–1686) seems to have been the first to have discussed how strata (layers of rock with similar compositions) are superimposed on one another, how they could have been laid down in different places at the same time but be different in appearance and how they could have become degraded and distorted. This was followed by detailed observational work that was undertaken by James Hutton and a few others in the late 18th century. The link between rocks and fossils was, however, made by William Smith soon after his realization that the various strata in a local coalminc in Somerset, U.K., could be identified by the fossils that they contained. In 1799, he published a geological map marking the various rock strata in the area around the city of Bath in Somerset and in 1815, a much larger geological map of most of England and Wales. This was a considerable achievement as rocks of the same age and even the same basic type can look very different in different places because of their local mineral content and the various weathering and other forces to which they have been subjected over very long periods.

Smith's work stimulated others to analyse strata and identify them by their type and fossil content (where possible), and specific types were often named after the places where they were first recognised. It is ironic that Sedgwick, who identified Cambrian rock in the 1830s and called it after the Latin name for Wales, never accepted Darwin's ideas on evolution, even though it turned out that the fossils in Cambrian rocks represent the time when multicellular organisms from the great majority of the phyla evolved. A list of the major strata, their periods and some major fossil features are given in Table 3.1.

What was particularly interesting to early geologists about each layer was that many of its constituent fossils bore little resemblance to those above or beneath it. Around 1800, Cuvier wrongly took this as evidence of major extinctions that were followed by periods of new creation (Appendix 2). This was because he failed to appreciate two important factors. First, although the extinctions were always important and sometimes catastrophic, a proportion of organisms always survived, and these provided base species for future diversification. Second, the nature of the surface in a particular region could change because of the effects of tectonic forces and evolutionary events: a region that was once sea could later become land and *vice versa*. It was for such reasons that very different organisms might be found in adjacent layers.

Common examples of invertebrate families with shells or robust exoskeletons that survived through at least several extinctions include ammonites, graptolites, and trilobites, although the last of these died out in the great Permian extinction (~252 Mya). Many such organisms evolved slowly, making small but clear incremental changes over time. Our knowledge of evolutionary history is now so good that we can link the detailed morphologies of particular organisms with the time intervals in which they lived. The presence of such *index fossils* in any rocks is a strong indication of when the original organisms were alive and provides a minimum age for that rock.

How does one work out that one piece of rock is older than another in the absence of fossil data? The simplest way is to see if one came from a lower level than another and, in principle, one could just take a drill and remove a core of rock from a particular place. One would then rank the various pieces on the basis of their depth from the surface: the further down, the older. Such relative measurements are, however, of limited use for three key reasons. First, even if one can identify distinct layers in the core, it is very hard to know how long each took to be laid down. Second, one cannot be sure that depth reflects age because the movements of tectonic plates and the enormous forces that occur when plates collide can lead to major foldings in the rock to the extent that older rock can be forced on top of younger. Third, such relative ageing doesn't help when one wants to compare rocks from different places. The only useful way to age rocks is to use physical measuring techniques to date them; only then can one integrate this information with that from the index and other fossils that they contain and get a full picture of the history of that rock level.

AGEING ROCKS (GEOCHRONOLOGY)

The best-known approach to measuring absolute time is radiometric dating and is based on radioactive decay: many elements have several isotopes (the nuclei have the same number of charged protons, but each isotope has its own number of neutrons), with only one or two being stable. As a result of radioactive decay, unstable isotopes can lose one or more neutrons or protons and so decay to either another, lighter isotope of the element or a different element with a lower atomic weight, a process that continues until a stable isotope forms. Each element of this decay process has a well-defined half-life (the time taken for half the original amount of the element to decay). In places where the amounts of original and decay constituents can be determined (either by measuring radioactive emission or by direct counting of each atomic type using a mass spectrometer), it is now relatively straightforward to work out how long ago the decay process started in a particular region of rock.

The two most important radiometric pairs for dating older rocks are potassium/argon and uranium/lead. Potassium-argon dating depends on the fact that K^{40} decays to Ar^{40} with a half-life of 1.3 By, and the system depends on the fact that argon only starts to accumulate once the rock has solidified. The amount of argon is tiny, but, if the rock is heated to 400°C, it can be released and measured. The technique can be used for rocks aged more than 100 Ky. The decay of uranium to lead occurs in a series of steps that is often measured in the mineral zircon ($ZrSiO_4$): uranium, unlike lead, is soluble in this mineral so any lead present in zircon is a result of uranium decay. Uranium exists as two isotopes, U^{235} decays to Pb^{207} with a half-life of about 700 MY while U^{238} decays to Pb^{206} with a half-life of about 4.5 By, which is the approximate age of the earth. Because the different isotopes decay to lead via different routes with each decay step having its own half-life, measurement of the relative amounts of all the elements and isotopes in the two decay sequences in a specimen gives a very good measure of its age. In practice and provided that the rock contains uranium, this technique can be used to age any material that acquired its uranium from more than 4.5 Bya to around 1 Mya.

There is also a range of nonradiometric techniques that can be used to date rocks, particularly surface ones, with perhaps the most important being magnetostratigraphy. This capitalises on the magnetic properties of the few magnetisable minerals, the most important of which is magnetite (Fe_3O_4), originally known as lodestone. This is found in many igneous and metamorphic rocks and has the property that, as the molten rock solidifies, its magnetic polarity is set along the local direction of the earth's magnetic field at that time. The rock then maintains this polarity, even when subject to later distortion or changes to the local magnetic field. Analysis of its magnetic remanence today thus gives the direction of the earth's magnetic field when the molten rock solidified. The resulting analysis has shown that, since the oldest magnetisable rocks were deposited, the polarity of the earth's magnetic field has changed direction many times, albeit for reasons that are still not properly understood. Such changes can be observed in the different layers of rock taken from a single location or in areas of slow rock deposition.

Perhaps the most striking examples of these reversals come from analysing the seabed. Igneous rock is continually being extruded in a molten state at oceanic ridges (such as the mid-Atlantic ridge) and cools to form basalt rock, which moves laterally away from either side of the ridge under the pressure of further extrusions, so incidentally driving local tectonic plate movement. As the rock solidifies, the magnetite within it aligns to the earth's current magnetic field. The most important investigation of the phenomenon was made using a magnetometer towed by a ship sailing directly away from the mid-Atlantic ridge. Measurements of the inherent magnetism of the sea floor showed that, as one moved away from the ridge, sea-floor rock comprises a series of parallel stripes of variable width marked by successive reversals in magnetic polarity. As the rate of flow of rock from the ridge appears roughly constant, we now have a clock linking polarity with time that extends back to about 180 Mya, the longest period between seafloor rock emerging from suboceanic ridges and being lost due to subduction when it eventually meets and then moves below continental plates.

The next step was to correlate the magnetic analysis with absolute time using radiometric dating. Measurement showed that most chrons (the period between magnetic reversals) are in the range 0.1–1 My, with the time of the transition usually being around 1 K–10 K years, although it can be faster. Over the last five million years, there have been nine major periods during which the direction of the earth's magnetism has been reversed (two of them including short periods of reversal). Going back further in time, there have been two superchrons, or chrons lasting more than 10 MY, the Cretaceous Normal from about 118–84 Mya and the Kiaman from about 362–212 Mya (the late Carboniferous Period to the late Permian Period). The net result of all this work is that we now have the Geomagnetic Polarity Time Scale that has had 184 polarity reversals over the last 83 MY.

PALEOCLIMATOLOGY

There is further information to be gained from rocks and fossils and that is indications of the local climate when they were laid down: some of this information comes from trace gases trapped in rocks and ancient ice and some from weather-affected structural details in fossils, such as the relative spacing of growth rings in trees (dendroclimatology) and corals. Other information comes from the nature of those fossilised organisms found, in particular, sediments: some, for example, flourished in particular climates.

Oxygen is a particularly important indicator here for two reasons. First, the absolute amount, which can be measured in trapped air, particularly in ice, is an indicator of cyanobacterium and then plant activity. Second, the ratio of normal O^{16}

to the rarer but heavier O^{18} in trapped water is an indicator of temperature. This is because, as the temperature rises, the former evaporates more quickly than the latter. In periods of cold when large amounts of water are stored in ice, the ratio is high (even MIS numbers), while warm, relatively ice-free periods have low ratios (odd MIS marine isotope stage numbers). There are now ~100 MIS periods (we are now in MIS 1) extending back ~2.5 My; samples over the last 40 Ky correlate well with carbon dating (below) and older ones correlate with magnetic polarity records (Wright, 2000). Ice cores are turning out to be particularly important for studying relatively recent evolution because some of the ice in Antarctica is more than a million years old (Knowlton et al., 2013).

HOW FOSSILS FORM

In most cases, dead organisms are rapidly lost because their tissues are eaten by vertebrates, invertebrates, and microbes, as well as being degraded by their own enzymes (autolysis; the study of the events taking place after death is known as taphonomy). The net result is that, within a year or so, all that is left of even a large vertebrate are the bones and teeth, the hardest of its tissues. As bones are held together by the softer connective tissue of joints and ligaments, these hard, skeletal elements may rapidly separate and even disperse if subjected to weathering, scavenging or water flow that breaks down connective tissue. Small invertebrates, other than those with a hard carapace, and nonwoody plants will also be rapidly lost, while even trees will eventually break down.

What this means is that the various decomposition and dispersal forces have to be kept at bay if an organism is to become fossilised. All organisms other than sulphate-reducing bacteria need oxygen and water to degrade tissues; the higher the temperature, the faster is the breakdown of soft tissue. If tissues are to be preserved, the key factors that, therefore, need to be minimized are the presence of oxygen, water, and high temperature, together with access by organisms, both prokaryote and eukaryote.

The processes of fossilisation and dispersal are thus most likely to take place if the dead organism is rapidly covered by sediment. This can occur at the bottom of the sea where it is also cool or by mud in a river; under these conditions, the dead organism is protected from the effects of oxygen and small degrading organisms—it is no surprise that marine fossils are particularly common. There are other ways of keeping the decomposition forces at bay: larger organisms can be desiccated by the hot dry winds of a desert or be buried in ice or tar pits, while small invertebrates can be embedded in resin that will become amber. Even under these conditions, the tissues most likely to become fossilised are hard materials that break down very slowly. Examples are the mineralized bones and teeth of vertebrates, the chitin exoskeletons of invertebrates, and the complex polysaccharides that produce wood in plants and the outer walls of pollen.

Fossilisation thus normally starts when dead tissue is rapidly buried in sediment, such as mud or sand that will eventually become rock. First, soft tissue is usually lost through autolysis or scavenging; this allows mineral-rich water to permeate slowly through the small spaces in nondegraded tissues (e.g. the hydroxyapatite in bone) so that mineral eventually replaces the organic material. Provided that the tissue structure does not break down before precipitation is complete, this process can allow very fine detail to be preserved.

In due course, the impregnated tissue and the surrounding sediment become rock but these fortunately remain different in appearance. This is because the constituents of the insoluble matrix surrounding the remnants of the organism will be different from those of the originally soluble inorganic constituents that penetrated the tissue. The result is that the fossil remains distinct within a wider rock matrix. The actual time that this takes depend on the local environment. The minimum age for a fossil is defined as 1,000 years, but this may not be enough to complete the process: the bones of *Homo floresiensis*, an extinct species of the genus *Homo*, sometime known as the "hobbit" man, which were dated to ~18 Kya, still remain unfossilised (Brown et al., 2004; Chapter 23).

The oldest fossils are stromatolites: these are layered mats of dense bacteria, some of which secreted carbonate, and trapped sediments; they date as far back as 3.5 Bya (Fig. 9.1). Perhaps surprisingly, living stromatolites can still be seen today in Australian and Brazilian hypersaline lakes where few other organisms can survive. Even small algal and bacterial assemblages can be fossilised if they have a strong coat, and it is these that tell us about pre-Ediacaran life. An interesting and important class of early fossils are spores and pollen. As they often have a tough coat, lack predators and contain no free water, they may well be preserved in ancient sediments as microfossils. Since their morphology is fairly species-specific, they are of particular importance in identifying early algae, fungi and plants, even if there is no other record of them.

Occasionally, soft-tissue structures are maintained long enough for them to become fossilised and the result has been some of our most wonderful examples of preserved organisms. Many such examples of both vertebrates and invertebrates have been found in areas that are known as conservation fossil *Lagerstätte* (this German word means storage place) and are detailed in Bottjer et al. (2002; see Websites). Perhaps the oldest known example is a cyst of several cells that has been preserved in phosphate and dates back >1 By (Strother et al., 2011). Fossils from the Ediacaran Period that are preserved in phosphate include those of the Doushantuo Formation, a site in Guizou Province in China (570 Mya, Chapter 12), which includes phosphatised organisms of groups of a few cells (Fig. 12.3). As these organisms seemed to show palintomic cleavage (cell division without growth), the fossils were thought to be of early invertebrate embryos, although this view was later doubted. The original view has now been confirmed by further analysis: more mature examples have an external epithelium, a key feature of all early embryos (Yin et al., 2007). Other fossilised organisms may, however, be nonmetazoan holozoans, and thus be more closely related to colonial unicellular organisms (Huldtgren et al., 2011).

Phosphate, because of its diffusion characteristics, can only preserve tiny organisms of millimeter size, but larger soft-bodied organisms have also been fossilised. The classic example is the fauna of the Burgess Shale, a series of shale layers in rock discovered in 1908 that is now some 2,500 m above sea level in the Rocky Mountains (Chapter 13). ~505 Mya (mid-Cambrian), however, these rocks were sand some 120 m below sea level, adjacent to a ledge ~10 m deep and at the base of a mountain. Mud and other debris would slide down the mountain and displace organisms living on the ledge pushing them off and down to the sea floor below where they were rapidly covered in silt and so died of suffocation. Such obrution conditions are ideal for rapid fossilisation and can, for example, result in the phosphate within the midgut glands of arthropods precipitating out so replacing the soft tissue with mineral and highlighting gut contents before the gut has been broken down by autolysis (Gaines et al., 2008).

As a result of such rapid, anoxic fossilisation, some of the Burgess Shale material show extraordinary detail of soft and hard tissues (Chapters 3 and 13). Further, even earlier soft-bodied fossils from the Cambrian have been found in the Chengjiang shales of China (<5018 Mya, Chapter 12). Such wonderful preservation is very rare in fossils later than the Cambrian period, perhaps because by then many types of marine worms and land invertebrates had evolved that would eat dead organisms (Brasier, 2010).

Fossils may form in other ways: the oxygen and nitrogen of plant tissues, such as leaves may be lost leaving only carbon films (compression fossils); and organisms may degrade completely, but slowly enough for the space that they had occupied to be filled by later minerals. In such cast fossils, the material that occupies the space is known as the *part* while the surface impression in the surrounding rock is called the *counterpart*; many of the Ediacaran organisms are represented today as such impression fossils. There are also trace fossils that represent evidence that an organism had been present: these include footprints, burrows, nests, and copralites (fossilised dung).

For vertebrates, the tissue that preserves better than any other is teeth; indeed, they are the sole remnants of some early Mesozoic mammaliaforms that were rodent-sized and whose small bones have generally not been preserved. Enamel, the surface coating of teeth is sufficiently tough to survive most degradative forces for a long time. A very great deal of work has now been done on fossilised teeth: their shape indicates their origin and function and, given this, their size is a measure of the original animal and palaeontological age, while the degree of wear is a measure of the age and diet of the organism (Gill et al., 2014).

What all this implies is that, although a dead animal is unlikely to be preserved as a fossil, it stands a much better chance if it is marine, has either a tough exoskeleton or bone skeleton and is rapidly covered with sediment after death, preferably as a result of suffocation. One the world's most common class of fossils comes from the calcareous protective plates of unicellular algal coccolithophorids that sank to the sea floor, were compressed and formed chalks made of calcite (e.g. the white cliffs of Dover). Such minerals would become marble in places if they were later subject to heating, compression, and metamorphosis. Plants stand a better chance of being fossilised if they possess tough polysaccharide coats or woody tissues, as the world's supply of coal bears witness.

DATING ANCIENT ORGANISMS

The techniques mentioned above can, of course, be used to age fossils which are themselves rock, but there are some specific techniques that can be used to date ancient organic material that has not yet been fossilised. The main method for unfossilised organic material is based on radioactive carbon decay: this makes use of the fact that cosmic radiation leads to normal carbon with an atomic weight of 12 (^{12}C) being activated at high altitudes to form radioactive ^{14}C; this reacts with oxygen to form $^{14}CO_2$ which can enter the food cycle. As ^{14}C breaks down to ^{12}C with a half-life of 5,730 years, there is a continuous but slow cycling of carbon between its two forms and the net result is that about one billionth of the carbon in CO_2 is ^{14}C. This cycling breaks down if the carbon is ingested and incorporated into organisms as it becomes inaccessible to high-energy cosmic radiation; as a result, the ^{14}C slowly decays to ^{12}C without being replenished.

Although the practical details for measuring radioactive carbon accurately are complicated, modern techniques, which, for example, allow for variation in the cosmic ray background, can use the ratio of ^{14}C to ^{12}C to date a sample of organic material within a window that extends from ~60 Kya to quite recently. (For further details, see Grün (2006) who has reviewed the methodologies of carbon dating, uranium-series dating and other such techniques as they have been applied to human bones).

To date older organisms, it is now possible to use marine oxygen isotope staging (MIS) which, as discussed above, is based on the ratio of O^{18} to O^{16} in material, such as calcite shells, high ratios indicating periods of cold and low ratios of warmth. MIS datings thus not only indicate the climate conditions when organisms died but allow rough datings from other techniques to be hardened up. Douze et al. (2015), for example, were able to use MIS datings to assign the earliest stage of the habitation of *H. sapiens* in the Blombos cave in South Africa to MIS 5c-5b or 105–90 Kya (Chapter 25).

Finally, it should also be mentioned that traces of organic material associated with rock-embedded fossils can now be analysed and visualised with a wide range of physical and chemical techniques (e.g. mass spectroscopy) to explore detail and chemical competition. For an example of what can be done, see Box 14.1 which considers the so-called Tully monster.

What all this analysis says is that, given a piece of rock, we can, as a result of two centuries of detailed analysis, usually work out approximately when it was laid down and what organisms have become fossilised within it. This is initially done on the basis of its inherent radioactivity and location, but we can then use residual magnification and other physical properties together with the presence of any index fossils within it to confirm and improve the accuracy of the original date.

FURTHER READINGS

Allègre CJ (2008) *Isotope Geology*. Cambridge University Press.

Benton J, Harper DAT (2009) *Basic Palaeontology: Introduction to Paleobiology and the Fossil Record* (3rd ed). Blackwell.

Hyman A (ed) (2017) *Principles of Paleoclimatology*. Columbia University Press.

McElhinny MW, McFadden PL (2000) *Paleomagnetism: Continents and Oceans*. Academic Press.

WEBSITES

- The Wikipedia entries on *Fossilization, Lagerstätte* and on *Radiometric Dating*.

- The Nature page on the dating of rocks and fossils: https://www.nature.com/scitable/knowledge/library/dating-rocks-and-fossils-using-geologic-methods-107924044/

Constructing Molecular Phylogenies

This appendix summaries the various ways in which phylograms (inheritance trees) can be constructed for a range of organisms on the basis that they share genes whose sequences are homologous (e.g. every animal with eyes includes a Pax6 gene in its genome). There are several ways of doing this, but the most important use maximum likelihood and Bayesian approaches. The former constructs a phylogenetic tree on the basis that the most likely ancestor sequences are those from which all the daughter sequences can be derived with the minimum number of mutations. Bayesian methods uses a population genetics model and Monte Carlo and Markov-chain approaches (the former emphasises random choices and the latter that history plays no role in future mutation) to explore all possible choices and to point to the most likely. The methods, of course, give very similar results

A full understanding of computational phylogenetics requires a fair amount of knowledge about probability theory and population genetics, together with some informatics and statistical skills (see Further Readings at the end of the appendix). The treatment here is intended to provide a qualitative explanation of the key principles behind the various types of analysis, many of which are hidden within software. This appendix thus supplements the information given in Chapter 7.

Phylogenetic trees are constructed computationally on the basis of sequence information available from online sequence databases; there are a host of programs that will do such computations, some of which are mentioned below (each, of course, has its own website). All molecular phylogenies are based on a set of homologous sequences for one or more species that may include sequence variants both across and within species (e.g. the paralagous Hox genes in the mouse genome—Chapter 7). Phylogenetic trees can be constructed on the basis of three different approaches that use increasing amounts of sequence information.

PHYLOGENIES BASED ON SHARED/ ABSENT SEQUENCES

Where a set of sequences is distributed across a group of species with each having only part of the set, it is straightforward to work out a phylogeny of the groups on the basis of that distribution. Such a tree is actually more a cladogram than a molecular phylogram, as inherited features (specific sequence names) are being analysed rather than sequence details. The insertion of a new sequence in the genome is really an apomorphy, and its inheritance by a set of daughter species represents a shared, derived characteristic or a synapomorphy. Because the number of sequence variants is small, constructing a tree is computationally simple. The best-known example of this sort of analysis was the use of retrotransposons to unpick the evolutionary line of whales and ungulates and so establish that their closest living relatives are hippopotamuses (Nikaido et al., 1999, Chapter 16).

PHYLOGENIES BASED ON DISTANCE MATRICES

The simplest approach for analysing sequence data is to choose a group of taxa and investigate the differences among their homologous genes. The first step in constructing a phylogenetic tree for **N** species for this sequence is to make a distance matrix. Its terms are numbers that reflect the genetic distance between each pair of homologous sequences across the group. For this, the sequence of one species (S_i) is aligned as closely as possible against each other sequence in turn. The differences between each pair are then estimated (by a computer program) and reduced to a single number. These can be expressed as a matrix whose terms are of the form D_{ij} for the difference between the sequence in species **i** and species **j**.

This process would be relatively straightforward if all gene sequences for a particular gene were the same length and mutations restricted to simple base changes. Unfortunately, possible mutational changes include deletions, insertions, secondary mutations (both backwards and forwards) and even duplications; the net result is that homologous sequences vary both in detail and in length. It is, therefore, unusual for there to be a unique way of producing this sequence alignment; indeed, one would not expect there to be because the effects of random mutation and selection over many millions of years are being compared and there is no logic linking the genomes of different organisms that have mutated independently.

The program has to cope with all these possibilities and usually starts with a progressive sequence alignment that first considers sequence elements where the differences are small and then adds those that are more and more distinct (this often means including gaps which are usually of different sizes and may be in different locations). It then assigns a number to each pair of sequences (**ij**) based on the number of mismatches between them (a gap counts as a mismatch). The most straightforward way of constructing the genetic difference between a pair of aligned sequences: (D_{ij}) is to use the Hamming distance

$$D^H = (\text{No. of sites that differ in a sequence})/(\text{Sequence length})$$

The difficulties here are that no account is taken of secondary mutation and that all differences are given equal weight so that, for example, gaps are equated with simple mutations; the net result is that D^H is always an underestimate of the correct distance. Two sorts of corrections are usually applied to D^H: the first makes allowance for the possibility that successive mutations can occur at a site, the simplest correction here being the Jukes-Cantor model whereby:

$$D^{JC} = -3/4 \ln\left(1 - 4/3 D_H\right)$$

The second correction gives more weight to related changes where the nature of the mutation is clear than to complex changes where it is not. Here, the aligned sequences are first broken up into clusters for each of which a D^{JC} is calculated and this figure is then modified using the Fitch–Margoliash method: because assigning a distance to closely related sequences is more accurate than to those that are very different, this method uses a least squares optimization algorithm that assigns more weight to the former than the latter in computing **D**.

The net result of all distance methods for analysing sequences is an **NxN** distance matrix where the components are the corrected D_{ij} and that is symmetric (the distance measure of sequence in species **X** and **Y** is the same as that for species **Y** and **X**). Once this matrix has been produced, the second step is to compute a tree that links sequences on the basis that the lower the D_{ij} for a pair of sequences the more closely related they are. There are a host of programs (e.g. Clustal and Mega6) that will construct such phylogenetic trees from distance matrices using procedures, such as *neighbour-joining*. In practice, and because different methods produce slightly different alignments, it is normal to analyse the whole dataset with several techniques, one of which is usually maximum-likelihood (below).

It is also important to note that this tree is calculated only on the basis of D_{ij} values and so has no explicit root or origin; it cannot, therefore, localise the position of the last-common-ancestor sequence. There are two ways of rooting such phylogenetic trees. The first is to include in the analysis a sequence only distantly related to those being studied. The analysis will show where this outlier links to the hierarchy, and its node should be with the root of the tree. In the case of the AmphiLim genes (Fig. 7.2), the outliers were the Amphioxus genes while, for example, the obvious outlier for a phylogenetic tree of the vertebrate Pax6 genes, would be the *Drosophila* Pax6 gene.

The other way of rooting a phylogenetic tree is to use implicit timing on the basis that the length of a branch represents the number of mutations between nodes and is thus a measure of the time taken for divergence. If this is so, the root should be at the midpoint of the longest distance between two most distant terminal nodes (e.g. two contemporary species from different phyla that have homologous sequences for some gene). Such rooting techniques generally give plausible candidate organisms, but sometimes fail. We still know little about the details of the LECA, the last eukaryotic ancestor from which all subsequent eukaryotic organisms derived.

PHYLOGENIES BASED ON TREE-SEARCHING METHODS

Difference matrix methods have the advantage that they are computationally fast but, because they analyse sequence differences rather than compare the sequences themselves, they are not very accurate. Rather richer and more commonly used are the more sophisticated techniques that analyse complete sequences and the details of their mutational differences, even though they are computationally more complex (see below).

Such methods are not only more accurate because they are based on more detailed information, but they also provide additional information. They can, for example, suggest likely ancestor sequences for each node in the phylogenetic hierarchy on the basis that mutations in each will lead to sets of downstream sequences in the most parsimonious way. It is this methodology, therefore, that is most commonly used today, although several approaches are generally used for any analysis. These different methods would be expected to give very similar results but any differences in their predictions may need to be averaged to construct the most-likely consensus tree.

Using sequences to make phylogenies is known as **tree-searching** because the procedure involves searching for the tree that generates all the current sequences on the basis of mutation from an ancestral root sequence. As there are many possible such trees, the procedure generally incorporates the test criterion that the optimal tree is that which requires the least number of mutations for its construction. A tree-searching program, in principle at least, thus looks at every possible tree and identifies the most likely nodes and associated sequences. There are three approaches to the computational analysis: maximum parsimony, maximum likelihood and Bayesian statistics. It should also be mentioned that such approaches are also used in coalescence analysis, which integrates molecular phylogenetics with population genetics to investigate population histories (Chapters 21 and 24).

Maximum parsimony

This approach sets out to find the most likely phylogenetic tree on the basis of identifying those ancestral sequences that generate contemporary ones with the minimum number of mutations. The MP approach is thus to search every possible scenario and see which is the most efficient. As the number of possibilities increases exponentially with the number of taxa, so does the amount of computational time required to identify the most parsimonious tree. In practice, computational algorithms have been devised that explore more likely solutions and reject less likely ones. Although the confidence that one has of the resulting tree can be improved by bootstrapping (see below), MP methods are not much used today because they do not include an evolutionary model and so exclude any information about the relative likelihoods of mutation possibilities.

Maximum likelihood

These methods also search through all possible trees, but incorporate a model of evolutionary change based, for example, on the details of mutation. Thus, it is known that: 1) the rates of base mutations from purine to purine (A->G, G->A) or pyrimidine to pyrimidine (T->C, C->T) are higher than those for transversions (e.g. A->T) and 2) that mutations in the final base of a codon that do not change an amino acid are more likely to be inherited than those that do. This is because the latter change may alter the final phenotype and so be lost as a result of natural selection. Any analysis that includes such information about mutation will be more accurate than one that does not. Maximum-likelihood methods use a program, such as IQ-TREE or RAxML to examine all plausible trees. The programs assign likelihoods to each and, as it were, recommend that tree with the highest likelihood of predicting the data, together with a statistical analysis of the confidence that one can have in it. This and Bayesian methods are the two approaches commonly used today.

One question that immediately arises from such an analysis is the degree of confidence that one can have in the result, and one way of answering this is to use the technique known as *bootstrap resampling*. Here, the tree-generating programs are repeated many times, with each reflecting the replacement of a different part of the data. Nodes that are robust are only rarely affected if a small amount of the data is altered, while nodes for which the support is weak immediately show up with a low bootstrap number. The general view is that a bootstrap figure of 70% is the absolute minimum for a reasonable hypothesis, while a figure of 95% or above is very unlikely to be wrong. (Note that bootstrap resampling is now more common than the alternative *jackknife resampling* technique which systematically removes a small part of the data.) The tree shown in Fig. 7.2, which gives the phylogenetic tree for the AmphiLim genes, was the result of 1,000 such simulations and only one link has a bootstrap value of less than 85% (albeit that putting exact confidence limits on bootstrap figures is difficult; Buckley & Cunningham, 2002). In practice, bootstrapping can be used with both distance-matrix and tree-searching approaches.

Bayesian methods

This approach aims to generate the set of hypothetical phylogenies that are supported by the data, and assign probabilities to each, implicitly recommending that with the greatest. They are thus subtly different from maximum-likelihood methods that essentially look for the phylogeny that best generates the data. As the methodology is based on Bayesian statistics, which in essence use existing information to refine expectations, it can also incorporate a model of evolution whose rules help generate that set (these rules are known as the prior probabilities). Neither the theory nor the methodology is simple as it is based on sophisticated statistical theory, but the practice is straightforward. The user submits to a program (e.g. MrBayes or BEAST) a set of homologous sequences that perhaps includes a likely outgroup sequence to help root the phylogeny together with a model of mutation (which, if it includes rate constants or other clock models, can generate timings).

The aim of such a program is to explore on the basis of the prior probabilities (this is the evolution model) the full space of trees that are suggested by the data and find that which is the most likely. The program uses Monte-Carlo and Markov-Chain methods; it starts by constructing a possible tree from the data, and then uses a random number (the Monte Carlo component) to mutate the sequences and construct another tree solely on the basis of the new sequences alone; this is the Markov criterion which essentially says that history does not influence the future. If the new tree is better on some criterion, it becomes the starting tree for the next simulation (or link in the Markov Chain); if worse, it only has some assigned probability of replacing the original tree for the next simulation. The program runs for as many as 1,000,000 Monte Carlo simulations (ignoring perhaps the first 25% as they might be dependent on the start tree), building up a frequency table for the various possible trees. One way of knowing when to stop is to run two simulations in parallel with different start trees. When their distributions are the same to within some predefined but strong criterion, one can be reasonably sure that the simulation is complete.

What emerges from the simulation is a set of trees, each of which could generate the data and with each having a frequency based on the number of times that the simulation "visited" that tree as it explored tree space. The branch lengths within the tree reflect the number of mutations required to make the link and, if a mutation rate was included in the original model, this provides a measure of how long a branch took to form. Integrating the set of simulation outputs gives a set of posterior probabilities that each tree is correct and is based on mathematically rigorous methodology.

It should, however, be said that there are some minor concerns about Bayesian trees. First, the distribution depends on both the prior probabilities and the inheritance model used, so one needs to test the quality of these and the robustness of the tree by using other priors and models. Second, Bayesian phylogenies turn out to be more sensitive to the choice of algorithms for measuring genetic differences between sequences than are other methods. Third, some doubt has been cast as to whether it is entirely appropriate here to use the Markov criterion that it is only the immediate state that determines the subsequent one (Cartwright et al., 2011).

PHYLOGENY CHOICES

For all the very large amounts of computation needed to construct phylogenies for large datasets, it is not easy to decide which mode of analysis is the most appropriate for determining a particular phylogeny. Hall (2004), therefore, constructed an interesting experiment. He took an *E. coli* protein-coding DNA sequence and generated successive generations of that sequence using a program that generated random mutations. The net result was a tree with 7 generations and >100 nodes, with a precise sequence associated for each, and with successive generations having increasing amounts of randomly generated deletions, mutations, and substitution. He then used the full range of phylogenetic techniques on the final-generation DNA sequences to reconstruct estimates of the original sequences, the intermediate sequences and the complete tree. To improve the quality of the analysis, he produced the sequences in three formats: first the nucleotide sequences, then the corresponding amino acid sequences, and finally the DNA sequences organised so that common codons were aligned (DNA-CA) as this format facilitates the handling of gaps.

His initial results illustrated the relative speeds of the various methods: neighbour-joining on the basis of distance matrices took less than a second for all formats; tree-searching methods using the same computer took much longer, with analyses on DNA-CA sequences giving more accurate results than the equivalent DNA or AA sequences. Here, MP, ML, and BI methods, respectively, took 5 min (without bootstrapping and about 1 h with), 6 h 42 min and 7 h 43 min (these figures would be much shorter today because computing speeds are faster). As to accuracy, tree methods were, as expected, better than those based only on distance matrices with, on average, maximum-likelihood methods giving the most accurate results. While Bayesian techniques could give good results, they turned out to be too sensitive to alignment choices to make them as reliable as maximum-likelihood methods, for this analysis at least. Hall, therefore, recommended standard distance-matrix methods for a quick analysis and maximum-likelihood methods for reliable phylogenetic trees. It should, however, be said that Bayesian methods have since been improved and many people see them as their first choice.

FURTHER READINGS

Hall BG (2017) *Phylogenetic Trees Made Easy: A How-To Manual*. Oxford University Press.

Lemey P, Salemi M, Vandamme A-M (eds) (2009) *The Phylogenetic Handbook: A Practical Approach to Phylogenetic Analysis and Hypothesis Testing* (2nd end). Cambridge University Press.

WEBSITES

- The Wikipedia entry on *Computational Phylogenetics*, which includes full referencing.
- The basics of phylogenetic analysis: https://bip. weizmann.ac.il/education/course/introbioinfo/03/lect12/ phylogenetics.pdf
- Reading a phylogenetic tree: https://www.nature.com/ scitable/topicpage/reading-a-phylogenetic-tree-the-meaning-of-41956/

Model organisms are key to studying evo-devo and several are important (Table A6.1). This book particularly focuses on the three of these: the mouse, a vertebrate, *Drosophila melanogaster*, an invertebrate fly, and humans. The first two two have been the key experimental animals for studying both evo-devo and the molecular mechanisms of evolutionary change. This appendix reviews the basic features of their development and also briefly considers the use of humans as a model organism for evolutionary studies. Humans have a particular significance here, partly because of the insights that human disease have given on the nature of mutation and partly because the extent of the detail now known about human fossils and their artifacts is so much greater than for any other organisms (Section 5). More detail on the mechanisms by which change is achieved in these organisms is given in Appendix 7.

The study of development has always focused on a few model organisms chosen both for their ease of use in one or another specific context and for their relative availability and cheapness (Table A6.1). Since the early 1980s, two of these model organisms have been particularly important in linking gene mutation to evolutionary change: the fruit fly, *Drosophila melanogaster*, and the mouse, *mus musculis*. *Drosophila* is frequently used in genetic studies as its breeding time is about two weeks, and it is both easy and cheap to grow individual strains in large numbers, while the mouse is a model system for mammalian development and also for the study of human genetic disorders; this is because there are many molecular and other homologies between the two organisms. As mutations are now so straightforward to create and study in these two organisms, they have also been used for experimental evolutionary biology.

Once a gene has been discovered in humans (or any other organism), its homologue in the mouse can easily be identified, and its developmental and other roles established. Similarly, any mutation in that human gene can be replicated in the mouse using gene technology, so that its effect on the phenotype can be followed in detail. Such parallel studies of humans and mice have not only established how development can go wrong if particular genes are mutated but has also established the roles of the unmutated genes in normal development. It is also straightforward to identify of homologues of mammalian genes in *Drosophila* and use genetic technology to produce flies with specific mutations. As a result, we can now study the roles of homologous genes in this very different organism, and such comparisons have provided a great deal of evo-devo knowledge that can be expanded to other organisms. The classic example of the role of the Pax6 gene in the development of all animal eyes is discussed in Chapter 8.

The next two sections describe how *Drosophila* and the mouse go through early development, producing in their profoundly different ways core body plans typical of an arthropod protostome and a mammalian deuterostome. These anatomical descriptions provide a framework for discussing the molecular mechanisms by which new tissues are generally produced in embryos, and how homologous mechanisms can sometimes underpin the development of very different organisms (Appendix 7). There is little point in providing such developmental information for humans as very

TABLE A6.1 COMMON MODEL ORGANISMS

Organism	Class	Use
Dictyostelium discoideum	slime mould	Origins of multicellularity, cell movement
Arabidopsis thaliana	Angiosperm	Genetics, molecular genetics and mutation analysis
Caenorhabditis elegans	Nematode	Genetics, molecular genetics and mutation analysis
Drosophila melanogaster	arthropod	Genetics, molecular genetics, developmental mechanisms
Sea urchin	Echinoderm	morphogenesis, molecular genetics
Danio rerio (zebrafish)	Fish	Genetics, mutation analysis & morphogenesis
Xenopus laevis	Amphibian	Experimental manipulation
Chick & quail	Birds	Experimental manipulation
Mus musculis (mouse)	Mammal	Molecular genetics, model for disease
Homo sapiens	Mammal	Identifying genetic abnormalities (+ evolution)

little experimental work is done on their embryos. For those interested, however, early human development is initially slightly different from that of the mouse, but very similar after neurulation (~E9 in the mouse and ~E28 in humans).

DROSOPHILA DEVELOPMENT

The fertilized *Drosophila* egg is a small ovoid structure ~0.5 mm long that is laid by the mother who takes no further part in its development. This egg consists of a developing oocyte together with nurse cells that produce maternal RNA that will pattern the cytoplasm and follicle cells that provide yolk and lay down the outer vitelline layer. The fertilised egg initially contains a single central nucleus but this soon undergoes rapid mitosis; after about seven rounds of division, the ~128 nuclei migrate to lie under the external cell membrane (the embryo is a syncytium). Those located in the posterior region will become the pole or future germ cells and are the first to cellularise. After about 12 divisions and some 3 h 30 m at 25C after fertilization, the blastoderm cell layer forms as the outer membrane of the embryo extends inwards to surround each nucleus with its allocation of cytoplasm, so forming an ovoid epithelial sheet underlying an acellular vitelline membrane (this is meroblastic cleavage).

After about 12 h, cell division ceases and will not restart until the larval stages; indeed, once the cell-cycle networks have been turned off, most of these cells will never divide again. The several thousand cells that have by now formed then start to produce embryonic tissues. The pole cells (the future germ cells) first segregate at the posterior end of the embryo, lying between the epithelium and the external membrane, and once the reproductive system has formed, migrate to the gonad (Fig. A6.1). The remainder of the epithelium soon segregates into the three germ layers: 1) the ectoderm that will form the outermost layer and the nervous system; 2) the endoderm that will form the gut; and 3) newly differentiated mesoderm that will particularly form muscles and other tissues. Two sets of stem cells form tissues, which will mainly develop during pupation, are also determined at around this time. These are the small groups of ectodermal cells (imaginal discs) and of mesenchymal cells (histoblast nests); once in place, they are quiescent until needed.

Gastrulation in *Drosophila* involves two separate events. First, regions of anterior and posterior endodermal epithelium invaginate, extend, and eventual meet to form the gut (in this highly evolved embryo, the original way of making the gut in protostomes has changed). Second, a furrow develops along the already-patterned ventral region of the embryo, with the cells inside the furrow breaking away to form mesoderm that extends around the body. Three rows of ventral ectodermal cells

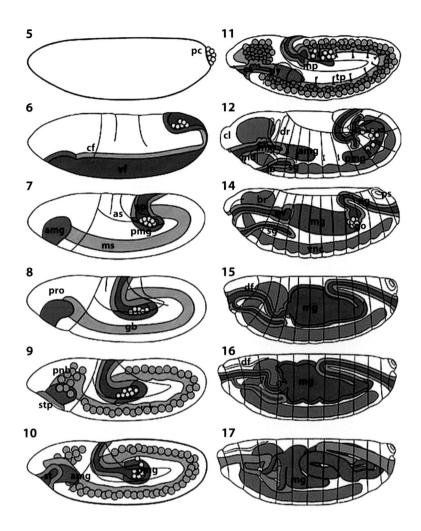

Figure A6.1 **Some important stages in the development of the Drosophila egg from just before gastrulation (stage 5) to hatching (stage 17).** A,D,V,P: anterior, posterior, dorsal, and ventral. amg: anterior midgut rudiment; br: brain; cf: cephalic furrow; cl: clypeolabrum; df: dorsal fold; dr: dorsal ridge; es: esophagus; gb: germ band; go: gonads; hg: hindgut; lb: labial bud; md: mandibular bud; mg: midgut; mg: Malpighian (nephric) tubules; mx: maxillary bud; pc: pole cells; pmg: posterior midgut rudiment; pnb: procephalic neuroblasts; pro: procephalon; ps: posterior spiracle; po: proventriculus; sg: salivary gland; stp: stomodeal plate; st: stomodeum; tp: tracheal pits; vf: ventral furrow; vnb: ventral neuroblasts; vnc: ventral nerve.

(From www.sdbonline.org/sites/fly/atlas/52. htm. Courtesy of Volker Hartenstein.)

delaminate from the furrow region, moving internally and becoming neuroblasts; they will become the central nervous system.

Soon after this, as the culmination of a series of molecular events that started soon after fertilisation, the ectoderm becomes segmented, and this is the most obvious external feature of the embryo. (It should be pointed out that, whereas most arthropods are short-germ-band [the segments are added sequentially], *Drosophila* is one of the rarer long-germ-band insects in which all the segments form at around the same time.) As this happens, the imaginal discs cells segregate as small, paired internalized buds in most ectodermal segments; these will become adult tissues during pupation. Simultaneously, the mesenchymal tissue differentiates to give muscle, somatic, visceral, and cardiac tissues, together with the set of small histoblast nests (typically four per segment). About 24 h after fertilization, embryogenesis is over, and a functioning and independent larva hatches from an egg that has not changed in size since fertilization.

Over the next 4 or so days, the organism goes through three larval stages that mainly involve growth, with successive layers of external cuticle being sloughed off as the larva enlarges. Pupation then starts and, over the next few days, results in a complete reorganization of the organism. The larval ectoderm is broken down and replaced by a new set of external adult structures that form from the imaginal discs (imago is another word for the adult insect), while the cells of the histoblast nests replace the embryonic tissues within which they are located with adult ones. The development of each disc depends on its initial segment location: anterior ones will form head structures including the antennae, the eyes, and mouth, while those from body segments form three pairs of legs, the wings and the halteres (balancing organs that evolved from the

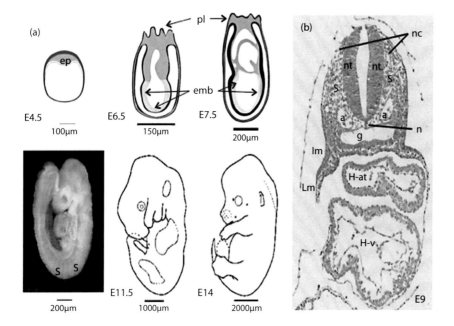

Figure A6.2 **Mouse development.** (a): **Diagrams of the early stages. E4.5**: Just before implantation, the blastocyst has an epiblast (ep), and primitive endoderm. **E6.5**: the embryo is a small cup of cells within the egg cylinder linked to the yolk sac (ys) extraembryonic membranes, such as and the early placenta (pl). **E.7.5**: Gastrulation is taking place, the embryo (emb) is lengthening, and the extraembryonic membranes and early placenta are developing. **E9**: a 3D reconstruction of the embryo based on sections; this shows the early brain, heart, eye placode, and the segmented somites (s). **E11.5**: The limb buds are beginning to develop and the jaws are forming from the maxillary and mandibular branches of the first pharyngeal arch; the second arch is much smaller. Internally, the metanephric kidney and reproductive system are beginning to develop. **E.14**: The embryonic limb is developing autopods, all the basic organ systems are in place, and the embryo is growing rapidly. (b): **A mid-thoracic transverse section of an E9 mouse embryo showing some of the major features**. (nt: neural tube; nc: neural crest cells; s: somites; g: gut; a: dorsal aorta; im: intermediate mesenchyme; lm lateral mesenchyme; n: notochord; H-at: heart common atrial chamber; H-v: heart common ventricular chamber. Distance bars show microns. (From the EMAP eMouse Atlas Project http://www.emouseatlas.org; Richardson et al., 2014).

(Courtesy of EMAP eMouse Atlas Project (http://www.emouseatlas.org; see also Richardson L, Venkataraman K, Stevenson P, et al. (2014) *Nucleic Acids Res.* 42:D835–44.)

rear wings of an ancestor fly), with the most posterior ones developing into the external genitalia. Some four days after pupation starts, or about 9 days after fertilisation and development at 25C, the adult fly emerges. (For further details, see Websites.)

The molecular basis of many of these events is now understood. The core principles are given in Appendix 7, while molecular and other details are accessible from Flybase (see Websites).

MOUSE DEVELOPMENT

Embryogenesis in the mouse could not be more different from that in the fly. The very small and almost yolkless egg (~75 µm in diameter) is internally fertilised and is never laid; it spends its first 4 days floating in the uterine tube and dividing to become the blastocyst. This consists of an external trophoblast layer that contains the inner cell mass and the fluid-filled blastocoel space. At around embryonic day (E) 4.5, the blastocyst implants itself in the wall of the uterus where the trophoblast cells cause a local vascular and proliferative response, the first stage of placentation. Around then, the inner cell mass segregates into the epiblast and the primitive endoderm layer and, over the next two days, the former will give rise to the early embryo while the latter, together with some of the trophoblast cells will form the extraembryonic membranes, the placenta, and the umbilical cord. At this stage, the extraembryonic part of the conceptus is very much larger than the embryo that it surrounds (Fig. A6.2).

At E6.5, this embryo is just a small, cup-shaped and apparently featureless structure made of a few hundred epithelial cells, but which has now acquired an antero-posterior polarity. Embryonic development now gathers momentum as gastrulation starts: a small part of the anterior ventral region, known as the node, moves posteriorly, taking about 48 h to reach the tail region. As it moves, it signals to local cells and these respond by reorganising themselves to form an overlying ectoderm, internal mesoderm, and ventral endoderm, with anterior regions continuing to develop while posterior regions await the gastrulating signal. Particularly interesting here are the derivatives of the mesoderm: it separates into four different regions, each of which has its own fate (see Fig. A6.3).

The key event that next takes place is neurulation: during this process, the dorsal ectoderm forms a furrow that folds over, with the internalised cells forming the neural tube and the cells on the borders between this tube, and the adjacent ectoderm forming the neural crest. Soon, the constituent cells migrate away from the neural crest region and differentiate into a wide variety of cell types, with their final fate depending partly on their original location and partly on their final one. Neural crest cells particularly form the enteric (gut) and autonomic nervous systems, pigment cells, much of the head tissues and skull bones, and the adrenal cortex. The

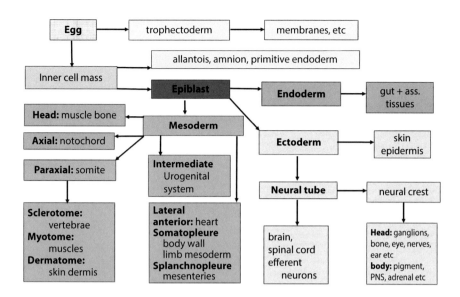

Figure A6.3 **Lineage in the mouse embryo: some major pathways of tissue development.**

importance of this tissue is seen from a comparison of craniates and *Amphioxus*, a cephalochordate that lacks neural crest (as do all invertebrates): *Amphioxus* has no eyes, skull or pigmented cells and has only limited sense organs.

The next major event is segmentation: for this. the paraxial mesoderm, which is adjacent to the neural tube, sequentially forms >60 distinct somites that extend from the hindbrain region to the tail (Fig. A6.2a). The anterior part at least of the neural tube reflects this patterning by developing segmental indentations. At the molecular level, the molecular profile of both sets of segmented tissue is defined by the expression of sets of segment-specific *Hox* genes which determine their future fates. Accompanying all this activity is considerable cell proliferation and growth (at its fastest, embryo cell numbers double in less than 10 h). The main lines of development and some early features in the mouse are laid out in Fig. A6.2, while the lineage of the primary germ layers and their descendants is given in Fig. A6.3.

By about E9, the essential features of the embryo are in place: the head region includes the neuroepithelium of the future brain and proximal part of the neural tube, together with eye and ear rudiments, while the body includes the neural tube (the future spinal cord), somites, gut, and a heart that is now beating, so circulating blood round the vascular system that links to the placenta via the developing umbilical cord. By E11.5, the brain and nervous system are beginning to form, while the branchial (or pharyngeal) arch system that will form the lower head and neck tissues is now present, with the jaws forming from the maxillary and mandibular components of the first branchial arch (Fig. A6.2a). In the body, the limbs are starting to extend and the musculoskeletal system is beginning to develop—the embryo has become a fetus. By E14.5, almost all of the major tissues and organ rudiments are in place (Fig. 10.5) and the remaining five or so days until birth are occupied by filling in anatomical detail and by growth.

SIMILARITIES AND DIFFERENCES

Although their development is profoundly different, there are a few basic similarities between *Drosophila* and mice. Both have bilateral symmetry, together with very similar cell types, such as neurons, muscles, tendon mesenchyme, and epithelia. At the anatomical level they each have an outer ectodermal layer, musculature, a gut, a nerve cord, and brain, although the cord is dorsal to the gut in Drosophila and ventral to it in mice (see Chapter 8).

The anatomical differences between them are, however, far more obvious, as examination of the head shows. Most of the *Drosophila* head derives from ectoderm, but the various tissues and sense organs of the mouse are each built from several of the derivatives of the mesoderm, neural crest, ectoderm, endoderm, and the anterior neural tube (the future brain). This is, in particularly, seen in the eye. The compound eye of *Drosophila* comprises about 800 omatidia, each of which has a set of optical components; all of it develops exclusively from the epithelium of the appropriate imaginal disc. This is in sharp contrast to the camera eye of the mouse which includes a lens and cornea from ectoderm, retina and pigmented epithelium that are an outgrowth from the brain, blood vessels from mesoderm and internal optic muscles, choroid, and sclera from neural crest.

There are further substantial differences between the limbs of mice and *Drosophila*. Those of the former derive from ectoderm (epidermis and motor nerves), neural crest cells (pigment-producing cells and sensory nerves) and four different mesoderm derivatives that separately give muscles, bones and tendons, blood vessels, and dermis. The limbs of *Drosophila*, in contrast, almost entirely derive from ectoderm apart from the small amounts of muscle that comes from mesoderm.

There are other differences between the two organisms. Many of their functional requirements are carried out by different organs and in different ways: respiration in *Drosophila*, for example, is handled by tracheoles (oxygen-carrying tubes that extend internally from the ectoderm) rather than lungs which derive from mesoderm and endoderm, while the urinary system in Drosophila is based on endoderm-derived Malpighian tubules as opposed to the mesoderm-derived kidney system in mice.

As there are such profound differences in the ways that the cells of these two organisms are organized into tissues, in their respective internal and external skeletons and in the mechanisms that handle that basic physiological functions, it had been hard until relatively recently to envisage that they could ever have had a common ancestor, other than a very primitive organism with just a few cell types. Some 30 years of evo-devo work has, however, shown that this pessimism was misplaced: comparative molecular anatomy has identified a range of homologous proteins in these two organisms that mediate homologous functions in their very different modes of development. The molecular similarities show that the two species share a distant common but primitive ancestor from which each has evolved in its own distinct way. This triumph of evo-devo research is discussed in Chapter 12.

Homo sapiens

Although we don't usually think of humans as a model system for studying *Drosophila*, *H. sapiens* has become an important model for both evolutionary biology and genetics. Through the power of molecular genetics technology and the importance of human disease, medical research has led to the identification of very large numbers of genetic abnormalities whose molecular basis can now be readily identified (the data is held in the OMIM database). These abnormalities not only provide pointers to the general effects of mutation but highlight the normal functions of the unmutated genes. (Note: Gene locations, types and functions are detailed in the Amigo database and structured on the basis of the Gene Ontology—see Websites.)

In addition, and because of a natural interest in our own species, more is known about the evolution of humans than of other species. This is partly because of our knowledge of the molecular basis of genetic disorders, partly because of its detailed fossil record, which is more extensive than for any other living organism (Chapter 23) and partly because we have been able to follow the migrations of the various human populations across the world over past few hundred thousand years (Chapter 24). This last conclusion is based partly on analysing group-specific mutations and genetic markers (particularly from mitochondrial and Y-chromosomal DNA) in today's humans and partly through the use of coalescent theory which gives information on past population numbers.

FURTHER READINGS

Gilbert SF, Barresi MJF (2016) *Developmental Biology* (11th ed). Oxford: Sinauer Press.
Kaufman MA, Bard JBL (1999) *The Anatomical Basis of Mouse Development*. Academic press.

Wilkins AS (2002) *The Evolution of Developmental Pathways*. Sunderland, MA: Sinauer Press.
Wolpert L et al. (2014) *Principles of Development*. Oxford University Press.

WEBSITES

- Each model organism has its own database, with many including large amounts of genetic and other data.

- Flybase: this *Drosophila* resource covers much of what is known about *Drosophila*: https://flybase.org

- Drosophila development https://www.ncbi.nlm.nih.gov/books/NBK10081/
- Human genetic disorders: these are given in the Online Mendelian Inheritance in Man (OMIM) database https://omim.org—the data is also available at https://www.ncbi.nlm.nih.gov/omim/
- Mouse Atlas website: this gives details of mouse developmental morphology https://www.emouseatlas.org/emap/home.html
- Gene data is given in the Amigo database which uses the Gene Ontology: http://amigo.geneontology.org/amigo
- Mouse gene expression (GXD): this particularly links gene expression developing tissues https://www.emouseatlas.org/emap/home.html
- Wormbase: this site includes detailed information about worms, and particularly *C. elegans*, the model nematode worm https://wormbase.org//#012-34-5

Some Principles of Animal Developmental Biology

Developmental biology plays a key role in understanding evolution for two reasons. First, some of the core evidence for evolution comes from showing that homologous proteins have very similar functions in embryos from different phyla (this is evo-devo, Chapter 8). Second, almost all anatomical change in adults depends on mutation modifying the developmental mechanisms that drive tissue formation in embryos (Chapter 18). This appendix sets out to explain how these mechanisms work. It should be emphasised that the treatment of the molecular basis of development given here is light, with the aim being to provide a framework to help understand the processes of normal embryogenesis and tissue construction rather than to provide detail. The reader who needs to know more about any aspect of developmental biology is referred to Barresi and Gilbert (2019) or Wolpert et al. (2019).

Appendix 6 described the development of the mouse and *Drosophila* embryos but made no attempt to explain the molecular basis of developmental change. This appendix looks at animal embryogenesis in more detail, focusing on the mechanisms of developmental change that take place as a fertilised egg progresses through its embryonic stages to become an independent organism. The dynamics here can seem bewilderingly complicated: in the mouse embryo, for example, around a hundred new named tissues form in a coordinated way during the 8th day of development, when the embryo is only ~3 mm long. In such embryos, several independent example of differentiation and morphogenesis can occur simultaneously within distances of only 100 µm—early embryos are very busy places! Understanding what is going on is not easy and the subject is made more complex as each group of organisms has its own anatomical features and distinct ways of making them. There are, however, some underlying similarities in the various modes of embryogenesis and this appendix discusses them.

All animal embryos start as a fertilised egg that divides to form a ball of cells (the blastula) that then undergoes reorganisation through a series of cell movements (gastrulation) to produce the gut. In protostomes, the first gut opening that forms becomes the mouth and the second becomes the anus. In deuterostomes, a phylum whose major clades are the chordates and echinoderms, it is the other way around. It should, however, be pointed out that gut formation in amniotes has been modified because the early embryo receives nourishment from the yolk (fishes, reptiles, birds, and monotremes) or through a placenta via an umbilical cord (most mammals) rather than by ingestion of food through the mouth. Similarly, other changes have occurred in some invertebrates to modify the details of their gut formation. Gastrulation is followed by neurulation, and at this point there is a further noticeable difference between protostomes and deuterostomes: in the former, the neural tube forms ventrally to the gut; in the latter, it is dorsal so that the neural tube is above the gut (Chapter 8).

More than a hundred years of experimental manipulation has shown that, in every embryo, the developmental fate of the cells in a particular tissue or region at

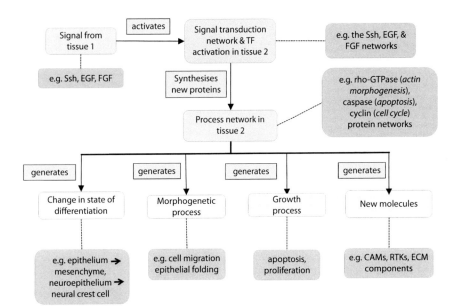

Figure A7.1 A diagram summarising the various processes that drive developmental change. The blocks in grey linked by dotted lines represent examples. Note that this diagram is actually a mathematical graph based on triplet facts (Appendix 1).

a particular stage depends partly on their *lineage* and partly on *molecular signaling*, usually from neighbouring tissues. The balance of importance of these influences varies among tissues and may also be species-specific. It is really only since the 1980s, however, that we have had a clear idea of the molecular nature of those signals and how their reception causes cells to change their properties on the basis of their lineage.

In chordates, the early cells are not determined, and it generally takes several rounds of intercellular signaling and successive rounds of developmental change before basic lineage is in place. At this point, a cell is generally part of a tissue and any further developmental activity before a tissue is functionally and anatomically complete tends to reflect lineage effects. In protostomes, lineage relationships tend to be determined rather earlier in development than in deuterostomes.

It is now known that, at any given time, the key effects of lineage in a cell at a particular development stage are to make it competent to secrete particular signals or to express both the appropriate signals-receptors (usually in its membrane) and the specific set of transcription factors (TFs) that are to be activated as a result of the reception of an external signal. Once activated, these TFs initiate transcription that leads to a set of proteins needed for the cell's future development (Fig. A7.1). Sometimes, a cell can express two or more sets of receptors and transcription factors so that it may have several developmental options. In the mouse, for example, this happens in the early mesoderm-derived somite regions, individual cells of which will become chondrocytes, muscles or mesenchyme according to the local signals that each somite domain receives from its neighbouring tissues (for further details on these molecular constituents, see Box A7.1).

The relative importance of signaling and lineage in deciding the future fate of a tissue is highlighted by the behaviour of the sonic hedgehog signal (SHH) in early mouse development. SHH protein is secreted by the notochord, a long thin cylinder that lies immediately below the neural tube; this signal diffuses both to the overlying ectoderm-derived neural tube, an epithelium at this stage, and to the adjacent mesoderm-derived somites (see Fig. A6.2). The result of SHH signaling is that the local neural-tube cells become motor neurons, while the adjacent somite cells enter the pathway that leads to them forming vertebrae. The effect of SHH signaling thus depends entirely on the lineage of the cells to which it binds.

SIGNALING

The main way in which signals in an early vertebrate embryo reach developing cells is through diffusion from a nearby neighbour; these signals then bind to cell-surface receptors (retinoic acid, an important developmental paracrine signal is an exception as it binds directly to receptors in the cell nucleus). This is known as *paracrine* or short-range signaling and well-known signals are members of the Hedgehog, Wnt and the various growth-factor families; homologues of these signals are found in many organisms. There are also cases where signal-receptor pairings work through *direct* intercellular contact (*juxtacrine signalling*) as both are located in cell membranes: two examples here are Notch-Delta (Fig. 8.6) and Eph-Ephrin signal-receptor families; the former is particularly involved in neuronal differentiation, while the latter facilitates the establishment of boundaries between tissues. The best-known form of signaling in adult vertebrates, endocrine (hormone) signaling via the bloodstream, is rare in early embryonic development because most endocrine tissues form relatively late in development.

Developmental change is sometimes initiated by a single signal; in other cases, however, signaling is better seen as a complex conversation between two tissues with signaling going both ways, and a well-studied example here is the development of the mouse kidney (below, Bard, 2002). Of particular importance, however, are those cases where the strength of the signal varies spatially and there is a concentration-dependent reaction across the responding cells. This mechanism is known as *pattern formation* and underpins much of embryonic development. Examples of particular evolutionary significance are the development of the limbs from sarcopterygians fins and the subsequent evolution of jointed digits in tetrapods (Chapter 6), together with the pigmentation patterns seen in butterfly wings and vertebrate skins (Chapters 11 and 12).

SIGNAL RECEPTORS

These receptors (e.g. receptor tyrosine kinases), which are often paired, are usually located in the cell membrane and have three domains: an external receptor part to which a signal binds, an intermediate region which locates it in the membrane, and an internal domain. Once that signal is bound, the internal domain undergoes a conformational change that switches its status from passive to active. This intracellular change activates a signal transduction pathway (e.g. Fig. A7.2), and there are at least ten of these in the mouse. There can often be several such receptors in a single cell and they and their associated pathway can be activated independently (see Table A7.1), so allowing several different events to happen at the same time or providing alternate developmental pathways for the cell to follow.

SIGNAL TRANSDUCTION

The main role of the signal transduction pathway is to translocate a transcription activator from the cytoplasm to the nucleus where it interacts with an appropriate transcription complex around a specific transcription factor that is already set up (this is part of the cell being competent). A few such transduction pathways, such as for Notch (Fig. 8.6) are relatively simple and involve only a few proteins, but most are more complex.

An important example here is the network activated when the epidermal growth factor (EGF) signal binds to the external domain of its ErbB-1 cell-surface receptor (a receptor tyrosine kinase). Kinase activation here triggers the activity of a complex network of ~60 proteins that includes enzymes, activators, transferases, and kinases and that has a range of functions that are lineage dependent (Fig. A1.1). This and other signaling pathways have homologues across a wide variety of organisms (and these are archived in the KEGG database, see Websites).

TRANSCRIPTION FACTORS

The net result of a signal-transduction network is that a transcription activator enables the transcription factor, now in the nucleus and already part of a transcription complex, to initiate gene expression and the synthesis of one or more proteins. Here, it is again worth noting that the presence of several signal transduction pathways in a single cell together with an associated set of transcription factors and complexes enables that cell either to go down alternate pathways or to undertake several sets of gene expression at once.

NEW GENE EXPRESSION

These newly synthesised proteins have a range of roles in development that include making new structures (e.g. collagen, actin, and myosin), enzyme activity, and network activation. Particularly important, however, are the synthesis of new signals, membrane receptors, and transcription factors that lead to further rounds of developmental change. It is those that are involved in forming or activating networks that are particularly important because it is network activity that drives the formation of new tissues and new physiological capabilities.

All of this is illustrated by the epidermal growth factor (EGF) activated pathway (Fig. A1.1). This has a range of functional outputs that depend on the lineage of the cell and include proliferation, protein degradation, implantation in mammals, and neural-cell patterning, particularly in *Drosophila*. This example shows that the same signal and signal-transduction network can be used in many different contexts because the downstream response always depends on which transcription factors are in place, and these are lineage-dependent (as the sonic hedgehog example above shows).

DRIVING DEVELOPMENTAL CHANGE

Although each animal develops uniquely, at the later stages at least, there are some unifying features across all of embryogenesis, and the most important is the set of processes that drive developmental change (for details, see Table A7.1 and websites). The processes include *proliferation* and *apoptosis* (Fig. A7.2; Baehrecke, 2000), that determine local and overall cell numbers, *differentiation* that causes changes in cell phenotypes and *morphogenesis* that leads to cell movements and shape changes. Each of these processes represents the output of the activities of a complex set of proteins that cooperate within a *network*, the fine details of which are usually taxon-specific.

Processes need to be activated and this is achieved through external *signaling* and *patterning* and or through self-signaling (a lineage effect). What makes each species unique are the ways, times, and locations that these processes are called into play: sometimes, one activity simply stimulates the next, but others require timing networks to initiate them. In the context of evolution, anatomical variants arise when mutations modulate the outputs of these networks in some way during development.

Most of these networks are complicated and may involve up to a hundred or more proteins (e.g. Fig. A7.2; see also Figs. 1.1, 8.6, 25.5, A1.1, and A1.2). In such cases, our understanding of the internal workings of these networks is very thin to the extent that even the functional role of most of their constituent proteins remains unknown. Investigating the details of how such complex networks operate is part of systems biology (Appendix 1), but one does not need to understand all the details of a network to appreciate that the key role of each is to produce a specific functional output, be it transcription, differentiation or apoptosis.

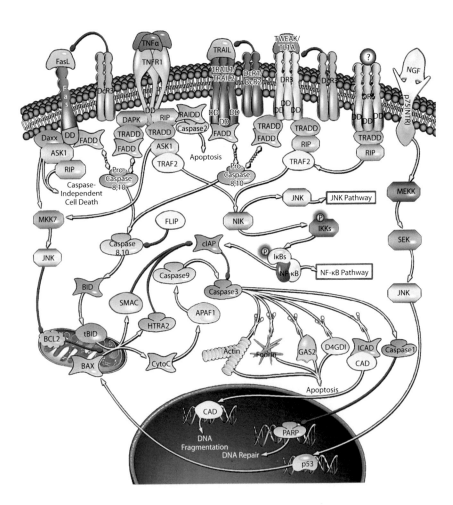

Figure A7.2 The apoptosis network. Ligand-activation of the death receptors of some cell initiates caspase activity that results in chromosome condensation, the inhibition of membrane activity and other events that lead to rapid cell death. Note that the network includes a second cell pathway: here, damage to mitochondria releases cytochrome C which activates caspase activity.

(From https://proteinlounge.com/pathway.php. Courtesy of ProteinLounge.)

TABLE A7.1 SOME MAJOR NETWORKS WHOSE OUTPUTS DRIVE DEVELOPMENT

SIGNALING NETWORKS		PROCESSES NETWORKS
Signaling pathways (signal-receptor names)	**Differentiation to**	**Morphogenetic events**
EGF (ERK/MAPK)	haematopoiesis lineage	boundary formation
FGF	erythroid lineage	(eph-ephrin)
Hippo	lymphocyte lineage	epithelium
JAK/STAT	myeloid lineage	branching
Notch-Delta	epithelium – monolayers	folding
HH	multilayer	migration
TGFβ	ep-mesenchyme	rearrangement
VEGF	mesenchyme	mesenchyme
Wnt	chondrocyte	adhesion
Patterning networks	fibroblast	migration
Notch oscillator system	muscle	**Proliferation**
Signaling gradients (e.g.	osteoblast	Cyclin + cell division
SHH)	mesothelium	**Apoptosis**
Patterning systems	neuron	*Extrinsic*: fas/TNF
Hox & RTK	neuron-support cell	*Intrinsic*: mitochondrial
Timing networks	pigment-producing cell	breakdown
These are not yet known		

In addition to the process networks that drive developmental change, organisms need the activities of many others protein networks if they are to thrive. Examples are those that maintain the organism's physiological abilities, such as the heartbeat cycle, which starts at E8–E8.5 in the mouse, the circadian clock cycle, neuronal activity, and muscle contraction, together with biochemical pathways, such as the Krebs cycle that is present in the mitochondria of all eukaryotic cells. Very often, more than one of these new processes operate simultaneously as a tissue develops. Table A7.2 gives some indication of the wide range of processes involved in making a mouse forelimb, a model system considered in more detail in Chapter 14. The ways in which mutation might modulate process outputs and so change developmental anatomy are particularly discussed in Chapter 18.

HOW TISSUES FORM

Organogenesis normally starts when some external signal binds to a receptor on the cell membranes of a small rudiment that is competent by virtue of its lineage (it expresses the appropriate receptors and transcription factors). These signals result in the activation of process networks within the rudiment that cause the tissue to enlarge, produce new cell types (differentiate) change shape and perhaps interact

TABLE A7.2 SOME OF THE PROCESSES INVOLVED IN MOUSE FORELIMB DEVELOPMENT

Patterning	Proximal-distal: long bone and muscle specification
	Dorso-ventral: e.g. the positioning of claws and paw pads
	Lateral-medial: patterning of digits
Differentiation	New chondroblasts, osteoblasts, osteoclasts, muscle cells, etc.,
Morphogenesis	Distal migration of mesenchymal cells and axon extension
	Distal extension and bifurcation of blood vessels
Proliferation	Everywhere, particularly in the distal mesenchyme (ZPA)
Apoptosis	Loss of mesenchyme & epithelium between digits

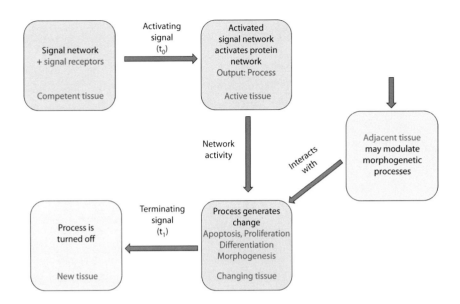

Figure A7.3 **A systems view of the events that take place in a single tissue as it changes its developmental state during embryogenesis.** The "Adjacent tissue" box represents the effect of the local environment.

with other tissues before the network activity is turned off (Figs. A7.1 and A7.3). In short, process networks drive anatomical development under the direction of signaling networks; they thus represent the executive arm of the genome (Davidson, 2010; Bard, 2013b; Fig. A7.5).

In the context of tissue formation, there is an interesting difference between morphogenesis, which includes the processes that organise the early embryo and shape later tissues, and the other activities downstream of signaling and patterning. Apoptosis, cell division and differentiation are all cell-autonomous: these properties reflect the internal responses of individual cells and the external context is of secondary importance. Morphogenetic responses, other than for individual cell migration, involve groups of cooperating cells (as in boundary formation, the folding of epithelial sheets, and the formation of condensations of mesenchymal cells). Such morphogenetic activity is often constrained by the features in the local environment, such as the boundaries provided by the external basal laminae of adjacent epithelia (Fig. A7.3 has an *Adjacent tissue* box to cover this possibility).

The actual production of a tissue from a primitive rudiment involves the coordination of many processes and tissue interactions and there are few if any cases where all these details are known. The example of the kidney, chosen because it highlights several important aspects of organogenesis, in discussed in Box A7.2.

TISSUE MODULES

It is worth pointing out that, during development, some transcription factors can be seen as being more powerful than others. While most cause a limited number of proteins to be synthesized and so perhaps activate a single network, there is a higher level of transcription factors whose activation leads to the production of a tissue. Such TFs lead to the expression of proteins amongst whose roles is the activation of further tiers of downstream process networks whose collective activity can lead to the formation of whole organs. An example of an extremely powerful, high-level transcription factor is Pax6 whose initial activation can lead to the production and transcription activity of a host of secondary transcription factors. This, in turn, leads to the production of a wide set of proteins, the activation of a range of process networks and the eventual development of a either a camera or a compound eye (Gehring, 1996; Chapter 8).

The compound eye is particularly interesting in this context as is may contains hundreds of ommatidia, each an identical optic fibre. The omatidium is an example of a *tissue module* that is repeated many times, and there are numerous examples of these across the world of multicellular organisms. In vertebrates, there are, in addition to the nephrons discussed above, somites, teeth, and the more than a hundred long bones, most of which have a synovial joint at each end, that form the majority of the vertebrate skeleton. In many invertebrates, tissue modules include segments, the ommatidia of compound eyes mentioned above, and often many limbs (e.g. in

BOX A7.2 THE DEVELOPMENT OF THE MOUSE KIDNEY

A typical example of vertebrate organogenesis is the development of the metanephric (adult) kidney in the mouse (Fig. A7.4): this starts to form in the 11.5-day embryo as a result of reciprocal signaling between the small metanephric mesenchyme (MM) condensation and the ureteric bud, an epithelial spur off the main nephric (or Wolffian) duct, both of which originally derive from the intermediate mesenchyme (Fig. A6.3). Organogenesis starts when the MM secretes glial-derived neural growth factor (GDNF). This signal diffuses to the nearby inducing duct inducing the epithelial ureteric bud to form and then extend towards the MM, soon invading it. As a result of further reciprocal signaling between the MM and the ureteric bud, the former differentiates to become nephron stem cells, and undergoes proliferation, while the proximal part of the ureter starts to bifurcate, producing branches throughout the MM. Meanwhile, the distal parts of the ureteric bud and the nephric duct together form the ureter which links to the newly formed bladder

Small groups of MM stem cells form condensations near the tips of a ureteric branch and each will develop to form a nephron, the basic functional unit of the kidney. Condensations first epithelialise (an example of a mesenchyme-to-epithelium transition) and then extends to make the early nephric tubule. One end of the tubules connects to nearby blood capillaries and will form the glomerulus, the other end connects to the local branch tip and the intervening length extends to form a hairpin

shape. As the branches continue to extend and bifurcate, new nephrons form, while earlier ones start to become functional. These soon produce urine that drains down the ducting system to the ureter and into the bladder from which it is excreted into the amniotic sac. In the mouse kidney, early nephrons start to produce urine at about E15.5, while the last of its many hundreds of nephrons are not formed until well after birth.

Every step in these processes involves signaling (Bard, 2002) with the responses depending on location and timing. The cell-autonomous process networks activated in the ureteric bud as a result of this signaling include that for proliferation, while those activated within the MM are for proliferation, for the mesenchyme-to-epithelium transition and for the differentiation of the various specialised epithelial types that subsequently form in nephrons. Also activated are the networks that direct the morphogenesis of the various components of the system: the ureteric-bud derivatives require the process that leads to epithelial bifurcation, while the MM requires the interactions that enable mesenchyme to form condensations that will epithelialise, and for the early tubule to extend and produce the characteristic nephron geometry.

Although there is now a large amount of molecular information about the instructions and processes that direct nephrogenesis in the mouse (Takasato & Little, 2015), much of the detail on the networks that drive individual aspects of its development still remains to be elucidated.

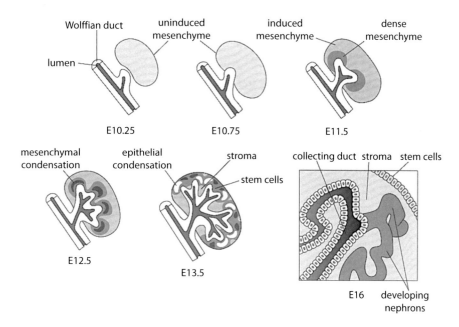

Figure A7.4 **Diagram showing some key events in the development of the mouse metanephric kidney over the period embryonic days (E) 10.25–16 (From Davies & Bard, 1998).** The kidney forms as a result of reciprocal signaling between metanephric mesenchyme (MM) and the Wolffian duct that leads to the production of the ureteric bud. MM become nephron stem cells that interact with capillaries within the early kidney to produce the glomerular system. The proximal part of the bud arborises within the mesenchyme to become the urine-collection system while the distal part lengthens to become the ureter.

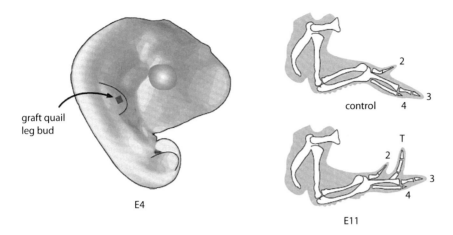

Figure A7.5 **The relative effects of lineage and environment.** A piece of quail limb-bud mesenchyme destined to become the femur is grafted into the apical tip of a chick wing bud from a 4-day embryo (E4). Because quail cells carry a nucleolar marker, the fate of each cell in the graft, and its surrounding tissues can be checked (arrowed region); histology on E11 embryos shows that the limb tissue changes its fate and becomes a toe (T) rather than a forelimb digit. The rudiment thus carries its hindlimb history but responds according to its new position. (Drawing based on data in Krabbenhoft & Fallon, 1989.)

millipedes). Plants, in particular, provide dramatic examples of structure repetition: they can produce thousands of leaves, flowers, fruits, and seeds, all essentially the same. Evidence to support the view that such modules are produced by a simple initiation system comes from the study of developmental mutants that affect the numbers of such modules: ectopic insect eyes (Chapter 8) have all their cell types, and extra digits in mammalian limbs have joints and nails (but note the interesting exception of sesamoid-derived digits in Chapter 18).

It should be pointed out that, in some cases, these modules are essentially templates in the sense that, although some may be identical replicates (e.g. hairs, leaves or fruits), others can be modulated in ways that depend on their locale (e.g. the different-sized digits on a forelimb are partially at least dependent on the level of a sonic hedgehog gradient; Sanz-Ezquerro & Tickle, 2003). In the invertebrate world, the hundreds of legs on a millipede have roughly the same morphology and size, but the limbs on a lobster vary greatly in size, even though their parts are much the same in terms of anatomical features. Here, as in vertebrates, segment details are regulated by Hox expression (Martin et al., 2016).

THE ORIGINS OF ANATOMICAL DIFFERENCES

As most organisms within a phylum carry developmental networks with very similar functional outputs, two immediate questions arise; these concern first the origins of the minor differences among individuals within a species and second the wider differences between related species that reflect more than scale (e.g. the right mouse lung has four lobes whereas the right human lung has three lobes). The minor differences among the individuals within a single species are mainly quantitative and reflect gender and individual mutations within an individual's genome. That all animals within a species are proportionate and roughly the same in size almost certainly reflects the systemic effects of endocrine signals such, for example, as growth hormone in mammals.

Superimposed on any variation in amounts of hormone and systemic intensity of response within a population are secondary effects which often reflect variation in the activity of local growth networks. An example here is face detail: all normal human faces have the same form and what marks one from another are minor differences, such as those in the relative size of the nose, the breadth of the forehead and the length of the chin together with the degree of melanisation in skin epithelium. These differences reflect the effects of secondary mutations in the local growth and pigmentation pathways that operate as the face develops. Such normal variation is discussed further in Chapters 17, 18, and 25.

The details of the origins of the larger differences across species within a taxon are rarely known, but probably start with the activities of the signaling pathways that underpin patterning and growth. These are, for instance, responsible for determining that the mouse left lung has a single lobe while the right one has four, or why all mammals have seven cervical vertebrae while the number in the birds and reptiles is highly species-dependent (Burke et al., 1995). It is not known whether this difference reflects natural selection on the basis of function or the random effects of genetic drift in early amniotes.

The differences in the details of process pathways, which go beyond the effects of the minor mutations seen in variants, probably reflect either different proteins in the networks or that a particular key protein has one rate constant in species A while its homologue has a different constant in species B. One interesting possible variable here is that a protein with alternative splice forms may express one or the other in a network. Livi and Davidson (2006) have found examples of proteins where the choice of which alternative splice form should be employed within a network depends on the cell's lineage.

Changes in the behaviour of a network can also be due to the developmental context in which they operate. Here, the detail of the differentiation of the vertebrate ectoderm provides a good example: it can be thick or thin in different parts of the body and the same basic patterning system can support hairs, feathers or scales depending both on the species and the body region as the classic work of Dhouailly and Sengel (1973) has shown. The effects of patterning can also be species-dependent and even individual-specific, as demonstrated by the external pigmentation patterns across the butterflies, giraffes and zebras or the patterning of teeth across the vertebrates makes clear. Many of these subtle differences derive from differences in the fine details of the behavior of individual proteins and reflect opportunities for mutation-derived variation. Other differences, particularly those that make individual pigmentation patterns unique, probably derive from random events, such as the precise geometry of local tissue domains and minor fluctuations in protein and metabolite concentrations (e.g. Bard, 1981).

THE ROLE OF THE GENOME IN DEVELOPMENT

Once the nature of DNA coding had been elucidated in the 1960s, it was initially thought that the genome drove embryogenesis, acting much like a computer program. Occasionally, this metaphor is helpful: the *C. elegans* worm develops from an egg to an adult with ~1000 cells, the fate of almost every cell is determined by its parent cell on the basis of lineage—there are few examples of fate flexibility in this highly determined system (Han, 1997). Things are, however, very different in more complex invertebrates, vertebrates, and plants, as a century of experimental work has shown. If, for example, a group of cells from a mid-stage vertebrate embryo is transplanted to a new location, the subsequent behavior of those cells can be highly influenced by the local signaling in their new environment to the extent that this can change its fate.

This is clearly shown in the classic experiment in which a fragment of mesenchyme (recognizable because of its nucleolar morphology) that would normally form a femur in a developing quail hindlimb was transplanted to the apical bud (the distal tip) of an early developing chick forelimb; this tissue would normally form wing digits (Fig. A7.5; see legend for details). The transplanted tissue did not form, however, a femur but mainly produced toes: it had "remembered" that it was fated to be a hindlimb but responded to the spatial patterning signals in its new location that instructed the tissue to display a distal phenotype (Krabbenhoft & Fallon, 1989). This emphasizes the point made earlier, that the behaviour of a tissue depends partly on its lineage (transcription factor expression), and partly on its environment (signaling).

Such experiments show that the future state of a cell within a tissue is determined by the demands that the current cell, and its environment make on the genome rather than being dictated to by the state of the genome. The main line of instruction is thus from the phenotype back to the genotype via the transcription factors in place, rather than the other way around. Perhaps the clearest way of seeing this comes from work on stem cells: it is now possible to take a differentiated cell expected to go down one pathway of development, modify its set of transcription factors and totally change its developmental fate (Pereira et al., 2012; Zakrzewski et al., 2019).

The idea that the genome acts as a program really arose because of the apparent inevitability of embryonic development, but it was one of those clever ideas that are disproved by simple experiments. In the context of development, it is now much more helpful to view the genome not as a program, but as a database of resources that is available on demand to a cell. Under normal circumstances, of course, the options that are available to a developing deuterostome cell, once the early embryo is past neurulation, are quite restricted (Fig. A6.3). Most changes to the state of differentiation are minor, although mesenchymal cells can occasionally, for example, become epithelia and *vice versa* (e.g. in the development of the epithelial nephrons from metanephric mesenchyme and the detachment of neural crest cells from their neural epithelium and their subsequent differentiation into a wide range of cell types that include mesenchymal cells and neurons).

There are, however, two areas where the computing metaphor is helpful. The first is the many networks that are used repeatedly in development; these can be viewed as subroutines with variable parameters. The second, which are the repeated tissue modules already mentioned, reflects a higher level of encoding. Here, it also seems appropriate to view the instructions for making these modules as reflecting the activity of genomic subroutines, although it is not clear how such a module might be coded or implemented. The challenge, therefore, is to identify a particular example that is both experimentally and genomically accessible; one possible tissue that is relatively simple in this context is the synovial joint (Decker et al., 2014).

KEY POINTS

- The general features of an organism develop early in development and provide the foundations for later, more specific characters (von Baer's laws).

- Developmental change can result either as a result of lineage (cell parentage) or as a result of signaling from other tissues. The latter is particularly important in deuterostomes.

- When signaling molecules from one tissue bind to surface receptors on another, conformational changes in the receptor activate a signal induction pathway that leads to activation of an existing transcription-factor complex on the genome (because the responding cell has both the receptor and transcription complex in place, it is said to be *competent*). The result of this activity is the synthesis of new proteins that may include further transcription factors.

- Such signaling may be uniform across a tissue, so causing uniform change, or it may vary in strength, so leading to a pattern of responses.

- The proteins that are produced as a result of signaling have many developmental roles:

  - They may have functional importance (e.g. as enzymic and structural proteins)

  - They may contribute to or activate one or more process networks

  - They may prepare the cell for future change (e.g. new signals, receptors, and transcription factors).

- Developmental change in animals and plants is driven by four main classes of process: 1) differentiation; 2) proliferation; 3) apoptosis (which together modulate growth); and 4) morphogenesis. The first three are cell-autonomous, while the last may involve groups of cells.

- A cell's lineage (or parentage) determines these possible future states through the expression of cell-surface receptors and primed transcriptional complexes on promoter regions of those genes that it will need. If there are alternative possibilities of change, there will be two sets of these proteins and the choice will be made by signals that bind to one or another receptor.

- A developing cell's current state of differentiation also includes the signals it can send to the cells of other tissues proteins and the proteins that it will need for possible future change.

- The results of morphogenetic changes, such as movement, adhesion, and folding depend on the local geometry of the embryo as well as on the properties of the participating cells.

- The control of development is thus distributed, with a cell's lineage and all aspects of its environment playing roles in effecting change.

- In development, the genome acts as a resource rather than a program, available on demand, as required by the state of a cell within a particular tissue at a particular time. This resource includes the genes and regulatory sequences for proteins, many of which participate in networks and tissue modules.

FURTHER READINGS

Gilbert SF, Barresi MJF (2016) *Developmental Biology* (11th ed). Oxford: Sinauer Press.
Wilkins AS (2002) *The Evolution of Developmental Pathways.* Sunderland, MA: Sinauer Press.

Wolpert L et al. (2014) *Principles of Development.* Oxford University Press.

WEBSITES

- A comprehensive collection of pathways: https://proteinlounge.com/pathway.php
- KEGG comparative pathway database and other bioinformatics resources: www.genome.jp

- A set of protein networks is downloadable from ProteinLounge: https://proteinlounge.com/pathway.php

Evolution and Creationism

Creationists who believe that the world was created by God or that, at the least, the creation of the human brain required divine intervention (the so-called theory of intelligent design) serve a useful role for evolutionary biologists: they make claims and criticisms that have to be refuted, something that can requires serious analysis and clear explanation. The first part of this appendix provides a set of questions to test creationism together with possible answers—a major aim here is to understand how creationist think about evolution. The second part provides a set of questions that creationists might ask to test evolution, and tries to answer them. The intention is to explore the implications of the two sets of views and to test them.

The focus of this appendix is on the Abrahamic faiths (Judaism, Christianity, and Islam) rather than on the faiths of other cultures. This is mainly because it is some members of these faiths, rather than others with creation myths, who particularly to want to deny that life today is the result of evolutionary events and changes that started almost four billion years ago. Readers who wish to widen their knowledge on this topic are recommended to browse through the fascinating Wikipedia file on creation myths.

Note: very little literature is cited here: interested readers should explore and enjoy the wide variety of web resources on this debate.

> All things bright and beautiful,
> All creatures great and small,
> All things wise and wonderful,
> The Lord God made them all.

This lovely, traditional children's hymn that Cecil Alexander wrote in 1848 summarises in the simplest terms the theistic view of creation, although it excludes all things dull, ugly, stupid, and mundane. If one takes this view, all the evidence for evolution given in this book is irrelevant, as it explicitly denies any role for God as creator. Some creationist groups, such as the Institute of Creation Research (websites) take what they view as a scientific approach to support their views and are prepared to consider experimentation, to argue about the facts and to ask hard questions about both theory and data; others feel obliged to attack evolution simply because it conflicts with the biblical view of creation which, in their view, needs no defense.

Arguments opposing evolution are important because they force evolutionary biologists to articulate answers and to test and even expand their understanding. Some questions, of course, cannot be answered fully yet because, as this book has emphasised, there are still gaps in knowledge about evolution; examples are details of the origins of life and the detailed evolution of the human brain over the last few million years. Nevertheless, it is to be regretted that creationist websites focus far more on querying the fine detail of evolutionary research than on explaining the data that supports evolution or finding new data that might contradict the evidence for evolution and test the Darwinian mechanism of evolution.

Public argument between evolution and creationism has been going on since the famous 1860 debate in Oxford between Thomas Huxley and Bishop Wilberforce (Appendix 2). Earlier evolutionary publications, such as Lamarck's *Zoologie Philosophique* (1809) seem not to have led to major theological arguments, probably because they were viewed as interesting but technical, although Chamber's *Vestiges of the Natural History of Creation* (1844) which was written for the wider public was publicly criticised on both theological and scientific grounds. An example of a more recent argument was the infamous Scopes trial in 1925 when John Scopes was prosecuted by the State of Tennessee for teaching evolution; he eventually won on a technicality rather than on the merits of the case.

There was also the more academic argument in 1949 between two New Zealanders, Dewar, a geologist, and Davies, a biologist, with Haldane, a major British evolutionary biologist and population geneticist. The former represented what they considered to be the "rationalist" anti-evolution approach, while the latter defended evolution. The debate took the form of question-and-answer letters and was published in the book *Is evolution a myth?* As this debate took place before anything was known about genomics and very little about developmental mechanisms, it inevitably focused on the fossil record, which was, of course, less detailed then than now. Haldane probably won but only on points: he at least tried to answer the questions posed but often could not do so completely because the information was lacking. The New Zealanders, in contrast, were far more interested in posing difficult questions about gaps in the fossil record than in providing detailed non-evolutionary answers to Haldane's questions.

The arguments between these two positions always focus on two distinct questions: did evolution happen and, if it did, how? The first asks whether organisms arose from pre-existing ones or were independently created; the second asks whether natural selection, Darwin's mechanism for producing new species, is correct. Most discussion focus on the first question as the second is more technical. Evolutionary biologists find overwhelming amounts of evidence that show that evolution happened with new organisms arising from old through descent with modification, while creationists deny that new organisms derive from old and believe that all were created *de novo* by a deity, perhaps allowing for a little change in organisms through mutation (e.g. breeding pigeon variants). Similarly, evolutionary biologists have discovered a great deal of evidence to identify how evolutionary change happens. Creationists do not, however, accept that any such evidence is relevant to what happened during creation and that it particularly does not apply to humans.

It would be optimistic, to put it mildly, to believe that anything written here will have much effect on the argument. Nevertheless, and in the hope of at least clarifying the battleground and explaining in detail their different perspectives, this appendix takes the form of a question-and-answer session between the two sides. The first part asks critical questions of creationism and the second asks critical questions of evolutionary science. It is to be hoped that both sides will feel that the criticisms, if not the responses, represent their views. Box A8.1 summarises the two sets of criticisms.

The following sections present the evolution (E) versus creationism (C) debate in an interview format. For each side, the procedure is first to give a brief summary of the basis for its position, then pose a series of questions (in bold, italic font) with some elaboration if needed (in italic font), followed by a paragraph or two for the response (normal font). Much of the content of these answers, of course, reflects points made earlier in the book. While each response might deserve further comeback comments from the other side, this would lead to long, tedious arguments, and it is left to the reader to decide on the merits of each case.

CREATIONISM

The claim

The living world today did not arise through simple science-based mechanisms but resulted from divine input. The theory of creationism exists in two forms: the full version is that each organism was created separately through divine activity and either that there has been no change with time or that only minor changes can occur as, for example, in breeding experiments (such as for new pigeon or dog varieties). Either way, new species cannot form other than by creation, but extant ones can, of course, die out (and form fossils). A further, perspective here comes from Rutman (2014) who suggests that God created the laws of nature in such a way that life and eventually humans were inevitable. In other words, God's role in human evolution

SOME EVOLUTIONARY CRITICISMS OF CREATIONISM
General criticism

- Creationism is a belief not a theory: if gives no explanations and makes no predictions.
- There are dozens of creationist myths from a wide range of cultures. They cannot all be right, but each one will be wrong if it cannot explain life today.

The data on the history of life

The creationist view cannot explain

- Fossil dating evidence and why dinosaur and human bones are not found together.
- The molecular data on mutation and its effects.
- The deep but imperfect similarities between very different organisms (evo-devo).
- The detailed history of life buttressed by vast amounts of fossil, sequence, and developmental data.

The theory of evolutionary change

- The bible, with its two very different accounts of creation, is inconsistent and just gives two more fables to add to a long list of world creation myths.
- Creationism cannot explain the large amounts of evidence in favour of the mechanism of selection (variation, selection, and speciation).
- Creationism involves magic to explain the origin of life, and its subsequent evolution.
- Human cognition is not fundamentally different from that of primates, but builds on their abilities.
- Intelligent design requires directing mutation at the single base level to micro-tune the necessary changes to make a primate brain human; such microtuning is incompatible with quantum mechanics because of the uncertainty principle.
- Finally, it is worth asking why does creationism not accept the fundamental tenet of natural theology: believers should seek to explore the nature of God through understanding his work in the natural world?

SOME CREATIONIST CRITICISMS OF EVOLUTION
General criticisms

- The perfect adaptation of organisms to their environment is evidence for creationism—no other explanation is required.
- Evolution is just a theory and has no more status than any other theory.
- The theory of evolution cannot make testable predictions and is, therefore, not a proper scientific theory.

The data on the history of life

- The fossil evidence is inadequate: too few fossils bridge morphologies across species.
- The fossil data is better explained by alternating major extinctions and periods of creation which were followed by minor tinkering than by continuous speciation.
- As one can never prove that hypothesized mating behaviours actually occurred in the deep past, computational analysis of ancient sequence data reflects no more than speculation.

The theory of evolutionary change

- Species, by definition, cannot interbreed, and the formation of new species is, therefore, impossible.
- The survival of the fittest is a meaningless tautology.
- Sophisticated organs are too complex to have evolved simply by selection acting on the effects of random mutation, particularly if they will not function until they have fully formed.
- Evolution can explain neither the human brain, and its abilities nor human social behaviour.
- Evolution cannot explain the origin of life.

would thus have been some 14 By after the birth of the universe or far earlier than others have suggested, albeit that the deistic time scale is not our time scale. The softer version of creationism says that Darwinian evolution can occur but is inadequate to account for the evolution of sophisticated anatomical features, such as the vertebrate eye and particularly the human brain. Here, a divine helping hand was required, and this was through **i**ntelligent design.

The mechanisms by which species were originally made is through God's undefined activity which either created organisms *de novo* or just sometimes helped them on their way. The basis for this view comes from Genesis. The logic to support it comes from perceived inadequacies in the scientific data and theory.

Evolutionary criticisms of creationism

General criticisms

E: *Creationism is a belief not a theory: it gives no explanations and makes no predictions.*
C: Creationism is not a theory but a correct description of the origins of life and the creation of biological diversity, based on revealed truth. All of life could have formed by an alternative method, such as evolution, but it didn't. The theory doesn't need to make predictions as what happened reflected divine will and not our logic.

Data criticisms

E: *The creationist view cannot explain evidence of the datings of fossils nor why dinosaur and human bones are not found together.*

C: There is no great problem with the very incomplete fossil evidence as this just consists of the remains of dead organisms. The presence of individual fossils gives no indication as to when their species was created or when it died out, while the fact that fossilised organisms that appear more evolved than those at lower levels merely reflects sampling problems. George Cuvier, the leading anatomist of the early 19th century, who discovered the great extinctions (and we can include Noah's flood here) by analysing layers of fossils, showed that each extinction was followed by a new bout of creation. Chateaubriand and Goss pointed out at around the same time that it was perfectly possible that God had created fossils, even coprolites (fossilised dung), much as, when he created man, he gave him a navel (the *omphalos* hypothesis). In short, the fossil record says nothing substantial, one way or the other, about evolution. As to scientific timings based on radioactive decay, we really do not know how God established the mineral details of rocks when He created the world. Scientific timings are essentially meaningless,

There is a further point here. There is no serious evidence of change over time, just differences among organisms. Indeed, there is no direct evidence of inheritance as the fossils reflect individuals not populations. As to timing, Genesis makes it clear that the sun was not created for several days after the "beginning" and what the bible refers to as days in those times obviously refers to undefined periods. The detailed times of creation given in the bible reflect God's timings not ours.

The reason why dinosaur and human bones are not found together is obvious: in those early days soon after creation, there were many more dinosaurs than humans because the former had been created earlier so the latter would have been much rarer and harder to find. Moreover, as they would not naturally have lived together, no reasonable person would expect their bones to be in the same place.

E: *The creationist view cannot explain the molecular data on mutation and its effects*

C: Minor mutations are no problem. Consider the creation of the Burgess Shale fossils: all the basic phyla are there and, while some got lost, few new ones emerged before the Cambrian extinction and the next round of creation. As to the similarities among the DNA sequence data, this merely means that God used a fairly standard template in creating related species. Any similarities reflect coincidences, and the genomic data prove nothing.

E: *The creationist view cannot explain the great but imperfect similarities between very different organisms (evo-devo).*

C: Analysis of contemporary organisms just shows that there is a repertoire of possible developmental mechanisms and appropriate ones were chosen for each organism; and if there has been a little mutation since their creation, it means nothing.

Theoretical criticisms

E: *The bible has two very different accounts of creation and so is inconsistent. It just adds two more fables to a very long list of creation myths (see the Wiki on creation myths).*

There are two versions of the creation of the world in Genesis. The first (1:1–2:4) is based around the events of a week ("And it was the evening and the morning ......") and has God creating Adam and Eve together on the sixth day. The second (2:5–25) starts with the creation of the world and of man, includes the Garden of Eden and ends with the creation of woman from the rib of Adam. Version one creates man after mice and flies and version two before.

Detailed academic analysis (e.g. Friedman, 1987) provides strong evidence that the second version of the creation was written in about the 9th century BCE in Judea, and the first version was written in about the 6th century BCE by the temple priests. Moses, himself, could not have written the Hebrew bible as he died in the 13th century BCE some 300 years before the Hebrew alphabet was invented and some 2,500 years after the rabbis believed that the world was created (5,776 years ago). All this suggests that the versions of creation told in the bible are just stories invented to make sense of a mystery.

C: The bible reports on what happened and there are indeed two versions of creation. The fact that they were written down much later merely reflects that difference. The bible does not, of course, say how God created each species and perhaps some scientific knowledge is helpful here, and one can see that this ambiguity is bound to excite differences in interpretation. We just do not know how much of these writings reflects signal and how much just noise (Silver, 2012). As to the creation myths of other cultures, they are simply wrong.

The rabbis are interesting on the two creation versions and have discussed them in depth. Rabbi Hertz asserts strongly that the two versions tell the same story, with the second amplifying the first. He also points out that Moses should not have been expected to be up to date with modern science. A Hasidic view is that the former story refers to the earthly creation of man and the latter to his spiritual creation—and, implicitly, that the creation stories should, therefore, be seen as metaphor. Rabbi Lieber says in the Aitz Haim edition of the Five Books of Moses that "It is a book of morality not cosmology" and hence should not be taken literally. In general, the modern orthodox view is that evolution merely reflects the slow way that God drives the world and is not particularly relevant to what happens on a day-to-day basis. Far more important to the orthodox rabbis are the laws that are contained in the five books of Moses, albeit that they still believe in creationism.

There is a deeper point here that scientists don't understand. The words of holy texts were divinely inspired and the details of how the text were actually written are not important. The words are as the words are and we accept them as coming from God. What is not clear is the meaning of the words, and it is this that we are obliged to find. It is the answers that we find in that task that tell us what we should believe and how we should behave. You focus on secondary and not very interesting matters.

E: *The creationist view cannot explain the evidence in favour of natural selection (variation, selection and speciation).*
C: We do not need to do so as there is no reason to suppose that modern research has anything to say on what God did all those centuries ago.
E: *The creationist view uses magic to explain the origin of life.*
C: It is insulting to say that acts of God are just magic. Creationists have little problem with minor DNA mutation and tinkering that lead to variants (such as the various dog breeds) but believe that the major organism groupings were created at around the same time. Here, it is worth noting that there are no antecedents to the Ediacaran fossils that vanished and to the later rich fauna of the Burgess Shale with their complete range of organisms and cell types. Creating the diversity of flora and fauna was clearly complicated and there may well have been, as Cuvier first suggested two centuries ago, a series of catastrophic destructions which were followed by new acts of creation. The evidence is very thin, and the jury is still out on this.
E: *Human cognition is not fundamentally different from primates but builds on their abilities.*
C: Of course, human cognition is fundamentally different from that of apes! The latter cannot talk and have no sense of history or culture or indeed cultural inheritance. Most important, and what distinguishes us from animals is their complete lack of morality (Matsuzawa, 2013). All of the Abrahamic faiths see humans as so profoundly different from all other animals that it is inconceivable that their presence can be accounted for in any way other than by the hand of God.
E: *Intelligent design or directing mutation at the single-base level to micro-tune evolution (e.g. to account for human cognition) to generate the power of the human brain is incompatible with quantum mechanics*. *Genomic manipulation of individual bases which involves interactions at the level of a few angstroms (10^{-8} cm) cannot be done with the necessary degree of precision because the uncertainty principle operates at such distances. In other words, intelligent design is impossible.*
C: There are two points here. First, we do not know just how God works and there is no reason to suppose that the laws of physics apply to Him as they apply to us. But suppose you are right that there was bush of early primate species (Chapter 15); this does no more than prove that God needed several efforts before he got *homo sapiens* "right" — no one thinks that this was an easy thing to do!

E: Finally, Why is creationism so against science? There is longstanding tradition of natural theology whereby believers seek to explore the nature of God through understanding his work in the natural world.

C: This is, of course, correct, but any understanding of the world today cannot be extrapolated back to explain the creation. You don't understand it and nor do we, and why bother? It is irrelevant to us now.

EVOLUTION

The claims

Evolution claims that the living world today derived through small changes that date back to a very early bacterium, with novel organisms arising through natural selection. The experimental data to support evolution comes from the vast amount of research work done in three separate and independent areas of research (Chapters 5–8).

1. *The fossil record*: Small differences between ancient organisms allow their relationship to be established on the basis of common descent with modification. This analysis gives trees of organisms (cladograms) going back in time to a common ancestor. The validity of this hierarchy is buttressed by the independent chronological information derived from the geological record and radioactive dating.

2. *DNA sequence analysis*: Comparisons of tens of thousands of sequences in different living animals whose proteins have similar sequences and functions shows that these sequences can be traced back to a common ancestor sequence. This work gives trees of descent on an evidential basis entirely different from that of cladograms. It is a prediction that phylograms and cladograms show should group living organisms in the same way. This prediction has been confirmed across the living world, wherever it has been examined.

3. *Protein function*: Comparisons of the functional mechanisms underpinning the development of anatomical features (e.g. eyes) in very different organisms show that they use homologous proteins (e.g. Pax6) for very similar purposes. This means that, although these protein sequences are now different as a result of mutation, each can be traced back to a common-parent sequence in an organism that may have lived millions of generations ago.

Such work allows relationships among distantly related organisms (e.g. mammals and insects) to be elucidated.

The theory of how evolutionary change happens (natural selection) is built on four successive stages, each supported by experimental evidence. (1) *Variation*: In any population of a single organism, each individual is distinct because its genome is a roughly random mix of the genes from its parents slightly modified by new mutations; the population thus reflects a spectrum of genotypes and phenotypes. (2) *Isolation*: A subgroup of an existing species becomes reproductively isolated in a novel environment; this small population will have, due to random sampling, a spectrum of phenotypes and genotypes slightly different from that of the parent population that will, due to genetic drift and further mutation, become amplified over time. As a result, modified proteins and networks with atypical properties will lead to the population having novel phenotypic variants. (3) *Selection*: Some of these variants may do well in their new environment and in due course become the dominant and eventually the unique phenotype. (4) *Speciation*: Further mutation combined with chromosomal change will eventually result in this population being unable to breed with the original population—it will then be recognized as a new species. As to timing, stages 1–3 can be relatively rapid, but the final stage is very slow. Thus, after ~4 Mya of separation, dogs and wolves can still interbreed because they still have the same number of chromosomes. Speciation can, however, be recognized at earlier stages if the morphological differences are sufficiently pronounced (the criterion for fossil species) or if organisms refuse to mate outside of the group.

The evidence to support this framework comes from many sources but is more limited than that showing that evolutionary change has occurred. This is because there is no simple way of studying all of these events in a single set of experiments. To study even aspects of speciation: the events of tens of thousands, even millions of generations have to be compressed into a few years through artificially amplifying mutation and selection pressures (Chapters 20–22).

Perhaps the greatest strengths of the history of life and theory of evolutionary change are that there is no evidence that is incompatible with it and that the predictions that it makes have been conformed rather than be proven false.

The creationists' criticisms of evolution

General criticisms

C: *Evolution is just a theory and has no more status than any other theory.*

E: There are three points here. First, theories are not intrinsically inadequate. Anyone who flies bets his or her life on three very different theories being not only correct in principle but quantitatively right too: the theory of mechanics, the theory of gravity, and the theory of thermodynamics. Second, the amount of evidence to support evolution (Chapters 6, 7, and 11) is so great as to be overwhelming—it is not theory but fact and there is no evidence to support any other view of the origins of today's biodiversity. Third, while much of the evidence comes from analysing descent with modification, actual timings depend on radioactive decay analysis. We know that this is correct because it is the same area of science that underpins atomic and nuclear bombs—and they work exactly as predicted. Perhaps more important here is that, even where there are gaps in the detail of our knowledge about evolution, no fact has emerged that contradicts the view that all of life evolved from very primitive beginnings ~4 Bya and then diverged. The sequential timing and the evidence of slow change refute ideas of creationism which does not predict gradual change.

C: *Neither the theory of evolution nor the mechanism of natural selection can make testable predictions; on their own terms, neither is a proper scientific theory.*

E: This statement is just wrong. It is, for example, a simple prediction of evolution that, if we analyse very different and independent types of data (e.g. contemporary DNA sequences, the anatomy of ancient fossils, and protein function), they should predict the same taxonomic groupings — and they do. It is also a prediction that the cladistics analysis of fossil anatomy should predict the approximate order for when those organisms lived. In addition, it is a prediction that no organisms will be found whose cladistic status clashes with the established history of life (e.g. there should be no teleost fish with a hand or gymnosperms with flowers). No such organisms have so far been found. All three predictions have so far been confirmed together with many more. The onus is on creationists to find data that contradict the theory of evolution and to make testable predictions that, if realised, would contradict the theory of evolution.

The second part of the criticism suggests that it is impossible to test the idea that speciation derives from the gradual accumulation of small differences. This prediction is proved possible by the natural experiment known as ring species: these are a few families of related organisms that originated in one place and migrated left and right, eventually meeting up to form a ring (Chapter 22). A nice example is the ring of greenish warbler varieties that surrounds the Himalayas. If one examines mating behaviours between neighbours, almost every group is found to mate with both of its adjacent groups. The exception occurs at a region at the north where there is a discontinuity: the group that evolved after a westerly migration cannot breed with that which evolved after an easterly migration. By the definition of failed intergroup breeding, a new species has formed by a series of small changes in accordance with prediction. Other experimentation (Chapter 14) has shown that selection acting on isolated groups can lead to novel phenotypes and reproductive isolation, the key step on the way to speciation.

Data criticisms

C: *The fossil data is better explained by alternating major extinctions and periods of creation that were followed by minor tinkering than by continuous evolution.*

E: This idea, first put forward by Cuvier at the beginning of the 19th century was wrong because it was based on inadequate data. The fossil evidence now shows that, for each extinction event, there was always continuity of some organism families; we can even follow a few transitons. The ancestors of today's mammals, for example, were the stem mammals that dominated the late Paleozoic era but were mainly lost in the Permian extinction. A few, however, survived (we know this because we have the fossils), becoming small rodent-like creatures that slowly evolved to become small monotremes, marsupials and placental mammals during the Mesozoic Era. All survived the K-T extinction to radiate during the Cenozoic era, so producing the diversity of the mammalian taxa that we see today.

C: *The fossil evidence is inadequate: too few fossils bridge the morphology between species.*

E: While it is true that the fossil record, depending as it does on dead organisms being preserved, is limited, it is still remarkably good, and getting better. Indeed, in a few cases, such as the Burgess Shale taxa, even soft tissue is preserved. As to the lack of transitional fossilised organisms, many are known and described in Chapter 6. There are, for example, a group of feathered dinosaurs that bridge the transition from dinosaurs and birds as well as a series of fossil fish showing how limbs evolved from bony fins (see Table 14.10). A full list of the several hundred currently known transitional fossils is readily available online (Websites).

C: *As one can never prove that particular mating behaviours actually occurred in the deep past, the computational analysis of sequence data reflects no more than speculation.*

E: The first part of the criticism is, of course, correct. The evidence here comes partly from analysing anatomical relationships among groups of similar organisms to identify lines of descent through mating and partly from showing that similar organisms have DNA sequences that clearly descended from an original ancestor sequence through mutation.

The perfect adaptation of organisms to their environment is evidence for creationism. *It is hard to see how a new species can perfectly adapt to its environment in a relatively short time.*

On the contrary, the perfection of adaptation is a very strong argument in favour of evolution and against creationism because the latter does not allow for modification in a population as the environment changes. In practice, adaptation reflects an ongoing but slow selection process as minor variants in a population due to mutation and genetic drift are continually being subject to natural selection. This is because an environment is only rarely static: it is always altering due to changes in the weather, food availability or the types of predators. Hence, those with slightly fitter phenotypes, defined as those that leave slightly more fertile offspring in the changed environment (not as being stronger!), will flourish at the expense of others and will eventually become the new norm. The perfection of adaptation reflects a dynamic process, continually tuning phenotype to the environment, as the data on how human skin pigmentation levels vary with latitude demonstrates (Chapter 24). A phenotype that seems static just reflects a current moment in a very slow process of change.

There is a second point: Nature is not as perfect as it might see as adaptations build on pre-existing tissues and organ. As a result, this process leads to far less "perfect" solutions than one might suppose. A classic example of how evolution works in an opportunistic way rather than by the achievement of perfection is provided by the panda's paw (Davis, 1964). Among the bears, and indeed most land-based vertebrates, the forelimb paw starts off in the embryo with five digits; only the panda has five digits together with an opposed thumb that it uses for shredding bamboo shoots. This extra thumb does not, however, form through a repatterning of the basic digit format. Instead, it forms by extension of one of the small wrist bones (Chapter 12). What strengthens this evidence is that other unrelated organisms with an extra sixth digit (e.g. (moles and elephants) show exactly the same adaptation.

There is an extra wrinkle here: although the panda only uses its front paw for eating bamboo, the alteration affects the hind paw too: the panda not only has six fingers but also has six toes, even though the extra toe has no function (a counter example to the perfection-of-nature idea). The mutation that causes this unusual morphology clearly leads to a minor change in the patterning of all limbs. The evidence shows that this adaptation has nothing to do with creation or intelligent design, but just represents a simple adaptation based on extra growth in a bone that is present in all mammalian limbs. It is hard to think of an example where the opportunistic nature of evolution is more explicitly shown.

Theoretical criticisms

C: *Species, by definition, cannot interbreed and the formation of new species from an older one is therefore impossible.*

E: As detailed above, speciation occurs very slowly when a subgroup of an existing species becomes isolated. In due course, differential mutation and eventually chromosomal change in the two populations make them first only able to produce sterile hybrids and then unable to produce any offspring at all. Examples are particularly provided by the breeding behavior of the different species of *Equidae* and the ring species just mentioned (Chapter 22). A failure of interbreeding is the culmination of a series of separation events.

C: *The survival of the fittest is a meaningless tautology.*

E: If fittest meant strongest, it would be meaningless, but it doesn't. In the context of evolution, fittest has the specific meaning of leaving the most offspring capable of reproduction that would themselves produce fertile offspring. Thus, if there is a subgroup of organisms whose offspring show better fitness in an environment than those of a parent group, the subgroup will eventually become the dominant group because its members will leave more fertile offspring than the parent group.

C: *Sophisticated organs are too complex to have evolved simply by selection acting on the effects of random mutation, particularly if they will not function until they have fully formed.*

E: The difficulty with discussing the evolution of sophisticated organs is that, because they are usually comprised of soft tissues, they are very rarely preserved in fossils. Where they reflect hard tissues, such as skeletal changes, the fossil record is often very good at demonstrating the steps of change. As to soft tissues, the evolution of the camera eye has, from Darwin onwards, been viewed as the classic problem in this context. We now understand the evolution of the mollusc camera-type eye (Chapter 20).

The evidence for this comes from two very different areas. First, zoological work has shown that there is a range of mollusc eyes that range from little more than a curved dish of photoreceptors to the squid's fully formed camera eye (Salvini-Plawen & Mayr, 1977; Chapter 19; Fig. 20.2). It is thus easy to see how the camera eye could have formed from a few surface photoreceptors through a sequence of small changes, with each having a small selective advantage. Second, an evolutionary genetics analysis of the evolution of the camera eye has been undertaken by Nilsson and Pelger (1994) on the assumption that a change in eye morphology that gave a 1% improvement in resolution would give a beneficial advantage to its host organism. Their analysis showed that it would take about ~350,000 generations, and probably the same number of years, for a minimal eye based on a set of flat photoreceptors to become a full camera eye; this is a trivial amount of time on an evolutionary timescale.

Second, complex organs can evolve from simpler ones that might already be fully functional for another role—this is known as exaptation. Thus, Clack (2012) has described how the lung system of land-based animals evolved from the swim bladders of fish and how walking limbs evolved from swimming fins. Similarly, wings evolved from forelimbs and flight feathers in birds evolved from dinosaur feathers whose original role had been insulation and then probably for courtship (as it still is for ground birds). There is no great conceptual problem in the evolution of complex organs and tissues over long periods provided that they each step can be shown to fulfil some function—these functions, however, can change.

C: *Evolution can explain neither the human brain, and its abilities nor social behaviour.*

E: The evolution of the unique features of the human brain, particularly those associated the cerebral cortex, is still unclear. The increase in size was not particularly rapid, taking perhaps two million years (~100,000 generations) to evolve from the 600cc brain of *Homo habilis* to the 1500cc brain of modern man. Even the major increase in the size of the cerebral cortex over this period probably involves only about 2–3 extra neuronal cell divisions.

More interesting is the increase in function, particularly with respect to language, an area about which we still know little. First, it is clear that human neuroanatomy both developmentally and physiologically builds on that seen in more neurologically simpler vertebrates, and in particular, nonhuman primates (Bystron et al., 2008). The two unique features are: (1) the additional cortical volume, particularly in the parietal region and (2) the functional area associated with language. This does not seem to reflect an unusually fast degree of evolutionary change for the ~6.5 Mya since the origin of the branch of primates leading to humans, particularly given that chimps, our closest relation here, can learn the basics of language if they are taught it.

As to social behaviour, it is hard to see any major differences between human and other primates that cannot be explained by the acquisition of language and our remarkable facility for cultural inheritance that builds on it. One can imagine how primitive modern humans would appear if they grew up with no adults to teach them language and social behaviour. Darwin (1871 and 1872) devoted considerable efforts to showing that human intellectual and social abilities have built on those to be found in the animal world, and modern research has confirmed this, particularly in chimpanzees (Matsuzawa, 2013; Melis, 2013). In short, while the details of the evolution of the human brain are not understood, its size and functions do not pose problems that are very much greater than understanding the brains of lesser primates.

C: *Evolution cannot explain the origin of either simple or complex life*. *Evolution has two problems here: first it cannot account for the origin of what it claims was the original organism, probably a bacterium; second it cannot account for the sudden richness of the Cambrian fauna that, it is claimed, suddenly appeared ~520 Mya. Such organisms, with their rich and novel complement of cell types and tissues have no antecedents and are obvious candidates for creation.*

E: Evolution as a subject makes no claim to explain the origin of life and has never done so; it considers how life forms change over time. That said, clues as to how life began are beginning to be discovered: much is known about how the complex biochemical molecules that underpin proteins and nucleic acids, both RNA and DNA, can form from very simple molecules; experimentation has shown how lipid drops can form and can contain complex molecules (Chapter 9). Whatever its inadequacies, this explanation makes far more sense than one based on either divine creation (How were all these complex biochemical networks assembled?) or through directed panspermia, the idea that basic life forms came from somewhere else in the universe and merely evolved here, a notion than does no more than relocate the question of the origin of life.

As to the suggestion that the Cambrian life forms emerged, as it were from nowhere, around 530 Mya, this is simply not so. There are early multicellular organisms dating back to >1 Bya, Ediacaran organisms (fronds and the like) dating to about ~580 Mya and fossil and trace evidence that sponges, and cnidarians (sea anemones, jellyfish, etc.) together with worm-like organisms, such as *Dicksonia* and particularly *Yilingia spiciformi* were all present well before the beginning of the Cambrian (Chapter 12). Although the details of how the Cambrian organisms evolved remain opaque, mainly because small, soft-bodied organisms rarely make good fossils, the evidence of these earlier organisms shows that the Cambrian fauna did not just suddenly appear as if they had been newly created.

In short, the data to hand shows that all the creationist objections to evolution can easily be answered.

CONCLUSIONS

Where does this question-and-answer session leave the two approaches? The evolutionary position remains extremely strong and nothing that the creationists say contradicts it. That said, there are still some gaps in our understanding of evolution: more information is still needed, for example, on how the many phyla of multicellular organisms with their various cell types evolved from unicellular and simple algae. Little is understood about the speed with which change can occur in new niches and after extinctions, although insights from systems biology on the effects of mutation on networks together with the possibility of soft evolutionary inheritance (e.g. via RNA sequences—Chapter 17) are giving some insight here. It is also true that our understanding of how the human cerebral cortex evolved is inadequate, although this may not be as hard a problem to solve as once seemed (Chapter 25). The evidence to support the science of evolution is overwhelming but not complete. Perhaps more important, there is no evidence to contradict it.

The crux of the creationist case is that, although subspecies may form by mutation and selective breeding, the species themselves were created independently rather than from ancestors. The evidence shows that this view is scientifically untenable as it cannot explain how the very detailed analysis of the mutational differences in the DNA sequences of similar genes in related organisms shows heritable relationships. This data is so robust that it is stretches credulity beyond breaking point even to suggest that these relationships arose by chance. Arguments against the scientific data on timing are specious—unless one takes the view that, when God set all the atomic clocks ticking, He made the time of creation location-dependent. Similarly, the creationist criticisms of the fossil data have become progressively weaker as more have been discovered. As to the mechanisms by which speciation takes place, the lovely example of the greenish warbler ring species shows how small changes can lead to new species forming and directly counters creationism.

There is, however, a different sort of problem with creationism, which is that, because it holds that God was responsible for all of life and that creation happened in the dim and distant past, its subscribers do not feel obliged to defend their views in any scientific way. It is enough to say that God created it all, even if He did it in several steps and perhaps allowed a little mutation to help Him; and, anyway, Who are we to question Him?

Any scientist who tries to explain creationist approaches inevitably feels like a defense lawyer desperately looking for something, indeed anything, to get their client off the hook while that client keeps saying "why bother, I know, and any reasonable jury who knew their bible would see that right is on my side." It does, however, it seems slightly sad to subject, which is as emotionally rich for humans, as God to this sort of scientific dissection when it is simply not necessary. Perhaps it is wiser to view a deity as an ideal that guides our moral actions rather than a "God of the gaps" someone who micromanages nature and becomes decreasingly significant as the gaps that this God is required to fill become ever smaller as science advances.

As this appendix shows, any evolutionary biologist should enjoy the challenge of trying to work out how creationists think and to provide the evidence that counters their views. It is to be hoped that any creationists who read this will feel that their criticisms have been adequately represented. One might even hope that a few creationists might even change sides, but that is unlikely not only because the mindsets of the two communities are so different but because books like this are hardly bedside reading for creationists! One side looks carefully at data and the other starts with the belief that the first and second chapters of Genesis reflect historical reality, even though they were clearly written well after the events that their authors believed they were describing. The former trust the well-proven methodology of science while the latter feel that, wherever else it that science might explain, evolution is a special case where science fails.

WEBSITES

- The Wikipedia entries on *Creationism*, the *List of Creation Myths* and the *List of Transitional Fossils*.

- Institute of Creation Research: www.icr.org

Glossary

Adaptation A phenotypic feature whose evolution reflects the selective pressure exerted by some aspect of the environment (in the widest sense).

Afrotheria A clade of placental mammals that evolved in Africa after the breakup of Gondwanaland.

Agnathan A taxon of early-evolving fish that lacked jaw bones.

Allele An alternate form of a gene. For a gene defined by a DNA sequence, it represents a variant sequence. For a traitgene, it represents the heritable component of a variant phenotype.

Allopatric speciation Speciation that occurs when a subpopulation becomes geographically isolated from its parent population.

Altricial Young organisms who require parental care after birth (e.g., mammals).

Altruism A form of behaviour whereby an organism performs some task that benefits its siblings or group at the expense of its own future reproductive capacity.

Amniotic egg A vertebrate egg for in which the embryo makes a series of protective membranes that ensure it does not need to develop in water. It is a synapomorphy of all amniote clades.

Amphibia A taxon of early vertebrates that evolved from sarcopterygian fish, retaining a fish-like larval form. Their modern descendants are the Lissamphibia.

Anagenesis (otherwise known as phyletic transformation) The process whereby a species changes its form over time to the extent that it comes to be considered as a distinct species.

Anapsids A taxon of sauropsid descendants that lost their temporal foramina and that now includes turtles and tortoises.

Anatomically modern humans (AMH) This is the standard name for all members of *H. sapiens*, a species that originally evolved about 200 Kya in south-east Africa (early European examples are known as Cro-Magnon man or stone-age man).

Ancient DNA DNA from organisms that died some thousands of years ago but whose DNA is still sequenceable, in part at least.

Anteriposterior patterning A key event in early development that establishes the head-to-tail organization of the tissues.

Anthropocene The current epoch of time that began when humans started to have a significant effect on Earth's ecosystem through their numbers and other effects on the environment. It dates back to 1945, a time marked by the presence in the rocks of the radioactive signature of atomic bombs.

Archaebateria The minor taxon of bacteria whose synapomorphies include cell membranes mainly made of L-glycerol-ether lipids with pseudomurein coats, histones, and intron-processing. They were initially identified on the basis of the stability of various clade to temperature, salt concentrations, etc.

Archaeplastida A taxon of eukaryotic phyla whose members contain chloroplasts with a double membrane and that originally, at leas, possessed two flagella (they are bikonts) The taxon includes plants, red and green algae, and glaucophytes (unicelled eukaryotes).

Archosaurs A taxon of diapsid reptiles characterized by several synapomorphies that include a preorbital fossa, socketed teeth, and a fourth trochanter (a ridge on the femoral). They included the dinosaurs, crocodiles, pterosaurs, and their descendants.

Artificial speciation Attempts to produce subgroups of a species that show the properties of separate species. So far, advances have been restricted to an unwillingness to mate.

Avemetarsalia A taxon of archosaurs that included dinosaurs and birds. They all had a flexible ankle joint.

Balanced polymorphism A rare genetic situation where the homozygotes of two alleles of a gene are each less fit than the heterozygote.

Bayesian tree construction This approach constructs a likely phylogenetic tree on the basis of as much data as possible and then using Monte-Carlo and Markov-chain methods to generate more trees. After a million or so simulations, the program identifies the tree with the greatest probability of representing the true phylogeny.

Bikont A very high-level taxon of phyla that carry two flagella and includes plants and algae.

Bilateria The taxon of animal phyla that show essential mirror symmetry, at least during their early stages (i.e., all Metazoa except the Coelenterata, which show rotational symmetry, and the Placozoa [sponges]).

Biramous limbs The double limb possessed by some arthropods. The inner limb is for walking while the outer is sometimes a gill.

Bootstrap resampling A means of validating phylogenetic analyses by using various subsets of the data to check the robustness of the analysis but adding resamples of the original data so that the original sample size is unaffected.

Boroeutheria A clade of placental mammals that evolved in the supercontinent of Laurasia after the breakup of Pangaea. Laurasia would later fragment into North America, Europe, and Asia.

Bottleneck A dramatic reduction in the size of a breeding population as a result of a catastrophe or as a result of separation of a subpopulation from a larger one (*see* Founder population). Protomammals in the early Mesozoic Era underwent a nocturnal bottleneck which ensured that only species able to live in the dark and underground survived.

Cambrian explosion The Period (541–485 Mya) during which the great majority of contemporary animal phyla were first identified.

Chordates A clade of deuterostome organisms characterized by possession, in the early developmental stage at least, of a long cartilaginous cylinder (or chord) ventral to the nerve cord.

Cis-acting element A DNA regulatory sequence that controls gene expression and that works through direct interaction rather than through an RNA or protein intermediary.

Cladistics A tissue classification that is based on evolutionary lineage and is based on rules of the type *Species P derives with modification from a last common ancestor Q.*

Cladogram A hierarchical tree based on cladistics analysis that details the evolutionary history of some taxon.

Co-evolution This occurs when two phenotypic features evolve to benefit one another. It can occur both in a single species when a giraffe's neck elongated and its heart became modified to cope with the new pressure requirements and in a pair of species when a parasite and a host adapt to one another.

Coalescent theory The approach to reconstructing the evolutionary history of an organism on the basis of sequence variants and a model of population genetics. Simulations are done to work out when in the past the different sequences would have coalesced back onto that of a last common ancestor. A strength of the approach is that it can also estimate the population history of the species and other such parameters.

Coalescent time An estimate of the number of generations required for the backward running of mutation to result in two variant DNA sequences to coalesce onto a most recent common ancestor sequence.

Competence A term describing the ability of an embryonic tissue to partake in a future developmental event. It usually means that the tissue has the appropriate cell-surface receptors to accept an initiating signal and transcription factors to respond to that signal by activating the requisite mRNA transcription.

Crurotarsi A taxon of archosaurs that includes crocodiles and is characterised by a semi-flexible ankle joint.

Cultural inheritance The transmission of knowledge across generations by communication or imitation rather than by genetic inheritance.

Cumulative cultural inheritance The ability to add some facility to an organism's cultural inheritance and pass on both to the next generation.

Cyanobacterium A phylum of diderm eubacteria that obtains energy through photosynthesis.

Deep homologies A term invented by Shubin et al. (2009) and used in evo-devo work to describe contemporary molecular homologies that reflect events in very ancient ancient ancestor. The best known examples typically lived before the protostome and deuterostome clades separated (c. 540 Mya) and were thus shown by *Urbilateria*.

Deuterostome An embryo whose initial gut cavity becomes the anus and whose second gut opening becomes the mouth (or that evolved from such an animal). This taxon includes all chordates and echinoderms.

Diapsids A major taxon of reptile species characterized by their skull having two temporal foramina posterior to the orbit on each side, or that descended from such animals. The taxon includes all birds, archosaurs (dinosaurs, pterosaurs, etc.) and contemporary reptiles.

Diderm A prokaryote with a cell membrane composed of two bilayers separated by polysaccharide layer.

Diploblast A taxon of animal organisms that includes the Cnidaria and the Ctenophera and whose early embryo has two germ layers, an ectoderm and an endoderm, both of which mainly give rise to epithelia. These organisms show radial symmetry.

Distance matrix A matrix whose components measure the genetic difference between specific pairs of regions in a set of homologous sequences.

Ecological separation A form of speciation in which reproductive isolation derives from specific ecological conditions (e.g., water depth, foliage type, and saline concentration).

Ectopic tissue An extra tissue.

Ediacaran Period The geological period (635–0541 Mya) before the Cambrian Period during which the first large organisms evolved. Early Ediacaran taxa were lost in the extinction at the end of the period leaving only the Metazoa and Algae as multicellular organisms of any size.

Endocrine signaling A form of intercellular signaling in which signals are transported between two different cell types through the vascular system.

Endosymbiosis The process whereby one organism engulfs another and, instead of breaking it down completely, develops a symbiotic relationship with it. The mechanism is thought to account for the origins of eukaryotic nuclei, chloroplasts, mitochondria, together with a great deal of molecular machinery.

Epigenetic inheritance Inheritance that is not through a DNA sequence but through secondary modifications such as methylation.

Epistasis A term from population genetics that reflects the interactions needed between trait-alleles to account for the observed phenotype distribution.

Eubacteria The major taxon of bacteria whose synapomorphies include cell membranes mainly made of R-glycerol-ester lipids with murein coats.

Eutherian mammals Those mammals with a full chorioallantoic placenta that give birth to newborn that are often capable of walking.

Evo-devo The subject of comparative embryology as interpreted through homologous proteins from very different organism having similar functions.

Evogram A phylogram or a cladogram with additional information (e.g., pictures, timing, anatomical details, and clade descriptions).

Evolutionary stable state A genetic state of a population that is stable to small changes.

Evolutionary stable strategy A term that derives from game-theory simulations and represents a behavioural dynamic for a population that is particularly stable over time.

Exaptation A property that initially serves as an adaptation in one context and then as a different adaptation in another (e.g., feathers were first used for warmth and later for flying).

Excavata A major supergroup of unicellular eukaryotic phyla that have diminished mitochondria and phylum-dependent numbers of flagella. Some, such as *Dictyostelium discoidium*, a slime mold, include a social period in their life cycle.

FECA The First Eukaryotic Common Ancestor; this was the first organism to acquire a nucleus by endosymbiosis and, under the standard assumption that this very rare event only happened once, the parent species of all other eukaryotic species.

Fitness An evolutionary measure of reproductive success essentially measured in terms of the likelihood of genes being passed on or in the number of offspring produced that are fertile (*see also* Inclusive fitness).

Fossils Preserved remnant of dead organisms, implicitly meaning that the organic material has been replaced by stone or some other substance.

Amber fossils Small organisms preserved in tree resin.

Cast or impression fossils The shape of the organism is maintained in the surface of a rock, but all evidence of the organic matter has been lost.

Compression fossils The organic matter of the organisms (perhaps ferns) has been compressed to a thin film that has been replaced by rock.

Index fossils Fossils of common species whose morphology and age are well known. Their presence can be used to date sediments.

Soft fossil These are rare fossils whose normal tissues, as opposed to their skeletal and other hard tissues, have been preserved. For this to happen, the normal decay processes of physical destruction, oxygenation, and animal predation have to be severely inhibited (*see* Obrution condition).

Trace fossils Evidence, normally in rock, of an organism's behavior. Examples include footprints, burrows, and copralites (fossil feces).

Transitional fossil A fossil whose anatomy has features intermediate between those of an ancestor species (e.g., the fin of a fish) and a later, and very different taxon (e.g., the limb of a reptile).

Founder effect The loss of genetic variability when a subgroup from a parent population colonises a new habitat.

Founder flush effect The rapid expansion of a founder population that colonises a new environment in which there are few if any predators.

Founder population A subpopulation that becomes isolated from its parent population when it colonises or expands into a new territory or environment.

FUCA The First Universal Common Ancestor; this was the first organism capable of reproducing itself and so gave rise to every other organism. It represents the origin of life.

Gastrulation An early set of embryonic events that drives the change in geometry from a form that is essentially spherical to one in which there is a nerve cord, a gut and appropriately distributed mesoderm.

Gene A word with two very different meanings. The molecular definition is that it represents a DNA sequence with a defined function. The Mendelian or traditional meaning is that it represents the heritable component of an observed phenotype. In this book, the latter is referred to as a **trait gene**.

Genetic assimilation The process whereby a feature that can be experimentally induced can, through very strong selection, eventually come to be part of the heritable phenotype. This is not Lamarckian inheritance, but the result of the selection of several alleles that were in distributed with low frequency across the wild-type genome and that together can generate the abnormal phenotype.

Genetic background This is the complex set of allele variants that distinguishes one subspecies from another. It is important because the phenotype produced by one protein often depends on others and different backgrounds give rise to subspecies-dependent genetic phenomena.

Genetic drift The process by which gene alleles are distributed across a population through recombination in meiosis and random breeding. The smaller the breeding population, the faster this happens.

Girdle The ring of bones in a vertebrate that supports the forelimbs (pectoral girdle) and hindlimbs (pelvic girdle) enabling them to lift the body without splaying.

Grade A group of organisms within some clade that are united by anatomical and physiological traits and that exludes other members of the clade.

Hardy-Weinberg equilibrium The distribution across a population of allele or gene distributions, each of which has its own frequency. If breeding is random and there is no selection, these frequencies remain constant from generation to generation and so are in equilibrium.

Hermaphrodite An organism that produces both male and female zygotes. It may or may not be capable of self fertilization.

Heterochony An evolutionary change in the timing at which homologous events happen in different species.

Heterokont A group of phyla (often unicellular algae)that carry two different flagella: one is typically for movement and the other facilitates feeding.

Homeothermy The maintenance of a constant body temperature. This can result from biochemical activity or simply be a function of size: large organisms have relatively less surface area through which they can lose heat.

Hominins (Hominini) A group that includes all *Homo* taxa together with extinct taxa closer to humans than any other living vertebrate.

Homologues Two genes or tissues are homologous is they derive through common descent from a common ancestor.

Homoplasy A phenotypic similarity that arose separately in different taxa as a response to selection rather than through common descent.

Hopeful monster An organism with a heritable mutation that led to a major change in anatomical phenotype that also increased fitness. Such a possibility was put forward by Goldschmidt (1940) to help explain complex evolutionary change. The nearest examples that have been found are mutations that affect embryonic patterning.

Horizontal gene transfer A means of acquiring lengths of DNA through donation from another organisms rather than through inheritance.

Hybrid The offspring of mating between a male and female from different taxa. If the parents are from different species, the hybrid should be sterile, but if the species are on the species-separation boundary a hybrid may occasionally be fertile.

Imperfect penetrance This term reflects the inability of an allele to display the expected Mendelian distribution of phenotypes: typically, there are too few heterozygotes and or homozygotes. Molecular genetics views imperfect penetrance as deriving from secondary, genetic-background effects on the relevant protein that may mask its effect.

Inclusive fitness The type of fitness use in discussing kin selection as it includes the likelihood of orphans being adopted.

Intelligent design The belief that, in spite of evidence to the contrary, the human brain is so sophisticated and unique that it could only have evolved with divine help.

Jackknife resampling A means of validating phylogenetic analysis by using various subsets of the data to check the robustness of the analysis.

Juxtacrine signaling A form of intercellular signaling based on contact between two different cell types.

K strategy A reproductive strategy that emphasizes small numbers of offspring and considerable amounts of parental care.

Kin selection Selection based on an individual's behavior that benefits its kin rather than itself.

Lagerstätte A site in which superbly preserved fossils, often with soft tissues, are found.

Lamarckian inheritance The inheritance of acquired adaptations; this that reflected Lamarck's view (1809) that organisms had the heritable ability to become more complex and to adapt to their environment. It was generally accepted until Weisman's work on the continuity of the germplasm at the end of the 19th century.

Last common ancestor In a cladogram, this is the first species in which some synapomorphy was present. As it cannot be identified, it simply has to be viewed is the ancester of all subsequent species carrying that synapomorphy.

LECA The Last Eukaryotic Common Ancestor; this descendant of the FECA had acquired a mitochondrion by endosymbiosis, and was the parent species of all other eukaryotes.

Lineage The term describes the developmental history of a tissue within an organism.

Linnaean taxonomy (classification) A means of classifying groups of species on the basis of shared properties. Higher classes incorporate lower classes on the basis of further and more basic (often developmentally earlier) shared properties. The classification is based on set or class membership with rules such as *species P has the anatomical properties that enable it to be a member of the class of taxon B*. This rule that does not include evolutionary data or any reference to time.

LUCA The Last Universal Common Ancestor; this organism is generally but not universally considered to be the parent organism of the eubacteria and archaebacteria.

Magical trait A phenotypic trait that is involved in both adaptation to an environment and in enhancing reproductive likelihood. Such traits have very high fitness coefficients.

Magnetic remanence The residual magnetism left in a ferromagnetic material once the original cause of the magnetism has been lost.

Magnetostratigraphy The study of the geographical distribution of magnetic remanence.

Mammaliaform The taxon of post-cynodont transitional synapsids (stem mammals) of the early Mesozoic Era that later evolved into mammals.

Mathematical graph A graphical representation of linked facts of the general form <term> <relationship> <term>. Such graphs include cladograms, phylograms, and networks. The most intuitively obvious example is a railways map where the facts are of the form *<station A> < is next to> <station B>*. More formally, it is a set of nodes (or vertices) connected by edges.

Maximum likelihood A computational means of constructing phylogenetic trees on the basis of homologous sequences, using knowledge of the significance and probabilities of various mutational alternative to construct all possible trees, and identifying that which is most likely.

Maximum parsimony A computational means of constructing phylogenetic trees through examining all possible sequence trees and constructing the most likely, without using any additional information. This method is rarely used today.

Meroblastic cleavage This occurs in fertilised eggs with considerable amounts of yolk: nuclei divide but the cell membranes only partially surround each during the early stages of embryogenesis leaving part of the cells directly exposed to yolk.

Metatherian mammals Marsupial mammals with an initial yolk sac placenta that is later supplemented by a choriovitelline placenta and that give birth to offspring still in an early fetal state of development.

Modern Evolutionary Synthesis A theory of evolution that built on Neo-Darwinism (the incorporaton of Mendelian genetics into Darwinian evolution) by including the mathematical field of population genetics as modified to handle first selection and later genetic drift.

Molecular clock A means of measuring time through the amount of mutational change that has occurred. It is only approximately reliable, except when used to analyse small numbers of similar organisms over an evolutionary short time.

Monoderm A prokaryote with a cell membrane composed of a single lipid bilayer with an external polysaccharide coat.

Monophyletic group A taxon containing an ancestral species and all its descendants.

Morphogenesis The processes involving cell movement and cooperative cell behavior that give shape and structural detail to embryos.

Morphospace A theoretical concept describing the multidimensional space within which every organism has a place. The interest in this concept derives from considering those parts of the space that are occupied and those that are not.

Mutualism The evolutionary process by which two organisms come to increase one another's fitness.

Natural extinction rate The typical period for which a species existed before going extinct. This is not easy to define because in general one cannot be certain when a taxon first appeared and when it was lost, unless it was at the time of a major extinction. The rate is typically 5–15 million years and probably a similar number of generations.

Natural selection The pressures from the environment in its widest sense that determine whether an organism will be reproductively successful (or fit).

Natural theology A means of investigating God through studying the evidence that comes from the natural world. In times past, it was an important justification for scientific investigation.

Neighbourhood joining A method for constructing phylogenetic trees on the basis of a distance matrix.

Neo-Darwinism A theory of evolution developed in the 1930s that combined Darwinian natural selection with Mendelian genetics.

Neoteny The slowing down of a developmental event so that its development takes longer than expected.

Nephrozoa The major clade of bilaterian organisms with nephridia or nephrons (excretory tubules). Its sister group is the Xenacoelomorpha.

Neural crest cells (NCCs) A population of cells that migrate away from the neural crest, a region within the vertebrate ectoderm at the interface between future neuronal tissues and the future epithelium that will cover the organisms. NCCs have a wide variety of fates that includes much of the peripheral nervous system, the face, the eye, the skull, pigments cells, and the adrenal medulla.

Non-disjunction The inability of chromosomes to separate properly during mitosis or meiosis.

Nonsynonymous mutation A mutation in a protein-coding DNA sequence that changes the encoded amino acid.

Obrution conditions A term used to describe the rapid covering of a dead organism so as to protect is from decay.

Opisthokont A very high-level taxon of phyla that carry a single posterior flagellum; it includes the animals and the fungi as well as some unicellular phyla.

Opportunism An informal term that reflects the ability of species to adapt through modifications rather than innovation.

Orthologous Homologous sequences are said to be orthologous if the sequence are from different organisms.

Paedomorphosis The retention in adults of an anatomical form of physiological ability normally seen in younger organisms and so is a form of **Heterochrony**.

Palynology The study of pollen spores that may be alive, dead, or even fossilised.

Pangenome The complete set of genes present in a species. It includes all variants.

Paracrine signaling A form of intercellular communication based on a molecular signal diffusing from one cell to another nearby.

Paralogous Homologous sequences are said to be paralagous if the sequences are from the same organism.

Parthenogenesis The ability of an organism to reproduce itself from an unfertilized egg.

Pattern formation The mechanism by which spatial patterns are set up within embryonic domains. It usually occurs through differential paracrine signaling.

Phenocopy A property of a wild-type organism that can be experimentally or environmentally manipulated to mimic a heritable mutation.

Phenotype The identifiable properties of an organism (typically a set of traits and proteins).

Phyletic gradualism The idea that evolutionary change and novel speciation occur at a relatively slow and uniform rate.

Phylogenetic tree A hierarchy of organisms based on evolutionary descent. Cladograms are based on anatomical homologies and molecular phylograms are based on sequence homologies.

Phylogenomics The evolutionary and comparative study of whole genomes so as to clarify the details of descent trees, horizontal gene transfer, and other such parameters.

Phylogeny An evolutionary history of a species or taxon visualized as a tree.

Phylogram A hierarchical tree (a mathematical graph) of a set of DNA or protein sequences, that connects genetic data of a group of existing sequences back in time to a series of shared last common ancestral sequences.

Plesiomorphy An ancestral innovation seen in all members of a clade and its sister clade.

Prezygotic isolation An external pressure that inhibits interbreeding between two groups.

Precocial Young organisms who can be independent from birth (e.g., most fish).

Protein network A group of protein that cooperate to produce an output such as proliferation or a physiological property such as a timing cycle. Their activity is normally initiated by a signal binding to the first protein, a receptor, whose response is to activate the network behavior.

Proto-mammals (mammal-like reptiles) These terms are out of date and both have been replaced by *stem mammal*. These synapsid tetrapods were dominant in the Permian period but became minor species during the Mesozoic Era, the age of the archosaurs, during which they evolved into mammals.

Protostome An embryo whose initial gut cavity becomes the mouth and whose second gut opening becomes the anus.

Prototherian mammals Monotreme mammals that lay eggs and have an essentially reptilian reproductive system.

Pterosaurs A clade of flying archosaurs that differed from dinosaurs in many anatomical features, particularly their possession of an elongated digit 4 and a pteroid bone that was part of the wrist and helped support the wing membrane, together with other adaptations for flight.

Punctuated equilibrium The idea that, although evolutionary change and novel speciation often occur at a relatively slow and uniform rate, it can sometimes be surprisingly rapid.

r strategy A reproductive strategy that emphasizes large numbers of offspring and little if any parental care.

Radiata An informal name for th Coelenterata: diploblast organisms with radial symmetery (e.g., jellyfish).

Radiometric dating Dating rock on the basis of the state of radioactive decay of isotopes that were laid down after the rock had formed.

Random drift The stochastic (and hence unpredictable) changes in the gene frequencies that arise in a finite population as a result of recombination in meiosis and random breeding. The smaller the population, the greater is the effect of drift.

Reinforcement (the Wallace effect) The process whereby a parent and daughter population in adjacent niches increase reproductive isolation through the loss of hybrids that are less fit in each niche than the parent or daughter population.

Retrotransposons These are potentially mobile interspersed DNA elements that may be short or long (SINEs or LINEs) that can be viewed as synapomorphies for phylogenetic purposes.

Ring species A group of species that spread from a founder population to surround an inhospitable domain and eventually met up. A property of the group is that neighbours can interbreed because there are only small genetic differences between them except at the meeting-up point. This is because the cumulative differences between the two populations there are too great for interbreeding to be possible.

Saltation The view that evolutionary changes to aspects of the phenotype do not arise by a series of small changes but can occasionally arise through a major heritable change (*see* Hopeful monster).

Sauropsid The earliest clade of reptiles with two temporal foramina (postorbital spaces in bones). Their descendants included the diapsids (e.g., snakes, reptiles, and archosaurs) and anapsids (e.g., turtles).

Selection pressure A measure of the effect of a feature of the environment on the ability of members of a species to reproduce.

Selective breeding A form of selection in which humans organise the deliberate breeding of organisms on the basis of their phenotypes.

Sexual selection A mode of selection that depends on traits that facilitate mating and breeding.

Somatic mutation A mutation that occurs in the DNA of a cell during mitosis. Although this mutation will then be carried by the subsequent lineage of that cell, it will not be transmitted to the next generation because it is not present in germ cells. An interesting example here is epigenetic change.

Somite A mesenchymal segment in the early embryo of chordates. As a result of local signalling, it partitions into a **dermatome** that will produce the dermis of the skin, a **sclerotome** that will make vertebrae and ribs, and a **myotome** that will give rise to muscles.

Species A core taxon ideally defined by its inability to breed with other species. Where this test cannot be done, it is usually defined by a unique morphology.

Species flock A group of species that formed from a founder population invading a new territory and that became diverse through the very different selection pressures in the various locales of that new territory.

Stromatolite A mound comprised of layers of bacteria and debris of sediment and biomaterials. Stromatolites are important because their fossils from ~3.5 Bya are the oldest evidence of life.

Survival of the fittest The idea that those organisms within a population that flourish are those that leave the most offspring that go on to reproduce. *Fittest* is a technical word in evolution and does **not** mean strongest.

Sympatric speciation The type of speciation that occurs when a subpopulation becomes reproductively isolated in some way from its parent population while still occupying the same territory (e.g., through a difference in eating habits, or light preferences).

Synapomorphy A term from cladistics which means a character possessed by a clade of evolutionarily related organism that is both shared and derived.

Synapsids A major taxon of reptiles characterized by their skull having a single temporal foramen posterior to each orbit. It includes all mammals and their Permian stem-mammal ancestors.

Synonymous mutation A mutation in a protein-coding DNA sequence that does not change the encoded amino acid.

Systematics The area of biology that covers species diversification and evolutionary history.

Systems biology The area of biology that studies biological complexity and how the proteins cooperate within networks produce their output (e.g., signaling, differentiation, and growth).

Taphonomy The study of organism death and decay.

Taxon A species or a group of species or any other rank, defined by common possession of some feature.

Teleological explanation An explanation based on purpose rather than a causative mechanism.

Therapsids A taxon of late Permian synapsid stem mammals whose descendants evolved into cynodonts and eventually mammals.

Tinkering (bricolage) A term introduced by Francois Jacob (1977) to describe the microevents downstream of mutation in the context of how evolutionary change occurs.

Tissue module A structural unit that is repeated may times in an organism. Examples include sperm, fruits, vertebrae, leaves, and nephrons. Their synthesis is often activated by a signal that sets in train a standard set of genomic, network, and other events.

Transcription factor A protein that is associated with a transcription complex that it locates to a specific promoter site on the genome. Activation of the complex leads to the synthesis of mRNA downstream of the site.

Tree-searching A computational method of identifying the most likely phylogenetic tree for some data through exploring the likelihood of every possible tree.

Triploblast A taxon of animal organisms whose early embryo has three germ layers, an ectoderm and an endoderm, both of which mainly give rise to epithelia, and an intervening mesoderm which particularly gives rise to muscles and connective tissue. It includes all Metazoa except coelenterates and sponges.

Urbilateria The last common ancestor of all multicellular organisms having an embryo with three germ layers and bilateral symmetry. It lived towards the end of the Ediacaran Period (~550 Mya) or a little earlier.

Urmetazoan The last common ancestor of all Metazoa and hence the first animal. It lived well before the end of the Ediacaran Period.

Unikont A unicellular eukaryote (protist) with a single flagellum.

Venus A term used for a figurine of a woman made over the period 40–10 Kya.

Wahlund effect This is a property of a daughter population that reflects the likely possibility that its distribution of alleles will be different from that in the parent population due to random drift.

The result is a loss of heterozygosity and an increase in the frequencies of heterozygotes in the daughter population.

Xenarthra A clade of placental mammals that evolved in South America after the fragmentation of Gondwanaland.

Xenocoelomorpha The simplest, worm-like members of the Bilateria and the sister group to the Nephrozoa. They have no coelom, through gut, excretory tubules, or gills.

Xenologue A DNA sequence in species A is a xenologue of a sequence in species B if an ancestor of species A originally acquired the sequence through horizontal gene transfer from an ancestor of species B.

References

1000 Genomes Project consortium (2010) A map of human genome variation from population-scale sequencing. *Nature*; 467:1061-1073. doi: 10.1038/nature09708.

1000 Genomes Project Consortium, Auton A, Brooks LD et al. (2015) A global reference for human genetic variation. *Nature*; 526:68-74. doi: 10.1038/Nature15393.

Aanen DK, Eggleton P, Rouland-Lefevre C, Guldberg-Froslev T, Rosendahl S, Boomsma JJ (2002) The evolution of fungus-growing termites and their mutualistic fungal symbionts. *Proc Natl Acad Sci*; 99:14887-14892. doi: 10.1073/pnas.222313099.

Abbot P, Abe J, Alcock J et al. (2011) Inclusive fitness theory and eusociality. *Nature*; 471:E1-E10. doi: 10.1038/Nature09831.

Abzhanov A (2010) Darwin's galapagos finches in modern biology. *Philos Trans*; 365B:1001-1007. doi: 10.1111/j.1558-5646.2011.01385.x.

Abzhanov A, Kuo WP, Hartmann C (2006) The calmodulin pathway and evolution of elongated beak morphology in Darwin's finches. *Nature*; 442:563-567. doi: 10.1038/nature04843.

Adl SM et al. (2012) The revised classification of eukaryotes. *J Eukaryot Microbiol*; 59:429-493. doi: 10.1111/j.1550-7408.2012.00644.x.

Aerts P (1998) Vertical jumping in *Galago senegalensis:* The quest for an obligate mechanical power amplifier. *Philos Trans R Soc*; 353:1607-1620.

Agić H, Moczydłowska M, Yin L (2017) Diversity of organic-walled microfossils from the early Mesoproterozoic Ruyang Group, North China Craton—A window into the early eukaryote evolution. *Precamb Res*; 297:101-130. doi: 10.1016/j.precamres.2017.04.042.

Ahlberg PE (2019) Follow the footprints and mind the gaps: A new look at the origin of tetrapods. *Trans R Soc Edin (Earth Env Sci)*; 109:115-137. doi.org/10.1017/S1755691018000695.

Ahlberg P, Trinajstic K, Johanson Z, Long J (2009) Pelvic claspers confirm chondrichthyan-like internal fertilization in arthrodires. *Nature*; 460:888-889. doi: 10.1038/Nature08176.

Aiello LC (2010) Five years of *Homo floresiensis*. *Am J Phys Anthr*; 142:167-169. doi: 10.1002/ajpa.21255.

Akçay E, Van Cleve J (2016) There is no fitness but fitness, and the lineage is its bearer. *Phil Trans Roy Soc*; 371B: 20150085. doi.org/10.1098/rstb.2015.0085.

Akkuratov EE, Gelfand MS, Khrameeva EE (2018) Neanderthal and Denisovan ancestry in Papuans: A functional study. *J Bioinform Comput Biol*; 16:1840011. doi: 10.1142/S0219720018400115.

Al-Qattan MM, Abou Al-Shaar H (2015) Molecular basis of the clinical features of Holt-Oram syndrome resulting from missense and extended protein mutations of the *TBX5* gene as well as *TBX5* intragenic duplications. *Gene*; 560:129-136. doi: 10.1016/j.gene.2015.02.017.

Albalat R, Martí-Solans J, Cañestro C (2012) DNA methylation in amphioxus: From ancestral functions to new roles in vertebrates. *Brief Funct Genomics*; 11:142-155. doi: 10.1093/bfgp/els009.

Alcaide M, Scordato ES, Price TD, Irwin DE (2014) Genomic divergence in a ring species complex. *Nature*; 511:83-85. doi: 10.1038/nature13285.

Alibardi L (2003) Adaptation to the land: The skin of reptiles in comparison to that of amphibians and endotherm amniotes. *J Exp Zool B Mol Dev Evol*; 298:12-41.

Alibardi L (2006) Structural and immunocytochemical characterization of keratinization in vertebrate epidermis and epidermal derivatives. *Int Rev Cytol*; 253:177-259.

Alibardi L (2008) Microscopic analysis of lizard claw morphogenesis and hypothesis on its evolution. *Acta Zoo Morph Evo*; 89:169-178.

Alibardi L (2012) Perspectives on hair evolution based on some comparative studies on vertebrate cornification. *J Exp Zool B (Mol Dev Evol)*; 318:325343. doi: 10.1002/jez.b.22447.

Alibardi L, Dalla Valle L, Nardi A, Toni M (2009) Evolution of hard proteins in the sauropsid integument in relation to the cornification of skin derivatives in amniotes. *J Anat*; 214:560-586. doi: 10.1111/j.1469-7580.2009.01045.x.

Alix K, Gérard PR, Schwarzacher T, Heslop-Harrison JSP (2017) Polyploidy and interspecific hybridization: Partners for adaptation, speciation and evolution in plants. *Ann Bot*; 120:183-194. doi: 10.1093/aob/mcx079.

Allègre CJ (2008) *Isotope Geology*. Cambridge University Press.

Allin EF (1975) Evolution of the mammalian middle ear. *J Morph*; 147: 403-437. doi:10.1002/jmor.1051470404.

Almén MS, Lamichhaney S, Berglund J et al. (2016) Adaptive radiation of Darwin's finches revisited using whole genome sequencing. *Bioessays*; 38:14-20. doi: 10.1002/bies.201500079.

Amborella Genome Project (2013) The Amborella genome and the evolution of flowering plants. *Science*; 342:1241089. doi: 10.1126/science.1241089.

Amemiya CT, Alföldi J, Lee AP et al. (2013) The African coelacanth genome provides insights into tetrapod evolution. *Nature*; 496:311-6. doi: 10.1038/nature12027.

Anamika K, Martin J, Srinivasan N (2008) Comparative kinomics of human and chimpanzee reveal unique kinship and functional diversity generated by new domain combinations. *BMC Genomics*; 9:625. doi: 10.1186/1471-2164-9-625.

Anderson E, Hill RE (2014) Long range regulation of the *sonic Hedgehog* gene. *Curr Opin Genet Dev*; 27:54-59. doi: 10.1016/j.gde.2014.03.011.

Anderson G, Beischlag TV, Vinciguerra M, Mazzoccoli G (2013) The circadian clock circuitry and the AHR signaling pathway in physiology and pathology. *Biochem Pharmacol*; 85:1405-1416. doi: 10.1016/j.bcp.2013.02.022.

Anderson JT, Wagner MR, Rushworth CA, Prasad KV, Mitchell-Olds T (2014) The evolution of quantitative traits in complex environments. *Heredity*; 112:4-12. doi: 10.1038/hdy.2013.33.

Anderson PS, Westneat MW (2007) Feeding mechanics and bite force modelling of the skull of *Dunkleosteus terrelli*, an ancient apex predator. *Biol Lett*; 3:76-79. doi: 10.1098/rsbl.2006.0569.

Andrews P (2020) Last common ancestor of apes and humans: Morphology and environment. *Folia Primatol (Basel)*; 91:122-148. doi: 10.1159/000501557.

Andrews M, Andrews ME (2017) Specificity in Legume-Rhizobia Symbioses. *Int J Mol Sci*; 18:705. doi: 10.3390/ijms18040705.

Antcliffe JB, Callow RH, Brasier MD (2014) Giving the early fossil record of sponges a squeeze. *Biol Rev Camb Phil Soc*; 89:972-1004. doi: 10.1111/brv.12090.

Anthwal N, Joshi L, Tucker AS (2012) Evolution of the mammalian middle ear and jaw: Adaptations and novel structures. *J Anat*; 222:147-160. doi: 10.1111/j.1469-7580.2012.01526.x.

Aplin LM (2019) Culture and cultural evolution in birds: A review of the evidence. *Anim Behav*; 147:179-187. doi.org/10.1016/j.anbehav.2018.05.001.

Acquaah A (2021) *Principles of Plant Genetics and Breeding* (3rd ed). Wiley Blackwell.

Archibald JM (2015) Endosymbiosis and eukaryotic cell evolution. *Curr Biol*; 25:R911-21. doi: 10.1016/j.cub.2015.07.055.

Ardelean CF, Becerra-Valdivia L, Pedersen MW, et al. (2020) Evidence of human occupation in Mexico around the Last Glacial Maximum. *Nature*; 584:87-92. doi: 10.1038/s41586-020-2509-0.

Arendt D, Tessmar-Raible K, Snyman H (2004) Ciliary photoreceptors with a vertebrate-type opsin in an invertebrate brain. *Science*; 306:869871.

Aria C, Caron JB (2019) A middle Cambrian arthropod with chelicerae and proto-book gills. *Nature*; 573:586-589. doi: 10.1038/s41586-019-1525-4.

Arsuaga JL, Martínez I, Arnold LJ (2014) Neandertal roots: Cranial and chronological evidence from Sima de los Huesos. *Science*; 344:1358-1363. doi: 10.1126/science.1253958.

Asfaw B, White T, Lovejoy O, Latimer B, Simpson S, Suwa G (1999) *Australopithecus garhi*: A new species of early hominid from Ethiopia. *Science*; 284: 629-35. doi: 10.1126/Science.284.5414.629.

Atkins JB, Reisz RR, Maddin HC (2019) Braincase simplification and the origin of lissamphibians. *PLoS One*; 14(3):e0213694. doi: 10.1371/journal.pone.0213694.

Atsumi T, McCarter L, Imae Y (1992) Polar and lateral flagellar motors of marine vibrio are driven by different ion-motive forces. *Nature*; 355:182–184.

Aubert M, Brumm A, Ramli M et al. (2014) Pleistocene cave art from Sulawesi, Indonesia. *Nature*; 514:223–227. doi: 10.1038/nature13422.

Aubineau J et al. (2018) Unusual microbial mat-related structural diversity 2.1 billion years ago and implications for the Francevillian biota. *Geobiology*; 16:476–497. doi: 10.1111/gbi.12296.

Aulehla A, Pourquié O (2010) Signaling gradients during paraxial mesoderm development. *Cold Spring Harb Perspect Biol*; 2:a000869. doi: 10.1101/cshperspect.a000869.

Ayala FJ (1999) Molecular clock mirages. *BioEssays*; 21:71–75.

Ayala FJ (2007) Darwin's greatest discovery: Design without designer. *Proc Natl Acad Sci U S A*; 104(Suppl 1):8567–73. doi: 10.1073/pnas.0701072104.

Ayala FJ, Avise JC (eds, 2014) *Essential Readings in Evolutionary Biology*. Johns Hopkins University Press.

Babbs C, Furniss D, Morriss-Kay GM, Wilkie AO (2008) Polydactyly in the mouse mutant *Doublefoot* involves altered *Gli3* processing and is caused by a large deletion in *cis* to *Indian hedgehog*. *Mech Dev*; 125:517–526. doi: 10.1016/j.mod.2008.01.001.

Bada JL (2013) New insights into prebiotic chemistry from Stanley Miller's spark discharge experiments. *Chem Soc Rev*; 42:2186–2196. doi: 10.1039/c3cs35433d.

Badyaev AV, Potticary AL, Morrison ES (2017) Most colorful example of genetic assimilation? Exploring the evolutionary destiny of recurrent phenotypic accommodation. *Am Nat*; 190:266–280. doi: 10.1086/692327.

Baedke J, Fábregas-Tejeda A, Nieves Delgado A (2020) The holobiont concept before Margulis. *J Exp Zool B Mol Dev Evol*; 334:149–155. doi: 10.1002/jez.b.22931.

Baehrecke EH (2000) How death shapes life during development. *Nat Rev Mol Cell Biol*; 3:779–787.

Bailleul AM, O'Connor J, Schweitzer MH (2019) Dinosaur paleohistology: Review, trends and new avenues of investigation. *PeerJ*; 7:e7764. doi: 10.7717/peerj.7764.

Bajpai S, Thewissen JGM, Sahni A (2009) The origin and early evolution of whales: Macroevolution documented on the Indian subcontinent. *J Biosci*; 34:673–686.

Balaresque P, King TE (2016) Human phenotypic diversity: An evolutionary perspective. *Curr Top Dev Biol*; 119:349–90. doi: 10.1016/bs.ctdb.2016.02.001.

Ballesteros JA, Sharma PS (2019) A critical appraisal of the placement of Xiphosura (Chelicerata) with account of known sources of phylogenetic error. *Syst Biol*; 68:896–917. doi.org/10.1093/sysbio/syz011.

Bar-On YM, Phillips R, Milo R (2018) The biomass distribution on earth. *Proc Natl Acad Sci*; 115:6506–6511. doi: 10.1073/pnas.1711842115.

Bar-Yosef Mayer DE, Groman-Yaroslavski I, Bar-Yosef O et al. (2020) On holes and strings: Earliest displays of human adornment in the middle Palaeolithic. *PLoS One*; 15(7):e0234924. doi: 10.1371/journal.pone.0234924.

Bard J (2002) Growth and death in the developing mammalian kidney: Signals, receptors and conversations. *Bioessays*; 24:72–82.

Bard J (2011) A systems biology formulation of developmental anatomy. *J Anat*; 218:591–599. doi: 10.1111/j.1469-7580.2011.01371.x.

Bard J (2013a) Systems biology—The broader perspective. *Cells*; 2(2):413–431.

Bard J (2013b) Driving developmental and evolutionary change: A systems biology view. *Prog Biophys Mol Biol*; 11:83–91. doi: 10.1016/j.pbiomolbio.2012.09.006.

Bard JBL (1977) A unity underlying the different zebra striping patterns. *J Zoology*; 183:527–539.

Bard JBL (1981) A model generating aspects of zebra and other mammalian coat patterns. *J Theor Biol*; 93:363–385.

Bard JBL (2018) Tinkering and the origins of heritable anatomical variation in vertebrates. *Biology (Basel)*; 26;7(1):20 doi: 10.3390/biology7010020.

Barreda VD, Cúneo NR, Wilf P, Currano ED, Scasso RA, Brinkhuis H (2012) Cretaceous/Paleogene floral turnover in Patagonia: Drop in diversity, low extinction, and a Classopollis spike. *PLoS One*; 7(12):e52455. doi: 10.1371/journal.pone.0052455.

Barresi MJF, Gilbert SF (2019) *Developmental Biology* (12th ed). Oxford University Press.

Barrio RA, Varea C, Aragón JL, Maini PK (1999) A two-dimensional numerical study of spatial pattern formation in interacting turing systems. *Bull Math Biol*; 61:483–505.

Bartels A, Han Q, Nair P et al. (2018) Dynamic DNA methylation in plant growth and development. *Int J Mol Sci*; 19:2144. doi: 10.3390/ijms19072144.

Barton NH (2010) Genetic linkage and natural selection. *Philos Trans R Soc Lond B Biol Sci*; 365:2559–69. doi: 10.1098/rstb.2010.0106.

Barton NH, Briggs DEG, Eisen JA (2007) *Evolution*. Cold Spring Harbor Laboratory Press.

Bass D, Czech L, Williams BAP et al. (2018) Clarifying the relationships between microsporidia and cryptomycota. *J Eukaryot Microbiol*; 65:773–782. doi: 10.1111/jeu.12519.

Bauernfeind AL, Soderblom EJ, Turner ME et al. (2015) Evolutionary divergence of gene and protein expression in the brains of humans and chimpanzees. *Genome Biol Evol*; 7: 2276–2288. doi: 10.1093/gbe/evv132.

Baujat G, Le Merrer M (2007) Ellis-van Creveld syndrome. *Orphanet J Rare Dis* 7; 2:27. doi: 10.1186/1750-1172-2-27.

Baum D (1992) Phylogenetic species concepts. *Trends Ecol Evol*; 7:1–3.

Bayrakdar M (1983) Al-Jahiz and the rise of biological evolution. *Islamic Quart* 1983;3:149.

Beall CM (2013) Human adaptability studies at high altitude: Research designs and major concepts during fifty years of discovery. *Am J Hum Biol*; 25:141–147. doi: 10.1002/ajhb.22355.

Beerling DJ, Osborne CP, Chaloner WG (2001) Evolution of leaf-form in land plants linked to atmospheric CO_2 decline in the Late Palaeozoic era. *Nature*; 410:352–354. doi: 10.1038/35066546.

Bejder L, Hall BK (2002) Limbs in whales and limblessness in other vertebrates: Mechanisms of evolutionary and developmental transformation and loss. *Evol Dev*; 4:445–458.

Bellwood P (2018) The search for ancient DNA heads east. *Science*; 36:31–32. doi: 10.1126/Science.aat8662.

Bengtson S, Belivanova V, Rasmussen B, Whitehouse M (2009) The controversial "Cambrian" fossils of the Vindhyan are real but more than a billion years older. *Proc Natl Acad Sci U S A*; 106:7729–7734. doi: 10.1073/pnas.0812460106.

Bengtson S, Rasmussen B, Ivarsson M et al. (2017a) Fungus-like mycelial fossils in 2.4-billion-year-old vesicular basalt. *Nature Ecol Evol*; 1:0141. doi: 10.1038/s41559-017-0141.

Bengtson S, Sallstedt T, Belivanova V, Whitehouse M (2017b) Three-dimensional preservation of cellular and subcellular structures suggests 1.6 billion-year-old crown-group red algae. *PLoS Biol*; 15:e2000735. doi: 10.1371/journal.pbio.2000735.

Benito-Kwiecinski S, Giandomenico SL, Sutcliffe M et al. (2021) An early cell shape transition drives evolutionary expansion of the human forebrain. *Cell*; 184:2084–2102.e19. doi: 10.1016/j.cell.2021.02.050.

Bennett GM, Moran NA (2015) Heritable symbiosis: The advantages and perils of an evolutionary rabbit hole. *Proc Natl Acad Sci*; 112:10169–76. doi: 10.1073/pnas.1421388112.

Bentham J et al. (2016) A century of trends in adult human height. *Elife*; 5:pii:e13410. doi: 10.7554/eLife.13410.

Benton J (2005) *Vertebrate Palaeontology*. Wiley-Blackwell.

Benton J, Harper DAT (2009) *Basic Palaeontology: Introduction to Paleobiology and the Fossil Record* (3rd ed). Blackwell.

Benton MJ, Dhouailly D, Jiang B, McNamara M (2019) The early origin of feathers. *Trends Ecol Evol*; 34:856–869. doi: 10.1016/j.tree.2019.04.018.

Beppu H, Malhotra R, Beppu Y (2000) BMP type II receptor is required for gastrulation and early development of mouse embryos. *Dev Biol*; 221:249–258.

Beran MJ, Heimbauer LA (2015) A longitudinal assessment of vocabulary retention in symbol-competent chimpanzees (*Pan troglodytes*). *PLoS One*; 10:e0118408. doi: 10.1371/journal.pone.0118408.

Becerra-Valdivia L, Higham T (2020) The timing and effect of the earliest human arrivals in North America. *Nature*; 584:93–97. doi: 10.1038/s41586-020-2491-6.

Beckers OM, Kijimoto T, Moczek AP (2017) Doublesex alters aggressiveness as a function of social context and sex in the polyphenic beetle *Onthophagus taurus*. *Anim Behav*; 132:261–269. doi: 10.1016/j.anbehav.2017.08.011.

Bennett MR, Bustos, D, Pigati JS, et al. (2021) Evidence of humans in North America during the Last Glacial Maximum. *Science*; 373:1528–1531. doi: 10.1126/science.abg7586

Berger LR, Hawks J, de Ruiter DJ et al. (2015) *Homo naledi*, a new species of the genus *Homo* from the Dinaledi Chamber, South Africa. *eLife*; 4:e09560. doi.org/10.7554/eLife.09560.

Berger LR, Hawks J, Dirks PH, Elliott M, Roberts EM (2017) *Homo naledi* and Pleistocene hominin evolution in subequatorial Africa. *Elife*; 6:e24234. doi: 10.7554/eLife.24234.

Berlocher SH (1998) Origins: A Brief History of Research on Speciation. In *Endless Forms: Species and Speciation* (Howard DJ, Berlocher SH, eds), pp. 3–15. Oxford University Press.

Berlocher SH, Feder JL (2002) Sympatric speciation in phytophagous insects: Moving beyond controversy? *Annu Rev Entomol*; 47:773–815.

Berna F, Goldberg P, Horwitz LK, Brink J, Holt S, Bamford M, Chazan M (2012) Microstratigraphic evidence of in situ fire in the Acheulean strata of Wonderwerk Cave, Northern Cape province, South Africa. *Proc Natl Acad Sci U S A*; 109:E1215–20. doi: 10.1073/pnas.1117620109.

Berner RA (1999) Atmospheric oxygen over phanerozoic time. *Science*; 96:10955–10957.

Bernstein M (2006) Prebiotic materials from on and off the early earth. *Philos Trans R Soc*; 61B:1689–1700.

Bernt M, Braband A, Schierwater B, Stadler PF (2013) Genetic aspects of mitochondrial genome evolution. *Mol Phylogenet Evol*; 69:328–338. doi: 10.1016/j.ympev.2012.10.020.

Berta A (2017) *The Rise of Marine Mammals: 50 Million Years of Evolution.* Johns Hopkins University Press.

Berto S, Nowick K (2018) Species-specific changes in a primate transcription factor network provide insights into the molecular evolution of the primate prefrontal cortex. *Genome Biol Evol*; 10:2023–2036. doi: 10.1093/gbe/evy149.

Besser K, Malyon GP, Eborall WS et al. (2018) Hemocyanin facilitates lignocellulose digestion by wood-boring marine crustaceans. *Nat Commun*; 9:5125. doi: 10.1038/s41467-018-07575-2.

Betts HC, Puttick MN, Clark JW, Williams TA, Donoghue PCJ, Pisani D (2018) Integrated genomic and fossil evidence illuminates life's early evolution and eukaryote origin. *Nat Ecol Evol*; 2:1556-1562. doi: 10.1038/s41559-018-0644.

Bhattacharyya T, Gregorova S, Mihola O (2013) Mechanistic basis of infertility of mouse intersubspecific hybrids. *Proc Natl Acad Sci U S A*; 110:468–477. doi: 10.1073/pnas.1219126110.

Biscotti MA, Canapa A, Forconi M, Barucca M (2014) Hox and ParaHox genes: A review on molluscs. *Genesis*; 52:935–945. doi: 10.1002/dvg.22839.

Blackburn DG, Flemming AF (2012) Invasive implantation and intimate placental associations in a placentotrophic African lizard *Trachylepis ivensi* (Scincidae). *J Morphol*; 273:137–159. doi: 10.1002/jmor.11011.

Blair JM, Webber MA, Baylay AJ, Ogbolu DO, Piddock LJ (2015) Molecular mechanisms of antibiotic resistance. *Nat Rev Microbiol*; 1:42–51. doi: 9.1038/nrmicro3380.

Blanco MJ, Misof BY, Wagner GP (1998) Heterochronic differences of Hoxa-11 expression in *Xenopus* fore and hind limb development: Evidence for lower limb identity of the Anuran ankle bones. *Dev Genes Evol*; 208:175–187.

Blatt H, Jones RL (1975) Proportions of exposed igneous metamorphic and sedimentary rocks. *Geol Soc Am Bull*; 86:1085–1088.

Blunden J, Arndt DS (eds) (2018) State of the climate in 2018. *Bull Am Met Soc*; 100:Si–S306 (journals.ametsoc.org/view/journals/bams/100/9/2019bamsstateoftheclimate.1.xml).

Bobola N, Merabet S (2017) Homeodomain proteins in action: Similar DNA binding preferences, highly variable connectivity. *Curr Opin Genet Dev*; 43:1–8. doi: 10.1016/j.gde.2016.09.008.

Böhm M, Boness D, Fantisch E et al. (2019) Channelrhodopsin-1 phosphorylation changes with phototactic behavior and responds to physiological stimuli in chlamydomonas. *Plant Cell*; 31:886–910. doi: 10.1105/tpc.18.00936.

Boisvert CA, Joss JM, Ahlberg PE (2013) Comparative pelvic development of the axolotl (*Ambystoma mexicanum*) and the Australian lungfish (*Neoceratodus forsteri*): Conservation and innovation across the fish-tetrapod transition. *Evo Devo*; 4:3. doi: 10.1186/2041-9139-4-3.

Boisvert CA, Mark-Kurik E, Ahlberg PE (2008) The pectoral fin of panderichthys and the origin of digits. *Nature*; 456:636–8. doi: 10.1038/nature07339.

Bonneville S et al. (2020) Molecular identification of fungi microfossils in a Neoproterozoic shale rock. *Sci Adv*; 6:eaax7599. doi: 10.1126/sciadv.aax7599.

Bordenstein SR, Theis KR (2015) Host biology in light of the microbiome: Ten principles of holobionts and hologenomes. *PLoS Biol*; 13:e1002226. doi: 10.1371/journal.pbio.1002226.

Bošković A, Rando OJ (2018) Transgenerational epigenetic inheritance. *Ann Rev Genet*; 52:21–41. doi: 10.1146/annurev-genet-120417-031404.

Bottjer DJ, Etter W, Hagadorn JW, Tang CM (eds) (2002) *Exceptional Fossil Preservation: A Unique View on the Evolution of Marine Life.* Columbia University Press.

Bourke AF (2014) Hamilton's rule and the causes of social evolution. *Phil Trans R Soc*; 369B:20130362. doi: 10.1098/rstb.2013.0362.

Bourke AF (2015a) Social evolution: Uneasy lies the head. *Curr Biol*; 25:R1077–R1079. doi: 10.1016/j.cub.2015.09.071.

Bourke AF (2015b) Sex investment ratios in eusocial Hymenoptera support inclusive fitness theory. *J Evol Biol*; 28:2106–2111. doi: 10.1111/jeb.12710.

Bouzouggar A, Barton N, Vanhaeren M (2007) 82000-year-old shell beads from North Africa and implications for the origins of modern human behavior. *Proc Natl Acad Sci U S A*; 104:9964–9969.

Box NF, Larue L, Manga P, Montoliu L, Spritz RA, Filipp FV (2018) The triennial international pigment cell conference (IPCC). *J Transl Med*; 16:252. doi: 10.1186/s12967-018-1609-1.

Boyle EA, Li YI, Pritchard JK (2017) An expanded view of complex traits: From polygenic to omnigenic. *Cell*; 169(7):1177–1186. doi: 10.1016/j.cell.2017.05.038.

Braddy SJ, Poschmann M, Tetlie OE (2008) Giant claw reveals the largest ever arthropod. *Biol Lett*; 4:106–109. doi: 10.1098/rsbl.2007.0491.

Brasier M (2010) *Darwin's Lost World: The Hidden History of Animal Life.* Oxford University Press.

Brawand D, Wagner C, Li Y et al. (2014) The genomic substrate for adaptive radiation in African cichlid fish. *Nature*; 513:375–381. doi: 10.1038/Nature13726.

Brawley SH, Blouin NA, Ficko-Blean E et al. (2017) Insights into the red algae and eukaryotic evolution from the genome of *Porphyra umbilicalis* (Bangiophyceae, Rhodophyta). *Proc Natl Acad Sci*; 114:E6361–E6370. doi: 10.1073/pnas.1703088114.

Brazeau MD, Friedman M (2014) The characters of Palaeozoic jawed vertebrates. *Zool J Linn Soc*; 170:779–821.

Brazeau MD, Giles S, Dearden RP et al. (2020) Endochondral bone in an early Devonian "placoderm" from Mongolia. *Nat Ecol Evol*; 4:1477–1484. doi: 10.1038/s41559-020-01290-2.

Briggs DE, Plint AG, Pickerill RK (1984) "Arthropleura trails from the Westphalian of eastern Canada" (PDF). *Palaeontology*; 27:843–855.

Brinkmann H, Venkatesh B, Brenner S, Meyer A (2004) Nuclear protein-coding genes support lungfish and not the coelacanth as the closest living relatives of land vertebrates. *Proc Natl Acad Sci*; 101:4900–4905.

Briscoe SD, Ragsdale CW (2018) Homology, neocortex, and the evolution of developmental mechanisms. *Science*; 362:190–193. doi: 10.1126/Science.aau3711.

Britt BB, Dalla Vecchia FM, Chure DJ, Engelmann GF, Whiting MF, Scheetz RD (2018) *Caelestiventus hanseni* gen. et sp. nov. extends the desert-dwelling pterosaur record back 65 million years. *Nat Ecol Evol*; 2:1386–1392. doi: 10.1038/s41559-018-0627-y.

Brown DD, Cai L (2007) Amphibian metamorphosis. *Dev Biol*; 306:20–33. doi: 10.1016/j.ydbio.2007.03.021.

Brown P, Sutikna T, Morwood MJ (2004) A new small-bodied hominin from the Late Pleistocene of Flores Indonesia. *Nature*; 431:1055–1061.

Brown RC, Lemmon BE (2011) Dividing without centrioles: Innovative plant microtubule organizing centres organize mitotic spindles in bryophytes, the earliest extant lineages of land plants. *AoB Plants*; 2011:plr028. doi: 10.1093/aobpla/plr028.

Brown RM, Siler CD, Oliveros CH (2013) The amphibians and reptiles of Luzon Island, Philippines, VIII: The herpetofauna of Cagayan and Isabela Provinces, northern Sierra Madre Mountain Range. *ZooKeys*; 266:1–120. doi: 10.3897/zookeys.266.3982.

Browne J (1995, 2002) *Charles Darwin.* Vols. 1, 2. Jonathan Cape, London.

Bruce WA, Wrensch DL (1990) Reproductive potential sex ratio and mating efficiency of the straw itch mite (Acari:Pyemotidae) *J Econ Entomol* 83:384–391.

Bruner E, De La Cuétara JM, Holloway R (2011) A bivariate approach to the variation of the parietal curvature in the genus *Homo. Anat Rec*; 294:1548–1556. doi: 10.1002/ar.21450.

Bruner E, Lozano M, Malafouris L et al. (2014) Extended mind and visuo-spatial integration: Three hands for the Neandertal lineage. *J Anthropol Sci*; 92:273–280. doi: 10.4436/JASS.92009.

Brunet T, Bouclet A, Ahmadi P et al. (2013) Evolutionary conservation of early mesoderm specification by mechanotransduction in bilateria. *Nat Commun*; 4:2821. doi: 10.1038/ncomms3821.

Brunet T, King N (2017) The origin of animal multicellularity and cell differentiation. *Dev Cell*; 43:124–140. doi: 10.1016/j.devcel.2017.09.016.

Brusatte SL, O'Connor JK, Jarvis ED (2015) The origin and diversification of birds. *Curr Biol*; 25:R888–R898. doi: 10.1016/j.cub.2015.08.003.

Buchmann R, Rodrigues T (2019) The evolution of pneumatic fenestrae in pterosaur vertebfenestrae. *An Acad Bras Cienc*; 91 (Suppl. 2):e20180782. doi: 10.1590/0001-3765201920180782.

Bücking H, Cox MP, Hudjashov G, Saag L, Sudoyo H, Stoneking M (2019) Archaic mitochondrial DNA inserts in modern day nuclear genomes. *BMC Genomics*; 20:1017. doi: 10.1186/s12864-019-6392-8.

Buckley M, Walker A, Ho SY et al. (2008) Comment on "Protein sequences from mastodon and *Tyrannosaurus rex* revealed by mass spectrometry" *Science*; 319:33. doi: 10.1126/science.1147046.

Buckley TR, Cunningham CW (2002) The effects of nucleotide substitution model assumptions on estimates of nonparametric bootstrap support. *Mol Biol Evol*; 19:394–405.

Budin I, Szostak JW (2010) Expanding roles for diverse physical phenomena during the origin of life. *Annu Rev Biophys*; 39:245–263.

Bui ET, Bradley PJ, Johnson PJ (1996) A common evolutionary origin for mitochondria and hydrogenosomes. *Proc Natl Acad Sci*; 93: 9651–9656. doi: 10.1073/pnas.93.18.9651.

Bulik-Sullivan B, Finucane HK, Anttila V et al. (2015) An atlas of genetic correlations across human diseases and traits. *Nat Genet*; 47:1236–41. doi: 10.1038/ng.3406.

Bullara D, De Decker Y (2015) Pigment cell movement is not required for generation of turing patterns in zebrafish skin. *Nat Commun*; 6:6971. doi: 10.1038/ncomms7971.

Bulmer MG (2003) *Francis Galton: Pioneer of Heredity and Biometry.* Johns Hopkins University Press.

Buneman P (1971) The Recovery of Trees from Measures of Dissimilarity. In *Mathematics in the Archaeological and Historical Sciences* (Hodson FR, Kendall DG, Tautu PT, eds), pp 387–395. Edinburgh University Press.

Burke AC, Nelson CE, Morgan BA, Tabin C (1995) Hox genes and the evolution of vertebrate axial morphology. *Development*; 12:333–346.

Burkhardt RW Jr. (2013) Lamarck, evolution, and the inheritance of acquired characters. *Genetics*; 194(4):793–805. doi: 10.1534/genetics.113.151852.

Burki F, Roger AJ, Brown MW, Simpson AGB (2020) The new tree of eukaryotes. *Trends Ecol Evol*; 35:43–55. doi: 10.1016/j.tree.2019.08.008.

Burki F, Shalchian-Tabrizi K, Pawlowski J (2008) Phylogenomics reveals a new "megagroup" including most photosynthetic eukaryotes. *Biol Lett*; 4:366–369. doi: 10.1098/rsbl.2008.0224.

Burnside RD, Molinari S, Botti C et al. (2018) Features of Feingold syndrome 1 dominate in subjects with 2p deletions including *MYCN*. *Am J Med Genet A*; 176:1956–1963. doi: 10.1002/ajmg.a.40355.

Burrow CJ, Rudkin D (2014) Oldest near-complete acanthodian: The first vertebrate from The Silurian Bertie Formation Konservat-Lagerstätte, Ontario. *PLoS One*; 9(8):e104171. doi: 10.1371/journal.pone.0104171.

Busiek KK, Margolin W (2015) Bacterial actin and tubulin homologs in cell growth and division. *Curr Biol*; 25:R243–R254. doi: 10.1016/j.cub.2015.01.030.

Butterfield NJ (2000) *Bangiomorpha pubescens* n. gen., n. sp.: Implications for the evolution of sex multicellularity and the Mesoproterozoic/Neoproterozoic radiation of eukaryotes. *Paleobiology*; 26:386–404. doi: 10.1666/0094-8373(2000)026<0386:BPNGNS>2.0.CO;2.

Bystron I, Blakemore C, Rakic P (2008) Development of the human cerebral cortex: Boulder Committee revisited. *Nat Rev Neurosci*; 9:110–122. doi: 10.1038/nrn2252.

Cabrera VM, Marrero P, Abu-Amero KK, Larruga JM (2018) Carriers of mitochondrial DNA macrohaplogroup L3 basal lineages migrated back to Africa from Asia around 70,000 years ago. *BMC Evol Biol*; 18:98. doi: 10.1186/s12862-018-1211-4.

Caetano-Anollés G (ed) (2010) *Evolutionary Genomics and Systems Biology*. John Wiley.

Caforio A, Siliakus MF, Exterkate M et al. (2018) Converting *Escherichia coli* into an archaebacterium with a hybrid heterochiral membrane. *Proc Natl Acad Sci U S A*; 115:3704–3709. doi: 10.1073/pnas.1721604115.

Calarco JA, Xing Y, Cáceres M et al. (2007) Global analysis of alternative splicing differences between humans and chimpanzees. *Genes Dev*; 21:2963–75. doi: 10.1101/gad.1606907.

Callery EM, Fang H, Elinson RP (2001) Frogs without polliwogs: Evolution of anuran direct development. *Bioessays*; 23:233–241. doi: 10.1002/1521-1878(200103)23:3<233:AID-BIES1033>3.0.CO;2-Q.

Campinho MA (2019) Teleost metamorphosis: The role of thyroid hormone. *Front Endocrinol*; 10:383. doi: 10.3389/fendo.2019.00383.

Cannell AER (2018) The engineering of the giant dragonflies of the Permian: Revised body mass, power, air supply, thermoregulation and the role of air density. *J Exp Biol*; 221:jeb185405. doi: 10.1242/jeb.185405.

Cannon JT, Vellutini BC, Smith J 3rd, Ronquist F, Jondelius U, Hejnol A (2016) Xenacoelomorpha is the sister group to Nephrozoa. *Nature*; 530:89–93. doi: 10.1038/Nature16520.

Cannon ME, Mohlke KL (2018) Deciphering the emerging complexities of molecular mechanisms at GWAS loci. *Am J Hum Genet*; 103:637–653. doi: 10.1016/j.ajhg.2018.10.001.

Caporale AL, Gonda CM, Franchini LF (2019) Transcriptional enhancers in the FOXP2 locus underwent accelerated evolution in the human lineage. *Mol Biol Evol*; Jul 29:msz173. doi: 10.1093/molbev/msz173.

Cappellini E et al. (2018) Ancient biomolecules and evolutionary inference. *Ann Rev Biochem*; 87:1029–1060. doi: 10.1146/annurev-biochem-062917-012002.

Capra F, Luisi PL (2014) *The Systems View of Life*. Cambridge University Press.

Cargnin F, Kwon JS, Katzman S, Chen B, Lee JW, Lee SK (2018) FOXG1 orchestrates neocortical organization and cortico-cortical connections. *Neuron*; 100:1083–1096.e5. doi: 10.1016/j.neuron.2018.10.016.

Carmona FD, Glösmann M, Ou J, Jiménez R, Collinson JM (2010) Retinal development and function in a "blind" mole. *Proc Biol Sci*; 277:1513–22. doi: 10.1098/rspb.2009.1744.

Caro T, Argueta Y, Briolat ES et al. (2019) Benefits of zebra stripes: Behaviour of tabanid flies around zebras and horses. *PLoS One*; 14:e0210831. doi: 10.1371/journal.pone.0210831.

Caron F, d'Errico F, Del Moral P, Santos F, Zilhão J (2011) The reality of Neandertal symbolic behavior at the Grotte du Renne, Arcy-sur-Cure, France. *PLoS One*; 6(6):e21545. doi: 10.1371/journal.pone.0021545.

Carr M, Leadbeater BS, Hassan R, Nelson M, Baldauf SL (2020) Molecular phylogeny of choanoflagellates, the sister group to Metazoa. *Proc Natl Acad Sci U S A*; 105:16641–6. doi: 10.1073/pnas.0801667105.

Carrasco J, Preston GM (2020) Growing edible mushrooms: A conversation between bacteria and fungi. *Environ Microbiol*; 22:858–872. doi: 10.1111/1462-2920.14765.

Carroll RL (1988) *Vertebrate Paleontology and Evolution*. WH Freeman & Co.

Carroll SB (1995) Homeotic genes and the evolution of arthropods and chordates. *Nature*; 376:479–485. doi: 10.1038/376479a0.

Cartwright P, Halgedahl SL, Hendricks JR et al. (2007) Exceptionally preserved jellyfishes from the Middle Cambrian. *PLoS One*; 2:e1121. doi: 10.1371/journal.pone.0001121.

Cartwright RA, Lartillot N, Thorne JL (2011) History can matter: Non-markovian behavior of ancestral lineages. *Syst Biol*; 60:276–90. doi: 10.1093/sysbio/syr012.

Carvalho-Santos Z, Azimzadeh J, Pereira-Leal JB, Bettencourt-Dias M (2011) Evolution: Tracing the origins of centrioles, cilia, and flagella. *J Cell Biol*; 194:165–175. doi: 10.1083/jcb.201011152.

Cary GA, Hinman VF (2017) Echinoderm development and evolution in the post-genomic era. *Dev Bio*; 427:203–211. doi: 10.1016/j.ydbio.2017.02.003.

Casillas S, Barbadilla A (2017) Molecular population genetics. *Genetics*; 205:1003–1035. doi: 10.1534/genetics.116.196493.

Cassidy LM, Martiniano R, Murphy EM et al. (2016) Neolithic and Bronze Age migration to Ireland and establishment of the insular Atlantic genome. *Proc Natl Acad Sci U S A*; 113:368–373. doi: 10.1073/pnas.1518445113.

Castellano S, Parra G, Sánchez-Quinto FA (2014) Patterns of coding variation in the complete exomes of three Neandertals. *Proc Natl Acad Sci U S A*; 111:6666–6671. doi: 10.1073/pnas.1405138111.

Catacchio CR et al. (2018) Inversion variants in human and primate genomes. *Genome Res*; 28:910–920. doi: 10.1101/gr.234831.118.

Cavalier-Smith T (1993) Kingdom Protozoa and its 18 phyla. *Microbiol Rev*; 57:953–994.

Cavalier-Smith T (2006) Rooting the tree of life by transition analyses. *Biol Direct*; 1:19.

Cavalier-Smith T (2010) Origin of the cell nucleus mitosis and sex: Roles of intracellular coevolution. *Biol Direct*; 5:7. doi: 10.1186/1745-6150-5-7.

Cavalier-Smith T (2014) The neomuran revolution and phagotrophic origin of eukaryotes and cilia in the light of intracellular coevolution and a revised tree of life. *Cold SpringHarb Perspect Biol*; 6:a016006.

Cavalier-Smith T (2018) Kingdom Chromista and its eight phyla: A new synthesis emphasising periplastid protein targeting, cytoskeletal and periplastid evolution, and ancient divergences. *Protoplasma*; 255:297–357. doi: 10.1007/s00709-017-1147-3.

Cavalier-Smith T, Chao EE (2003) Molecular phylogeny of centrohelid heliozoa a novel lineage of bikont eukaryotes that arose by ciliary loss. *J Mol Evol*; 56:387–396.

Chan YF, Marks ME, Jones FC (2010) Adaptive evolution of pelvic reduction in sticklebacks by recurrent deletion of a *Pitx1* enhancer. *Science*; 327:302–305. doi: 10.1126/science.1182213.

Chang S, Zhang L, Clausen S, Bottjer DJ, Feng F (2019) The Ediacaran-Cambrian rise of siliceous sponges and development of modern oceanic ecosystems. *Precambrian Res*; 333:105438. doi.org/10.1016/j.precamres.2019.105438.

Charlesworth D (2002) Plant sex determination and sex chromosomes. *Heredity*; 88:94–101. https://doi.org/10.1038/sj.hdy.6800016.

Charmantier A, Perrins C, McCleery RH, Sheldon BC (2006) Quantitative genetics of age at reproduction in wild swans: Support for antagonistic pleiotropy models of senescence. *Proc Natl Acad Sci U S A*; 103:6587–92. doi: 10.1073/pnas.0511123103.

Chatterjee S (2015) *The Rise of Birds: 225 Million Years of Evolution*. Johns Hopkins University Press.

Chaudhary R, Burleigh JG, Fernández-Baca D (2013) Inferring species trees from incongruent multi-copy gene trees using the Robinson-Fauld distance. *Alg Mol Biol*; 8:28–37. doi: 10.1186/1748-7188-8-28.

Chen CF, Foley J, Tang PC (2015) Development, regeneration, and evolution of feathers. *Annu Rev Anim Biosci*; 3:169–195. doi: 10.1146/annurev-animal-022513-114127.

Chen F et al. (2019) A late Middle Pleistocene Denisovan mandible from the Tibetan Plateau. *Nature*; 569:409–412. doi.org/10.1038/s41586-019-1139-x.

Chen J-Y (2009) The sudden appearance of diverse animal body plans during the Cambrian explosion. *Int J Dev Biol*; 53:733–751. doi: 10.1387/ijdb.072513cj.

Chen L, Xiao S, Pang K, Zhou C, Yuan X (2014) Cell differentiation and germ-soma separation in Ediacaran animal embryo-like fossils. *Nature*; 516:238–241. doi: 10.1038/Nature13766.

Chen Z, Zhou C, Yuan X, Xiao S (2019) Death march of a segmented and trilobate bilaterian elucidates early animal evolution. *Nature*; 573:412–415. doi: 10.1038/s41586-019-1522-7.

Chew KY, Shaw G, Yu H (2014) Heterochrony in the regulation of the developing marsupial limb. *Dev Dyn*; 243:324–338. doi: 10.1002/dvdy.24062.

Chiarenza AA, Farnsworth A, Mannion PD et al. (2020) Asteroid impact, not volcanism, caused the end-Cretaceous dinosaur extinction. *Proc Natl Acad Sci U S A*; 117:17084–17093. doi: 10.1073/pnas.2006087117.

Chiatante G, Giannuzzi G, Calabrese FM, Eichler EE, Ventura M (2017) Centromere destiny in dicentric chromosomes: New insights from the evolution of human chromosome 2 ancestral centromeric region. *Mol Biol Evol*; 34:1669–1681. doi: 10.1093/molbev/msx108.

Childebayeva A, Jones TR, Goodrich JM et al. (2019) LINE-1 and EPAS1 DNA methylation associations with high-altitude exposure. *Epigenetics*; 14:1–15. doi: 10.1080/15592294.2018.1561117.

Chisholm RL, Firtel RA (2004) Insights into morphogenesis from a simple developmental system. *Nat Rev Mol Cell Biol*; 5:531–41. doi: 10.1038/nrm1427.

Chittka L, Osorio D (2007) Cognitive dimensions of predator responses to imperfect mimicry. *PLoS Biol*; 5(12):e339. doi: 10.1371/journal. pbio.0050339.

Choe CP, Miller SC, Brown SJ. (2006) A pair-rule gene circuit defines segments sequentially in the short-germ insect *Tribolium castaneum*. *Proc Natl Acad Sci U S A*; 103:6560–4. doi: 10.1073/pnas.0510440103.

Choi J, Kim SH (2017) A genome tree of life for the Fungi Kingdom. *Proc Natl Acad Sci U S A*;114(35):9391–9396. doi: 10.1073/ pnas.1711939114.

Choudhury A, Aron S, Sengupta D, Hazelhurst S, Ramsay M (2018) African genetic diversity provides novel insights into evolutionary history and local adaptations. *Hum Mol Genet*; 27:R209–R218. doi: 10.1093/hmg/ddy161.

Chow RL, Altmann CR, Lang RA, Hemmati-Brivanlou A (1999) Pax6 induces ectopic eyes in a vertebrate. *Development*; 126:4213–4222.

Chown SL, Gaston KJ (2010) Body size variation in insects: A macroecological perspective. *Biol Rev Camb Philos Soc*; 85:139–169. doi: 10.1111/j.1469-185X.2009.00097.x.

Cisneros JC, Abdala F, Rubidge BS (2011) Dental occlusion in a 260-million-year-old therapsid with saber canines from the Permian of Brazil. *Science*; 33:1603–1605. doi: 10.1126/ science.1200305.

Clack J (2012) *Gaining Ground: The Origin and Evolution of the Tetrapods* (2nd ed). Indiana University Press.

Clack JA (2002) An early tetrapod from 'Romer's Gap. *Nature*; 418:72–76.

Claessens LP, O'Connor PM, Unwin DM (2009) Respiratory evolution facilitated the origin of pterosaur flight and aerial gigantism. *PLoS One*; 4(2):e4497. doi: 10.1371/journal. pone.0004497.

Clark ED, Peel AD, Akam M (2019) Arthropod segmentation. *Development*; 146:pii:dev170480. doi: 10.1242/dev.170480.

Clark-Hachtel CM, Tomoyasu Y (2020) Two sets of candidate crustacean wing homologues and their implication for the origin of insect wings. *Nat Ecol Evol*; 4:1694–1702. doi: 10.1038/ s41559-020-1257-8.

Clarkson C, Jacobs Z, Marwick B et al. (2017) Human occupation of northern Australia by 65,000 years ago. *Nature*; 547:306–310. doi: 10.1038/Nature22968.

Clarkson E, Levi-Setti R, Horváth G (2006) The eyes of trilobites: The oldest preserved visual system. *Arthropod Struct Dev*; 35:247–259. doi: 10.1016/j.asd.2006.08.002.

Clarkson E, Twitchet R, Smart C (2014) *Invertebrate Palaeontology and Evolution*. Wiley-Blackwell.

Cleber A, Trujillo ES, Rice NK et al. (2021) Reintroduction of the archaic variant of *NOVA1* in cortical organoids alters neurodevelopment. *Science*; 371:eaax2537. doi:10.1126/science.aax2537.

Clements T, Dolocan A, Martin P, Purnell MA, Vinther J, Gabbott SE (2016) The eyes of *Tullimonstrum* reveal a vertebrate affinity. *Nature*; 532:500–503. doi: 10.1038/ Nature17647.

Cloutier R, Clement AM, Lee MSY et al. (2020) Elpistostege and the origin of the vertebrate hand. *Nature*; 579:549–554. doi: 10.1038/ s41586-020-2100-8.

Coates MI (1994) The origin of vertebrate limbs. *Dev Suppl*; 1994:169–80.

Coates MI (1996) The Devonian tetrapod *Acanthostega gunnari* Jarvik: Postcranial anatomy, basal tetrapod interrelationships and patterns of skeletal evolution. *Trans Royal Soc Edin (Earth Sciences)*; 87:363–421. doi:10.1017/S0263593300006787.

Cobb A, Cobb S (2019) Do zebra stripes influence thermoregulation? *J Nat Hist*; 53:863–879. doi.org/10.1080/00222933.2019. 1607600.

Cohen PA, Riedman LA (2018) It's a protist-eat-protist world: Recalcitrance, predation, and evolution in the tonian-cryogenian ocean. *Emerg Top Life Sci*; 2:173–180. doi: 10.1042/ETLS20170145.

Cohn MJ, Tickle C (1999) Developmental basis of limblessness and axial patterning in snakes. *Nature*; 399:474–479.

Coiffard C, Kardjilov N, Manke I, Bernardes-de-Oliveira MEC (2019) Fossil evidence of core monocots in the Early Cretaceous. *Nat Plant*; 5:691–696. doi: 10.1038/ s41477-019-0468-y.

Collins D (1996) The "evolution" of *Anomalocaris* and its classification in the arthropod class Dinocarida (Nov.) and order Radiodonta (Nov.). *J Paleontol*; 70:280–293.

Compant S, Samad A, Faist H, Sessitsch A (2019) A review on the plant microbiome: Ecology, functions, and emerging trends in microbial application. *J Adv Res*; 19:29–37. doi: 10.1016/j. jare.2019.03.004.

Conard NJ, Delagnes A (eds, 2010) *Settlement Dynamics of the Middle Paleolithic and Middle Stone Age*; Vol 3. Kerns Verlag, Tübingen.

Condamine FL, Clapham ME, Kergoat GJ (2016) Global patterns of insect diversification: Towards a reconciliation of fossil and molecular evidence? *Sci Rep*; 6:19208. doi: 10.1038/srep19208.

Conine CC, Sun F, Song L, Rivera-Pérez JA, Rando OJ (2018) Small RNAs gained during epididymal transit of sperm are essential for embryonic development in mice. *Dev Cell*; 46:470–480.e3. doi: 10.1016/j. devcel.2018.06.024.

Conway Morris S (1977) Fossil Priapulid Worms.

Cook LM, Saccheri IJ (2013) The peppered moth and industrial melanism: Evolution of a natural selection case study. *Heredity*; 10:207–212. doi: 10.1038/ hdy.2012.92.

Coots PS, Seifert AW (2015) Thyroxine-induced metamorphosis in the axolotl (*Ambystoma mexicanum*). *Methods Mol Biol*; 1290:141–5. doi: 10.1007/978-1-4939-2495-0_11. PMID:25740483.

Cordaux R, Gilbert C (2017) Evolutionary significance of *Wolbachia*-to-animal horizontal gene transfer: Female sex determination and the *f* element in the isopod *Armadillidium vulgare*. *Genes*; 8: pii:E186. doi: 10.3390/genes8070186.

Cordes EE, Arthur MA, Shea K, Arvidson RS, Fisher CR (2005) Modeling the mutualistic interactions between tubeworms and microbial consortia. *PLoS Biol*; 3(3):e77. doi: 10.1371/journal.pbio.0030077.

Cordingley JE, Sundaresan SR, Fischhoff IR, Shapiro B, Ruskey J, Rubenstein DI (2009) Is the endangered Grevy's zebra threatened by hybridization? *Animal Conservation*; 12:505–513. doi.org/10.1111/j.1469-1795. 2009.00294.x.

Costanzo E, Trehin C, Vandenbussche M (2014) The role of *WOX* genes in flower development. *Ann Bot*; 114:1545–1553. doi: 10.1093/aob/ mcu123.

Cox CJ (2018) Land plant molecular phylogenetics: A review with comments on evaluating incongruence among phylogenies. *Crit Rev Plant Sci*; 37:1130127. doi.org/10.1080/0735268 9.2018.1482443.

Coyne JA, Orr HA (1989) Two Rules of Speciation. In *Speciation and Its Consequences* (Otte D, Endler JA, eds), pp 180–207. Sinauer Press.

Coyne JA, Orr HA (2004) *Speciation*. Sinauer Press.

Cresko WA, Armores A, Wilson C (2004) Parallel genetic basis for repeated evolution of armor loss in Alaskan threespine stickleback populations. *Proc Natl Acad Sci U S A*; 101:6050–6055.

Crews D (2003) Sex determination: Where environment and genetics meet. *Evol Dev*; 5:50–55.

Crick AP, Babbs C, Brown JM, Morriss-Kay GM (2003) Developmental mechanisms underlying polydactyly in the mouse mutant Doublefoot. *J Anat*; 202:21–6. doi: 10.1046/j.1469-7580.2003.00132.x.

Cruz YP (1997) Mammals. In *Embryology: Constructing the Organism* (Gilbert SF, Raunio AM eds), pp 459–492. Sinauer Press.

CSaAC (Chimpanzee Sequencing and Analysis Consortium) (2005) Initial sequence of the chimpanzee genome and comparison with the human genome. *Nature*; 437:69–87. doi: 10.1038/Nature04072.

Culyba MJ, Mo CY, Kohli RM (2015) Targets for combating the evolution of acquired antibiotic resistance. *Biochemistry*; 54:3573–3582. doi: 10.1021/ acs.biochem.5b00109.

Cunningham JA, Varga, K, Yin Z, Bengtson S, Donoghue PC (2017) The Weng'an Biota (Doushantuo Formation): An Ediacaran window on soft-bodied and multicellular microorganisms. *J Geol Soc*; 174:793–802. doi.org/10.1144/jgs2016-142.

Curini-Galletti M, Artois T, Delogu V et al. (2012) Patterns of diversity in soft-bodied meiofauna: Dispersal ability and body size matter. *PLoS One*; 7(3):e33801. doi: 10.1371/journal. pone.0033801.

Curnoe D (2010) A review of early *Homo* in Southern Africa focusing on cranial, mandibular and dental remains, with the description of a new species (*Homo gautengensis* sp. nov). *Homo*; 6:151–177. doi: 10.1016/j.jchb.2010.04.002.

Curnoe D (2010) A review of early *Homo* in Southern Africa focusing on cranial mandibular and dental remains with the description of a new species (*Homo gautengensis* sp. Nov). *J Comp Hum Biol*; 61:151–177. doi: 10.1016/j.jchb.2010.04.002.

Curnoe D, Thorne A (2003) Number of ancestral human species: A molecular perspective. *Homo*; 53:201–224. doi: 10.1078/0018-442x-00051.

Cyran KA, Kimmel M (2010) Alternatives to the Wright-Fisher model: The robustness of mitochondrial eve dating. *Theor Popul Biol*; 78:165–172. doi: 10.1016/j. tpb.2010.06.001.

D'Anastasio R, Wroe S, Tuniz C et al. (2013) Micro-biomechanics of the kebara 2 hyoid and its implications for speech in Neanderthals. *PLoS One*; 8:e82261. doi: 10.1371/journal.pone.0082261.

D'Mello SA, Finlay GJ, Baguley BC, Askarian-Amiri ME (2016) Signaling pathways in melanogenesis. *Int J Mol Sci*; 17:1144. doi: 10.3390/ijms17071144.

da Silveira JC, de Ávila ACFCM, Garrett HL, Bruemmer JE, Winger QA, Bouma GJ (2018) Cell-secreted vesicles containing microRNAs as regulators of gamete maturation. *J Endocrinol*; 236:R15–R27. doi: 10.1530/JOE-17-0200.

Dabney J, Knapp M, Glocke I (2013) Complete mitochondrial genome sequence of a Middle Pleistocene cave bear reconstructed from ultrashort DNA fragments. *Proc Natl Acad Sci U S A*; 110:15758–15763. doi: 10.1073/pnas.1314445110.

Dahl TW, Hammarlund EU, Anbar AD et al. (2010) Devonian rise in atmospheric oxygen correlated to the radiations of terrestrial plants and large predatory fish. *Proc Natl Acad Sci U S A*; 107:17911–17915. doi: 10.1073/pnas.1011287107.

Dailey L (2015) High throughput technologies for the functional discovery of mammalian enhancers: New approaches for understanding transcriptional regulatory network dynamics. *Genomics*; 106:151–158. doi: 10.1016/j.ygeno.2015.06.004.

Daley AC, Antcliffe JB (2019) Evolution: The Battle of the first animals. *Curr Biol*; 29:R257–R259. doi: 10.1016/j.cub.2019.02.031.

Danley PD, Husemann M, Ding B (2012) The impact of the geologic history and paleoclimate on the diversification of East African cichlids. *Int J Evol Biol*; 2012:574851. doi: 10.1155/2012/574851.

Darroch SAF, Smith EF, Laflamme M, Erwin DH. (2018) Ediacaran extinction and Cambrian explosion. *Trends Ecol Evol*; 33:653–663. doi: 10.1016/j.tree.2018.06.003.

Darwin CR (1855) I - Extract from an unpublished work on species, by C. Darwin, Esq., consisting of a portion of a Chapter entitled, "On the variation of organic beings in a state of nature; on the natural means of selection; on the comparison of domestic races and true species." *Proc. Linn Soc*; 3:46–53. http://darwin-online.org.uk/converted/pdf/1858_species_F350.pdf.

Darwin CR (1859) On the Origin of Species by Means of Natural Selection or the Preservation of Favoured Races in the Struggle for Life. John Murray. http://darwin-onlineorguk/converted/pdf/1861_OriginNY_F382.pdf.

Darwin CR (1871) *The Descent of Man and Selection in Relation to Sex*. John Murray.

Darwin CR (1872) *The Expression of the Emotions in Man and Animals*. John Murray.

Darwin E (1796) *Zoonomia or The Laws of Organic Life* (2nd ed.). J Johnson. http://www.gutenberg.org/files/15707/15707-h/15707-h.htm

Daum B, Gold V (2018) Twitch or swim: Towards the understanding of prokaryotic motion based on the type IV pilus blueprint. *Biol Chem*; 399:799–808. doi.org/10.1515/hsz-2018-0157.

Dave AJ, Godward MB (1982) Ultrastructural studies in the Rhodophyta. I. Development of mitotic spindle poles in *Apoglossum ruscifolium*, Kylin. *J Cell Sci*; 58:345–62.

David B, Mooi R (2014) How Hox genes can shed light on the place of echinoderms among the deuterostomes. *Evodevo*; 5:22. doi: 10.1186/2041-9139-5-22.

Davidson AJ (2009) Mouse kidney development. *StemBook*. www.ncbi.nlm.nih.gov/pubmed/20614633.

Davidson EH (2010) Emerging properties of animal gene regulatory networks. *Nature*; 468:911–920. doi: 10.1038/nature09645.

Davidson EH, Erwin DH (2009) An integrated view of Precambrian Eumetazoan evolution. *Cold Spring Harb Symp Quant Biol*; 74:65–80. doi: 10.1101/sqb.2009.74.042.

Davies JA, Bard J (1998) The development of the kidney. *Curr Top Dev Biol*; 39:245–301.

Davis BM (2011) Evolution of the tribosphenic molar pattern in early mammals with comments on the "dual-origin" hypothesis. *J Mammal Evol*; 18:227–244.

Davis DD (1964) *The Giant Panda a Morphological Study of Evolutionary Mechanisms Fieldiana Zoology Memoirs*, pp 120–124. Chicago: Chicago Natural History Museum.

Davis MC, Shubin N, Daeschler EB (2004) A new specimen of *Sauripterus taylori* (*Sarcopterygii, Osteichthyes*) from the Fammenian Catskill Formation of North America. *J Vert Paleont*; 24:26–40.

Dawkins R (1976) *The Selfish Gene*. Oxford University Press.

Dawkins R (1996) The Blind Watchmaker. W W Norton & Co.

de Bakker MA, Fowler DA, den Oude K (2013) Digit loss in archosaur evolution and the interplay between selection and constraints. *Nature*; 500:445–448. doi: 10.1038/nature12336.

de Heinzelin J, Clark JD, White T et al. (1999) Environment and behavior of 2.5-million-year-old Bouri hominids. *Science*; 284:625–629. doi: 10.1126/Science.284.5414.625.

de Herder WW (2012) Acromegalic gigantism, physicians and body snatching. Past or present? *Pituitary*; 15:312–318. doi: 10.1007/s11102-012-0389-5.

De León LF, Podos J, Gardezi T et al. (2014) Darwin's finches and their diet niches: The sympatric coexistence of imperfect generalists. *J Evol Biol*; 27:1093–1104. doi: 10.1111/jeb.12383.

de Queiroz K, Cantino P (2020) *PhyloCode: A Phylogenetic Code of Biological Nomenclature*. CRC Press.

De Robertis EM (2008) Evo-devo: Variation on ancestral themes. *Cell*; 132:185–195. doi: 10.1016/j.cell.2008.01.003.

De Robertis EM, Sasai Y (1996) A common plan for dorsoventral patterning in bilateria. *Nature*; 380:37–44.

de Vries J, Gould SV (2018) The monoplastidic bottleneck in algae and plant evolution. *J Cell Sci*; 131:jcs203414. doi: 10.1242/jcs.203414.

de Vries J, de Vries S, Slamovits CH, Rose LE, Archibald JM (2017) How embryophytic is the biosynthesis of phenylpropanoids and their derivatives in streptophyte algae? *Plant Cell Physio*; 58:934–945. doi: 10.1093/pcp/pcx037.

Dean P, Hirt RP, Embley TM (2016) Microsporidia: Why make nucleotides if you can steal them? *PLoS Pathog*; 12:e1005870. doi: 10.1371/journal.ppat.1005870.

Decelle J et al. (2018) Worldwide occurrence and activity of the reef-building coral symbiont symbiodinium in the open ocean. *Curr Biol*; 28:3625–3633.e3. doi: 10.1016/j.cub.2018.09.024.

Decker RS, Koyama E, Pacifici M (2014) Genesis and morphogenesis of limb synovial joints and articular cartilage. *Matrix Biol*; 39:5–10. doi: 10.1016/j.matbio.2014.08.006.

del Pino EM (2018) The extraordinary biology and development of marsupial frogs (Hemiphractidae) in comparison with fish, mammals, birds, amphibians and other animals. *Mech Dev*; 154:2–11. doi: 10.1016/j.mod.2017.12.002.

Del-Bem LE (2018) Xyloglucan evolution and the terrestrialization of green plants. *New Phyto*; 219:1150–1153. doi: 10.1111/nph.15191.

Delezene LK (2015) Modularity of the anthropoid dentition: Implications for the evolution of the hominin canine honing complex. *J Hum Evol*; 86:1–12. doi: 10.1016/j.jhevol.2015.07.001.

De Menocal PB (2004) African climate change and faunal evolution during the Pliocene-Pleistocene. *Earth Planet Sci Lett*; 220:3–24.

Demircan T, İlhan AE, Aytürk N, Yıldırım B, Öztürk G, Keskin İ (2016) A histological atlas of the tissues and organs of neotenic and metamorphosed axolotl. *Acta Histochem*; 118:746–759. doi: 10.1016/j.acthis.2016.07.006.

Denes AS, Jékely G, Steinmetz PR (2007) Molecular architecture of annelid nerve cord supports common origin of nervous system centralization in bilateria. *Cell*; 129:277–288.

Dennis MY, Nuttle X, Sudmant PH (2012) Evolution of human-specific neural SRGAP2 genes by incomplete segmental duplication. *Cell*; 149:912–922. doi: 10.1016/j.cell.2012.03.033.

Denton D, Aung-Htut MT, Kumar S (2013) Developmentally programmed cell death in *Drosophila*. *Biochim Biophys Acta*; 1833:3499–3506. doi: 10.1016/j.bbamcr.2013.06.014.

Der Sarkissian C, Allentoft ME, Ávila-Arcos MC et al. (2015) Ancient genomics. *Philos Trans R Soc Lond B Biol Sci*; 370:20130387. doi: 10.1098/rstb.2013.0387.

Derelle R, Lopez P, Le Guyader H, Manuel M (2007) Homeodomain proteins belong to the ancestral molecular toolkit of eukaryotes. *Evol Dev*; 9:212–219. doi: 10.1111/j.1525-142X.2007.00153.x.

Desalle R, Rosenfeld J (2012) *Phylogenomics*. Garland Press.

Deshpande O, Batzoglou S, Feldman MW, Cavalli-Sforza LL (2009) A serial founder effect model for human settlement out of Africa. *Proc Biol Sci*; 276:291–300. doi: 10.1098/rspb.2008.0750.

Détroit F, Mijares AS, Corny J et al. (2019) A new species of *Homo* from the Late Pleistocene of the Philippines. *Nature*; 568:181–186. doi: 10.1038/s41586-019-1067-9.

Dewar D, Merson Davies L, Haldane JBS (1949) *Is Evolution a Myth*. CA Watts & The Paternoster Press.

Deryckere A, Styfhals R, Vidal EAG et al. (2020) A practical staging atlas to study embryonic development of Octopus vulgaris under controlled laboratory conditions. *BMC Dev Biol*; 20:7. doi: 10.1186/s12861-020-00212-6.

Dhouailly D (2009) A new scenario for the evolutionary origin of hair feather and avian scales. *J Anat*; 214:587–606. doi: 10.1111/j.1469-7580.2008.01041.x.

Dhouailly D, Sengel P (1973) Morphogenic interactions between reptilian epidermis and birds or mammalian dermis. *C R Acad Sci Hebd Seances Acad Sci D*; 277:1221–1224.

Diamond J (1997) *Guns, Germs and Steel*. Vintage Books, London.

Diaz GA, Gelb BD, Risch N et al. (2000) Gaucher disease: The origins of the Ashkenazi Jewish N370S and 84GG acid beta-glucosidase mutations. *Am J Hum Genet*; 66:1821–32. doi: 10.1086/302946.

Dietz L, Dömel JS, Leese F, Mahon AR, Mayer C (2019) Phylogenomics of the longitarsal colossendeidae: The evolutionary history of an Antarctic sea spider radiation. *Mol Phylogenet Evol*; 136:206–214. doi: 10.1016/j.ympev.2019.04.017.

Dines JP, Otárola-Castillo E, Ralph P et al. (2014) Sexual selection targets cetacean pelvic bones. *Evolution*; 68:3296–3306. doi: 10.1111/evo.12516.

Diogo R (2008) *The Origin of Higher Clades: Osteology, Myology, Phylogeny and Evolution of Bony Fishes and the Rise of Tetrapods*. CRC Press.

Diogo R, Ziermann JM (2015) development, metamorphosis, morphology, and diversity: The evolution of chordate muscles and the origin of vertebrates [published correction appears in dev dyn. 2015 Sep;244(9):1179]. *Dev Dyn*; 244:1046–1057. doi: 10.1002/dvdy.24245.

Dirks PH et al. (2015) Geological and taphonomic context for the new Hominin species *Homo naledi* from the Dinaledi Chamber, South Africa. *Elife*; 4:e09561. doi: 10.7554/eLife.09561.

Dirks PH et al. (2017) The age of *Homo naledi* and associated sediments in the Rising Star Cave, South Africa. *Elife*; 6:pii:e24231. doi: 10.7554/eLife.24231.

Doan RN, Bae BI, Cubelos B et al. (2016) Mutations in human accelerated regions disrupt cognition and social behavior. *Cell*; 167:341–354.e12. doi: 10.1016/j.cell.2016.08.071.

Dobzhansky T (1937) *Genetics and the Origin of Species*. Columbia University Press.

Dobzhansky T (1973) Nothing in biology makes sense except in the light of evolution. *Am Biol Teacher*; 35:125–129.

Dohrmann M, Wörheide G (2017) Dating early animal evolution using phylogenomic data. *Sci Rep*; 7:3599. doi: 10.1038/s41598-017-03791-w.

Dolgova O, Lao O (2018) Evolutionary and Medical Consequences of Archaic Introgression into Modern Human Genomes. *Genes (Basel)*; 9:358. doi: 10.3390/genes9070358.

Donoghue PC, Sansom IJ (2002) Origin and early evolution of vertebrate skeletonization. *Microsc Res Tech*; 59:352–372.

Donoghue PC, Forey PL, Aldridge RJ (2000) Conodont affinity and chordate phylogeny. *Biol Rev Camb Philos Soc*; 75:191–251.

Donoghue PC, Yang Z (2016) The evolution of methods for establishing evolutionary timescales. *Philos Trans R Soc*; 371B:20160020. doi: 10.1098/rstb.2016.0020.

Dor Y, Cedar H (2018) Principles of DNA methylation and their implications for biology and medicine. *Lancet*; 392:777–786. doi: 10.1016/S0140-6736(18)31268-6.

Dorard C, Vucak G, Baccarini M (2017) Deciphering the RAS/ERK pathway in vivo. *Biochem Soc Trans*; 45:27–36. doi: 10.1042/BST20160135. PMID:28202657.

Douka K, Slon V, Jacobs Z et al. (2019) Age estimates for hominin fossils and the onset of the upper palaeolithic at denisova cave. *Nature*; 565:640–644. doi: 10.1038/s41586-018-0870-z.

Douze K, Wurz S, Henshilwood CS. (2015) Techno-cultural characterization of the MIS 5 (c. 105–90 ka) lithic industries at Blombos Cave, Southern Cape, South Africa. *PLoS One*; 10(11):e0142151. doi: 10.1371/journal.pone.0142151.

Drongitis D, Aniello F, Fucci L, Donizetti A (2019) Roles of transposable elements in the different layers of gene expression regulation. *Int J Mol Sci*; 20:pii:E5755. doi: 10.3390/ijms20225755.

Droser ML, Gehling JG (2015) The advent of animals: The view from the Ediacaran. *Proc Natl Acad Sci U S A*; 112:4865–4870. doi: 10.1073/pnas.1403669112.

Drumm P (2018) Australian fantasy revisited. *Hist Psych*; 21:295–296. https://doi.org/10.1037/hop0000099.

Duboule D, Tarchini B, Zàkàny J, Kmita M (2007) Tinkering with constraints in the evolution of the vertebrate limb anterior-posterior polarity. *Novartis Found Symp*; 284:130–141.

DuBuc TQ, Ryan JF, Martindale MQ (2019) "Dorsal-ventral" genes are part of an ancient axial patterning system: Evidence from trichoplax adhaerens (Placozoa). *Mol Biol Evol*; 36:966–973. doi: 10.1093/molbev/msz025.

Duda P, Zrzavý J (2013) Evolution of life history and behavior in hominidae: Towards phylogenetic reconstruction of the chimpanzee-human last common ancestor. *J Hum Evo*; 65:424–446. doi: 10.1016/j.jhevol.2013.07.009.

Dudley R (2000) The evolutionary physiology of animal flight: Paleobiological and present perspectives. *Ann Rev Physiol*; 62:135–155. doi: 10.1146/annurev.physiol.62.1.135.

Dunn FS, Liu AG, Donoghue PCJ. Ediacaran developmental biology (2018) *Biol Rev Camb Philos Soc*; 93:914–932. doi: 10.1111/brv.12379.

Dunne EM, Close RA, Button DJ et al. (2018) Diversity change during the rise of tetrapods and the impact of the "Carboniferous rainforest collapse." *Proc Biol Sci*; 285:20172730. doi: 10.1098/rspb.2017.2730.

Dunning-Hotopp JC (2011) Horizontal gene transfer between bacteria and animals. *Trends Genet* 27:157–163. doi: 10.1016/j.tig.2011.01.005.

Durand-Smet P, Chastrette N, Guiroy A et al. (2014) A comparative mechanical analysis of plant and animal cells reveals convergence across kingdoms. *Biophys J*; 107:2237–44. doi: 10.1016/j.bpj.2014.10.023.

Dziecio AJ, Mann S (2012) Designs for life: Protocell models in the laboratory. *Chem Soc Rev* 41:79–85. doi: 10.1039/c1cs15211d.

Dzik J (2003) Anatomical information content in the Ediacaran fossils and their possible zoological affinities. *Integr Comp Biol*; 43:114–126. doi: 10.1093/icb/43.1.114.

Eames BF, Allen N, Young J, Kaplan A, Helms JA, Schneider RA (2007) Skeletogenesis in the swell shark cephaloscyllium ventriosum. *J Anat*; 210:542–554. doi: 10.1111/j.1469-7580.2007.00723.x.

Edel KH, Marchadier E, Brownlee C, Kudla J, Hetherington AM. (2017) The evolution of calcium-based signalling in plants. *Curr Biol*; 27: R667–R679. doi: 10.1016/j.cub.2017.05.020.

Edwards CJ, Suchard MA, Lemey P (2011) Ancient hybridization and an Irish origin for the modern polar bear matriline. *Curr Biol*; 21:12511258. doi: 10.1016/j.cub.2011.05.058.

Edwards M, Cha D, Krithika S, Johnson M, Cook G, Parra EJ (2016) Iris pigmentation as a quantitative trait: Variation in populations of European, East Asian and South Asian ancestry and association with candidate gene polymorphisms. *Pigment Cell Melanoma Res*; 29:141–62. doi: 10.1111/pcmr.12435.

Ehrenfreund P, Rasmussen S, Cleaves J, Chen L (2006) Experimentally tracing the key steps in the origin of life: The aromatic world. *Astrobiology*; 6:490–520. doi: 10.1089/ast.2006.6.490.

El Albani A, Bengtson S, Canfield DE et al. (2010) Large colonial organisms with coordinated growth in oxygenated environments 2.1 Gyr ago. *Nature*; 466:100–104. doi: 10.1038/nature09166.

El Albani A, Bengtson S, Canfield DE (2014) The 2.1 Ga old Francevillian biota: Biogenicity, taphonomy and biodiversity. *PLoS One*; 9:e99438. doi: 10.1371/journal.pone.0099438.

Eldredge N, Gould SJ (1972) Punctuated Equilibria: An Alternative to Phyletic Gradualism. In *Models in Paleobiology* (Schopf TJM ed), pp 82–115. Freeman Cooper. http://www.nileseldredge.com/uploads/3/8/1/1/38114941/punctuated_equilibria_eldredge_gould_1972.pdf

Elgorriaga A, Escapa IH, Rothwell GW et al. (2018) Origin of Equisetum: Evolution of horsetails (Equisetales) within the major euphyllophyte clade Sphenopsida. *Am J Bot*; 105:1286–1303. doi: 10.1002/ajb2.1125.

Eliason CM, Hudson L, Watts T et al. (2017) Exceptional preservation and the fossil record of tetrapod integument. *Proc Biol Sci*; 284:20170556. doi: 10.1098/rspb.2017.0556.

Elinson RP, Beckham Y (2002) Development in frogs with large eggs and the origin of amniotes. *Zoology (Jena)*; 105(2):105–117. doi: 10.1078/0944-2006-00060.

Ellegren H (2011) Sex-chromosome evolution: Recent progress and the influence of male and female heterogamety. *Nat Rev Genet*; 12:157–66 doi: 10.1038/nrg2948.

Ellis NA, Roe AM, Kozloski J, Proytcheva M, Falk C, German J (1994) Linkage disequilibrium between the FES, D15S127, and BLM loci in Ashkenazi Jews with Bloom syndrome. *Am J Hum Genet*; 55:453–60.

Elzanowski A, Ostell J (2013) The genetic codes. www.ncbi.nlm.nih.gov/Taxonomy/Utils/wprintgc.cgi

Eme L, Spang A, Lombard J, Stairs CW, Ettema TJG (2017) Archaea and the origin of eukaryotes. *Nat Rev Microbiol*; 15:711–723. doi: 9.1038/nrmicro.2017.133.

Emelyanov VV (2001) Rickettsiaceae, rickettsia-like endosymbionts, and the origin of mitochondria. *Biosci Rep*; 21:1–17. doi: 10.1023/a:1010409415723.

Epifanova E, Babaev A, Newman AG, Tarabykin V (2019) Role of Zeb2/Sip1 in neuronal development. *Brain Res*; 1705:24–31. doi: 10.1016/j.brainres.2018.09.034.

Ernst OP, Lodowski DT, Elstner M, Hegemann P, Brown LS, Kandori H (2014) Microbial and animal rhodopsins: Structures, functions, and molecular mechanisms. *Chem Rev*; 114:126–63. doi: 10.1021/cr4003769.

Erwin DH (2015) Early metazoan life: Divergence, environment and ecology. *Phil Trans R Soc*; 370B. pii:20150036. doi: 10.1098/rstb.2015.0036.

Erwin DH, Davidson EH (2002) The last common bilaterian ancestor. *Development*; 129:3021–3032.

Erzurumlu RS, Gaspar P (2012) Development and critical period plasticity of the barrel cortex. *Eur J Neurosci*; 35:1540–1553. doi: 10.1111/j.1460-9568.2012.08075.x.

Evans DD, Hughes IV, Gehling JG, Droser ML (2020) Discovery of the oldest bilaterian from the Ediacaran of South Australia. *Proc. Nat. Acad. Sci*; 202001045. doi: 10.1073/pnas.2001045117.

Ewart JC (1897) *A Critical Period in the Development of the Horse*. A & C Black.

Ewart JC (1899) *The Pennycuik Experiments*. A & C Black.

Ewen WJ (2010) *Mathematical Population Genetics: I. Theoretical Introduction*. New York: Springer.

Ewens W (1972) The sampling theory of selectively neutral alleles. *Theoret Pop Biol*; 3:87–112.

Ezcurra MD (2014) The osteology of the basal archosauromorph tasmaniosaurus triassicus from the lower triassic of Tasmania, Australia. *PLoS One*; 9(1):e86864. doi: 10.1371/journal.pone.008686.

Falcon-Lang HJ, Benton MJ, Stimson M (2007) Ecology of the earliest reptiles inferred from basal Pennsylvanian trackways. *J Geol Soc*; 164:1113–1118.

Fan HT, Zhang M, Zhan P, Yang X, Tian WJ, Li RW (2016) Structural chromosomal abnormalities in couples in cases of recurrent spontaneous abortions in Jilin Province, China. *Genet Mol Res*; 15:10.4238/gmr.15017443 doi: 10.4238/gmr.15017443.

Farbstein R, Radić D, Brajković D, Miracle PT (2012) First epigravettian ceramic figurines from Europe (Vela Spila, Croatia). *PLoS One.*2; 7:e41437. doi: 10.1371/journal.pone.0041437.

Fedonkin MA, Waggoner BM (1997) The late Precambrian fossil *Kimberella* is a mollusc-like bilaterian organism. *Nature*; 388:868–871.

Feduccia A (2014) Avian extinction at the end of the Cretaceous: Assessing the magnitude and subsequent explosive radiation. *Cretaceous Research*; 50:1–15. doi: 10.1016/j.cretres.2014.03.009.

Fei X, Shi J, Liu Y, Niu J, Wei A (2019) The steps from sexual reproduction to apomixis. *Planta*; 249:1715–1730. doi: 10.1007/s00425-019-03113-6.

Fenchel T (2003) *The Origin and Early Evolution of Life*. Oxford University Press.

Ferner K, Mess A (2011) Evolution and development of fetal membranes and placentation in amniote vertebrates. *Respir Physiol Neurobiol*; 178:39–50. doi: 10.1016/j.resp.2011.03.029.

Ferris JP, Ertem G (1992) Oligomerization of ribonucleotides on montmorillonite: Reaction of the 5-phosphorimidazolide of adenosine. *Science*; 257:1387–1389.

Fesel PH1, Zuccaro A (2016) β-glucan: Crucial component of the fungal cell wall and elusive MAMP in plants. *Fungal Genet Biol*; 90:53–60. doi: 10.1016/j.fgb.2015.12.004.

Fiddes IT, Pollen AA, Davis JM, Sikela JM (2019) Paired involvement of human-specific Olduvai domains and *NOTCH2NL* genes in human brain evolution. *Hum Genet*; 138:715–721. doi: 10.1007/s00439-019-02018-4.

Fisher RA (1918a) The correlation between relatives on the supposition of mendelian inheritance. *Trans R Soc Edinb*; 52:399–433.

Fisher RA (1918b) The causes of human variability. *Eugen Rev*; 10:213–20.

Fisher RA (1930) *The Genetical Theory of Natural Selection*. Oxford Clarendon Press.

Flannery DT, Walter MR (2012) Archean tufted microbial mats and the great oxidation event: New insights into an ancient problem. *Aust J Earth Sci*; 59:1–11. doi: 10.1080/08120099.2011.607849.

Fleming PA, Verburgt L, Scantlebury M et al. (2009) Jettisoning ballast or fuel? Caudal autotomy and locomotory energetics of the Cape dwarf gecko *Lygodactylus capensis* (Gekkonidae). *Physiol Biochem Zool*; 82:756–765. doi: 10.1086/605953.

Ford DP, Benson RBJ (2020) The phylogeny of early amniotes and the affinities of Parareptilia and Varanopidae. *Nat Ecol Evol*; 4:57–65. doi.org/10.1038/s41559-019-1047-3.

Forterre P (2015) The universal tree of life: An update. *Front Microbiol*; 6:717. doi: 10.3389/fmicb.2015.00717.

Fortey R (2011) *Survivors: The Animals and Plants That Time Left Behind*. Harper Press.

Fothergill T, Donahoo AL, Douglass A (2014) Netrin-DCC signaling regulates corpus callosum formation through attraction of pioneering axons and by modulating Slit2-mediated repulsion. *Cereb Cortex*; 24:1138–1151. doi: 10.1093/cercor/bhs395.

Fouts RS, Fouts DH, Van Cantfort TE (1989) The infant Loulis learns signs from cross-fostered chimpanzees. In *Teaching Sign Language to Chimpanzees*. (Gardner RA, Gardner BT, Van Cantfort TE, eds). Albany: State University of New York Press; 8:280–2922.

Franch-Marro X, Martín N, Averof M, Casanova J (2006) Association of tracheal placodes with leg primordia in Drosophila and implications for the origin of insect tracheal systems. *Development*; 133:785–790. doi: 10.1242/dev.02260.

Frank SA (1995) George Price's contributions to evolutionary genetics. *J Theor Biol*; 175:373–388.

Frank SA (2012) Natural selection IV: The price equation. *J Evol Biol*; 25:1002–1019. doi: 10.1111/j.1420-9101.2012.02498.x.

Freyer C, Zeller U, Renfree MB (2003) The marsupial placenta: A phylogenetic analysis. *J Exp Zool*; 299A:59–77.

Friedlaender JS et al. (2008) The genetic structure of pacific islanders. *PLoS Genet*; 4:e19. doi: 10.1371/journal.pgen.0040019.

Friedman JR, Nunnari J (2014) Mitochondrial form and function. *Nature*; 505:335–343. doi: 10.1038/nature12985.

Friedman RE (1987) *Who Wrote the Bible?* Summit Books.

Frieling J, Gebhardt H, Huber M, Adekeye OA, Akande SO, Reichart GJ, Middelburg JJ, Schouten S, Sluijs A (2017) Extreme warmth and heat-stressed plankton in the tropics during the Paleocene-Eocene Thermal Maximum. *Sci Adv*; 3:e1600891. doi: 10.1126/sciadv.1600891.

Friis EM, Pedersen KR, Crane PR (2016) The emergence of core eudicots: New floral evidence from the earliest Late Cretaceous. *Proc R Soc*; B283:pii:20161325. doi: 10.1098/rspb.2016.1325.

Fritzenwanker JH, Uhlinger KR, Gerhart J, Silva E, Lowe CJ (2019) Untangling posterior growth and segmentation by analyzing mechanisms of axis elongation in hemichordates. *Proc Natl Acad Sci U S A*; 116:8403–8408. doi: 10.1073/pnas.1817496116.

Frumkin A, Bar-Yosef O, Schwarcz HP (2011) Possible paleohydrologic and paleoclimatic effects on hominin migration and occupation of the Levantine Middle Paleolithic. *J Hum Evol*; 60:437–451. doi: 10.1016/j.jhevol.2010.03.010.

Fry JD (2003) Multilocus models of sympatric speciation: Bush versus Rice versus Felsenstein. *Evolution*; 57:1735–1746.

Fry JD (2009) Laboratory Experiments on Speciation. In *Experimental Evolution: Concepts, Methods, and Applications of Selection Experiments* (Garland T, Rose MR, eds). pp 631–656. doi: 10.1525/california/9780520247666.003.0020.

Fryer G (2001) On the age and origin of the species flock of haplochromine cichlid fishes of Lake Victoria. *Proc Biol Sci*; 268:1147–1152.

Fu Q et al. (2013) A revised timescale for human evolution based on ancient mitochondrial genomes. *Curr Biol*; 23:553–559. doi: 10.1016/j.cub.2013.02.044.

Fu Q et al. (2018) An unexpected noncarpellate epigynous flower from the Jurassic of China. *Elife*; 7:pii:e38827. doi: 10.7554/eLife.38827.

Fuerst JA, Sagulenko E (2011) Beyond the bacterium: Planctomycetes challenge our concepts of microbial structure and function. *Nat Rev Microbiol*; 9:403–413. doi: 10.1038/nrmicro2578.

Fuerst JA, Sagulenko E (2012) Keys to eukaryality: Planctomycetes and ancestral evolution of cellular complexity. *Front Microbiol*; 3:167. doi: 10.3389/fmicb.2012.00167.

Fuss J, Spassov N, Begun DR, Böhme M (2017) Potential hominin affinities of graecopithecus from the Late Miocene of Europe. *PLOS One*; 12(5):e0177127. doi.org/10.1371/journal.pone.0177127.

Futuyma DJ (2013) *Evolutionary Biology*. Sinauer Press.

Gabora L, Steel M. (2020) A model of the transition to behavioural and cognitive modernity using reflexively autocatalytic networks. *J R Soc Interface*; 171:20200545. doi: 10.1098/rsif.2020.0545.

Gabr A, Grossman AR, Bhattacharya D (2020) Paulinella, a model for understanding plastid primary endosymbiosis. *J Phycol*; 56:837–843. doi: 10.1111/jpy.13003.

Gagat P, Sidorczuk K, Pietluch F, Mackiewicz P (2020) The photosynthetic Adventure of *Paulinella* spp. *Results Probl Cell Differ*; 69:353–386. doi: 10.1007/978-3-030-51849-3_13.

Gage PJ, Rhoades W, Prucka SK, Hjalt T (2005) Fate maps of neural crest and mesoderm in the mammalian eye. *Invest Ophthalmol Vis Sci*; 46:4200–4208.

Gaines RR, Briggs DEG, Yuanlong Z (2008) Cambrian Burgess Shale-type deposits share a common mode of fossilization. *Geology*; 36:755–758.

Gañan Y, Macias D, Basco RD (1998) Morphological diversity of the avian foot is related with the pattern of msx gene expression in the developing autopod. *Dev Biol*; 196:33–41.

Gardner A (2020) Price's equation made clear. *Philos Trans R Soc Lond*; 375B:20190361 doi: 10.1098/rstb.2019.0361.

Garg SG, Martin WF. (2016) Mitochondria, the cell cycle, and the origin of sex via a syncytial eukaryote common ancestor. *Genome Biol Evol*; 8:1950–1970. doi: 10.1093/gbe/evw136.

Garm A, Ekström P (2010) Evidence for multiple photosystems in jellyfish. *Int Rev Cell Mol Biol*; 280:41–78. doi: 10.1016/S1937-6448(10)80002-4.

Garwood RJ, Edgecombe GD (2011) Early terrestrial animals, evolution, and uncertainty. *Evo Edu Outreach*; 4:489–501. doi: 10.1007/s12052-011-0357-y.

Gaston KJ, Spicer JI (2004) *Biodiversity: An Introduction*. Blackwell Science.

Gates TA, Organ C, Zanno LE (2016) Bony cranial ornamentation linked to rapid evolution of gigantic theropod dinosaurs. *Nat Commun*; 7:12931. doi: 10.1038/ncomms12931.

Gatesy J, Geisler JH, Chang J et al. (2013) A phylogenetic blueprint for a modern whale. *Mol Phylogenet Evol*; 66:479–506. doi: 10.1016/j.ympev.2012.10.012.

Gaunt SJ, Paul YL (2012) Changes in cis-regulatory elements during morphological evolution. *Biology (Basel)*; 1:557–574. doi: 10.3390/biology1030557.

Gaunt SJ. (2018) Hox cluster genes and collinearities throughout the tree of animal life. *Int J Dev Biol*; 62:673–683. doi: 10.1387/ijdb.180162sg.

Gauthier C, Peck LS (1999) Polar gigantism dictated by oxygen availability. *Nature*; 399:114115.

Gavelis GS, Keeling PJ, Leander BS (2017) How exaptations facilitated photosensory evolution: Seeing the light by accident. *Bioessays*; 39:10.1002/bies.201600266. doi: 10.1002/bies.201600266.

Gehring WJ (1996) The master control gene for morphogenesis and evolution of the eye. *Genes Cells*; 1:11–15.

Gehring WJ, Kloter U, Suga H (2009) Evolution of the Hox gene complex from an evolutionary ground state. *Curr Top Dev Biol*; 88:35–61. doi: 10.1016/S0070-2153(09)88002-2.

Geisler JH (2019) Whale evolution: Dispersal by paddle or fluke. *Curr Biol*; 29:R294–R296. doi: 10.1016/j.cub.2019.03.005.

George D, Blieck A (2011) Rise of the earliest tetrapods: An early Devonian origin from marine environment. *PLoS One*; 6(7):e22136. doi: 10.1371/journal.pone.0022136.

Gerkema MP, Davies W, Foster RG et al. (2013) The nocturnal bottleneck and the evolution of activity patterns in mammals. *Proc Biol Sci*; 280:20130508. doi: 10.1098/rspb.2013.0508.

Gibbons A (2018) How islands shrink people. *Science*; 361:439. doi: 10.1126/Science. 361.6401.439.

Gierlinski GD, Niedźwiedzki G, Lockley MG et al. (2017) Possible hominin footprints from the Late Miocene (c. 5.7 Ma) of Crete? *Proc Geol Ass*; 697–710. doi.org/10.1016/j.pgeola.2017.07.006.

Gilbert SF (2019) Developmental symbiosis facilitates the multiple origins of herbivory. *Evol Dev*; 22:e12291. doi: 10.1111/ede.12291.

Gilbert SF, Epel D (2015) *Ecological Developmental Biology* (2nd ed). Sinauer Press.

Gilbert SF, Bosch TC, Ledón-Rettig C (2015) Eco-evo-devo: Developmental symbiosis and Developmental plasticity as evolutionary agents. *Nat Rev Genet*; 16:611–622. doi: 10.1038/nrg3982.

Giles S, Friedman M, Brazeau MD (2015) Osteichthyan-like cranial conditions in an early Devonian stem gnathostome. *Nature*; 520:82–85. doi: 10.1038/nature14065.

Gill GG, Purnell MA, Crumpton N (2014) Dietary specializations and diversity in feeding ecology of the earliest stem mammals. *Nature*; 512:303–305. doi: 10.1038/nature13622.

Gillis JA, Alsema EC, Criswell KE (2017) Trunk neural crest origin of dermal denticles in a cartilaginous fish. *Proc Natl Acad Sci*; 114:13200–13205. doi: 10.1073/pnas.1713827114.

Gingerich PD, Haq Mu, Zalmout IS, Khan IH, Malkani MS (2001) Origin of whales from early artiodactyls: Hands and feet of eocene protocetidae from Pakistan. *Science*; 293:2239–2242. doi: 10.1126/science.1063902.

Giribet G (2008) Assembling the lophotrochozoan (spiralian) tree of life. *Philos Trans R Soc Lond B Biol Sci*; 363B:1513–1522. doi: 10.1098/rstb.2007.2241.

Giribet G, Edgecombe GD (2012) Reevaluating the arthropod tree of life. *Annu Rev Entomol*; 57:167–186. doi: 10.1146/annurev-ento-120710-100659.

Giribet G, Distel DL, Polz M (2000) Triploblastic relationships with emphasis on the acoelomates and the position of Gnathostomulida, Cycliophora, Plathelminthes, and Chaetognatha: A combined approach of 18S rDNA sequences and morphology. *Syst Biol*; 49:539–562.

Giribet G, Edgecombe GD (2019) The phylogeny and evolutionary history of arthropods. *Curr Biol*; 29:R592–R602. doi: 10.1016/j.cub.2019.04.057.

Gissis SB, Jablonka E (eds) (2011) *Transformations of Lamarckism*. MIT Press.

Glansdorff N, Xu Y, Labedan B (2008) The last universal common ancestor: Emergence constitution and genetic legacy of an elusive forerunner. *Biol Direct*; 3:29. doi: 10.1186/1745-6150-3-29.

Glikson A (2016) Cenozoic mean greenhouse gases and temperature changes with reference to the Anthropocene. *Glob Chang Biol*; 22:3843–3858. doi: 10.1111/gcb.13342.

Godefroit P, Cau A, Hu D-Y, Escuillié F et al. (2013) A Jurassic avialan dinosaur from China resolves the early phylogenetic history of birds. *Nature*; 498:359–362. doi: 10.1038/nature12168.

Goebel T, Waters MR, O'Rourke DH (2008) The Late Pleistocene dispersal of modern humans in the Americas. *Science*; 319:1497–1502. doi: 10.1126/Science.1153569.

Goedert J, Lécuyer C, Amiot R et al. (2018) Euryhaline ecology of early tetrapods revealed by stable isotopes. *Nature*; 558:68–72. doi: 10.1038/s41586-018-0159-2.

Goldberg A, Mychajliw AM, Hadly EA (2016) Post-invasion demography of prehistoric humans in South America. *Nature*; 532:232–235. doi: 10.1038/Nature17176.

Goldschmidt RB (1940) *The Material Basis of Evolution*. Yale University Press.

Goldschmidt T (1996) *Darwin's Dreampond: Drama in Lake Victoria*. MIT Press.

Gokhman D, Mishol N, de Manuel M et al. (2019) Reconstructing Denisovan anatomy using DNA methylation maps. *Cell*; 179:180–192.e10. doi: 10.1016/j.cell.2019.08.035.

Gokhman D, Nissim-Rafinia M, Agranat-Tamir L et al. (2020) Differential DNA methylation of vocal and facial anatomy genes in modern humans. *Nat Commun*; 4;11:1189. doi: 10.1038/s41467-020-15020-6.

Gomez B, Daviero-Gomez V, Coiffard C (2015) *Montsechia*, an ancient aquatic angiosperm. *Proc Natl Acad Sci*; 112:10985–10988. doi: 10.1073/pnas.1509241112.

Good BH, McDonald MJ, Barrick JE, Lenski RE, Desai MM (2017) The dynamics of molecular evolution over 60,000 generations. *Nature*; 551:45–50. doi: 10.1038/Nature24287.

Goodenough U, Heitman J (2014) Origins of eukaryotic sexual reproduction. *Cold Spring Harb Perspect Biol*; 6:pii:a016154. doi: 10.1101/cshperspect.a016154.

Goodman MB, Sengupta P (2019) How caenorhabditis elegans senses mechanical stress, temperature, and other physical stimuli. *Genetics*; 212:25–51. doi: 10.1534/genetics.118.300241.

Goodwin AF, Chen CP, Vo NT, Bush JO, Klein OD (2020) YAP/TAZ regulate elevation and bone formation of the mouse secondary palate. *J Dent Res*; 99:1387–1396. doi: 10.1177/0022034520935372.

Gornik SG, Bergheim BG, Morel B, Stamatakis A, Foulkes NS, Guse A (2020) Photoreceptor diversification accompanies the evolution of anthozoa. *Mol Biol Evol:msaa* 304. doi: 10.1093/molbev/msaa304.

Gorrell JC, McAdam AG, Coltman DW et al. (2010) Adopting kin enhances inclusive fitness in asocial red squirrels. *Nat Commun*; 1:1–4. doi: 10.1038/ncomms1022.

Goswami A, Milne N, Wroe S (2011) Biting through constraints: Cranial morphology disparity and convergence across living and fossil carnivorous mammals. *Proc R Soc*; 278B:1831–1839. doi: 10.1098/rspb.2010.2031.

Gould SJ (1977) *Ontogeny and Phylogeny*. Belknap Press.

Gould SJ (2000) A Tree Grows in Paris: Lamarck's Division of Worms and Revision of Nature. In *The Lying Stones of Marrakech*: *Penultimate Reflections in Natural History*. pp 115–114. Harmony Books. http://www.facstaff. bucknell.edu/sdjordan/PDFs/Gould_Lamarck'sTree.pdf

Gould SJ, Eldredge N (1972 [1993]) Punctuated equilibrium comes of age. *Nature*; 366:223–227.

Gould SJ, Eldredge N (1972) Punctuated Equilibria: The Tempo and Mode of Evolution Reconsidered. *Paleobiol*; 3:115–151.

Grabowski M, Hatala KG, Jungers WL (2018) Body mass estimates of the earliest possible hominins and implications for the last common ancestor. *J Hum Evol*; 122:84–92. doi: 10.1016/j.jhevol.2018.05.001.

Grady JM, Enquist BJ, Dettweiler-Robinson E et al. (2014) Dinosaur physiology. Evidence for mesothermy in dinosaurs. *Science* 344:1268–1272. doi: 10.1126/science.

Graham A, Shimeld SM (2013) The origin and evolution of the ectodermal placodes. *J Anat*; 222:32–40. doi. org/10.1111/j.1469-7580.2012.01506.x.

Graham JB, Wegner NC, Miller LA et al. (2014) Spiracular air breathing in polypterid fishes and its implications for aerial respiration in stem tetrapods. *Nat Commun*; 5:3022. doi: 10.1038/ncomms4022.

Grant BR, Grant PR (2008) Fission and fusion of Darwin's finches populations. *Philos Trans R Soc Lond B Biol Sci*; 63B:2821–2829. doi: 10.1098/rstb.2008.0051.

Grasby SE, Sanei H, Beauchamp B (2011) Catastrophic dispersion of coal fly ash into oceans during the latest Permian extinction. *Nat Geosci*; 4:104–107. doi: 10.1038/NGEO1069.

Gray MW (2012) Mitochondrial evolution. *Cold Spring Harb Perspect Biol*; 4:a011403. doi: 10.1101/cshperspect.a011403.

Gray MW, Burger G, Lang BF (2001) The origin and early evolution of mitochondria. *Genome Biol*; 2:1–5.

Gredler ML, Larkins CE, Leal F et al. (2014) Evolution of external genitalia: Insights from reptilian development. *Sex Dev*; 8:311–26. doi: 10.1159/000365771.

Green RE, Krause J, Brigs AW et al. (2010) A draft sequence of the Neandertal genome. *Science*; 328:710–722. doi: 10.1126/science.1188021.

Greenfield A, Bard J (2015) The Reproductive System. In *Supplement to Kaufman's Atlas of Mouse Development* (Baldock R, Bard J, Davidson D, Morriss-Kay G, eds). pp 121–132. Academic Press.

Gribaldo S, Brochier-Armanet C (2006) The origin and evolution of Archaea: A state of the art. *Philos Trans R Soc Lond B Biol Sci*; 361:1007–1022. doi: 10.1098/rstb.2006.1841.

Gribaldo S, Poole AM, Daubin V (2010) The origin of eukaryotes and their relationship with the Archaea: Are we at a phylogenomic impasse? *Nat Rev Microbiol*; 8:743–752. doi: 10.1038/nrmicro2426.

Grimaldi D, Engel MS (2005) *Evolution of Insects*. Cambridge University Press.

Grindley JC, Hargett LK, Hill RE et al. (1997) Disruption of PAX6 function in mice homozygous for the Pax6Sey-1Neu mutation produces abnormalities in the early development and regionalization of the diencephalon. *Mech Dev*; 64:111–126.

Grine FE, Wurz S, Marean CW (2017) The Middle Stone Age human fossil record from Klasies River main site. *J Hum Evol*; 103:53–78. doi: 10.1016/j.jhevol.2016.12.001.

Gritli-Linde A (2007) Molecular control of secondary palate development. *Dev Biol*; 301:309–326.

Groenewald GH, Welman J, Maceachern JA (2001) Vertebrate burrow complexes from the early Triassic cynognathus zone (Driekoppen Formation Beaufort Group) of the Karoo Basin South Africa. *Palaios*; 16:148–160.

Grossniklaus U, Kelly WG, Kelly B, Ferguson-Smith AC, Pembrey M, Lindquist S (2013 Mar) Transgenerational epigenetic inheritance: How important is it? *Nat Rev Genet*; 14(3):228–235. doi: 10.1038/nrg3435.

Grossnickle DM, Smith SM, Wilson GP (2019) Untangling the multiple ecological radiations of early mammals. *Trends Ecol Evol*; 34:936–949. doi: 10.1016/j.tree.2019.05.008.

Groucutt HS et al. (2015) Rethinking the dispersal of *Homo sapiens* out of Africa. *Evol Anthropol*; 24:149–64. doi: 10.1002/evan.21455.

Grube M, Wedin M (2016) Lichenized fungi and the evolution of symbiotic organization. *Microbiol Spectr*; 4:10.1128/microbiolspec. FUNK-0011-2016. doi: 10.1128/microbiolspec. FUNK-0011-2016.

Grün R (2006) Direct dating of human fossils. *Am J Phys Anthr*; 49:2–48.

Grün R, Pike A, McDermott F et al. (2020) Dating the skull from Broken Hill, Zambia, and its position in human evolution. *Nature*; 580:372–375. doi: 10.1038/s41586-020-2165-4.

Grün R, Stringer C, McDermott F (2005) U-series and ESR analyses of bones and teeth relating to the human burials from skhul. *J Hum Evol*; 49:316–334.

Gu H, Goodale E, Chen J (2015) Emerging directions in the study of the ecology and evolution of plant-animal mutualistic networks: A review. *Zoo Res*; 36:65–71.

Guernsey MW, Chuong EB, Cornelis G, Renfree MB, Baker JC (2017) Molecular conservation of marsupial and eutherian placentation and lactation. *Elife*. 2017;6:e27450. doi: 10.7554/eLife.27450.

Guerrero G (2008) The session that did not shake the world (The Linnaean Society, 1st July 1858). *Int Microbiol*; 11:209–212.

Guerrero R, Margulis L, Berlanga M (2013) Symbiogenesis: The holobiont as a unit of evolution. *Int Microbiol*; 16:133–143. doi: 10.2436/20.1501.01.188.

Guo R, Chen LH, Xing C, Liu T (2019) Pain regulation by gut microbiota: Molecular mechanisms and therapeutic potential. *Br J Anaesth*; 123:637–654. doi: 10.1016/j.bja.2019.07.026.

Gupta RS (2011) Origin of diderm (gramnegative) bacteria: Antibiotic selection pressure rather than endosymbiosis likely led to the evolution of bacterial cells with two membranes. *Antonie Van Leeuwenhoek* 100:171–82. doi: 10.1007/s10482-011-9616-8.

Haas PA, Goldstein RE (2018) The flipping and peeling of elastic lips. *Phys Rev E*; 98:052415. doi: 10.1103/PhysRevE.98.052415.

Hagen IJ, Donnellan SC, Bull CM (2012) Phylogeography of the prehensile-tailed skink *Corucia zebrata* on the Solomon Archipelago. *Ecol Evol*; 2:1220–1234. doi: 10.1002/ece3.84.

Hajkova Erhardt S, Lane N (2002) Epigenetic reprogramming in mouse primordial germ cells. *Mech Dev*; 117:15–23.

Haldane JBS (1927) A mathematical theory of natural and artificial selection, part v: Selection and mutation. *Proc Camb Philos Soc*; 23:838–844.

Halder G, Callaerts P, Gehring WJ (1995) Induction of ectopic eyes by targeted expression of the eyeless gene in *Drosophila*. *Science*; 267:17881792.

Halfmann R, Jarosz DF, Jones SK, Chang A, Lancaster AK, Lindquist S (2012) Prions are a common mechanism for phenotypic inheritance in wild yeasts. *Nature*; 482:363–8. doi: 10.1038/Nature10875.

Hall BG (2004) Comparison of the accuracies of several phylogenetic methods using protein and DNA sequences. *Mol Biol Evol*; 22:792–802.

Hallett E, Marean, CW, Steele TE, et al. (2021) A worked bone assemblage from 120,000–90,000 year old deposits at Contrebandiers Cave, Atlantic Coast, Morocco. *iScience*; September:102988. doi: 10.1016/j.isci.2021.102988.

Hamilton MB (2009) *Population Genetics*. Wiley Blackwell.

Hamilton WD (1964) The genetical evolution of social behavior. I. *J Theoret Biol*; 7:1–16.

Hamilton WD (1967) Extraordinary sex ratios. *Science*; 156:477–488.

Han B, Weiss LM (2017) Microsporidia: Obligate intracellular pathogens within the fungal Kingdom. *Microbiol Spectr*; 5:10.1128/microbiolspec.FUNK-0018-2016. doi: 10.1128/microbiolspec.FUNK-0018-2016.

Han G, Mao F, Bi S, Wang Y, Meng J (2017) A Jurassic gliding euharamiyidan mammal with an ear of five auditory bones. *Nature*; 551:451–456. doi: 10.1038/Nature24483.

Han M (1997) Gut reaction to Wnt signaling in worms. *Cell*; 90:581–584.

Han TM, Runnegar B (1992) Megascopic eukaryotic algae from the 21-billion-year-old negaunee iron-formation Michigan. *Science*; 257:232–235.

Hanson IM, Fletcher JM, Jordan T (1994) Mutations at the PAX6 locus are found in heterogeneous anterior segment malformations including Peters' anomaly. *Nat Genet*; 6:168–173.

Hardy IC (ed) (2002) *Sex Ratios: Concepts and Research Methods*. Cambridge University Press.

Harmand S, Lewis JE, Feibel CS et al. (2015) 3.3-million-year-old stone tools from Lomekwi 3, West Turkana, Kenya. *Nature*; 521:310–315. doi: 10.1038/nature14464. PMID: 25993961.

Harmston N, Lenhard B (2013) Chromatin and epigenetic features of long-range gene regulation. *Nucleic Acids Res*; 41:7185–7199.

Harris WA (1997) Pax-6: Where to be conserved is not conservative. *Proc Natl Acad Sci*; 94:2098–2100.

Harrison CJ, Morris JL (2018) The origin and early evolution of vascular plant shoots and leaves. *Phil Trans*; 373B:20160496. doi: 10.1098/rstb.2016.0496.

Harrison JF (2015) Handling and use of oxygen by pancrustaceans: Conserved patterns and the evolution of respiratory structures. *Integr Comp Biol*; 55:802–15. doi: 10.1093/icb/icv055.

Hartstone-Rose A, Dickinson E, Boettcher ML, Herrel A (2019) A primate with a Panda's thumb: The anatomy of the pseudothumb of *Daubentonia madagascariensis*. *Am J Phys Anthropol*; 1–9. doi.org/10.1002/ajpa.23936.

Harvati K et al. (2019) Apidima Cave fossils provide earliest evidence of *Homo sapiens* in Eurasia. *Nature*; 571:500–504. doi: 10.1038/s41586-019-1376-z.

Hastie ND (2017) Wilms' tumour 1 (WT1) in development, homeostasis and disease. *Development*; 144:2862–2872. doi: 10.1242/dev.153163.

Hayes B (2013) Overview of statistical methods for genome-wide association studies (GWAS). *Methods Mol Biol*; 1019:149–169. doi: 10.1007/978-1-62703-447-0_6.

He D, Fiz-Palacios O, Fu CJ, Fehling J, Tsai CC, Baldauf SL (2014) An alternative root for the Eukaryote tree of life. *Curr Biol*; 24:465–470. doi: 10.1016/j.cub.2014.01.036.

Hedrick PW (2010) *Genetics of Populations*. Jones & Bartlett.

Heger TJ, Edgcomb VP, Kim E (2014) A resurgence in field research is essential to better understand the diversity, ecology, and evolution of microbial eukaryotes. *J Eukaryot Microbiol*; 61:214–223. doi: 10.1111/jeu.12095.

Hein J, Schierup M, Carten W (2005) *Gene Genealogies, Variation and Evolution: A Primer in Coalescent Theory*. Oxford University Press.

Heinz T, Pala M, Gómez-Carballa A, Richards MB, Salas A (2017) Updating the African human mitochondrial DNA tree: Relevance to forensic and population genetics. *Forensic Sci Int Genet*; 27:156–159. doi: 10.1016/j.fsigen.2016.12.016.

Held LI (2014) *How the Snake Lost Its Legs: Curious Tales from the Frontier of Evo-Devo*. Cambridge University Press.

Held LI Jr. (2017) *Deep Homology? Uncanny Similarities of Humans and Flies Uncovered by Evo-Devo*. Cambridge University Press.

Held LI Jr. (2021) *Animal Anomalies*. Cambridge University Press.

Hellmann JK, Hamilton IM (2019) Intragroup social dynamics vary with the presence of neighbors in a cooperatively breeding fish. *Curr Zool*; 65:21–31. doi: 10.1093/cz/zoy025.

Hendry AP, Nosily P, Rieseberg LH (2007) The speed of ecological speciation. *Funct Ecol*; 21:455–464.

Henn BM, Cavalli-Sforza LL, Feldman MW (2012) The great human expansion. *Proc Natl Acad Sci U S A*; 109:17758–17764. doi: 10.1073/pnas.1212380109.

Hennig W (1966) *Phylogenetic Systematics*. University of Illinois Press.

Henry RJ, Furtado A, Rangan P (2020) Pathways of photosynthesis in non-leaf tissues. *Biology (Basel)*; 9:438. doi: 10.3390/biology9120438.

Henshilwood CS, d'Errico F, van Niekerk KL et al. (2011) A 100,000-year-old ochre-processing workshop at Blombos Cave, South Africa. *Science*; 334:219–222. doi: 10.1126/Science.1211535.

Henshilwood CS, d'Errico F, Marean CW (2001) An early bone tool industry from the Middle Stone Age at Blombos Cave, South Africa: Implications for the origins of modern human behaviour, symbolism and language. *J Hum Evol*; 41:631–678.

Henshilwood CS, d'Errico F, Yates R (2002) Emergence of modern human behavior: Middle Stone age engravings from South Africa. *Science*; 295:1278–1280.

Herron JC, Freeman S (2014) *Evolutionary Analysis* (5th ed). Pearson: England.

Hershberger WK, Myers JM, Iwamoto RN, Mcauley WC, Saxton AM (1990) Genetic changes in the growth of coho salmon (*Oncorhynchus kisutch*) in marine net-pens, produced by ten years of selection. *Aquaculture*; 85:187–197. doi.org/10.1016/0044-8486(90)90018-I.

Hershkovitz I et al. (2018) The earliest modern humans outside Africa. *Science*; 359:456–459. doi: 10.1126/Science.aap8369.

Hershkovitz I, May H, Sarig R et al. (2021) A Middle Pleistocene *Homo* from Nesher Ramla, Israel. *Science*; 372:1424–1428. Doi: 10.1126/science.abh3169.

Hertweck KL, Kinney MS, Stuart SA et al. (2015) Phylogenetics, divergence times and diversification from three genomic partitions in monocots. *Bot J Linn Soc*; 178:375–393. doi.org/10.1111/boj.12260.

Hervella M, Svensson EM, Alberdi A, et al. (2016) The mitogenome of a 35,000-year-old Homo sapiens from Europe supports a Palaeolithic back-migration to Africa. *Sci Rep*; 6:25501. doi: 10.1038/srep25501.

Hickford D, Frankenberg S, Renfree MB (2009) The tammar wallaby, *Macropus eugenii*: A model kangaroo for the study of developmental and reproductive biology. *Cold Spring Harb Protoc*; 2009:pdb.emo137. doi: 10.1101/pdb.emo137.

Higgs PG, Lehman N (2015) The RNA world: Molecular cooperation at the origins of life. *Nat Rev Genet*; 6:7–17. doi: 10.1038/nrg3841.

Higham T, Douka K, Wood R (2014) The timing and spatiotemporal patterning of Neanderthal disappearance. *Nature*; 512:306–309. doi: 10.1038/nature13621.

Hill RE, Favor J, Hogan BL et al. (1991) Mouse small eye results from mutations in a paired-like homeobox-containing gene. *Nature*; 354:522–525. doi: 10.1038/354522a0.

Hill WG (2010) Understanding and using quantitative genetic variation. *Philos Trans R Soc Lond*; 365B:73–85. doi: 10.1098/rstb.2009.0203.

Holland LZ, Holland ND (2001) Evolution of neural crest and placodes: amphioxus as a model for the ancestral vertebrate? *J Anat*; 199:85-98. doi: 10.1046/j.1469-7580.2001.

Ho NTT, Kutzner A, Heese K (2017) Brain plasticity, cognitive functions and neural stem cells: A pivotal role for the brain-specific neural master gene |-SRGAP2-FAM72-|. *Biol Chem*; 399:55-61. doi: 10.1515/hsz-2017-0190.

Ho SY, Duchêne S (2014) Molecular-clock methods for estimating evolutionary rates and timescales. *Mol Ecol*; 23:5947-5965. doi: 10.1111/mec.12953.

Hochuli PA, Feist-Burkhardt S (2013) Angiosperm-like pollen and afropollis from the Middle Triassic (Anisian) of the Germanic Basin. (Northern Switzerland). *Front Plant Sci*; 4:344. doi: 10.3389/fpls.2013.00344.

Hodgson JA, Mulligan CJ, Al-Meeri A, Raaum RL (2014) Early back-to-Africa migration into the Horn of Africa. *PLoS Genet*; 10:e1004393. doi: 10.1371/journal.pgen.1004393.

Hoffmann DL et al. (2018) U-th dating of carbonate crusts reveals Neandertal origin of Iberian Cave art. *Science*; 359:912-915. doi: 10.1126/Science.aap7778.

Hohagen J, Jackson DJ (2013) An ancient process in a modern mollusc: Early development of the shell in *Lymnaea stagnalis*. *BMC Dev Biol*; 13:27 doi: 10.1186/1471-213X-13-27.

Höhn S, Hallmann A (2011) There is more than one way to turn a spherical cellular monolayer inside out: Type b embryo inversion in *Volvox globator*. *BMC Biol*; 9:89. doi: 10.1186/1741-7007-9-89.

Holland LZ (2015) Evolution of basal deuterostome nervous systems. *J Exp Biol*; 218:637-645. doi: 10.1242/jeb.109108.

Holland ND (2005) Chordates. *Curr Biol*; 22:R911-R914.

Holland PWH (2013) Evolution of homeobox genes. *Wiley Interdiscip Rev Dev Biol*; 2:31-45. doi: 10.1002/wdev.78.

Holland PWH, Takahashi T (2005) The evolution of homeobox genes: Implications for the study of brain development. *Brain Res Bull*; 66:484-490.

Holleley CE, O'Meally D, Sarre SD et al. (2015) Sex reversal triggers the rapid transition from genetic to temperature-dependent sex. *Nature*; 523:79-82. doi: 10.1038/nature14574.

Holmes G (2012) The role of vertebrate models in understanding craniosynostosis. *Childs Nerv Syst*; 28:1471-1481. doi: 10.1007/s00381-012-1844-3.

Hopson JA (1950) The origin of the mammalian middle ear. *Am Zool*; 6:437-450.

Hooper LV, Gordon JI (2001) Commensal host-bacterial relationships in the gut. *Science* 292; 1115-1118.

Horie R, Hazbun A, Chen K, Cao C, Levine M, Horie T (2018) Shared evolutionary origin of vertebrate neural crest and cranial placodes. *Nature*; 560:228-232. doi: 10.1038/s41586-018-0385-7.

Hornbruch A, Wolpert L (1970) Cell division in the early growth and morphogenesis of the chick limb. *Nature*; 226:764-766.

Horner JR, Varricchio DJ, Goodwin MB (1992) Marine transgressions and the evolution of Cretaceous dinosaurs. *Nature*; 358:59-61.

Hoyal Cuthill JF, Guttenberg N, Budd GE (2020) Impacts of speciation and extinction measured by an evolutionary decay clock. *Nature*; 588:636-641. doi: 10.1038/s41586-020-3003-4.

Hrdy SB (2009) *Mothers and Others: The Evolutionary Understanding*. Harvard: Belknap Press.

Huang JD, Motani R, Jiang DY et al. (2019, Sep 9) The new ichthyosauriform *Chaohusaurus brevifemoralis* (Reptilia, Ichthyosauromorpha) from Majiashan, Chaohu, AnHui Province, China. *PeerJ*. 2019;7:e7561. doi: 10.7717/peerj.7561.

Hubaud A, Pourquié O (2014) Signalling dynamics in vertebrate segmentation. *Nat Rev Mol Cell Biol*; 15:709-721. doi: 10.1038/nrm3891.

Huber SK, De León LF, Hendry AP (2007) Reproductive isolation of sympatric morphs in a population of Darwin's finches. *Proc Biol Sci*; 274:1709-1714.

Hublin JJ et al. (2017) New fossils from Jebel Irhoud, Morocco and the pan-African origin of *Homo sapiens*. *Nature*; 546:289-292. doi: 10.1038/Nature22336.

Hudson C (2016) The central nervous system of ascidian larvae. *Wiley Interdiscip Rev Dev Biol*; 5:538-61. doi: 10.1002/wdev.239.

Huerta-Sánchez E, Jin X et al. (2014) Altitude adaptation in Tibetans caused by introgression of Denisovan-like DNA. *Nature*; 512:194-197. doi: 10.1038/nature13408.

Hughes JF, Rozen S (2012) Genomics and genetics of human and primate y chromosomes. *Ann Rev Genomics Hum Genet*; 3:83-108. doi: 10.1146/annurev-genom-090711-163855.

Hulbert AJ, Else PL (1989) Evolution of mammalian endothermic metabolism: Mitochondrial activity and cell composition. *Am J Physiol*; 256:R63-R69.

Huldtgren T, Cunningham JA, Yin C, Stampanoni M (2011) Fossilized nuclei and germination structures identify Ediacaran "animal embryos" as encysting protists. *Science*; 334:1696-1699. doi: 10.1126/science.1209537.

Hutchinson JR, Delmer C, Miller CE (2011) From flat foot to fat foot structure, ontogeny, function, and evolution of elephant sixth toes. *Science*; 34:1699-1703. doi: 10.1126/science.1211437.

Huysseune A, Sire JY, Witten PE (2009) Evolutionary and developmental origins of the vertebrate dentition. *J Anat*; 21:465-476. doi: 10.1111/j.1469-7580.2009.01053.x.

Ijdo JW, Baldini A, Ward DC et al. (1991) Origin of human chromosome 2: An ancestral telomere-telomere fusion. *Proc Natl Acad Sci U S A*; 88:9051-9055.

Ikuta T (2011) Evolution of invertebrate deuterostomes and Hox/ParaHox genes. *Genomics Proteomics Bioinformatics*; 9:77-96. doi: 10.1016/S1672-0229(11)60011-9.

Imachi H et al. (2020) Isolation of an archaeon at the prokaryote-eukaryote interface *Nature*. 577:519-525. doi: 9.1038/s41586-019-1916-6.

Ioannidis AG, Blanco-Portillo J, Sandoval K, et al. (2021). Paths and timings of the peopling of Polynesia inferred from genomic networks. *Nature*; 597:522-526. doi: 10.1038/s41586-021-03902-8.

Iqbal K, Tran DA, Li AX (2015) Deleterious effects of endocrine disruptors are corrected in the mammalian germline by epigenome reprogramming. *Genome Biol*; 16:59. doi: 10.1186/s13059-015-0619-z.

Irwin DE, Irwin JH, Price TD (2001) Ring species as bridges between microevolution and speciation. *Genetica*; 112-113:223-243.

Irwin G (1994) *The Prehistoric Exploration and Colonisation of the Pacific*. Cambridge University Press.

Ishida N, Oyunsuren T, Mashima S, Mukoyama H, Saitou N (1995) Mitochondrial DNA sequences of various species of the genus Equus with special reference to the phylogenetic relationship between Przewalskii's wild horse and domestic horse. *J Mol Evol*; 41:180-8. doi: 10.1007/BF00170671.

Islam S, Bučar DK, Powner M (2017) Prebiotic selection and assembly of proteinogenic amino acids and natural nucleotides from complex mixtures. *Nature Chem*; 9:584-589. doi.org/10.1038/nchem.2703.

Ivantsov AY, Fedonkin MA (2002) Conulariid-like fossil from the vendian of Russia: A metazoan clade across the Proterozoic/Palaeozoic boundary. *Palaeontology*; 45:1219-1229. doi: 10.1111/1475-4983.00283.

Jablonski NG, Chaplin G (2000) The evolution of skin coloration. *J. Hum. Evol.* 39:57-106. doi: 10.1006/jhev.2000.0403.

Jablonski NG, Chaplin G (2010) Human skin pigmentation as an adaptation to UV radiation. *Proc Natl Acad Sci U S A*; 107(Suppl. 2):8962-8968. doi: 10.1073/pnas.0914628107.

Jacob F (1977) Evolution and tinkering. *Science*; 196:1161-1166.

Jacobs PA, Browne C, Gregson N, Joyce C, White H (1992) Estimates of the frequency of chromosome abnormalities detectable in unselected newborns using moderate levels of banding. *J Med Genet*; 29:103-108. doi: 10.1136/jmg.29.2.103.

Jacobs Z, Li B, Shunkov MV et al. (2019) Timing of archaic hominin occupation of Denisova Cave in Southern Siberia. *Nature*; 565:594-599. doi: 10.1038/s41586-018-0843-2.

James TY, Berbee ML (2012) No jacket required-new fungal lineage defies dress code: Recently described zoosporic fungi lack a cell wall during trophic phase. *Bioessays*; 34:94-102. doi: 10.1002/bies.201100110.

Jandzik D, Garnett AT, Square TA (2015) Evolution of the new vertebrate head by co-option of an ancient chordate skeletal tissue. *Nature*; 518:534-537. doi: 10.1038/nature14000.

Janik VM (2014) Cetacean vocal learning and communication. *Curr Opin Neurobiol*; 28:60-5. doi: 10.1016/j.conb.2014.06.010.

Janvier P (1996) *Early Vertebrates*. Clarendon Press.

Janvier P (2015) Facts and fancies about early fossil chordates and vertebrates. *Nature*: 520:483-489. doi.org/10.1038/nature14437.

Jarman AP (2000) Developmental genetics: Vertebrates and insects see eye to eye. *Curr Biol*; 10:R857-8599.

Jarne P, Auld JR (2006) Animals mix it up too: The distribution of self-fertilization among hermaphroditic animals. *Evolution*; 60:1816-1824.

Jarvis P, López-Juez E (2013) Biogenesis and homeostasis of chloroplasts and other plastids. *Nat Rev Mol Cell Biol*; 14:787-802. doi: 10.1038/nrm3702.

Javaux EJ (2007) The early eukaryotic fossil record. *Adv Exp Med Biol*; 607:1-19.

Javaux E, Marshall C, Bekker A (2010) Organic-walled microfossils in 3.2-billion-year-old shallow-marine siliciclastic deposits. *Nature*; 463:934-938. doi: 10.1038/nature08793.

Jaruga A, Hordyjewska E, Kandzierski G, Tylzanowski P (2016) Cleidocranial dysplasia and RUNX2-clinical phenotype-genotype correlation. *Clin Genet*; 90:393-402. doi: 10.1111/cge.12812.

Jensen MP, Allen CD, Eguchi T et al. (2018) Environmental warming and feminization of one of the largest sea turtle populations in the world. *Curr Biol*; 28:154-159.e4. doi: 10.1016/j.cub.2017.11.057.

Jensen PE, Leister D (2014) Chloroplast evolution structure and functions. *F1000Prime Rep*; 6:40. doi: 10.12703/P6-40.

Jeong C, Di Rienzo A (2014) Adaptations to local environments in modern human populations. *Curr Opin Genet Dev*; 29:1–8. doi: 10.1016/j.gde.2014.06.011.

Jeong H, Mendizabal I, Berto S et al. (2021) Evolution of DNA methylation in the human brain. *Nat Commun*; 12:2021. doi.org/10.1038/s41467-021-21917-7.

Ji Q, Luo ZX, Yuan CX, Tabrum AR (2006) A swimming mammaliaform from the Middle Jurassic and ecomorphological diversification of early mammals. *Science*; 311:1123–1127.

Jiggins FM (2017) The spread of *Wolbachia* through mosquito populations. *PLoS Biol*; 15:e2002780. doi: 10.1371/journal.pbio.2002780.

Jobling M, Hollox E, Hurles M et al. (2013) *Human Evolutionary Genetics* (2nd ed). Garland Press.

Jodar M, Sendler E, Krawetz SA (2016) The protein and transcript profiles of human semen. *Cell Tiss Res*; 363:85–96.

Johansson S (2014) Neanderthals did speak, but FOXP2 doesn't prove it. *Behav Brain Sci*; 37:558–9. doi: 10.1017/S0140525X13004068.

Johnson WE, Eizirik E, Pecon-Slattery J et al. (2006) The late miocene radiation of modern felidae: A genetic assessment. *Science*; 311:73–77.

Johnson WE, O'Brien SJ (1997) Phylogenetic reconstruction of the felidae using 16S rRNA and NADH-5 mitochondrial genes. *J Mol Evol*; 44(Suppl 1):S98–S116. doi: 10.1007/pl00000060.

Jones TR (1997) Quantitative aspects of the relationship between the sickle-cell gene and malaria. *Parasitol Today*; 13:107–111.

Jorde LB, Fineman RM, Martin RA (1983) Epidemiology and genetics of neural tube defects: An application of the Utah Genealogical Database. *Am J Phys Anthropol*; 62:23–31.

Jude E, Johanson Z, Kearsley A, Friedman M (2014) Early evolution of the lungfish pectoral-fin endoskeleton: evidence from the Middle Devonian (Givetian) Pentlandia macroptera. *Front. Earth Sci*; 2:18. doi: 10.3389/feart.2014.00018.

Just J, Kristensen RM, Olesen J (2014) Dendrogramma, new genus, with two new non-bilaterian species from the marine bathyal of southeastern Australia (Animalia, Metazoa incertae sedis) – With similarities to some medusoids from the Precambrian ediacara. *PLoS One*; 9:e102976. doi: 10.1371/journal.pone.0102976.

Kaas JH (2011) Neocortex in early mammals and its subsequent variations. *Ann N Y Acad Sci*; 1225:1225–1236. doi: 10.1111/j.1749-6632.2011.05981.x.

Kaltenpoth M, Flórez LV (2020) Versatile and dynamic symbioses between insects and burkholderia bacteria. *Ann Rev Entomol*; 65:145–170. doi: 10.1146/annurev-ento-011019-025025.

Kamenz C, Dunlop JA, Scholtz G, Kerp H, Hass H (2008) Microanatomy of early Devonian book lungs. *Biol Lett*; 4: 212–215. doi: 10.1098/rsbl.2007.0597.

Kanehisa M, Goto S, Sato Y et al. (2012) KEGG for integration and interpretation of large-scale molecular data sets. *Nucleic Acids Res*; 40:D109–D114. doi: 10.1093/nar/gkr988.

Kardong KV (2014) *Vertebrates: Comparative Anatomy, Function, Evolution* (7th ed). McGrawHill.

Karmin M, Saag L, Vicente M et al. (2015) A recent bottleneck of y chromosome diversity coincides with a global change in culture. *Genome Res*; 25:459–66. doi: 10.1101/gr.186684.114.

Karnkowska A (2016) A eukaryote without a mitochondrial organelle. *Curr Biol*; 26:1274–84. doi: 10.1016/j.cub.2016.03.053.

Kathman ND, Fox JL (2018) Representation of haltere oscillations and integration with visual inputs in the fly central complex. *J Neurosci*; 39:4100–4112. doi: 10.1523/JNEUROSCI.1707-18.2019.

Katz A (2012) Origin and diversification of eukaryotes. *Annu Rev Microbiol*; 66:411–427. doi: 10.1146/annurev-micro-090110-102808.

Katz MJ (1983) Comparative anatomy of the tunicate tadpole. *Ciona Intestinalis. Biol Bull*; 164:1–27. www.jstor.org/stable/1541186.

Kaufman AJ (2018) A Resource-Based Hypothesis for the Rise and Fall of the Ediacara Biota. In *Chemostratigraphy across Major Chronological Boundaries* (Sial AN, Gaucher C, Ramkumar M, Pinto Ferreira V, eds). pp 115–142. American Geophysical Union. doi.org/10.1002/9781119382508.ch7.

Kaufman M (2017) *Black Tudors: The Untold Story*. Oneworld Publications.

Kaucka M, Adameyko I (2019) Evolution and development of the cartilaginous skull: From a lancelet towards a human face. *Semin Cell Dev Biol*. doi: 10.1016/j.semcdb.2017.12.007.

Kawakami Y, Capdevila J, Büscher D (2001) WNT signals control FGF-dependent limb initiation and AER induction in the chick embryo. *Cell*; 104:891900. doi: 10.1016/S0092-8674(01)00285-9.

Kawashima T, Takahashi H, Ogasawara M et al. (2014) On a possible evolutionary link of the stomochord of hemichordates to pharyngeal organs of chordates. *Genesis*; 52:925–34. doi: 10.1002/dvg.22831.

Kawamoto Y, Nakajima YI, Kuranaga E (2016) Apoptosis in cellular society: Communication between apoptotic cells and their neighbors. *Int J Mol Sci*; 17:2144. doi: 10.3390/ijms17122144.

Kay T, Keller L, Lehmann L (2020) The evolution of altruism and the serial rediscovery of the role of relatedness. *Proc Natl Acad Sci U S A*; 202013596. doi: 10.1073/pnas.2013596117.

Keating JN, Marquart CL, Donoghue PC (2015) Histology of the heterostracan dermal skeleton: Insight into the origin of the vertebrate mineralised skeleton. J Morphol; 276:657–80. doi: 10.1002/jmor.20370.

Keeling PJ (2013) The number, speed, and impact of plastid endosymbioses in eukaryotic evolution. *Annu Rev Plant Biol*; 64:583–607. doi: 10.1146/annurev-arplant-050312-120144.

Keeling PK, Palmer JD (2008) Horizontal gene transfer in eukaryotic evolution. *Nat Rev Genet*; 9:605–618. doi: 10.1038/nrg2386.

Kehrer-Sawatzki H, Cooper DN (2007) Understanding the recent evolution of the human genome: Insights from human-chimpanzee genome comparisons. *Hum Mutat*; 28:99–130. doi: 10.1002/humu.20420.

Keith A (1910) Abnormal ossification of Meckel's cartilage. *J Anat Physiol*; 44:151–152.

Keller CB, Husson JM, Mitchell RN et al. (2019) Neoproterozoic glacial origin of the Great Unconformity. *Proc Natl Acad Sci U S A*; 116:1136–1145. doi: 10.1073/pnas.1804350116.

Keller I, Wagner CE, Greuter L (2013) Population genomic signatures of divergent adaptation, gene flow and hybrid speciation in the rapid radiation of Lake Victoria cichlid fishes. *Mol Ecol*; 22:2848–2863. doi: 10.1111/mec.12083.

Kellner AW, Wang X, Tischlinger H, de Almeida Campos D, Hone DW, Meng X (2010) The soft tissue of jeholopterus (*Pterosauria anurognathidae, batrachognathinae*) and the structure of the pterosaur wing membrane. *Proc Biol Sci*; 277:321–329. doi: 10.1098/rspb.2009.0846.

Kelly JK (2011) The breeder's equation. *Nat Educ Knowl*; 4(5):5. www.nature.com/scitable/knowledge/library/the-breeder-s-equation-24204828/.

Kemp TS (2005) *The Origin and Evolution of the Mammals*. Oxford University Press.

Keyte AL, Smith KS (2010) Developmental origins of precocial forelimbs in marsupial neonates. *Development*; 137:4283–4294. doi: 10.1242/dev.049445.

Kielan-Jaworowska Z, Cifelli RL, Luo Z-X (2004) *Mammals from the Age of Dinosaurs: Origins, Evolution, and Structure*. Columbia University Press.

Kikuchi Y, Nakatsukasa M, Tsujikawa H et al. (2018) Sexual dimorphism of body size in an African fossil ape, nacholapithecus kerioi. *J Hum Evol*; 123:129–140. doi: 10.1016/j.jhevol.2018.07.003.

Kim HK, Pham MHC, Ko KS, Rhee BD, Han J (2018) Alternative splicing isoforms in health and disease. *Pflugers Arch*; 470:995–1016. doi: 10.1007/s00424-018-2136-x.

Kim J, Ahn Y, Adasooriya D et al. (2019) Shh plays an inhibitory role in cusp patterning by regulation of Sostdc1. *J Dent Res*; 98:98–106. doi: 10.1177/0022034518803095.

Kim KM, Caetano-Anollés G (2011) The proteomic complexity and rise of the primordial ancestor of diversified life. *BMC Evol Biol*; 11:140. doi: 10.1186/1471-2148-11-140.

Kim KM, Park JH, Bhattacharya D, Yoon HS (2014) Applications of next-generation sequencing to unravelling the evolutionary history of algae. *Int J Syst Evol Microbiol*; 64(Pt 2):333–345. doi: 10.1099/ijs.0.054221-0.

Kimbel WH, Rak Y (2010) The cranial base of *Australopithecus afarensis*: New insights from the female skull. *Philos Trans R Soc*; 365B:3365–3376. doi: 10.1098/rstb.2010.0070.

Kimura I, Miyamoto J, Ohue-Kitano R et al. (2020) Maternal gut microbiota in pregnancy influences offspring metabolic phenotype in mice. *Science*; 367(6481):eaaw8429. doi: 10.1126/Science.aaw8429.

Kimura M, Ohta T (1969) The average number of generations until fixation of a mutant gene in a finite population. *Genetics*; 61:763–771.

Kimura M (1968) Evolutionary rate at the molecular level. *Nature*; 217:624–626. doi: 10.1038/217624a0.

Kimura M (1991) The neutral theory of molecular evolution: A review of recent evidence. *Jap J Gen*; 66:367–386.

King N (2004) The unicellular ancestry of animal development. *Dev Cell*; 7:313–325. doi: 10.1016/j.devcel.2004.08.010.

King N, Hittinger CT, Carroll SB (2003) Evolution of key cell signaling and adhesion protein families predates animal origins. *Science*; 301:361–363.

Kingman JF (2000) Origins of the coalescent 1974–1982. *Genetics*; 156:1461–1463.

Kingman JFC (1982) The coalescent. *Genetics*; 156:1461–1463.

Kirkilionis M (2010) Exploration of cellular reaction systems. *Brief Bioinform*; 11:153–178. doi: 10.1093/bib/bbp062.

Kirsch R, Gramzow L, Theißen G et al. (2014) Horizontal gene transfer and functional diversification of plant cell wall degrading polygalacturonases: Key events in the evolution of herbivory in beetles. *Insect Biochem Mol Biol*; 52:33–50. doi: 10.1016/j.ibmb.2014.06.008.

Kitching IJ, Forey PL, Humphries CJ, Williams D (1998) *Cladistics: Theory and Practice of Parsimony Analysis* (2nd ed). Oxford University Press.

Kleisner K, Ivell R, Flegr J (2010) The evolutionary history of testicular externalization and the origin of the scrotum. *J Biosci*; 35:27–37. doi: 10.1007/s12038-010-0005-7.

Knoll AH (2004) *Life on a Young Planet: The First Three Billion Years of Evolution on Earth*. Princeton University Press.

Knoll AH (2015) Paleobiological perspectives on early eukaryotic evolution. *Cold Spring Harb Perspect Biol*; 7:a018093. doi: 10.1101/cshperspect.a018093.

Knoll AH, Javaux EJ, Hewitt D, Cohen P (2006) Eukaryotic organisms in proterozoic oceans. *Philos Trans R Soc B*; 361:1023–1038. doi: 10.1098/rstb.2006.1843.

Knowlton C, Veerapaneni R, D'Elia T, Rogers SO (2013) Microbial analyses of ancient ice core sections from Greenland and Antarctica. *Biology (Basel)*; 2:206–232. doi: 10.3390/biology2010206.

Kocot KM, Poustka AJ, Stöger I, Halanych KM, Schrödl M (2020) New data from Monoplacophora and a carefully-curated dataset resolve molluscan relationships. *Sci Rep*; 10:101. doi: 10.1038/s41598-019-56728-w.

Koenig KM, Sun P, Meyer E, Gross JM (2016) Eye development and photoreceptor differentiation in the cephalopod *Doryteuthis pealeii*. Development; 143:3168-81. doi: 10.1242/dev.134254.

Koga R, Meng XY, Tsuchida T, Fukatsu T (2012) Cellular mechanism for selective vertical transmission of an obligate insect symbiont at the bacteriocyte-embryo interface. *Proc Natl Acad Sci*; 109:E1230. doi: 10.1073/pnas.1119212109.

Kohno S, Vang D, Ang E, Brunell AM, Lowers RH, Schoenfuss HL (2020) Estrogen-induced ovarian development is time-limited during the temperature-dependent sex determination of the American alligator. *Gen Comp Endocrinol*; 291:113397. doi: 10.1016/j.ygcen.2020.113397.

Koonin EV, Galperin MY (2003) *Sequence - Evolution - Function: Computational Approaches in Comparative Genomics*. Kluwer Academic. www.ncbi.nlm.nih.gov/books/NBK20260/

Kopp RE, Kirschvink JL, Hilburn IA, Nash CZ (2005) The paleoproterozoic snowball earth: A climate disaster triggered by the evolution of oxygenic photosynthesis. *Proc Natl Acad Sci U S A*; 102:11131-6. doi: 10.1073/pnas.0504878102.

Kosmidis EK, Moschou V, Ziogas G et al. (2014) Functional aspects of the EGF-induced MAP kinase cascade: A complex self-organizing system approach. *PLoS One*; 9:e111612. doi: 10.1371/journal.pone.0111612.

Koumandou VL, Wickstead W, Ginger ML (2013) Molecular paleontology and complexity in the last eukaryotic common ancestor. *Crit Rev Biochem Mol Biol*; 48:373–396. doi: 10.3109/10409238.2013.821444.

Kozmik Z, Daube M, Frei E (2003) Role of Pax genes in eye evolution: A cnidarian PaxB gene uniting Pax2 and Pax6 functions. *Dev Cell*; 5:773–785. doi: 10.1016/s1534-5807(03)00325-3.

Kozmik Z (2005) Pax genes in eye development and evolution. *Curr Opin Genet Dev*; 15:430–438. doi: 10.1016/j.gde.2005.05.001.

Krabbenhoft KM, Fallon JF (1989) The formation of leg or wing specific structures by leg bud cells grafted to the wing bud is influenced by proximity to the Apical Ridge. *Dev Biol*; 131:373-82.

Kramer J, Meunier J (2016) Kin and multilevel selection in social evolution: A never-ending controversy? *F1000Res*; 5:eCollection 2016. doi: 10.12688/f1000research.8018.1.

Kramerov DA, Vassetzky NS (2011) SINEs. *Wiley Interdiscip Rev RNA*; 2:772–786. doi: 10.1002/wrna.91.

Krause WJ, Cutts JH (1985) Placentation in the opossum, *Didelphis virginiana*. *Acta Anat*; 123:156–171.

Krause WJ, Cutts JH (1986) Scanning electron microscopic observations on developing opossum embryos: Days 9 through 12. *Anat Anz*; 161:11–21.

Krings M, Harper CJ, Taylor EL (2018) Fungi and fungal interactions in the Rhynie chert: A review of the evidence, with the description of *Perexiflasca tayloriana* gen. et sp. nov. *Philos Trans R Soc*; 373B:20160500. doi: 10.1098/rstb.2016.0500.

Krissansen-Totton J, Arney GN, Catling DC (2018) Constraining the climate and ocean pH of the early Earth with a geological carbon cycle model. *Proc Natl Acad Sci U S A*; 115:4105–4110. doi: 10.1073/pnas.1721296115.

Kropotkin P (1902 [2009]) *Mutual Aid: A Factor of Evolution*. New York University Press. www.complementarycurrency.org/ccLibrary/Mutual_Aid-A_Factor_of_Evolution-Peter_Kropotkin.pdf

Kuhlemeier C (2017) Phyllotaxis. *Curr Biol*; 27:R882–R887. doi: 10.1016/j.cub.2017.05.069.

Kumar JP, Moses K (2001) Eye specification in *Drosophila*: Perspectives and implications. *Sem Cell Dev Biol*; 12:469–474. doi: 10.1006/scdb.2001.0270.

Kumar S (2005) Molecular clocks: Four decades of evolution. *Nat Rev Genet*; 6:654–662. doi: 10.1038/nrg1659.

Kun Á, Szilágyi A, Könnyű B (2015) The dynamics of the RNA world: Insights and challenges. *Ann N Y Acad Sci*; 1341:75–95. doi: 10.1111/nyas.12700.

Kundaje A and the Roadmap Epigenomics Consortium (2015) Integrative analysis of 111 reference human epigenomes. *Nature*; 518:317–30. doi: 10.1038/Nature14248. PMID:25693563.

Kusuhashi N (2013) A new Early Cretaceous eutherian mammal from the Sasayama Group Hyogo Japan. *Proc Biol Sci*; 280:20130142. doi: 10.1098/rspb.2013.0142.

Kvenvolden K, Lawless J, Pering K et al. (1970) Evidence for extraterrestrial amino-acids and hydrocarbons in the Murchison meteorite. *Nature*; 228:923–926. doi: 10.1038/228923a0.

Lafuente E, Beldade P (2019) Genomics of developmental plasticity in animals. *Front Genet*; 10:720. doi: 10.3389/fgene.2019.00720.

Laird MK, McShea H, Murphy CR et al. (2018) Non-invasive placentation in the marsupials *Macropus eugenii* (Macropodidae) and *Trichosurus vulpecula* (Phalangeridae) involves redistribution of uterine Desmoglein-2. *Mol Reprod Dev*; 85:72–82. doi: 10.1002/mrd.22940.

Laland K, Uller T, Feldman M, Sterelny K et al. (2014) Does evolutionary theory need a rethink? *Nature*; 514:161–4. doi: 10.1038/514161a.

Lallemand T, Leduc M, Landès C, Rizzon C, Lerat E (2020) An overview of duplicated gene detection methods: Why the duplication mechanism has to be accounted for in their choice. *Genes (Basel)*; 11:1046. doi: 10.3390/genes11091046.

Lalueza-Fox C, Gilbert MTP (2011) Paleogenomics of archaic hominins. *Curr Biol*; 21:R1002–R1009. doi: 10.1016/j.cub.2011.11.021.

Lamichhaney S, Han F, Webster MT, Andersson L, Grant BR, Grant PR (2018) Rapid hybrid speciation in Darwin's finches. *Science*; 359:224–228. doi: 10.1126/Science.aao4593.

Lamsdell JC, Braddy SJ (2010) Cope's rule and Romer's theory: Patterns of diversity and gigantism in eurypterids and Palaeozoic vertebrates. *Biol Lett*; 6:265–269. doi: 10.1098/rsbl.2009.0700.

Lane CS, Chorn BT, Johnson TC (2013) Ash from the Toba supereruption in Lake Malawi shows no volcanic winter in East Africa at 75 ka, *Proc Natl Acad Sci U S A*; 110:8025–8029. doi: 10.1073/pnas.1301474110.

Langeland JA, Holland LZ, Chastain RA, Holland ND (2006) An amphioxus LIM-homeobox gene *AmphiLim1/5* expressed early in the invaginating organizer region and later in differentiating cells of the kidney and central nervous system. *Int J Biol Sci*; 2:110–116.

Langmore NE et al. (2011) Visual mimicry of host nestlings by cuckoos. *Proc Biol Sci*; 278:2455–63. doi: 10.1098/rspb.2010.2391.

Lanna E (2015) Evo-devo of non-bilaterian animals. *Genet Mol Biol*; 38:284–300. doi: 10.1590/S1415-475738320150005.

Lappin TR, Grier DG, Thompson A, Halliday HL (2006) HOX genes: Seductive science, mysterious mechanisms. *Ulster Med J*; 75:23–31. Erratum in: Ulster Med J. 2006 May; 75:135.

Laraba I, Kim HS, Proctor RH (2020) *Fusarium xyrophilum*, sp. nov., a member of the *Fusarium fujikuroi* species complex recovered From pseudoflowers on yellow-eyed grass (*Xyris* spp.) from Guyana. *Mycologia*; 112:39–51. doi: 10.1080/00275514.2019.1668991.

Larsen CS (2003) Equality for the sexes in human evolution? Early hominid sexual dimorphism and implications for mating systems and social behavior. *Proc Natl Acad Sci*; 100:9103–9104. doi: 10.1073/pnas.1633678100.

Lau AN, Peng L, Goto H et al. (2009) Horse domestication and conservation genetics of Przewalski's horse inferred from sex chromosomal and autosomal sequences. *Mol Biol Evol*; 26:199–208. doi: 10.1093/molbev/msn239.

Laumer CE, Gruber-Vodicka H, Hadfield MG et al. (2018) Support for a clade of Placozoa and Cnidaria in genes with minimal compositional bias. *Elife*; 7:e36278. doi: 10.7554/eLife.36278.

Lautenschlager S, Gill P, Luo ZX, Fagan MJ, Rayfield EJ (2017) Morphological evolution of the mammalian jaw adductor complex. *Biol Rev Camb Philos Soc*; 92:1910–1940. doi: 10.1111/brv.12314.

Lautenschlager S, Gill PG, Luo ZX, Fagan MJ, Rayfield EJ (2018) The role of miniaturization in the evolution of the mammalian jaw and middle ear. *Nature*; 561:533–537. doi: 10.1038/s41586-018-0521-4.

Lawton JH, May RM (2005) *Extinction Rates*. Oxford University Press.

Le PT, Pontarotti P, Raoult D (2014) Alphaproteobacteria species as a source and target of lateral sequence transfers. *Trends Microbiol*; 22:147–156. doi: 10.1016/j.tim.2013.

Lebedev OA, Coates MI (1995) The postcranial skeleton of the Devonian tetrapod *Tulerpeton curtum* Lebedev. *Zoo J Linn Soc*; 114 (3):307–348. doi: 10.1111/j.1096-3642.1995.tb00119.x.

LeBlanc AR, Reisz RR (2014) New postcranial material of the Early Caseid *Casea broilii* Williston, 1910 (Synapsida: Caseidae) with a review of the evolution of the sacrum in Paleozoic non-mammalian synapsids. *PLoS One*; 9:e115734. doi: 10.1371/journal.pone.0115734.

Leclère L, Röttinger E (2017) Diversity of cnidarian muscles: Function, anatomy, development and regeneration. *Front Cell Dev Biol*; 4:157. doi: 10.3389/fcell.2016.00157.

Lee MH, Jeon HS, Kim SH et al. (2019) Lignin-based barrier restricts pathogens to the infection site and confers resistance in plants. *EMBO J*; 2:38:e101948. doi: 10.15252/embj.2019101948.

Lee MS, Cau A, Naish D, Dyke GJ (2014) Dinosaur evolution. Sustained miniaturization and anatomical innovation in the dinosaurian ancestors of birds. *Science*; 345:662–566. doi: 10.1126/science.1252243.

Lee RT, Thiery JP, Carney TJ (2013) Dermal fin rays and scales derive from mesoderm not neural crest. *Curr Biol*; 6:R336–R337. doi: 10.1016/j.cub.2013.02.055.

Lefebvre V (2019) Roles and regulation of SOX transcription factors in skeletogenesis. *Curr Top Dev Biol*; 133:171–193. doi: 10.1016/bs.ctdb.2019.01.007.

Lemey P, Salemi M, Vandamme A-M (eds) (2009) *The Phylogenetic Handbook: A Practical Approach to Phylogenetic Analysis and Hypothesis Testing* (2nd end). Cambridge University Press.

Leonard WE (2008) On the Nature of Things (Part 5): Translation of Titus Lucretius Carus (d. 55 BCE) *De Rerum Natura*. http://www.gutenberg.org/files/785/785-h/785-h.htm#link2H_4_0027

Levis NA, Pfennig DW (2019) Plasticity-led evolution: A survey of developmental mechanisms and empirical tests. *Evol Dev*. doi: 10.1111/ede.12309.

Lewis-Williams, D (2002) *The Mind in the Cave*. Thames and Hudson.

Lewsey MG, Hardcastle TJ, Melnyk CW et al. (2016) Mobile small RNAs regulate genome-wide DNA methylation. *Proc Natl Acad Sci*; 113(6):E801–10. doi: 10.1073/pnas.1515072113.

Leys SP, Nichols SA, Adams ED (2009) Epithelia and integration in sponges. *Integr Comp Biol*; 49:167–177. doi: 10.1093/icb/icp038.

Li H, Durbin R (2011) Inference of human population history from individual whole-genome sequences. *Nature*; 475:493–496. doi: 10.1038/nature10231.

Li J, Chen X, Kang B, Liu M (2015) Mitochondrial DNA genomes organization and phylogenetic relationships analysis of eight anemonefishes (Pomacentridae: Amphiprioninae). *PLoS One*; 10(4):e0123894. doi: 10.1371/journal.pone.0123894.

Li JZ, Absher DM, Tang H (2008a) Worldwide human relationships inferred from genome-wide patterns of variation. *Science*; 319:1100–1104. doi: 10.1126/science.1153717.

Li Q, Gao KQ, Vinther J (2010) Plumage color patterns of an extinct dinosaur. *Science*; 327:13691372. doi: 10.1126/science.1186290.

Li R, Lockley MG, Makovicky PJ et al. (2008b) Behavioral and faunal implications of Early Cretaceous deinonychosaur trackways from China. *Naturwissenschaften*; 95:185–191. doi: 10.1007/s00114-007-0310-7.

Li Z-Y et al. (2017) Late Pleistocene archaic human crania from Xuchang, China. *Science*; 355:969–972. doi: 10.1126/Science.aal2482.

Lichtneckert R, Reichert H (2005) Insights into the urbilaterian brain: Conserved genetic patterning mechanisms in insect and vertebrate brain development. *Heredity*; 94:465–477.

Lieberman D (2011) *Evolution of the Human Head*. Harvard University Press.

Lieberman D (2013) *The Story of the Human Body: Evolution Health and Disease*. Pantheon Press.

Liebers R, Rassoulzadegan M, Lyko F (2014) Epigenetic regulation by heritable RNA. *PLoS Genet*; 10:e100429. doi: 10.1371/journal.pgen.1004296.

Ligrone R, Duckett JG, Renzaglia KS (2012) Major transitions in the evolution of early land plants: A bryological perspective. *Ann Bot*; 109:851–871. doi: 10.1093/aob/mcs017.

Lin BD, Willemsen G, Abdellaoui A et al. (2016) The genetic overlap between hair and eye color. *Twin Res Hum Genet*; 19:595–599. doi: 10.1017/thg.2016.85.

Linde AM, Eklund DM, Kubota A et al. (2017) Early evolution of the land plant circadian clock. *New Phytol*; 216:576–590. doi: 10.1111/nph.14487.

Lipscombe D (1998) Basics of cladistics analysis. www.gwu.edu/~clade/faculty/lipscomb/Cladistics.pdf

Liu AG, Matthews JJ, Menon LR, McIlroy D, Brasier MD (2014) Haootia quadriformis n. gen., n. sp., interpreted as a muscular cnidarian impression from the Late Ediacaran period (approx. 560 Ma). *Proc Biol Sci*; 281:20141202. doi: 10.1098/rspb.2014.1202.

Liu H, Weisz DA, Zhang MM et al. (2019a) Phycobilisomes harbor FNRL in cyanobacteria. *mBio*; 10:e00669-19. doi: 10.1128/mBio.00669-19.

Liu H, Zartman RE, Ireland TR, Sun WD (2019b) Global atmospheric oxygen variations recorded by Th/U systematics of igneous rocks. *Proc Natl Acad Sci*; 116:18854–18859. doi: 10.1073/pnas.1902833116.

Liu L, Yu L, Kubatko L (2009) Coalescent methods for estimating phylogenetic trees. *Mol Phylogenet Evol*; 53:320–328.

Liu Y, Xiao S, Shao T, Broce J, Zhang H (2014) The oldest known priapulid-like scalidophoran animal and its implications for the early evolution of cycloneuralians and ecdysozoans. *Evol Dev*; 16:155–65. doi: 10.1111/ede.12076.

Liu YJ, Hodson MC, Hall BD (2006) Loss of the flagellum happened only once in the fungal lineage: Phylogenetic structure of Kingdom fungi inferred from RNA polymerase II subunit genes. *BMC Evol Biol*; 6:74. doi: 10.1186/1471-2148-6-74.

Liu Z, Lavine KJ, Hung IH, Ornitz DM (2007) FGF18 is required for early chondrocyte proliferation hypertrophy and vascular invasion of the growth plate. *Dev Biol*; 302:80–91.

Liu ZJ, Wang X (2016) A perfect flower from the Jurassic of China. *Hist Biol*; 28:707–719. doi: 10.1080/08912963.2015.1020423.

Livi CB, Davidson EH (2006) Expression and function of blimp1/krox, an alternatively transcribed regulatory gene of the sea urchin endomesoderm network. *Dev Biol*; 293:513–525.

Lloyd GT, Davis KE, Pisani D et al. (2008) Dinosaurs and the Cretaceous Terrestrial Revolution. *Proc Biol Sci*; 275(1650):2483–2490. doi: 10.1098/rspb.2008.0715.

Lomiguen C, Vidal L, Kozlowski P, Prancan A, Stern R (2018) Possible role of chitin-like proteins in the etiology of Alzheimer's disease. *J Alzheimers Dis*; 66:439–444.

Long JA (2011) *The Rise of Fishes: 500 Million Years of Evolution* (2nd ed). Johns Hopkins University Press.

Long JA, Trinajstic K, Young GC, Senden T (2008) Live birth in the Devonian period. *Nature*; 453:650–652. doi: 10.1038/nature06966.

Long JA, Young GC, Holland T et al. (2006) An exceptional Devonian fish from Australia sheds light on tetrapod origins. *Nature*; 444:199–202.

Long M, Van Kuren NW, Chen S, Vibranovski MD (2013) New gene evolution: Little did we know. *Annu Rev Genet*; 47:307–333. doi: 10.1146/annurev-genet-111212-133301.

Longrich NR, Scriberas J, Wills MA (2016) Severe extinction and rapid recovery of mammals across the Cretaceous-Palaeogene boundary, and the effects of rarity on patterns of extinction and recovery. *J Evol Biol*; 29:1495–512. doi: 10.1111/jeb.12882.

Longrich NR, Tokaryk T, Field DJ (2011) Mass extinction of birds at the Cretaceous-Paleogene (K-pg) boundary. *Proc Natl Acad Sci*; 108:15253–15257. doi: 10.1073/pnas.1110395108.

Longrich NR, Vinther J, Meng Q et al. (2012) Primitive wing feather arrangement in *Archaeopteryx lithographica* and *Anchiornis huxleyi*. *Curr Biol*; 22:2262–2267. doi: 10.1016/j.cub.2012.09.052.

Loomis WF (2015) Genetic control of morphogenesis in *Dictyostelium*. *Dev Biol*; 402:146–161. doi: 10.1016/j.ydbio.2015.03.016.

Lordkipanidze D, Ponce de León MS, Margvelashvili A (2013) A complete skull from Dmanisi, Georgia, and the evolutionary biology of early *Homo*. *Science*; 342:326–331. doi: 10.1126/science.1238484.

Loron, CC. François C, Rainbird RH, Turner EC, Borensztajn S, Javaux EJ (2019) Early fungi from the Proterozoic era in Arctic Canada. *Nature*; 570:232–23. doi: 10.1038/s41586-019-1217-0.

Louchart A, Viriot L (2011) From snout to beak: The loss of teeth in birds. *Trends Ecol Evol*; 26:663–673. doi: 10.1016/j.tree.2011.09.004.

Lovegrove BG (2017) A phenology of the evolution of endothermy in birds and mammals. *Biol Rev Camb Philos Soc*; 92:1213–1240. doi: 10.1111/brv.12280.

Lowe CB, Clarke JA, Baker AJ, Haussler D, Edwards SV (2015a) Feather development genes and associated regulatory innovation predate the origin of Dinosauria. Mol Biol Evol; 32:23–8. doi: 10.1093/molbev/msu309.

Lowe CB, Clarke JA, Baker AJ et al. (2015b) Feather development genes and associated regulatory innovation predate the origin of Dinosauria. *Mol Biol Evol*; 32:23–28. doi: 10.1093/molbev/msu309.

Lowe CJ, Clarke DN, Medeiros DM (2015) The deuterostome context of chordate origins. *Nature*; 520:456–465. doi: 10.1038/nature14434.

Lowery RK, Uribe G, Jimenez EB et al. (2013) Neanderthal and denisova genetic affinities with contemporary humans: Introgression versus common ancestral polymorphisms. *Gene*; 530:83–94. doi: 10.1016/j.gene.2013.06.005.

Luo S-H, Schatten H, Sunac Q-Y (2013) Sperm mitochondria in reproduction: Good or bad and where do they go? *J Genet Genomics*; 40:549–556. doi: 10.1016/j.jgg.2013.08.004.

Luo Z-X, Ji Q, Wible JR, Yuan C-X (2003) An early Cretaceous tribosphenic mammal and metatherian evolution. *Science*; 302:1934–1940.

Luo ZX (2007) Transformation and diversification in early mammal evolution. *Nature*; 450:1011–1019. doi: 10.1038/Nature06277.

Luo ZX, Cifelli RL, Kielan-Jaworowska Z (2001) Dual origin of tribosphenic mammals. *Nature*; 409:53–7. doi: 10.1038/35051023.

Luo ZX, Ruf I, Schultz JA, Martin T (2011a) Fossil evidence on evolution of inner ear cochlea in Jurassic mammals. *Proc Biol Sci*; 278:28–34. doi: 10.1098/rspb.2010.1148.

Luo ZX, Yuan CX, Meng QJ, Ji Q (2011b) A Jurassic eutherian mammal and divergence of marsupials and placentals. *Nature*; 476:442–445. doi: 10.1038/nature10291.

Lyson TR, Bever GS, Scheyer TM (2013) Evolutionary origin of the turtle shell. *Curr Biol*; 23:1113-1119. doi: 10.1016/j.cub.2013.05.003.

Lyson TR, Sperling EA, Heimberg AM (2012) MicroRNAs support a turtle + lizard clade. *Biol Lett*; 8:104-107. doi: 10.1098/rsbl.2011.0477.

Macaulay V, Hill C, Achilli A (2005) Single rapid coastal settlement of Asia revealed by analysis of complete mitochondrial genomes. *Science*; 308:1034-1036.

Macdonald DW (1979) "Helpers" in fox society. *Nature*; 282:69-71.

Macdonald DW, Moehlman, PD (1982) Cooperation, altruism, and restraint in the reproduction of carnivores. *Persp Ethol*; 5:433-467.

Maciver SK (2016) Asexual amoebae escape Muller's ratchet through polyploidy. *Trends Parasitol*; 32:855-862. doi: 10.1016/j.pt.2016.08.006.

MacNaughton RB, Cole JM, Dalrymple RW (2002) First steps on land: Arthropod trackways in Cambrian-Ordovician eolian sandstone southeastern Ontario, Canada. *Geology*; 30:391-394.

Maddison WP (1997) Gene trees in species trees. *Syst Biol*; 46:523-536.

Mah JL, Christensen-Dalsgaard KK, Leys SP (2014) Choanoflagellate and choanocyte collar-flagellar systems and the assumption of homology. *Evol Dev*; 16:25-37. doi: 10.1111/ede.12060.

Malaspinas AS, Lao O, Schroeder H (2014) Two ancient human genomes reveal polynesian ancestry among the indigenous botocudos of Brazil. *Curr Biol*; 24:R1035-R1037. doi: 10.1016/j.cub.2014.09.078.

Malfait F, De Paepe A (2014) The Ehlers-Danlos syndrome. *Adv Exp Med Biol*; 802:129-143. doi: 10.1007/978-94-007-7893-1_9.

Malik S (2013) Polydactyly phenotypes genetics and classification. *Clin Genet*; 85B:203-212. doi: 10.1111/cge.12276.

Mallarino R, Grant PR, Grant BR et al. (2011) Two developmental modules establish 3D beak-shape variation in Darwin's finches. *Proc Natl Acad Sci U S A* 108:4057-4062. doi: 10.1073/pnas.1011480108.

Mallick CB et al. (2013) The light skin allele of SLC24A5 in South Asians and Europeans shares identity by descent. *PLoS Genet*; 9:e1003912. doi: 10.1371/journal.pgen.1003912.

Malo AF, Roldan ERS, Garde J, Soler AJ, Gomendio M (2005) Antlers honestly advertise sperm production and quality. *Proc Biol Sci*; 272B:149-57. doi: 10.1098/rspb.2004.2933.

Mangalam M, Fragaszy DM (2015) Wild bearded capuchin monkeys crack nuts dexterously. *Curr Biol* 25:1334-1339. doi: 10.1016/j.cub.2015.03.035.

Manjrekar J, Shah H (2020) Protein-based inheritance. *Semin Cell Dev Biol*; 97:138-155. doi: 10.1016/j.semcdb.2019.07.007.

Mann J, Patterson EM (2013) Tool use by aquatic animals. *Philos Trans R Soc Lond B Biol Sci*; 368:20120424. doi: 10.1098/rstb.2012.0424. PMID:24101631.

Manolio TA, Collins FS, Cox NJ et al. (2009) Finding the missing heritability of complex diseases. *Nature*; 461:747-53. doi: 10.1038/Nature08494.

Mansuit R, Clément G, Herrel A et al. (2020) Development and growth of the pectoral girdle and fin skeleton in the extant coelacanth latimeria chalumnae. *J Anat*; 236:493-509. doi: 10.1111/joa.13115.

Mansy SS, Schrum JP, Krishnamurthy M (2008) Template-directed synthesis of a genetic polymer in a model protocell. *Nature*; 454:122-125. doi: 10.1038/nature07018.

Marcelis CL, Hol FA, Graham GE et al. (2008) Genotype-phenotype correlations in MYCN-related Feingold syndrome. *Hum Mutat*; 29:1125-32. doi: 10.1002/humu.20750.

Margolin W (2014) Sculpting the bacterial cell. *Curr Biol*; 19:R812-R822. doi: 10.1016/j.cub.2009.06.033.

Marin B, Nowack EC, Glöckner G, Melkonian M (2007) The ancestor of the *Paulinella* chromatophore obtained a carboxysomal operon by horizontal gene transfer from a *Nitrococcus*-like gamma-proteobacterium. *BMC Evol Biol*; 7:85. doi: 10.1186/1471-2148-7-85.

Marjanović D, Laurin M (2009) The origin(s) of modern amphibians: A commentary. *Evol Biol*; 36:336-338.

Markov AV, Kaznacheev IS (2016) Evolutionary consequences of polyploidy in prokaryotes and the origin of mitosis and meiosis. *Biol Direct*; 1128. doi: 10.1186/s13062-016-0131-8.

Marsicano CA, Irmis RB, Mancuso AC et al. (2016) The precise temporal calibration of dinosaur origins. *Proc Natl Acad Sci U S A*; 113:509-513. doi: 10.1073/pnas.1512541112.

Martill DM, Tischlinger H, Longrich NR (2015) A four-legged snake from the Early Cretaceous of Gondwana. *Science*; 349:416-419. doi: 10.1126/science.aaa9208.

Martin A, Serano JM, Jarvis E et al. (2016) CRISPR/Cas9 mutagenesis reveals versatile roles of Hox genes in crustacean limb specification and evolution. *Curr Biol*; 26(1):14-26. doi: 10.1016/j.cub.2015.11.021.

Martin LD, Czerkas SA (2000) The fossil record of feather evolution in the Mesozoic. *Am Zool*; 40:687-694.

Martin MW, Grazhdankin DV, Bowring SA (2000) Age of Neoproterozoic bilatarian body and trace fossils, White Sea, Russia: Implications for metazoan evolution. *Science*; 288:841-845.

Martinez RN, Sereno PC, Alcober OA et al. (2011) A basal dinosaur from the dawn of the dinosaur era in southwestern Pangaea. *Science*; 331:206-210. doi: 10.1126/science.1198467.

Martínez-Abadías N, Motch SM, Pankratz TL (2013) Tissue-specific responses to aberrant FGF signaling in complex head phenotypes. *Dev Dyn*; 242:80-94. doi: 10.1002/dvdy.23903.

Martino RL, Greb S (2009) Walking trails of the giant terrestrial arthropod arthropleura from the upper carboniferous of Kentucky. *J Paleaont*; 83:140-146. doi: 10.1666/08-093R.1.

Martins AB, de Aguiar MA, Bar-Yam Y (2013) Evolution and stability of ring species. *Proc Natl Acad Sci U S A*; 110:5080-5084. doi: 10.1073/pnas.1217034110.

Martone PT, Estevez JM, Lu F et al. (2009) Discovery of lignin in seaweed reveals convergent evolution of cell-wall architecture. *Curr Biol*; 19:169-175. doi: 10.1016/j.cub.2008.12.031.

Masel J (2011) Genetic drift – A quick guide. *Curr Biol*; 21:R837-R838. doi: 10.1016/j.cub.2011.08.007.

Mason J (2006) Apollo 11 cave in southwest Namibiasome observations on the site and its rock art. *S African Archeol Bull*; 61:76-89.

Masson-Boivin C, Sachs JL (2018) Symbiotic nitrogen fixation by rhizobia-the roots of a success story. *Curr Opin Plant Biol*; 44:7-15. doi: 10.1016/j.pbi.2017.12.001.

Mat W-K, Hong Xue H, Wong JT-F (2008) The genomics of LUCA. *Front Biosci*; 13:5605-5613.

Mateo JM (2002) Kin-recognition abilities and nepotism as a function of sociality. *Proc R Soc Lond B*; 269:721-727.

Mateo JM (2017) The ontogeny of kin-recognition mechanisms in Belding's ground squirrels. *Physiol Behav*; 173:279-284. doi: 10.1016/j.physbeh.2017.02.024.

Mathieson I et al. (2015) Genome-wide patterns of selection in 230 ancient Eurasians. *Nature*; 528:499-503. doi: 10.1038/Nature16152.

Matsui T, Yamamoto T, Wyder S (2009) Expression profiles of urbilaterian genes uniquely shared between honey bee and vertebrates. *BMC Genomics*; 10:17. doi: 10.1186/1471-2164-10-17.

Matsuzawa T (2013) Evolution of the brain and social behavior in chimpanzees. *Curr Opin Neurobiol*; 23:443-449. doi: 10.1016/j.conb.2013.01.012.

Maynard Smith J, Burian R, Kauffman S (1985) Developmental constraints and evolution. *Q Rev Biol*; 60:265-287.

Mayr E (1982) *The Growth of Biological Thought*. Belknap Press.

McBrearty S, Brooks AS (2000) The revolution that wasn't: A new interpretation of the origin of modern human behaviour. *J Hum Evol*; 39:453-563.

McCourt RM, Delwiche CF, Karol KG (2004) Charophyte algae and land plant origins. *Trends Ecol Evol*; 19:661-666.

McCoy VE, Saupe EE, Lamsdell JC et al. (2016) The "Tully monster" is a vertebrate. *Nature*; 532:496-499. doi: 10.1038/Nature16992.

McCoy VE, Wiemann J, Lamsdell JC et al. (2020) Chemical sig of soft tissues distinguish between vertebrates and invertebrates from the Carboniferous Mazon Creek Lagerstätte of Illinois. *Geobiol*. doi: 10.1111/gbi.12397.

McDougall I, Brown FH, Fleagle JG (2005) Stratigraphic placement and age of modern humans from Kibish, Ethiopia. *Nature*; 433:733-736.

McElhinny MW, McFadden PL (2000) *Paleomagnetism: Continents and Oceans*. Academic Press.

McFadden DE, Friedman JM (1997) Chromosome abnormalities in human beings. *Mutat Res*; 396:129-140. doi: 10.1016/s0027-5107(97)00179-6.

McFadden GI (2001) Primary and secondary endosymbiosis and the origin of plastids. *J Phycol*; 37:951-959.

McFadden GI, Gilson PR, Hofmann CJ, Adcock GJ, Maier UG. (1994) Evidence that an amoeba acquired a chloroplast by retaining part of an engulfed eukaryotic alga. *Proc Natl Acad Sci*; 91:3690-3694. doi: 10.1073/pnas.91.9.3690.

McFall-Ngai M, Hadfield MG, Bosch TC et al. (2013) Animals in a bacterial world, a new imperative for the life sciences. *Proc Natl Acad Sci U S A*; 110:3229-36. doi: 10.1073/pnas.1218525110.

McGhee G (2011) *Convergent Evolution*. MIT Press.

McGuire JA, Dudley R (2011) The biology of gliding in flying lizards (genus *Draco*) and their fossil and extant analogs. *Integr Comp Biol;* 51:983-990. doi: 10.1093/icb/icr090.

McIntyre B, Asahara T, Alev C (2020) Overview of basic mechanisms of notch signaling in development and disease. *Adv Exp Med Biol*; 227:9-27. doi: 10.1007/978-3-030-36422-9_2.

McKenna MC (1975) Toward a Phylogenetic Classification of the Mammalia. In *Phylogeny of the Primates* (Luckett WP, Szalay ES, eds), pp 21-46. Plenum Press.

McKenna Kelly E, Sears KE (2011) Limb specialization in living marsupial and eutherian mammals: Constraints on mammalian limb evolution. *J Mamm*; 92:1038–1049. doi: 10.1644/10-MAMM-A-425.1.

McLysaght A, Hokamp K, Wolfe KH (2002) Extensive genomic duplication during early chordate evolution. *Nat Genet*; 31:200–4. doi: 10.1038/ng884.

McPherron SP, Alemseged Z, Marean CW et al. (2010) Evidence for stone-tool-assisted consumption of animal tissues before 3.39 million years ago at Dikika, Ethiopia. *Nature*; 466:857–860. doi: 10.1038/Nature09248.

McQueen C, Towers M (2020) Establishing the pattern of the vertebrate limb. *Development*; 147:dev177956. doi: 10.1242/dev.177956.

Medina L, Abellán A (2009) Development and evolution of the pallium. *Semin Cell Dev Biol*; 20:698–711. doi: 10.1016/j.semcdb.2009.04.008.

Medland SE, Nyholt DR, Painter JN et al. (2009) Common variants in the trichohyalin gene are associated with straight hair in Europeans. *Am J Hum Genet*; 85:750–755. doi: 10.1016/j.ajhg.2009.10.009.

Megrian D, Taib N, Witwinowski J, Beloin C, Gribaldo S (2020) One or two membranes? Diderm Firmicutes challenge the Gram-positive/Gram-negative divide. *Mol Microbiol*; 113:659–671. doi: 10.1111/mmi.14469.

Meinhardt H (1984) Models for positional signalling the threefold subdivision of segments and the pigmentation pattern of segments. *J Embryol Exp Morphol*; 83(Suppl.):289–311.

Melis AP (2013) The evolutionary roots of human collaboration: Coordination and sharing of resources. *Ann N Y Acad Sci*; 1299:68–76. doi: 10.1111/nyas.l2263.

Mendel G (1866) Experiments in plant hybridization. English translation: http://www.esp.org/foundations/genetics/classical/gm-65.pdf

Mendez FL, Poznik GD, Castellano S, Bustamante CD (2016) The divergence of Neandertal and modern human Y chromosomes. *Am J Hum Genet*; 98:728–34. doi: 10.1016/j.ajhg.2016.02.023.

Meng L, Fan Z, Zhang Q et al. (2018) BEL1-LIKE HOMEODOMAIN 11 regulates chloroplast development and chlorophyll synthesis in tomato fruit. *Plant J*; 94:1126–1140. doi: 10.1111/tpj.13924.

Meng J, Hu Y, Wang Y et al. (2006) A Mesozoic gliding mammal from northeastern China. *Nature*; 444:889–893.

Meredith P, Riesz J (2004) Radiative relaxation quantum yields for synthetic eumelanin. *Photochem Photobiol*; 79:211–6. doi: 10.1562/0031-8655(2004)079<0211:rcrqyf>2.0.co;2.

Mesoudi A, Thornton A (2018) What is cumulative cultural evolution? *Proc Biol Sci*; 285:20180712. doi: 10.1098/rspb.2018.0712.

Metspalu M, Kivisild T, Metspalu E et al. (2004) Most of the extant mtDNA boundaries in south and southwest Asia were likely shaped during the initial settlement of Eurasia by anatomically modern humans. *BMC Genet*; 5:26. doi: 10.1186/1471-2156-5-26.

Metz JAJ (2011) Thoughts on the Geometry of Meso-evolution: Collecting Mathematical Elements for a Postmodern Synthesis. In *The Mathematics of Darwin's Legacy* (Chalub FACC, Rodrigues JF, eds), pp 193–229. Springer Basel AG. doi: 10.1007/978-3-0348-0122-5_11.

Meunier FJ, Laurin M (2012) A microanatomical and histological study of the fin long bones of the Devonian sarcopterygian *Eusthenopteron foordi*. *Acta Zool*; 93:88–97. doi: 10.1111/j.1463-6395.2010.00489.x.

Meyer M, Fu Q, Aximu-Petri A (2014) A mitochondrial genome sequence of a hominin from Sima de los Huesos. *Nature*; 505:403–406. doi: 10.1038/nature12788.

Meyerowitz EM (2002) Plants compared to animals: The broadest comparative study of development. *Science*; 295:1482–1485.

Michod RE, Bernstein H, Nedelcu AM (2008) Adaptive value of sex in microbial pathogens. *Infect Genet Evol*; 8:267–85. doi: 10.1016/j.meegid.2008.01.002.

Miga KH (2017) Chromosome-specific centromere sequences provide an estimate of the ancestral chromosome 2 fusion event in hominin genomes. *J Hered*; 108:45–52. doi: 10.1093/jhered/esw039.

Mikhailov KV, Konstantinova AV, Nikitin MA et al. (2009) The origin of Metazoa: A transition from temporal to spatial cell differentiation. *Bioessays*; 3:758–768. doi: 10.1002/bies.200800214.

Mikkelsen MD, Harholt J, Ullvskov P (2014) Evidence for land plant cell wall biosynthetic mechanisms in charophyte green algae. *Ann Bot*; 114:1217–1236. doi: 10.1093/aob/mcu171.

Miller RF, Cloutier R, Turner S (2003) The oldest articulated chondrichthyan from the early Devonian period. *Nature*; 425:501–4.

Miller SL (1953) Production of amino acids under possible primitive earth conditions. *Science*; 117:528–529.

Mills RE, Bennett EA, Iskow RC, Devine SE (2007) Which transposable elements are active in the human genome? *Trends Genet*; 23:183–91. doi: 10.1016/j.tig.2007.02.006.

Minelli A (2018) *Plant Evolutionary Developmental Biology: The Evolvability of the Phenotype.* Cambridge University Press.

Minelli A, Fusco G (2010) Developmental plasticity and the evolution of animal complex life cycles. *Philos Trans R Soc*; 365B:631–640. doi: 10.1098/rstb.2009.0268.

Mirabeau O, Joly J-S (2013) Molecular evolution of peptidergic signalling systems in bilaterians. *Proc Natl Acad Sci U S A* 110:E2028–E2037. doi: 10.1073/pnas.1219956110.

Misof B, Liu S, Meusemann KB (2014) Phylogenomics resolves the timing and pattern of insect evolution. *Science*. 346:763–767. doi: 10.1126/science.1257570.

Mitchell D.R (2007) The evolution of eukaryotic cilia and flagella as motile and sensory organelles. *Adv. Exp. Med. Biol*; 607:130–140. doi: 10.1007/978-0-387-74021-8_11.

Mitchell FL, Lasswell J (2005) *A Dazzle of Dragonflies.* Texas A&M University Press.

Mitchell G, Skinner JD (2009) An allometric analysis of the giraffe cardiovascular system. *Comp Biochem Physiol A Mol Integr Physiol*; 154:523–529. doi: 10.1016/j.cbpa.2009.08.013.

Mitchison GJ (1977) Phyllotaxis and the fibonacci series. *Science*; 196:270–275. doi: 10.1126/Science.196.4287.270.

Mitgutsch C, Richardson MK, Jiménez R (2012) Circumventing the polydactyly "constraint": The mole's 'thumb.' *Biol Lett*; 8:74–77. doi: 10.1098/rsbl.2011.0494.

Miyagi R, Terai Y (2013) The diversity of male nuptial coloration leads to species diversity in Lake Victoria cichlids. *Genes Genet Syst*; 88:145153.

Miyashita T, Gess RW, Tietjen K et al. (2021) Non-ammocoete larvae of Palaeozoic stem lampreys. *Nature*; 591:008–412. doi. org/10.1038/s41586-021-03305-9.

Mizukami Y, Hayashi R, Tsuruta D, Shimomura Y, Sugawara K (2018) Novel splice site mutation in the LIPH gene in a patient with autosomal recessive woolly hair/

hypotrichosis: Case report and published work review. *J Dermatol*; 45:613–617. doi: 10.1111/1346-8138.14257.

Mohammed H, Carroll JS (2013) Approaches for assessing and discovering protein interactions in cancer. *Mol Cancer Res*; 11:1295–302. doi: 10.1158/1541-7786. MCR-13-0454.

Mokkonen M, Lindstedt C (2015) The evolutionary ecology of deception. *Biol Rev Camb Philos Soc* In Press. doi: 10.1111/brv.12208.

Molnar JL, Johnston PS, Esteve-Altava B, Diogo R (2017) Musculoskeletal anatomy of the pelvic fin of polypterus: Implications for phylogenetic distribution and homology of pre- and postaxial pelvic appendicular muscles. *J Anat*; 230:532–541. doi: 10.1111/joa.12573.

Molnar MJ, Kovacs GG (2017) Mitochondrial diseases. *Handb Clin Neurol*; 145:147–155. doi: 10.1016/B978-0-12-802395-2.00010-9. PMID: 28987165.

Molnár Z, Kaas JH, de Carlos JA et al. (2014) Evolution and development of the mammalian cerebral cortex. *Brain Behav Evol*; 83:126–139. doi: 10.1159/000357753.

Momigliano P, Jokinen H, Fraimout A, Florin AB, Norkko A, Merilä J (2017) Extraordinarily rapid speciation in a marine fish. *Proc Natl Acad Sci*; 114:6074–6079. doi: 10.1073/pnas.1615109114.

Monk D (2015) Germline-derived DNA methylation and early embryo epigenetic reprogramming: The selected survival of imprints. *Int J Biochem Cell Biol*; 67:128–138. doi: 10.1016/j.biocel.2015.04.014.

Monnier G. (2012) Neanderthal behavior. *Nature Education Knowledge* 3(10):11. www.nature.com/scitable/knowledge/library/neanderthal-behavior-59267999/

Moore PB, Steiz TA (2005) The Role of RNA in the Synthesis of Proteins. In *The RNA World*, (3rd ed) (Gesteland RF, Cech TR, Atkins JF, eds), pp 257–285. Cold Spring Harbor Laboratory Press.

Mora C, Tittensor DP, Adl S (2011) How many species are there on earth and in the ocean? *PLoS Biol*; 9:e1001127. doi: 10.1371/journal. pbio.1001127.

Moroz LL, Kocot KM, Citarella MR et al. (2014) The ctenophore genome and the evolutionary origins of neural systems. *Nature*; 510:109–114. doi: 10.1038/Nature13400.

Morris JL, Puttick MN, Clark JW et al. (2018) The timescale of early land plant evolution. *Proc Natl Acad Sci*; 115:E2274–E2283. doi: 10.1073/pnas.1719588115.

Morris SC, Caron JB (2012) *Pikaia gracilens* Walcott, a stem-group chordate from the Middle Cambrian of British Columbia. *Biol Rev Camb Philos Soc*; 87:480–512. doi: 10.1111/j.1469-185X.2012.00220.x.

Morriss GM (1975) Placental Evolution and Embryonic Nutrition. In *Comparative Placentation* (Dawes GS ed). pp. 87–107, Academic Press.

Morriss-Kay GM (2010) The evolution of human artistic creativity. *J Anat*; 216:158–176. doi: 10.1111/j.1469-7580.2009.01160.x.

Morriss-Kay GM (2011) A New Hypothesis on the Creation of the Hohle Fels "Venus" Figurine. In *L'art Pléicistocène Dans Le Monde* (Clotte J ed), pp 65–66. Société préhistorique de l'Ariège.

Morriss-Kay GM, Wilkie AO (2005) Growth of the normal skull vault and its alteration in craniosynostosis: Insights from human genetics and experimental studies. *J Anat*; 207:637–653.

Morse D, Daoust P, Benribague S. (2016) A transcriptome-based perspective of cell cycle regulation in dinoflagellates. *Protist*; 167:610–621. doi: 10.1016/j.protis.2016.10.002.

Motani R, Jiang DY, Chen GB et al. (2015) A basal ichthyosauriform with a short snout from the Lower Triassic of China. *Nature*; 517:485–488. doi: 10.1038/Nature13866.

Muensterer OJ, Berdon W, McManus C et al. (2013) Ellis-van Creveld syndrome: Its history. *Pediatr Radiol*; 43:1030–6. doi: 10.1007/s00247-013-2709-y.

Müller J, Berman DS, Henrici AC, Martens T, Sumida TS (2006) The basal reptile thuringothyris mahlendorffae (Amniota: Eureptilia) from the lower Permian of Germany. *J Paleont*; 80:726–739. doi. org/10.1666/0022-3360(2006)80[726:TBRTMA]2.0.CO;2.

Müller J, Reisz RB (2005) An early captorhinid reptile (amniota eureptilia) from the upper carboniferous of Hamilton Kansas. *J Vert Pal*; 25:561–568.

Müller WE (2001) Review: How was metazoan threshold crossed? The hypothetical Urmetazoa. *Comp Biochem Physiol A Mol Integr Physiol*; 129:433–60. doi: 10.1016/s1095-6433(00)00360-3.

Munns SL, Owerkowicz T, Andrewartha SJ, Frappell PB (2012) The accessory role of the diaphragmaticus muscle in lung ventilation in the estuarine crocodile *Crocodylus porosus*. *J Exp Biol*; 215:845–852. doi: 10.1242/jeb.061952.

Mus F, Alleman AB, Pence N, Seefeldt LC, Peters JW (2018) Exploring the alternatives of biological nitrogen fixation. *Metallomics*; 10:523–538. doi: 10.1039/c8mt00038g.

Mus MM, Palacios T, Jensen S (2008) Size of the earliest mollusks: Did small helcionellids grow to become large adults? *Geology*; 36:175–178. doi.org/10.1130/G24218A.1.

Muscatine L, Falkowski PG, Porter W, Dubinsky Z (1984) Fate of photosynthetic fixed carbon in light- and shade-adapted colonies of the symbiotic coral *Stylophora pistillata*. *Proc R Soc*; 222B:181–202.

Muscente AD, Bykova N, Boag TH et al. (2019) Ediacaran biozones identified with network analysis provide evidence for pulsed extinctions of early complex life. *Nat Commun*; 10:911. doi: 10.1038/s41467-019-08837-3.

Nabavizadeh A, Weishampel DB (2016) The Predentary Bone and Its Significance in the Evolution of Feeding Mechanisms in Ornithischian Dinosaurs. *Anat Rec*; 299(10):1358–1388. doi: 10.1002/ar.23455.

Nachman MW, Crowell SL (2000) Estimate of the mutation rate per nucleotide in humans. *Genetics*; 156:297–304.

Næraa T, Scherstén A, Rosing MT et al. (2012) Hafnium isotope evidence for a transition in the dynamics of continental growth 3.2 Gyr ago. *Nature*; 485:627-30. doi: 10.1038/nature11140.

Nagle N, Ballantyne KN, van Oven M et al. (2017) Mitochondrial DNA diversity of present-day Aboriginal Australians and implications for human evolution in Oceania. *J Hum Genetr*; 62:343–353. doi: 10.1038/jhg.2016.147.

Nakano M, Miwa N, Hirano A et al. (2009) A strong association of axillary osmidrosis with the wet earwax type determined by genotyping of the ABCC11 gene. *BMC Genet*; 10:42. doi: 10.1186/1471-2156-10-42.

Nance RD, Murphy JB (2018) Supercontinents and the Case for Pannotia. *Geological Society London Special Publications*. doi: 10.1144/SP470.5.

Nanglu K, Caron JB, Cameron CB (2020) Cambrian tentaculate worms and the origin of the hemichordate body plan. *Curr Biol*; 30(21):4238-4244.e1. doi: 10.1016/j.cub.2020.07.078.

Nara T, Hashimoto T, Aoki T (2000) Evolutionary implications of the mosaic pyrimidine-biosynthetic pathway in eukaryotes. *Gene*; 257:209–222. doi: 10.1016/s0378-1119(00)00411-x.

Naranjo-Ortiz MA, Gabaldón T (2019) Fungal evolution: Major ecological adaptations and evolutionary transitions. *Biol Rev Camb Philos Soc*; 94:1443–1476. doi: 10.1111/brv.12510.

Nastou KC, Tsaousis GN, Papandreou NC, Hamodrakas SJ (2016) MBPpred: Proteome-wide detection of membrane lipid-binding proteins using profile Hidden Markov Models. *Biochim Biophys Acta*; 1864:747–54. doi: 10.1016/j.bbapap.2016.03.015.

Needham J (1931) *Chemical Embryology*, Cambridge University Press.

Neher RA, Shraiman BI (2011) Genetic draft and quasi-neutrality in large facultatively sexual populations. *Genetics*; 188:975–96. doi: 10.1534/genetics.111.128876.

Nei M, Suzuki Y, Nozawa M (2010) The neutral theory of molecular evolution in the genomic era. *Ann Rev Gen Hum Genet*; 11:265–289. doi: 10.1146/annurev-genom-082908-150129.

Nesbitt SJ (2011) The early evolution of archosaurs: Relationships and the origin of major clades. *Bull Am Mus Nat Hist*; 352:1–292.

Nesbitt SJ, Butler RJ, Ezcurra MD et al. (2017) The earliest bird-line archosaurs and the assembly of the dinosaur body plan. *Nature*; 544:484–487. doi: 10.1038/Nature22037.

Neves WA, Araujo AG, Bernardo DV, Kipnis R, Feathers JK (2012) Rock art at the pleistocene/holocene boundary in Eastern South America. *PLoS One*; 7(2):e32228. doi: 10.1371/journal.pone.0032228.

Neves WA, Hubbe M, Piló LB (2007) Early Holocene human skeletal remains from Sumidouro Cave, Lagoa Santa, Brazil: History of discoveries, geological and chronological context, and comparative cranial morphology. *J Hum Evol*; 52:16–30.

Newman JH (1973) Letter to J. Walker of Scarborough, May 22, 1868. In *The Letters and Diaries of John Henry Newman*, pp 77–78. Clarendon Press.

Nichols DJ, Johnson KR (2008) *Plants and the K-T Boundary*. Cambridge University Press.

Nielsen C (2018) Origin of the trochophore larva. *R Soc Open Sci*; 5:180042. doi: 10.1098/rsos.180042.

Nielsen C (2019) Early animal evolution: A morphologist's view. *R Soc Open Sci*; 6:190638. doi: 10.1098/rsos.190638.

Nielsen R, Akey JM, Jakobsson M, Pritchard JK, Tishkoff S, Willerslev E (2017) Tracing the peopling of the world through genomics. *Nature*; 541:302–310. doi: 10.1038/Nature21347.

Nieuwenhuis BP, James TY (2016) The frequency of sex in fungi. *Philos Trans R Soc Lond B Biol Sci*; 19; 371(1706):20150540. doi: 10.1098/rstb.2015.0540.

Nikaido M, Rooney AP, Okada N (1999) Phylogenetic relationships among cetartiodactyls based on insertions of short and long interspersed elements: Hippopotamuses are the closest extant relatives of whales. *Proc Natl Acad Sci U S A*; 96:10261–10266.

Nikolov LA, Runions A, Das Gupta M, Tsiantis M (2019) Leaf development and evolution. *Curr Top Dev Biol*; 131:109–139. doi: 10.1016/bs.ctdb.2018.11.006.

Nilsson DE, Pelger S (1994) A pessimistic estimate of the time required for an eye to evolve. *Proc Biol Sci*; 256B:53–58.

Nisbet EG, Nisbet REN (2008) Methane oxygen photosynthesis rubisco and the regulation of the air through time. *Proc R Soc B*; 363:2745–2754. doi: 10.1098/rstb.2008.0057.

Niu XM et al. (2019) Transposable elements drive rapid phenotypic variation in *Capsella rubella*. *Proc Natl Acad Sci*; 116:6908–6913. doi: 10.1073/pnas.1811498116.

Niwa N, Akimoto-Kato A, Niimi T, Tojo K, Machida R, Hayashi S. (2010) Evolutionary origin of the insect wing via integration of two developmental modules. *Evol Dev*; 12:168–76. doi: 10.1111/j.1525-142X.2010.00402.x.

Noble D (2008) *The Music of Life: Biology Beyond Genes*. Oxford University Press.

Noble D (2011) Successes and failures in modeling heart cell electrophysiology. *Heart Rhythm*; 8:1798–1803. doi: 10.1016/j.hrthm.2011.06.014.

Noble D (2012) A theory of biological relativity: No privileged level of causation. *Interface Focus*; 2:55–64. doi: 10.1098/rsfs.2011.0067.

Nobles DR, Romanovicz DK, Brown RM (2001) Cellulose in cyanobacteria. Origin of vascular plant cellulose synthase? *Plant Physiol*; 127:529–542.

Nochomovitz YD, Li H (2006) Highly designable phenotypes and mutational buffers emerge from a systematic mapping between network topology and dynamic output. *Proc Natl Acad Sci U S A*; 103:4180–4185.

Noffke N, Christian D, Wacey D, Hazen RM (2013) Microbially induced sedimentary structures recording an ancient ecosystem in the ca. 3.48 billion-year-old Dresser Formation, Pilbara, Western Australia. *Astrobiology*; 13:1103–1124. doi: 10.1089/ast.2013.1030.

Noonan JP, Coop G, Kudaravalli S et al. (2006) Sequencing and analysis of Neanderthal genomic DNA. *Science*; 314:1113–1118. doi: 10.1126/science.1131412.

Nord AS (2015) Learning about mammalian gene regulation from functional enhancer assays in the mouse. *Genomics*; 106:178–184. doi: 10.1016/j.ygeno.2015.06.008.

Nosil P (2012) *Ecological Speciation*. Oxford University Press.

Nosil P, Feder JL (2012) Genomic divergence during speciation: Causes and consequences. *Philos Trans R Soc*; 367B:332–342. doi: 10.1098/rstb.2011.0263.

Nosil P, Schluter D (2011) The genes underlying the process of speciation. *Trends Ecol Evol*; 26:160167. doi: 10.1016/j.tree.2011.01.001.

Novacek MJ, Rougier GW, Wible JR (1997) Epipubic bones in eutherian mammals from the late Cretaceous of Mongolia. *Nature*; 389:483–486.

Nowack EC, Melkonian M, Glöckner G (2008) Chromatophore genome sequence of Paulinella sheds light on acquisition of photosynthesis by eukaryotes. *Curr Biol*; 18:410–418. doi: 10.1016/j.cub.2008.02.051.

Nowak H, Schneebeli-Hermann E, Kustatscher E (2019) No mass extinction for land plants at the Permian-Triassic transition. *Nat Commun*. 10:384. doi: 10.1038/s41467-018-07945-w.

Nowak M, Tarnita, C, Wilson E (2010) The evolution of eusociality. *Nature*; 466:1057–1062. https://doi.org/10.1038/Nature09205.

Nowak MA, McAvoy A, Allen B, Wilson EO (2017) The general form of Hamilton's rule makes no predictions and cannot be tested empirically. *Proc Natl Acad Sci U S A*; 114:5665–5670. doi: 10.1073/pnas.1701805114.

Nowitzki U, Flechner A, Kellermann J (1998) Eubacterial origin of nuclear genes for chloroplast and cytosolic glucose-6-phosphate isomerase from spinach: Sampling eubacterial gene diversity in eukaryotic chromosomes through symbiosis. *Gene*; 214:205–213.

Nozaki H, Matsuzaki M, Takahara M (2003) The phylogenetic position of red algae revealed by multiple nuclear genes from mitochondria-containing eukaryotes and an alternative hypothesis on the origin of plastids. *J Mol Evol*; 56:485–497.

Nyakatura JA, Allen VR, Lauströer J et al. (2015) A Three-Dimensional Skeletal Reconstruction of the Stem Amniote *Orobates pabsti* (Diadectidae): Analyses of Body Mass, Centre of Mass Position, and Joint Mobility. *PLoS One*; 10(9):e0137284. doi: 10.1371/journal.pone.0137284.

O'Connell JF, Allen J (2015) The process, biotic impact, and global implications of the human colonization of sahul about 47,000 years ago. *J Arch Sci*; 56:73–84. doi: 10.1016/j.jas.2015.02.020.

O'Connell JF, Allen J, Williams MAJ et al. (2018) When did *Homo sapiens* first reach Southeast Asia and Sahul? *Proc Natl Acad Sci U S A*; 115:8482–8490. doi: 10.1073/pnas.1808385115.

O'Day DH (1979) Cell differentiation during fruiting body formation in polysphondylium pallidum. *J Cell Sci*; 35:203–15.

O'Hara TD, Hugall AF, MacIntosh H, Naughton KM, Williams A, Moussalli A (2016) Dendrogramma is a siphonophore. *Curr Biol*; 26:R457–R458. doi: 10.1016/j.cub.2016.04.051.

O'Rand MG (1988) Sperm-egg recognition and barriers to interspecies fertilization. *Gamete Res*; 19:315–28. doi: 10.1002/mrd.1120190402. PMID:3058566.

Oakenfull RJ, Davis SJ (2017) Shining a light on the Arabidopsis circadian clock. *Plant Cell Environ*; 40:2571–2585. doi: 10.1111/pce.13033.

Oborník M (2019) Endosymbiotic evolution of algae, secondary heterotrophy and parasitism. *Biomolecules*; 9:pii:E266 doi: 10.3390/biom9070266.

Oakley TH, Wolfe JM, Lindgren AR, Zaharoff AK (2013) Phylotranscriptomics to bring the understudied into the fold: Monophyletic ostracoda, fossil placement, and pancrustacean phylogeny. *Mol Biol Evol*; 30:215–33. doi: 10.1093/molbev/mss216.

Offield MF, Jetton TL, Labosky PA et al. (1996) PDX-1 is required for pancreatic outgrowth and differentiation of the rostral duodenum. *Development*; 122:983–95.

Oftedal OT (2002) The mammary gland and its origin during synapsid evolution. *J Mammary Gland Biol Neoplasia*; 7:225–252.

Oftedal OT (2012) The evolution of milk secretion and its ancient origins. *Animal*; 6:355–368. doi: 10.1017/S1751731111001935.

Ohta T (1973) Slightly deleterious mutant substitutions in evolution. *Nature*; 246:96–98. doi: 10.1038/246096a0.

Ohtomo Y, Kakegawa T, Ishida A (2014) Evidence for biogenic graphite in early Archaean Isua metasedimentary rocks. *Nature Geosci*; 7:25–28.

Oisi Y, Ota K, Kuraku S et al. (2013) Craniofacial development of hagfishes and the evolution of vertebrates. *Nature*; 493:175–180. https://doi.org/10.1038/nature11794

Olsen PE, Shubin NH, Anders MH (1987) New early Jurassic tetrapod assemblages constrain Triassic-Jurassic tetrapod extinction event. *Science*; 237(4818):1025–1029. doi: 10.1126/Science.3616521.

One Thousand Plant Transcriptomes Initiative (2019) One thousand plant transcriptomes and the phylogenomics of green plants. *Nature*; 574:679–685. doi: 10.1038/s41586-019-1693-2.

Oppenheimer S (1998) *Eden in the East: The Drowned Continent of Southeast Asia*. Phoenix Press.

Oppenheimer S (2003) *Out of Eden: The Peopling of the World*. Constable.

Oppenheimer S (2003) *Out of Eden: The Peopling of the World*. Robinson: London.

Oppenheimer S (2009) The great arc of dispersal of modern humans: Africa To Australia. *Quat Int*; 202:2–13.

Oppenheimer S (2012) Out-of-Africa, the peopling of continents and islands: Tracing uniparental gene trees across the map. *Philos Trans R Soc Lond B Biol Sci*; 367B:770–784. doi: 10.1098/rstb.2011.0306.

Oren A, Garrity GM (2014) Then and now: A systematic review of the systematics of prokaryotes in the last 80 years. *Antonie Van Leeuwenhoek* 106:43–56. doi: 10.1007/s10482-013-0084-1.

Orlando L, Ginolhac A, Zhang G (2013) Recalibrating *Equus* evolution using the genome sequence of an early Middle Pleistocene horse. *Nature*; 499:74–78. doi: 10.1038/nature12323.

Orlando L, Male D, Alberdi MT, Prado JL, Prieto A, Cooper A, Hänni C (2008) Ancient DNA clarifies the evolutionary history of American Late Pleistocene equids. *J Mol Evol*; 66:533-8. doi: 10.1007/s00239-008-9100-x.

Ornitz DM, Legeai-Mallet L (2017) Achondroplasia: Development, pathogenesis, and therapy. *Dev Dyn*; 46:291–309. doi: 10.1002/dvdy.24479. Epub 2017 Mar 2. PMID:27987249; PMCID:PMC5354942.

Orr HA (2009) Fitness and its role in evolutionary genetics. *Nature Rev Genetics*; 10:531–539. doi.org/10.1038/nrg2603.

Orr HA (2010) The population genetics of beneficial mutations. *Phil Trans R Soc*; 365B:1195–1201. doi: 10.1098/rstb.2009.0282.

Oster G, Wang H (2003) Rotary protein motors. *Trends Cell Biol*; 13:114–121. doi: 10.1016/S0962-8924(03)00004-7.

Ota KG, Fujimoto S, Oisi Y, Kuratani S (2011) Identification of vertebra-like elements and their possible differentiation from sclerotomes in the hagfish. *Nat Commun*; 2:373. doi: 10.1038/ncomms1355.

Otsuna H, Shinomiya K, Ito K (2014) Parallel neural pathways in higher visual centers of the *Drosophila* brain that mediate wavelength- specific behavior. *Front Neural Circuits*; 8:8. doi: 10.3389/fncir.2014.00008.

Otto SP (2018) Adaptation, speciation and extinction in the Anthropocene. *Proc Biol Sci*; 285:20182047. doi: 10.1098/rspb.2018.2047.

Pagani L, Schiffels S, Gurdasani D et al. (2015) Tracing the route of modern humans out of Africa by using 225 human genome sequences from Ethiopians and Egyptians. *Am J Hum Genet*; 96:986–991. doi: 10.1016/j.ajhg.2015.04.019.

Page RDM, Holmes EC (1998) *Molecular Evolution: A Phylogenetic Approach*. Wiley-Blackwell.

Pan Y-H, Sha J-G, Zhou Z-H, Fürsich FT (2013) The Jehol Biota: Definition and distribution of exceptionally preserved relicts of a continental early Cretaceous ecosystem. *Cretaceous Res*; 44:30–38. doi: 10.1016/j.cretres.2013.03.007.

Pannell JR (2017) Plant sex determination. *Curr Biol*; 27:R191–R197. doi: 10.1016/j.cub.2017.01.052.

Panyutina AP, Korzun LP, Kuznetsov AN (2013) *Flight of Mammals: From Terrestrial Limbs to Wings*. Springer.

Pardini AT, O'Brien PC, Fu B et al. (2007) Chromosome painting among proboscidea, hyracoidea and sirenia: Support for paenungulata (Afrotheria, mammalia) but not tethytheria. *Proc Biol Sci*; 274:1333–1340. doi: 10.1098/rspb.2007.0088.

Parés JM, Pérez-González A, Weil AB, Arsuaga JL (2000) On the age of the hominid fossils at the Sima de los Huesos Sierra de Atapuerca Spain:paleomagnetic evidence. *Am JPhys Anthropol*; 111:451–461.

Parfrey LW, Lahr DJ, Knoll AH, Katz LA (2011) Estimating the timing of early eukaryotic diversification with multigene molecular clocks. *Proc Natl Acad Sci U S A*; 108:13624–13629. doi: 10.1073/pnas.1110633108.

Parker ET, Cleaves HJ, Dworkin JP et al. (2011) Primordial synthesis of amines and amino acids in a 1958 miller H2S-rich spark discharge experiment. *Proc Natl Acad Sci U S A*; 108:5526–31. doi: 10.1073/pnas.1019191108.

Pascual-Anaya J, DAniello S, Kuratani S, Garcia-Fernández J (2013) Evolution of Hox gene clusters in deuterostomes. *BMC Dev Biol*; 13:26. doi: 10.1186/1471-213X-13-26.

Patané JSL, Martins J Jr, Setubal JC (2018) Phylogenomics. *Methods Mol Biol*; 1704:103–187. doi: 10.1007/978-1-4939-7463-4_5.

Paton RL, Smithson TR, Clack JA (1999) An amniote-like skeleton from the early carboniferous of Scotland. *Nature*; 398:508–513. doi: 10.1038/19071.

Pavlopoulos A, Akam M (2011) Hox gene *Ultrabithorax* regulates distinct sets of target genes at successive stages of *Drosophila* haltere morphogenesis. *Proc Natl Acad Sci U S A*; 108:2855–2860. doi: 10.1073/pnas.1015077108.

Payer LM, Burns KH (2019) Transposable elements in human genetic disease. *Nat Rev Genet*; 20:760–772. doi: 10.1038/s41576-019-0165-8.

Pereira CF, Lemischka IR, Moore K (2012) Reprogramming cell fates: Insights from combinatorial approaches. *Ann N Y Acad Sci*; 1266:7–17. doi: 10.1111/j.1749-6632.2012.06508.x.

Pereira J, Johnson WE, O'Brien SJ (2014) Evolutionary genomics and adaptive evolution of the Hedgehog gene family (*Shh ihh and dhh*) in vertebrates. *PLoS One*; 9:e74132. doi: 10.1371/journal.pone.0074132.

Perini, FA, Russo CAM, Schrago CG (2010) The evolution of South American endemic canids: A history of rapid diversification and morphological parallelism. *J Evol Biol*; 23:311–322. doi: 10.1111/j.14209101.2009.01901.x.

Persons WS, Currie PJ (2019) Feather evolution exemplifies sexually selected bridges across the adaptive landscape. *Evolution* 73:1686–1694. doi: 10.1111/evo.13795.

Peter J, De Chiara M, Friedrich A et al. (2018) Genome evolution across 1,011 saccharomyces cerevisiae isolates. *Nature*; 556:339–344. doi: 10.1038/s41586-018-0030-5.

Peters J (2014) The role of genomic imprinting in biology and disease: an expanding view. *Nat Rev Genet*; 15:517-30. doi: 10.1038/nrg3766.

Petersen KK, Hørlyck A, Ostergaard KH (2013) Protection against high intravascular pressure in giraffe legs. *Am J Physiol Regul Integr Comp Physiol*; 305:R1021–R1030. doi: 10.1152/ajpregu.00025.2013.

Peterson KJ, Eernisse DJ (2001) Animal phylogeny and the ancestry of bilaterians: Inferences from morphology and 18S rDNA gene sequences. *Evol Dev*; 3:170–205.

Peterson RE et al. (2019) Genome-wide association studies in ancestrally diverse populations: Opportunities, methods, pitfalls, and recommendations. *Cell*; pii:S0092-8674(19)31002-5. doi: 10.1016/j. cell.2019.08.051.

Peterson E, Kau K (2018) Antibiotic resistance mechanisms in bacteria: Relationships between resistance determinants of antibiotic producers, environmental bacteria, and clinical pathogens. *Front Microbiol*. 2018; 9:2928. doi: 10.3389/fmicb.2018.02928.

Petroutsos D, Amiar S, Abida H (2014) Evolution of galactoglycerolipid biosynthetic pathways —From cyanobacteria to primary plastids and From primary to secondary plastids. *Prog Lipid Res*; 54:68–85. doi: 10.1016/j. plipres.2014.02.001.

Pfennig DW, Mullen SP (2010) Mimics without models: Causes and consequences of allopatry in batesian mimicry complexes. *Proc Biol Sci*; 277:2577-2585. doi: 10.1098/ rspb.2010.0586.

Philippe H, Brinkmann H, Copley RR (2011) Acoelomorph flatworms are deuterostomes related to Xenoturbella. *Nature*; 470:255–258. doi: 10.1038/nature09676.

Philippe H, Lartillot N, Brinkmann H. (2005) Multigene analyses of bilaterian animals corroborate the monophyly of ecdysozoa, lophotrochozoa, and protostomia. *Mol Biol Evol*; 22:1246–53. doi: 10.1093/molbev/msi111.

Phillips MJ, Fruciano C (2018) The soft explosive model of placental mammal evolution. *BMC Evol Biol*; 18:104. doi: 10.1186/s12862-018-1218-x.

Piazza P, Jasinski S, Tsiantis M. (2005) Evolution of leaf developmental mechanisms. *New Phytt*; 167:693–710. https://doi.org/10.1111/ j.1469-8137.2005.01466.x.

Pickett-Heaps JD, Staehelin LA (1975) The ultrastructure of scenedesmus (Chlorophyceae). II. Cell division and colony formation. *J Phycol*. 11:186–202. doi: 10.1111/ j.1529-8817.1975.tb02766.x.

Pierce SE, Clack JA, Hutchinson JR (2012) Three-dimensional limb joint mobility in the early tetrapod ichthyostega. *Nature*; 486:523–526. doi: 10.1038/nature11124.

Pigliucci M, Murren CJ, Schlichting CD (2006) Phenotypic plasticity and evolution by genetic assimilation. *J Exp Biol*; 209:2362–7. doi: 10.1242/jeb.02070.

Pilder SH, Olds-Clarke P, Phillips DM, Silver LM (1993) Hybrid sterility-6: A mouse t complex locus controlling sperm flagellar assembly and movement. *Dev Biol*; 159:631–642.

Piovesan A, Caracausi M, Antonaros F, Pelleri MC, Vitale L (2016) GeneBase 1.1: A tool to summarize data from NCBI gene datasets and its application to an update of human gene statistics. *Database (Oxford)*; 2016:baw153. doi:10.1093/database/baw153.

Pisani D, Poling LL, Lyons-Weiler M, Hedges SB (2004) The colonization of land by animals: Molecular phylogeny and divergence times among arthropods. *BMC Biol*; 2:1.

Pittman M, Gatesy SM, Upchurch P, Goswami A, Hutchinson JR (2013) Shake a tail feather: The evolution of the theropod tail into a stiff aerodynamic surface. *PLoS One*; 8(5):e63115. doi: 10.1371/journal.pone.0063115.

Plard F, Fay R, Kéry M, Cohas A, Schaub M (2019) Integrated population models: Powerful methods to embed individual processes in population dynamics models. *Ecology*; 100(6):e02715. doi: 10.1002/ecy.2715.

PLoS One 8:e82261. doi: 10.1371/journal. pone.0082261.

Plummer T (2004) Flaked stones and old bones: Biological and cultural evolution at the dawn of technology. *Am JPhys Anthropol*; 39(Suppl.):118–164.

Polaszek A (ed) (2010) *Systema Naturae 250—The Linnaean Ark*. CRC Press.

Pollock FJ, Lamb JB, van de Water JAJM, Smith HA, Schaffelke B, Willis BL, Bourne DG (2019) Reduced diversity and stability of coral-associated bacterial communities and suppressed immune function precedes disease onset in corals. *R Soc Open Sci*; 6:190355. doi: 10.1098/rsos.190355.

Ponce-Toledo RI, Deschamps P, López-García P, Zivanovic Y, Benzerara K, Moreira D (2017) An early-branching freshwater cyanobacterium at the origin of plastids. *Curr Biol*; 27:386–391. doi: 10.1016/j.cub.2016.11.056.

Ponce-Toledo RI, Moreira D, López-García P, Deschamps P (2018) Secondary plastids of euglenids and chlorarachniophytes function with a mix of genes of red and green algal ancestry. *Mol Biol Evol*; 35:2198–2204. doi: 10.1093/molbev/msy121.

Popper K (1959) *The Logic of Scientific Discovery*. Routledge.

Porter SM (2020) Insights into eukaryogenesis from the fossil record. *Interface Focus*; 10(4):20190105. doi: 10.1098/rsfs.2019.0105.

Postler TS, Ghosh S (2017) Understanding the holobiont: How microbial metabolites affect human health and shape the immune system. *Cell Metab*; 26:110–130. doi: 10.1016/j. cmet.2017.05.008.

Potter BA, Baichtal JF, Beaudoin AB et al. (2018) Current evidence allows multiple models for the peopling of the Americas. *Sci Adv*; 84:eaat5473. doi: 10.1126/sciadv.aat5473.

Pradel A, Sansom IJ, Gagnier PY et al. (2007) The tail of the Ordovician fish *Sacabambaspis*. *Biol Lett*; 3:72–75.

Prado JL, Alberdi MT (1996) A cladistic analysis of the horses of the tribe equine. *J Paleontol*; 39:663–680.

Prinzinger R, Preßmar A, Schleucher E (1991) Body temperature in birds. *Comp Biochem Phys*; 99A:499–506.

Prokop J, Pecharová M, Nel A, Hörnschemeyer T, Krzemińska E, Krzemiński W, Engel MS (2017) Paleozoic nymphal wing pads support dual model of insect wing origins. *Curr Biol*; 27:263–269. doi: 10.1016/j. cub.2016.11.021.

Prüfer K, Racimo F, Patterson N et al. (2014) The complete genome sequence of a Neanderthal from the Altai mountains. *Nature*; 505:43–49. doi: 10.1038/nature12886.

Prum RO, Brush AH (2002) The evolutionary origin and diversification of feathers. *Q Rev Biol*; 77:261–295.

Pugach I, Delfin F, Gunnarsdóttir E (2013) Genome-wide data substantiate holocene gene flow from India to Australia. *Proc Natl Acad Sci U S A*; 110:1803–1808. doi: 10.1073/ pnas.1211927110.

Pushkarev A, Inoue K, Larom S et al. (2018) A distinct abundant group of microbial rhodopsins discovered using functional metagenomics. *Nature*; 558:595–599. doi: 10.1038/s41586-018-0225-9.

Pyron RA (2011) Divergence time estimation using fossils as terminal taxa and the origins of Lissamphibia. *Syst Biol*; 60:466–481. doi: 10.1093/sysbio/syr047.

Qin P, Stoneking M (2015) Denisovan ancestry in East Eurasian and native American populations. *Mol Biol Evol*; 32:2665–2674. doi: 10.1093/molbev/msv141.

Quach H, Rotival M, Pothlichet J et al. (2016) Genetic adaptation and Neandertal admixture shaped the immune system of human populations. *Cell*; 167:643–656.e17. doi: 10.1016/j.cell.2016.09.024.

Rader B, McAnulty SJ, Nyholm SV (2019) Persistent symbiont colonization leads to a maturation of hemocyte response in the Eupsrymna scolopes/Vibrio fischeri symbiosis. *Microbiologyopen*; e858. doi: 10.1002/mbo3.858.

Ramírez-Chaves HE, Weisbecker V, Wroe S, Phillips MJ (2016) Resolving the evolution of the mammalian middle ear using bayesian inference. *Front Zool*; 13:39. doi: 10.1186/ s12983-016-0171-z.

Rands CM, Darling A, Fujita M (2013) Insights into the evolution of Darwin's finches from comparative analysis of the geospiza magnirostris genome sequence. *BMC Genomics*; 12:1495. doi: 10.1186/1471-2164-14-95.

Rangel de Lázaro G, de la Cuétara JM, Pišová H et al. (2016) Diploic vessels and computed tomography: Segmentation and comparison in modern humans and fossil hominids. *Am J Phys Anthropol*; 159:313–324. doi: 10.1002/ ajpa.22878.

Rauhut OW, Foth C, Tischlinger H, Norell MA (2012) Exceptionally preserved juvenile megalosauroid theropod dinosaur with filamentous integument from the Late Jurassic of Germany. *Proc Natl Acad Sci U S A*; 109:1174611751. doi: 10.1073/ pnas.1203238109.

Ravel A, Marivaux L, Tabuce R et al. (2011) The oldest African bat from the early Eocene of El Kohol (Algeria). *Naturwissenschaften*; 98:397–405. doi: 10.1007/s00114-011-0785-0.

Raven JA, Edwards D (2014) Roots: Evolutionary origins and biogeochemical significance. *J Exp Bot*; 52(Suppl.):381–401.

Redecker D, Kodner R, Graham LE (2000) Glomalean fungi from the ordovician. *Science*; 289:1920–1921. doi: 10.1126/ Science.289.5486.1920.

Reich D, Green RE, Kircher M et al. (2010) Genetic history of an archaic hominin group from Denisova Cave in Siberia. *Nature*; 468:1053– 1060. doi: 10.1038/Nature09710.

Reich D, Patterson N, Kircher M et al. (2011) Denisova admixture and the first modern human dispersals into southeast asia and oceania. *Am J Hum Genet*; 89:516–528. doi: 10.1016/j. ajhg.2011.09.005.

Reidenberg JS (2007) Anatomical adaptations of aquatic mammals. *Anat Rec (Hoboken)*; 290:507–513.

Reisz RR (1977) Petrolacosaurus, the oldest known diapsid reptile. *Science*; 196:1091–1093.

Reisz RR, Fröbisch J (2014) The oldest caseid synapsid from the late Pennsylvanian of Kansas, and the evolution of herbivory in terrestrial vertebrates. *PLoS One*; 9(4):e94518. doi: 10.1371/journal. pone.0094518.

Renard E, Leys SP, Wörheide G, Borchiellini C (2018) Understanding animal evolution: The added value of sponge transcriptomics and genomics: The disconnect between gene content and body plan evolution. *Bioessays*; 40:e1700237/ doi: 10.1002/ bies.201700237.

Renfree MB (2010) Marsupials: Placental mammals with a difference. *Placenta*; 31:S21–S26. doi: 10.1016/j. placenta.2009.12.023.

Renne PR et al. (2013) Time scales of critical events around the Cretaceous-Paleogene boundary. *Science*. 2013; 339:684–687. doi: 10.1126/Science.1230492.

Retallack GJ (2013) Ediacaran life on land. *Nature*; 493:89–92. doi: 10.1038/nature11777.

Rice R, Kallonen A, Cebra-Thomas J, Gilbert SF (2016) Development of the turtle plastron, the order-defining skeletal structure. *Proc Natl Acad Sci U S A*; 113:5317–5322. doi: 10.1073/pnas.1600958113.

Rice WR, Hostert EE (1993) Laboratory experiments on speciation: What have we learned in forty years? *Evolution*; 47:1637–1653.

Rice WR, Salt GW (1990) The evolution of reproductive isolation as a correlated character under sympatric conditions: Experimental evidence. *Evolution*; 44:1140–1152.

Rice WR, Chippindale AK (2002) The evolution of hybrid infertility: Perpetual coevolution between gender-specific and sexually antagonistic genes. *Genetica*; 116(2–3):179–188.

Rich SS, Bell AE, Wilson SP (1979) Genetic drift in small populations of tribolium. *Evolution*; 33(2):579–584. doi: 10.1111/j.1558-5646.1979.tb04711.x.doi: 10.1111/j.1558-5646.1979.tb04711.x.

Richardson L, Venkataraman S, Stevenson P et al. (2014) EMAGE mouse embryo spatial gene expression database: (2014 update). *Nucleic Acids Res*; 42:D835–D844. doi: 10.1093/nar/gkt1155.

Richmond S, Howe LJ, Lewis S, Stergiakouli E, Zhurov A. (2018) Facial genetics: A brief overview. *Front Genet*; 9:462. doi: 10.3389/fgene.2018.00462.

Richter DJ, King N (2013) The genomic and cellular foundations of animal origins. *Ann Rev Genet*; 47:509–537. doi: 10.1146/annurev-genet-111212-133456.

Ridley M (2004) *Evolution*. Wiley-Blackwell.

Riedel-Kruse IH, Müller C, Oates AC (2007) Synchrony dynamics during initiation failure and rescue of the segmentation clock. *Science*; 317:1911–1915.

Riga BJG, Lamanna M, Ortiz L, Coria JP (2016) A gigantic new dinosaur from Argentina and the evolution of the sauropod hind foot. *Sci Rep*; 6:19165. doi: 10.1038/srep19165

Rincón Baron EJ, Rolleri CH, Alzate Guarin F, Dorado Gálvez JM (2014) Ontogeny of the sporangium, spore formation and cytochemistry in Colombian lycopodials (Lycopodiaceae)]. *Rev Biol Trop*; 62:273–298. PMID:24912358 (Spanish, English abstract).

Rinehart LF, Lucas SG (2013) Tooth form and function in Temnospondyl amphibians: Relationship of shape to applied stress. *New Mexico Mus Nat Hist Sci Bull*; 61:533–542.

Risch N, Tang H, Katzenstein H, Ekstein J (2003) Geographic distribution of disease mutations in the Ashkenazi Jewish population supports genetic drift over selection. *Am J Hum Genet*; 72:812–22. doi: 10.1086/373882.

Rito T, Vieira D, Silva M et al. (2019) A dispersal of *Homo sapiens* from Southern to Eastern Africa immediately preceded the out-of-Africa migration. *Sci Rep*; 9:4728. doi: 10.1038/s41598-019-41176-3.

Roach JC, Glusman G, Smit AF et al. (2010) Analysis of genetic inheritance in a family quartet by whole-genome sequencing. *Science*; 328:636–9. doi: 10.1126/science.1186802.

Robinson A (2007) *The Story of Writing Alphabets, Hieroglyphs and Pictograms*. Thames & Hudson.

Robson SL, Wood B (2008) Hominin life history: Reconstruction and evolution. *J Anat*; 212:394–425. doi: 10.1111/j.1469-7580.2008.00867.x.

Rodrigues JA, Zilberman D (2015) Evolution and function of genomic imprinting in plants. *Genes Dev*; 29:2517–2531. doi:10.1101/gad.269902.115.

Roebroeks W, Sier MJ, Nielsen TK (2012) Use of red ochre by early Neandertals. *Proc Natl Acad Sci U S A*; 109:1889–1894. doi: 10.1073/pnas.1112261109.

Roff DA (2017) A centennial celebration for quantitative genetics. *Evolution*; 61:1017–32. doi: 10.1111/j.1558-5646.2007.00100.x.

Roger AJ, Muñoz-Gómez SA, Kamikawa R (2017) The origin and diversification of mitochondria. *Curr Biol*; 27(21):R1177–R1192. doi: 10.1016/j.cub.2017.09.015.

Roger AJ, Simpson AG (2009) Evolution: Revisiting the root of the eukaryote tree. *Curr Biol*; 19(4):R165–7. doi: 10.1016/j.cub.2008.12.032.

Rogers CS, Astrop TI, Webb SM, Ito S, Wakamatsu K, McNamara ME (2019) Synchrotron x-ray absorption spectroscopy of melanosomes in vertebrates and cephalopods: Implications for the affinity of *Tullimonstrum*. *Proc Biol Sci*; 286:20191649. doi: 10.1098/rspb.2019.1649.

Roh SW, Kim KH, Nam YD, Chang HW, Park EJ, Bae JW (2010) Investigation of archaeal and bacterial diversity in fermented seafood using barcoded pyrosequencing. *ISME J*; 4:1–16. doi: 10.1038/ismej.2009.83.

Rohner PT, Macagno ALM, Moczek AP (2020) Evolution and plasticity of morph-specific integration in the bull-headed dung beetle *Onthophagus taurus*. *Ecol Evol*; 10:10558–10570. doi: 10.1002/ece3.6711.

Rolland-Lagan AG, Paquette M, Tweedle V, Akimenko MA (2012) Morphogen-based simulation model of ray growth and joint patterning during fin development and regeneration. *Development*. 139:1188–1197. doi: 10.1242/dev.073452.

Romanes GJR (1910) *Darwin and after Darwin*. Open Court Publishing.

Romeo, A (2008) *Bacterial Biofilms*. Berlin: Springer.

Rosenberg NA, Nordborg M (2002) Genealogical trees, coalescent theory and the analysis of genetic polymorphisms. *Nat Rev Genet*; 3:380–390. doi: 10.1038/nrg795.

Ross J, Westaway K, Travers M, Morwood MJ, Hayward J (2016) Into the past: A step towards a robust Kimberley rock art chronology. *PLoS One*; 11(8):e0161726. doi: 10.1371/journal.pone.0161726.

Ross P, Mayer R, Benziman M (1991) Cellulose biosynthesis and function in bacteria. *Microbiol Rev*; 55:35–58.

Rota-Stabelli O, Daley AC, Pisani D (2013) Molecular timetrees reveal a Cambrian colonization of land and a new scenario for ecdysozoan evolution. *Curr Biol*; 23:392–398. doi: 10.1016/j.cub.2013.01.026.

Rots V, Van Peer P (2006) Early evidence of complexity in lithic economy: Core-axe production hafting and use at Late Middle Pleistocene site 8-b-11, Sai Island (Sudan). *J Archaeol Sci*; 33:360–371.

Rowe TB, Macrini TE, Luo ZX (2011) Fossil evidence on origin of the mammalian brain. *Science*; 332:955–957. doi: 10.1126/science.1203117.

Rowntree LG, Clark TH, Hanson AM (1935) Accruing acceleration in growth and development in five successive generations of rats under continuous treatment with thymus extract. *Arch Int Med*; 56:1–29.

Rücklin M, Donoghue PC, Johanson Z, Trinajstic K et al. (2012) Development of teeth and jaws in the earliest jawed vertebrates. *Nature*; 491:748–751. doi: 10.1038/nature11555.

Rücklin M, Donoghue PC (2015) Romundina and the evolutionary origin of teeth. *Biol Lett*; 11:20150326. doi: 10.1098/rsbl.2015.0326.

Ruff C (2002) Variation in human body size and shape. *Ann Rev Anthrop*; 31:211–232. doi: 10.1146/annurev.anthro.31.040402.085407.

Ruhfel BR, Gitzendanner MA, Soltis PS, Soltis DE, Burleigh JG. (2014) From algae to angiosperms-inferring the phylogeny of green plants (viridiplantae) from 360 plastid genomes. *BMC Evol Biol*; 14:23. doi: 10.1186/1471-2148-14-23.

Ruijtenberg S, van den Heuvel S (2015) G1/S inhibitors and the SWI/SNF complex control cell-cycle exit during muscle differentiation. *Cell*; 162:300–313. doi: 10.1016/j.cell.2015.06.013.

Ruse M, Travis J (eds) (2009) *Evolution: The First Four Billion Years*. Harvard University Press.

Rutman JY (2014) *Why Evolution Matters: A Jewish View*. Valentine Mitchell.

Ryan JF, Baxevanis AD (2007) Hox Wnt and the evolution of the primary body axis: Insights from the early-divergent phyla. *Biol Direct*; 2:37.

Ryan JF, Chiodin M (2015) Where is my mind? How sponges and placozoans may have lost neural cell types. *Phil Trans R Soc*; 370B:pii:20150059. doi: 10.1098/rstb.2015.0059.

Ryder OA, Chemnick LG, Bowling AT, Benirschke K (1985) Male mule foal qualifies as the offspring of a female mule and jack donkey. *J Hered*; 76:379–381.

Saastamoinen M et al. (2018) Genetics of dispersal. *Biol Rev Camb Philos Soc*; 93(1):574–599. doi: 10.1111/brv.12356.

Sabeti PC, Reich DE, Higgins JM et al. (2002) Detecting recent positive selection in the human genome from haplotype structure. *Nature*; 419:832–7. doi: 10.1038/Nature01140.

Sacerdot C, Louis A, Bon C, Berthelot C, Roest Crollius H (2018) Chromosome evolution at the origin of the ancestral vertebrate genome. *Genome Biol*; 19:166. doi: 10.1186/s13059-018-1559-1.

Saez AG, Probert I, Geisen M, Quinn P, Young JR, Medlin LK (2003) Pseudo-cryptic speciation in coccolithophores. *Proc Natl Acad Sci*; 100:7163–7168. doi: 10.1073/pnas.1132069100.

Sagulenko E, Morgan GP, Webb RI (2014) Structural studies of planctomycete *Gemmata obscuriglobus* support cell compartmentalization in a bacterium. *PLoS One*; 9:e91344. doi: 10.1371/journal.pone.0091344.

Saitou N (2013) *Introduction to Evolutionary Genetics*. Springer.

Sakamaki K, Imai K, Tomii K, Miller DJ (2015) Evolutionary analyses of caspase-8 and its paralogs: Deep origins of the apoptotic signaling pathways. *Bioessays*; 37:767–776. doi: 10.1002/bies.201500010.

Sakamoto M, Benton MJ, Venditti C (2016) Dinosaurs in decline tens of millions of years before their final extinction. *Proc Natl Acad Sci U S A*; 113:5036–5040. doi: 10.1073/pnas.1521478113.

Salgado S (2021) *Amazônia*. Taschen GmbH.

Sallan et al. (2017) The "Tully monster" is not a vertebrate: Characters, convergence, and taphonomy in paleozoic problematic animals. *Palaeont*; 60:149–157. doi: 10.1111/pala.12282.

Salvini-Plawen LV, Mayr E (1977) On the evolution of photoreceptors and eyes. *Evol Biol*; 10:207–263.

Samuels ME, Regnault S, Hutchinson JR (2017) Evolution of the patellar sesamoid bone in mammals. *PeerJ*; 5:e3103. doi: 10.7717/peerj.3103.

Samonds KE, Godfrey LR, Ali JR et al. (2013) Imperfect isolation: factors and filters shaping Madagascar's extant vertebrate fauna. *PLoS One;* 8(4):e62086. doi: 10.1371/journal.pone.0062086.

San Mauro D (2010) A multilocus timescale for the origin of extant amphibians. *Mol Phylogenet Evol;* 56:554–561. doi: 10.1016/j.ympev.2010.04.019.

Sankararaman S, Mallick S, Dannemann M et al. (2014) The genomic landscape of neanderthal ancestry in present-day humans. *Nature;* 507:354–357. doi: 10.1038/nature12961.

Sankararaman S, Mallick S, Patterson N, Reich D (2016) The combined landscape of denisovan and neanderthal ancestry in present-day humans. *Curr Biol;* 26:1241–7. doi: 10.1016/j.cub.2016.03.037.

Sansom IJ, Donoghue PCJ, Albanesi G (2005) Histology and affinity of the earliest armoured vertebrate. *Biol Lett;* 1:446–449. doi: 10.1098/rsbl.2005.0349.

Santarella-Mellwig R, Franke J, Jaedicke A (2010) The compartmentalized bacteria of the Planctomycetes-Verrucomicrobia-Chlamydiae superphylum have membrane coat-like proteins. *PLoS Biol;* 8:e1000281. doi: 10.1371/journal.pbio.1000281.

Sanz-Ezquerro JJ, Tickle C (2003) Digital development and morphogenesis. *J Anat;* 202:51–58.

Sarkar S, Jolly DJ, Friedmann T, Jones OW (1984) Swimming behavior of X and Y human sperm. *Differentiation;* 27:120–125. doi: 10.1111/j.1432-0436.1984.tb01416.x.

Sartorius GA, Nieschlag E (2010) Paternal age and reproduction. *Hum Reprod Update;* 16:65–79. doi: 10.1093/humupd/dmp027.

Sasakura Y, Hozumi A (2018) Formation of adult organs through metamorphosis in ascidians. *Wiley Interdiscip Rev Dev Biol;* 7(2). doi: 10.1002/wdev.304.

Sassa T (2013) The role of human-specific gene duplications during brain development and evolution. *J Neurogenet;* 27:86–96.

Sato A, Tichy H, O'hUigin C (2001) On the origin of Darwin's finches. *Mol Biol Evol;* 18:299–311.

Satoh N (2003) The ascidian tadpole larva: Comparative molecular development and genomics. *Nat Rev Genet;* 4:285–295. doi: 10.1038/nrg1042.

Satoh N (2016) *Chordate Origins and Evolution: The Molecular Evolutionary Road to Vertebrates.* Academic Press.

Satoh N, Rokhsar D, Nishikawa T (2014) Chordate evolution and the three-phylum system. *Proc. R. Soc;* 281B:20141729. doi: 10.1098/rspb.2014.1729.

Sauquet H, von Balthazar M, Magallón S et al. (2017) The ancestral flower of angiosperms and its early diversification. *Nat Commun;* 8:16047. https://doi.org/10.1038/ncomms16047

Saxena A, Towers M, Cooper KL (2017) The origins, scaling and loss of tetrapod digits. *Philos Trans R Soc Lond B Biol Sci;* 372:20150482. doi: 10.1098/rstb.2015.0482.

Saylo MC, Escoton CC, Saylo MM (2011) Punctuated equilibrium vs. phyletic gradualism. *Int J Biosci Biotech;* 3:27–42. www.sersc.org/journals/IJBSBT/vol3_no4/3.pdf.

Schedel FDB, Musilova Z, Schliewen UK (2019) East African cichlid lineages (Teleostei: Cichlidae) might be older than their ancient host lakes: New divergence estimates for the East African Cichlid radiation. *BMC Evol Biol;* 19:94. doi: 10.1186/s12862-019-1417-0.

Schierwater B, de Salle R (2021) *Invertebrate Zoology: A Tree of Life Approach.* CRC Press.

Schilthuizen M, Giesbers MC, Beukeboom LW (2011) Haldane's rule in the 21st century. *Heredity;* 107:95–102. doi: 10.1038/hdy.2010.170.

Schluter D, Conte GL (2009) Genetics and ecological speciation. *Proc Natl Acad Sci U S A;* 106 (Suppl. 1):9955–9962. doi: 10.1073/pnas.0901264106.

Schmid MW, Heichinger C, Coman Schmid D et al. (2018) Contribution of epigenetic variation to adaptation in Arabidopsis. *Nat Commun;* 9:4446. doi: 10.1038/s41467-018-06932-5.

Schneider H, Sampaio I (2013) The systematics and evolution of new world primates—a review. *Mol Phylogenet Evol;* 82B:348–357. doi: 10.1016/j.ympev.2013.10.017.

Schneider I, Aneas I, Gehrke AR, Dahn RD, Nobrega MA, Shubin NH (2011) Appendage expression driven by the Hoxd Global Control Region is an ancient gnathostome feature. *Proc Natl Acad Sci U S A;* 108:12782–6. doi: 10.1073/pnas.1109993108.

Schneider NY (2011) The development of the olfactory organs in newly hatched monotremes and neonate marsupials. *J Anat;* 219:229–242.

Schoch RR, Sues HD (2016) The diapsid origin of turtles. *Zoology (Jena);* 119:159–161. doi: 10.1016/j.zool.2016.01.004.

Schoenfelder S, Fraser P (2019) Long-range enhancer-promoter contacts in gene expression control. *Nat Rev Genet;* 20:437–455. doi: 10.1038/s41576-019-0128-0.

Scholtz G, Wolff C (2013) Arthropod Embryology: Cleavage and Germ Band Development. In. *Arthropod Biology and Evolution* (Minelli A, Boxshall G, Fusco G, eds), pp 63–85. Springer.

Schraiber JG, Akey JM (2015) Methods and models for unravelling human evolutionary history. *Nat Rev Genet;* 16:727–740. doi: 10.1038/nrg4005.

Schubbert S, Bollag G, Shannon K (2007) Deregulated ras signaling in developmental disorders: New tricks for an old dog. *Curr Opin Genet Dev;* 17:15–22.

Schulte P, Alegret L, Arenillas I et al. (2010) The chicxulub asteroid impact and mass extinction at the Cretaceous-Paleogene boundary. *Science;* 327:1214–8. doi: 10.1126/science.1177265. PMID: 20203042.

Schultz JA, Martin T (2014) Function of pretribosphenic and tribosphenic mammalian molars inferred from 3D animation. *Naturwissen;* 101:771–781. doi: 10.1007/s00114-014-1214-y.

Schwartz JH, Tattersall I, Chi Z (2014) Comment on "A complete skull from Dmanisi, Georgia, and the evolutionary biology of early *Homo.*" *Science;* 344:360. doi: 10.1126/science.1250056.

Sebé-Pedrós A, de Mendoza A1, Lang BF, Degnan BM, Ruiz-Trillo (2011) Unexpected repertoire of metazoan transcription factors in the unicellular holozoan Capsaspora owczarzaki. *Mol Biol Evol;* 28:1241–1254. doi: 10.1093/molbev/msq309.

Sebé-Pedrós A, Degnan BM, Ruiz-Trillo I (2017) The origin of Metazoa: A unicellular perspective. *Nat Rev Genet;* 18:498–512. doi: 10.1038/nrg.2017.21.

Secor DH (2015) *Marine Ecology of Marine Fishes.* Johns Hopkins University Press.

Seehausen O (2002) Patterns in fish radiation are compatible with Pleistocene desiccation of Lake Victoria and 14600 year history for its cichlid species flock. *Proc Biol Sci;* 269:491–497.

Seilacher A, Bose PK, Pfluger F (1998) Triploblastic animals more than 1 billion years ago: Trace fossil evidence from India. *Science;* 282:80–83.

Seipel K, Schmid V (2005) Evolution of striated muscle: Jellyfish and the origin of triploblasty. *Dev Biol;* 282:14–26.

Self S, Blake S, Sharma K, Widdowson M, Sephton S (2008) Sulfur and chlorine in late Cretaceous Deccan magmas and eruptive gas release. *Science;* 319:1654–1657. doi: 10.1126/Science.1152830.

Sella G, Barton NH (2019) Thinking about the evolution of complex traits in the era of genome-wide association studies. *Ann Rev Genomics Hum Gene;* 20:461–493. doi: 10.1146/annurev-genom-083115-022316.

Sellers WI, Margetts L, Coria RA, Manning PL (2013) March of the titans: The locomotor capabilities of sauropod dinosaurs. *PLoS One;* 8(10):e78733. doi: 10.1371/journal.pone.0078733.

Selwood L, Johnson MH (2006) Trophoblast and hypoblast in the monotreme, marsupial and eutherian mammal: Evolution and origins. *Bioessays;* 28:128–145. doi: 10.1002/bies.20360.

Semon R (1901) Book – Normal Plates of the Development of Vertebrates 3. https://embryology.med.unsw.edu.au/embryology/index.php/Book_-_Normal_Plates_of_the_Development_of_Vertebrates_3

Senji Laxme RR, Suranse V, Sunagar K (2018) Arthropod venoms: Biochemistry, ecology and evolution. *Toxicon;* 158:84–103. doi: 10.1016/j.toxicon.2018.11.433.

Serbet R, Rothwell GW (1992) Characterizing the most primitive seed ferns. I. A reconstruction of Elkinsia polymorpha. *Int J Plant Sci;* 153:602–621.

Serre C, Busuttil V, Botto JM (2018) Intrinsic and extrinsic regulation of human skin melanogenesis and pigmentation. *Int J Cosmet Sci;* 40:328–347. doi: 10.1111/ics.12466.

Seymour M, Räsänen K, Kristjánsson BK (2019) Drift versus selection as drivers of phenotypic divergence at small spatial scales: The case of Belgjarskógur threespine stickleback. *Ecol Evol;* 9:8133–8145. doi: 10.1002/ece3.5381.

Seymour RS, Bennett-Stamper CL, Johnston SD et al. (2004) Evidence for cndothermic ancestors of crocodiles at the stem of archosaur evolution. *Physiol Biochem Zool;* 77:1051–1067.

Shapiro BJ, Leducq JB, Mallet J (2016) What is speciation? *PLoS Genet;* 12:e1005860. doi: 10.1371/journal.pgen.1005860.

Shapiro JA (2011) *Evolution: A View from the 21st Century.* FT Press.

Sharma PP, Tarazona OA, Lopez DH et al. (2015) A conserved genetic mechanism specifies deutocerebral appendage identity in insects and arachnids. *Proc Biol Sci;* 282:20150698. doi: 10.1098/rspb.2015.0698.

Shattock SG (1880) A new bone in human anatomy together with an investigation into the morphological significance of the so-called internal lateral ligament of the human lower jaw. *J Anat Physiol;* 14:201–204.

Shear WA, Edgecombe GD (2010) The geological record and phylogeny of the Myriapoda. *Arthropod Struct Dev;* 39:174–190. https://doi.org/10.1016/j.asd.2009.11.002.

Sheehan S, Harris K, Song YS (2013) Estimating variable effective population sizes from multiple genomes: A sequentially Markov conditional sampling distribution approach. *Genetics;* 194:647–662. doi: 10.1534/genetics.112.149096.

Shen B, Dong L, Xiao S, Kowalewski M (2008) The avalon explosion: Evolution of Ediacara morphospace. *Science;* 319:81–84. doi: 10.1126/science.1150279.

Sheppard CE, Marshall HH, Inger R et al. (2018) Decoupling of genetic and cultural inheritance in a wild mammal. *Curr Biol*; 28:1846–1850.e2. doi: 10.1016/j.cub.2018.05.001.

Sherman PW (1977) Nepotism and the evolution of alarm calls. *Science*; 97:1246–1253.

Shi Y, Yokoyama S (2003) Molecular analysis of the evolutionary significance of ultraviolet vision in vertebrates. *Proc Natl Acad Sci U S A*; 100:83088313.

Shichida Y, Matsuyama T (2009) Evolution of opsins and phototransduction. *Philos Trans R Soc*; 364B:2881–2895. doi: 10.1098/rstb.2009.0051.

Shigeno S, Sasaki T, Moritaki T, Kasugai T, Vecchione M, Agata K (2008) Evolution of the cephalopod head complex by assembly of multiple molluscan body parts: Evidence from nautilus embryonic development. *J Morphol*; 269:1–17. doi. 10.1002/jmor.10564.

Shih PM, Matzke NJ (2013) Primary endosymbiosis events date to the later proterozoic with cross-calibrated phylogenetic dating of duplicated ATPase proteins. *Proc Natl Acad Sci U S A*; 110:12355–12360. doi: 10.1073/pnas.1305813110.

Shimada H, Yamagishi A (2011) Stability of heterochiral hybrid membrane made of bacterial sn-G3P lipids and archaeal sn-G1P lipids. *Biochemistry*; 50:4114–4120R. doi: 10.1021/bi200172d.

Shin HY (2018) Targeting super-enhancers for disease treatment and diagnosis. *Mol Cells*; 41:506–514. doi: 10.14348/molcells.2018.2297.

Shoshani J (1998) Understanding proboscidean evolution: A formidable task. *Trends Ecol Evol*; 13:4 80–487. doi: 10.1016/s0169-5347(98)01491-8.

Shu DG, Morris SC, Han J et al. (2003a) Head and backbone of the Early Cambrian vertebrate haikouichthys. *Nature*; 421: 526–529. doi: 10.1038/nature01264.

Shu D, Morris SC, Zhang ZF (2003b) A new species of Yunnanozoan with implications for deuterostome evolution. *Science*; 299:1380–1384.

Shu D, Zhang X, Chen L (1996) Reinterpretation of Yunnanozoon as the earliest known hemichordate. *Nature*; 380:428–430.

Shubin N (2007) *Your Inner Fish*. Random House.

Shubin N, Tabin C, Carroll S (2009) Deep homology and the origins of evolutionary novelty. *Nature*; 457:818–823. doi: 10.1038/nature07891.

Shubin NH, Daeschler EB, Jenkins FA Jr (2014) Pelvic girdle and fin of *Tiktaalik roseae*. *Proc Natl Acad Sci U S A*; 111:893–899. doi: 10.1073/pnas.1322559111.

Sibbald SJ, Archibald JM (2017) More protist genomes needed. *Nat Ecol Evol*; 1:145. doi: 10.1038/s41559-017-0145.

Silver N (2012) *The Signal and the Noise*. Penguin Press.

Simons RS, Brainerd EL (1999) Morphological variation of hypaxial musculature in salamanders (Lissamphibia: caudata). *J Morphol*; 241:153–164.

Singer A, Poschmann G, Mühlich C et al. (2017) Massive protein import into the early-evolutionary-stage photosynthetic organelle of the Amoeba Paulinella chromatophora. *Curr Biol*; 25;27(18):2763–2773.e5. doi: 10.1016/j.cub.2017.08.010.

Skoglund P, Mathieson I (2018) Ancient genomics of modern humans: The first decade. *Ann Rev Genomics Hum Genet*; 19:3:81–404. doi: 10.1146/annurev-genom-083117-021749.

Skoglund P, Thompson JC, Prendergast ME et al. (2017) Reconstructing Prehistoric African Population Structure. *Cell*; 171:59–71.e21. doi: 10.1016/j.cell.2017.08.049.

Skvortsova K, Iovino N, Bogdanović O (2018) Functions and mechanisms of epigenetic inheritance in animals. *Nat Rev Mol Cell Biol*; 19:774–790. doi: 10.1038/s41580-018-0074-2.

Sleep NH, Bird DK, Pope EC (2011) Serpentinite and the dawn of life. *Philos Trans R Soc Lond B Biol Sci*; 366:2857-69. doi: 10.1098/rstb.2011.0129.

Slon V, Hopfe C, Weiß CL et al. (2017) Neandertal and Denisovan DNA from Pleistocene sediments. *Science*; 356:605–608. doi: 10.1126/science.aam9695.

Slon V, Mafessoni F, Vernot B et al. (2018) The genome of the offspring of a Neanderthal mother and a Denisovan father. *Nature*; 561:113–116. doi: 10.1038/s41586-018-0455-x.

Smith J, Coop G, Stephens M, Novembre J (2018) Estimating time to the common ancestor for a beneficial allele. *Mol Biol Evol*; 35:1003–1017. doi: 10.1093/molbev/msy006.

Smith JN, Goldizen AW, Dunlop RA, Noad MJ (2008a) Songs of male humpback whales, megaptera novaeangliae, are involved in intersexual interactions. *Animal Behav*; 76:467–477. doi: 10.1016/j.anbehav.2008.02.013.

Smith MR, Ortega-Hernández J (2014) Hallucigenia's onychophoran-like claws and the case for tactopoda. *Nature*; 514:363–366. doi: 10.1038/nature13576.

Smith MM, Cruz Smith L, Cameron RA, Urry LA (2008b) The larval stages of the sea urchin, Strongylocentrotus purpuratus. *J Morphol*; 269:713–33. doi: 10.1002/jmor.10618.

Smith MR, Caron JB (2015) Hallucigenia's head and the pharyngeal armature of early ecdysozoans. *Nature*; 523:75–78. doi: 10.1038/Nature14573.

Snell-Rood EC, Cash A, Han MV, Kijimoto T, Andrews J, Moczek AP (2011) Developmental decoupling of alternative phenotypes: Insights from the transcriptomes of horn-polyphenic beetles. *Evolution*; 65:231–245. doi: 10.1111/j.1558-5646.2010.01106.x.

Soares P, Alshamali F, Pereira V (2012) The expansion of mtDNA haplogroup L3 within and out of Africa. *Mol Biol Evol*; 29:915–927. doi: 10.1093/molbev/msr245.

Soares P, Ermini L, Thomson N (2009) Correcting for purifying selection: An improved human mitochondrial molecular clock. *Am J Hum Genet*; 84:740–759. doi: 10.1016/j.ajhg.2009.05.001.

Sokoloff DD, Remizowa MV, El ES, Rudall PJ, Bateman RM (2020) Supposed Jurassic angiosperms lack pentamery, an important angiosperm-specific feature. *New Phytol*; 228:420–426. doi: 10.1111/nph.15974.

Solovyeva VV, Shaimardanova AA, Chulpanova DS, Kitaeva KV, Chakrabarti L, Rizvanov AA. (2018) New approaches to Tay-Sachs disease therapy. *Front Physiol*; 9:1663. doi: 10.3389/fphys.2018.01663.

Soltis DE, Bell CD, Kim S, Soltis PS (2008) Origin and early evolution of angiosperms. *Ann N Y Acad Sci*; 1133:3–25. doi: 10.1196/annals.1438.005.

Soltis PS, Marchant DB, Van de Peer Y, Soltis DE (2015) Polyploidy and genome evolution in plants. *Curr Opin Genet Dev*; 35:119–125. doi: 10.1016/j.gde.2015.11.003.

Solymosi K, Lethin J, Aronsson H (2018) Diversity and plasticity of plastids in land plants. *Methods Mol Biol*; 1829:55–72. doi: 10.1007/978-1-4939-8654-5_4.

Soressi M, McPherron SP, Lenoir M et al. (2013) Neandertals made the first specialized bone tools in Europe. *Proc Natl Acad Sci U S A*; 110:14186–14190. doi: 10.1073/pnas.1302730110.

Soucy SM, Huang J, Gogarten JP (2015) Horizontal gene transfer: Building the web of life. *Nat Rev Genet*; 16:472–82. doi: 10.1038/nrg3962.

Soya O (ed) (2012) *Evolutionary Systems Biology*. Springer.

Spang A, Saw JH, Jørgensen SL (2015) Complex archaea that bridge the gap between prokaryotes and eukaryotes. *Nature*; 521:173–179. doi: 10.1038/nature14447.

Spatafora JW, Aime MC, Grigoriev IV, Martin F, Stajich JE, Blackwell M (2017) The fungal tree of life: From molecular systematics to genome-scale phylogenies. *Microbiol Spectr*; Sep; 5(5). doi: 10.1128/microbiolspec.FUNK-0053-2016.

Spec papers. Palaeontol Paleaeol Soc. No 20.

Speijer D (2016) What can we infer about the origin of sex in early eukaryotes? *Philos Trans R Soc*; 371B; 20150530. doi: 10.1098/rstb.2015.0530.

Sperling EA, Stockey RG (2018) The temporal and environmental context of early animal evolution: Considering all the ingredients of an "explosion". *Integr Comp Biol*; 58:605–622. doi: 10.1093/icb/icy088.

Spoor F (2015) Palaeoanthropology: The middle pliocene gets crowded. *Nature*; 521:432–433. doi: 10.1038/521432a.

Springer MS, Gatesy J (2018) Pinniped diphyly and bat triphyly: More homology errors drive conflicts in the mammalian tree. *J Hered*; 109:297–307. doi: 10.1093/jhered/esx089.

Standen EM, Du TY, Larsson HC (2014) Developmental plasticity and the origin of tetrapods. *Nature*; 513:54–58. doi: 10.1038/Nature13708.

Stappenbeck TS, Hooper LV, Gordon JI (2002) Developmental regulation of intestinal angiogenesis by indigenous microbes via paneth cells. *Proc Nat Acad Sci*; 99:15451–15455. doi: 10.1073/pnas.202604299.

Stechmann A, Cavalier-Smith T (2003) The root of the eukaryote tree pinpointed. *Curr Biol*. 13:R665–6. doi: 10.1016/s0960-9822(03)00602-x.

Stein K, Prondvai E, Huang T, Baele JM, Sander PM, Reisz R (2019) Structure and evolutionary implications of the earliest (Sinemurian, Early Jurassic) dinosaur eggs and eggshells. *Sci Rep*; 9:4424. doi: 10.1038/s41598-019-40604-8.

Steinfeld L, Vafaei A, Rösner J, Merzendorfer H (2019) Chitin prevalence and function in bacteria, fungi and protists. *Adv Exp Med Biol*; 1142:19–59. doi: 10.1007/978-981-13-7318-3_3.

Stern R (2017) Go fly a chitin: The mystery of chitin and chitinases in vertebrate tissues. *Front Biosci*; 22:580–595. doi: 10.2741/4504.

Stevens J, Last PR, Paxton JR, Eschmeyer WN (eds) (1998) *Encyclopedia of Fishes*. Academic Press.

Steventon B, Mayor R, Streit A (2014) Neural crest and placode interaction during the development of the cranial sensory system. *Dev Biol*; 389:28–38. doi: 10.1016/j.ydbio.2014.01.021.

Stewart TA, Lemberg JB, Taft NK, Yoo I, Daeschler EB, Shubin NH (2020) Fin ray patterns at the fin-to-limb transition. *Proc Natl Acad Sci U S A*; 117:1612–1620. doi: 10.1073/pnas.1915983117.

Stolfi A, Ryan K, Meinertzhagen IA, Christiaen L (2015) Migratory neuronal progenitors arise from the neural plate borders in tunicates. *Nature*; 527:371–374. doi: 10.1038/Nature15758.

Stott R (2012) *Darwin's Ghosts: In Search of the First Evolutionists*. Bloomsbury.

Stringer C (2011) *The Origin of Our Species*. Penguin Books.

Stringer C (2016) The origin and evolution of *Homo sapiens*. *Phil Trans R Soc, 5B*; 371(1698). pii:20150237. doi: 10.1098/rstb.2015.0237.

Strother PK, Battison L, Brasier MD, Wellman CH (2011) Earth's earliest non-marine eukaryotes. *Nature*; 473:505–509. doi: 10.1038/nature09943.

Strzyz P (2021) Shaping the human brain. *Nat Rev Mol Cell Biol*. doi.org/10.1038/s41580-021-00364-8.

Sturm RA (2009) Molecular genetics of human pigmentation diversity. *Hum Mol Genet*; 18(R1):R9–R17. doi: 10.1093/hmg/ddp003.

Suárez R, Gobius I, Richards LJ (2014) Evolution and development of interhemispheric connections in the vertebrate forebrain. *Front Hum Neurosci*; 8:497. doi: 10.3389/fnhum.2014.00497.

Subirana L, Farstey V, Bertrand S, Escriva H (2020) Asymmetron lucayanum: How many species are valid? *PLoS One*. 2020;15(3):e0229119. doi: 10.1371/journal.pone.0229119.

Subramanian M (2019) Anthropocene now: Influential panel votes to recognize Earth's new epoch. *Nature*. doi: 10.1038/d41586-019-01641-5.

Sues H-D (2019) *The Rise of Reptiles: 320 Million Years of Evolution*. Johns Hopkins University Press.

Sulej T, Niedźwiedzki G (2019) An elephant-sized late triassic synapsid with erect limbs. *Science*; 36:78–80. doi: 10.1126/science.aal4853.

Summerhayes GR, Leavesley M, Fairbairn A (2010) Human adaptation and plant use in highland New Guinea 49000 to 44000 years ago. *Science*; 330:78–81. doi: 10.1126/science.1193130.

Sun G, Dilcher DL, Wang H, Chen Z (2011) A eudicot from the Early Cretaceous of China. *Nature*; 471:625–8. doi: 10.1038/Nature09811.

Sun J, Liu X, McKenzie EHC et al. (2019) Fungicolous fungi: Terminology, diversity, distribution, evolution, and species checklist. *Fungal Diversity*; **95**, 337–430. doi.org/10.1007/s13225-019-00422-9.

Sutton VR (2002) Tay-Sachs disease screening and counseling families at risk for metabolic disease. *Obstet Gynecol Clin North Am*; 29:287–96. doi: 10.1016/s0889-8545(01)00002-x.

Suwa G, Kono RT, Simpson SW, Asfaw B, Lovejoy CO, White TD (2009) Paleobiological implications of the ardipithecus ramidus dentition. *Science*; 326:94–99 (PMID:19810195).

Suzuki IK, Gacquer D, Van Heurck R et al. (2018) Human-specific NOTCH2NL genes expand cortical neurogenesis through Delta/Notch regulation. *Cell*; 173:1370–1384.e16. doi: 10.1016/j.cell.2018.03.067.

Swarup R, Bhosale R (2019) Developmental roles of AUX1/LAX auxin influx carriers in plants. *Front Plant Sci*, 28;10:1306. doi: 10.3389/fpls.2019.01306.

Szöllősi GJ, Tannier E, Daubin V, Boussau B (2015) The inference of gene trees with species trees. *Syst Biol*; 64:e42–e62. doi: 10.1093/sysbio/syu048.

Szövényi P, Waller M, Kirbis A (2019) Evolution of the plant body plan. *Curr Top Dev Biol*; 131:1–34. doi: 10.1016/bs.ctdb.2018.11.005.

Tabin CJ, Carroll SB, Panganiban G (1999) Out on a limb: Parallels in vertebrate and invertebrate limb patterning and the origin of appendages. *Am Zool*; 39:650–666.

Tajsharghi H, Oldfors A (2013) Myosinopathies: Pathology and mechanisms. *Acta Neuropathol*; 125:3–18. doi: 10.1007/s00401-012-1024-2.

Takarada T, Hinoi E, Nakazato R (2013) An analysis of skeletal development in osteoblast-specific and chondrocyte-specific runt-related transcription factor-2 (Runx2) knockout mice. *J Bone Miner Res*; 28:2064–2069. doi: 10.1002/jbmr.1945.

Takasato M, Little MH (2015) The origin of the mammalian kidney: Implications for recreating the kidney in vitro. *Development*. 142:1937–1947. doi: 10.1242/dev.104802.

Takechi M, Kuratani S (2010) History of studies on mammalian middle ear evolution: A comparative morphological and developmental biology perspective. *J Exp Zool Mol Dev Evol*; 314:417–433.

Takeichi T, Akiyama M (2016) Inherited ichthyosis: Non-syndromic forms. *J Dermatol*; 43:242–251. doi: 10.1111/1346-8138.13243.

Talbert PB, Meers MP, Henikoff S (2019) Old cogs, new tricks: The evolution of gene expression in a chromatin context. *Nat Rev Genet*; 20:283–297. doi: 10.1038/s41576-019-0105-7.

Tanaka K, Zelenitsky DK, Therrien F, Kobayashi Y (2018) Nest substrate reflects incubation style in extant archosaurs with implications for dinosaur nesting habits. *Sci Rep*; 8:3170. doi: 10.1038/s41598-018-21386-x.

Tang Q, Pang K, Yuan X, Xiao S (2020) A one-billion-year-old multicellular chlorophyte. *Nat Ecol Evol*; 4:543–549. doi: 10.1038/s41559-020-1122-9.

Tang Q, Wan B, Yuan X, Muscente AD, Xiao S (2019) Spiculogenesis and biomineralization in early sponge animals. *Nat Commun*; 10:3348. doi: 10.1038/s41467-019-11297-4.

Tang WJ, Fernandez J, Sohn JJ, Amemiya CT (2015) Chitin is endogenously produced in vertebrates. *Curr Biol*; 30;25:897–900. doi: 10.1016/j.cub.2015.01.058

Tapaltsyan V, Charles C, Hu J (2016) Identification of novel Fgf enhancers and their role in dental evolution. *Evol Dev* 18:31–40 doi: 10.1111/ede.12132.

Tarver JE, Dos Reis M, Mirarab S et al. (2016) The interrelationships of placental mammals and the limits of phylogenetic inference. *Genome Biol Evol*; 8:330–344. doi: 10.1093/gbe/evv261.

Tassi F, Ghirotto S, Mezzavilla M (2015) Early modern human dispersal from Africa: Genomic evidence for multiple waves of migration. *Investig Genet*; 6:13. doi: 10.1186/s13323-015-0030-2.

Taylor DW, Li H (2018) Did flowering plants exist in the Jurassic period? *Elife*; 7:e43421. doi: 10.7554/eLife.43421.

Teeling EC, Springer MS, Madsen O (2005) A molecular phylogeny for bats illuminates biogeography and the fossil record. *Science*; 307:580–584.

Templeton AR (2008) The reality and importance of founder speciation in evolution. *BioEssays*; 30:470–479. doi: 10.1002/bies.20745.

Tennant JP, Mannion PD, Upchurch P, Sutton MD, Price G (2017) Biotic and environmental dynamics through the Late Jurassic-Early Cretaceous transition: Evidence for protracted faunal and ecological turnover. *Biol Rev Camb Philos Soc*; 92:776–814. doi: 10.1111/brv.12255.

Tesson SVM, Skjøth CA, Šantl-Temkiv T, Löndahl J (2016) Airborne microalgae: Insights, opportunities, and challenges. *Appl Environ Microbiol 22*;82(7):1978–1991. doi: 10.1128/AEM.03333-15.

Thieme H (1997) Lower palaeolithic hunting spears from Germany. *Nature*; 385:807–810.

Thompson JN (1989) Concepts of coevolution. *Trends Ecol Evol*; 4:179–83. doi: 10.1016/0169-5347(89)90125-0.

Thornton JW (2004) Resurrecting ancient genes: Experimental analysis of extinct molecules. *Nat Rev Genet*; 5:366–375.

Thulborn RA (1980) The ankle joints of archosaurs. *Aust J Palaeont*; 4:241–261. doi: 10.1080/03115518008558970.

Tickle C (2006) Developmental cell biology: Making digit patterns in the vertebrate limb. *Nat Rev Mol Cell Biol*; 7:45–53.

Tikhonenkov DV, Hehenberger E, Esaulov AS et al. (2020) Insights into the origin of metazoan multicellularity from predatory unicellular relatives of animals. *BMC Biol*; 18:39. doi: 10.1186/s12915-020-0762-1.

Tissir F, Goffinet AM (2013) Shaping the nervous system: Role of the core planar cell polarity genes. *Nat Rev Neurosci*; 14:525–535. doi: 10.1038/nrn3525.

Tokita M, Iwai N (2010) Development of the pseudothumb in frogs. *Biol Lett*; 6:517–520. doi: 10.1098/rsbl.2009.1038.

Tokita M (2015) How the pterosaur got its wings. *Biol Rev Camb Philos Soc*; 90:1163–1178. doi: 10.1111/brv.12150.

Tolker-Nielsen T (2015) Biofilm development. *Microbiol Spectr*; 3:MB-0001-2014. doi: 10.1128/microbiolspec.MB-0001-2014.

Towers M, Mahood R, Yin Y, Tickle C (2008) Integration of growth and specification in chick wing digit-patterning. *Nature*; 452:882–886. doi: 10.1038/nature06718.

Tucci S et al. (2018) Evolutionary history and adaptation of a human pygmy population of Flores island, Indonesia. *Science*; 361:511–516. doi: 10.1126/Science.aar8486.

Turetsky MR et al. (2019) Permafrost collapse is accelerating carbon release. *Nature*; 569:32–34. doi: 10.1038/d41586-019-01313-4.

Turing AM (1952) The chemical basis of morphogenesis. *Philos Trans R Soc B*; 237:37–72.

Turner EC (2021) Possible poriferan body fossils in early Neoproterozoic microbial reefs. *Nature*. doi.org/10.1038/s41586-021-03773-z.

Turner GF, Seehausen O, Knight ME (2001) How many species of cichlid fishes are there in African lakes? *Mol Ecol*; 10:793–806.

Twitchett RJ (2007) The Lilliput effect in the aftermath of the end-Permian extinction event. *Palaeogeography, Palaeoclimatology, Palaeoecology*; 252:132–144. doi: 10.1016/j.palaeo.2006.

Tyndale-Biscoe CH, Renfree MB (1987) *Reproductive Physiology of Marsupials*. Cambridge University Press.

Uhen MD (2007) Evolution of marine mammals: Back to the sea after 300 million years. *Anat Rec*; 290:514–522.

Uhl EW, Warner NJ (2015) Mouse models as predictors of human responses: Evolutionary medicine. *Curr Pathobiol Rep*; 3:219–223. doi:10.1007/s40139-015-0086-y.

Uitterlinden AG (2016) An introduction to genome-wide association studies: GWAS for dummies. *Semin Reprod Med*; 34:196–204. doi: 10.1055/s-0036-1585406.

Ungar PS (2010) *Mammal Teeth: Origin Evolution and Diversity*. Johns Hopkins Press.

Urbanek AK, Rymowicz W, Mirończuk AM (2018) Degradation of plastics and plastic-degrading bacteria in cold marine habitats. *Appl Microbiol Biotechnol*; 102:7669–7678. doi: 10.1007/s00253-018-9195-y.

Vaelli PM, Theis KR, Williams JE, O'Connell LA, Foster JA, Eisthen HL. (2020) The skin microbiome facilitates adaptive tetrodotoxin production in poisonous newts. *Elife*; 9:e53898. doi: 10.7554/eLife.53898.

Valas RE, Bourne PE (2011) The origin of a derived superkingdom: How a Gram-positive bacterium crossed the desert to become an archaeon. *Biol Direct*; 6:16. doi: 10.1186/1745-6150-6-16.

Valenzuela N, Adams DC (2011) Chromosome number and sex determination coevolve in turtles. *Evolution*; 65:1808–13. doi: 10.1111/j.1558-5646.2011.01258.x.

Van Dyke JU, Griffith OW (2018) Mechanisms of reproductive allocation as drivers of developmental plasticity in reptiles. *J Exp Zool A Ecol Integr Physiol*; 329:275–286. doi: 10.1002/jez.2165.

Van Grouw K (2018) *Unnatural Selection.* Princeton University Press.

van Tuinen M, Blair Hedges S (2001) Calibration of avian molecular clocks. *Mol Biol Evol*; 18:206–213.

Van Valen LM (1973) A new evolutionary law. *Evol Theory*; 1:1–30, (https://ebme. marine.rutgers.edu/HistoryEarthSystems/ HistEarthSystems_Fall2010/VanValen%20 1973%20Evol%20%20Theor%20.pdf)

Van Valkenburgh B (2007) Deja vu: The evolution of feeding morphologies in the carnivora. *Integr Comp Biol*; 47:147–163. doi: 10.1093/icb/icm016.

Van Valkenburgh B, Jenkins I (2002) Evolutionary patterns in the history of permo-triassic and cenozoic synapsid predators. *Paleont Soc Papers*; 8:267–288. doi:10.1017/ S1089332600001121

van Veelen M, Allen B, Hoffman M, Simon B, Veller C (2017) Hamilton's rule. *J Theor Biol*; 414:176–230. doi: 10.1016/j.jtbi.2016.08.019.

Vandermeersch B, Bar-Yosef O (2019) The Paleolithic Burials at Qafzeh Cave, Israel. *Paleo*; 30:256–275. doi: 10.4000/paleo.4848.

Vandiver PB, Soffer O, Klima B, Svoboda J (1989) The origins of ceramic technology at Dolni Vecaronstonice, czechoslovakia. *Science*; 246:1002–8.

Varga T, Krizsán K, Földi C et al. (2019) Megaphylogeny resolves global patterns of mushroom evolution. *Nat Ecol Evol*; 3:668–678. doi: 10.1038/s41559-019-0834-1.

Varki A, Altheide TK (2005) Comparing the human and chimpanzee genomes: Searching for needles in a haystack. *Genome Res*; 15:1746–58. doi: 10.1101/gr.3737405.

Varriale A (2014) DNA methylation epigenetics and evolution in vertebrates: Facts and challenges. *Int J Evol Biol*; 2014:475981. doi: 10.1155/2014/475981.

Vasco A, Moran RC, Ambrose BA (2013) The evolution, morphology, and development of fern leaves. *Front Plant Sci*; 4:345. doi: 10.3389/fpls.2013.00345.

Vaškaninová V, Chen D, Tafforeau P et al. (2020) Marginal dentition and multiple dermal jawbones as the ancestral condition of jawed vertebrates. *Science*; 369:211–216. doi: 10.1126/science.aaz9431.

Vaziri SH, Majidifard MR, Laflamme M (2018) Diverse assemblage of Ediacaran fossils from Central Iran. *Sci Rep8*; 8:5060. doi: 10.1038/s41598-018-23442-y.

Veeramah KR, Hammer MF (2014) The impact of whole-genome sequencing on the reconstruction of human population history. *Nat Rev Genet*; 15:149–162. doi: 10.1038/nrg3625.

Venkatesh B, Lee AP, Ravi V et al. (2014) Elephant shark genome provides unique insights into gnathostome evolution. *Nature*; 505:174–179. doi: 10.1038/Nature12826.

Venkatesh B, Erdmann MV, Brenner S (2001) Molecular synapomorphies resolve evolutionary relationships of extant jawed vertebrates. *Proc Natl Acad Sci U S A*; 98:11382–7. doi: 10.1073/pnas.201415598.

Ventura GT, Kenig F, Reddy CM (2007) Molecular evidence of Late Archaen archaea and the presence of a subsurface hydrothermal biosphere. *Proc Natl Acad Sci U S A*; 104:14260–14265.

Verbrugge LM (1982) Sex differentials in health. *Public Health Rep*; 97:417–437.

Vereecken NJ, Wilson CA, Hötling S et al. (2012) Pre-adaptations and the evolution of pollination by sexual deception: Cope's rule of specialization revisited. *Proc Biol Sci*; 279:4786–4794. doi: 10.1098/rspb.2012.1804.

Via S (2009) Natural selection in action during speciation. *Proc Natl Acad Sci*; 106(Suppl 1):9939–9946. doi: 10.1073/pnas.0901397106.

Visick KL1, Fuqua C (2005) Decoding microbial chatter: Cell-Cell communication in bacteria. *J Bacteriol*; 187:5507–19. doi: 10.1128/JB.187.16.5507-5519.2005.

von Koenigswald WV (2000) Two Different Strategies in Enamel Differentiation: Marsupialia Versus Eutheria. In *Development Function and Evolution of Teeth* (Teaford MF, Meredith-Smith M, Ferguson MWJ, eds), pp 107–118. Cambridge University Press.

Von Konrat M, Shaw AJ, Renzaglia KS (2010) A special issue of phytotaxa dedicated to bryophytes: The closest living relatives of early land plants. *Phytotaxa*; 9. doi: 10.11646/phytotaxa.9.1.3.

von Petzinger G, Nowell A (2014) A place in time: Situating Chauvet within the long chronology of symbolic behavioral development. *J Hum Evol*; 74:37–54. doi: 10.1016/j.jhevol.2014.02.022.

Waddington CH (1953) Genetic assimilation of the bithorax phenotype. *Evolution*; 10:1–13.

Waddington CH (1961) Genetic assimilation. *Adv Genet*; 10:257–293.

Wagner DL, Grames EM, Forister ML, Berenbaum MR, Stopak D (2021) Insect decline in the Anthropocene: Death by a thousand cuts. *Proc Natl Acad Sci U S A*; 118:e2023989118. doi: 10.1073/pnas.2023989118.

Wahl LM (2011) Fixation when N and S vary: Classic approaches give elegant new results. *Genetics*; 188:783–785. doi: 10.1534/genetics.111.131748.

Wakeley J (2010) Natural Selection and Coalescent Theory. In *Evolution since Darwin: The First 150 Years* (Bell MA, Futuyma DJ, Eanes WF, Levinton JS, eds), pp 119–149. Sinauer Press.

Wall JD, Slatkin M (2012) Paleopopulation genetics. *Annu Rev Genet*; 46:635–649.

Wallace AR (1858) On the tendency of varieties to depart indefinitely from the original type. *Proc. Linn Soc*; 3:53–62. http://people.wku.edu/charles.smith/wallace/S043.htm

Wallace IJ, Shea, JJ (2006) Mobility patterns and core technologies in the middle paleolithic of the levant. *J. Arch. Sci*; 33:1293–1309. doi: 10.1016/j.jas.2006.01.005.

Wallace RVS, Martínez R, Rowe T (2019) First record of a basal mammaliamorph from the early Late Triassic Ischigualasto Formation of Argentina. *PLoS One*; 14(8):e0218791. doi: 10.1371/journal.pone.0218791.

Walter Anthony K, Schneider von Deimling T, Nitze I et al. (2018) 21st-century modeled permafrost carbon emissions accelerated by abrupt thaw beneath lakes. *Nat Commun*; 9:3262. doi: 10.1038/s41467-018-05738-9.

Wang B, Sullivan TN (2017) A review of terrestrial, aerial and aquatic keratins: The structure and mechanical properties of pangolin scales, feather shafts and baleen plates. *J Mech Behav Biomed Matter*; 76:4–20. doi: 10.1016/j.jmbbm.2017.05.015.

Wang DY, Kumar S, Hedges SB (1999) Divergence time estimates for the early history of animal phyla and the origin of plants, animals and fungi. *Proc Biol Sci*; 266:163–171. doi: 10.1098/rspb.1999.0617.

Wang H, Meng J, Wang Y. (2019a) Cretaceous fossil reveals a new pattern in mammalian middle ear evolution. *Nature*; 576:102-105. doi: 10.1038/s41586-019-1792-0.

Wang H, Meng J, Wang Y (2019b) Cretaceous fossil reveals a new pattern in mammalian middle ear evolution. *Nature*; 576:102–105. doi: 10.1038/s41586-019-1792-0.

Wang, J, Santiago E, Caballero A (2016) Prediction and estimation of effective population size. *Heredity*; 117:193–206. doi.org/10.1038/hdy.2016.43.

Wang M, Li Z, Zhou Z (2017) Insight into the growth pattern and bone fusion of basal birds from an Early Cretaceous enantiornithine bird. *Proc Natl Acad Sci*; 114:11470–11475. doi: 10.1073/pnas.1707237114.

Wang M, O'Connor JK, Xu X, Zhou Z (2019a) A new Jurassic scansoriopterygid and the loss of membranous wings in theropod dinosaurs. *Nature*; 569:256–259 doi: 10.1038/s41586-019-1137-z.

Wang M, O'Connor JK, Xu X, Zhou Z (2019b) A new Jurassic scansoriopterygid and the loss of membranous wings in theropod dinosaurs. *Nature*; 569:256–259. doi: 10.1038/s41586-019-1137-z.

Wang X, Nudds RL, Palmer C, Dyke GJ (2012) Size scaling and stiffness of avian primary feathers: Implications for the flight of Mesozoic birds. *J Evol Biol*; 25:547–555. doi: 10.1111/j.14209101.2011.0 2449.x.

Wanninger A, Wollesen T (2019) The evolution of molluscs. *Biol Rev*; 94:102–115. doi: 10.1111/brv.12439.

Ward CV, Kimbel WH, Harmon EH, Johanson DC (2012) New postcranial fossils of *Australopithecus afarensis* from Hadar Ethiopia (1990–2007). *J Hum Evol*; 65:1–51. doi: 10.1016/j.jhevol.2011.11.012.

Wayne L (1988) International committee on systematic bacteriology: Announcement of the report of the ad hoc committee on reconciliation of approaches to bacterial systematics. *Zentralb Bakteriol Hyg A*; 268:433–434.

Weaver LN, Varricchio DJ, Sargis EJ, Chen M, Freimuth WJ, Wilson Mantilla GP (2021) Early mammalian social behaviour revealed by multituberculates from a dinosaur nesting site. *Nat Ecol Evol*; 5:32–37. doi: 10.1038/s41559-020-01325-8.

Weaver TD (2012) Did a discrete event 200,000–100,000 years ago produce modern humans? *J Hum Evol*; 63:121–126. doi: 10.1016/j.jhevol.2012.04.003.

Wedel MJ (2012) A monument of inefficiency: The presumed course of the recurrent laryngeal nerve in sauropod dinosaurs. *Act Palaentol Pl*; 57:251–256. doi: 10.4202/app.2011.0019.

Wei Y et al. (2019) Genetic mapping and evolutionary analysis of human-expanded cognitive networks. *Nat Commun*; 10:4839. doi: 10.1038/s41467-019-12764-8.

Weir JT, Schluter D (2008) Calibrating the avian molecular clock. *Mol Ecol*; 17:2321–2328. doi: 10.1111/j.1365-294X.2008.03742.x.

Weismann A (1889) *The Continuity of the Germ- Plasm as the Foundation of a Theory of Heredity. In Essays on Heredity.* Oxford Clarendon Press. www.esp.org/books/weismann/germ-plasm/facsimile/

Weiss MC, Sousa FL, Mrnjavac N et al. (2016) The physiology and habitat of the last universal common ancestor. *Nat Microbiol*; 1:16116. doi: 10.1038/nmicrobiol.2016.116.

Weiss-Lehman C, Tittes S, Kane NC, Hufbauer RA, Melbourne BA (2019) Stochastic processes drive rapid genomic divergence during experimental range expansions. *Proc Biol Sci*; 286:20190231. doi: 10.1098/rspb.2019.0231.

Welker F, Ramos-Madrigal J, Gutenbrunner P et al. (2020) The dental proteome of *Homo antecessor*. *Nature*; 580:235–238. doi: 10.1038/s41586-020-2153-8.

Wellman CH, Gray J (2000) The microfossil record of early land plants. *Philos Trans R Soc Lond B Biol Sc*; 355:717–732. doi: 10.1098/rstb.2000.0612.

Wessenberg H, Antipa G (1970) Capture and ingestion of Paramecium by *Didinium nasutum*. *J Eukaryote Microbiol*; 17:250–270. doi: 10.1016/0169-5347(94)90202-X.

Westneat MW, Betz O, Blob RW (2003) Tracheal respiration in insects visualized with synchrotron x-ray imaging. *Science*; 299:558–560.

White D, Rabago-Smith M. (2011) Genotype-phenotype associations and human eye color. *J Hum Genet*; 56:5–7. doi: 10.1038/jhg.2010.126.

White TD, Asfaw B, Beyene Y, Haile-Selassie Y, Lovejoy CO, Suwa G, Wolde-Gabriel G (2009) Ardipithecus ramidus and the paleobiology of early hominids. *Science*; 326:75–86.

Whitehead H, Laland KN, Rendell L, Thorogood R, Whiten A (2019) The reach of gene-culture coevolution in animals. *Nat Commun*; 10:2405. doi: 10.1038/s41467-019-10293-y.

Wickett NJ, Mirarab S, Nguyen N et al. (2014) Phylotranscriptomic analysis of the origin and early diversification of land plants. *Proc Natl Acad Sci*; 111:E4859–E4868. doi: 10.1073/pnas.1323926111.

Wierzchos J, de los Ríos A, Ascaso C (2012) Microorganisms in desert rocks: The edge of life on earth. *Int Microbiol*; 15:173–83. doi: 10.2436/20.1501.01.170.

Wijesena N, Simmons DK, Martindale MQ (2017) Antagonistic BMP-cWNT signaling in the cnidarian *Nematostella vectensis* reveals insight into the evolution of mesoderm. *Proc Natl Acad Sci*; 114:E5608–E5615. doi: 10.1073/pnas.1701607114.

Wildman DE, Uddin M, Liu G et al. (2003) Implications of natural selection in shaping 99.4% nonsynonymous DNA identity between humans and chimpanzees: Enlarging genus *Homo*. *Proc Natl Acad Sci U S A*; 100:7181–7788.

Wiles AM, Doderer M, Ruan J (2010) Building and analyzing protein interactome networks by cross-species comparisons. *BMC Syst Biol* 4:36. doi: 10.1186/1752-0509-4-36.

Wilkins AS (2002) *The Evolution of Developmental Pathways*. Sinauer Press.

Wilkins AS (2007) Genetic networks as transmitting and amplifying devices for natural genetic tinkering. *Novartis Found Symp*; 284:71–86.

Will M, Parkington JE, Kandel AW, Conard NJ (2013) Coastal adaptations and the Middle Stone age lithic assemblages from Hoedjiespunt 1 in the Western Cape South Africa. *J Hum Evol*; 64:518–537. doi: 10.1016/j.jhevol.2013.02.012.

Wille M, Nägler TF, Lehmann B (2008) Hydrogen sulphide release to surface waters at the Precambrian/Cambrian boundary. *Nature*; 453:767–769. doi: 10.1038/nature07072.

Williams TA, Foster PG, Nye TMW (2012) A congruent phylogenomic signal places eukaryotes within the Archaea. *Proc Biol Sci*; 279:4870–4879. doi: 10.1098/rspb.2012.1795.

Williams TA, Foster PG, Cox CJ, Embley TM (2013) An archaeal origin of eukaryotes supports only two primary domains of life. *Nature*; 304:231–236. doi: 10.1038/nature12779.

Williams TL, Senft SL, Yeo J et al. (2019) Dynamic pigmentary and structural coloration within cephalopod chromatophore organs. *Nat Commun*; 10:1004. Published 2019 Mar 1. doi: 10.1038/s41467-019-08891-x.

Williamson TE, Brusatte SL, Wilson GP (2014) The origin and early evolution of metatherian mammals: The Cretaceous record. *Zookeys*; 465:1–76. doi: 10.3897/zookeys.465.8178.

Willis K, McElwain J (2013) *The Evolution of Plants*. Oxford University Press.

Willman S, Peel JS, Ineson JR, Schovsbo NH, Rugen EJ, Frei R (2020) Ediacaran Doushantuo-type biota discovered in Laurentia. *Commun Biol*; 3:647. doi: 10.1038/s42003-020-01381-7.

Wilson EO (2000) *Sociobiology: The New Synthesis*. Belknap Press.

Wilson NG, Rouse GW, Giribet G (2010) Assessing the molluscan hypothesis Serialia (Monoplacophora+Polyplacophora) using novel molecular data. *Mol Phylogenet Evol*; 54:187–193. doi: 10.1016/j.ympev.2009.07.028.

Witmer LM (1987) The nature of the antorbital fossa of archosaurs: Shifting the null hypothesis. In *Fourth Symposium on Mesozoic Terrestrial Ecosystesms* (Currie PJ, Koster EH, eds), pp 230–235. http://www.ohio.edu/people/witmerl/Downloads/1987_Witmer_Archosaur_antorbital_cavity.pdf

Witmer LM (1995) Homology of facial structures in extant archosaurs (birds and crocodilians), with special reference to paranasal pneumaticity and nasal conchae. *J Morphol*; 225:269–327. doi: 10.1002/jmor.1052250304.

Witton, MP (2013) *Pterosaurs: Natural History, Evolution, Anatomy*. Princeton University Press.

Witton MP (2015) Were early pterosaurs inept terrestrial locomotors? *PeerJ*; 3(3):e1018. doi: 10.7717/peerj.1018.

Witton MP, Habib M (2010) On the size and flight diversity of giant pterosaurs, the use of birds as pterosaur analogues and comments on pterosaur flightlessness. *PLoS One*; 5(11):e13982. doi: 10.1371/journal.pone.0013982.

Woese CR, Kandler O, Wheelis ML (1990) Towards a natural system of organisms: Proposal for the domains Archaea, Bacteria and Eucarya. *Proc Natl Acad Sci U S A*; 87:4576–4579.

Wolpert L, Tickle C, Jessell T (2019) *Principles of Development* (6th ed). Oxford University Press.

Wong M, Balshine S (2011) The evolution of cooperative breeding in the African cichlid fish, *Neolamprologus pulcher*. *Biol Rev Camb Philos Soc*; 86(2):511–530. doi: 10.1111/j.1469-185X.2010.00158.x.

Wood B (2019) *Human Evolution: A Very Short Introduction* (2nd ed). Oxford University Press.

Wood B, Constantino P (2007) *Paranthropus boisei*: Fifty years of evidence and analysis. *Am J Phys Anthropol*; 45(Suppl.):106–132.

Wood B, Elton S (eds) (2008) Human evolution: Ancestors and relatives. *J Anat* 4:335–562. (This is a set of symposium papers.)

Wood B, Harrison T (2011) The evolutionary context of the first hominins. *Nature*; 470:347–352. doi: 10.1038/nature09709.

Wood B, Lonergan N (2008) The hominin fossil record: Taxa, grades and clades. *J Anat*; 212:354–376. doi: 10.1111/j.1469-7580.2008.00871.x).

Wood B, Boyle EK (2016) Hominin taxic diversity: Fact or fantasy? *Am J Phys Anthropol*; 159 (Suppl 61):S37–78. doi: 10.1002/ajpa.22902.

Wood B, Doherty D, Boyle, E (2020) Hominin taxic diversity. *Oxf Res Encycl Anthropol*; doi: 10.1093/acrefore/9780190854584.013.194.

Wood TWP, Nakamura T (2018) Problems in Fish-to-Tetrapod Transition: Genetic Expeditions Into Old Specimens. *Front Cell Dev Biol*; 6:70. doi: 10.3389/fcell.2018.00070.

Woodhouse RM, Ashe A (2020) How do histone modifications contribute to transgenerational epigenetic inheritance in *C. elegans*? *Biochem Soc Trans*; 48(3):1019–1034. doi: 10.1042/BST20190944.

World Meteorological Organization (2020). WMO Statement on the State of the Global Climate in 2019. WMO-No. 1248. Geneva.

Wray GA (1997) Echinoderms. In *Embryology: Constructing the Organism* (Gilbert SF, Raunio AM, eds), pp 309–330. Sinauer Press.

Wright JD (2000) Global Climate Change in Marine Stable Isotope Records, In *Quaternary Geochronology: Methods and Applications* (Noller JS, Sowers JM, Lettis WR, eds), pp 427–433. American Geophysical Union. doi: 10.1029/RF004p0427.

Wright S (1955) Classification of the factors of evolution. *Cold Spring Harbor Symp Quant Biol*; 20:16–24. doi: 10.1101/SQB.1955.020.01.004.

Wu P, Ng CS, Yan J et al. (2015) Topographical mapping of α- and β-keratins on developing chicken skin integuments: Functional interaction and evolutionary perspectives. *Proc Natl Acad Sci U S A*; 112:E6770–9. doi: 10.1073/pnas.1520566112.

Wu P, Yan J, Lai YC et al. (2018) Multiple regulatory modules are required for scale-to-feather conversion. *Mol Biol Evol*; 35:417–430. doi: 10.1093/molbev/msx295.

Wu S et al. (2016) Genome-wide scans reveal variants at EDAR predominantly affecting hair straightness in han Chinese and Uyghur populations. *Hum Genet*; 135:1279–1286. doi: 10.1007/s00439-016-1718-y.

Wu XJ et al. (2019) Archaic human remains from Hualongdong, China, and Middle Pleistocene human continuity and variation. *Proc Natl Acad Sci*; 1 16:9820–9824. doi: 10.1073/pnas.1902396116.

Wunderling N, Willeit M, Donges JF, Winkelmann R (2020) Global warming due to loss of large ice masses and Arctic summer sea ice. *Nat Commun*; 11:5177. doi: 10.1038/s41467-020-18934-3.

Wurster-Hill DH, Gray CW (1973) Giemsa banding patterns in the chromosomes of twelve species of cats (Felidae). *Cytogenet Cell Genet*; 12:388–397.

Xiao S, Tang Q (2018) After the boring billion and before the freezing millions: evolutionary patterns and innovations in the Tonian Period. *Emerg Top Life Sci*; 2:161-171. doi: 10.1042/ETLS20170165.

Xing H, Mallon JC, Currie ML (2017) Supplementary cranial description of the types of Edmontosaurus regalis (Ornithischia: Hadrosauridae), with comments on the phylogenetics and biogeography of Hadrosaurinae. *PLoS One*; 12:e0175253. doi.org/10.1371/journal.pone.0175253.

Xu X (2012) A gigantic feathered dinosaur from the Lower Cretaceous of China. *Nature*; 484:92–95. doi: 10.1038/nature10906.

Xu X, Zhou Z, Dudley R et al. (2014) An integrative approach to understanding bird origins. *Science*; 346:1253293. doi: 10.1126/science.1253293.

Xue H, Tong KL, Marck C et al. (2003) Transfer RNA paralogs: Evidence for genetic code-amino acid biosynthesis coevolution and an archaeal root of life. *Gene*; 310:59–66.

Yahalomi D, Atkinson SD, Neuhof M et al. (2020) A cnidarian parasite of salmon (Myxozoa: *Henneguya*) lacks a mitochondrial genome. *Proc Natl Acad Sci U S A*; 117:5358–5363. doi: 9.1073/pnas.1909907117.

Yan W (2014) Potential roles of noncoding RNAs in environmental epigenetic transgenerational inheritance. *Mol Cell Endocrinol*; 398:24–30. doi: 10.1016/j.mce.2014.09.008.

Yang F, Fu B, O'Brien PC (2004) Refined genome-wide comparative map of the domestic horse, donkey and human based on cross-species chromosome painting: Insight into the occasional fertility of mules. *Chromosome Res*; 12:65–76.

Yang Y, Guillot P, Boyd Y et al. (1998) Evidence that preaxial polydactyly in the Doublefoot mutant is due to ectopic Indian Hedgehog signalling. *Development*; 125:3123–3132.

Yehuda R, Daskalakis NP, Bierer LM (2015) Holocaust exposure induced intergenerational effects on FKBP5 methylation. *Biol Psychiatry*; pii:S0006-3223(15)00652–6. doi: 10.1016/j.biopsych.2015.08.005.

Yew CW, Lu D, Deng L, Wong LP et al. (2018) Genomic structure of the native inhabitants of Peninsular Malaysia and North Borneo suggests complex human population history in southeast Asia. *Hum Genet*; 137:161–173. doi: 10.1007/s00439-018-1869-0.

Yin L, Meng F, Kong F, Niu C (2020) Microfossils from the Paleoproterozoic Hutuo Group, Shanxi, North China: Early evidence for eukaryotic metabolism. *Precamb Res*; 342:105650. doi.org/10.1016/j.precamres.2020.105650.

Yin L, Zhu M, Knoll AH, Yuan X, Zhang J, Hu J (2007) Doushantuo embryos preserved inside diapause egg cysts. *Nature*; 446:661–3. doi: 10.1038/nature05682.

Yin Z, Zhu M, Davidson EH, Bottjer DJ, Zhao F, Tafforeau P (2015) Sponge grade body fossil with cellular resolution dating 60 Myr before the Cambrian. *Proc Natl Acad Sci U S A*; 112:E1453–E1460. doi: 10.1073/pnas.1414577112.

Yoshida MA, Ogura A (2011) Genetic mechanisms involved in the evolution of the cephalopod camera eye revealed by transcriptomic and developmental studies. *BMC Evol Biol*; 11:180. doi: 10.1186/1471-2148-11-180.

Yuan CX, Ji Q, Meng QJ, Tabrum AR, Luo ZX (2013) Earliest evolution of multituberculate mammals revealed by a new Jurassic fossil. *Science*; 341:779–83. doi: 10.1126/science.1237970.

Yubuki N, Leander BS (2013) Evolution of microtubule organizing centers across the tree of eukaryotes. *Plant J*; 75:230–44. doi: 10.1111/tpj.12145.

Zakrzewski W, Dobrzyński M, Szymonowicz M, Rybak Z (2019) Stem cells: Past, present, and future. *Stem Cell Res Ther*; 10:68. doi: 10.1186/s13287-019-1165-5.

Zardoya R, Meyer A (1998) Complete mitochondrial genome suggests diapsid affinities of turtles. *Proc Natl Acad Sci*; 95:14226–14231.

Zhang F, Zhang Y, Lv X, Xu B, Zhang H, Yan J, Li H, Wu L (2019) Evolution of an X-Linked miRNA family predominantly expressed in mammalian male germ cells. *Mol Biol Evol*; 36:663–678. doi: 10.1093/molbev/msz001.

Zhang F, Zhou Z, Xu X (2008) A bizarre Jurassic maniraptoran from China with elongate ribbonlike feathers. *Nature*; 455:1105–1108. doi: 10.1038/nature07447.

Zhang QL et al. (2018) A phylogenomic framework and divergence history of Cephalochordata Amphioxus. *Front Physiol*; 9:1833. doi: 10.3389/fphys.2018.01833.

Zhang R, Nowack EC,2, Price DC, Bhattacharya D, Grossman AR (2017) Impact of light intensity and quality on chromatophore and nuclear gene expression in Paulinella chromatophora, an amoeba with nascent photosynthetic organelles. *Plant J*. 90:221–234. doi: 10.1111/tpj.13488.

Zhang Y, O'Connor J, Di L, Qingjin M, Sigurdsen T, Chiappe LM (2014) New information on the anatomy of the Chinese Early Cretaceous Bohaiornithidae (Aves: Enantiornithes) from a subadult specimen of Zhouornis hani. *PeerJ*; 2:e407. doi: 10.7717/peerj.407.

Zhao B, Tumaneng K, Guan KL (2011) The Hippo pathway in organ size control tissue regeneration and stem cell self-renewal. *Nat Cell Biol* 13:877–883. doi: 10.1038/ncb2303.

Zhao F, Bottjer DJ, Hu S (2013) Complexity and diversity of eyes in Early Cambrian ecosystems. *Nat Sci Rep*; 3:2751. doi: 10.1038/srep02751.

Zheng H, Perreau J, Powell JE et al. (2019) Division of labor in honeybee gut microbiota for plant polysaccharide digestion. *Proc Natl Acad Sci U S A*; 116:25909–25916. doi: 10.1073/pnas.1916224116.

Zheng X-T, You HL, Xu X, Dong ZH (2009) An Early Cretaceous heterodontosaurid dinosaur with filamentous integumentary structures. *Nature*; 458:333–336. doi: 10.1038/nature07856.

Zhong B, Xi Z, Goremykin VV, Fong R, McLenachan PA, Novis PM, Davis CC, Penny D (2014) Streptophyte algae and the origin of land plants revisited using heterogeneous models with three new algal chloroplast genomes. *Mol Biol Evol*; 3:177–83. doi: 10.1093/molbev/mst200.

Zhou Z, Zhang F (2005) Discovery of an ornithurine bird and its implication for Early Cretaceous avian radiation. *Proc Natl Acad Sci U S A*; 102(52):18998–19002. doi: 10.1073/pnas.0507106102.

Zhu M, Ahlberg PE (2004) The origin of the internal nostril of tetrapods. *Nature*; 432:94–7.

Zhu M, Yu X, Ahlberg PE (2013) A Silurian placoderm with osteichthyan-like marginal jaw bones. *Nature*; 502:188–193. doi: 10.1038/nature12617.

Zhu M, Ahlberg PE, Pan Z, Zhu Y, Qiao T, Zhao W, Jia L, Lu J (2016a) A Silurian maxillate placoderm illuminates jaw evolution. *Science*; 354:334–336. doi: 10.1126/science.aah3764.

Zhu R, Zhisheng AZ, Potts R, Hoffmand KA (2003) Magnetostratigraphic dating of early humans in China. *Earth Sci Rev*; 61:341–359.

Zhu X, Wang J, Chen Q, Chen G, Huang Y, Yang Z (2016b) Costs and trade-offs of grazer-induced defenses in scenedesmus under deficient resource. *Sci Rep*; 6:22594. doi: 10.1038/srep22594.

Zhuravlev AY, Wood RA (2018) The two phases of the Cambrian Explosion. *Sci Rep*; 8:16656. doi: 10.1038/s41598-018-34962-y.

Zilhão J et al. (2010) Symbolic use of marine shells and mineral pigments by Iberian Neandertals. *Proc Natl Acad Sci*; 107:1023–8. doi: 10.1073/pnas.0914088107.

Zmasek CM, Zhang Q, Ye Y, Godzik A (2007) Surprising complexity of the ancestral apoptosis network. *Genome Biol*; 8:R226.

Zumberge JA, Love GD, Cárdenas P et al. (2018) Demosponge steroid biomarker 26-methylstigmastane provides evidence for Neoproterozoic animals. *Nat Ecol Evol*; 2:1709–1714. doi: 10.1038/s41559-018-0676-2.

Index

Note: Locators in *italics* may represent species, genes, books, or figures; **bold** indicates tables in the text.

T - #0271 - 111024 - C536 - 280/210/25 - PB - 9780367357016 - Gloss Lamination